PRACTICAL METALLURGY AND MATERIALS OF INDUSTRY

PRACTICAL METALLURGY AND MATERIALS OF INDUSTRY

FIFTH EDITION

JOHN E. NEELY
Instructor, Retired
Machine Technology
Lane Community College
Eugene, Oregon

THOMAS J. BERTONE
Metallurgy Instructor
Fullerton Community College
Fullerton, California

Prentice Hall
Upper Saddle River, New Jersey *Columbus, Ohio*

Library of Congress Cataloging-in-Publication Data
Neely, John, 1920–
Practical metallurgy and materials of industry / John E. Neely,
Thomas J. Bertone. —5th ed.
p. cm.
Includes bibliographical references and index.
ISBN 0-13-064552-8
1. Metallurgy. 2. Materials. I. Bertone, Thomas J. II. Title.
TN665 .N3 2000
669—dc21
99–39264
CIP

Editor: Stephen Helba
Assistant Editor: Michelle Churma
Production Editor: Louise N. Sette
Production Supervision: Carlisle Publishers Services
Design Coordinator: Karrie Converse-Jones
Cover Designer: John Bertone
Cover Photo: Thomas J. Bertone
Production Manager: Matthew Ottenweller
Marketing Manager: Chris Bracken

This book was set in Times Roman by Carlisle Communications, Ltd. and was printed and bound by R.R. Donnelley & Sons Company. The cover was printed by Phoenix Color Corp.

Cover image: Material of Industry: Graphitized Coal Tar Pitch viewed through an optical microscope using crossed polarized light and a 1/4 wave plate.

Printed in the United States of America

10 9 8 7 6 5 4 3 2 1

ISBN: 0-13-624552-8

Prentice-Hall International (UK) Limited, *London*
Prentice-Hall of Australia Pty. Limited, *Sydney*
Prentice-Hall of Canada, Inc., *Toronto*
Prentice-Hall Hispanoamericana, S. A., *Mexico*
Prentice-Hall of India Private Limited, *New Delhi*
Prentice-Hall of Japan, Inc., *Tokyo*
Prentice-Hall (Singapore) Pte. Ltd., *Singapore*
Editora Prentice-Hall do Brasil, Ltda., *Rio de Janeiro*

Acknowledgments

I wish to express my thanks to Thomas J. Nordby, a former student, who took an interest in my work on the book and helped me in preparing lab specimens for photomicrography. I am grateful to all individuals who served as models for photographs.

J.E.N.

I wish to express my thanks and appreciation to the staff of Prentice Hall, especially Ms. Michelle Churma, who was very patient, understanding, and helpful in guiding the effort of producing this book.

I also wish to thank Mr. William James for his diligent updating and editing of Chapter 22, "Plastics and Elastomers"; and Mr. Mike Bastian of Accurate Steel Treating, South Gate, CA, for his help in providing heat treated samples for metallography along with photographs of their new "ECO–GLO" ion-plasma surface hardening processing equipment.

My thanks to Mr. and Mrs. Hod T. Finch for their confidence and inspiration.

T.J.B.

The following industries and people have been most helpful in supplying illustrations and technical information.

Accurate Steel Treating Inc., South Gate, California
American Chain & Cable Company, Inc., Wilson Instrument Division, Bridgeport, Connecticut
Aluminum Company of America, Pittsburgh, Pennsylvania
American Iron & Steel Institute, Washington, D.C.
American Society for Metals, Metals Park, Ohio
Armco Steel Corporation, Middletown, Ohio
Bethlehem Steel Corporation, Bethlehem, Pennsylvania
Bohemia Incorporated, Eugene, Oregon
Buehler, Ltd., Evanston, Illinois
Chilton Company, Radnor, Pennsylvania
W. C. Dillon & Company, Inc., Van Nuys, California

DoAll Company, Des Plaines, Illinois
Dover Publications, Inc., New York, New York
The Eugene Register-Guard, Eugene, Oregon
Hitchiner Manufacturing Company, Inc., Milford, New Hampshire
HPM Corporation, Mount Gilead, Ohio
John J. Hren, Department of Materials Science and Engineering, University of Florida, Gainesville, Florida
The International Nickel Company, Inc., New York, New York
Investment Casting Corporation, Springfield, New Jersey
John Wiley & Sons, Inc., New York, New York
Kennecott Copper Corporation, Salt Lake City, Utah
Lane Community College, Eugene, Oregon
Lindberg Sola Basic Industries, Chicago, Illinois
The Lincoln Electric Company, Cleveland, Ohio
Linn-Benton Community College, Albany, Oregon
Magnaflux Corporation, Chicago, Illinois
Megadiamond Industries, New York, New York
Mt. Hood Community College, Gresham, Oregon
Pacific Machinery & Tool Steel Company, Portland, Oregon
R. A. Parham of the Institute of Paper Chemistry, Appleton, Wisconsin
Republic Steel Corporation, Cleveland, Ohio
The Shore Instrument & Manufacturing Company, Inc., Jamaica, New York
Stork-MMA Laboratories, Huntington Beach, California
Tinius Olsen Testing Machine Company, Inc., Willow Grove, Pennsylvania
United States Steel Corporation, Pittsburgh, Pennsylvania
Universal Alloys, Anaheim, California
Wah Chang Albany, Albany, Oregon
Western Wood Products Association, Portland, Oregon
Weyerhaeuser Company, Springfield, Oregon

Contents

Preface

This textbook on metallurgy and materials is ideally used as an introductory book for both materials science and metallurgy courses and for students whose majors are highly related, such as quality control, machine tool technology, welding technology, and many others. *Practical Metallurgy and Materials of Industry,* 5th edition, now includes many of the latest industry processes which change the physical and mechanical properties of materials and is highly recommended as a "materials processing" reference handbook in support of design, process, electrical and chemical technicians, and engineers.

The book is intended to be easy to read. We make a special effort to explain the complicated metallurgical terms in basic shop language when possible. Although the terms used by the metallurgist can be baffling to the student, terms such as *austenitizing, retained austenite, allotropic transformation, eutectoid, passivation,* and hundreds more are necessary because they avoid extensive explanation when referring to the concept. Only those terms deemed necessary to convey the thought or idea are used, and we make an effort to explain each new word or term in the text; an extensive glossary is also included.

With a basic knowledge of the behavior, characteristics, and commonly used processing of metals and materials, our intent is to provide a solid platform of information for the responsible technician and/or engineer upon which to base materials decisions. Workers who handle the materials should understand their behavior and characteristics to avoid sometimes costly difficulties.

The highly visual approach in this book uses graphics, drawings, illustrations, and photographs of actual shop applications of metallurgy. These aids will help the reader to comprehend the many abstractions that must be used. In the study of metals, metallurgists and students observe the tiny grain structures that are in metals by sectioning and then polishing a surface. It is then etched with an acid or other chemical in order to make the grains visible. A microscope is used to enlarge the view of those structures. Photos are often taken. In this text, photomicrographs are often used to show the differences in metals when they are subjected to certain conditions such as heating, forming, or forging. The first few chapters deal with the extraction of metals from their ores, the identification of irons and steels, along with basic application of metallurgical processing, manufacturing and welding practice. The next chapters deal with the introduction and metallurgical processing of nonferrous metals and alloys, followed by chapters on precious metals, composites, adhesives, elastomers, wood and paper products, and failure analysis.

This revised edition of *Practical Metallurgy and Materials of Industry* was prepared for the purpose of providing a more comprehensive explanation of metals and materials processing by updating the text in view of changing technology. Many chapters have been reorganized and expanded. Color illustrations have been added. Service-related real-life problems and solutions have been integrated into the general text, and the case problems offered at the end of each chapter challenge the student to participate in problem solving as part of a self-evaluation process of learning by using information within the subject chapter. The case problem provides a practical application of the chapter information. Solutions to all the case problems are in the Instructor's Manual.

The chapters in this revised edition have been reorganized in a manner more closely aligned with the historical development of metallurgy, keeping ferrous metals somewhat separated from nonferrous metals. A reinforcement approach to instruction is used through the book by building on previously covered information and by encouraging the student to read the material, use the worksheets, read the case problems, and participate in the self-evaluation section at the end of each chapter. The format of the book is adaptable to either the conventional lecture, lab, or individual approach to instruction. Objectives, chapter text, self-tests, and worksheets are intended to help both the student and the instructor: All these are consistent; that is, testing is relevant to objectives and discussions. The self-test review questions are aimed toward helping the student understand the material. For this purpose, answers to the self-evaluation review questions are found in Appendix 3. Multiple-choice posttest questions and keys for most chapters are provided in the Instructor's Manual, which is available to any instructor who is using this textbook. Instructions are also given in the Instructor's Manual on how to set up a simple metallurgical laboratory and on the various methods of utilizing the book for establishing teaching programs.

Our hope for this book is that students of metallurgy, material science, and related technologies will gain a better understanding of the processes that influence the behavior of metals and materials with which they work.

John E. Neely/ Thomas J. Bertone

Introduction

The ability of human beings to make and use tools is the single most important reason for the tremendous progress leading to our present technological age. The first humans made tools of wood, bone, and stone (Figure I.1). However primitive and crude, the first discovery of metal most likely occurred with humans finding pieces of meteorites and with time, learning how to pound and sharpen meteorites into useful tools. The history of the seven Metals of Antiquity can trace origins back to ~6000 B.C. when humans first discovered gold, and then silver, copper, iron (iron-nickel meteorites), tin, and mercury in a nearly pure or native state. Copper, a red metal, was discovered ~4700 B.C. and widely used for armament and tools by the Mesopotamians, Egyptians, Greeks, Bolivians, Romans, Indians, and Chinese. Copper was melted and combined with beryllium by ancient Bolivians, using a long lost alloying process which scientists today have not been able to duplicate. Silver was discovered ~4000 B.C. and widely used with gold as old-world coinage. Lead was discovered ~3500 B.C. The Roman Empire used this abundant, easy-to-form metal to line drinking utensils, pipes, and aqueducts. Tin was discovered in ~1750 B.C. and was melted together, alloyed with copper for the Egyptians' decorative purposes to harden and strengthen copper. Scandinavians discovered rudimentary methods to extract iron from ores by chance observation, finding a puddle of metal at the bottom of campfire pits. Humans quickly associated metal-rich ores as the source of the puddle metal and cleverly observed how wind increased the temperature of fire and the quantity of metal. The addition of a bellows and dried wood to the fire allowed humans to reach the higher temperatures required for extraction of metal from their ore, which thus began the technology of extractive metallurgy. The process to reduce iron ore is believed to have been discovered by the Chinese about 2000 B.C. Mercury, discovered ~1600 B.C., was referred to by the ancients as "quicksilver." The discovery and useful implementation of the Metals of Antiquity provided mankind with the tools necessary for the development and growth of technology and creation of the modern world.

Nearly everything we need for our present-day civilization depends on metals. Vast quantities of steels, aluminum, titanium, copper, and nickel alloys are used for automobiles, ships, aircraft, spacecraft, bridges, and buildings and the machines required to produce them (Figure I.2). Almost all uses of electricity depend on copper. All around us we see the utilization of aluminum, copper, steels, and often new applications combining metals with plastics and composite materials. Some metals

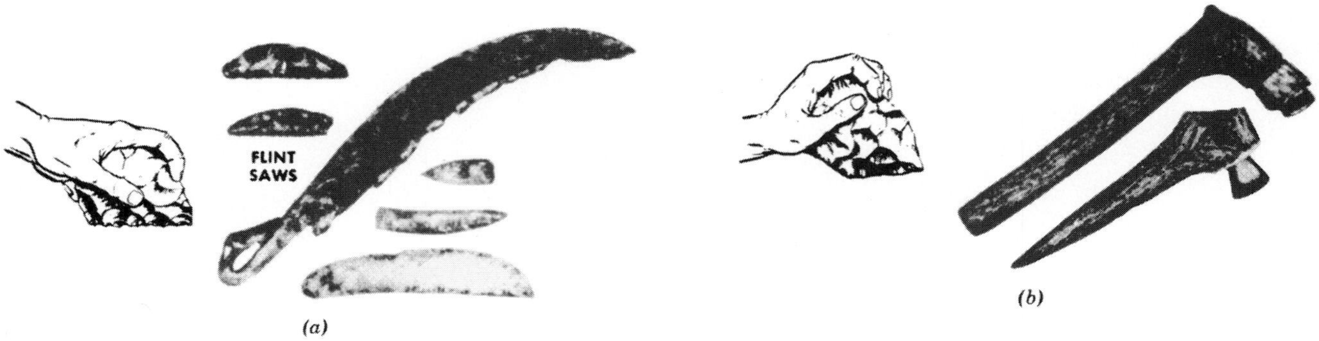

FIGURE I.1. (*a*) Stone saws and (*b*) stone hatchets. (*Courtesy of DoAll Company*)

FIGURE I.2. Large-scale production of metal products is achieved in modern steel mills. (*Bethlehem Steel Corporation*)

such as titanium and zirconium, impossible to smelt or extract from ores just a few years ago, are now used in large quantities and referred to as "space-age" metals. There are also hundreds of combinations of metals and nonmetals called alloys, and many new tool steels.

Indeed, because the use of metals in the present-day world is so heavy, many have been designated as "strategic metals" such as chromium, cobalt, manganese, and others due to their short supply. These metals are heavily traded, commanding high prices in the world marketplace. Iron and aluminum are more plentiful; however, even the production of these metals is affected by the increasing cost of energy that contrives to create man-made shortages. Aluminum requires a great deal more energy per ton to extract than steel, and titanium requires nearly 10 times more energy to produce than steel. Aluminum and titanium alloys have been substituted for steel in ever increasing applications, meeting the demands of high technology that require better corrosion resistance combined with a high mechanical strength to weight ratio.

Gold, copper, silver, and some other metals were known to the ancient people because these metals were often found near or at the surface of the ground as "native or pure" metals. Nuggets of soft pure metal could be easily shaped by hammering to produce ornaments and items of jewelry. Humans soon discovered the process known as work hardening since some metals become harder when hammered or plastically deformed and would hold an edge. Gold and silver are too soft to make cutting tools because they do not work harden. Native copper, however, could be hammered or forged into tool shapes, and as a result, copper became a primary armament metal for the Mesopotamians and Egyptians. The discovery of tin, along with enhanced mechanical properties and hardness when alloyed with copper, gave rise to the Bronze Age which lasted for many centuries to the time of the Roman Empire and the use of iron for tools.

Meteoric iron was probably the first ferrous metal used by the ancients. Modern chemical analysts have discovered meteoric iron to be composed of an iron-nickel alloy. Nearly all the early names for *iron* meant "stone from heaven" or "star metal." Iron is only rarely found as native iron, and since meteors are also rarely found, little use of iron was made until a means was discovered to extract it from ore by the Chinese about 2000 B.C. and in Asia Minor about 1500 B.C. Recent discoveries of ancient man-made iron artifacts in China and in India dating further in time than 3500 B.C. now have scientists revising the time frame in which humans are believed to have discovered how to reduce iron ore into usable items. In ancient times, iron was produced in primitive furnaces or forges. The white-hot heat required to reduce iron ore such as hematite or magnetite into iron was accomplished by forcing air through burning charcoal in close proximity to the ore. This process produced a product sometimes referred to as "sponge" iron by ancient historians, because the iron mass was not molten—although within the sponge, impurities called "gangue" (mostly silicon) were molten due to their lower melting point.

By mashing and squeezing the spongy mass, the molten gangue was nearly removed, leaving a white-hot mass of iron which could be further pounded or forged into a solid bar. The wrought iron which was produced contained a high percentage of silicon, giving the metal a tough fibrous quality. In combination with silicon, the very low carbon content of the wrought iron provides excellent corrosion resistance; in fact, some of the wrought iron nails used in Viking ships buried over a 1000 years ago remained intact when recently unearthed.

Humans have come a long way from the first discovery of copper and iron, into a world of industrial revolution, where the development and refinement of extractive processes created metals available for industrial application. Prior to the Industrial Revolution, for hundreds of years, humans made very limited quantities of steel weapons, processed by secret religious rites, such as those used to produce the Japanese samurai sword. The Industrial Revolution brought about a transition of metallurgy from art to science, with improvements in methods of ore processing and metal refinement. The first version of the modern rolling mill was invented by Cort in 1783. Wrought iron was the primary iron produced by humans until the invention of the Bessemer converter in 1855, making crude alloy steel available in support of the railroad industry. The "scientific" mankind began to study and expand on knowledge of metals with the development of the metallurgical microscope by Sorby in 1864. This new vision into the structure of metals provided humans with a tool for basic understanding of the microstructural changes which take place in metals during heat treatment and thermal fabrication.

Modern metallurgy stems from the ancient desire to understand fully the behavior of metals. Long ago, the art of the metal worker was enshrouded in mystery and folklore. Crude methods of making and heat treating small amounts of steel were discovered by trial and error. Unfortunately these methods were often forgotten and had to be rediscovered when the metal craft was not handed down to the next of kin (see Figure I.3). Our progress during the Industrial Revolution has led

FIGURE I.3. Just prior to the Industrial Revolution, iron working had become a highly skilled craft. (*Dover Publications, Inc.*)

us from those early open forges which produced 20 or 30 pounds of wrought iron a day to our modern production furnaces producing nearly 150 million tons of steel yearly in North America.

In this book you will learn how metallic ores are smelted into metals and how these metals are combined into alloys which are cast or wrought into the many diversified products needed for our society today and in the future. In addition, the metallurgical processes are presented to challenge the student of metallurgy and material science in a self-evaluation format with learning exercises and case problem evaluations with practical examples of failure analysis and resolution to service problems. When appropriate, safety precautions are given for the safe handling of materials and metallurgical equipment such as abrasive cutoff saws, acids, and other reagents used for etching specimens. Also, the toxicity of certain metals and materials are discussed.

The study of metallurgy is needed not only by students of metallurgy and material science, but also by the welding, machine shop, quality control, and industrial technology students who equally share responsibility for the design, development, and implementation of metals and materials processing in industry today. The chapters are presented in a developmental format intended to give you a greater understanding and awareness of the physical properties of metals, including the influence of metallurgical processes such as smelting, casting, forging, welding, machining, and heat treatment on the character and physical behavior of metals.

This textbook is intended to give you a greater understanding of metals and materials of industry and an awareness of their properties and behavior. This knowledge should help you become far more capable and efficient in your chosen profession.

Ductile Failure by Tensile Overload

CHAPTER 1

Extracting Metal from Ores

Iron is the fourth most plentiful element in the earth's crust. It is almost never found in its native or metallic state, but as part of various mineral compounds called ores. The separation of metals from their ores is known as extractive metallurgy. This chapter will describe how iron ores are processed and converted into pig iron in blast furnaces. It will also describe various methods of converting pig iron into steel. As with other technologies today, iron and steel making are undergoing rapid changes to compete in the world markets. Older methods of reduction from ores and purification of extracted metals are discussed, because many are still in use. Some of the newer processes are gaining in use, and new iron and steel plants are being built to accommodate them. Still others are in the experimental stage and are briefly mentioned in this chapter. Pig iron, cast iron, and steel are known as ferrous metals, and all other metals are known as nonferrous metals. A second section of this chapter is devoted to the description of the extraction of copper, aluminum, and other nonferrous metals from their ores.

OBJECTIVES **After completing this chapter, you should be able to:**

1. List the various steps, basic materials, and principles involved in making pig iron.
2. Identify various steel-making processes.
3. Explain several processes used in producing nonferrous metals.

MINING NATURAL ORES OF IRON

Iron ores are found all over the world, but in the past only certain deposits were considered rich enough in iron to be mined. Not many years ago, most iron and steel makers in the United States would not have considered mining an ore with an iron content of less than 30 percent, particularly if the mineral were difficult to process. Today, however, a mineral called taconite is a primary source for making pig iron in blast furnaces in this country. From Table 1.1, you can see that the iron content of taconite ranges from 25 to 35 percent. In the United States, for many years the Lake Superior Mesabi Range has produced hematite ore, which contained a high iron content up to 68 percent, but this source has since been depleted. Table 1.1 lists some minerals of iron and their chemical formulas. (*Note:* The chemical symbol for iron is Fe, oxygen is O, hydrogen is H, and carbon is C.)

These ores are removed by open-pit mining (Figures 1.1*a* and 1.1*b*) as contrasted with the more costly

TABLE 1.1. Some natural ores of iron

Name	Formula	Iron Content (Percent)
Magnetite	Fe_3O_4	72.4
Hematite	Fe_2O_3	68.0
Limonite	$2Fe_2O_3, 3H_2O$	59.8
Goethite	Fe_2O_3, H_2O	62.9
Siderite	$FeCO_3$	48.2
Taconite	Fe_3O_3	25–35

***FIGURE 1.1*a.** Open-pit mining showing a 12 1/4-in. rotary drill in the foreground. (*Kennecott Copper Corporation photograph by Don Green*)

***FIGURE 1.1*b.** A 25-yard shovel loading a 150-ton truck in an open-pit mine. (*Kennecott Copper Corporation photograph by Don Green*)

and dangerous underground mining method. When the ore has been removed from the mine, it is cleaned and separated from the gangue, or worthless rock, by a process called ore dressing. This process could be carried out by any of several methods such as flotation, agglomeration, and magnetic separation. By these processes, low-grade ores such as taconite are upgraded and pellitized before being shipped to the steel mill. See Appendix 1, Table 9, "Some Common Ores of Industrial Metals."

PRODUCTION OF PIG IRON

The iron ore is converted into pig iron in a blast furnace. Figure 1.2 shows the operation of a blast furnace. The three raw materials, **iron ore, coke,** and **limestone,** are put into the furnace alternately at intervals, thus making the process continuous. About 2 tons of ore, 1 ton of coke, and 1/2 ton of limestone are required to produce 1 ton of iron.

One of the three major ingredients in the production of pig iron is coke, a residue left after certain soft coals have been heated in the absence of air. When coal is heated in coke ovens and the resulting gases are driven off (Figure 1.3), coke is the result. Coke is a hard, brittle, and porous material containing from 85 to 90 percent carbon, together with some ash, sulfur, and phosphorus. An older type of coke oven, called a beehive oven because of its shape (Figure 1.4), is now obsolete because it wasted gases that were produced during the process. Many useful products are made from the coke oven gas that is driven off: fuel gas, ammonia, sulfur, oils, and coal tars. From the coal tars come many important products such as dyes, plastics, synthetic rubbers, perfumes, sulfa drugs, and aspirin.

Reduction is a process by which oxygen (O) is removed from a compound, such as an iron ore, and combined with the carbon (C). Hence, when iron ore and coke are put into the blast furnace, the metallic iron is released from its oxide state (formulas in Table 1.1) by reduction. The solid materials—coke, limestone, and ore—enter the blast furnace through the hopper at the top, and heated air is blasted in at the bottom. This air is heated in stoves composed of refractory brick before it enters the furnace. Refractory materials have a high melting point to enable them to retain their shape and strength at high temperatures.

In the furnace, the fuel burns near the bottom, and the heat rises to meet the descending charge of coke and ore. At very high temperatures (about 3000°F or 1649°C), the coke unites with the oxygen in the air blast and is converted to carbon monoxide gas. The carbon monoxide (CO) rises to near the top of the charge in the furnace where it unites with the iron oxide to produce iron (Fe) and carbon dioxide (CO_2). The carbon dioxide gas combines with some of the remaining coke to form CO again. The flux (limestone) decomposes to form lime and carbon dioxide ($CaCO_3 \rightarrow CaO + CO_2$). Lime (basic) combines with silica gangue (acid) to form the flux—calcium silicate $(CaO) \cdot SiO_2$).

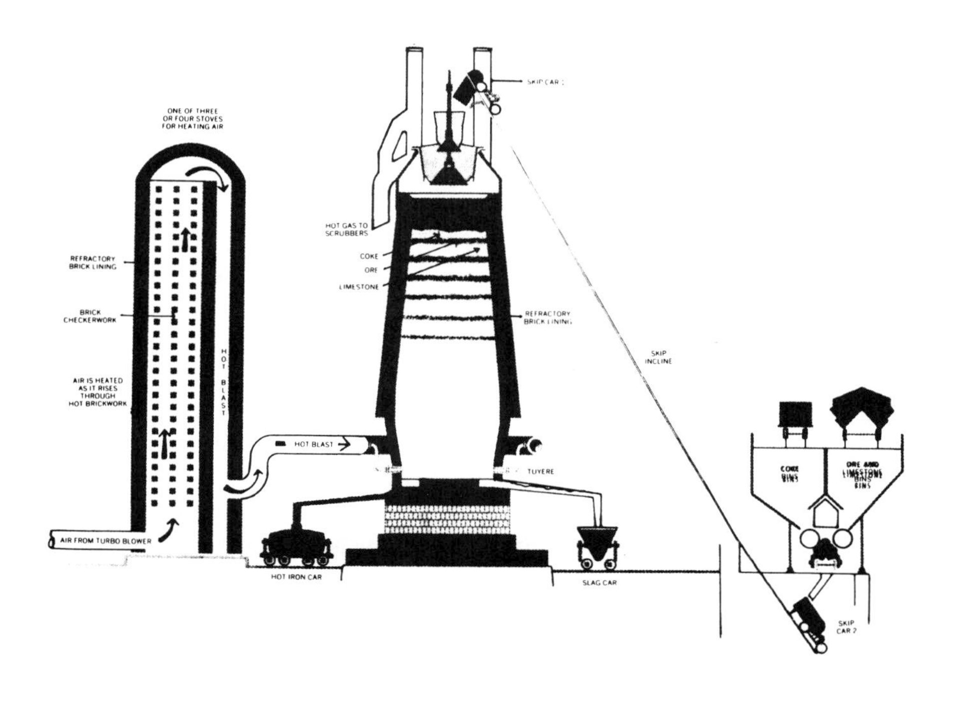

FIGURE 1.2. Blast furnace. Iron ore is converted into pig iron by means of a series of chemical reactions that take place in the blast furnace. (*Bethlehem Steel Corporation*)

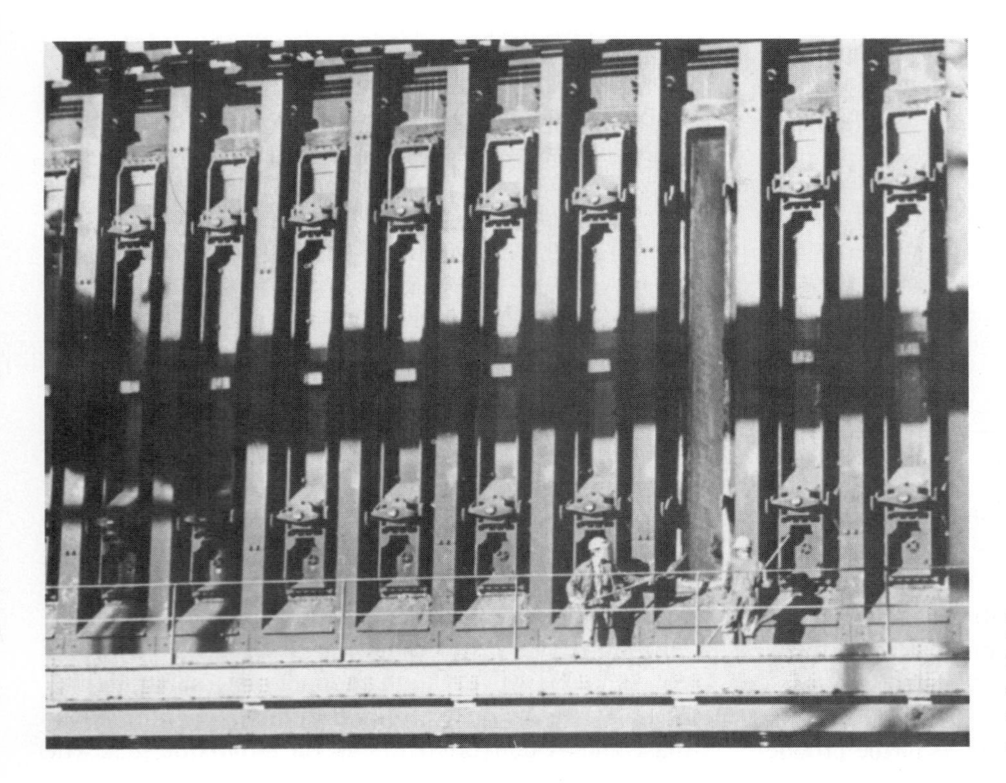

FIGURE 1.3. Between each pair of vertical dividers in this coke oven battery is an individual oven in which coal is heated and converted to coke. (*Courtesy of American Iron & Steel Institute*)

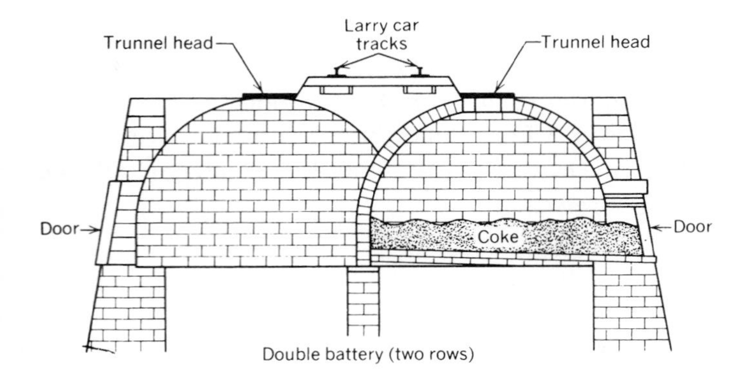

FIGURE 1.4. Beehive coke oven. The coal is charged through the trunnel head from which the gases escape. This method is obsolete because it was wasteful. The gases are not collected and utilized but are lost through the upper vent.

As the process continues, iron is released to the bottom of the furnace, where it remains a molten mass. Here the limestone is used to separate the impurities (mostly SiO_2) from the iron by combining to form a lower temperature melting compound. Because this waste slag is lighter in weight than the iron, it floats on top and is drawn off periodically to be hauled away in slag cars. The slag is sometimes ground into an aggregate that is used for asphaltic concrete roads and in concrete building blocks.

The molten iron at the bottom of the furnace contains from 3 to 4.5 percent carbon, 4 percent silicon, up to 1 percent manganese, and some other unwanted impurities such as phosphorus and sulfur. These are later removed to some extent by other refining processes. The reduction of iron ore into pig iron is actually an intermediate step in the manufacture of steel. The iron is tapped (drawn off) at intervals and collected in a transfer car, which is insulated so that the molten metal will stay hot (Figure 1.5). It is then moved to the steel furnaces and added to the charge of scrap steel, ore, and limestone. Sometimes the iron is not made into steel but is instead poured directly into molds. Before pig casting machines were developed, the iron was poured in open sand molds consisting of a groove or trough with many small molds on each side, reminding one of a sow and pigs, hence, the name "pig iron" (Figure 1.6). Iron pigs are remelted in cast iron foundries and in steel mills.

FIGURE 1.5. Tapping the blast furnace. Blast furnaces operate continuously. The molten iron is tapped every 5 to 6 hours. (*Bethlehem Steel Corporation*)

FIGURE 1.6. Some blast furnace iron is poured to solidify in molds in the automatic pig casting machine. Iron foundries are major users. (*Courtesy of American Iron & Steel Institute*)

SPONGE IRON AND POWDER

Iron powder has been produced directly from the ore since the 1920s and is used to produce small parts by forming them under high pressure and sintering them in a furnace, that is, heating them without melting. Many other systems operating today convert the ore directly into the pellet form (Figure 1.7).

FIGURE 1.7. Iron pellets produced by direct reduction.

The HyL, Midrex, and SL/RN systems are some of the many processes in which small pellets of sponge containing about 0.95 percent iron are produced. The pellets can be remelted in steel-making furnaces such as the basic oxygen furnace, but most of this iron is shipped to electric furnaces. Only a few million tons of sponge iron are produced yearly at present as compared with over 55 million tons of pig iron produced in blast furnaces in the United States. There is an increasing demand for sponge, and a rapid increase in its production can be expected in the near future. This is a result of the fact that iron can be mined in remote areas in the world and shipped economically as sponge pellets to industrial nations.

WROUGHT IRON

In ancient times, iron was produced by the process of direct reduction; that is, the ore was heated in a forge to a white heat to remove the impurities and to produce iron from the ore. The charcoal fire in the forge did not get hot enough to melt the iron; the result then was a pasty mass of "sponge" iron that was then hammered to remove the molten gangue or slag. The result was soft wrought iron containing little or no carbon. This method of smelting iron ore has not been used commercially since the Middle Ages.

Before the processes of modern steel making were known (over 100 years ago), wrought iron was used extensively for bars, plates, rails, and structural shapes for bridges, boilers, and many other uses. Wrought iron was made by a puddling process in which pig iron was melted in an open-hearth type of furnace. Iron oxide was then added to form a slag. The iron was slowly cooled to a pasty mass and puddled or stirred with "rabble arms" by hand. The carbon and other impurities were removed by the iron oxide slag. The mass of iron was removed and squeezed to remove the slag; however, much of it remained in the iron. The result was a very low carbon, fibrous (from the trapped slag), soft iron. This process is obsolete, but wrought iron is now produced by the Aston process in which molten pig iron and steel are poured into an open-hearth furnace in a prepared slag; this cools the steel to a pasty mass that is later squeezed in a hydraulic press. Wrought iron is used for ironwork such as on railings for stairs and for the production of pipe and other products subject to deterioration by rusting. Advantages of this metal are high ductility (can be deformed easily without splitting), good weldability, and corrosion resistance. Even though direct reduction of iron ore to produce wrought iron as it was practiced in ancient times had become obsolete, recently the direct reduction process has been greatly improved, and currently several processes are being used.

STEEL-MAKING PROCESSES

Because pig iron contains too many impurities to be useful, it must be refined to produce steel or cast iron of various types. Steel is simply an alloy of iron with most of the impurities removed; that is, it contains 0.05 to 2 percent carbon plus any other alloying elements. Over 90 percent of all steel produced is classified as plain carbon steel; its carbon content is controlled, usually under 1 percent. Different alloy steels comprise the remaining 10 percent of all the steels produced.

Small amounts of manganese are added to remove unwanted oxygen and to control the sulfur and other impurities that still remain. Sulfur is difficult to remove from steel, and iron sulfide (the form it naturally takes in iron) causes steel to be hot-short; that is, it tends to split when forged or rolled at forging temperatures because the stringers of iron sulfide are weak and do not adhere to the steel at that temperature. If manganese is present, it combines more readily with the sulfur, forming manganese sulfide, which has greater strength at forging temperatures. Thus, no steel is made without some manganese.

In some resulfurized steels, sulfur is deliberately added along with additional manganese. This is done to make a free-machining steel. Manganese sulfide not only lubricates the chip being cut off in a machine tool, but it also breaks up the chips—both of which make for ease of cutting. Lead is also sometimes added to steel to make it free-machining.

Steel can be given many different and useful properties by alloying the iron with other metals such as chromium, nickel, tungsten, molybdenum, vanadium, and titanium, and with nonmetals such as boron and silicon.

The major steel-making processes are **basic oxygen, open hearth, Bessemer,** and **electric.** The Bessemer converter (Figure 1.8) uses air to burn out excess carbon and some impurities. Although the converter is fast (20 minutes to a blow, that is, an air blast forced through the molten iron), very little steel (none in the United States) is made by this process today, as it produces a low-grade product. The Bessemer process is limited to using pig iron for steel making. Small, measured amounts of carbon are also added to the molten metal after the blow, thus producing a carbon steel. Although this process is obsolete, it is mentioned here because of its historical significance, showing the development toward modern steel-making processes.

A modern development that resembles the old Bessemer converter is the basic oxygen furnace (BOF) (Figures 1.9*a* to 1.9*c*). It is designed to make steel of high quality in a very short period as compared with the open-hearth process (discussed later in the chapter). The basic oxygen process uses a lance that blows oxygen down from the top of the furnace to burn out the impurities (Figure 1.10). It has the added advantage of being able to use fairly large amounts of scrap, unlike the old Bessemer process. In 1989, the total world production of raw steel was 862,600,000 tons. The United States'

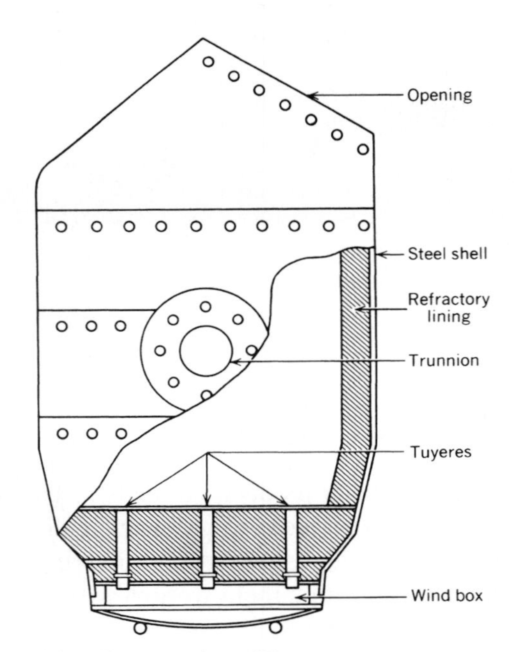

FIGURE 1.8. Cutaway view of Bessemer converter.

***FIGURE 1.9*a.** Molten iron from the blast furnace is charged into one of two basic oxygen furnaces at Bethlehem Steel Corporation's Bethlehem, Pa., plant. After the charge has been completed, the vessel will return to its upright position for the oxygen "blow," during which the blast furnace iron, mixed with scrap and selected additives, will be refined into steel. (*Bethlehem Steel Corporation*)

***FIGURE 1.9*b.** Basic oxygen furnace in Figure 1.9*a* is charged with scrap. The charging machine empties scrap into the vessel before the hot metal is added. In addition to positioning and charging the loaded boxes of scrap, the charging machine also positions the heat shield that is utilized for testing during furnace turndowns. (*Bethlehem Steel Corporation*)

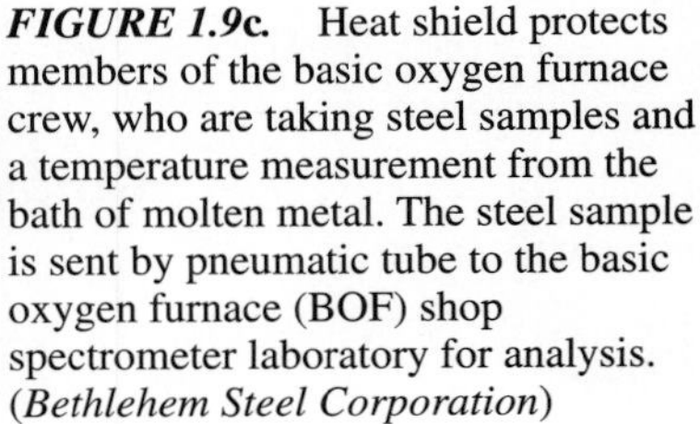

FIGURE 1.9c. Heat shield protects members of the basic oxygen furnace crew, who are taking steel samples and a temperature measurement from the bath of molten metal. The steel sample is sent by pneumatic tube to the basic oxygen furnace (BOF) shop spectrometer laboratory for analysis. (*Bethlehem Steel Corporation*)

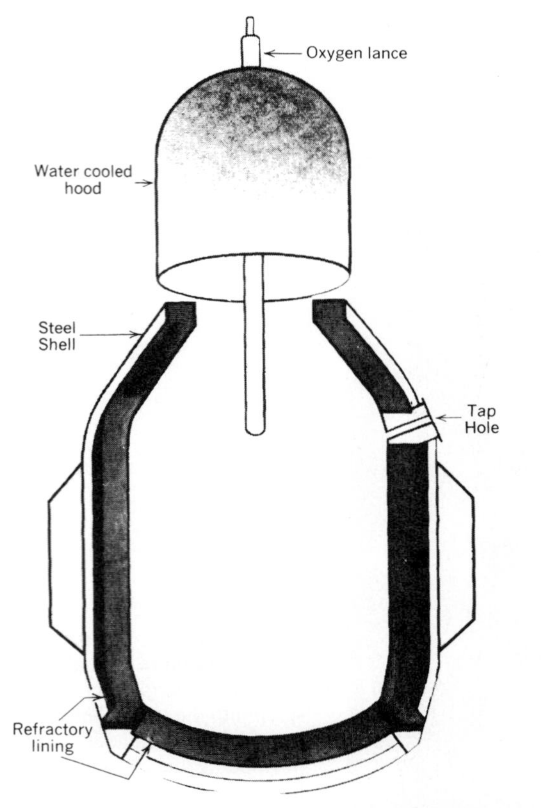

FIGURE 1.10. The basic oxygen furnace resembles the old Bessemer converter in appearance, but it is an entirely new development, designed expressly to get the best results by the use of oxygen in steel making. Steel of excellent quality can be made at a high rate of speed—about one heat an hour. (*Bethlehem Steel Corporation*)

total production was 11.4 percent of that amount. The basic oxygen process produced 53,300,000 tons of the United States' total 97,900,000 tons.

The basic oxygen furnace can turn out steel at the rate of about one 200-ton heat per hour. A heat is a quantity of metal produced in one melting period; alloying ingredients are added under closely controlled conditions. The furnace is charged with molten pig iron, iron ore, steel scrap, and fluxing materials such as limestone that react with the impurities and form a slag on top of the molten metal. The water-cooled lance is lowered to within a few feet of the charge and then it blows a stream of oxygen at more than 100 pounds per square inch (psi) on the surface of the bath. The oxidation of carbon and impurities causes a violent agitation of the molten bath, bringing all the metal into contact with the oxygen stream. The ladle is first tipped to remove the slag and then rotated to pour out the molten steel into a ladle. Carbon and other elements are added to produce the desired quality of steel.

Today, steel is still produced in the United States by the open-hearth process (Figure 1.11), which produces a steel of high quality. In 1989, 4,400,000 tons of steel were produced by the open-hearth process. Open-hearth furnaces produce 100 to 375 tons per heat (Figure 1.12), with each heat taking from 8 to 10 hours, unless an oxygen lance is used, in which case the time per heat is about 4 hours. During this time

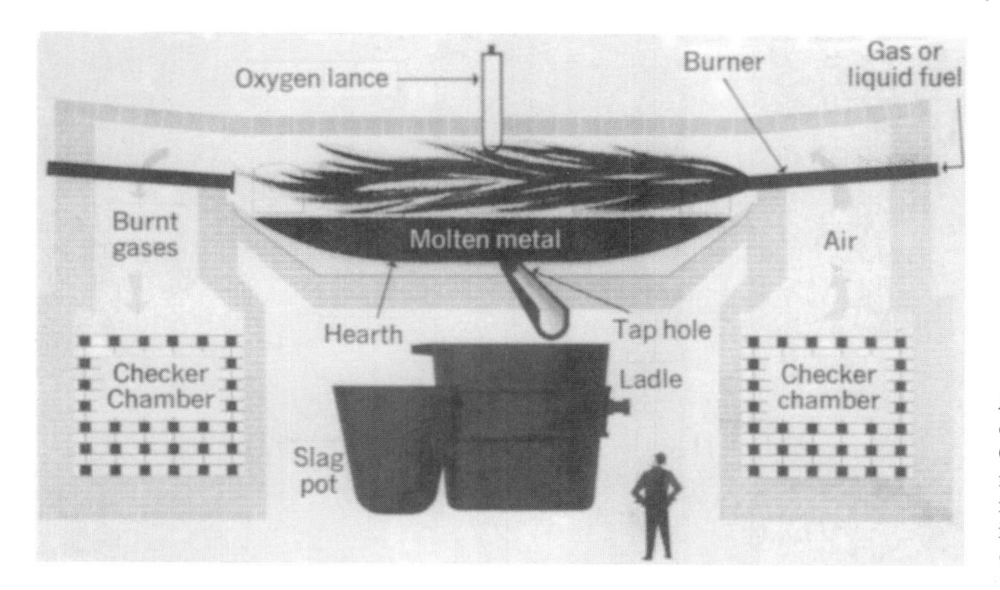

FIGURE 1.11. Simplified cutaway diagram of a typical open-hearth furnace. Oxygen may be injected through one or more lances. In some cases it is introduced through the burners to improve combustion. (*Bethlehem Steel Corporation*)

FIGURE 1.12. Open-hearth charging floor. Molten cast iron being poured from the ladle into the hearth furnace. (*Bethlehem Steel Corporation*)

molten pig iron, scrap, iron ore, and other elements such as manganese are added to control the condition of the melt or heat. Ferrosilicon is added if the steel is to be killed. A **killed steel** is one that has been sufficiently deoxidized in the ladle or ingot to prevent gas evolution in the ingot mold. This makes a uniform steel. If the heat is not killed, gas is formed in the ingot and the resultant holes remain in the ingot when the steel hardens. This is called **rimmed steel,** because the rim or outside surface of the ingot is free from defects and the flaws remain in the center. These flaws, however, are mostly removed by rolling processes, discussed in the next chapter.

When the contents in the melt have been controlled and the mixture and temperature are right, the furnace is tapped and the molten metal poured into a ladle (Figure 1.13). Adjustments in carbon content are made in the ladle, usually by adding pellets of anthracite coal. The slag on top of the molten steel overflows into the slag spout and into a smaller slag pot.

About 100 tons of molten steel are picked up by a crane and brought over a series of heavy cast iron ingot

FIGURE 1.13. Tapping an open-hearth furnace. Molten steel fills the ladle; the slag spills over into the slag pot. (*Bethlehem Steel Corporation*)

FIGURE 1.14. Teeming ingots. The steel is poured into molds, where it solidifies. (*Bethlehem Steel Corporation*)

molds. It is teemed (poured) into the molds by means of an opening in the bottom of the ladle (Figure 1.14). Aluminum is sometimes added in the mold to make the steel finer grained and less porous.

Electric furnace steel makes up about 35 percent of the total production in the United States. As with the open-hearth process, electric furnaces (Figures 1.15 and 1.16) use blast furnace iron, selected scrap, and other control elements. Where very little coal and iron ore are found, and where considerable steel scrap and cheap electricity are available, the electric furnace is a competitive producer of high-quality steel. Most of the special alloys such as stainless steels and tool steels are produced by this method. Steel foundries, where steel is melted down and remolded, use electric furnace steel to produce steel castings.

Electric furnaces produce up to 100 tons per heat. The time is determined by the amount of cold scrap and the amount of the furnace current. Scrap is charged through a door in the side or through the furnace top that swings to one side. The entire furnace tilts to pour the molten steel into the ladle.

The direct arc electric furnaces are most popular for making alloys of steel. In this type of furnace the current passes from one electrode through an arc to the metal charge and then through an arc to another electrode. Induction-heating furnaces are often used for remelting metals such as cast iron in an iron foundry. In this type a coil surrounds a crucible, and a high frequency electric current is passed through the coil, causing the charge to heat. Combinations of both direct arc and induction furnaces are also used.

A number of methods are used to purify the steel after it leaves the furnace, an operation especially

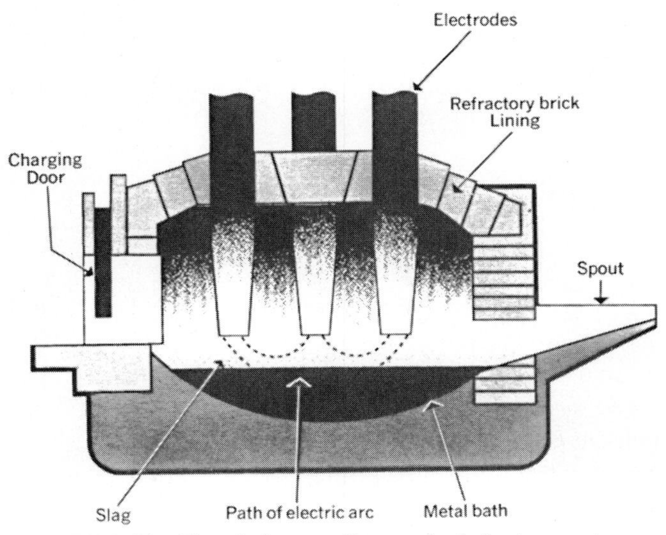

FIGURE 1.15. Electric furnace. Because both the temperature and atmosphere can be closely controlled in an electric furnace, it is ideal for producing steel to exacting specifications. The entire process takes from 4 to 12 hours, depending on the type of steel to be produced. (*Bethlehem Steel Corporation*)

FIGURE 1.16. Tapping an electric furnace, which is tilted to pour its steel into the huge ladle below. (*Bethlehem Steel Corporation*)

needed for tool steels and other high-grade steels. Two of the most important technologies are vacuum degassing and ladle injection metallurgy. Vacuum degassing is done by cycling molten steel into a vacuum chamber where unwanted gases such as oxygen, nitrogen, and hydrogen, which cause porosity in steel, are removed by the vacuum. The degassed steel is returned to the ladle and the cycle is repeated until all the steel has been exposed to the vacuum.

Exceptionally pure steels are also produced by injecting argon gas and powdered alloys and fluxes into a covered ladle that contains the molten steel. These processes not only purify the steel, lowering sulfur content and other impurities, but also increase homogeneity, improve machinability, and make possible specialized steel grades with highly predictable content and quality.

The production of iron and steel from the raw materials to the final finished products may be followed in the flowchart (Figure 1.17). The following chapter deals with steel making from the ingot stage to the formation of the many types of steel products.

RAW MATERIALS

The best iron ores have long since been depleted in the United States; but the United States has perhaps the world's greatest supply of two basic raw materials: coal and scrap metals. Coal is essential to most processes of iron smelting and it is a relatively low-cost fuel. It is estimated that there are 750 to 800 million tons of ferrous scrap lying around in the United States. It is abundant and inexpensive. However, there is one major drawback in recycling steel scrap: It must be carefully selected for use in making high-grade steel. The two most damaging contaminants in scrap steel come mostly from automobile scrap: copper and tin. Both of these metals tend to congregate in clusters during solidification and, when the metal is reheated, gather at grain boundaries, causing a weakness known as hot-shortness. Hot-short steel tends to split when hot forged or hot rolled. This happens even if only 0.2 percent or more copper is present; the copper content must be below 0.1 percent for drawing sheet steel. Copper and tin cannot be removed in the steel-making process at present, and so scrap must be

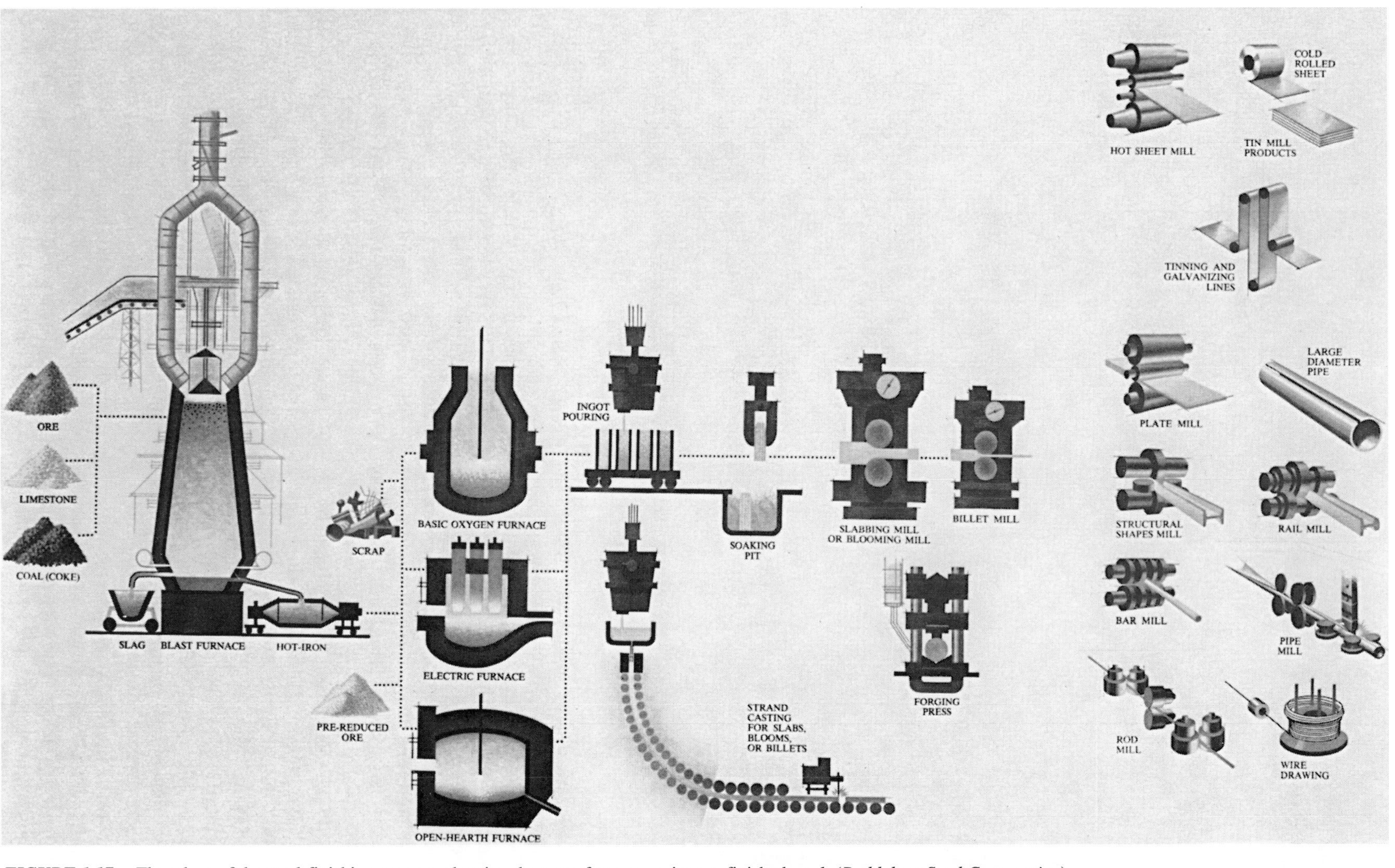

FIGURE 1.17. Flowchart of the steel-finishing process, showing the route from ore to iron to finished steel. (*Bethlehem Steel Corporation*)

carefully selected, which is a time-consuming process. If automobiles could be made without the use of copper, this scrap would be more easily recyclable. Aluminum would then be an acceptable substitute for copper, from a steel-making standpoint. For example, wire in alternators and electrical wiring could be aluminum instead of copper because aluminum conducts heat and electricity almost as well as copper.

In addition, there are environmental concerns which put restrictions on the recycling of automobiles. Containment and disposal of freon, oil, antifreeze, and other materials makes scrapping cars less profitable, possibly leading to the closing of many recycling yards.

SMELTING NONFERROUS METALS

Several kinds of furnaces are used for smelting nonferrous metals. The blast furnace, similar to but smaller than those used for iron smelting, is used for reducing oxide ores of nonferrous metals such as copper, tin, lead, and zinc. Some ores contain sulfides that are removed in roasting ovens that drive off the sulfur, leaving the oxide. Roasting is a process in which heat and air are applied to an ore or concentrate, converting it to an oxide or sulfate. Roasting is usually a preliminary operation prior to smelting many nonferrous metals. In the case of lead and zinc concentrates, removal of the sulfur is necessary before smelting. A reverberatory furnace, similar to a hearth furnace, is often used to smelt nonferrous metals. Carbon (charcoal or coke) is also used for reducing nonferrous metals, but unlike iron, it does not combine with the metal to alter its characteristics.

Another type of electric furnace in use today is the consumable electrode unit (Figure 1.18). Two basic designs of consumable electrode furnaces are the vacuum arc and vacuum induction models. High purity steels, titanium, zirconium, and various other metals are melted by these units. The metal used to charge the furnace is fashioned into long rods by methods such as briquetting sponge metal, as is done with titanium, and then welding these briquettes into an electrode. The vacuum in the furnace is constantly maintained to remove any gases before they can be absorbed by the

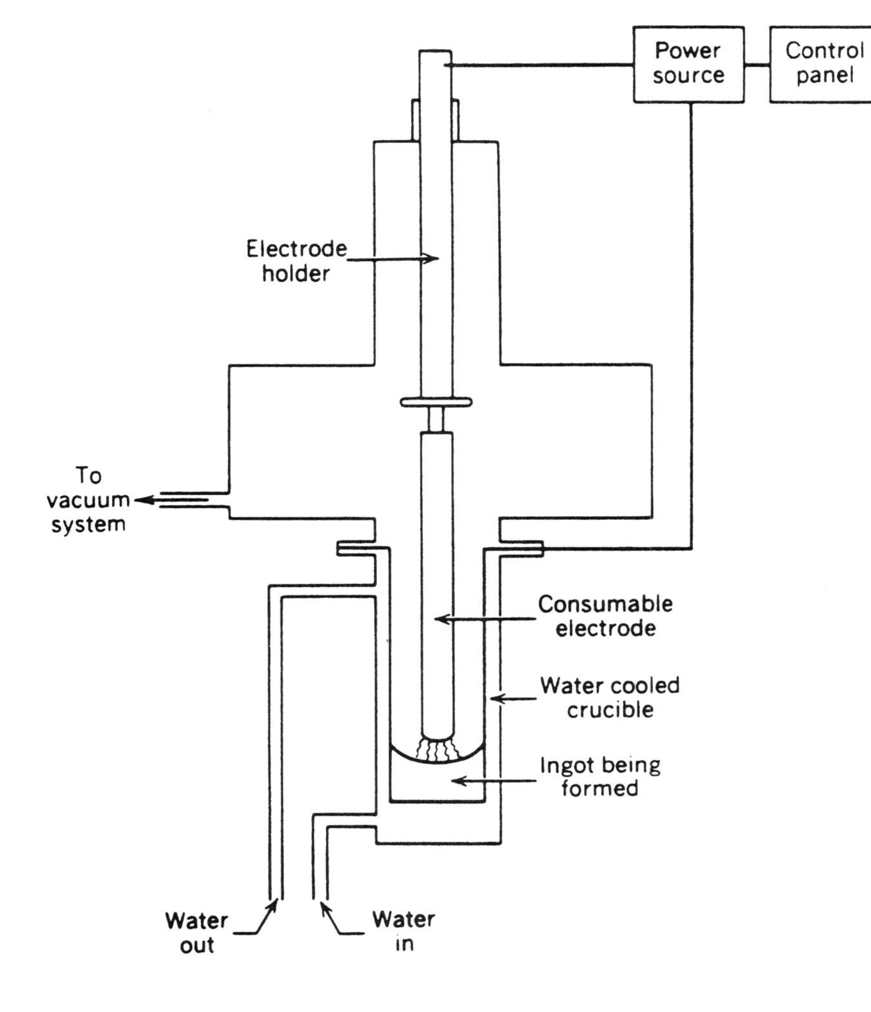

FIGURE 1.18. Vacuum arc furnace (consumable electrode).

***FIGURE 1.19*a.** Copper ores are first ground in the primary crusher. (*Kennecott Copper Corporation photograph by Don Green*)

***FIGURE 1.19*b.** The copper ore is transferred by conveyer from the primary crusher to the next step in the grinding of the ores. (*Kennecott Copper Corporation photograph by Don Green*)

metal. The metal electrode is fed into the melt at a controlled rate by which the arc length is automatically adjusted. By this process of melting in a water-cooled mold at the base of the furnace, an ingot is gradually formed. The finished ingot is removed from the mold and processed.

COPPER

The use of copper dates from prehistoric times. Neolithic communities of Eskimos and ancient dwellers in Turkey, Egypt, and North America hammered native copper into tools and ornaments. Copper is one of the

FIGURE 1.19c. The ore is finally ground in these ball mills. (*Kennecott Copper Corporation photograph by Don Green*)

FIGURE 1.20a. Copper converter. (*Kennecott Copper Corporation photograph by Don Green*)

comparatively few metals found in the "native" or metallic state. Most copper used today is extracted from various ores by smelting. Sulfide ores are ground (Figures 1.19*a* to 1.19*c*) and then require a roasting process to reduce sulfur and arsenic content. The fume given off in the process is destructive to the environment surrounding the smelter, so that collection of the fume is required. The copper oxide also contains gold, silver, and other impurities, which are later removed.

The copper ore is then smelted in a reverberatory or electric furnace to produce an impure alloy called matte. Air is then blown through the molten matte in a converter to remove any iron and remaining sulfur to produce a refined copper known as "blister" copper (Figures 1.20*a* and 1.20*b*). Further refinement is necessary to produce copper for electrical and other uses,

FIGURE 1.20b. Copper matte being charged into converter from a 50-ton ladle. The converter produces "blister" copper. (*Kennecott Copper Corporation photograph by Don Green*)

***FIGURE 1.21*a.** These 600-lb copper anodes are produced in the casting machine at the rear. These will later be placed in electrolytic cells. (*Kennecott Copper Corporation photograph by Don Green*)

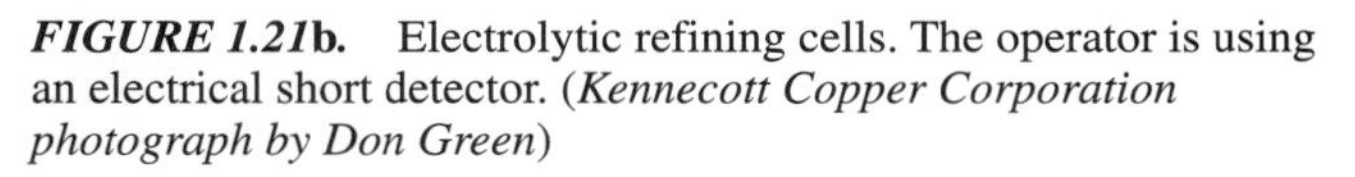

***FIGURE 1.21*b.** Electrolytic refining cells. The operator is using an electrical short detector. (*Kennecott Copper Corporation photograph by Don Green*)

FIGURE 1.21c. Copper microstructure showing equiaxed annealed grains. (*Courtesy Metals Handbook, 9th edition, Vol. 9, p. 410, picture 77, 1985. Metallography and MicroStructures, American Society for Metals.*)

which is done by electrolysis. It is at this stage that impurities such as gold and silver are collected at the bottom of the tanks and removed from the sludge by a separate process. Copper 99.9 percent pure is produced by the electrolytic process (Figures 1.21*a* and 1.21*b*). The microstructure is shown in Figure 1.21*c*.

ZINC

The chief ores of zinc are sulfides, carbonates, silicates, and oxides. One source of zinc ore in the United States is called franklinite, a zinc-iron-manganese oxide, an ore named after the Franklin furnace mines in New Jersey. Zinc ores are found principally in Missouri, Kansas, Oklahoma, Tennessee, and Idaho.

Zinc smelting begins with grinding or milling the ores to separate them from ores of other metals. The flotation process further separates out the zinc concentrates. Zinc ore may be reduced by burning off the zinc as an oxide by using the roasting process. The desulfurized ore may then be leached with acid and the metal obtained by electrolysis. The first method of obtaining zinc from the oxide, however, was by distillation since it had a low volatization point. In this process, the zinc vapor is produced in a heated, lower chamber and comes up through a small opening into a cooler, upper chamber where the distilled zinc vapor condenses on the sides, runs down, and collects to form slabs.

LEAD

Lead was smelted and used from early times for ornaments, pipes, weights, and other articles. The most common lead ore is the sulfite (PbS), known as galena. Galena is a dark, heavy mineral that is often associated with zinc and sometimes with silver.

The galena ore is ground to small pieces and the impurities are separated by flotation. One common method of smelting is the roasting and reduction process in which the sulfur is driven off by roasting and the lead oxide that remains is reduced in a blast furnace with coke to produce molten lead, which is drawn off at the bottom of the furnace.

ALUMINUM

Of all the structural metals used, such as iron, aluminum, and copper, aluminum is by far the most abundant in the earth's crust. Approximately 7.5 percent of the earth's crust is made up of aluminum. Many of the earth's ores contain aluminum, but the ore most used in the production of aluminum is bauxite, which is found in relatively few places.

The bauxite ore basically goes through two refining processes. First, a refining process breaks the ore down into its components: silicon, iron, titanium oxide, water, and aluminum oxide. This is done by crushing the bauxite to a powder and mixing it with caustic soda (sodium hydroxide, NaOH). The aluminum oxide is dissolved by the caustic soda, leaving the impurities. The aluminum hydroxide then settles to the bottom of large tanks to which it is pumped. It is then washed to remove the caustic soda. Heat is applied to drive off the water, leaving a white powder: aluminum oxide. The process of making aluminum can be seen in Figure 1.22.

In the second stage, the aluminum oxide (Al_2O_3), called alumina, is refined into metallic aluminum by the electrolytic reduction process. This is done in an electrolytic cell (Figure 1.23). Cryolite (sodium aluminum fluoride, Na_3AlF_6), a material mined in Greenland, is melted by an electric current in the cell and the alumina is placed in the cryolite bath.

The reaction that takes place in the electrolytic cell is quite simple. The electrolyte (molten cryolite)

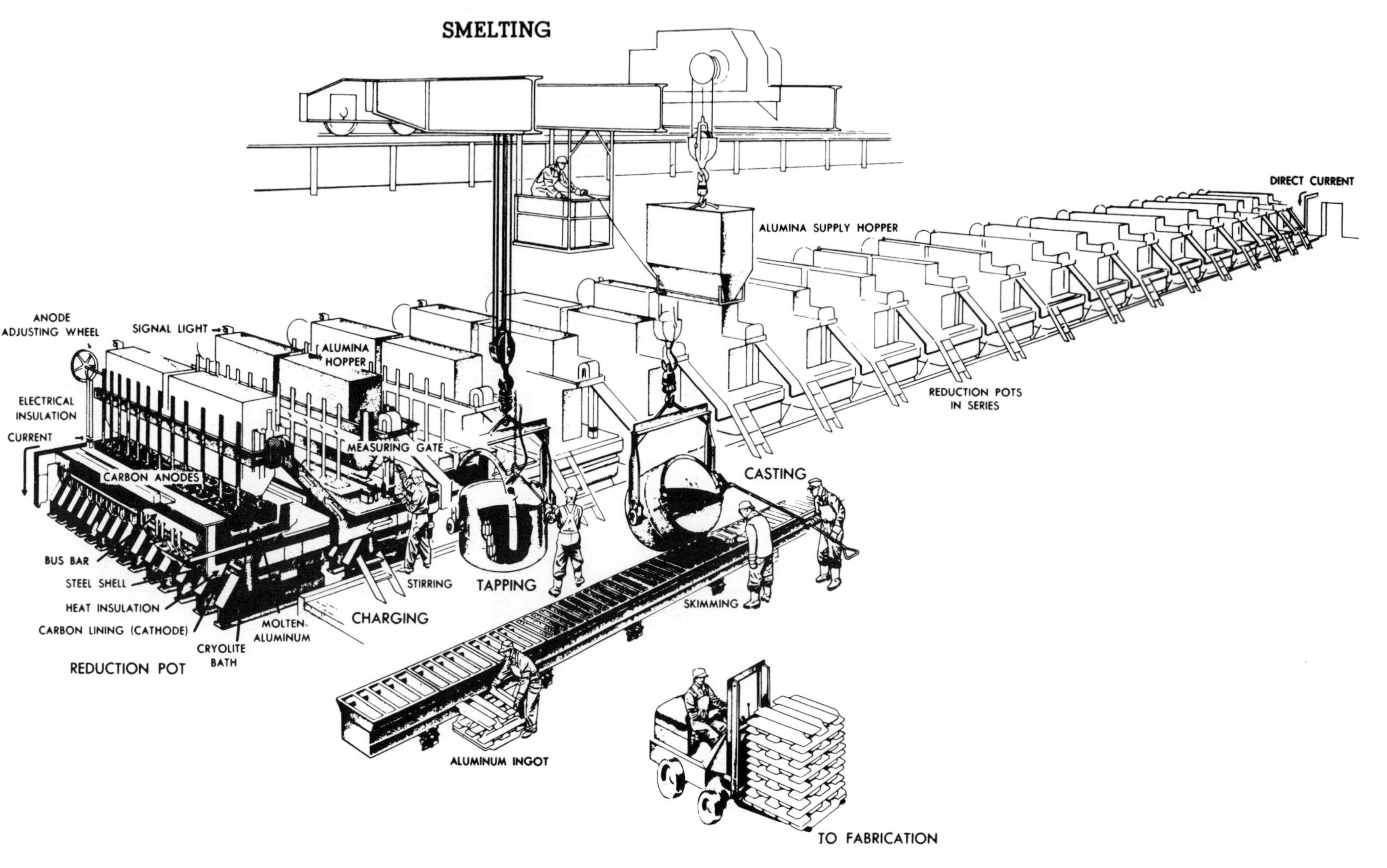

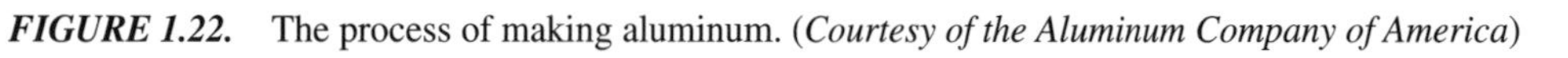

FIGURE 1.22. The process of making aluminum. (*Courtesy of the Aluminum Company of America*)

FIGURE 1.23. An electrolytic cell showing a cutaway view of the reduction pot and the carbon anodes. (*Courtesy of the Aluminum Company of America*)

dissolves the alumina. An electrical current is forced to flow from the cathode through the alumina to the anode. By the process of electrolysis, the alumina breaks down to form metallic aluminum and oxygen. The oxygen goes off as a gas, leaving metallic aluminum. The aluminum, because of its density, sinks to the bottom of the bath, where it is siphoned off at intervals and cast into ingots. Some typical microstructures of cast aluminum alloys are shown in Figures 1.24*a*, 1.24*b*, and 1.24*c*.

Many other complex processes are used to extract and process metals such as nickel, titanium, zirconium, tungsten, columbium, cerium, and many rare metals. Many publications including those of the Bureau of Mines that explain these processes in detail are available.

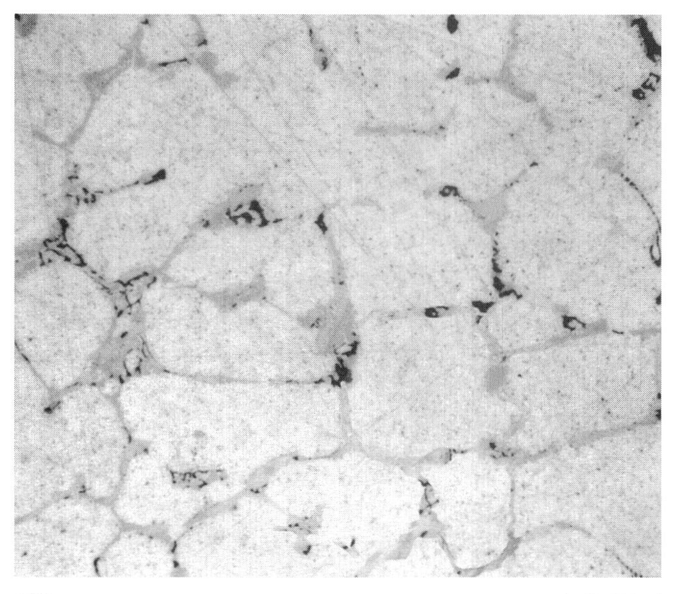

250× As-Polished

***FIGURE 1.24*a.** Investment cast aluminum alloy 356-F, with sodium modified ingot. Interdendritic structure: particles of silicon (dark gray), $FeMg_3Si_6Al_8$ (light gray script), $Fe_2Si_2Al_9$ (medium-gray blades) and Mg_2Si (black).

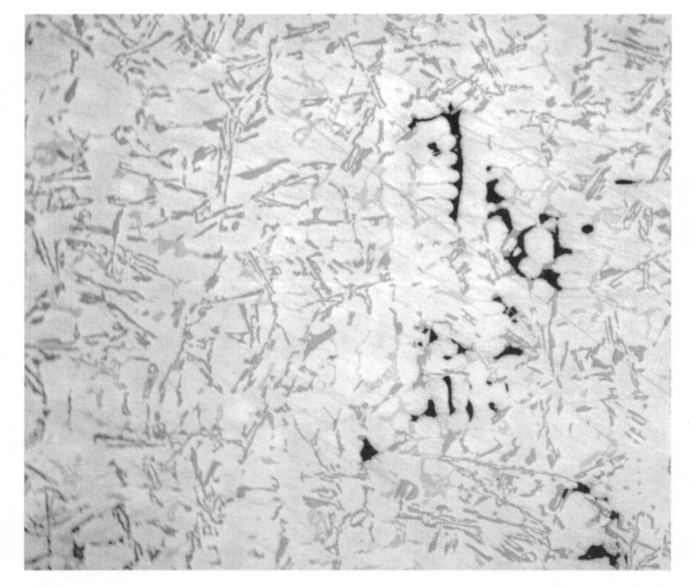

100× As-Polished

***FIGURE 1.24*b.** Sand cast aluminum alloy A356-F sand casting showing dendritic solidification with shrinkage porosity (black).

100× Keller's Etch

***FIGURE 1.24*c.** Permanent mold cast aluminum alloy 319-F, etched to show segregation (coring). Other constituents are interdendritic network of silicon (dark gray), rounded $CuAl_2$, and $(Fe_9Mn)_3SiAl_{12}$ script.

SELF-EVALUATION

1. What needs to be removed from the iron ore to produce pig iron? How is this done?
2. Is the pig iron made into useful articles? How is it used?
3. What ingredients are necessary to make pig iron?
4. Is the carbon content of pig iron the same as that of steel? Explain.
5. What is the purpose of ore dressing?
6. In what way does the steel-making furnace change pig iron to make steel? Are any other changes effected?
7. List some of the advantages of the electric furnace.
8. What is "killed" steel?
9. What happens to steel made of steel scrap containing considerable copper and tin?
10. What are the environmental concerns in recycling automobile scrap?
11. Name the heating process that sulfide ores of copper, lead, and zinc must first go through prior to oxidation reduction in a blast furnace. What is the product called after the oxidation reduction process?
12. By what process is copper finally refined to obtain 99.9 percent pure copper?
13. What is the name of the ore that is commonly used to produce aluminum?
14. The refined aluminum oxide (alumina) is placed in an electrolytic cell in a bath of molten ______.
15. By the process of electrolysis, oxygen is removed from the aluminum oxide. What happens to the free oxygen and aluminum?

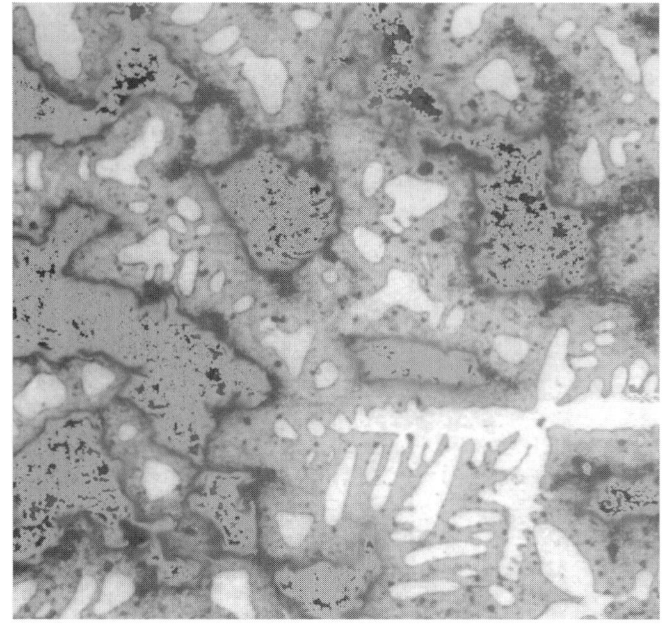

500× As-Polished

CHAPTER 2

Casting Processes

The casting of molten metals into molds to produce useful items is one of the oldest methods of metal forming. Cast ornaments and tools over 4000 years old have been found from ancient Egyptian, Assyrian, and Chinese cultures.

Molten metals such as iron, steel, and aluminum are cast into ingot molds and allowed to solidify before they are further processed. For the purpose of producing a needed or useful shape, casting is done by using many different techniques. These processes involve a large segment of the metals industry. The castings produced range from small, intricate precision parts to massive machinery sections weighing many tons. In this chapter, you will be introduced to the various casting methods used by the industry today, with the emphasis on the metallurgical condition of castings.

OBJECTIVES **After completing this chapter, you should be able to:**

1. Identify and list the various types of casting processes.
2. Describe each casting process.
3. Select the appropriate casting process for various manufactured products.

Almost any metal can be cast into molds from its molten state. Iron, steel, aluminum, brass, bronze and die cast metals are examples (see Chapter 13 for identification of these metals). The casting process requires a **pattern** (having the shape of the desired casting) and a **mold** made from the pattern. The mold must be made to withstand the heat of the molten metal, either of sand, plaster of paris, ceramic, or metal. Wood is most often used for the mold patterns, but metal, plastic, styrofoam, and wax are also used. The different methods of casting are as follows:

A. Sand casting.
  1. Green sand molding.
  2. Dry sand molding.
    a. Core sand molding.
    b. Shell molding.
    c. Vacuum molding.
  3. Chemically bonded sand molding.
B. Centrifugal casting.
C. Investment (lost wax) molding.
  1. Solid investment casting (plaster molding).
  2. Shell investment casting.
D. Permanent molds.
  1. Book type.
  2. Deep cavity.
E. Die casting.

SAND CASTING

One type of sand mold, **green sand molding,** consists of moist sand with small amounts of clay and other substances. Green refers to the moisture content rather than to the color of the sand. The second type of sand casting is **dry sand molding,** in which a resin bond is mixed with the sand. By far the greatest number of castings is made by using a sand aggregate (additives and various particle sizes). Silica sand is combined with cereal, moisture, and sometimes a carbonaceous substance such as pitch by a process called mulling. After sand has been used for casting, it is usually reclaimed by crushing, screening, magnetic separation, and impinging of sand grains against a wear plate at high velocity. The fines are vacuumed away and the clean sand is transported back to hoppers to be used again. A simple casting, such as the one shown in Figure 2.1, can be made by sand-casting processes. In these processes, the mold is made by packing or ramming sand around a pattern and removing the pattern when the sand is sufficiently hard (Figure 2.2*a*). Of course, most sand-casting processes for manufacturing small parts are automated. Large castings are still done by hand and are more labor intensive. However, for nonautomated sand castings, a jolt-squeeze machine (Figures 2.2*b* and 2.2*c*) is generally used. These machines quickly compact the sand mix around the pattern.

Patterns for green sand molding are usually made in two halves and are called **match plate patterns,** or **cope** (top half) (Figure 2.3*a*) and **drag** (bottom half) (Figure 2.3*b*). Typical patterns are shown in Figure 2.3*c*. Loose patterns are simply placed on the mold board with the drag half of the flask, and the sand is rammed into place and struck off (Figure 2.4). A mold board is placed on the drag half, which is then rolled over so that the cope half of the pattern can be put into position over the drag half. The cope pattern is placed and sand is rammed into the mold; one or more holes are provided for pouring the metal. This hole, called the sprue, is connected to the cavity by a gate after the cope

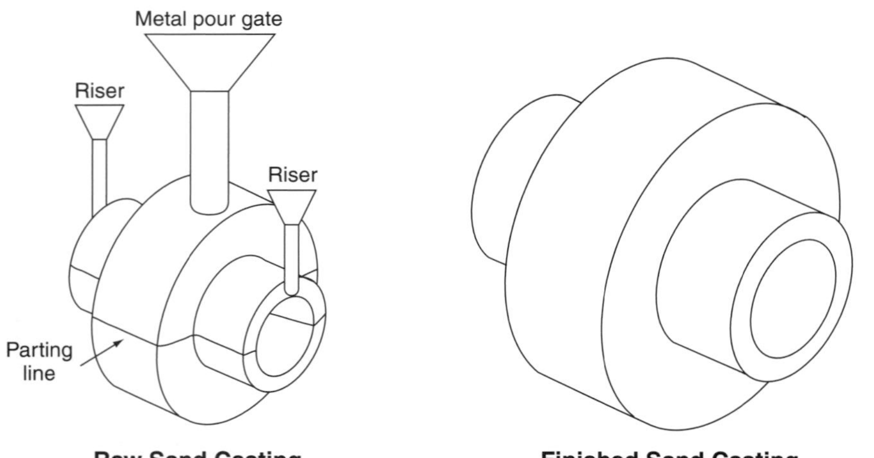

FIGURE 2.1. The "raw" sand casting must have the gate and risers removed in order to become a "finished" casting.

FIGURE 2.2a. Sand mold being rammed. A cope and sprue will be made in the finished mold and molten metal will be gravity cast into it. *(Photo courtesy of the Bureau of Mines)*

FIGURE 2.2b. Operator placing core print in place.

FIGURE 2.2c. Flask ready for assembly. It will be taken from here to the casting floor.

is removed. Other holes, called risers, are used (1) as reservoirs to feed liquid metal to the casting as it shrinks while solidifying, and (2) as vents for escaping gases. Both the cope and drag patterns must be removed before the flask is put back together. All patterns must have a draft or tapered shape so they can be removed easily from the sand without disrupting the mold. Patterns must also be larger than the required size to allow for shrinkage when the molten metal solidifies as it cools. Pattern makers allow for shrinkage, depending on what metal is used for the casting; for example, 1/8 in. per foot for cast iron. After the patterns are removed, and the new molds are put back together, the molten metal can be poured (Figure 2.5).

In dry sand molding, the specially prepared sand packs well when it is rammed into the mold, but it must be baked to drive off the excess moisture and harden before the molten metal can be poured. Core sand is similar to that used in dry sand molding and is used often when a core is needed to produce hollow parts. Core sand is clean, free of clay, and mixed with an organic binder such as linseed oil, cereal, or resins. After the cores are formed, they are hardened by baking in an oven. **Cores** are made by ramming sand into a simple core box (Figure 2.6), or by forcing sand into a mold using a core blowing machine. They are, like most sand molds, formed in halves and then put together to be baked (Figure 2.7). The tedious process of oven drying cores has been, for the most part, replaced by the use of **no-bake sands.** These sands are combined with a resin and a catalyst. Several types of resins are used such as furan and phenolic.

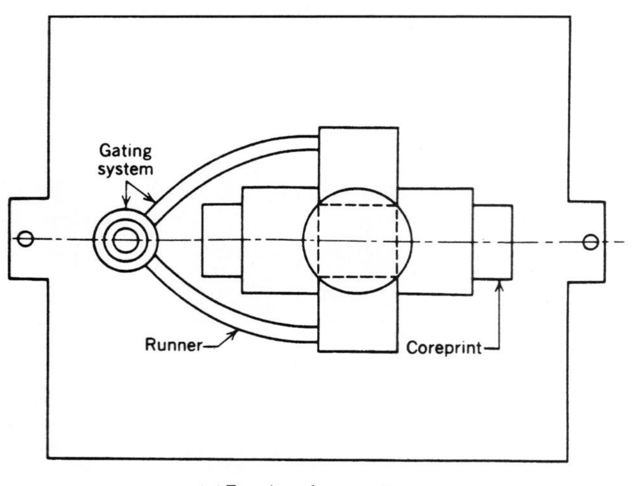

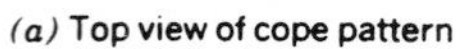

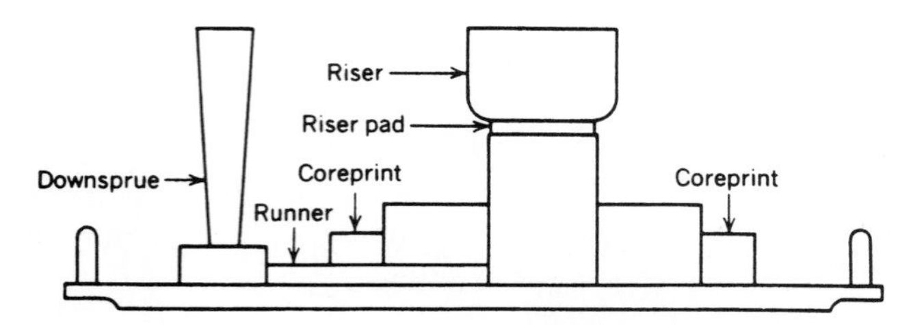

*FIGURE 2.3*a. Cope match plate pattern. (*a*) Top view of cope pattern. (*b*) Side view of cope pattern. The down sprue and riser patterns are removed from one side and the cope pattern is removed from the other side. *(Machine Tools and Machining Practices)*

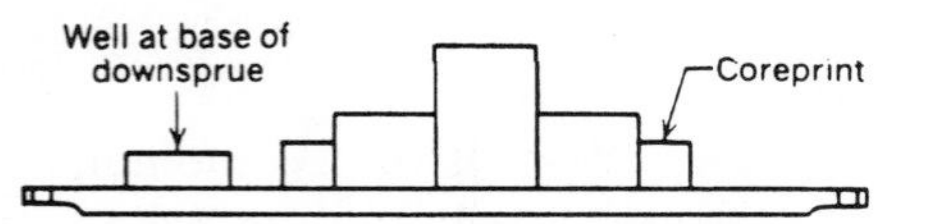

*FIGURE 2.3*b. Drag match plate pattern. *(Machine Tools and Machining Practices)*

*FIGURE 2.3*c. A number of metal match plate patterns ready to be used.

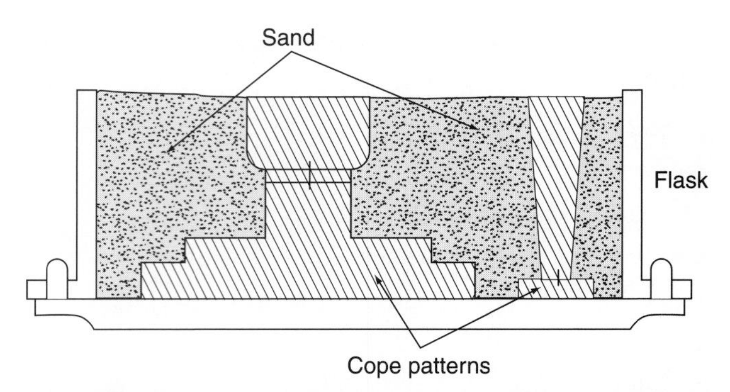

FIGURE 2.4. Sectional view of cope patterns in flask with the sand rammed in place and struck off.

FIGURE 2.5. Here molten aluminum has been poured into a number of molds. The one at the center has been broken open to reveal the casting.

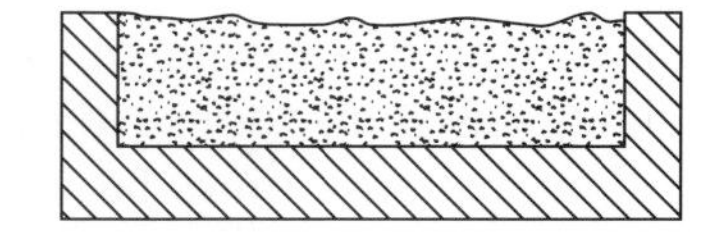
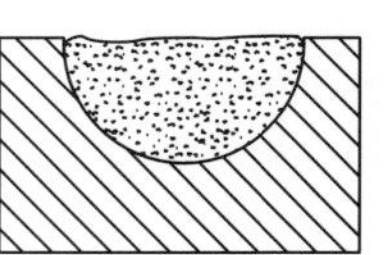

FIGURE 2.6. Sectional view of core box with the core.

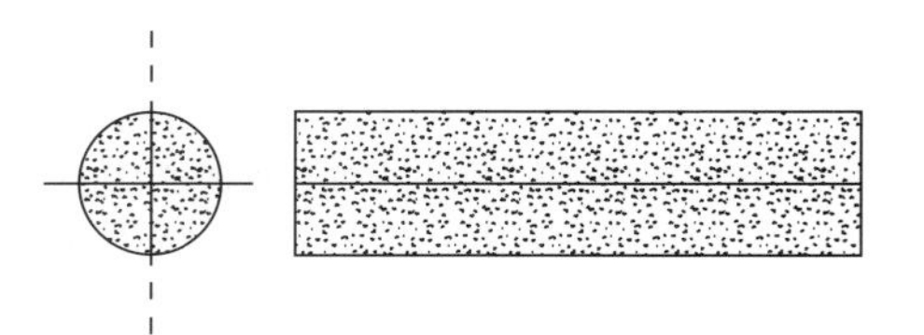

FIGURE 2.7. Two halves of core fastened together after being removed from core box and baked in oven.

Some of the requirements for sand cores are as follows:

1. They must be able to withstand high temperatures.
2. They must have a green bond strength so that they will hold together while damp.
3. They must have dry strength after baking.
4. They must have permeability to enable gases to escape through them.
5. They must be easily broken apart and removed from the finished casting (collapsibility).

When cores are positioned in a flask, they are held in place by a section of the pattern called the **core print.** The core print does not contribute to the final shape of the casting in any way. The cope and drag with core in place are shown in Figure 2.8, ready to make the casting.

Very little sand casting is performed by hand methods except for very large castings. Small part casting is mostly automated. The molds move along a continuous belt and are automatically cast with a minimum of labor.

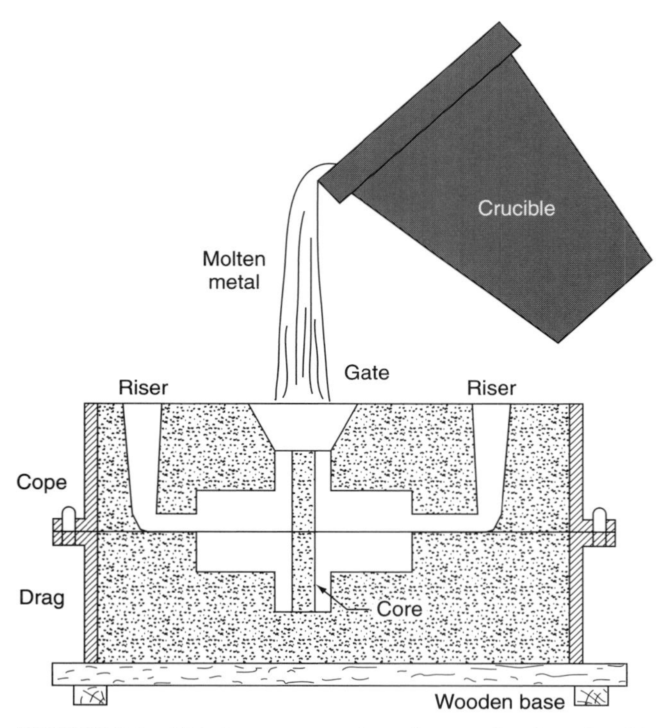

FIGURE 2.8. This is a cutaway view of a completed sand mold ready for metal pouring. The metal will flow into the mold through the gate, down the sprue, and into the main body of the casting. Liquid metal also fills the risers which help to remove air pockets and prevent cavities due to shrinkage when the metal solidifies. The gate and risers are removed upon cooldown.

FOUNDRY PRACTICE

The metallurgical properties of various cast irons are discussed in Chapter 7. Of all the types of cast iron, gray cast iron is most commonly used for casting purposes. However, large amounts of malleable, ductile, and some white cast irons are also produced. In most cases, molten cast iron is ladled (poured) into sand molds.

The **cupola furnace** (Figure 2.9) has been of primary importance in the melting of cast irons. Prior to the development of various kinds of electric furnaces, cast iron was melted in cupola furnaces that normally used coke for producing heat. When fossil fuels such as coke are burned, considerable fumes are released into the atmosphere; therefore in areas where air pollution is a critical problem, electric furnaces are most often used. Electric furnaces for melting cast iron can be either direct arc, electric induction, or electric resistance types.

The cupola furnace is a circular steel shell lined with a refractory material such as firebrick. It is equipped with a blower, air duct, and wind box with tuyeres for admitting air into the cupola. A coke bed is prepared and burned in with propane torches, and a charge of pig iron (or scrap cast iron) and steel scrap is placed in the furnace. Limestone is used for flux. The blower is turned on and the iron begins to melt at the top of the coke bed. The melting rate is determined by the diameter of the cupola. The molten iron is drawn off into a ladle after suitable metallurgical tests are made.

Metallurgical control of the molten cast iron is made at the furnace. Small wedges are poured in prepared sand molds and quickly cooled. The wedges are broken in cross section, revealing the chill area (white iron). The depth of the chill area reveals the amount of graphitization. The tests reveal whether sufficient inoculation of silicon or magnesium (for nodular iron) has taken place. Newer tests for metallurgical control include thermal analysis of the cooling curves during solidification.

Thin sections in castings likewise tend to cool more quickly than thick sections. Thinner areas in a casting will tend to be harder and contain less graphite (with smaller flakes) than the thicker sections.

The green sand molds are poured from a ladle that was filled at the spout of the cupola or electric furnace. A crane or monorail transports the ladle to the molds on the pouring floors.

Gray iron is an alloy easy to cast because it possesses good fluidity; this allows intricate designs to

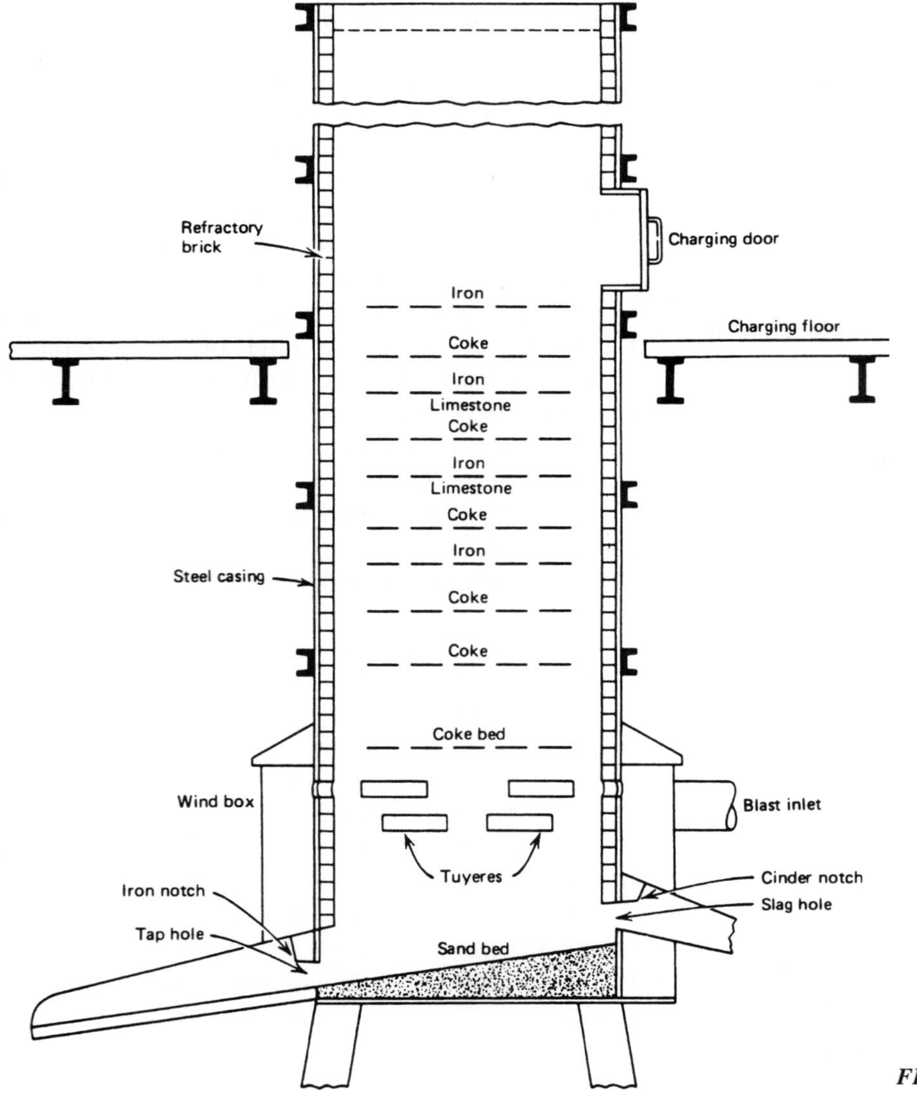

FIGURE 2.9. The cupola furnace.

be cast. Logical designs are necessary, however, to compensate for contraction, which can leave voids or shrinkage cavities. Large cross sections adjacent to thin sections should be avoided. A means of feeding extra molten metal into the mold as the iron shrinks is necessary to prevent shrinkage cavities. This is done with risers, which are reservoirs of molten metal. The pattern must also allow for progressive solidification toward the risers to produce sound castings. Impurities and gas pockets move to the last area of solidification that should be in the riser. Gray irons are more sensitive to differences in cooling rates than most alloys, and heavy sections of castings cool more slowly than lighter ones.

Chills are metal pieces placed in a mold to control the cooling rate of a casting. A wheel with a heavy rim section may crack in a flange or spoke during solidification if the rim section cools too slowly. A chill ring is therefore put into the mold to overcome this tendency. When cast iron is poured against a chill plate, the resultant product is a hard, abrasive-resistant iron that can be either partial or complete white iron.

Malleable iron is produced by prolonged heat treatment of white cast iron, but ductile (nodular) iron is made by inoculating a megnesium alloy into the ladle. Usually the magnesium alloy is placed in the ladle and the molten cast iron is poured over it. A nodular graphite is formed in the iron by this process in place of the flake graphite found in ordinary gray irons. The nodular graphite gives the iron more ductility. Nodular iron is replacing some malleable iron because of its lower cost. The microstructures of various cast irons are shown and discussed in Chapter 7.

STEEL CASTING

Although steel is somewhat more difficult to sand cast than gray cast iron, it is used in higher toughness and higher ductility applications, and for high-strength parts. These range in size from small parts to huge hydraulic press castings weighing many tons. Most steel castings are of a medium carbon steel, with some elements added to ensure soundness. Most important among these is manganese, since it has a favorable effect on the mechanical properties of cast steels. If a cast steel contains less than 1.0 percent manganese, it is a carbon steel; if more, it would be an alloy steel. Other alloying elements in carbon cast steel would be limited to residual amounts. Cast steels can contain carbon in the same range as wrought steel: low, medium, and high. Annealing low carbon cast steel will not produce much change except to toughen it in terms of the Izod impact test. Medium carbon steels are usually heat treated, either for hardness or refinement as in normalizing. High carbon steels are usually quenched and tempered for a specific hardness and tensile strength.

Alloy steel castings, especially massive ones that cool slowly, are usually heat treated by normalizing or are quenched and tempered. The mechanical properties of normalized steel castings approach those of corresponding wrought steels, but they do not have anisotropic properties. The ductility and toughness are much greater than those of cast iron. The welding of steel castings presents the same problems as those of wrought steels.

PROBLEMS IN CASTING

Since molten metal shrinks when it solidifies (as much as 6 percent in some metals), cracks, called hot tears, often occur at narrow sections that are restrained by the mold. Shrinkage cavities that weaken the casting are sometimes found in thicker sections. These problems can be avoided with proper part design and by designing risers large enough to supply molten metal during solidification and keeping heavy sections uppermost in the mold. Air and gases trapped in the metal can produce blow holes or porosity, which weaken the casting. Large, slowly cooled castings usually have coarse, irregular grains; but small, rapidly cooled castings usually are fine grained. As discussed in Chapter 6, fine-grained metals are stronger than coarse-grained metals. Alloy steel castings that are to be machined are usually normalized—reheated to recrystallize and refine the coarse grain to homogenize the structure and eliminate hard, difficult-to-machine areas.

An example of one difficulty seen in a medium carbon alloy steel casting was that of a repeated failure of a flange on a large casting (Figures 2.10 and 2.11). When a section of the flange was removed and examined with a microscope, the grain structure was found to be exceedingly coarse. This fact alone could explain the inherent weakness of the flange. Also, a considerable amount of oxide inclusions were found, and some porosity could be seen as well.

The **shell-molding process,** a type of sand mold casting, is often used for casting high-strength steel parts.

FIGURE 2.10. A section of a broken steel casting that failed after a short time in service. A section was cut out for microscopic examination.

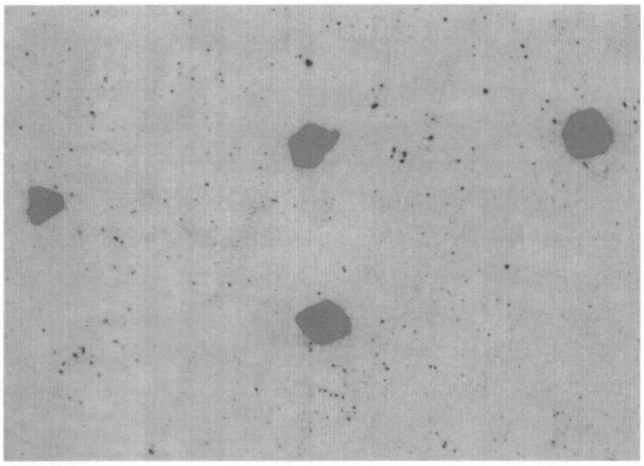

100× 2% Nital Etch

FIGURE 2.11. The nonmetallic inclusions (gray) were found in the microstructure of the fractured casting in Figure 2.10. The inclusions were identified as an aluminum-nitride compound and are the result of a poor casting practice. These inclusions were the cause of the casting failure. Mechanical properties, such as fatigue and impact strength, are severely compromised by inclusions and grain size. The actual grain size measured ASTM 2 (Figure 4.22). The use of *heat sinks* and *chills* can often resolve grain size variations by controlling and equalizing the solidification and cooling rate between thick and thin cast sections.

This method requires the use of metal patterns, which are heated to 450°F, coated with a silicon release agent, and put into a dump box. A prepared sand is then poured over them (Figure 2.12). The sand is mixed with a phenolic resin binder. Some of the sand adheres to the pattern and solidifies, forming a shell around it. The thickness of the shell is determined by the length of time the pattern is in contact with the sand, and ranges from 1/8 to 1/4 in. The dump box is turned over and the loose sand is removed (Figure 2.13). The pattern and the adhering sand are placed in an oven and heated to a temperature of 600°F for 1 or 2 minutes (Figure 2.14). The half shell is then removed from the pattern (Figure 2.15) and clamped together with its mating half to form the mold (Figure 2.16). When the mold is used for small or thin castings, the shells have sufficient strength to withstand the pressure of the molten metal. Heavier castings require the use of backup materials around the shells, usually sand, gravel, or metal shot. The advantages of shell molding over other forms of sand casting are that high precision, good finishes, and more complex shapes are possible, and less machining is needed.

CENTRIFUGAL CASTING

Centrifugal casting is a process in which molten metal is poured into a rapidly rotating mold. The liquid metal is forced outward by centrifugal forces to the mold

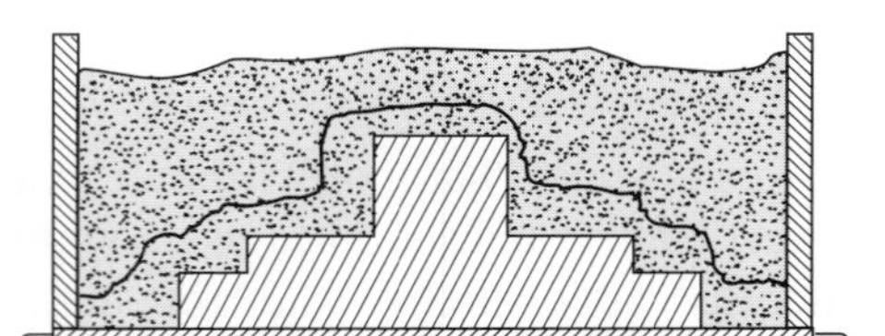

FIGURE 2.12. Heated metal pattern in dump box with sand.

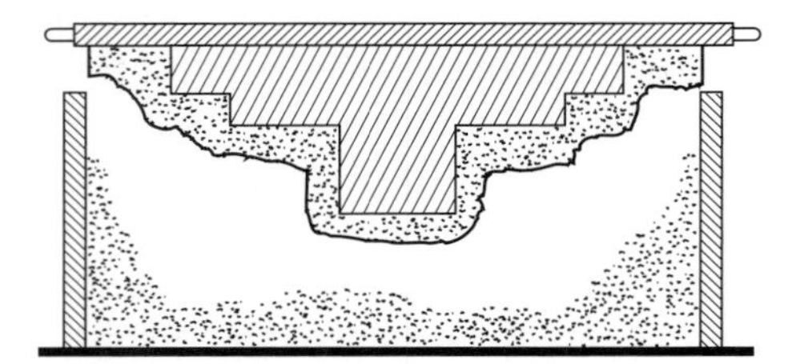

FIGURE 2.13. Dump box is turned, leaving some sand adhering to the pattern.

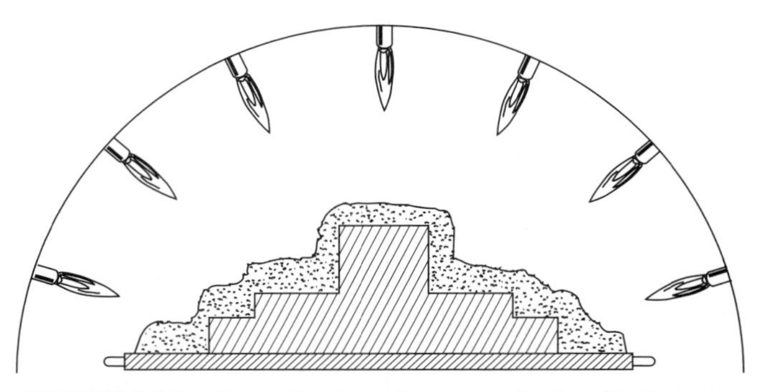

FIGURE 2.14. Pattern is placed in an oven for 1 or 2 minutes to harden it.

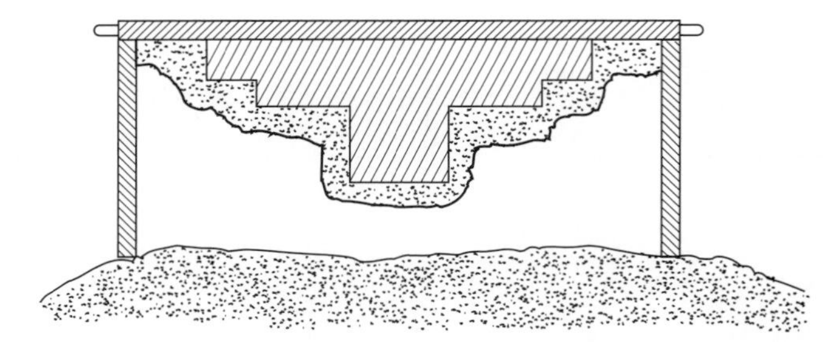

FIGURE 2.15. Half shell mold is stripped from pattern.

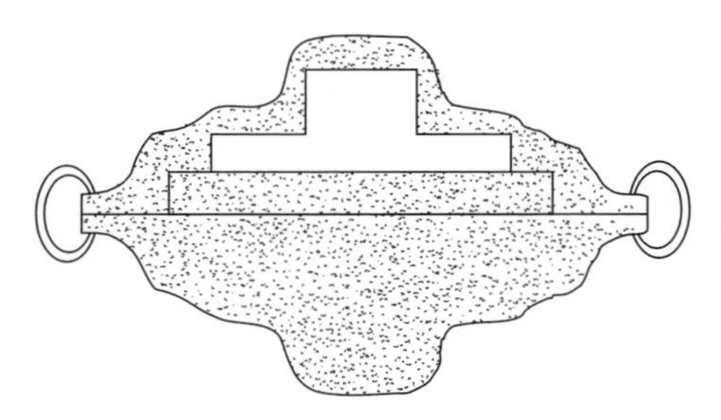

FIGURE 2.16. Mold halves are clamped together with core. The top half shell is shown as a sectional view.

cavity. Wheels, tubing, and pipe (Figure 2.17) are made by the centrifugal casting process. The centrifugal process also makes possible the casting of two dissimilar metals. For example, an outer surface of hard alloy can be poured, followed by an inner layer of softer metal. This would give the casting an outer wear surface while maintaining the machinability and weldability of the inner core.

Centrifuging of castings is a similar process, but with this method the entire mold or group of molds is rotated away from the center of rotation. Castings made by this process have superior mechanical properties and a uniform grain structure. The center area is the last to solidify so that the impurities work to the center where they are later removed by machining processes.

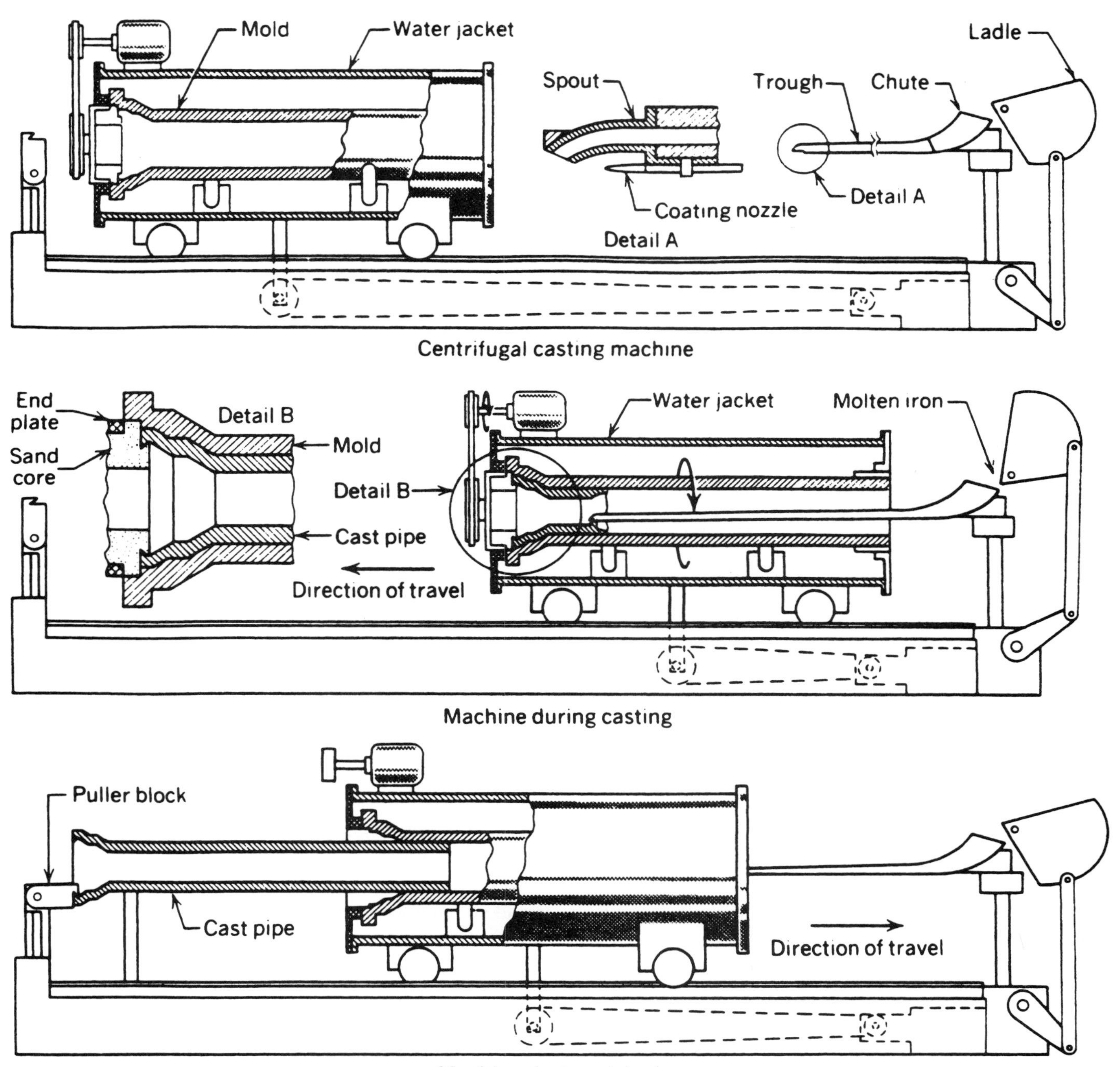

FIGURE 2.17. Machine for centrifugal casting of iron pipe in a rotating water-jacketed mold. *(By permission, from Metals Handbook, Volume 5, Copyright American Society for Metals, 1970)*

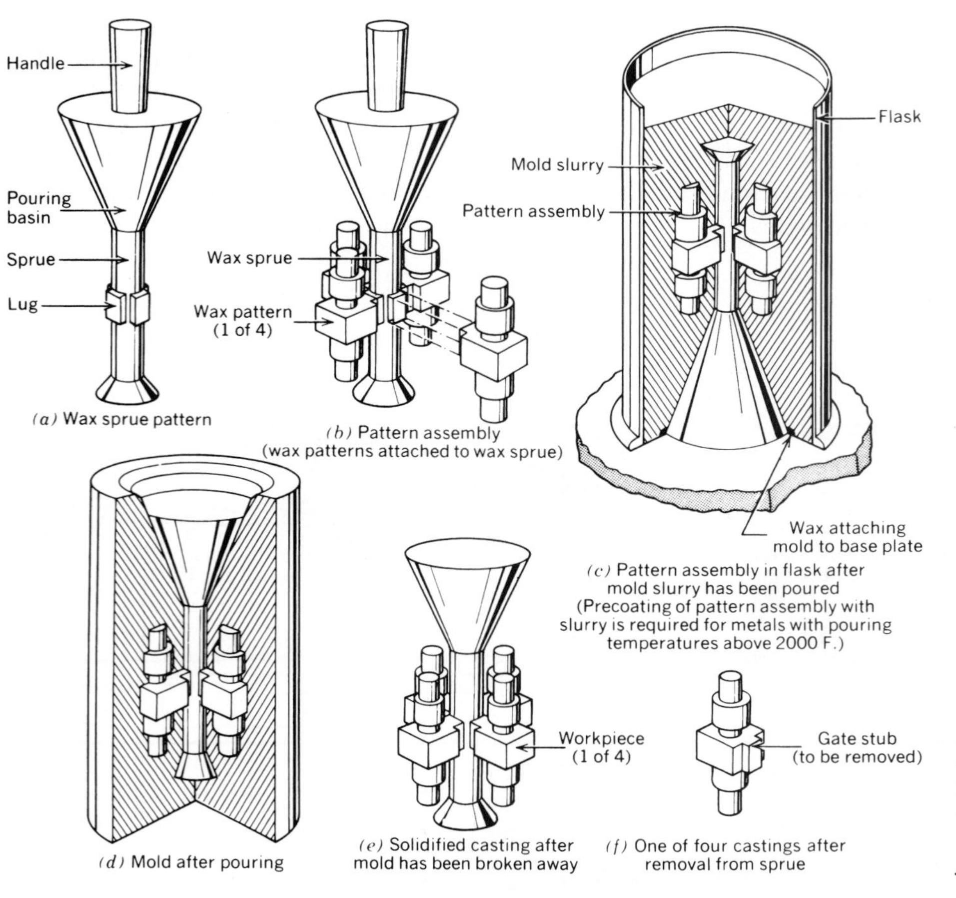

FIGURE 2.18. Steps in the production of a casting by the solid investment molding process, using a wax pattern. *(By permission, from Metals Handbook, Volume 5, Copyright American Society for Metals, 1970)*

INVESTMENT CASTING

One of the oldest methods of casting metals is that of **investment casting** or **lost wax process.** The pattern is made of wax and is used only once, but the wax is not really lost as it may be reused in another mold. Wax patterns, including their sprues and risers, are usually cast in metal molds (dies) or formed by injection molding. This method is used for industrial purposes, but unique (one-of-a-kind) patterns are often used for art metal castings.

Plaster of paris has long been used to make molds for solid investment castings (Figure 2.18). Plaster casting can only be used for the lower temperature metals such as aluminum, zinc, tin, and some bronzes. Metals having higher pouring temperatures such as steel and cast iron are precision cast in molds that are made by the **investment shell** process. In this process, the wax pattern is dipped into a slurry (thin liquid mixture of several substances) of refractory (resistant to high temperatures) material and dried. This process is repeated until a suitable shell up to 1/4 in. thick has been built up. This mold is then heated in an oven to melt out the wax, which is collected and reused. The shell is heated in a furnace to 1600°F and the metal is then poured. The mold cannot be reused—it must be destroyed to remove the casting. Figures 2.19*a* to 2.19*j* show the process from the wax pattern to the finished casting.

Investment casting is used for intricate shapes that would make the withdrawal of a normal pattern impossible. Great detail and high precision with no parting line or seam is typical of this method. Since nonmachinable alloys can be cast with this precision casting, the need for machining is almost eliminated.

Many alloys made into castings cannot be reasonably manufactured in the wrought form—they are classed as unworkable; that is, they cannot be hot or cold worked. Examples are some superalloys, wear-resistant alloys, many magnet alloys, and cast irons. Some of these alloys have a very high melting point and so cannot be produced with die casting or permanent mold techniques, and sand casting does not produce the near-net shape that is needed in some cases. Plaster of paris molds would not withstand the high temperatures. The investment shell process is ideal for these high-temperature melting alloys. Automotive and aerospace industries, for example, use the investment casting method for small parts (Figure 2.20) such as jet turbine blades and impellers for pumps.

***FIGURE 2.19*a.** First step in the production of a shell investment casting shows a wax replica of the lower half of valve assembly, together with die. Special wax is forced into the die under high pressure. *(Investment Casting Corporation)*

***FIGURE 2.19*b.** Wax replica of upper half of valve assembly, made in a separate mold to provide for interchangeability of various vane configurations. *(Investment Casting Corporation)*

***FIGURE 2.19*c.** Two-piece wax assembly shown prior to sealing. *(Investment Casting Corporation)*

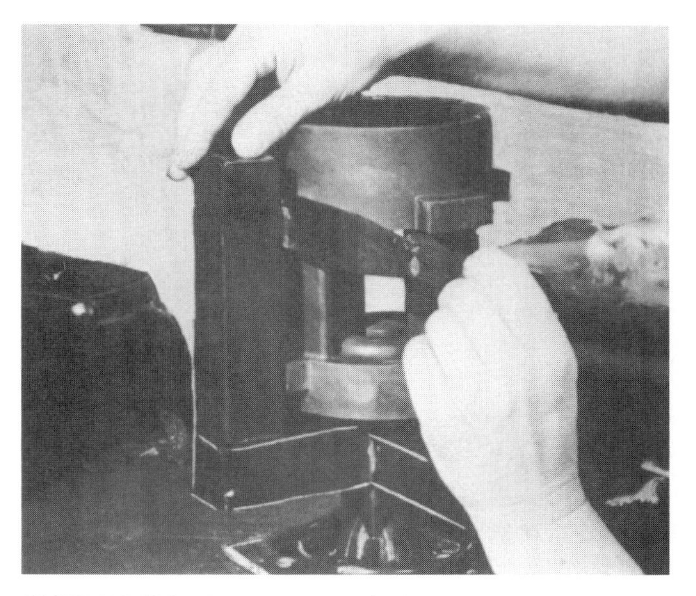

***FIGURE 2.19*d.** Runners are attached to wax assembly, which provide the necessary gating and risers for sound castings. *(Investment Casting Corporation)*

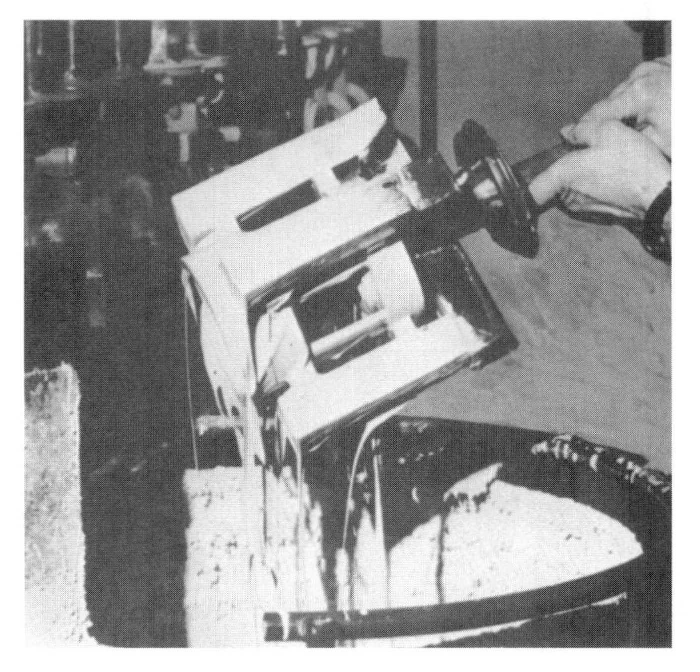

***FIGURE 2.19*e.** Wax assembly is coated with first coat of ceramic slurry. As many as eight coats are applied. *(Investment Casting Corporation)*

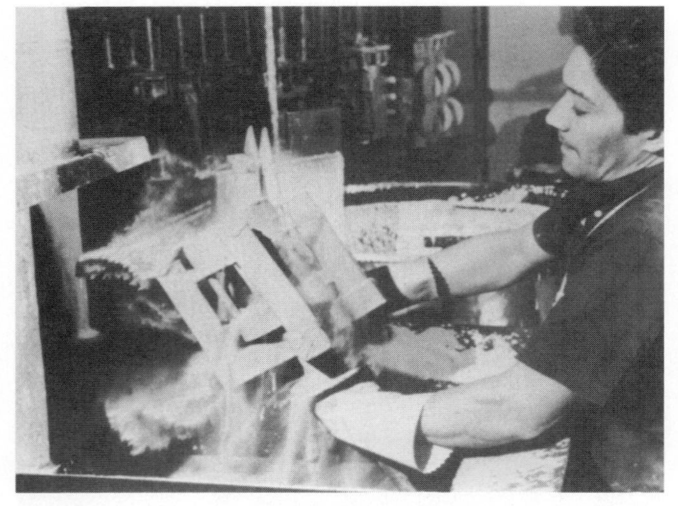

***FIGURE 2.19*f.** Slurry ceramic is stuccoed with zircon sand and allowed to dry. *(Investment Casting Corporation)*

***FIGURE 2.19*g.** Ceramic shells drying. *(Investment Casting Corporation)*

***FIGURE 2.19*h.** Ceramic slurries about to enter high-pressure steam autoclave to remove majority of wax present in mold. *(Investment Casting Corporation)*

***FIGURE 2.19*i.** Induction heat metal being ladled into ceramic shell. *(Investment Casting Corporation)*

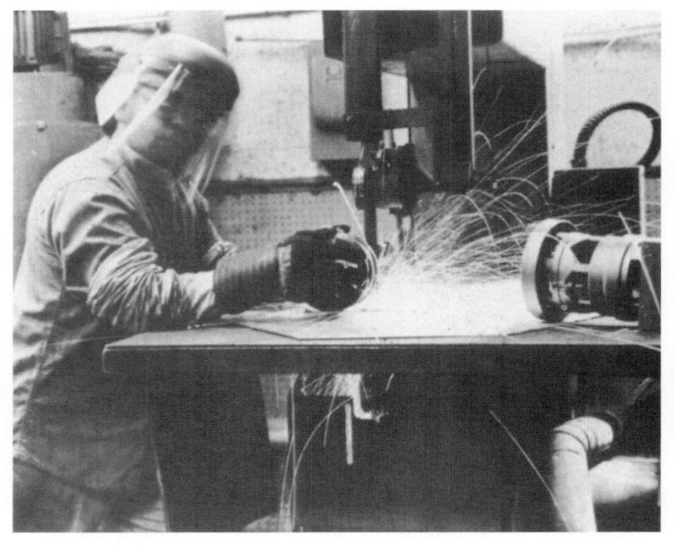

***FIGURE 2.19*j.** After shakeout, castings are cut off on friction saw, finish ground, and subjected to quality assurance checks prior to shipment. X-ray, Zyglo, and Magnaflux facilities are available if required, in addition to dimensional checks. *(Investment Casting Corporation)*

PERMANENT MOLDING

Permanent molds (Figures 2.21 and 2.22) are made of metal, usually gray cast iron or steel. Graphite and other refractory materials are sometimes used for steel casting. The molds are usually machined to a rough shape and hand finished. A refractory wash is applied to the mold prior to casting. As in sand casting, the cores are made of sand and are not reused. Molten metal is poured into the mold from a ladle as in sand casting. The mold must be heated to and maintained at about 700°F to produce good castings.

Permanent molding is a step between sand-casting and die-casting processes. Initial costs are low as compared with die casting. Fairly high precision can be achieved, thus eliminating considerable machining time when compared with a sand casting. Permanent molds are used mostly for limited production runs since they can be reused for only a few thousand castings before they have lost their shape and must be scrapped.

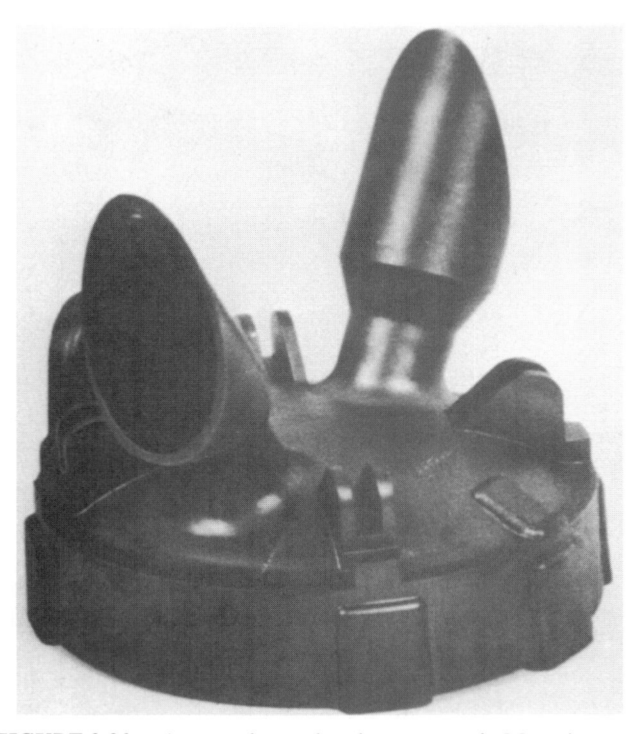

FIGURE 2.20. An experimental rocket part cast in Maraging steel by the investment casting process. *(Hitchiner Manufacturing Company, Inc.)*

DIE CASTING

Die casting differs from permanent molding and sand casting in that the metal is not poured into the mold but is injected under high pressures from 1000 to 100,000 psi. The two basic systems of die casting are the **hot-chamber** and the **cold-chamber** methods. Cold-chamber machines are used for casting aluminum, magnesium, copper-base alloys, and other high melting point alloys. Hot-chamber machines are used for casting zinc, tin, lead, and the low melting point alloys. Die-casting machines are heavy and massive (Figure 2.23), generally hydraulically operated, and capable of exerting the hundreds of tons of force needed to hold the die halves together. This high pressure is necessary to keep the injected molten metal from leaking at the parting line.

A typical hot-chamber machine with the submerged plunger or gooseneck injector is illustrated in Figure 2.24. This type has an oil- or gas-fired furnace with a cast iron pot for melting and holding the metal. The plunger and cylinder are submerged in the molten metal that is forced through the gooseneck and nozzle into the die.

Cold-chamber machines (Figure 2.25) have to a large extent replaced the gooseneck-type machines. One major difference is that cold-chamber machines have a separate melting and holding furnace. A sequence of steps of operation can be seen in Figure 2.26.

High production rates, from 100 to 500 cycles an hour, are made possible by using the die-casting process. Small, thin sections can be produced and very

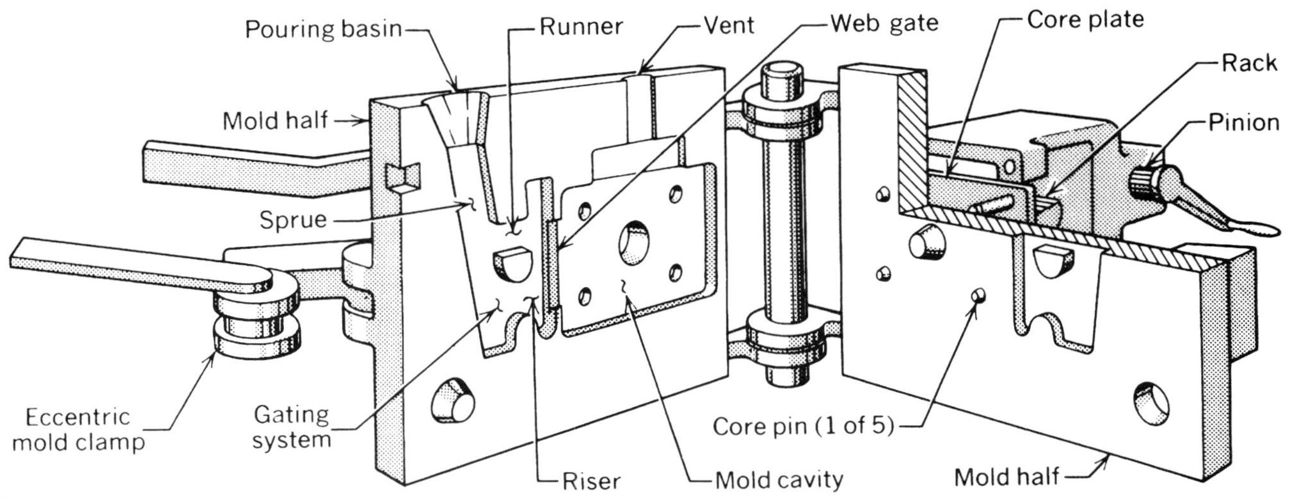

FIGURE 2.21. Book-type manually operated mold-casting machine, used principally with molds having shallow cavities. *(By permission, from Metals Handbook, Volume 5, Copyright American Society for Metals)*

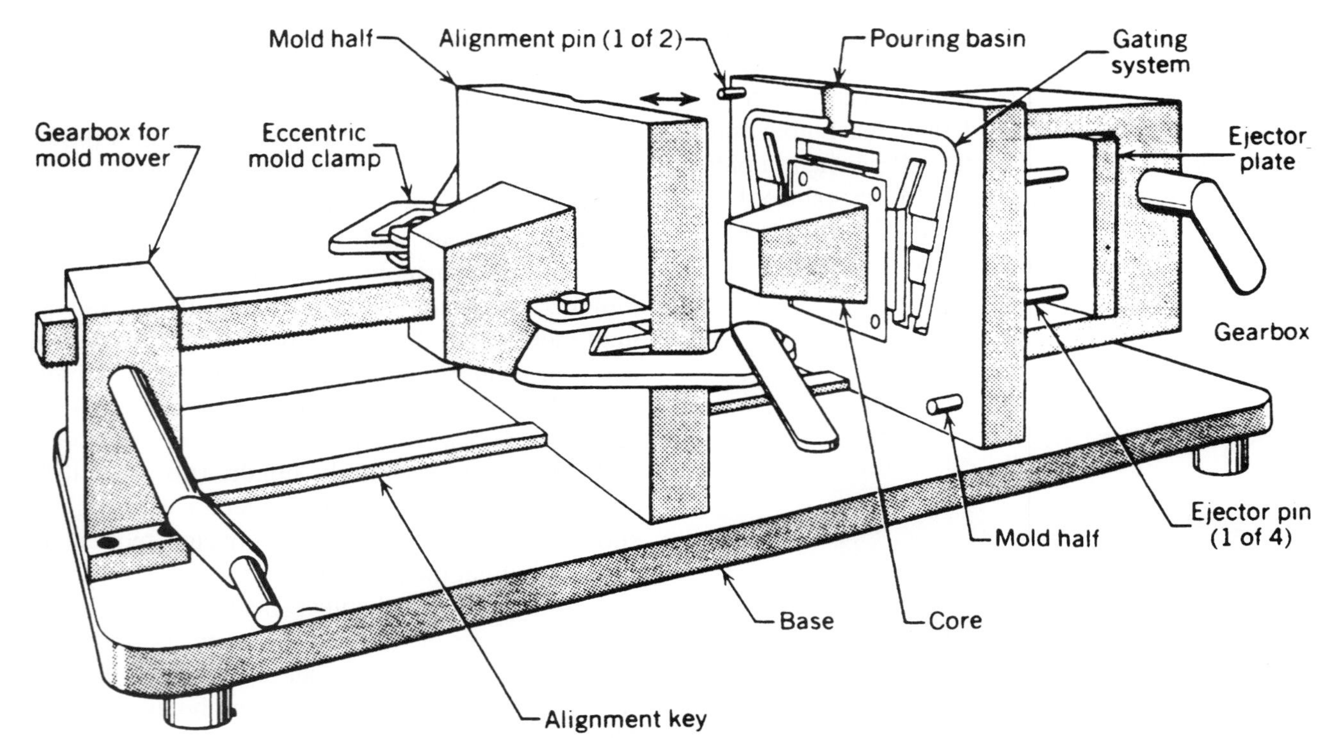

FIGURE 2.22. Manually operated permanent mold-casting machine with straight-line retraction; required for deep-cavity molds. *(By permission, from Metals Handbook, Volume 5, Copyright American Society for Metals)*

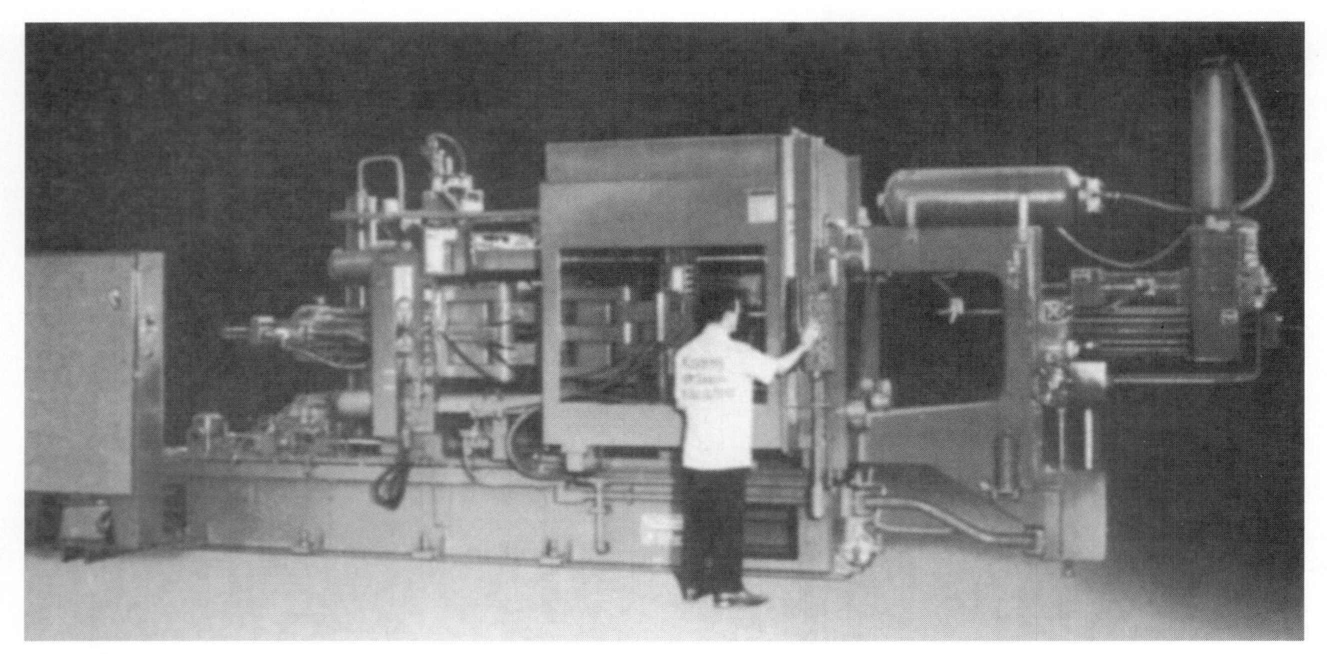

FIGURE 2.23. Die-casting machine. *(HPM Corporation)*

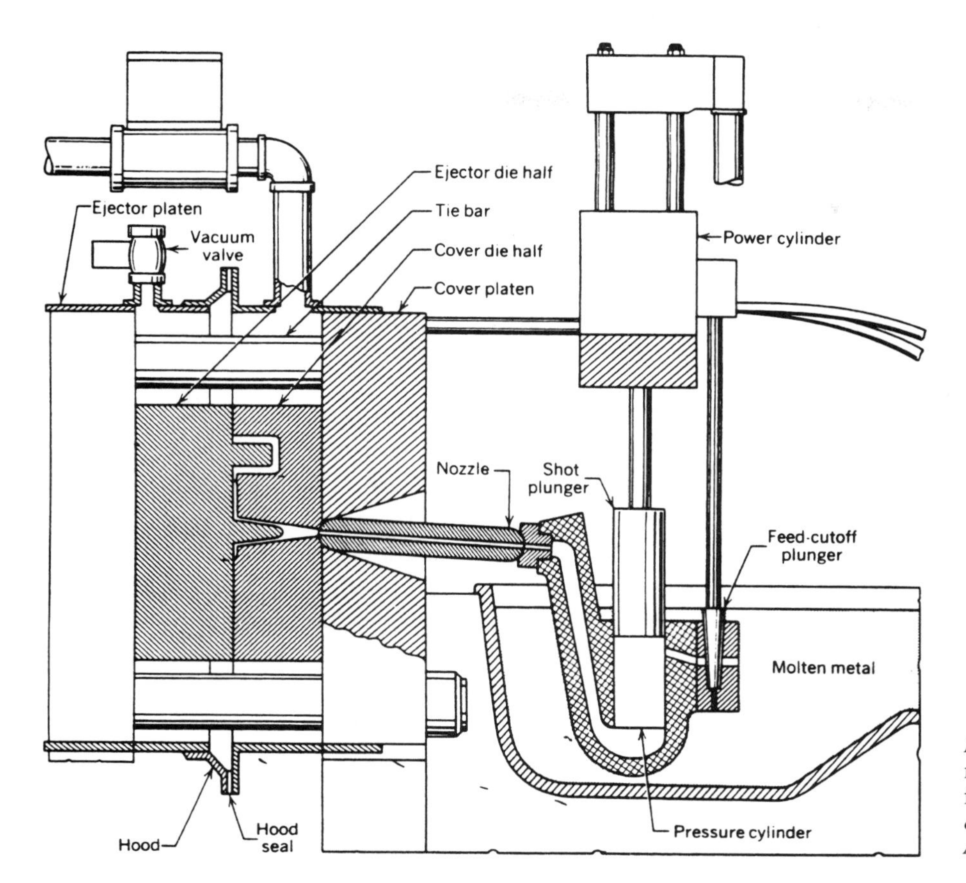

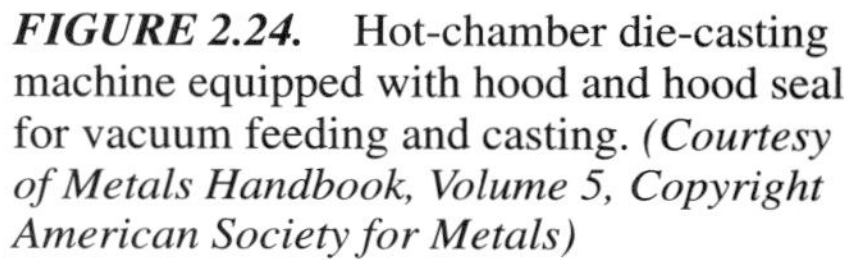
FIGURE 2.24. Hot-chamber die-casting machine equipped with hood and hood seal for vacuum feeding and casting. *(Courtesy of Metals Handbook, Volume 5, Copyright American Society for Metals)*

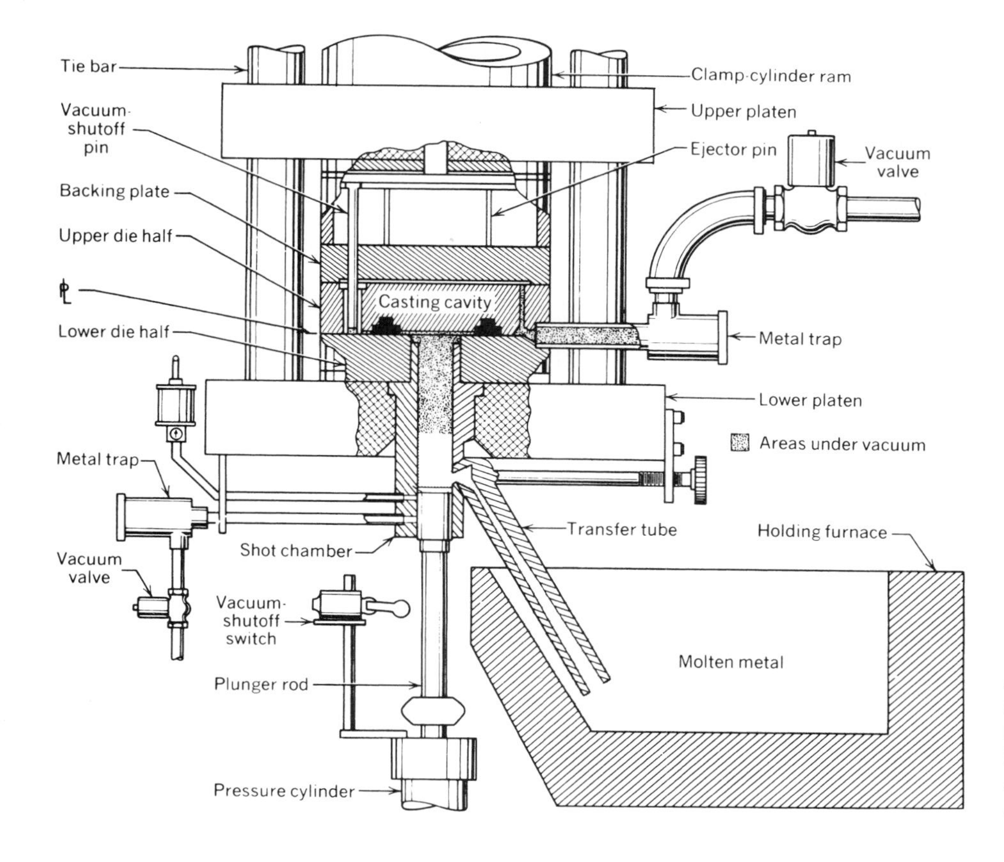

FIGURE 2.25. Principal components of a vertical cold-chamber die-casting machine with the die-parting line in the horizontal plane. *(Courtesy of Metals Handbook, Volume 5, Copyright American Society for Metals)*

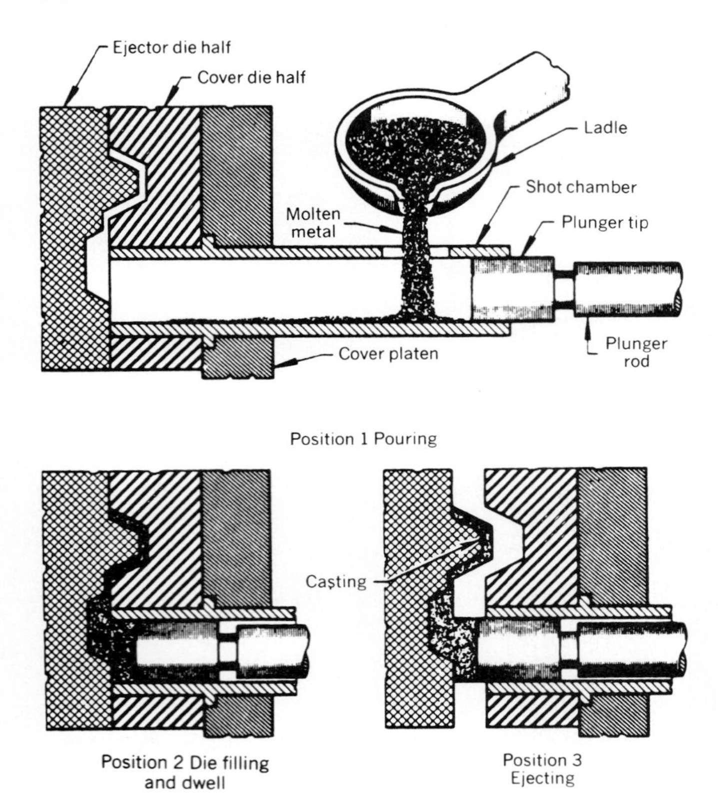

FIGURE 2.26. Operating cycle of a horizontal cold-chamber die-casting machine. *(By permission, from Metals Handbook, Volume 5, Copyright American Society for Metals, 1970)*

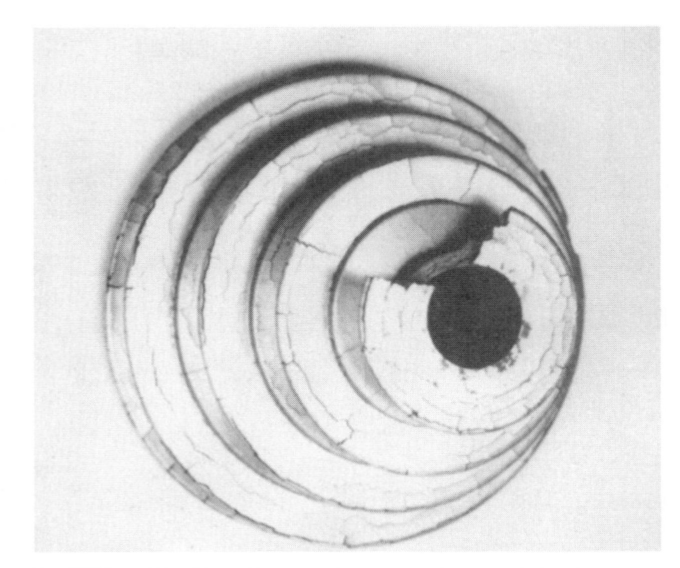

FIGURE 2.27. This die-cast step vee pulley was taken from a wood lathe that was in storage for several years and had never been used. The failure is typically intergranular corrosion; pieces will break off quite easily. This kind of failure is not very common today, as care is taken to prevent such occurrences.

high tolerances maintained. Surface finishes are usually so smooth that subsequent finishing or machining processes are not needed. The quality of the cast metal is better than that of sand castings. The metal mold or die cools the molten metal at a higher rate, thus producing a superior grain structure in the metal. However, zinc die-casting alloys must be of high purity to ensure good impact strength and to avoid chemical deterioration at the grain boundaries during the life of the casting. This phenomenon is known as intergranular corrosion (Figure 2.27). This problem can be accentuated by the presence of impurities such as lead, tin, and/or cadmium, but their detrimental effect may be offset by adding magnesium, copper, and/or nickel to the alloy. Intergranular corrosion may be observed as a swelling and subsequent disintegration of the casting. Die castings are of typically small, high-production parts (Figure 2.28). Larger parts cannot be made because thick sections will contain shrinkage holes having poorer quality.

FIGURE 2.28. Die-cast parts. *(Machine Tools and Machining Practices)*

SELF-EVALUATION

1. Sand casting may be divided into two types. Name them and explain the differences.
2. What kinds of materials are used to make patterns? Briefly describe the patterns used in sand molding.
3. Explain how cores are made and how they are used.
4. What is a *chill?* Explain.
5. Briefly list the steps used in the shell molding process. Explain the types of materials needed.
6. How does centrifugal casting work? Name two advantages. What is centrifuging?
7. Can you use a wood or metal pattern for investment casting? Explain. List the steps necessary to produce an investment casting.
8. Name two advantages and one disadvantage of permanent molding.
9. How does die casting differ from permanent molding?
10. What is the major difference between the hot-chamber and cold-chamber machines? Name two or more advantages of die casting.
11. If a zinc-based die-casting alloy contains a significant amount of impurity atoms, what can happen to the die-cast part?
12. How are investment and investment shell molds removed from the casting?

CASE PROBLEM: THE POLLUTING FURNACE

A small foundry in the Pacific Northwest had a lucrative business in sand casting. They produced sprocket wheels for mills, sewer manhole rings and covers, and other products. State authorities told them they would have to permanently shut down their cupola furnace because it was causing considerable air pollution. Did this mean they would have to close down their business and lay off 40 employees? Was there something that could be done so they could continue operating without polluting the atmosphere?

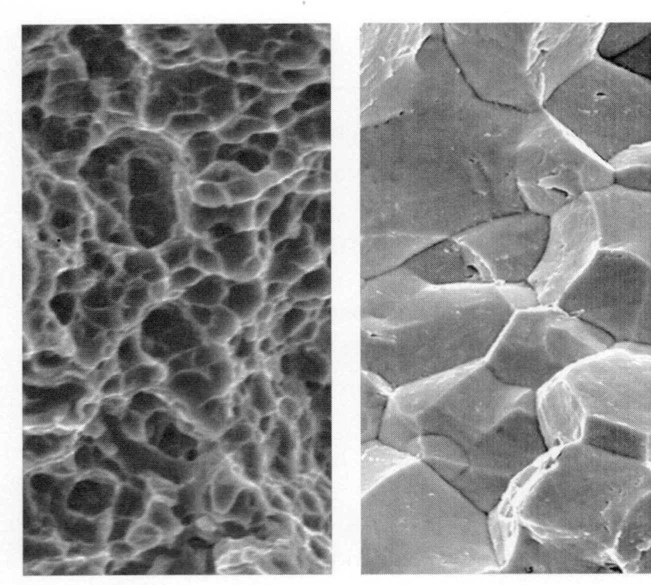

Ductile Failure by Tensile Overload Brittle Failure by Tensile Overload

CHAPTER 3

The Physical and Mechanical Properties of Metals

This chapter introduces the principles of material strength, elasticity, brittleness, plasticity, creep, and others. These *mechanical properties* of a material determine its usefulness for a particular job. An understanding of mechanical properties and how they are measured and their relevance will assist in the selection and design of materials for the shop. An understanding of these properties will help you to become more aware of the problems associated with severe applications.

Before reading this chapter, you should review in the Glossary the definitions of the following terms used when describing mechanical properties: *brittleness, ductility, strength, elasticity, hardness, elastic limit, yield point, tensile strength, fatigue strength, toughness,* and *creep.* (See Chapter 26 for hardness testing.)

OBJECTIVES **After completing this chapter, you should be able to:**

1. Correctly define and describe the mechanical properties of metals.
2. Understand the terms applied to mechanical testing.
3. Describe the various testing machines and their uses.
4. Prepare tensile and shear test specimens and perform these tests.
5. Calculate stress, elastic limit, yield point, ultimate tensile strength, percent elongation and percent reduction in area, using formulas provided.

STRENGTH

The **strength** of a metal is its ability to resist changing its shape or size when external forces are applied. There are four basic types of stresses: **tensile, compressive, shear,** and **torsion** (Figure 3.1). When we consider strength, the type of stress to which the material will be subjected must be known. Steel has equal compressive and tensile strength, but cast iron has low tensile strength and high compressive strength. Shear strength is less than tensile strength in virtually all metals. See Table 3.1.

The tensile test is used to obtain information about the mechanical properties of a material. These include ductility, tensile strength, proportional limit, elastic limit, modulus of elasticity, resilience, yield point, yield strength, ultimate strength, and breaking strength.

The tensile strength of a material in pounds per square inch can be determined by dividing the maximum load (in pounds) by the original cross-sectional area (in square inches) before testing. Thus,

$$\text{Tensile strength (psi)} = \frac{\text{maximum load (lb)}}{\text{original cross-sectional area (in.}^2)}$$

To put it another way, the strength of materials is expressed in terms of pounds per square inch. This is called **unit stress** (Figure 3.2). The unit stress equals the load divided by the total area.

$$\text{Unit stress} = \frac{\text{load}}{\text{area}}$$

When **stress** is applied to a metal, it changes shape. For example, a metal in compressive stress will shorten, and metal in tension is stretched longer. This change in shape is called **strain** and is expressed as inches of deformation per inch of material length. As stress increases,

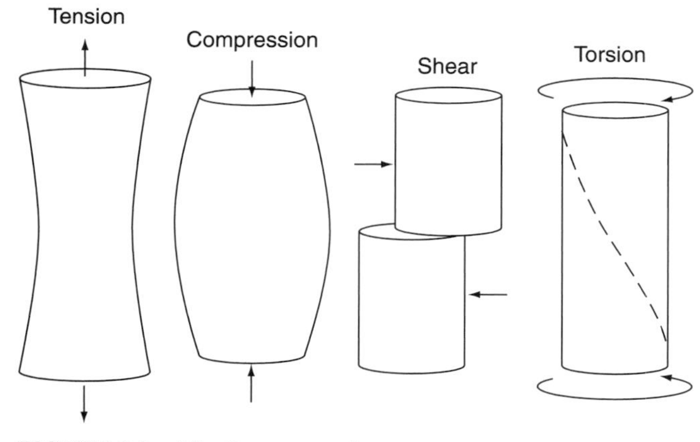

FIGURE 3.1. The four types of stresses.

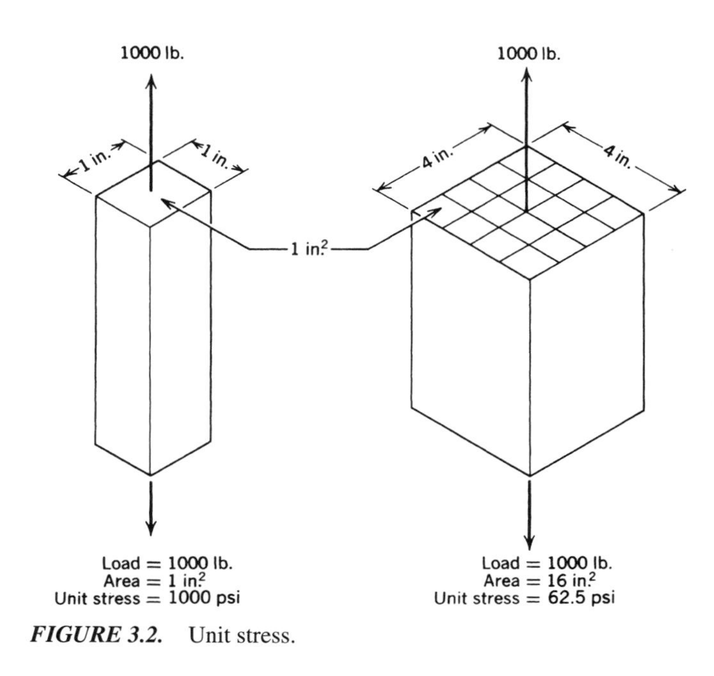

FIGURE 3.2. Unit stress.

TABLE 3.1. Material strength

Material	Modulus of Elasticity (psi)	Allowable Working Unit Stress			
		Tension	Compression	Shear	Extreme Fiber in Bending
Cast iron	15,000,000	3,000	15,000	3,000	
Wrought iron	25,000,000	12,000	12,000	9,000	12,000
Steel, structural	29,000,000	20,000	20,000	13,000	20,000
Tungsten carbide	50,000,000				

Material	Elastic Limit (psi)		Ultimate Strength (psi)		
	Tension	Compression	Tension	Compression	Shear
Cast iron	6,000	20,000	20,000	80,000	20,000
Wrought iron	25,000	25,000	50,000	50,000	40,000
Steel, structural	36,000	36,000	65,000	65,000	50,000
Tungsten carbide	80,000	120,000	100,000	400,000	70,000

strain increases by direct proportion within the elastic range. When the load is removed, the material returns to its original shape. This is known as *Hooke's law.* However, when a load is applied at or slightly above the elastic limit, the metal is permanently deformed.

Metals are "pulled" on a machine called a tensile tester (Figure 3.3). A specimen of known dimensions is placed in the machine and loaded until it breaks. Instruments are sometimes used to make a continuous record of the load and the amount of strain. This information is put on a graph called a stress-strain diagram. If a stress-strain recorder is not used, the tensile test can be performed by first marking the specimen with a prick punch.

To obtain a full compliment of data from the tensile test specimen, a gage length must be established. Within the established gage length, the nominal dimensions must be accurately measured in thousands of an inch to allow calculation of the original area (see example in Figure 3.4). Measurements are usually made with a micrometer. When a knife blade type extensiometer is used the extensiometer is often set to zero and used to mark the gage length directly on the specimen. A blue machinist's ink is used to delineate the marking.

A stress-strain diagram can be made for most metals and composites by preparing a specimen such as shown in Figure 3.5*a*. Metals exhibiting a hardness above 60 HRC, metal matrix composites, and ceramics are often tensile tested using a tensile machine with a load cell rather than an extensiometer, because it is difficult to determine the load at which the yield point oc-

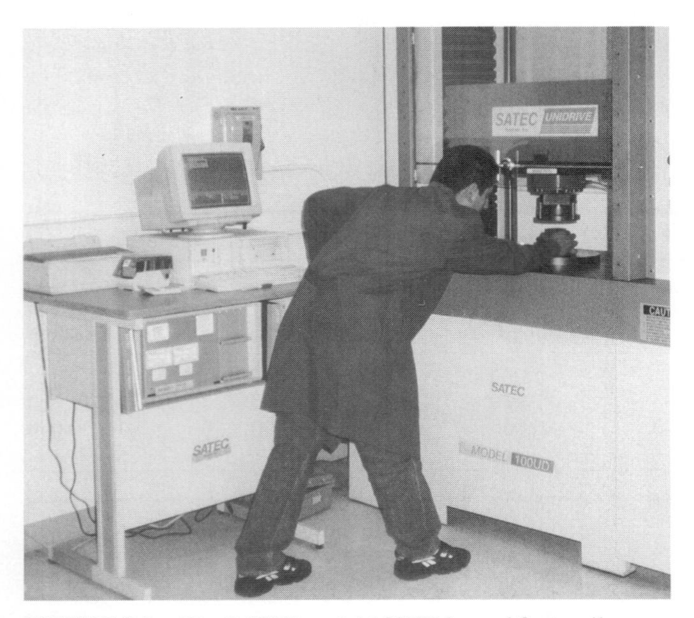

FIGURE 3.3. The SATEC model 100UD is used for tensile, compression and shear testing. Loading of the specimen is by computer control. (*Courtesy of Stork-MMA Laboratories, Huntington Beach, CA.*)

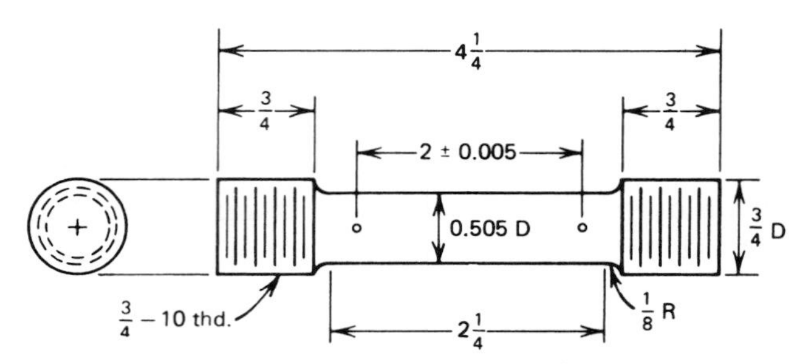

FIGURE 3.4. This standard test specimen has a cross-sectional area of 0.200 in. and a gage length of 2 in. Round numbers make calculations for percent reduction of area and percent elongation easier.

curs in these materials (see Figure 3.5*b*). Serious damage results to the expensive extensiometer if it remains on the specimen at the time breakage occurs.

When conducting a tensile test, the specimen is clamped securely in the upper and lower grips or jaws between the crosshead and the machine table (Figures 3.6*a* and 3.6*b*). In some machines, the table is part of a gigantic hydraulic cylinder and the crosshead is stationary. The opposite is true in other testing machines. In either case, once the specimen is securely held in the grips, a preload is applied to test the setup for specimen slippage. This load is released and the machine is rezeroed to show no applied load. Load is usually applied following a preset rate called *pacer,* throughout the entire test until breakage occurs. The rate of pacer is established by customer specification. During the test, the technician can observe the rate of load and resulting strain. If an extensiometer is used, the unit must be removed after the material exhibits a definite yielding past the indicated strain which is required for determining the yield point. The application of load continues while the extensiometer is removed. The maintenance of load rate, pacer, is important to establish the maximum load-carrying capability of the specimen being tested, with minimal influence resulting from irregular yielding due to operation of the tensile machine. Pacing also provides a means of test method standardization in the application of load since many materials are load-rate sensitive. Once the machine and test specimen are readied for the test, the load is applied until the specimen is broken.

Load is recorded during the entire tensile test. Usually the yield point must be determined by the 0.2 percent offset method because the theoretical dip in the load strain diagram only occurs in pure metals under very slow load rate application and the extensiometer must be removed immediately following the yield point and before the specimen breaks. The yield point dip occurs because during slow load application, soft metals will ex-

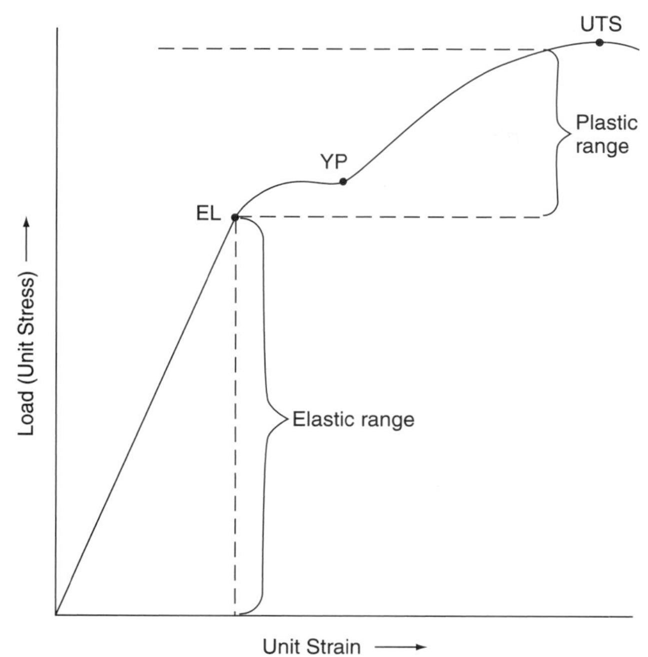

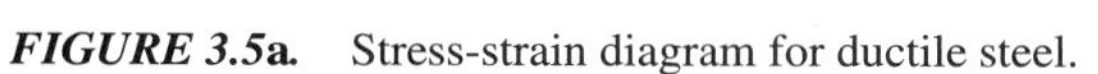

***FIGURE 3.5*a.** Stress-strain diagram for ductile steel.

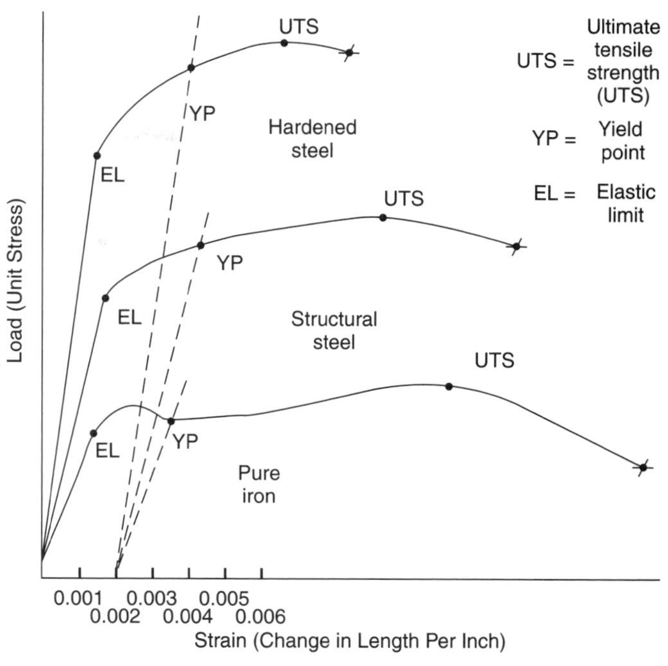

***FIGURE 3.5*b.** This "Load-Strain" diagram or "Stress-Strain" diagram compares the strength differences between Pure Iron, Structural Steel, and Hardened Steel (steel hardened by thermal treatment).

***FIGURE 3.6*a.** Servohydraulic mechanical testing system used for tensile and fatigue testing of materials. *(Photo courtesy of MTS Systems Corporation.)*

***FIGURE 3.6*b.** Hydraulic wedge rips are used to prevent specimen slippage during tensile and/or fatigue testing. An extensiometer is shown attached to the tensile specimen. *(Photo courtesy of MTS Systems Corporation.)*

hibit a faster rate of material stretching under the applied load. This dip is almost impossible to observe in alloys and hardened materials especially when load paced.

The cross-sectional area to be pulled is usually a standard diameter of which the area can easily be calculated in round numbers. The C-1 standard is the one most commonly used for test specimens; the diameter is 0.505 in., making the area 0.2 in.^{2} A specific gage length is used, in this case 2 in. ± 0.005 in. The specimen in the tensile testing machine is then slowly pulled. If a strain gage and an X-Y recorder are used, the diagram is automatically made. The X-Y recorders are gradually being replaced with computerized systems. If no recorder is available, the tensile testing machine may be stopped at intervals and the load, or stress, written down and the strain or distance between the center-punched marks measured and copied down on the same line. With some equipment, readings may be taken without stopping the machine. Later, these stress-strain increments can be plotted on a graph to produce the stress-strain diagram for that particular metal.

TORSION

The torsion test is useful for testing parts that are subjected to twisting loads such as shafts, axles, sockets, and rotating tools. A torsion testing machine (Figure 3.7) has a twisting head that grips the specimen and applies a twisting motion. The other end of the specimen is gripped in a measuring head, and a measuring device measures angular displacement of points on opposite ends of the specimen gage length. Stress-strain curves may be plotted from torsional data.

ELASTICITY

The ability of a material to strain under load and then return to its original size and shape when unloaded is called **elasticity.** The **elastic limit** (proportional limit) is the greatest load a material can withstand and still spring back to its original shape when the load is removed. Elastic hysterisis plays an important factor controlling the length of time required for materials to return to their original size and shape upon load removal. Materials are subject to **visco-elastic lag** which means the closer a material is loaded to the elastic limit, the longer it will take for the material to return to its original size and shape upon load removal.

The elastic limit is easy to determine on any stress-strain diagram. It is the end of the straight-line portion of the stress-strain curve in Figure 3.5*b*. When a metal is pulled in tension within the elastic range, the crystal lattice is elastically distorted and lengthened. At the same time, the specimen becomes thinner at right angles to the applied force. The ratio of lateral change to the change in length is called **Poisson's Ratio.**

YIELD POINT

Yielding or plastic flow occurs in materials when the elastic limit has been exceeded. The standardized

FIGURE 3.7. Torsion testing machine. These machines read torque in in-lb or Kg-cm.

method for determining the yield point in a material is by the 0.2 percent offset method. This means the yield point is determined on the stress-strain diagram by drawing a line parallel to the straight-line portion of the diagram starting at 0.002 in. per inch offset. Strain is measured in inches per inch (0.000 in./inch) increments on the diagram. By drawing a parallel line, starting at 0.002 in. offset from the straight elastic line, the yield point is determined where this line crosses the curved portion of the diagram. Most yield points are determined by this method because a dip in the curve does not occur for most alloyed metals.

PLASTICITY

Metals undergo plastic flow when stressed at or beyond their elastic limits. For this reason, the area of the stress-strain curve beyond the elastic limit in Figure 3.5*b* is called the **plastic range.** It is this property that makes metals so useful. When enough force is applied by rolling, pressing, or hammer blows, metals can be formed when hot or cold into useful shapes. Many metals tend to work harden when cold formed, which in most cases increases their usefulness. They must be annealed for further cold work when a certain limit of plastic deformation has been reached.

BRITTLENESS

A material that will not deform plastically under load is said to be **brittle.** Excessive cold working causes brittleness and loss of ductility. Cast iron does not deform plastically under a breaking load and is therefore brittle.

A very sharp "notch" that concentrates the load in a small area can also reduce plasticity (Figure 3.8). Notches are common causes of premature failure in parts. Weld undercut, sharp shoulders on machined shafts, and sharp angles on forgings and castings are examples of unwanted notches (stress raisers).

STIFFNESS (MODULUS OF ELASTICITY)

Stiffness is expressed by the **modulus of elasticity,** also called Young's modulus. Within the elastic range, if the stress is divided by the corresponding strain at any given point, the result will be the modulus of elasticity for that material. Therefore, the modulus of elasticity is represented by the slope of the stress-strain curve below the elastic limit.

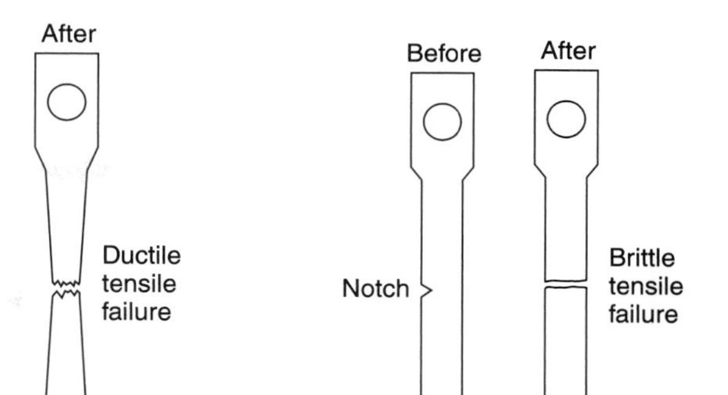

FIGURE 3.8. A sharp radius, a notch or gouge produced when machining or a fold or crack created when forging metal can cause a normally ductile metal to exhibit a brittle fracture mode.

$$\text{Modulus of elasticity in psi} = \frac{\text{stress}}{\text{strain}}$$

EXAMPLE

In Figure 3.5*b*, the stress at the elastic (or proportional) limit is 40,000 psi. Let us assume the strain is measured at that point as 0.0013 in./in. and the modulus of elasticity is represented by *E*.

$$s = \text{stress}$$
$$e = \text{strain}$$
$$E = \frac{s}{e} \text{ or } s = Ee$$

Thus,

$$E = \frac{40{,}000}{0.0013} = 30{,}769{,}230 \text{ psi}$$

And if the strain is measured at the 20,000 psi point of stress, it will be one-half that of the previous example because the elastic range is proportional. Thus,

$$E = \frac{20{,}000}{0.0065} = 30{,}769{,}230 \text{ psi}$$

The modulus of elasticity for some common materials is given in Table 3.1.

For an example of stiffness, two rods of equal dimensions are suspended horizontally on one end, with equal weights hanging on the other end. Of course, both rods will deflect the same amount if they are made of the same steel. Even if one rod were made of mild steel and the other of hardened tool steel, both would still deflect the same amount within the elastic range (Figure 3.9). The reason is that all steels have about the same modulus of elasticity. If one rod were made of tungsten carbide, the results would be quite different because the carbide rod would deflect much less than the steel since

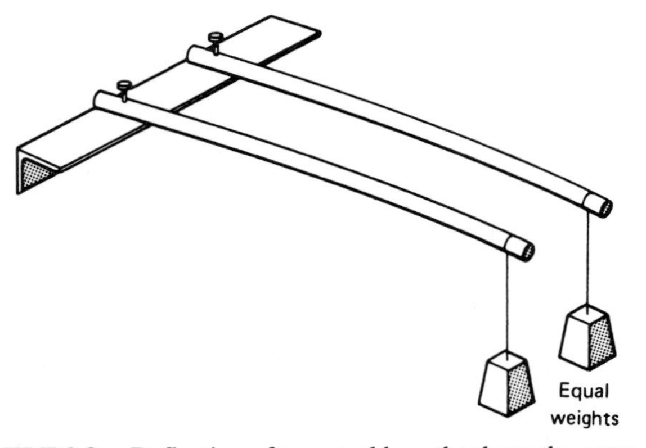

FIGURE 3.9. Deflection of two steel bars that have the same modulus of elasticity. One is made of hardened alloy steel and the other is of low carbon steel. Both bars deflect the same amount.

its modulus of elasticity is considerably higher than that of steel. The deflection as related to the modulus of elasticity also applies to torsion as measured on a torsion testing machine. Tension and compression within the elastic range can also be tested.

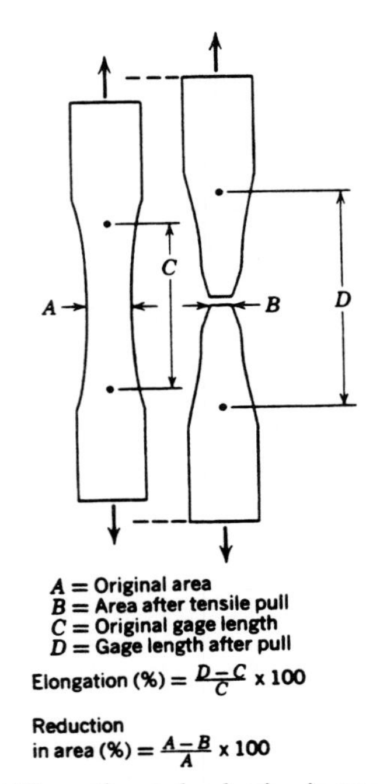

FIGURE 3.10. Elongation and reduction in area.

DUCTILITY

The property that allows a metal to deform permanently when loaded in tension is called **ductility.** Any metal that can be drawn into a wire is ductile. Iron, aluminum, gold, silver, and nickel are examples of ductile metals.

The tensile test is used to measure ductility. Tensile specimens are measured for area and length between gage marks before and after they are pulled. The **percent of elongation** (increase in length) and the **percent of reduction** in area (decrease of area at the narrowest point) are measures of ductility. The amount of elongation before the specimen breaks is an indication of the amount of plastic deformation (cold work) that can occur in that sample of metal. These data are useful for cold forming operations in manufacturing.

The percent of reduction in area refers to the necking down of the specimen and the difference in the original area and the area at the smallest point after the specimen breaks. This is a measure of the specimen's plasticity and/or brittleness. A small percentage of reduction (about 10 percent) would show very little deformation on the broken ends and would show mostly a brittle fracture, whereas an 80 percent reduction would show much necking down and very little brittle fracture. The method for calculating these values is shown in Figure 3.10.

Mechanical properties charts (Figure 3.11) for specific steels are provided by manufacturers. These are based on tensile tests and show the tensile strength, yield point, reduction of area, and elongation based on various hardnesses obtained with different tempering temperatures. As can be seen in Figure 3.11, these properties are often related to the hardness of the specimen. A very hard metal is usually brittle, having little ductility, and a soft metal is usually plastic, with very little tendency to be brittle under normal conditions. Impact testing is an excellent method for determining if a soft metal will exhibit brittle fracture.

MALLEABILITY

The ability of a metal to deform permanently when loaded in compression is called **malleability.** Metals that can be hammered or rolled into sheets are malleable. Most ductile metals are also malleable, but some very malleable metals such as lead are not very ductile and cannot be drawn into wire easily. Metals with low ductility, such as lead, can be extruded or pushed out of a die to form wire and other shapes. Some highly malleable pure metals are lead, tin, gold, silver, iron, and copper. Alloys usually tend to be less malleable and therefore require a softening heat treatment such as annealing to remove work hardening effects.

IMPACT STRENGTH/FRACTURE TOUGHNESS

Impact strength is a term used to describe the resistance of a material to fracture resulting from impact loading. Nearly all of the metallurgical properties play a role in

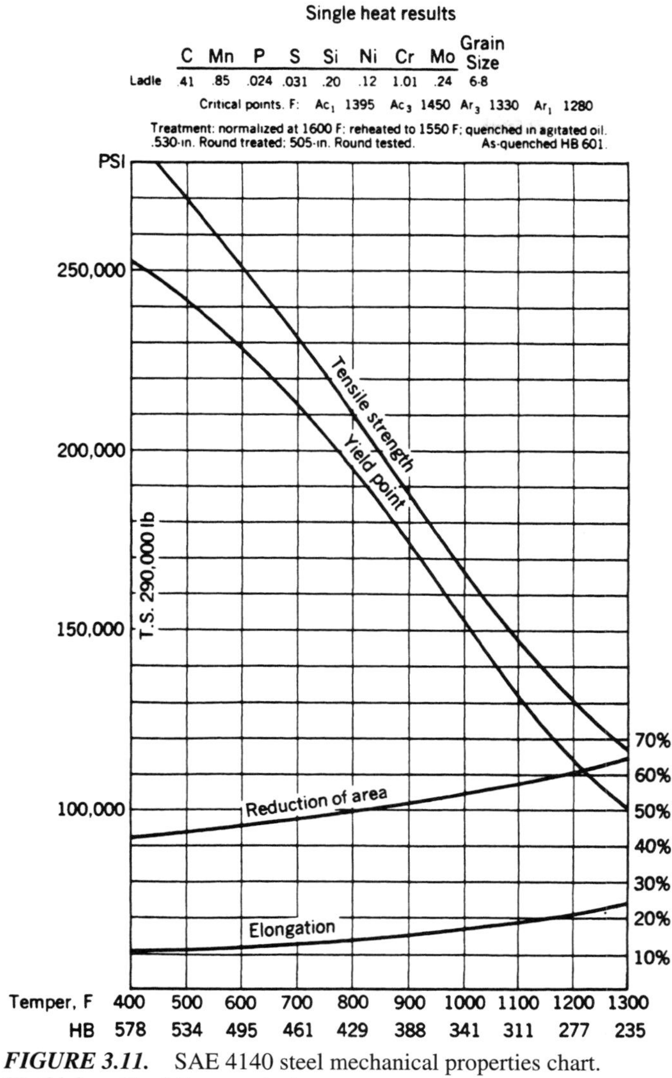

FIGURE 3.11. SAE 4140 steel mechanical properties chart. (*Bethlehem Steel Corporation*)

determining impact strength. A metal may exhibit high ductility and softness when tensile or hardness tested, but often will fracture in a brittle manner upon impact loading. The microstructure of gray cast iron contains flakes of graphite in a hard metal matrix; thus, it is the microstructure of gray cast iron that makes this alloy intolerant to shock loads. Low alloy steels, in the annealed condition, usually behave in a ductile fracture-resistant manner and would rather deform than fracture. As a rule, a large grain size is more malleable and ductile than a fine grain size in the same metal or alloy. The residual stresses and crystal structure produced by hardening heat treatments also influence fracture behavior. Cold can affect the fracture behavior of most steels, causing them to markedly lose fracture strength when the temperature is lowered. Many aerospace and government specifications now require material testing for impact strength to be conducted at subzero temperatures in addition to room temperature. A surface groove created during machining can serve as a source of fracture initiation. The design or shape of a part can greatly affect the ability to resist impact loading; for example, a notch or a sharp radius can lower the impact strength. Standardized test specimens, *Charpy* and *Izod,* were designed with a specific notch shape and dimensions to produce uniform test results when measuring impact strength.

Fracture toughness is a generic term for measures of resistance to extension of a crack (i.e., crack propagation or growth). This term is sometimes used to describe the results of an impact test, such as a test specimen exhibiting high test results in an impact test is said to have a high fracture toughness. The American Society for Testing and Materials (ASTM) publishes an annual book of ASTM standards that lists the terminology and practice common to fracture mechanics and mechanical testing. It is an excellent information source for standard test methods and analytical procedures. The ASTM procedure for notched bar impact testing is E-23.

As discussed, the device used to measure notch toughness is called the Izod-Charpy testing machine (Figure 3.12). The method of supporting and differences in specimens are what distinguish the Izod from the Charpy method (Figures 3.13*a* and 3.13*b*), but the results are practically the same. The base (Figure 3.14) has two leveling pads for adjusting the machine. The hammer straddles the anvil support and the striking bit is in the hammer. Standard test specimens (Figure 3.15) are used for either the Izod or the Charpy test. The testing machine consists of a vise where the test specimen is clamped. A weight on a swinging arm is allowed to drop (Figure 3.16). Note that the specimens are of standard geometry. The pendulum drops, strikes the specimen, and continues to swing forward; but it will not swing up as high as the starting position. The difference between the pendulum's beginning height and ending height indicates how much energy was absorbed in breaking the specimen. This energy is measured in foot-pounds (ft-lb). Tough metals absorb more foot-pounds of energy than brittle metals, and therefore the pendulum will not swing as far.

METALS AT LOW TEMPERATURES

As the temperature decreases, the strength, hardness, and modulus of elasticity increase for almost all metals. The effect of temperature drop on ductility separates metals into two groups: those that become brittle at low temperatures and those that remain ductile. Figure 3.17 shows examples of these two groups. Plastic deformation which occurs in Charpy and Izod test specimens is shown in Figure 3.18.

FIGURE 3.12. Izod-Charpy testing machine.

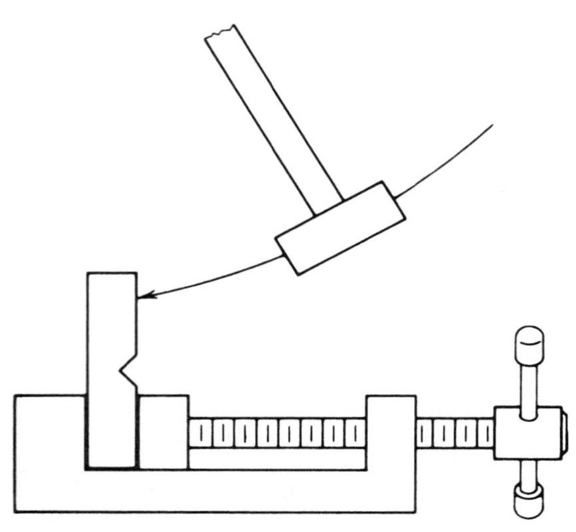

*FIGURE 3.13*a. The vertical mounting position for the Izod test specimen is shown as it is clamped in the vise.

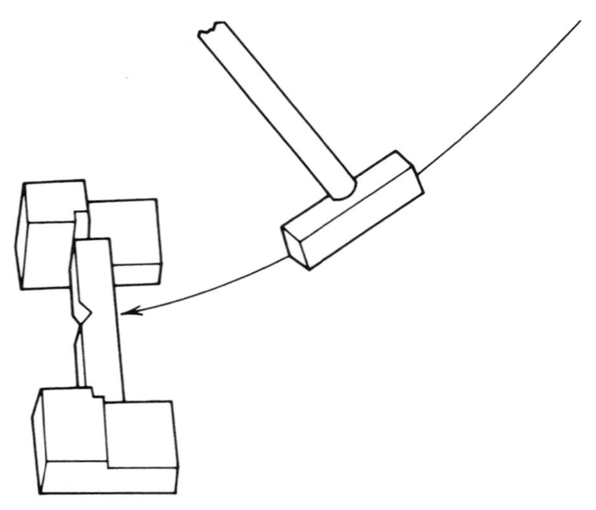

*FIGURE 3.13*b. The horizontal mounting position for the Charpy test specimen is shown as clamped in the vise.

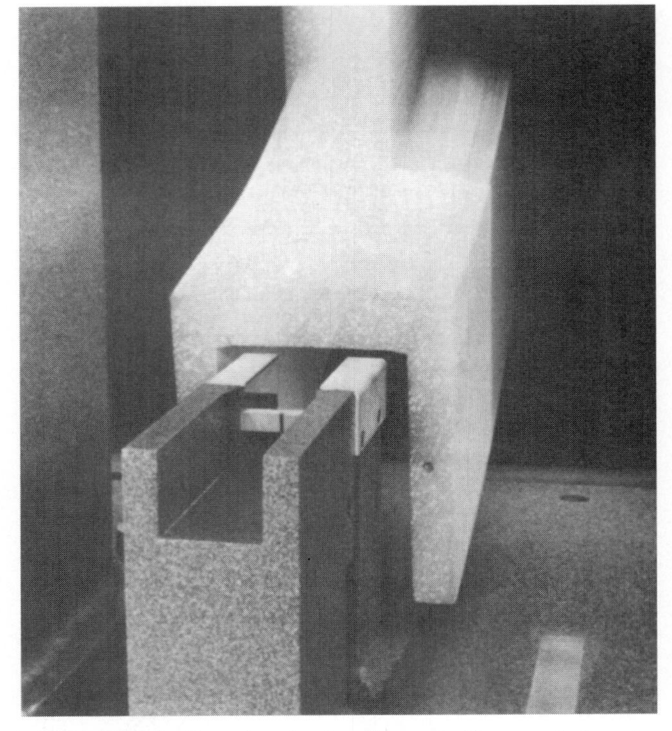

FIGURE 3.14. Base showing leveling pads. Hammer is dropping and about to strike the specimen.

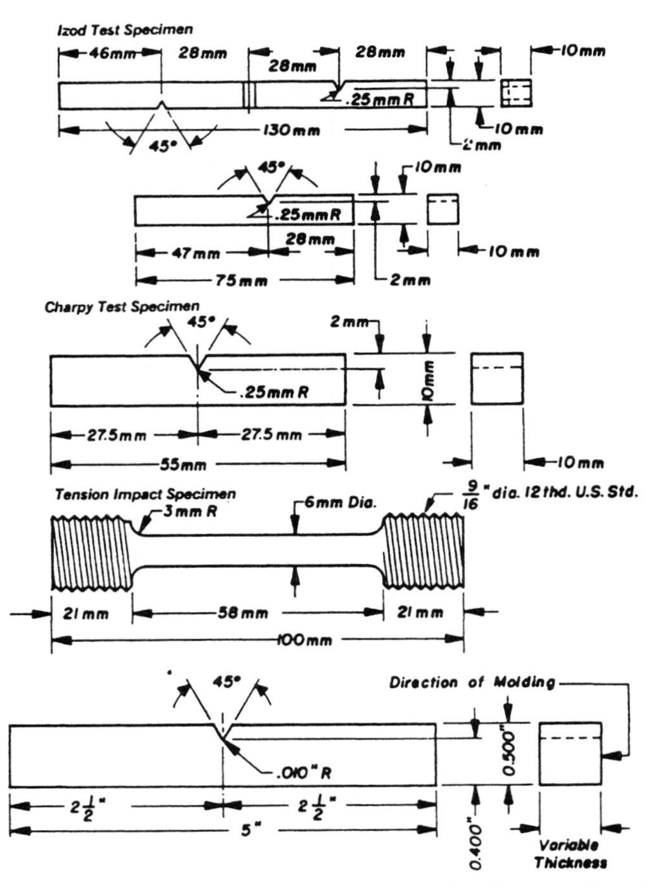

FIGURE 3.15. Test specimen specifications for Charpy and Izod tests. (*Tinius Olsen Testing Machine Company, Inc.*)

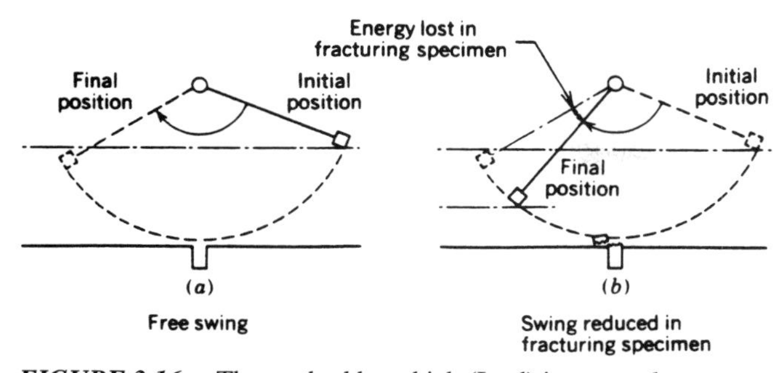

FIGURE 3.16. The method by which (Izod) impact values are determined. (*Machine Tools and Machining Practices*)

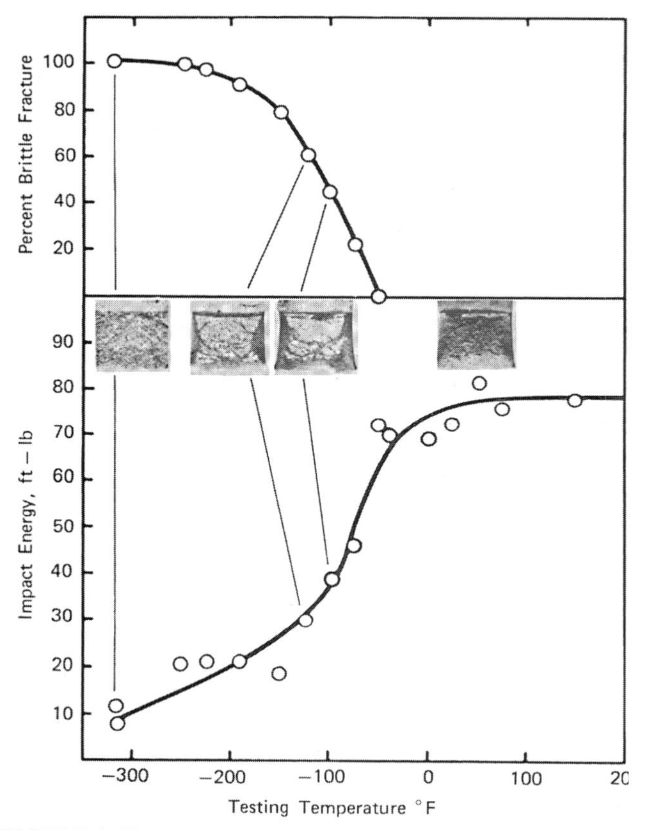

FIGURE 3.17. Appearance of Charpy V-notch fractures obtained at a series of testing temperatures with specimens of tempered martensite of hardness around Rockwell C 30. (*Courtesy of Republic Steel Corporation*)

Metals of the group that remain ductile show a slow, steady decrease in ductility with a drop in temperature. Metals in the group that become brittle at low temperatures show a temperature range where ductility and, most important, toughness drop rapidly. This range is called the transition zone. The Izod-Charpy impact test is the most common method used to determine the transition zone of metals. When the notch-bar specimens show half brittle failures and half ductile failures, the transition temperature has been reached. When parts for low-temperature service

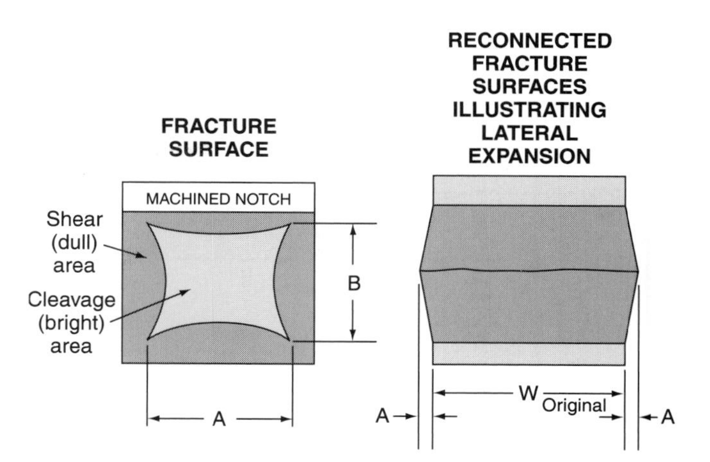

FIGURE 3.18. A and B indicate areas on the fracture surface where percent shear is determined. The fracture surfaces placed together illustrate where lateral expansion is measured.

are designed, the operating temperature should be well above the transition temperature. Nickel is one of the most effective alloying elements for lowering the transition temperature of steels. Following are some examples of operating temperatures for a few alloys and metals.

1. For operating temperatures as low as −50°F.
   a. Killed low carbon steel.
   b. Three percent nickel low carbon steel.
2. For operating temperatures as low as −150°F.
   a. Six percent nickel low carbon steel.
   b. Stainless steels with 8 percent nickel.
3. For operating temperatures below −150°F.
   a. Stainless steels with 8 percent nickel or more.
   b. Nine percent nickel steel.
   c. FCC metals such as aluminum or monel.

FATIGUE

When metal parts are subjected to repeated loading and unloading, they may fail at stresses far below their yield strength with no sign of plastic deformation. This is called a **fatigue failure.** When designing machine parts that are subject to vibration or cyclic loads, fatigue strength may be more important than ultimate tensile or yield strength.

Fatigue testing machines put specimens through many cycles of loading at a given stress. The results of repeated tests at different stresses can be plotted on a graph called a stress cycle diagram (Figure 3.19). The **fatigue limit,** or endurance limit, is the maximum load

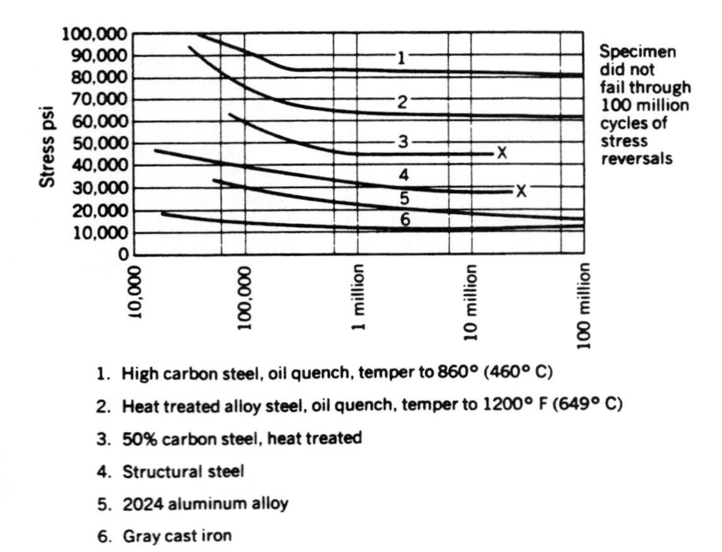

FIGURE 3.19. Relation between fatigue limit and tensile strength. The fatigue limit of steel is approximately 45 to 50 percent of its tensile strength up to about 200,000 lb. Repeated stresses in excess of this fatigue limit cause ultimate failure. (*Machine Tools and Machining Practices*)

in pounds per square inch that can be applied an infinite number of times without causing failure. Ten million loading cycles are usually considered enough to establish fatigue limits.

Fatigue life can be enhanced by smooth design. Avoiding undercut in welds, sharp corners and shoulders, and deep tool marks in machined parts will help eliminate stress raisers (notches) and thereby increase fatigue life.

To determine the material design limitations for parts used in fatigue applications, direct measurement of the mechanical properties must be conducted. The tensile specimen must be heat treated from the same lot of material and is often machined from an integral test coupon or an area designated as **stress critical.** The objective of the tensile test is to determine the actual **elastic limit.** Because fatigue is an elastic function with many million cycles of load fluctuation, elasticity is key in determining endurance limit. In most items designed for fatigue applications, the part is never cyclic loaded at more than half the stress determined by direct measurement of the elastic limit, especially in cases when the endurance limit is not determined. Modern fatigue testing machines use spectrum load data developed from direct measurements made when the part is in actual use. Aircraft manufacturers will apply strain gauges and accelerometers in key locations on the plane to measure the loads during actual flights and especially when firing weapons. The spectra data accumulated during flight testing are fed into the fatigue testing machine computer and applied to the fatigue test specimen (see Figure 3.20).

Fatigue testing is often conducted on standardized test coupons such as that shown in Figure 3.21. Fatigue crack growth or propagation studies are conducted on metals and alloys to determine the influence of microstructure, heat treatment, and physical defects on crack growth.

CREEP STRENGTH

Creep is a continuing, slow plastic flow at a stress below the elastic limit of a metal. Creep is usually associated with high temperatures but does occur to some ex-

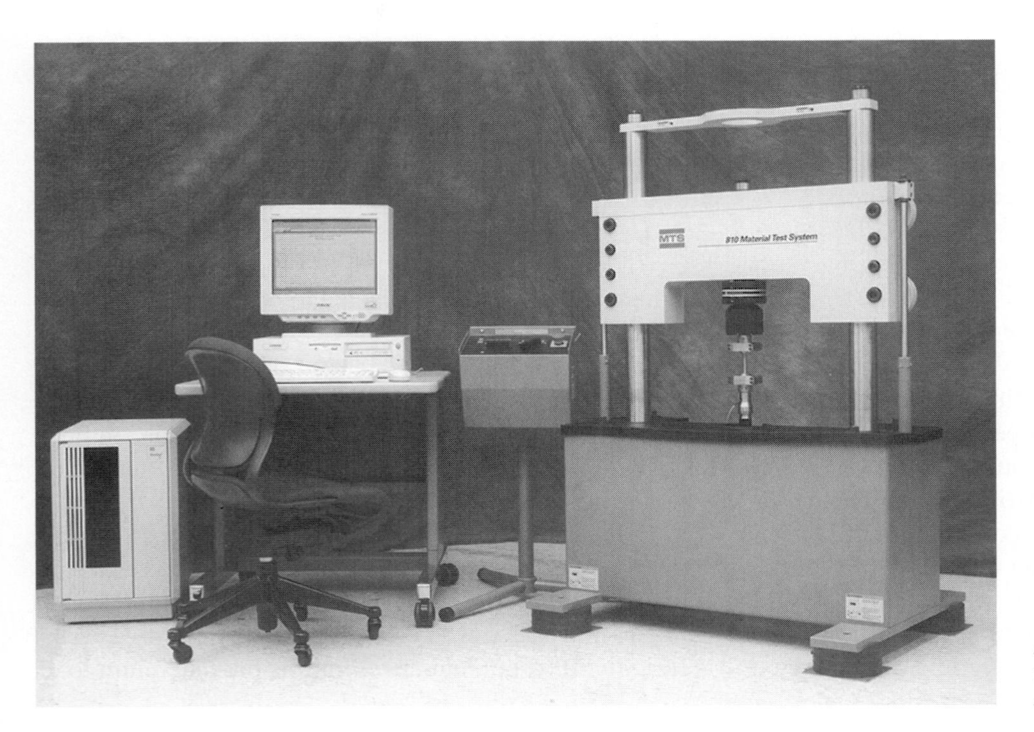

FIGURE 3.20. MTS fatigue testing machine. (*Photograph Courtesy of MTS Systems Corporation*)

FIGURE 3.21. A metal-foil gage is applied to the surface of a fracture propagation fatigue test specimen to monitor real crack growth. (*Photo courtesy of MTS Systems Corporation*)

FIGURE 3.22. One type of creep test machine. (*Photo courtesy of the Bureau of Mines*)

tent at normal temperatures. Low-temperature creep can take months or years to alter machine parts that are habitually left in a stressed condition. As the temperature increases, creep becomes more of a problem. Creep strength for a metal is given in terms of an allowable amount of plastic flow (creep) per 1000-hour period.

Table 3.2 gives creep strength for some alloys. Note that strengths are given for 1 percent creep per 10,000 hours at 800°F and for 1 percent creep per 100,000 hours at 1200°F. Stress to failure is also given.

Creep in metals at elevated temperatures is tested on machines such as the one shown in Figure 3.22. Metal is heated to a specified temperature, then a given load or stress is applied. A recorder makes a graph of the elongation movement of the specimen (creep) over a period of many hours.

TABLE 3.2. Creep strengths for several alloys

		Creep Strength (psi)		
Alloy	**70°F Tensile Strength (psi)**	**800°F—Stress for 1 Percent Elongation per 10,000 Hr**	**1200°F—Stress for 1 Percent Elongation per 100,000 Hr**	**1500°F—Stress to Failure**
0.20 percent carbon steel	62,000	35,100	200	1,500
0.50 percent molybdenum	64,000	39,000	500	2,600
0.08 percent to 20 percent carbon steel				
1.00 percent chromium	75,000	40,000	1,500	3,500
0.60 percent molybdenum				
0.20 percent C steel				
304 stainless steel	85,000	28,000	7,000	15,000
19 percent chromium				
9 percent nickel				

EXPANSION AND CONDUCTIVITY OF METALS (PHYSICAL PROPERTIES OF METALS)

Metals conduct heat better than nonmetals. Silver conducts heat the best of all metals. This ability to conduct heat and the ability to conduct electricity are related. Since silver is the best heat conductor, it is also the best electrical conductor. Figure 3.23 compares the thermal conductivity of some common metals and alloys. Note that in all cases the pure metals are better conductors than their alloys. Pure copper would be a better choice for electrical wiring than a copper alloy. A pure aluminum automobile radiator would conduct heat away from the water inside better than an aluminum alloy radiator would.

THERMAL EXPANSION

In almost all cases, solids become larger when heated and smaller when cooled. Each substance expands and contracts at a different rate. This rate is expressed in inches per inch per degree Fahrenheit and is called the coefficient of thermal expansion. Figure 3.24 compares the coefficient of thermal expansion of some common metals and alloys.

SOME CASES WHEN A KNOWLEDGE OF THERMAL EXPANSION HELPS

Knowing the thermal expansion coefficient of steel allows the engineer to calculate the sizes of the expansion joints in bridges and other steel structures. Heat treaters must be aware of differing expansion and contraction rates when heating and quenching steels. The internal expansion rate is often lower than the external rate when a piece of steel is heated rapidly. The stresses caused by uneven heating can cause cracking in metals with low ductility.

If a mechanic must remove a bronze bushing from a housing, or if heat is inadvertently applied to a bushing area such as by welding in the vicinity of the bushing, it may be loosened as a result of applying heat (Figure 3.25). The thermal expansion coefficient of bronze is almost twice that of steel. If the bushing and housing are heated, the bronze will try to expand at almost twice

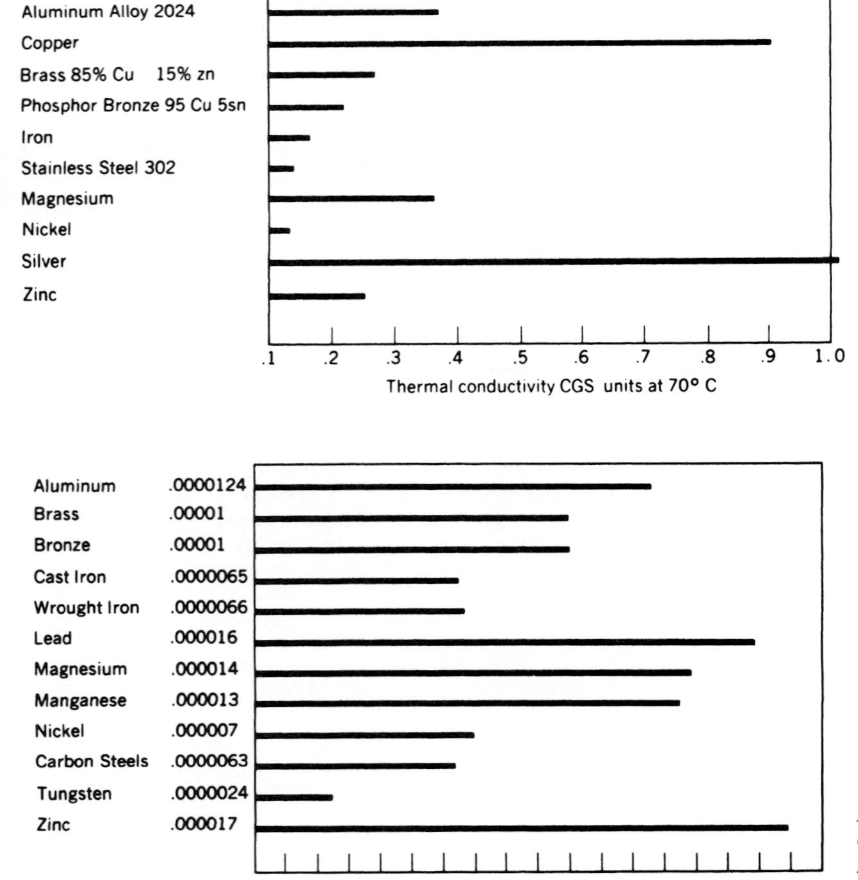

FIGURE 3.23. Comparison of thermal conductivity. (*Machine Tools and Machining Practices*)

FIGURE 3.24. Coefficient of thermal expansion per degree Fahrenheit per unit length. (*Machine Tools and Machining Practices*)

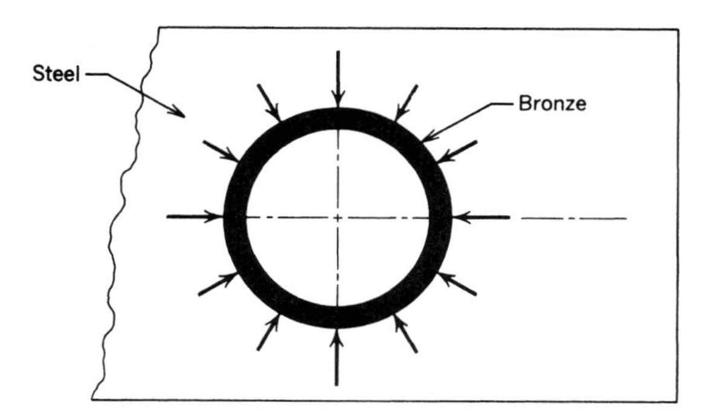

FIGURE 3.25. An application of thermal expansion of metals. (*Machine Tools and Machining Practices*)

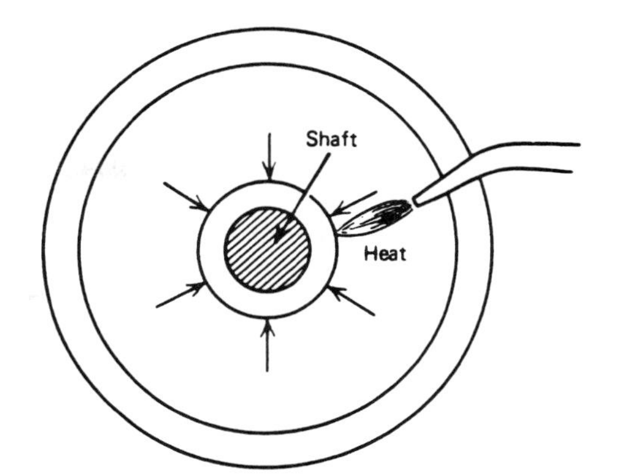

FIGURE 3.26. Heat applied to hub only causes it to expand. The restraint of the flange and rim makes the hub grip the shaft even more tightly.

the rate of steel for a given amount of heat. The steel restricts the bronze from expansion. The bronze then is stressed above its elastic limit and into the plastic range, where the bushing is deformed to a slightly smaller diameter. When both steel and bronze are cool, the bushing is now smaller than the steel bore, which makes it easy to remove from the hole.

A machinist has turned a bearing fit with a 0.0001-in. tolerance on a 4-in. steel shaft. The shaft is still hot from turning when the operator measures it. After a coffee break, he returns and checks his work to find that it is 0.0025 in. under size. What happened? The shaft had cooled down 100°F to room temperature, causing it to shrink in size. The following formula is used to calculate the amount of contraction after cooling to room temperature.

$$\text{Coefficient of expansion} \times \text{diameter} \times \text{rise in temperature (°F)} = \text{expansion}$$

or

$$0.0000063 \times 4\text{-in. dia} \times 100°\text{F} = 0.0025 \text{ in.}$$

If a 1 in. in diameter steel shaft would expand 0.0000063 in. for 1°F rise in temperature, it would expand 0.00063 in. for a 100°F rise. If the shaft were 4 in. in diameter and had a 100°F rise, the expansion would be 0.0025 in. If machined at that temperature, it would contract the same amount when cooled. All lathe operators are familiar with the lengthwise expansion of turned workpieces, which causes the dead center to tighten and heat up.

Temperature differences on workpieces, especially thin parts, can cause them to "crawl" on the milling machine table. The heat generated by milling with carbide cutters, when no coolant is used, is most likely to cause this problem.

Mechanics and machinists sometimes need to remove a sprocket or gear that has been pressed onto a shaft. Since the bore is often corroded and has no lubrication, the pressure required for removal is often many times greater than that required to press it on. When the available press is not capable of removing the gear or sprocket, heat is often applied to the hub to expand it so it will come off easily. The trouble with this process is that unless a high heat input rate is used, the shaft will heat along with the hub and it will not help. Even with a high heat input rate, if only the hub is heated on a large gear or sprocket with a solid flange or rim, the bore may not be expanded. In fact, just the reverse may happen: The bore is squeezed tighter than ever because of the constraint placed upon the hub by the cold flange and rim (Figure 3.26). The expansion of the hub from the heat is deflected inward instead of outward. The solution is to heat the entire outer portion of the gear or sprocket gently and then to heat the hub area quickly with a high heat input while pressing on the shaft.

Welders often experience expansion and contraction problems because of the heat involved in welding. In structural welding, large beams or weldments are sometimes preheated prior to welding. If a beam were 100 ft. long and preheated to 300°F, the amount of expansion could cause it to buckle if no allowance were made for expansion. Its total expansion would be

$$0.0000063 \times 100 \times 12 \times 300°\text{F} = 2.268 \text{ in.}$$

Steel structures such as bridges have expansion joints to allow for differences in temperature that can be as much as 150°F.

UNUSUAL PROPERTIES OF SOME METALS

Amorphous Metals

Metals, as we usually think of them, solidify from the molten state into crystals, called grains. When metals are fractured, these crystals can often be seen on the broken ends. The atoms are arranged in uniformly spaced rows within each grain and are oriented randomly from grain to grain. In metallic alloys, crystallization normally starts within a microsecond of the temperature's dip below the metal's melting point.

Amorphous metals do not form into crystals when solidifying, but instead, the atoms are in random locations, resembling the structure of glass. This is why they are often called metallic glasses. The phenomenon of amorphous metals is produced by extremely rapid quenching rates and with certain alloys that lend themselves to the process. Some of these alloys are Al-Ni, Al-Cu, Al-Fe, Ag-Cu, and Cu-Zn. One method of rapid quenching is by the "splat" method. Molten metal is directly shot from a "gun," from which the metal is thrown at a high velocity onto a water-cooled copper plate. The thin "splat" foil thus formed is an amorphous metal.

In the method known as melt spinning, a jet of molten metal alloy liquefied by radio-frequency energy is directed at the outside edge of a copper disk spinning at 6000 revolutions per second. The metal solidifies at a rate of between 10,000 to 1 million degrees per second, which is required to produce an amorphous metal. An amorphous solid ribbon forms on the edge of the copper disk. This method shows great promise for future manufacturing of these amorphous alloys in large quantities.

These amorphous metals possess a very high strength and are easily magnetized, a feature that makes them especially attractive to the electrical industry. This metal makes possible an advanced transformer technology that has the potential of reducing core losses as much as 80 percent. No doubt, many more uses for this unusual kind of metal will be discovered in the near future.

Metals with a Memory

Shape memory alloys are not new; certain brasses, alloys of copper and zinc, and other alloys were known long ago that could alter their shape with a change in temperature. However, some new alloys show this unusual property to a marked extent. Nitinol, so named because it is an alloy of nickel and titanium and because it was discovered at the U.S. Naval Ordinance Laboratory (NOL) in 1958 by metallurgist William Buehler, shows great promise for future applications.

When nitinol is heated to a high temperature, formed to a desired shape while hot, and then quenched, it has a martensite structure. This metal in this state is not hard but instead is unusually soft and pliable. If it is now reformed cold to a new shape, it will still "remember" its original shape. Now, if the soft metal is heated to its particular transition temperature, it will suddenly reshape itself to its original configuration. The transition temperature can be adjusted by the proportions of nickel and titanium, so that small differences of temperature can produce the memory response. A heat engine using nitinol wire loops was developed in 1973 by Ridgway Banks at the Lawrence Laboratory in Berkeley, California. While this engine was running it was noted that nitinol has a "double memory." When the remembered shape is again cooled in water, it begins to "learn" its second shape in the soft condition during repeated cycles as in a heat engine. This phenomenon is known as two-way memory. Even by the use of a single memory response, a spring can be used to pull the nitinol back to the original position when it is cooled from the transition temperature, since it becomes soft at the lower temperature.

Considerable experimentation is being done on this new metal, and one surprise is that after millions of cycles, nitinol shows no signs of fatigue or wearing out its shape memory response; indeed, it appears to get stronger with use. This metal has almost unlimited possibilities, such as in the areas of space technology, surgical implants, electrical power generation, and automobile engines.

SUPERCONDUCTING METALS

Silver, copper, and aluminum wires are all very good conductors of electricity, but not *perfect* conductors. Because of the resistance of wire to the passage of electricity, great losses in electrical transmission systems are experienced. Recently, several superconducting metals and ceramics have been developed that offer no resistance to the flow of electricity at very low (cryogenic) temperatures. Superconducting metals and alloys lose their electrical resistance only at very low temperatures. In 1962, the first superconducting alloy of niobium and zirconium (NbZr) was produced. This metal had to be cooled to 2 to 5 K. This was not a very ductile metal, and surface hardening occurred during wire-drawing operations. A niobium-titanium alloy (NbTi) was later developed and is still in use, since it is more ductile and multifilaments can be made of it. A niobium-tin alloy (NbSn) has the advantage of a higher critical temperature. However, to keep these alloys at the superconducting temperature, they must be immersed in liquid helium. A new sintered ceramic superconductor, composed of $Y_1Ba_2Cu_3O_7$, was developed in the laboratories of Teledyne Wah Chang. This ceramic material has the advantage of a higher transition temperature where it becomes a superconductor, about 90 K. Therefore, liquid nitrogen could be used instead of the more costly liquid helium.

Figure 3.27 shows a sintered ceramic superconductor hanging by a string near a magnet. It is being repelled by the magnet, showing the Meissner effect, the creation of diamagnetism within the sample, indicating its superconductivity. This new material is being considered for use in high-speed communications and will possibly be made into fibers or strands and encapsulated in a tube.

Other possible uses of superconducting materials are in the area of high-energy physics, where the greatest interest is displayed in superconductors because great field strengths are needed for high-energy particle accelerators. Superconductors may be used in the area of nuclear fusion, the same process that occurs in the core of the sun. Large confinement magnetic force fields are needed to contain the high-temperature plasma. Superconducting windings would greatly help this project.

FIGURE 3.27. Sintered ceramic superconductor (hanging from a string) repelling the magnetic force away from the magnet. This demonstrates that it is a superconductor at cryogenic temperatures. (*Teledyne Wah Chang Albany*)

SELF-EVALUATION

1. Does creep occur within the elastic or plastic range of steels?
2. Explain the relative rate of failure in creep. Are creep failures sudden or do they take a long time to develop?
3. What happens below the transition temperature of a metal?
4. What happens to the properties of hardness, strength, and modulus of elasticity with a decrease in temperature?
5. Name an alloying element that can be added to steel to lower its transition temperature.
6. Describe the three categories of hardness and how they can be measured.
7. Name the three basic stresses.
8. A 2-in. square bar in tension has a load of 40,000 lb. What is the unit stress?
9. Explain what is meant by ductility.
10. Explain what is meant by malleability.
11. In what ways can fatigue strength be improved?
12. What is the correlation between metals for electrical and thermal conductivity?
13. In what state does a metal conduct heat best, alloyed or unalloyed?
14. How is the rate of thermal expansion for a particular material expressed?
15. Metals tend to solidify into grain structures when cooled normally. Amorphous metals have no grain structure. How are they made?
16. What two things happen to certain metallic alloys when immersed in liquid helium?

CASE PROBLEM: RAILROAD TANK CAR FRAME COLLAPSE

A railroad tank car frame made of welded mild steel suddenly fractured in several places when liquid oxygen spilled onto the frame after the tank was overfilled. The steel contained 0.23 percent carbon and 0.39 percent manganese. It was rimmed steel in the as-rolled condition. What was the cause of this sudden failure? Can it be corrected? How? (Liquid oxygen can be approximately –300°F.)

WORKSHEET 1

Objectives
1. Learn to use the tensile testing machine.
2. Calculate elongation, reduction in area, and unit stress of a pulled specimen.

Materials A tensile testing machine and a prepared specimen.

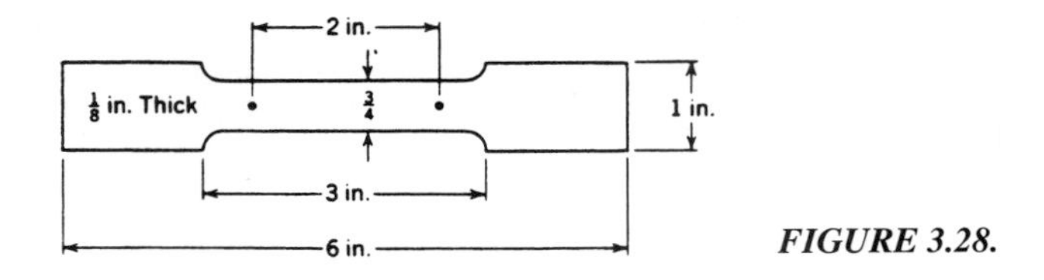

FIGURE 3.28.

Procedure

1. Prepare a specimen of mild steel as shown in Figure 3.28. Punch the gage marks exactly 2 in. apart.
2. With a micrometer, measure the width at the narrowest point and the thickness.
3. Record this information.
4. Set up the tensile tester with flat gripping jaws and a 0 to 10,000 lb dial if it is the interchangeable type.
5. Place the specimen in the tensile testing machine and pull it to rupture.
6. Record the yield point.
7. Record the ultimate load.
8. Record the breaking load.
9. Remove the specimen and fit the broken pieces together. Measure the width and the thickness of the specimen at the narrowest point; measure the length between gage marks and record the information.
10. Calculate the elongation and reduction in area using the formulas that are given in Figure 3.10.
11. Calculate the unit stress using the following formula (use ultimate load).

$$\text{Unit stress} = \frac{\text{load}}{\text{original area}}$$

Note The original area is equal to the width $\times$ the thickness at the narrowest point before pulling.

Conclusion Do you think this was a ductile metal? Why?

Data Tensile specimen

Length between gage marks =
Thickness =
Width =
Area =

Tensile specimen after pull
- Length between gage marks =
- Ultimate load =
- Breaking load =
- Yield point =
- Thickness =
- Width =
- Area =

Results
- Ultimate unit stress =
- Reduction in area, percent =
- Elongation, percent =

WORKSHEET 2

Objective Determine the relative notch toughness of a carbon steel in its annealed and hardened state.

Materials An Izod-Charpy testing machine and prepared specimens.

Procedure

1. Prepare two Izod or Charpy specimens of annealed or as-rolled high carbon steel, SAE 1080 to 1095 according to specifications in Figure 3.15 or 3.16.
2. Harden one specimen in a water quench from 1500°F and temper to 400°F.
3. Test both specimens for hardness. Record your results.
4. Test both specimens on the Izod-Charpy machine. Record your results.

Conclusion Which metal shows greater toughness? Which is more brittle?

Data

	Specimen 1 (Soft)	Specimen 2 (Hard)
Hardness		
Ft-lb		

WORKSHEET 3

Objective Demonstrate the difference in thermal conductivity between copper and stainless steel.

Materials A convenient heat source such as a Bunsen burner or propane torch, a strip of copper, and an equally shaped strip of stainless steel.

Procedure

1. Set up the burner so the flame is at the ends of the strips arranged as in Figure 3.29.
2. Mark the opposite ends with a 200°F temperature crayon. Instead of the crayon, a wooden match will work, as shown in Figure 3.29.
3. Note which crayon mark melts first or which match lights first.

Conclusion Which metal has the higher thermal conductivity?

Note You may demonstrate the difference between thermal conductivity of other metals such as steel and stainless steel if you use closely controlled conditions.

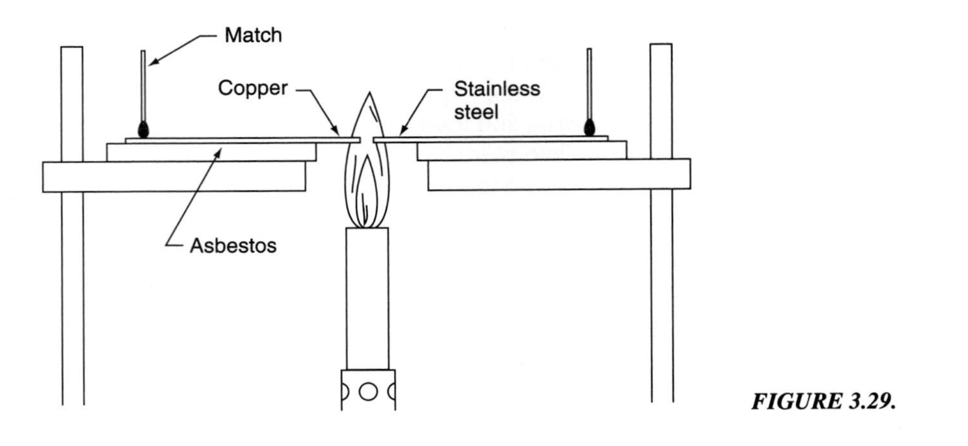

FIGURE 3.29.

WORKSHEET 4

Objective Determine the relative scaling characteristics of two metals.

Materials A furnace or a Bunsen burner, one piece of mild steel about 1⁄16 × 1 × 4 in., and one piece of stainless steel about 1⁄16 × 1 × 4 in.

Procedure

1. Place the two specimens over a Bunsen burner or in an electric furnace and allow them to remain at a yellow heat for 1 hour.
2. Check and record observations at 10-minute intervals.

Conclusion What happens to the mild steel?
What happens to the stainless steel?
Which one would you suggest for high-temperature service?

Face Centered Cubic Aluminum Crystals
(SEM Fracture-Surface)

CHAPTER 4

The Crystal Structure of Metals and Phase Diagrams

What are the forces that hold metals together? Why do metals behave as they do? These and other questions relating to the atomic and crystalline structure of metals will be discussed in this chapter.

Constitutional diagrams (sometimes called alloy or phase diagrams) are a useful means of explaining and understanding the behavior of metals. Many very complex diagrams for various alloys are used by metallurgists, but only several of the simple types will be used in this chapter. These alloy diagrams are the "road maps" of the metallurgist that help to develop new alloys. The iron-carbon diagram is basic to an understanding of heat-treating iron and steel.

OBJECTIVES **After completing this chapter, you should be able to:**

1. Describe the various phases of crystalline structures of metals.
2. Describe the various aspects of solid solutions.
3. Conduct a Metcalf experiment and determine the approximate grain size in steel samples.
4. Prepare metal specimens for microscopic study by polishing and etching.
5. Demonstrate an understanding of phase diagrams by recognizing their parts.
6. Establish relative carbon content by microscopic evaluation.
7. Identify various cast iron compositions by microscopic examination.

The great utility of metals is due to their elastic behavior to a certain level of stress followed by a plastic behavior at higher levels of stress. Along with ceramic materials that are brittle, or polymers such as wood or leather, metals play a unique and useful role in the economy.

The physical world is composed of matter and energy. **Matter** is defined as anything that occupies space; it exists in three forms: solid, liquid, and gas. Energy is the ability to do work. It can be either potential or kinetic. Potential energy may be chemical or physical; it is "stored" energy, for example, as water in a reservoir or as the power in explosives. Kinetic energy is energy in motion that is doing work. The laws of conservation of matter and energy state that matter can neither be created nor destroyed but may be changed from one form to another. This is true of the everyday world, but Einstein's discoveries leading to this atomic age revealed that matter and energy are related. A small amount of matter releases a tremendous amount of energy under certain conditions.

Matter is composed of atoms that are too small to be seen with the aid of ordinary microscopes, but the outline of molecules has been detected with such devices as the ion field emission microscope and the electron microscope (Figure 4.1). The **molecule** is defined as the smallest particle of any substance that can exist free and still exhibit all the chemical properties of that substance. A molecule can consist of one or more atoms. Atoms of different materials vary only in number and arrangement but not in the type of their parts. Matter composed of a single kind of atom is called an **element.** There are more than 100 elements, and new ones are being discovered. Those that have atomic numbers greater than 92 are not found in nature but are produced by atomic reactions. Elements are classed in two groups: metals and nonmetals. A **compound** is composed of two or more elements combined chemically. A **mixture** is two or more elements or compounds physically combined in the same way in which two fine powders are mixed together.

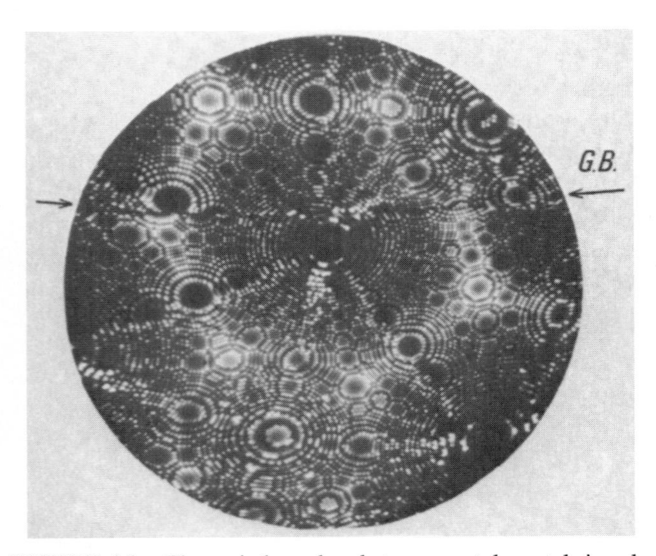

FIGURE 4.1. The grain boundary between metal crystals is only a few atoms wide, as seen in a field ion micrograph. Here the tip of a tungsten needle is enlarged several million diameters. Each bright dot represents a tungsten atom. The distance between dots is approximately 5 angstroms. (*John J. Hren*)

ATOM

An atom resembles a miniature solar system that has orbits in many planes. Its chief parts are shown in Figure 4.2. The nucleus of the atom consists of protons and neutrons. The protons have a positive electrical charge. Neutrons weigh essentially the same as a proton but are neutral in charge. Revolving at high speed around the nucleus are much smaller particles, called electrons. Electrons are negatively charged, which means that they are strongly attracted to the positively charged nucleus.

Each atom has preferred electron paths or orbits, called shells. The number, arrangement, and spin of the electrons in these shells in combination with the positively charged nucleus determine the kind of atom and its characteristics. Each shell contains a given number of electrons for each atom. The sum of the electrons in all shells equals the atomic number. Also, the number of protons in the nucleus equals the atomic number. For example, the atomic number of hydrogen is 1, and it contains one proton and one electron.

The combining power of an atom is called its **valence.** The electrons on the outer (and sometimes the second) shell, called the valence electrons, are the most important in determining chemical and physical properties. When the valence shell in a free atom has a full complement of electrons, it is said to have zero valence. When an atom has more or fewer electrons in the outer shell than in its uncombined or free state, it is called an ion and possesses an electrical charge. The charge is negative when extra electrons are present and positive when some are missing. Metal atoms are easily stripped of their valence electrons and thus form positive **ions.**

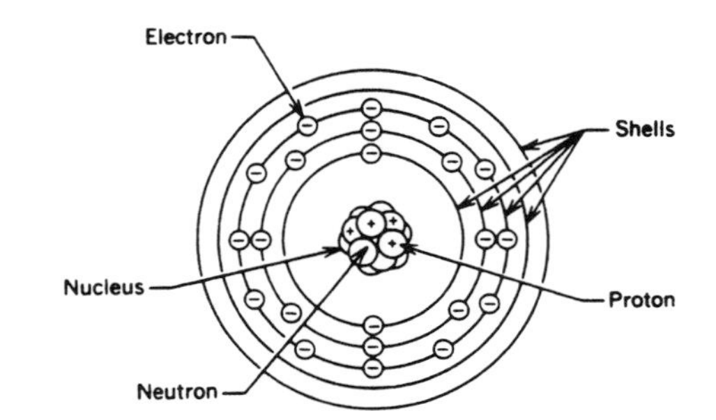

FIGURE 4.2. The atom. (*Machine Tools and Machining Practices*)

Each element is assigned a symbol that is derived from its name, such as O for oxygen and C for carbon. Some are derived from the Latin name. For example, the Latin for sodium is Natrium (Na), iron is Ferrum (Fe), silver is Argentum (Ag), lead is Plumbum (Pb), and gold is Aurum (Au).

The number of atoms or ions of each element in a compound is indicated by a subscript immediately following the symbol. H_2O indicates that in one molecule of pure water there are two atoms of hydrogen (H) and one of oxygen (O). A number placed in front of the symbol indicates the number of molecules of that substance. For example, $2CO_2$ indicates two molecules of carbon dioxide.

BONDING OF METALS

With this information, we can now go on to find out how metals are held together. There are four basic types of bonding arrangements that hold atoms together. They are ionic, covalent, metallic, and van der Waals forces. A molecule may be held together by a combination of several or all of these forces.

Ionic bonding holds dissimilar atoms together by the attraction of negative and positive ions. Sodium chloride (NaCl) is an example of ionic bonding (Figure 4.3), where a metal (sodium) loses its single valence electron to a nonmetal (chlorine) to complete its valence shell. The positively charged sodium atom and the negatively charged chlorine atom combine to become a neutral compound. The resultant structure of salt (Figure 4.4) is rather weak and brittle because the electrostatic attractions are very selective and directional.

Covalent bonding, or shared valence electrons, is a very strong atomic bond whose strength depends on the number of shared electrons. Covalent bonding is found primarily among nonmetallic elements such as

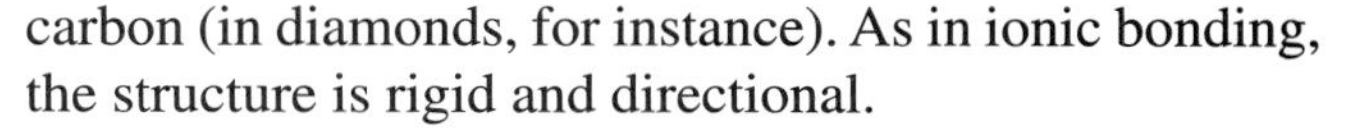

carbon (in diamonds, for instance). As in ionic bonding, the structure is rigid and directional.

For example, the oxygen atom has six electrons in the outer shell. This shell would like to be complete by having eight electrons. To make an oxygen molecule, two electrons are shared by each atom to provide a satisfactory arrangement (Figure 4.5).

The **metallic bond** is formed among similar metal atoms when some electrons in the valence shell separate from their atoms and exist in a cloud surrounding all the positively charged atoms. These positively charged atoms arrange themselves in a very orderly pattern. The atoms are held together because of their mutual attraction for the negative electron cloud (Figure 4.6). The free movement of electrons accounts for the high electrical and heat conductivity of metals and for their elasticity and plasticity. The metallic bond is very strong.

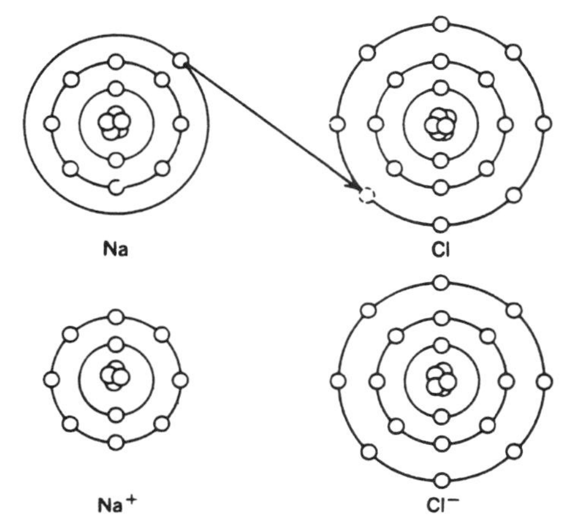

FIGURE 4.3. Ionic bonding. (*Machine Tools and Machining Practices*)

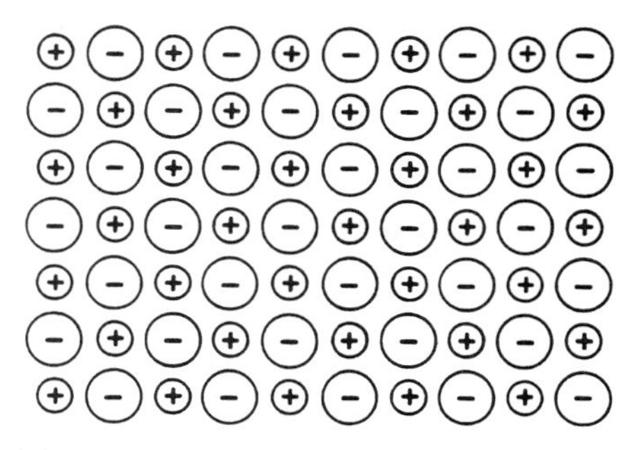

FIGURE 4.4. Lattice structure of salt. (*Machine Tools and Machining Practices*)

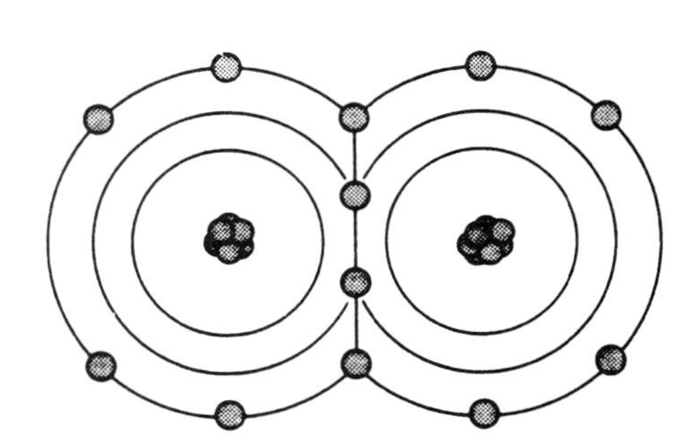

FIGURE 4.5. Oxygen molecule has a covalent bond. (*Machine Tools and Machining Practices*)

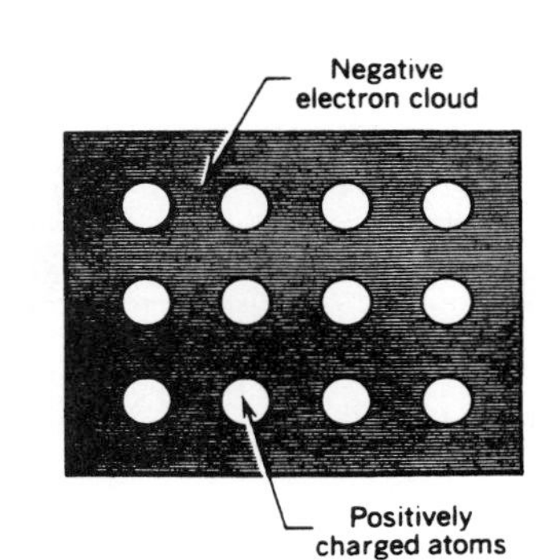

FIGURE 4.6. Metallic bond. (*Machine Tools and Machining Practices*)

Van der Waals bonding is found in neutral atoms such as the inert gases. There is only a very weak attractive force, and it is of importance in the study of metals only at very low temperatures.

METALS AND NONMETALS

Approximately three-quarters of all the elements are considered to be metals. Some of these are metalloids such as silicon or germanium. Some of the properties that an element must have to be considered a metal are as follows:

1. Ability to donate electrons and form a positive ion
2. Crystalline structure—grain structure
3. High thermal and electrical conductivity
4. Ability to be deformed plastically
5. Metallic luster or reflectivity

Metalloids, such as silicon, possess one or more of these properties, but they are not true metals unless they have all of the characteristics of metal.

CRYSTALLINE UNIT STRUCTURES

When metals solidify from the molten to the solid state, the atoms align themselves in orderly rows, which is an arrangement that is peculiar to that metal. This arrangement is called a **space lattice,** which can be considered a series of points in space. By means of the study of X-ray diffraction, the space lattice of the crystal has been determined for the different metals.

Metals solidify into six main lattice structures:

1. Body-centered cubic (BCC)
2. Face-centered cubic (FCC)
3. Hexagonal close-packed (HCP)
4. Cubic
5. Body-centered tetragonal
6. Rhombohedral

The three most common crystal patterns of unit cells are shown in Figure 4.7.

As solidification is taking place, the arrangement of the crystalline lattice structure takes on its characteristic pattern. Each unit cell builds on another to form crystalline needle patterns that resemble small pine trees. These structures are called **dendrites** (Figure 4.8).

The crystalline lattice structures begin to grow first by the formation of seed crystals or nuclei as the metal solidifies. The number of nuclei or grain starts formed determines the fineness or coarseness of the metal grain crystal structure. Slow cooling promotes large grains and fast cooling promotes smaller grains. The grain grows outward to form the dendrite crystal until it meets another dendrite crystal that is also growing. The places where these grains meet are called grain boundaries.

Figure 4.9 represents the growth of the dendrite from the nucleus to the final grain when metal is solidifying from the melt. The nucleus can be a small impurity particle or a unit cell of the metal. Iron, copper, silver, and other metals are composed of these **tiny grain structures,** which can be seen under a microscope when a specimen is polished and etched. This grain structure can also be seen with the naked eye as small crystals in the rough broken section of a piece of metal (Figure 4.10).

Grain Boundary

As the crystal structures grow in different directions, it can be seen that at the **grain boundaries,** the atoms are jammed together in a misfit pattern (Figure 4.11). Although the nature of the grain boundaries is not entirely understood, it is generally assumed to be an interlocking border in a highly strained condition. Grain boundaries are only about one or two atoms wide, but their strained condition causes them to etch differently; this means they may be observed with the aid of a microscope.

Body-Centered Cubic (BCC)

This cubic unit structure is made up of atoms at each corner of the cube and one in the very center. Steel un-

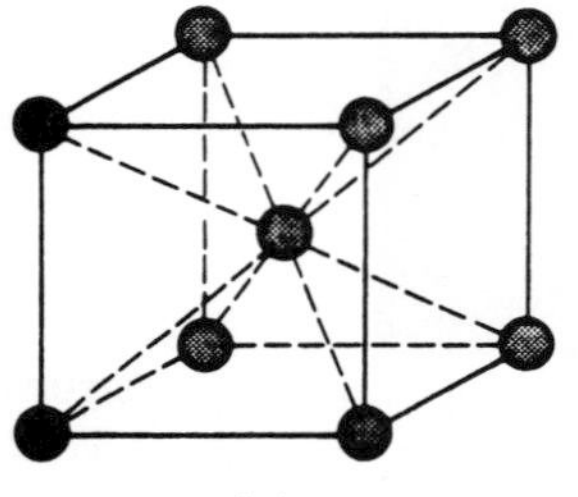

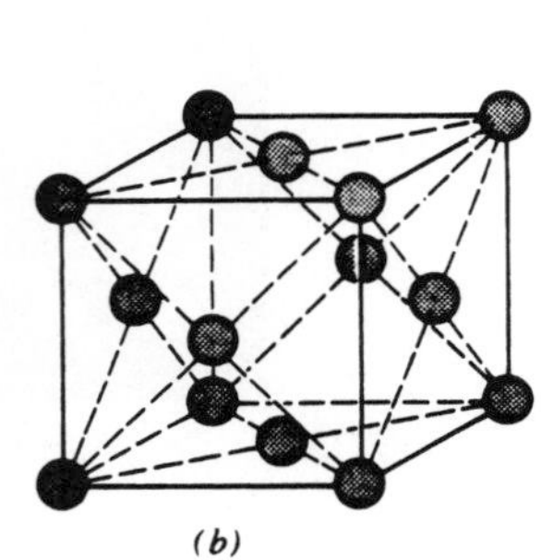

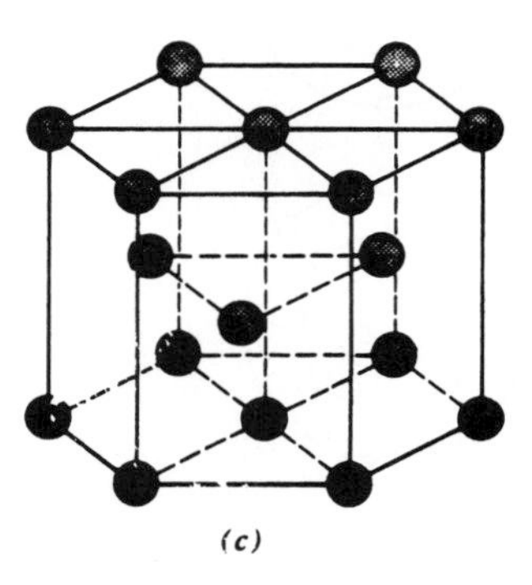

FIGURE 4.7. Three common unit cells of the space lattice are (*a*) body-centered cubic, (*b*) face-centered cubic, and (*c*) hexagonal close-packed.

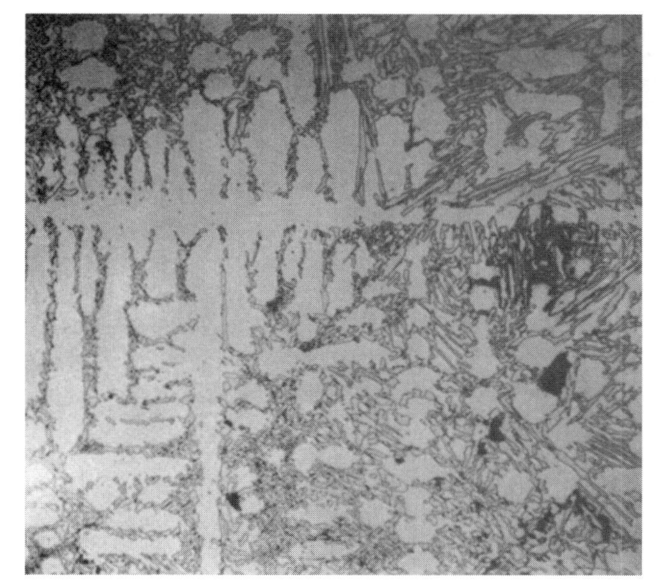

FIGURE 4.8. A typical dendrite formation is shown in this micrograph.

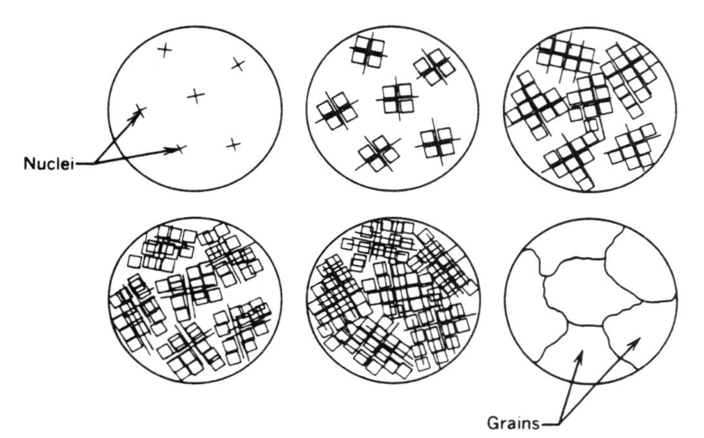

FIGURE 4.9. The formation of grains during solidification. (*Machine Tools and Machining Practices*)

der 1333°F has this arrangement, and it is called alpha iron or **ferrite.** Other metals such as chromium, columbium, barium, vanadium, molybdenum, and tungsten crystallize into this lattice structure. These cubes are identified within the lattice structure as seen in Figure 4.12. Body-centered cubic metals (see Figure 4.7*a*) show a lower ductility but a higher yield strength than face-centered cubic metals.

Face-Centered Cubic (FCC)

Atoms of calcium, aluminum, copper, lead, nickel, gold, platinum, and some other metals arrange themselves with an atom in each corner of the cube and one in the center of each cube face. When steel is above the upper critical temperature, it rearranges its atoms to this FCC structure and is called gamma iron or **austenite** (see Figure 4.7*b*).

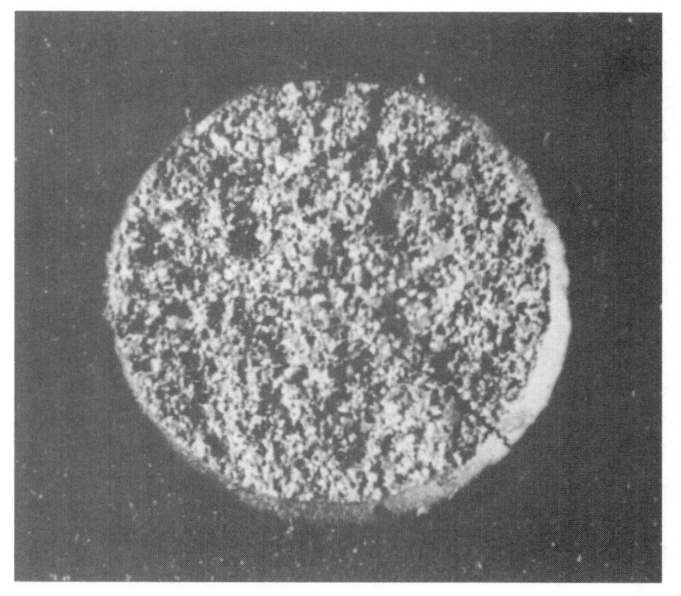

FIGURE 4.10. Broken section showing the crystalline structure. (*Machine Tools and Machining Practices*)

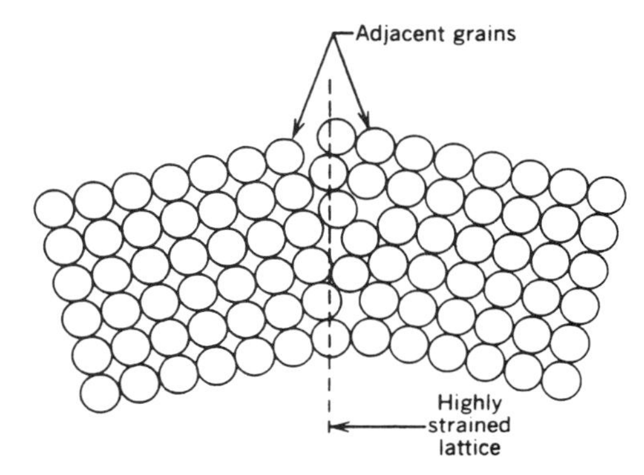

FIGURE 4.11. Grain boundary is highly strained. (*Machine Tools and Machining Practices*)

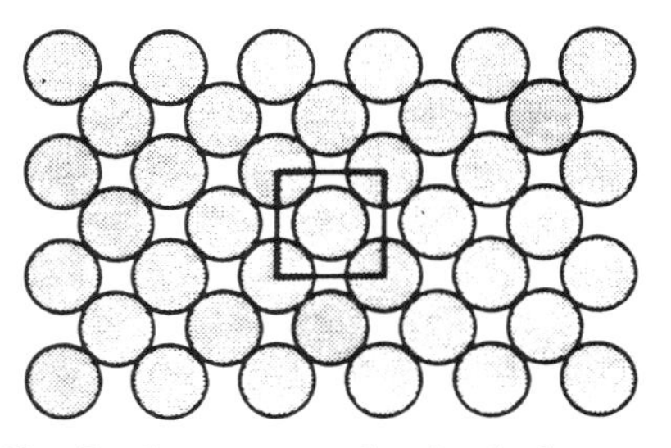

FIGURE 4.12. Lattice structure showing body-centered cubic formation. (*Machine Tools and Machining Practices*)

Hexagonal Close-Packed (HCP)

The hexagonal close-packed structure (see Figure 4.7*c*) is found in many of the least common metals. Beryllium, zinc, cobalt, titanium, magnesium, and cadmium are examples of metals that crystallize into this structure. Because of the spacing of the lattice structure, rows of atoms do not easily slide over one another in HCP.

For this reason, these metals have lower plasticity and ductility than cubic structures.

Other Structures

The metal manganese has a simple cubic structure. Manganese is used as an alloying element in steel. Antimony has a rhombohedral crystal structure and is used as an alloying element with zinc and tin.

When carbon steel that has been heated above the upper critical temperature is quenched, the FCC structure attempts to change to the BCC structure. Since there is a solid solution of carbon and iron at the quenching temperature, the lattice contains the smaller carbon atoms in the interstices (spaces between the iron atoms) and complete conversion to BCC is not possible. This is because the carbon atoms interfere with the conversion since there is not enough room in the BCC structure to hold all of them. The result is a structure that ranges from an elongated body-centered cubic crystal to a body-centered tetragonal crystal. This distortion of the lattice is what causes the hardness of **martensite,** which has a body-centered tetragonal unit structure (Figure 4.13). (Martensite is the hard structure that is formed when carbon steel is quenched from high temperatures.)

Table 4.1 lists some common metals and their chemical symbols and crystalline structures. See the periodic table of the elements in Appendix 1.

CRYSTALLINE CHANGES DURING HEATING

When metals are heated slowly to their melting points, certain changes occur. Most nonferrous metals, such as aluminum, copper, and nickel, do not change their crystalline lattice structure before becoming a liquid; however, this is not the case with iron. (The term *iron* refers to elemental or pure iron unless specified as wrought or cast.)

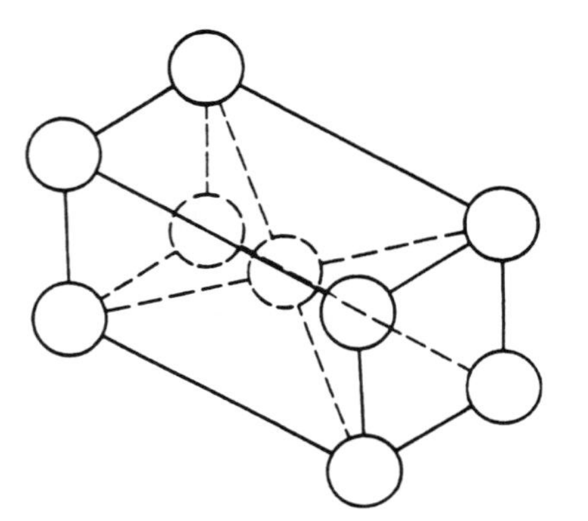

FIGURE 4.13. Body-centered tetragonal structure. This is the distorted cubic form of the unit cell.

TABLE 4.1. Crystal structures of some common metals

Symbol	Element	Crystal Structure
Al	Aluminum	FCC
Sb	Antimony	Rhombohedral
Be	Beryllium	HCP
Bi	Bismuth	Rhombohedral
Cd	Cadmium	HCP
C	Carbon (graphite)	Hexagonal
Cr	Chromium	BCC (above 26°C)
Co	Cobalt	HCP
Cu	Copper	FCC
Au	Gold	FCC
Fe	Iron (alpha)	BCC
Pb	Lead	FCC
Mg	Magnesium	HCP
Mn	Manganese	Cubic
Mo	Molybdenum	BCC
Ni	Nickel	FCC
Nb	Niobium (columbium)	BCC
Pt	Platinum	FCC
Si	Silicon	Cubic, diamond
Ag	Silver	FCC
Ta	Tantalum	BCC
Sn	Tin	Tetragonal
Ti	Titanium	HCP
W	Tungsten	BCC
V	Vanadium	BCC
Zn	Zinc	HCP
Zr	Zirconium	HCP

Iron is a special type of metal that undergoes three crystalline changes as it is heated from room temperature to a liquid state. Iron at room temperature is a body-centered cubic (BCC) (see Figure 4.14), but when heated to 1666°F, it changes to face-centered cubic (FCC). Upon reaching a temperature of 2554°F, pure iron changes back to BCC and remains as such until it becomes liquid at 2795°F. A material that can exist in more than one crystalline lattice structure is called **allotropic.** An allotropic material is able to exist in two or more forms having various properties without change in chemical composition.

COLD FORMING AND PLASTIC DEFORMATION IN METALS

As you learned in Chapter 3, the ability of a metal to undergo plastic deformation when stressed beyond its elastic limit is one of its most useful characteristics. Forging, drawing, forming, extruding, rolling, stamping, and pressing involve plastic deformation. A great deal of useful information regarding the behavior of metals in plastic deformation may be obtained by the study of the crystalline and lattice structures under stress. Deformation may occur

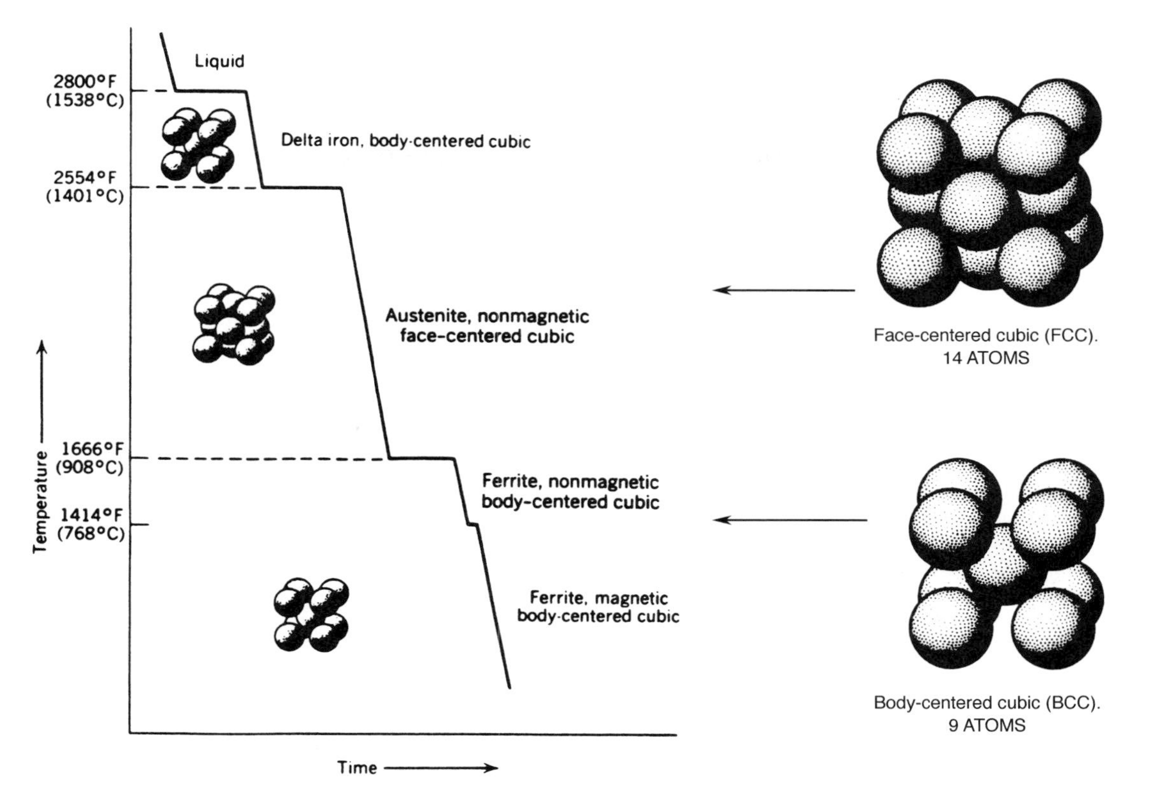

FIGURE 4.14. Cooling curve for pure iron. As iron is cooled slowly from the liquid phase, it undergoes these allotropic transformations.

by slip, twinning, or a combination of these. Some of the factors that influence slip and twinning in metals are precipitate particles or inclusions in the grains, impurity atoms added intentionally or unintentionally, vacancies where no atom exists in the lattice, atoms ordered and arranged in a pattern or disordered, interstitial atoms, dislocation or distortion of the lattice arrangement, and factors involving polycrystalline material (grains whose crystal axes are oriented at random).

In an amorphous material such as tar, plastic deformation begins immediately when a load is applied and the rate is proportional to the load. In metals, however, the plastic deformation rate is essentially zero until a given stress is reached (the yield point).

It is theorized that commercial metals do not have perfectly arranged lattice structures. It is believed that the irregularities of vacancies, impurity atoms, and dislocations are responsible for the plasticity of metals. Metallurgists have grown "whiskers" of pure metal with perfectly ordered lattice structures. These whiskers have tensile strengths many times that of the commercial metal. This is because these pure metals have few irregularities such as dislocations to allow slip to take place. These high-strength fibers are sometimes used in composite materials as reinforcers to impart higher tensile strengths to the base material.

When a dislocation is present, rows of atoms in the lattice can slip or slide until the vacancy or dislocation is blocked by impurity atoms or other irregularities (Figure 4.15). The dislocation moves one atom at a time across the lattice (Figure 4.16). When many planes of atoms are free to slide to a dislocation, a large percentage of elongation or deformation is possible. This is roughly analogous to a deck of cards, each card sliding a small amount, adding to the total movement (Figure 4.17).

Up to this point, the discussion about slip has been confined to the single crystal or grain. In the polycrystalline state of metal, there is no uniform order of lattice planes from one grain to another. Since each grain has a random orientation (direction of dendrite growth), no continuous slip line through the material is possible since resistance to slip is exhibited at the grain boundaries. This explains why large grain metals have a greater capacity to slip and have more plasticity than small grain metals of the same hardness. A fine grain steel is therefore stronger than a coarse-grain steel. Slip takes place along certain crystallographic planes (Figures 4.19*a* and 4.19*b*). BCC metals have 4 times as many possibilities or directions for slip as FCC metals and 16 times as many as HCP metals such as zinc, magnesium, and cadmium. A method of identifying slip planes used by metallurgists is called *Miller Indices.* Many good reference books are available for those who wish to make a further study in this area.

Twinning takes place along certain planes and results in a new lattice orientation or direction along the

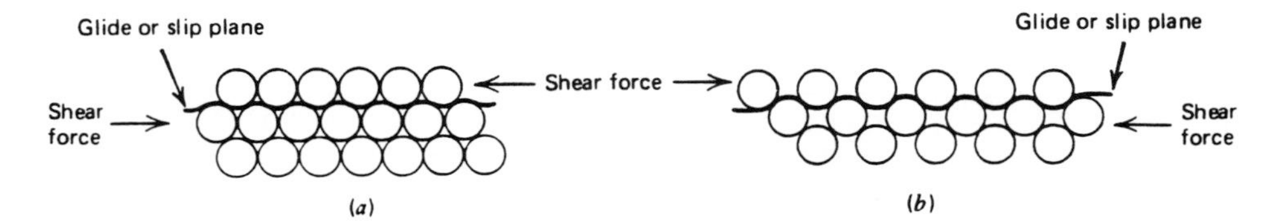

FIGURE 4.15. Rows of atoms can glide along slip planes as shown. BCC and FCC have rows of atoms closely spaced as in (*a*), making movement easier than in (*b*). HCP-type metals are less ductile since many slip planes as in (*b*) are spaced closer.

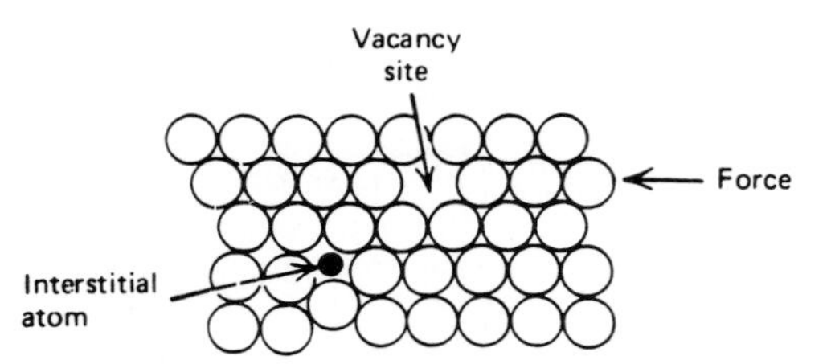

FIGURE 4.16. Rows of atoms move, one atom at a time, to fill in a vacancy or space created by dislocated atoms. Interstitial atoms jam up the slip planes of atoms, making the metal that contains them harder and less ductile.

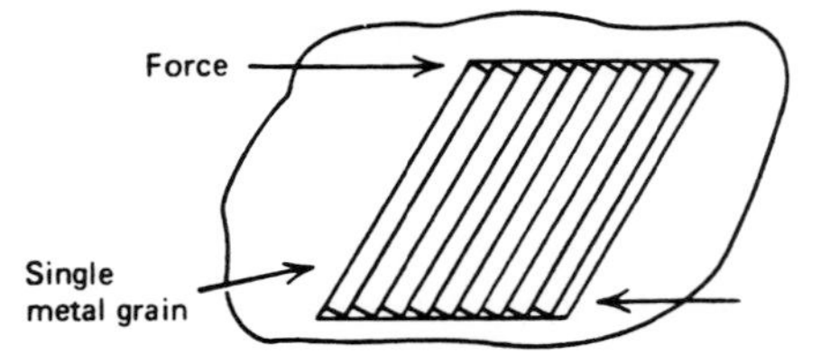

FIGURE 4.17. When stress is applied to a metal grain above its elastic range, slip occurs and the grain flattens and becomes elongated. Slip planes are rotated when stress is applied.

100× Battleship Etch

***FIGURE 4.18*a.** Slip planes in a work-hardened austenitic steel.

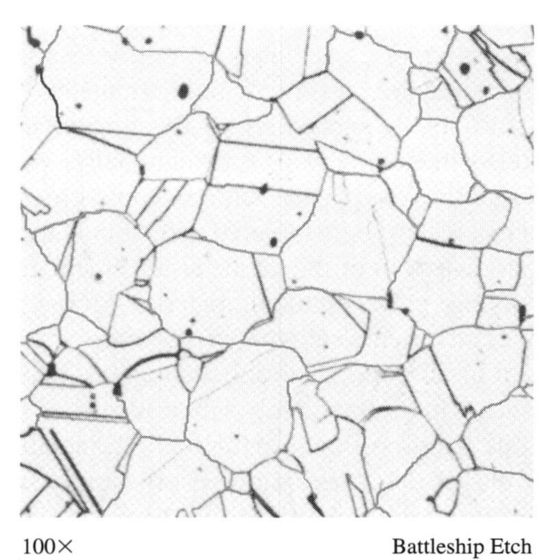

100× Battleship Etch

***FIGURE 4.18*b.** Same austenitic steel as shown in Figure 4.18*a*, except the steel has been heat treated to a full annealed condition. The grain structure formed is called "equiaxed."

twinning plane (Figure 4.20). The HCP metals are particularly susceptible to twinning when a shear stress is applied. Twinning forms a mirror image and occurs in pairs. It is easily observed by microscopic examination. If the surface steps are removed by additional polishing, the twin pattern can still be seen as wide bands. Slip lines, by contrast, are seen as lines that are changed or removed by additional polishing.

When any of these forms of deformation have reached their limit because of filled vacancies or jammed up dislocations resulting from impurity or alloying atoms, no more slip is possible and the material is no longer plastic. Continued stress results in fracture at this point.

GRAIN SIZE CONSIDERATION

The size of the grain has a great effect on the mechanical properties of the metal. The effects of **grain growth** caused by heat treatment are easily predictable. Temperatures, alloying elements, and soaking time all affect grain growth. Grain growth in hardened tool steel is especially damaging since it causes the steel to be weak and brittle. Tempering will not remove this brittleness.

In metals, a small grain is generally preferable to a large grain. The small grain metals have more tensile strength, greater hardness, and will distort less during quenching, as well as be less susceptible to cracking. Fine grain is best for tools and dies. However, in steels a large grain has increased hardenability, which is often preferable for carburizing. Large grains are better for steel that will be subjected to extensive cold working.

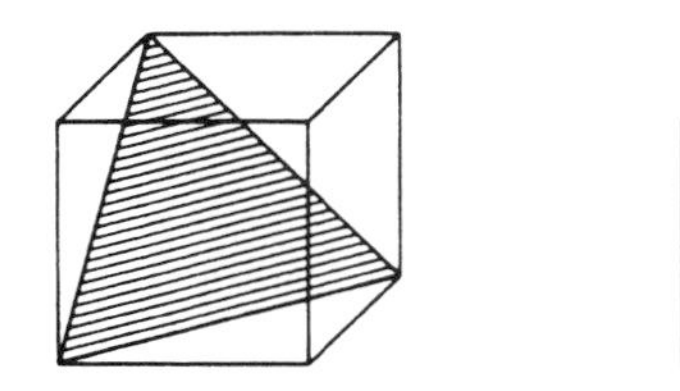

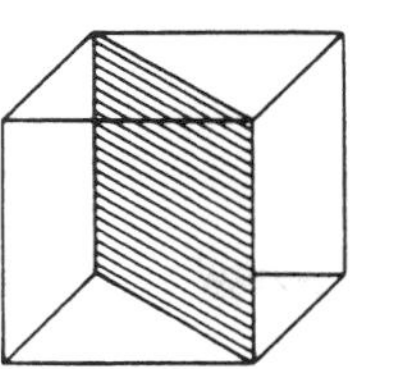

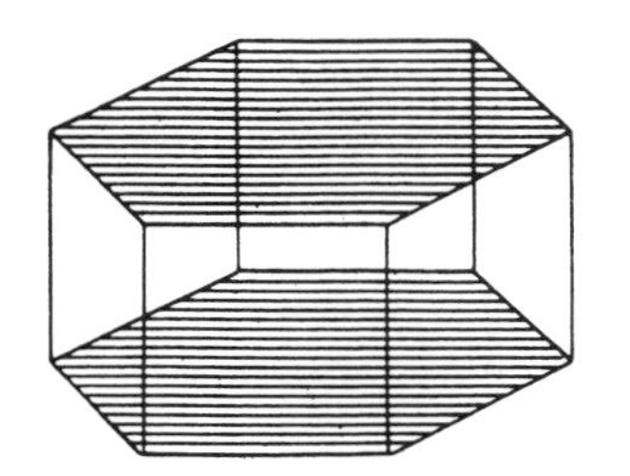

FIGURE 4.19a. Slip is confined to certain crystallographic planes (darkened area). Here, a few of the most important slip planes in the unit cell are shown; however, there are many others.

FIGURE 4.19b. When a metal specimen is being pulled in a tensile tester, slip bands can sometimes be seen on the surface of the metal moving progressively from one end of the test area to the other as the stress increases. These lines, called Eüler's bands, can be seen in the upper end of the specimen in the photograph.

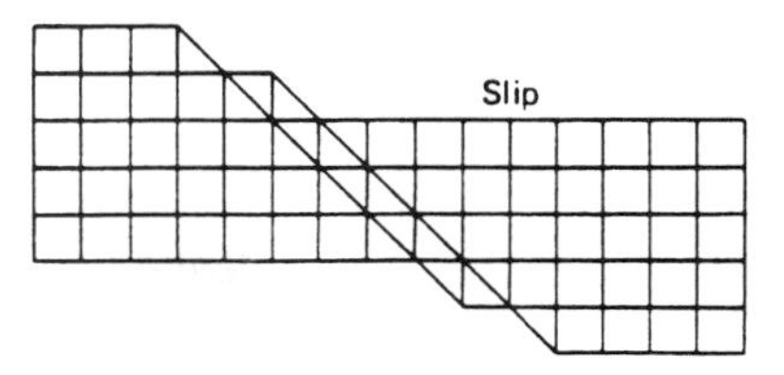

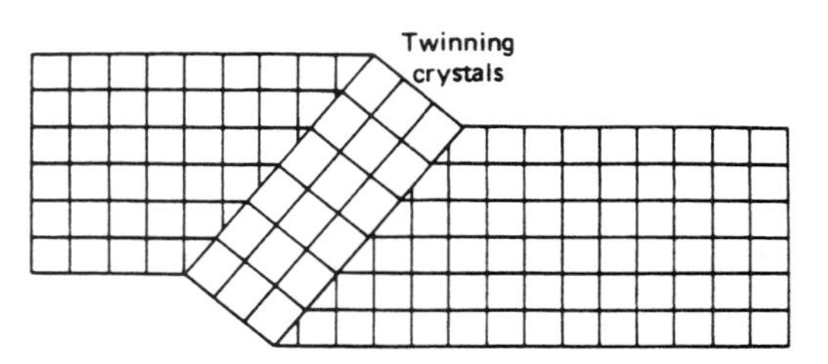

FIGURE 4.20. The difference between slip and twinning. Twinning is a kind of slip in which a uniform tilting of the twinning crystals results from shear force. Slip occurs along the planes of greatest atomic density.

All metals will experience grain growth at high temperatures (Figure 4.21). However, there are some steels that can reach relatively high temperatures (about 1800°F) with very little grain growth, but as the temperature rises, there is a rapid growth rate. These steels are referred to as fine grained. A wide range of grain sizes may be produced in the same steel.

GRAIN SIZE CLASSIFICATION

There are several methods of determining grain size, as seen under a microscope. The method explained here is one that is widely used by suppliers. The size of grain is determined by a count of the grains per square inch under 100× magnification. Figure 4.22 is a chart representing the actual size of the grains as they appear when magnified 100×. Specified grain size is generally the austenitic grain size. A properly quenched steel should show a fine grain.

In describing a piece of steel, one must often specify the grain size, which is done by comparing the specimen with the grain size classification chart. The chart includes eight different grain sizes. A steel is considered fine grained if it is predominantly 5 to 8 inclusive, and coarse grained if it is predominantly 1 to 5 inclusive. If 70 percent of the grain size falls into the given range, it is considered acceptable. Two size classifications may be necessary if there is a wide variation in a section of metal. When austenitic grain size is specified, as it is in most mechanical property tables, the accepted method of determining it is the McQuaid-Ehn test. This test

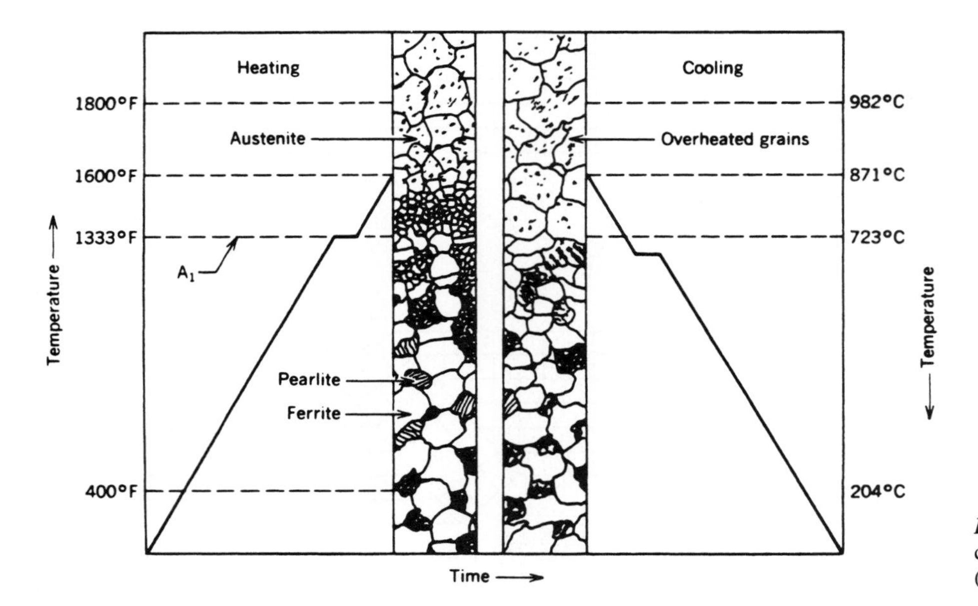

FIGURE 4.21. The effect of overheating carbon steel and the resultant grain growth. (*Machine Tools and Machining Practices*)

consists of carburizing a specimen at 1700°F, followed by slow cooling to develop a carbide network at the grain boundaries. The specimen is then polished and etched, and then compared with the grain size standards at 100× magnification.

Another method of determining grain size is by comparing a fractured surface of the metal with a fracture standard. A common standard is the Shepherd grain size standard (Figure 4.23). This comparison is made without any magnification.

PHASE DIAGRAMS

Matter may exist in three states: solid, liquid, or gaseous. Some substances are capable of changing within the solid state to other phases or crystalline structures. The ability to change into different phases is called allotropy. Iron is an allotropic element and changes from face-centered cubic to body-centered cubic during cooling. When a substance changes phases while cooling, there is a release of heat, which appears on the cooling curve graph as a straight or curved horizontal line (Figure 4.24). As the temperature rises, more energy is required to make the transformation from one structure to another. This means that here too there is an elapsed time at the point of transformation, shown in Figure 4.24 as a straight line. Heating and cooling transformation points are at slightly different temperatures.

When two metals are alloyed together, the temperature of phase changes will be different for every combination. The temperatures and compositions of phase changes can be graphed so that all possible combinations of two pure metals are represented.

SOLUTIONS, LIQUID AND SOLID

When two or more metals are heated to or above their melting points and combined, they usually become a solution that is considered an alloy. The metal composing the greater percentage is the solvent and the metal composing the smaller percentage is the solute. Some molten metals will not dissolve at all in other molten metals but will separate or form a mixture. We are accustomed to thinking of solutions in terms of liquids such as salt or sugar solutions. There are also limits to solubility; water will dissolve only a certain amount of salt or sugar. Oil *will* dissolve in water, but to a very limited extent. There is a similarity between liquid metal solutions and other liquid solutions in that some metals will only partially dissolve other metals.

Solutions may also be found in solid metals, but the changes are made within the lattice and grain structure in solids. The atoms are not quite so free to move about as they are in liquids. Atoms move only to a limited extent and much more slowly in solids. The rate of movement (diffusion) is dependent upon the temperature.

TYPES OF SOLUTION

The dissolving of one material into another can take place in two ways: **substitutional** (replacement of atoms by others) and **interstitial** (spaces between the atoms in the lattice), as illustrated in Figure 4.25.

Substitutional Solid Solutions

A substitutional solid solution is a solution of two or more elements with atoms that are nearly the same size.

FIGURE 4.22. Standard grain size numbers. The grain size per square inch at a magnification of 100×. (*Bethlehem Steel Corporation*)

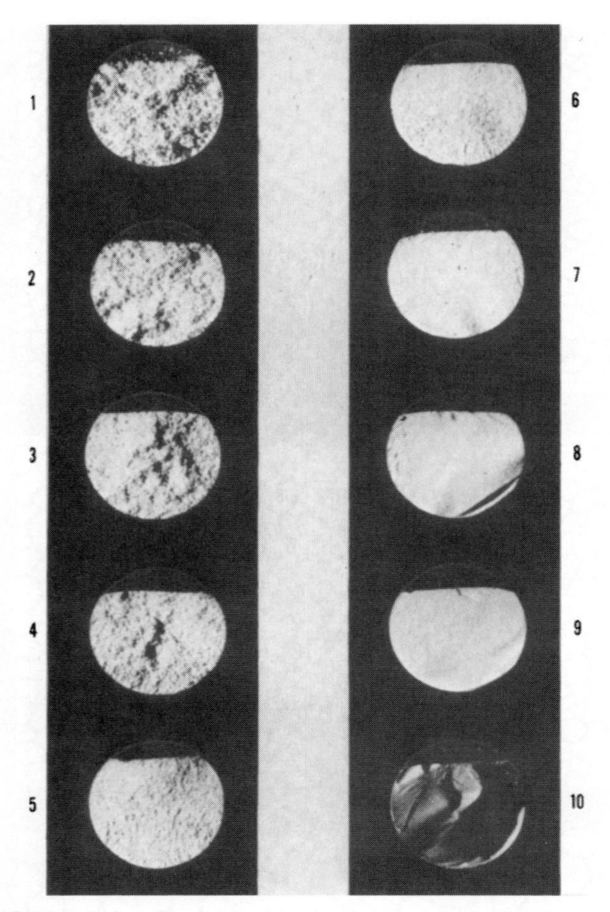

FIGURE 4.23. Shepherd grain size fracture standards. The fracture grain size test specimen can be compared visually with this series of 10 standards. Number 1 constitutes the coarsest and number 10 the finest fracture grain size. (*Courtesy Republic Steel Corporation*)

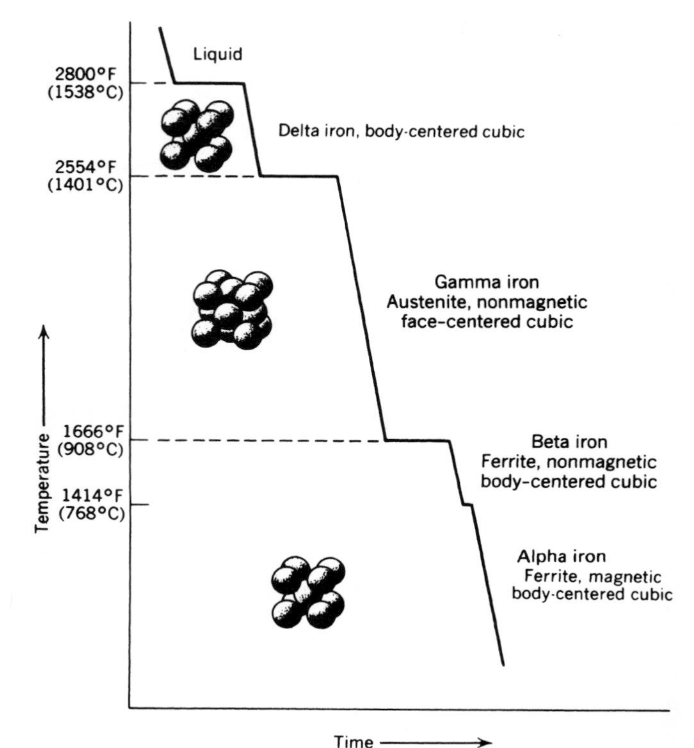

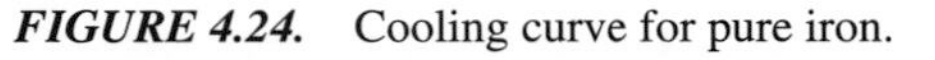

FIGURE 4.24. Cooling curve for pure iron.

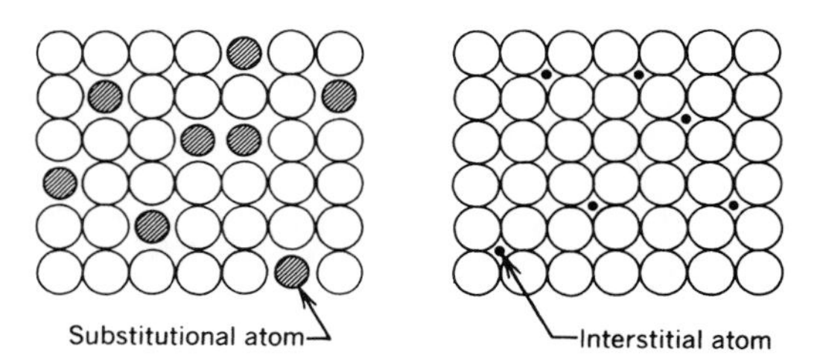

FIGURE 4.25. Substitutional solid solution and interstitial solid solution.

This requirement is necessary in that the alloying atoms need to replace the regular atoms in the lattice structure and not just fit in the spaces between the regular atoms as they do in interstitial solutions.

Interstitial Solid Solutions

Interstitial solid solutions are made up of alloying elements or atoms that differ greatly in size, as Figure 4.25 demonstrates. The alloying atoms must be small enough to fit within the lattice structure of the base material. It has been determined that the alloying atoms should be about one-half the size of the base atom. Common elements that are able to form interstitial solutions with iron are carbon, nitrogen, oxygen, hydrogen, and boron. Some of these elements are also important in their ability to combine chemically with the base metal to form compounds such as the combining of iron (Fe) and carbon (C) to form iron carbide (cementite) (Fe_3C). Iron carbide and other compounds such as iron nitrides and chromium carbides have the potential for greatly increasing the hardness, strength, heat resistance, and abrasion resistance of metals. The development of the carbides, nitrides, and borides in metals has greatly aided the aircraft and space industry in its building programs.

A third condition is the combining of substitutional and interstitial atoms with the same base metal. Figure 4.26 illustrates how this would be done. Many alloys are formed in this way. It makes possible the strength, hardness, and heat-treatment advantages of both types of lattice structures within the same material.

In both the substitutional and interstitial solid solutions, the strengthening of the material is accomplished through the distortion of the lattice structure caused by the alloyed atoms. The lattice distortion creates a strain along the slip planes and grains of the material, which results in the increase of strength and hardness, because it makes atom-by-atom slip along those planes more difficult.

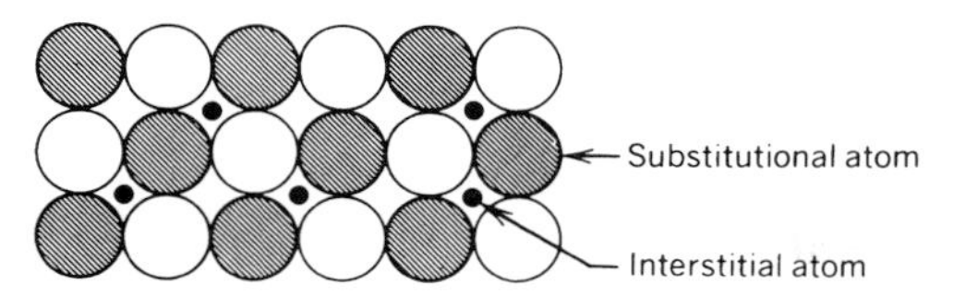

FIGURE 4.26. Substitutional and interstitial solid solution. (*Machine Tools and Machining Practices*)

TYPE I: SOLID SOLUTION ALLOYS

When a pure metal solidifies from the molten state, the change from liquid to solid takes place at a constant temperature, as shown by the cooling curves in Figure 4.27*a*. The curves on each end of the phase diagram describe the cooling rates of the pure metals, copper and nickel, showing no changes in composition while cooling.

Type I alloys are completely soluble in both the liquid and solid states, and their solid phase has a substitutional lattice. When one metal is completely soluble in another, such as copper and nickel, both metals must have the same lattice structure (distance between the atoms), atoms of the same relative size, and a chemical desire to combine. This type of solution is called a continuous solid solution. Figure 4.27*b* is a diagram showing how nickel and copper combine into a continuous solid solution. When the above factors vary, metals take on varying degrees of solubility. Copper-silver is one example. Their atom size differs somewhat more than copper-nickel, and their chemical desire to combine is much less. As a general rule, the more alike the two metals are chemically and physically, the more they tend to form continuous solid solutions. The following are necessary for a continuous solid solution to form.

1. The size of the atom of the alloying metals cannot differ by more than 15 percent.
2. The chemical characteristics should be similar.
3. The metals must crystallize in the same pattern, such as BCC, FCC, or HCP.

In Figure 4.27*b* the upper line indicates the beginning of solidification and the lower line the end of solidification. The area above the upper line is a homogenous liquid solution; the area below the lower line is a solid solution. Between the liquidus (upper) and solidus (lower) lines is a two-phase region consisting of a mixture of a liquid solution and a solid solution. As the alloy cools from any particular location in the mushy area between the lines, crystals begin to form, having an infinite variety of solid solutions. The result is a dendritic structure (Figure 4.28).

Since the composition of the alloy is constantly changing as it is cooling in the mushy area, a homogenous alloy is not possible. However, the slower the cooling rate is, the more diffusion takes place and the more uniform the alloy becomes. With fast-cooling rates, nonequilibrium conditions prevail, not allowing time for diffusion, and the final solid will consist of a "cored structure." Coring or dendritic segregation is caused by a higher melting central portion on a dendrite that solidifies first, followed by lower melting alloys, which are last to solidify as a shell. Dendritic and cored structures are often found in castings and welds. These are not strong structures and are usually refined by reheating to allow diffusion to take place, which produces a finer, more uniform structure.

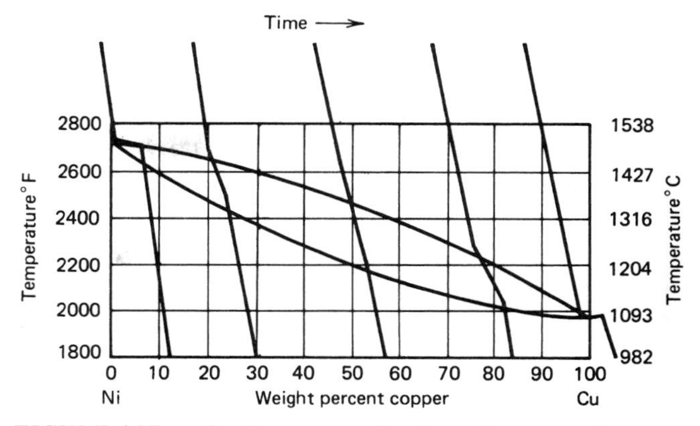

FIGURE 4.27a. Cooling curves of various alloy proportions as they might appear on a Type I phase diagram.

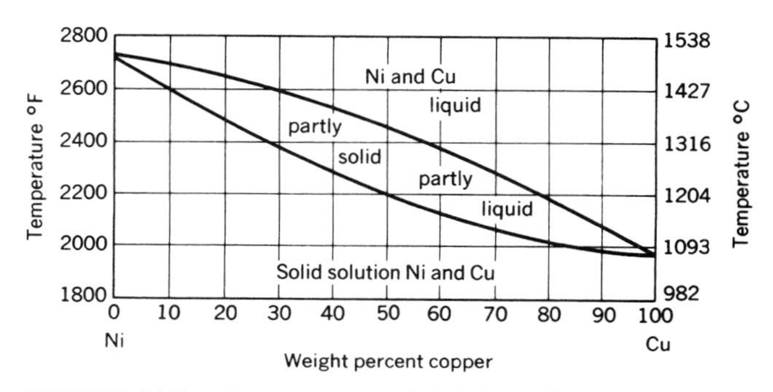

FIGURE 4.27b. Type I, copper-nickel phase diagram. (*Machine Tools and Machining Practices*)

TYPE II: EUTECTIC ALLOYS

Type II alloys are soluble in the liquid state but insoluble in the solid state (Figure 4.29). Alloys that are not solid solutions are usually mixtures of various forms. Compounds may also form between two metals (for example, copper and aluminum form copper aluminide) and between metals and nonmetals like iron and carbon to form iron carbide (Fe_3C). The following are some examples of mixtures in metals.

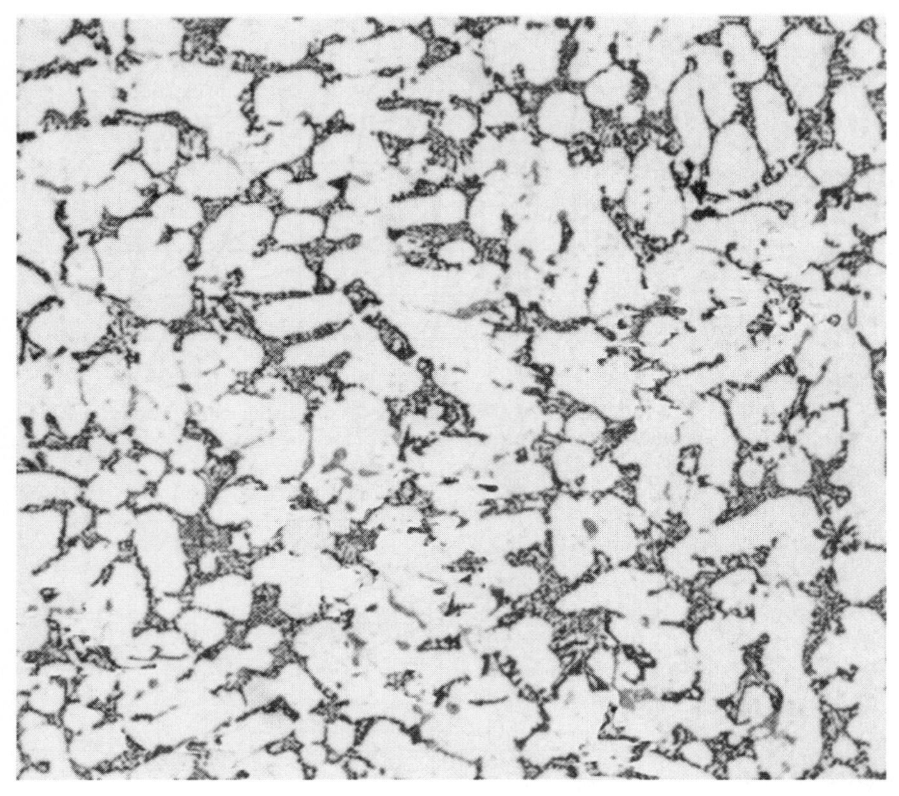

FIGURE 4.28. Dendritic microstructure of cast monel.

1. Solid solution and solid solution
2. Solid solution and compound
3. Compounds and compounds
4. Compound and pure material

Solid solutions that become insoluble below the transformation temperature often separate into a lamellar or rosette pattern microstructure formation. The eutectic composition is always the lowest melting point of the alloy system; other percentages in a two element alloy system produce some alloy A plus eutectic or some alloy B plus eutectic (see Figure 4.30).

The cooling curve for a eutectic is like that of a pure metal; there is no change in composition, only a slight constant temperature time delay. The eutectic is not a homogenous alloy but contains crystals of nearly pure metal A and nearly pure metal B. The eutectic for this alloy would always be 40 percent metal A and 60 percent metal B. Since these particular metals are completely insoluble in the solid state, it can be seen that the only solid that can form when freezing begins is a pure metal, not a solution. Then, when any alloy of the two metals is completely solidified, it must be a mixture of these two metals. The curve of the liquidus line indicates that starting from the left at 100 percent A, any percentage of A and B to the right will require progressively lower temperatures to begin to solidify. In Figure 4.29 at the 70–30 percent composition,

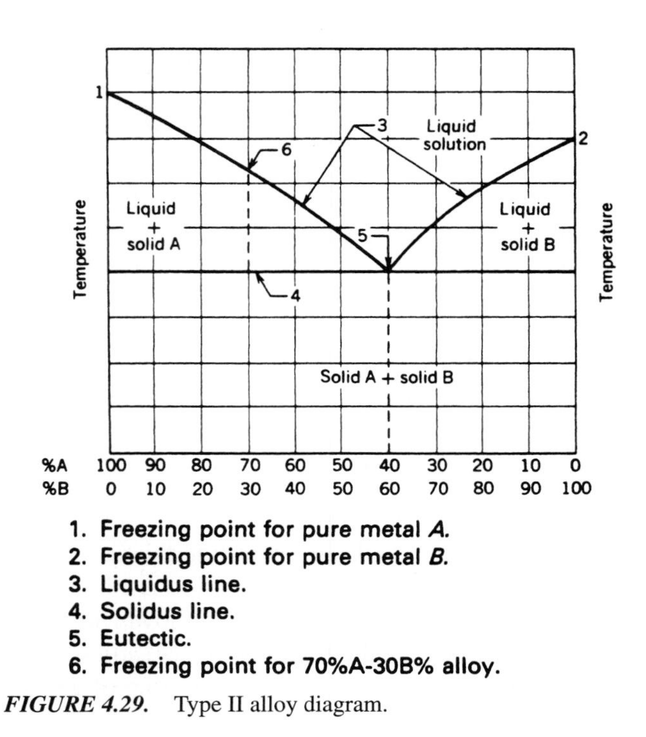

FIGURE 4.29. Type II alloy diagram.

shown at the dotted line at the left, some of the pure metal A can solidify first. The remaining liquid is now richer in metal B, so some of the eutectic mixture A plus B solidifies to restore this equilibrium (the 70–30 percent composition). This process continues to the solidus line where

100× Nital Etch

FIGURE 4.30. Lamellar pearlite as seen in a AISI 1080 carbon steel.

freezing is completed. The metal is now composed of grains of pure metal A and grains of the eutectic mixture.

A practical application of these changes in two-phase alloys was for autobody "wiping" solder, which is no longer in use to any great extent. It was necessary to have the metal remain somewhat fluid or "mushy" while it was cooling so it would not freeze on the car body before it was shaped. An alloy that solidified in the largest part of the mushy area (between the liquidus and solidus lines) was therefore used. Some welding rods for brazing in which rapid solidification was needed would have to be of eutectic composition; also, the eutectic produces a more uniform weld structure.

TYPE III: TERMINAL SOLID SOLUTION

A terminal solid solution consists of two metals that are soluble in the liquid state but only partly soluble in the solid state. It can be seen from Figure 4.31 that solubility is **continuous** at each end, but **insoluble** (a mixture) in the middle. These terminal solid solutions are found in solution heat treating and in the precipitation type of hardening used for some kinds of stainless steels and some nonferrous metals. This is further discussed in Chapter 15.

PHASE CHANGES OF IRON

Iron, being an allotropic element, can exist in more than one lattice unit structure, depending on temperature. When a substance goes through a phase change while cooling, it releases heat. When a phase change is reached while heating, the substance absorbs heat. This characteristic is used to construct cooling curve graphs. If a continuous record is kept of the temperature of cooling iron, we can construct a graph that will resemble Figure 4.24.

Each flat segment of the cooling curve represents a phase change. These flat portions are caused by the release of heat during phase changes. At 2800°F, iron changes to a solid, body-centered cubic structure (delta iron). All the remaining changes involve a solid of one lattice structure transforming into a solid of another lattice structure. At 2554°F BCC delta iron changes to FCC austenite (gamma iron). Austenite transforms to BCC ferrite at 1666°F. The next change is not a phase change at all, but it does give off heat. This is the change from nonmagnetic ferrite (beta iron) to magnetic ferrite (alpha iron).

THE IRON-CARBON DIAGRAM

In the preceding phase diagrams of alloy systems, none of the metals went through any solid state phase changes (transformation). There are some new lines that must be added to account for solid phase transformations (Figure 4.32). Many of the lines, the liquidus, solidus, and the eutectic point, are similar. This diagram differs from previous ones in that the diagram ends on the right at 6.67 percent carbon (C) instead of 100 percent carbon. The rest of the diagram from 6.67 to 100 percent carbon

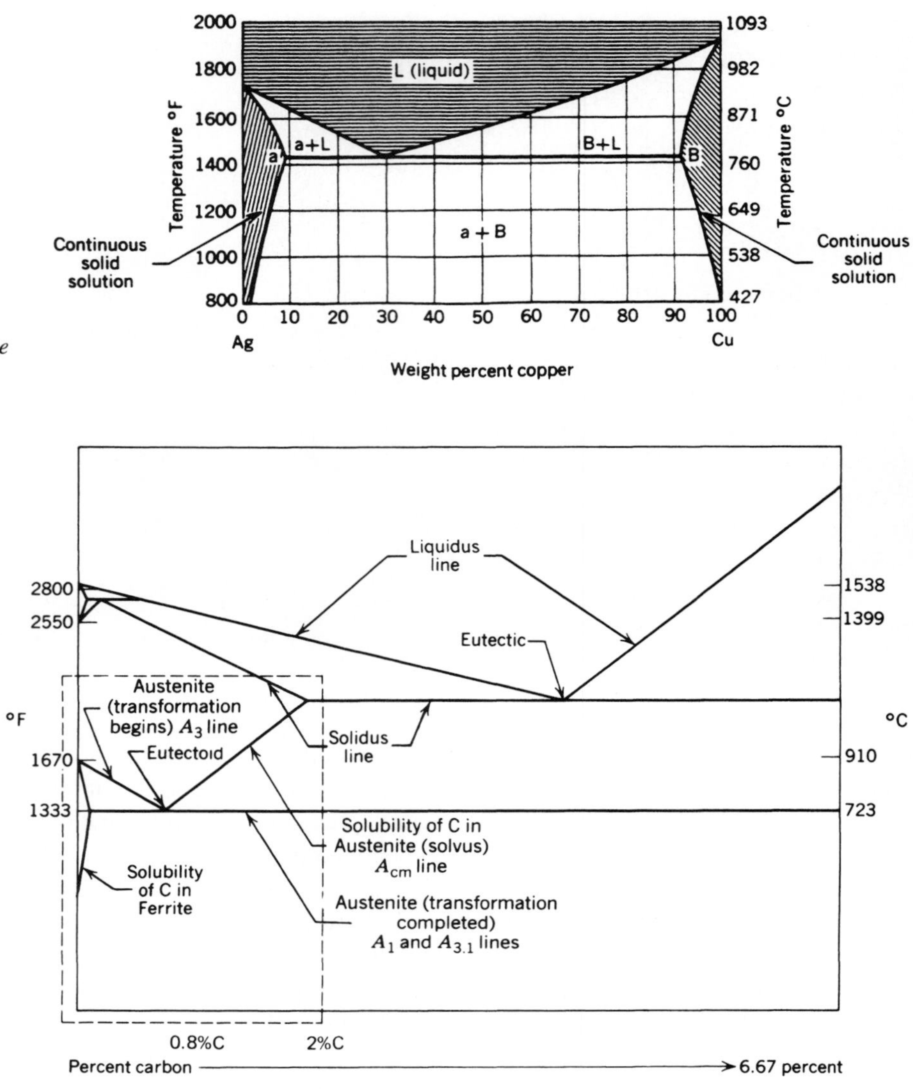

FIGURE 4.31. Silver-copper phase diagram represents the terminal solid solution. (*Machine Tools and Machining Practices*)

FIGURE 4.32. Simplified iron-carbon diagram in equilibrium or very slow cooling. (*Machine Tools and Machining Practices*)

would give no useful information about steels and cast irons. But a more important reason for ending the diagram at this point is that 6.67 is the percent of carbon by weight in the compound iron carbide (Fe_3C). The iron-carbon diagram is referred to as an equilibrium or phase diagram. The phase diagram indicates the transformations that take place during very slow cooling or near-equilibrium conditions.

THE DELTA IRON REGION

The area at the left of the diagram between 2800°F and 2554°F in Figure 4.32 describes the solidification and transformation of delta iron. This area has no commercial value in heat treating and therefore is only of passing interest.

THE STEEL PORTION

Study the portion of the diagram outlined by dashed lines in Figure 4.32 and you will see how it resembles phase diagrams that you have already studied. However, a little different terminology is used for the lines. Since the metal is now a solid, the terms *liquidus* and *solidus* do not apply. Since *eutectic* means low melting point, this term is not correct, either. The suffix "-oid" means similar but not the same. The **eutectoid** point appears like a eutectic on the diagram, but it is the lowest temperature transformation point of solid phases, while eutectic is the lowest freezing point of a liquid phase. Instead of a liquidus line, we have the line that shows the transformation from austenite to ferrite, called the A_3 line, and the line showing the amount of carbon that is soluble in austenite, called the A_{cm} line (Figure 4.33). Instead of a solidus line, we have the line that shows where austenite completes its transformation to ferrite and where pearlite is formed. This line is called A_1 to the left of the eutectoid point and $A_{3,1}$ to the right of the eutectoid point.

MICROSTRUCTURES

When etched carbon steel is viewed through a metallurgical microscope, the microstructure may be identified as ferrite, pearlite, iron carbide, martensite, retained austenite, or bainite with limitless variations dependent upon processing and heat treatment. The primary ASM reference books used to identify microstructures are *The Metals Handbook,* vol. 7, 8th edition and vol. 9, 9th edition; and the *Heat Treater's Guide,* 2nd edition. Metallographic sample preparation techniques are described in Chapter 27.

Austenite is a face-centered cubic iron. It has the ability to dissolve carbon interstitially to a maximum of 2 percent at 2065°F. It is therefore a solid solution, not a mixture (see Figure 4.32). The iron-carbon diagram shows the solubility range of carbon in austenite. Austenite is not stable below the lower transformation, or critical, temperature, the A_1-$A_{3,1}$ line, except in some alloy steels, but decomposes into pearlite, ferrite, and/or cementite.

Ferrite is a body-centered cubic iron. It will dissolve only 0.008 percent carbon at room temperature and a maximum of 0.025 percent carbon at 1333°F. Locate the narrow solid solution area at the left of the diagram as seen in Figure 4.33. To a very limited extent in this area, ferrite forms an interstitial solid solution with carbon. Ferrite is the softest structure that appears in the diagram. The color of ferrite is a bright white when viewed through the optical microscope.

Pearlite is the lamellar eutectoid mixture of ferrite (white) and cementite (dark) layers. The actual color of cementite is yellow when compared to ferrite, although due to the hardness and abrasive resistance of cementite, the layers become rounded during metallographic polishing. The rounded cementite layers protrude above the soft ferrite and scatter the illuminating microscope light rather than reflecting it, thus producing what appears to be a black image (see Figure 4.34).

THE SLOW COOLING OF 1020 CARBON STEEL

For a plain carbon steel less than eutectoid, such as AISI 1020 (or 0.20 percent carbon) steel during cooling, it can be seen in Figure 4.35, line 1 at a point *a* that the entire microstructure is austenite. After crossing the A_3 line (point *b*), ferrite begins to form along the austenite grain boundaries.

Further cooling causes more ferrite to form. Because the solubility of carbon in ferrite is so low, excess carbon must be contained by the remaining austenite. At point *c,* most of the austenite has become ferrite and the remaining austenite has a carbon content of 0.8 percent. As the steel crosses the A_1 line, the remaining carbon-rich austenite transforms to the eutectoid or pearlite and remains pearlite at point *d.* Steels with carbon content below 0.8 percent are called hypoeutectoid.

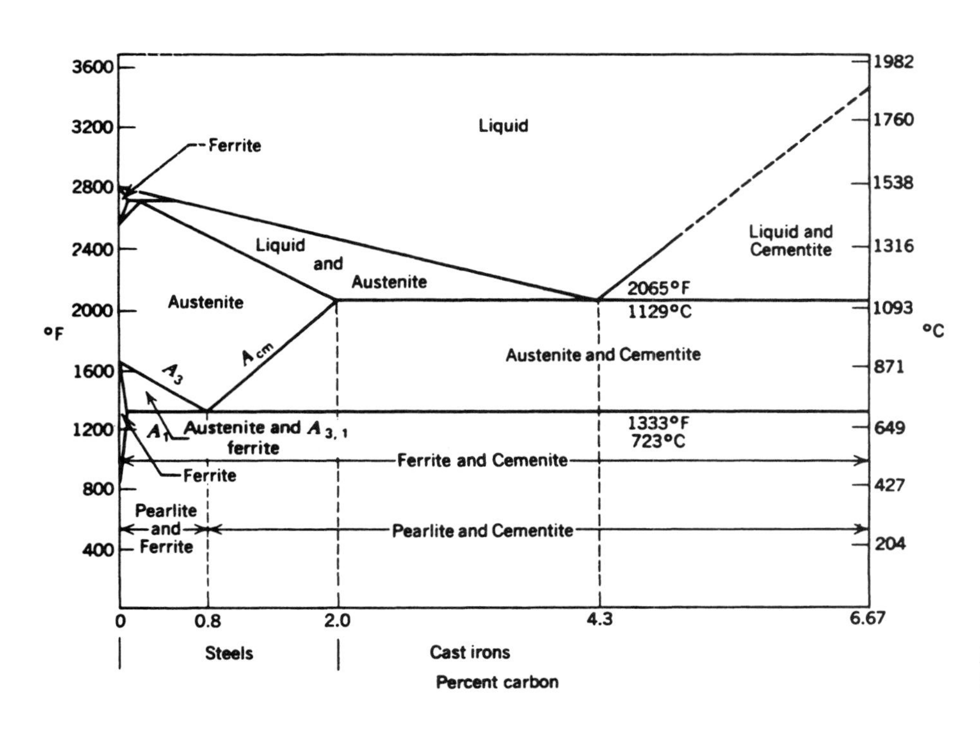

FIGURE 4.33. The iron-carbon diagram. (*Machine Tools and Machining Practices*)

400× Nital Etch

FIGURE 4.34. This is the microstructure of AISI 4340 alloy steel which formed as a result of slow cooling through allotropic transformation from FCC to BCC. The microstructure consists of fine pearlite grains (dark) and proeutectoid ferrite (white).

SLOW HEATING OF 1020 STEEL

When hypoeutectoid steels like AISI 1020 are slowly heated to the austenitic temperature range, microstructure transformation occurs in reverse to that of slow cooling. Remember, pearlite upon reaching the A_1 line undergoes allotropic transformation to austenite. As the temperature rises, carbon from the austenite grains migrates to the adjacent ferrite grains, causing them to become richer in carbon and undergo allotropic transform. This process continues as the temperature rises above the A_3 line, until all the ferrite has transformed to austenite. To homogenize the alloy and cause the carbon atoms to diffuse evenly throughout the austenite the metal must be heated 50° to 100°F above the A_3 line and maintained for a minimum of 30 minutes per inch of thickness. The process of homogenization plays a very important role in producing a quench hardened steel with consistent hardness and mechanical properties.

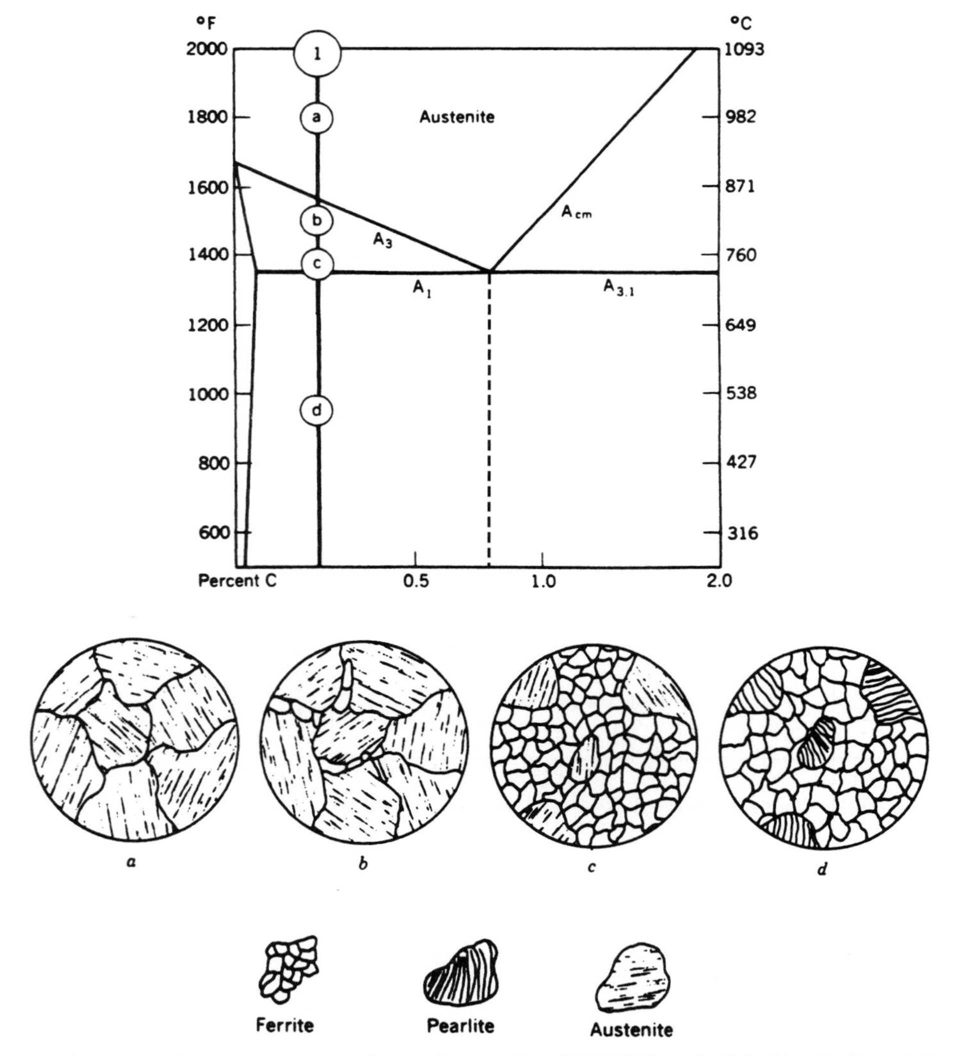

FIGURE 4.35. Microstructures at various temperatures of very slow cooling SAE 1020 steel. (*Machine Tools and Machining Practices*)

Grain size also changes as the metal becomes fully austenitic. Grain size has a marked effect on the strength, toughness, and plasticity of steels at room temperature. Large grained steels usually have lower strength and greater toughness than fine grain steels. There is an optimum grain size for every manufactured product. For example, when fastener heads are formed by upset process, if the grain size is too fine, the metal will work harden rapidly and will split rather than take the desired head shape. The same alloy with an ASTM 4 grain size will plastically deform much more readily and with less required force. When stretch-forming sheet steel, too large of a grain size can produce an "orange peel" effect.

When recommended processing temperatures are exceeded, grain growth can occur. Most steels can be normalized, which will cause refinement of the grains as long as alloy segregation has not occurred. An example of a low alloy steel with banded alloy segregation is shown in Figures 4.38*a*, 4.38*b*.

SLOW COOLING OF A 1095 CARBON STEEL

At point *a* (Figure 4.36), the AISI 1095 steel is entirely austenite and interstitially dissolved carbon. When the steel cools to point *b*, it has crossed the A_{cm} line. The A_{cm} line indicates the limit of carbon solubility in austenite. Since the austenite can no longer hold the entire amount of carbon in solution, cementite begins to appear along the grain boundaries. At point *c*, the solubility of the austenite has dropped even further, so that the additional cementite has formed a fairly continuous network between the austenite grains. The carbon content of the remaining austenite is now the eutectoid composition of 0.8 percent carbon. At point *d*, the steel has crossed the $A_{3,1}$ line and the remaining austenite transforms to pearlite. Steels with a carbon content above 0.8 percent are called hypereutectoid.

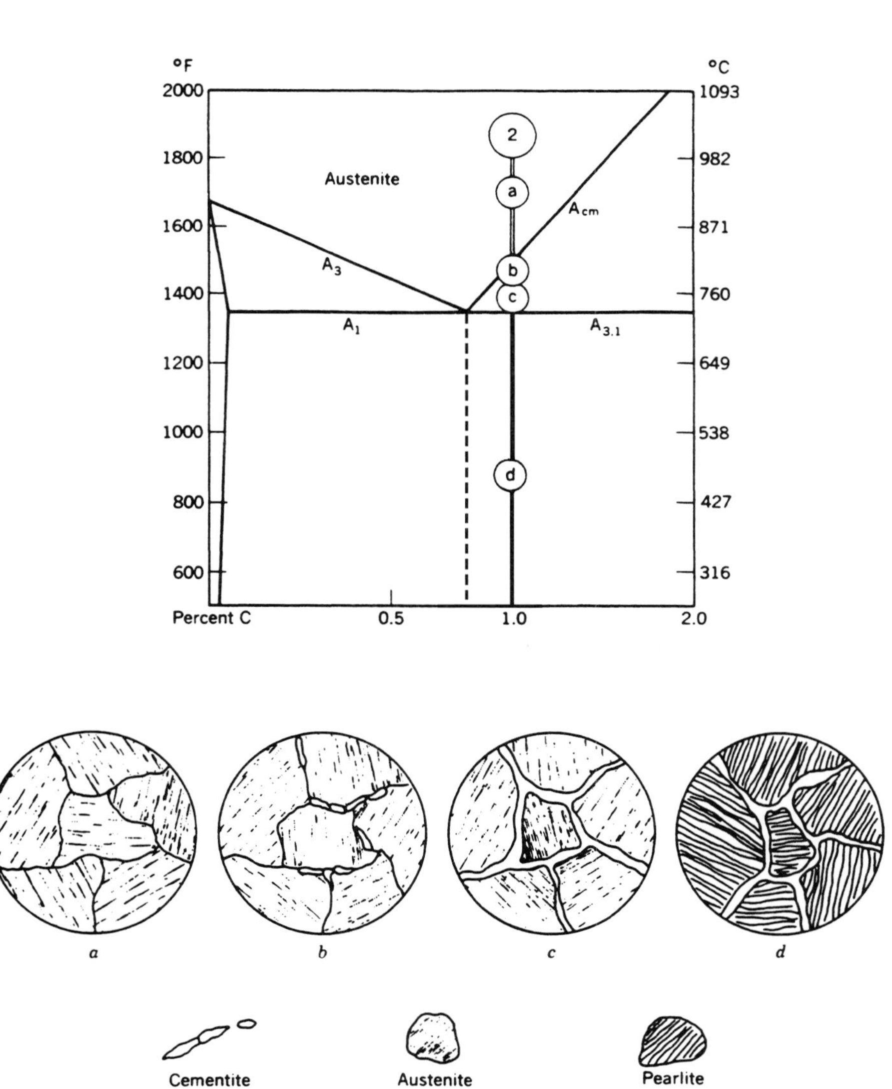

FIGURE 4.36. Microstructures at various temperatures when cooling SAE 1095 steel. (*Machine Tools and Machining Practices*)

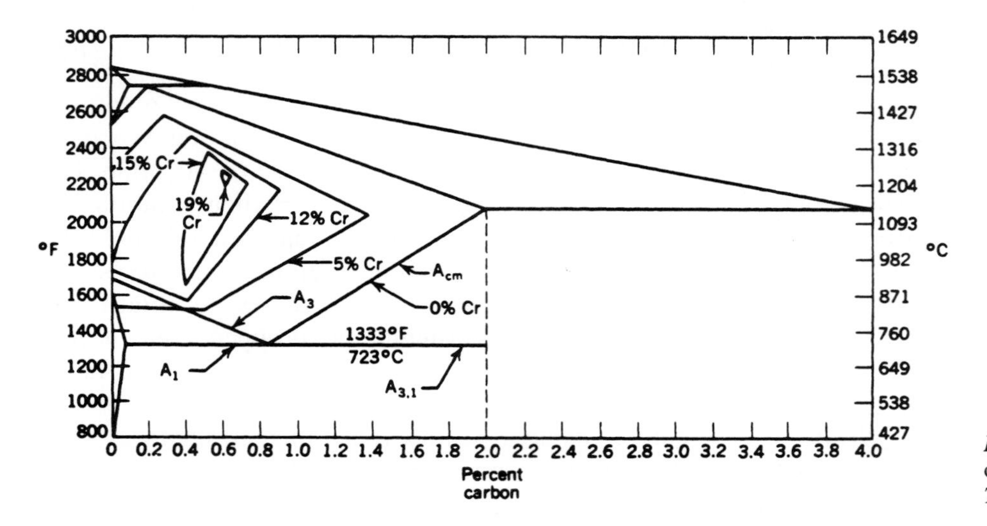

FIGURE 4.37. The effect of chromium on the austenite range of steel. (*Machine Tools and Machining Practices*)

ALLOYING ELEMENTS AND THE IRON-CARBON DIAGRAM

Alloying elements move the transformation lines of the iron-carbon diagram. A common alloying element is chromium. Considering the austenite area of the iron-carbon diagram, it can be seen that the effect of increasing chromium is to decrease the austenite range (Figure 4.37). This will increase the ferrite range. Many other alloying elements are ferrite promoters also, such as molybdenum, silicon, and titanium.

Nickel and manganese tend to enlarge the austenite range and lower the transformation temperature (austenite to ferrite). A large percentage of these metals will cause steels to remain austenitic at room temperature. Examples are 18.8 stainless steel and 14 percent manganese steel. The A_1-$A_{3,1}$ line is also lowered by rapid cooling. Any alloy addition will move the eutectoid point to the left (with a few exceptions such as copper) or, in other words, lower the carbon content of the eutectoid composition.

CAST IRON ALLOYS

Cast iron, essentially an alloy of iron, carbon, and silicon, is composed of iron and from 2 to 6.67 percent carbon, plus manganese, sulfur, and phosphorus. Commercial cast iron contains no more than 4 percent carbon. Cast iron is often alloyed with elements such as nickel, chromium, molybdenum, vanadium, copper, and titanium. Alloying elements toughen and strengthen cast irons.

Since the maximum amount of carbon that iron can contain in an austenitic solution is 2 percent, any amount of carbon in excess is in a supersaturated condition. The solubility of carbon in iron varies with the temperature. The supersaturated or excess carbon will either form graphite flakes as seen in gray cast iron, or iron carbide as often seen in white cast iron. Graphite may also take the form of spheres or nodules as found in nodular cast iron and malleable cast iron.

The types of cast irons are:

1. *White cast iron.* Most of the carbon is in the combined form of cementite (iron carbide) (Figure 4.39). This is a very hard, brittle material that often contains pearlite grains. It is essentially unweldable.
2. *Gray cast iron.* Much of the carbon is in the form of graphite. Other microstructures are pearlite and ferrite (see Figures 4.40*a* and 4.40*b*).
3. *Alloy cast iron.* This is cast iron to which alloys have been added to enhance certain characteristics, for example, an addition of nickel to retain austenite.
4. *Nodular or ductile cast iron.* The carbon is mostly graphite in the form of spheroids and is produced during solidification by inoculating the cast iron with an element such as magnesium while it is still in the ladle (see Figure 4.41).
5. *Malleable cast iron.* The carbon in malleable cast iron is also in the form of graphite spheroids but is formed as a result of lengthy heat treatment of white cast iron at high temperatures; see Figure 4.45.

The solubility of carbon in iron when slowly cooled may be observed in the iron-graphite phase diagram (Figure 4.42). Silicon is a graphitizer and about 2.5 to 3 percent is added to cast iron to promote the formation of graphite. A 3 percent carbon cast iron that contains silicon can dissolve 2 percent carbon. The remaining 1 percent carbon is free to change to graphite when solidification occurs at 2075°F. The solubility of carbon decreases as the temperature drops, and at 1360°F the solubility of carbon has dropped to 0.5 percent. Now a total of 1.5 percent carbon has been precipitated (pushed out) to form graphite because of the silicon content. The final structure, there-

100× 2% Nital Etch

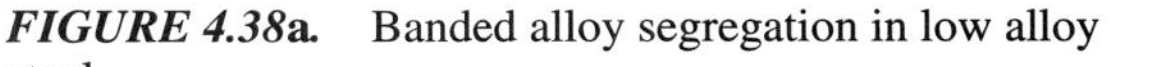

***FIGURE 4.38*a.** Banded alloy segregation in low alloy steel.

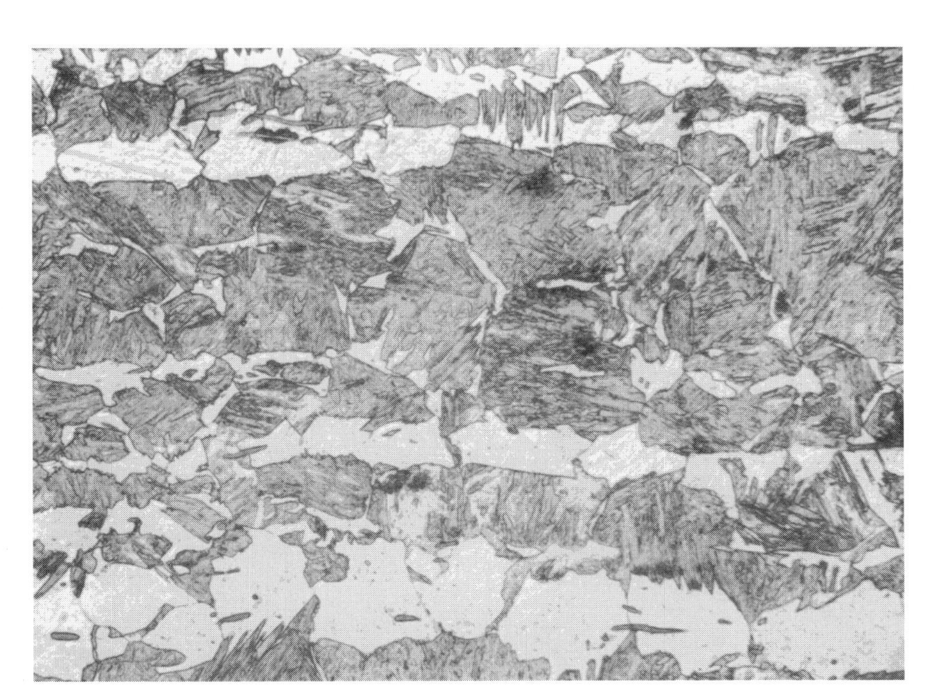

400× 2% Nital Etch

***FIGURE 4.38*b.** Banded alloy segregation in low alloy steel; martensite (dark), ferrite (white).

fore, will be ferrite and graphite flakes (ferritic gray cast iron). If less silicon were added, the final structure would probably be pearlite and graphite flakes (pearlitic gray cast iron). The rate of cooling also determines the type of final structure. Rapid cooling promotes cementite or massive white areas to form in combination with gray iron. Certain chemical compositions, in conjunction with cooling rate, can create a mixture of microstructures found in gray and white cast iron. This is called a **mottled** cast iron. Very rapid cooling can produce martensite. Figure 4.43 shows the composition limit for gray, mottled, and white cast irons. It can be seen in the graph that as the carbon increases, a smaller amount of silicon is required to form gray cast iron.

Pearlitic cast irons may be heat treated just as steel can. They can also be locally or surface hardened by flame or induction heating. Hardening operations do not affect the graphite constituent of the casting; they only change the pearlite to martensite.

Gray Cast Iron

Gray cast irons are classified according to their tensile strength, as Table 3.5 in Chapter 3 indicates. The Ameri-

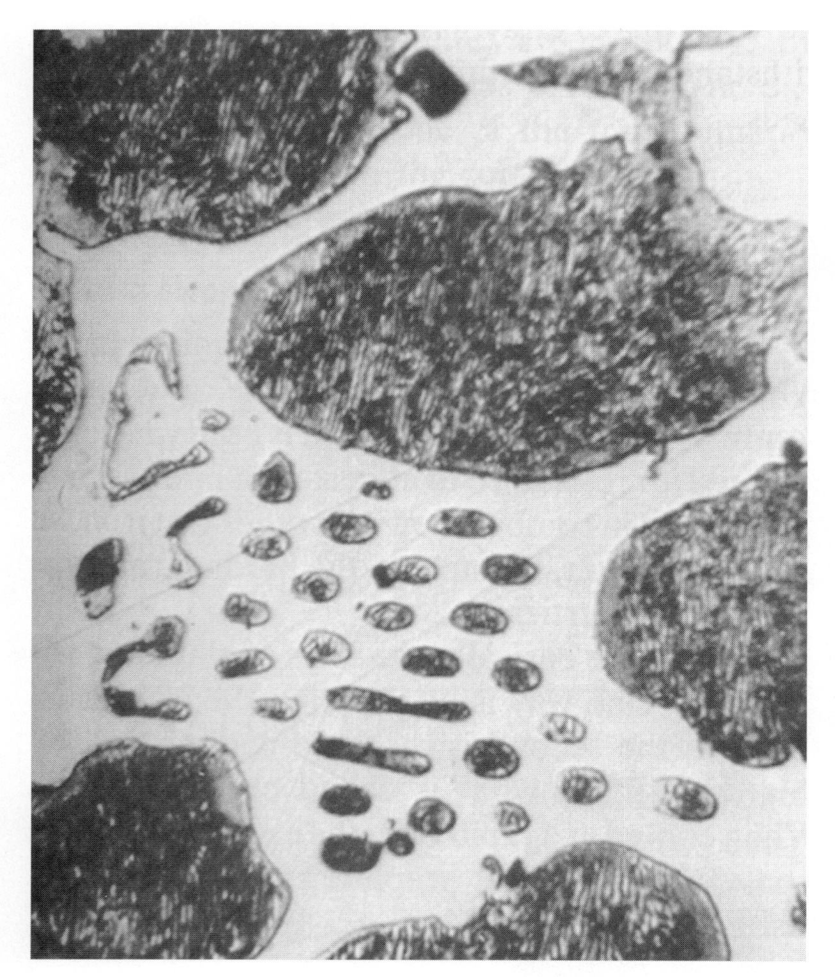

White cast iron, etched (500×).

FIGURE 4.39. White cast iron showing islands of pearlite grains in a matrix of cementite.

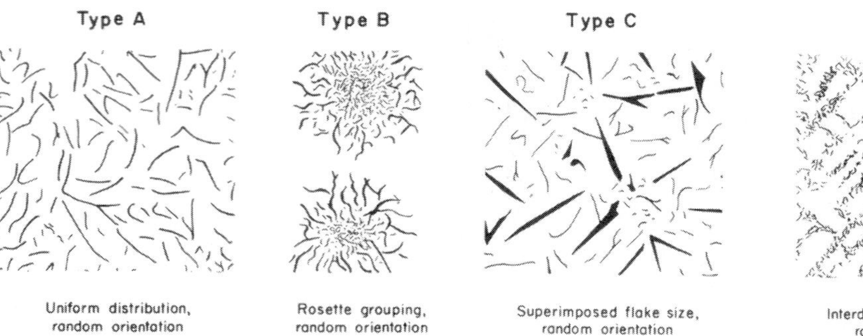

***FIGURE 4.40*a.** Types of graphite flakes in gray iron (per ASTM)

can Society for Testing and Materials (ASTM) classes run from 20 to 60. The number correlates to the tensile strength of the material in thousand pounds per square inch.

Gray cast iron is a relatively brittle material, mainly because of its long thin graphite flakes that are very weak. The types of graphite flake formations are shown in Figure 4.40*a*. Gray cast iron is a metal that will withstand large compressive loads but small tensile loads.

White Cast Iron

White cast iron is very hard, brittle, and virtually nonmachinable. In some cases it is used where there is a need for resistance to abrasion. White cast iron is often found in combination with other cast iron, such as gray cast iron, to improve the hardness and wear-resistant properties.

***FIGURE 4.40*b.** Pearlitic gray cast iron.

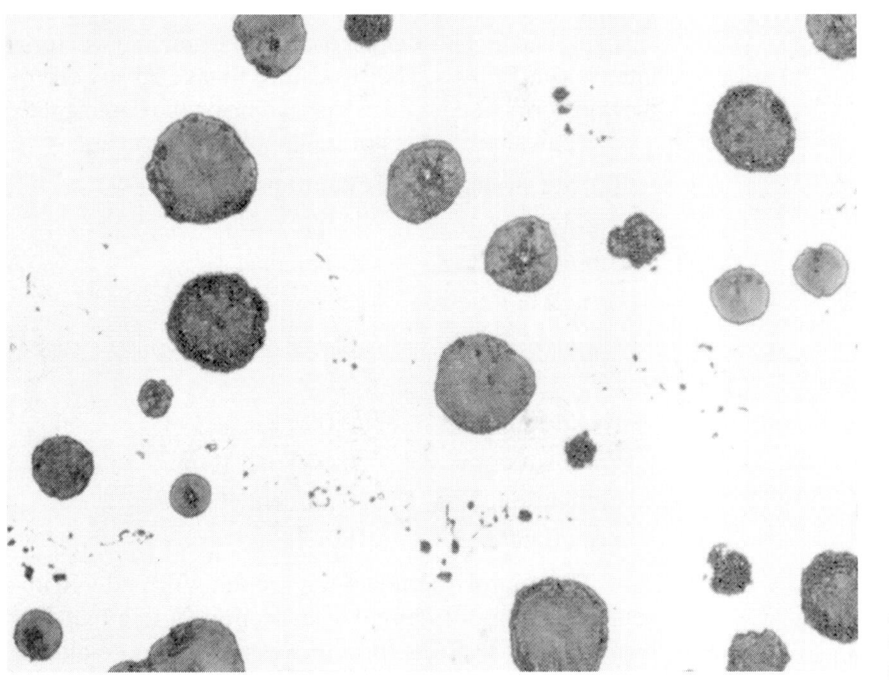

FIGURE 4.41. The microstructure of nodular or ductile cast iron consists of spheroidal graphite in a matrix of ferrite.

There are basically two ways of obtaining white cast iron. One way is by lowering the iron's silicon content; the second is by rapid cooling, which in this case yields what is called chilled cast iron. When cooled at a rapid rate, the excess carbon forms iron carbide and not graphite, thus making white cast iron.

Many times it is advantageous to have a hard, wear-resistant surface on the part, such as a bearing surface or outer rim. This is easily accomplished by putting chill plates in the mold so that the molten iron will cool faster in these localized areas, creating white cast iron.

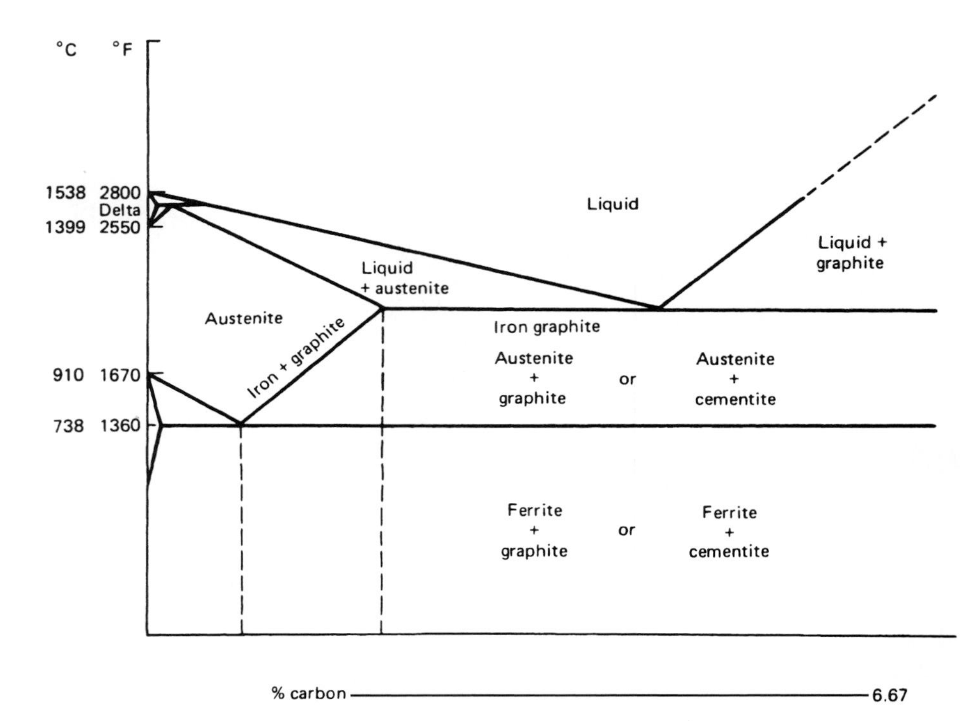

FIGURE 4.42. Iron-graphite equilibrium diagram.

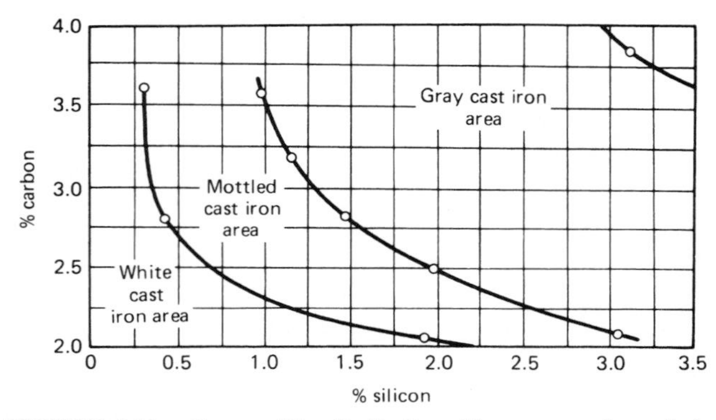

FIGURE 4.43. Composition limits for white, gray, and mottled cast irons.

High-chromium irons have been used for some time as an abrasive-resistant material. They are often used for plowshares that are used in sandy soils. The chromium carbides are dispersed in austenitic or hardenable matrix such as the martensitic irons.

Malleable Cast Iron

Malleable cast iron is noted for its strength, toughness, ductility, and machinability. In the process of making malleable cast iron, it is first necessary to begin with white cast iron. The white cast iron is then heat treated as follows.

1. Heat to about 1700°F.
2. Hold at this temperature for about 15 hours. This breaks down the iron carbide to austenite and graphite.
3. Slow cool to about 1300°F.
4. Hold at this temperature for approximately 15 hours.
5. Air cool to room temperature.

This process breaks down the iron carbide into additional austenite and graphite. Upon cooling, the graphite will form into clusters or balls. The austenite will take on any one of the transformation products, depending on the cooling rate.

Typical properties of malleable cast iron are:

Tensile strength	50,000 psi
Yield strength	35,000 psi
Elongation, 2 in.	15 percent
Brinell hardness	120

For pearlitic malleable cast iron, typical properties are:

Tensile strength	70,000 psi
Yield strength	50,000 psi
Elongation, 2 in.	20 percent
Brinell hardness	180

The microstructure of the graphite formed will appear as shown in Figure 4.40*a,* section C, in a ball-like form. Graphite in this form makes a much more ductile metal than the metal would be if the graphite were in flake form (Figure 4.44).

Nodular Cast Iron

Nodular cast iron is known by several names: nodular iron, ductile iron, and spheroidal graphite iron. It gets the names from the ball-like form of the graphite in the metal and the very ductile property it exhibits. Nodular

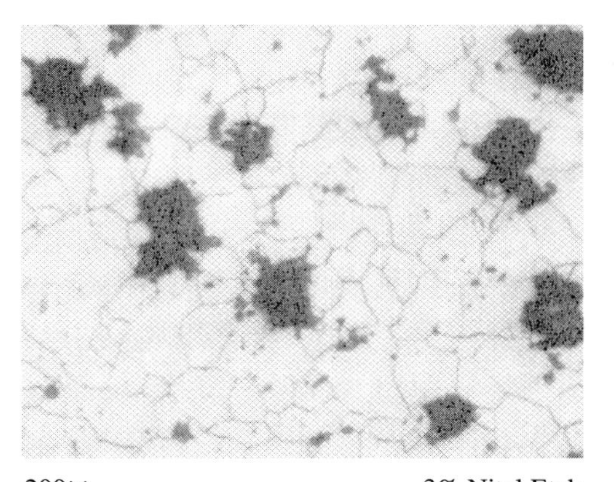

FIGURE 4.44. The microstructure of malleable cast iron consisting of irregular shaped temper-carbon graphite in aferrite matrix. Compare the shape of this temper-carbon with the graphite in nodular cast iron (Figure 4.41).

cast iron combines many of the advantages of cast iron and steel. Its advantages include good castability, toughness, machinability, good wear resistance, weldability, low melting point, and hardenability.

The formation of the graphite into a ball form is accomplished by adding certain elements such as magnesium and cerium to the melt just prior to casting. The vigorous mixing reaction caused by adding these elements results in a homogenous spheroidal or ball-like structure of the graphite in the cast iron. The iron matrix or background material can be heat treated to form any one of the microstructures associated with steels, such as ferrite, bainite (as in austempered ductile iron), pearlite, or martensite.

Ductile iron is increasingly replacing gray cast iron and steel castings for many applications, because it can be produced at less cost and with excellent mechanical qualities.

SELF-EVALUATION

1. Briefly describe the structure of an atom and the importance of the valence electrons.
2. How does the metallic bond work, and what effect does it have on metals?
3. Name five crystalline unit structures found in metals.
4. The growth from the nucleus of the grain that resembles small pine trees is called a ____________________.
5. Are the grain boundaries a continuation of the regular lattice structure from one grain to another? Explain.
6. Locate the following on the phase diagram (Figure 4.45):
   a. Eutectic point
   b. Liquidus line
   c. Solidus line
   d. Area of liquid and solid
   e. Area of 100 percent solubility

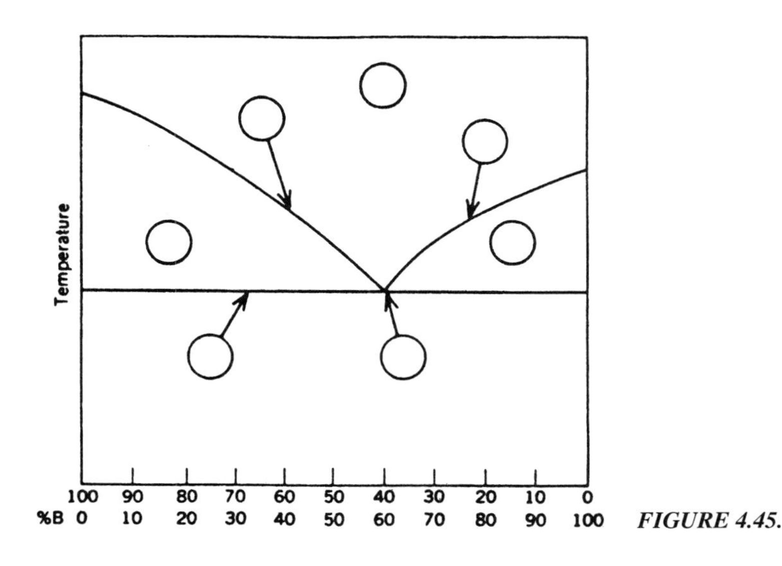

FIGURE 4.45.

7. When a substance changes phases while cooling, there is a release of heat. How does this appear on a cooling curve graph?
8. What does eutectoid mean? How does this differ from eutectic?
9. What does the A_3 line indicate? The A_1? The A_{cm}?
10. What is the hardest structure found in the iron-carbon alloy system? How would it appear in pearlite under the microscope?

CASE PROBLEM: RAPID WEAR IN A GRAVEL CHUTE

At a rock quarry, crushed rock was sent down a chute made of mild steel plate. It quickly wore holes through the bottom, so gray cast iron plates were made and bolted to the bottom. But they quickly wore out as well. However, the second time they were replaced, they showed very little wear after months of use. The rock crusher operator noticed the last set of plates had a whiter appearance than the first ones, so he looked up the requisition for the foundry that had made them. The buyer had added the words "wear-resistant plates." What kind of cast iron had he ordered that last time? Two things had been done to the cast iron in the foundry to make this change. What were they?

WORKSHEET 1

Objective Determine carbon content and heat treatment by metallurgical observation.

Materials Metallurgical equipment for encapsulating, polishing, and etching; two unknown samples of carbon steel, one of relatively low carbon and one of very high carbon content; metallurgical microscope (100× to 500× magnification).

Procedure

1. Heat both specimens to above 1650°F and slowly cool them in the furnace (anneal).
2. Prepare the two specimens for microscopic study as explained in Chapter 27 and Appendix 1. Identify the capsule by marking on the bottom or side.
3. Observe each specimen. Estimate the percentage of pearlite and ferrite. Compare your observations with the following:

   100 percent pearlite = 0.8 percent C
   75 percent pearlite = 0.6 percent C
   50 percent pearlite = 0.4 percent C
   25 percent pearlite = 0.2 percent C

Conclusion

1. How much carbon do you estimate for each sample?
2. Draw an iron-carbon diagram and locate each numbered specimen on the diagram for carbon content by drawing a vertical line.
3. Turn in your drawings and conclusions to your instructor.

WORKSHEET 2

Objective Learn to identify the microstructures of various forms of cast iron.

Materials Metallurgical equipment for microscopic study; encapsulated and polished specimens of gray, white, malleable, and nodular cast irons.

Procedure

1. Polish the specimens only. Do not etch.
2. View each specimen and make a drawing of each, showing what you see. Identify each microstructure with an arrow and label.
3. Etch the specimens and view each in the microscope. Draw four more sketches and identify the new features that are revealed by the etch.

Conclusion

1. What are the differences between the unetched and etched specimens?
2. Are the cast irons ferritic or pearlitic?
3. How can you identify gray cast iron?
4. Turn in your sketches and conclusions to your instructor.

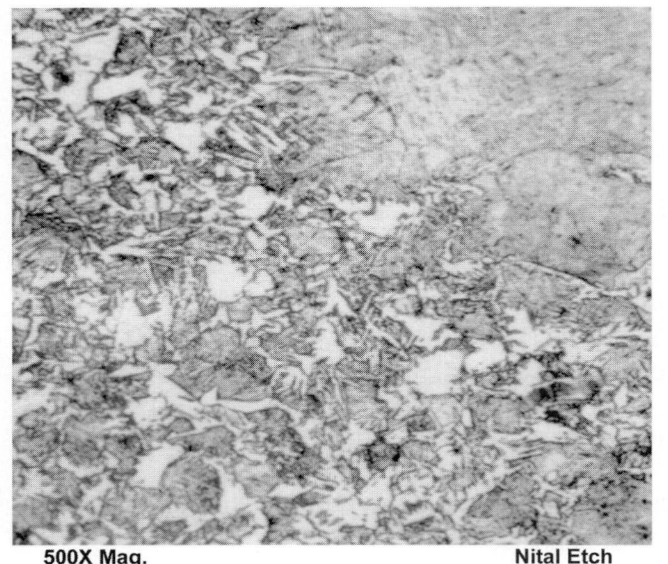

What Alloy Steel ?

CHAPTER 5

Identification and Selection of Iron Alloys

When the village smithy plied his trade, there was only wrought iron and plain carbon steel for making tools, implements, and horseshoes; so this made the task of separating metals relatively simple. As industry began to need more alloy steels and special metals, they were gradually developed. Today many hundreds of these metals are in use. Without some means of reference or identification, machine design and work in welding and machine shops would be confusing. Therefore, this chapter will introduce you to several systems used for identifying steels and to some ways of choosing between them.

OBJECTIVES **After completing this chapter, you should be able to:**

1. Identify different types of ferrous metals by various means of shop testing.
2. Select several commercial methods for determining AISI chemical analysis.

Classification of irons and steels makes it easier for the producer and user to identify the hundreds of different compositions available today. A few of the more common classification systems are given for plain carbon and low alloy steels, stainless steels, and cast irons.

Color coding is used as one means of identifying a particular type of steel. Its main disadvantage is that there is no universal color coding system. Manufacturers and local shops all have different codes.

PLAIN CARBON AND LOW ALLOY STEELS

The most common systems in the United States used to classify steels by chemical composition were developed by the Society of Automotive Engineers (SAE) and the American Iron and Steel Institute (AISI). The SAE and AISI systems use a four- or five-digit number (Table 5.1). The first number indicates the type of steel. Carbon, for instance, is denoted by the number 1, 2 is a nickel steel, 3 is a nickel-chromium steel, and so on. The second digit indicates the approximate percentage of the predominant alloying element. The third and fourth digits, represented by x, always denote the percentage of carbon in hundredths. For plain carbon steel it could be anywhere from 0.08 to 1.70 percent. For example, SAE 1040 denotes plain carbon steel with 0.40 percent carbon; SAE 4140 denotes a chromium-molybdenum steel containing 0.40 percent carbon and about 1.0 percent of the major alloy (molybdenum).

The AISI numerical system is basically the same as the SAE system, with certain capital letter prefixes. These prefixes designate the process used to make the steel.

The AISI prefixes are as follows:

B—Acid Bessemer, carbon steel

C—Basic open-hearth carbon steel

CB—Either acid Bessemer or basic open-hearth carbon steel at the option of the manufacturer

D—Acid open-hearth carbon steel

E—Electric furnace alloy steel

The AISI suffix H is used when hardenability is a major requirement.

TABLE 5.1. SAE-AISI numerical designation of alloy steels

Carbon Steels	
Plain carbon	10xx
Free-cutting; resulfurized	11xx
Manganese Steels	13xx
Nickel Steels	
0.50% nickel	20xx
1.50% nickel	21xx
3.50% nickel	23xx
5.00% nickel	25xx
Nickel-Chromium Steels	
1.25% nickel, 0.65% chromium	31xx
1.75% nickel, 1.00% chromium	32xx
3.50% nickel, 1.57% chromium	33xx
3.00% nickel, 0.80% chromium	34xx
Corrosion and Heat-Resisting Steels	303xx
Molybdenum Steels	
Chromium	41xx
Chromium-nickel	43xx
Nickel	46xx and 48xx
Chromium Steels	
Low chromium	50xx
Medium chromium	51xx
High chromium	52xx
Chromium-Vanadium Steels	6xxx
Tungsten Steels	7xxx
Triple Alloy Steels	8xxx
Silicon-Manganese Steels	9xxx
Steels	11Lxx (example)

(x represents percent of carbon in hundredths)

STAINLESS STEEL

It is the element chromium (Cr) that makes stainless steels stainless. Steel must contain a minimum of about 11 percent chromium to gain resistance to atmospheric corrosion. Higher percentages of chromium make steel even more resistant to corrosion at high temperatures. Nickel is added to improve ductility, corrosion resistance, and other properties.

There are three basic types of stainless steel: the martensitic and ferritic types of the 400 series and the austenitic types of the 300 series. Among these are the precipitation hardening types that harden over a period of time after solution heat treatment.

Since the martensitic, hardenable type has a carbon content up to 1 percent, it can be hardened by heating to a high temperature and then quenching (cooling) in oil or air. The grades of stainless steel for cutlery are to be found in this group. The ferritic type contains little or no carbon. It is essentially soft iron that has 12 percent or more chromium content. It is the least expensive of the stainless steels and is used for building trim, pots, and pans, for example. Both the ferritic and the martensitic types are magnetic.

Austenitic stainless steel contains chromium and nickel and little or no carbon, and cannot be hardened by quenching, but it readily work hardens while retaining much of its ductility. For this reason it can be work hardened until it is almost as hard as hardened martensitic steel. Austenitic stainless steel is somewhat magnetic in its work hardened condition but is nonmagnetic when annealed or soft.

TABLE 5.2. Classification of stainless steels

Alloy Content	Metallurgical Structure	Ability to Be Heat Treated
	Martensitic	Hardenable (Types 410, 416, 420)
Chromium types		Nonhardenable (Types 405, 14 SF)
	Ferritic	Nonhardenable (Types 430, 442, 446)
	Austenitic	Nonhardenable (except by cold work) (Types 301, 302, 304, 316)
		Strengthened by aging (Types 314, 17-14 CuMo, 22-4-9)
	Semiaustenitic	Precipitation hardening (PH 15-7 Mo, 17-7 PH)
Chromium-nickel types		Precipitation hardening (17-4 PH, 15-5 PH)

Source: Armco Steel Corporation, Middletown, Ohio, *Armco Stainless Steels,* 1966. The following are registered trademarks of Armco Steel Corporation: 17-4 PH, 15-5 PH, 17-7 PH, and PH 15-7 Mo.

Table 5.2 illustrates the method of classifying the wrought stainless steels. Only a very few of the basic types are given here. Cast stainless steels often use a separate system. Consult a manufacturer's catalog for further information.

TOOL STEELS

Special carbon and alloy steel, called tool steels, have their own classification. There are seven major tool steels, for which one or more letter symbols have been assigned.

1. Water-hardening tool steels
   W—high carbon steels
2. Shock-resisting tool steels
   S—Medium carbon, low alloy
3. Cold-work tool steels
   O—Oil-hardening types
   A—Medium alloy air-hardening types
   D—High carbon, high chromium types
4. Hot-work tool steels
   H—H10 to H19, inclusive, chromium base types
   H20 to H39, inclusive, tungsten base types
   H40 to H59, inclusive, molybdenum base types
5. High-speed tool steels
   T—Tungsten base types
   M—Molybdenum base types
6. Special-purpose tool steels
   L—Low alloy types
   F—Carbon tungsten types
7. Mold steels
   P—P1 to P19, inclusive, low carbon types
   P20 to P39, inclusive, other types

Several metals can be classified under each group, so an individual type of tool steel will also have a suffix number that follows the letter symbol of its alloy group. The carbon content is given only in those cases when it is considered an identifying element of that steel. In Appendix 1, Table 1, tool steels are listed according to manufacturers' names and heat treatments.

For example, the classification of some types of steel is as follows:

Water hardening:	
Straight carbon tool steel	W1, W2, W4
Manganese, chromium, tungsten:	
Oil-hardening tool steel	O1, O2, O6
Chromium (5.0%):	
Air-hardening die steel	A2, A5, A10
Silicon, manganese, molybdenum:	
Punch steel	S1, S5
High-speed tool steel	M2, M3, M30 T1, T5, T15

UNIFIED NUMBERING SYSTEM

A new numbering system, called the Unified Numbering System for Metals and Alloys (UNS), provides a designation system for all present and future metals and alloys. The system was published by the SAE in 1975. Both SAE and ASTM (American Society for Testing and Materials) will always use these numbers. The system will also be proposed to the ISO (International Standards Organization).

The Unified Numbering System (Table 5.3) establishes 15 series of numbers for metals and alloys. Each

TABLE 5.3. Unified numbering system for metals and alloys

Ferrous Metals	
D00001-D99999	Specified mechanical properties steels
F00001-F99999	Cast irons: Gray, malleable, pearlitic malleable, ductile (nodular)
	Carbon steel castings, low alloy steel castings
G00001-G99999	AISI-SAE carbon and alloy steels
H00001-H99999	AISI-"H" steels
K00001-K99999	Miscellaneous steels and ferrous alloys
S00001-S99999	Stainless steels
T00001-T99999	Tool steels
Nonferrous Metals	
A00001-A99999	Aluminum and aluminum alloys
C00001-C99999	Copper and copper alloys
E00001-E99999	Rare earth and rare earthlike metals
L00001-L99999	Low melting metals
M00001-M99999	Miscellaneous metals
N00001-N99999	Nickel and nickel alloys
P00001-P99999	Precious metals
R00001-R99999	Reactive and refractory metals

Source: Reprinted by John E. Neely from *Iron Age,* December 16, 1974, Chilton Company, 1976.

UNS number consists of a single letter prefix followed by five digits. In most cases, the letter is suggestive of the family of metals identified; for example, A for aluminum and P for precious metals.

When feasible, identification numbers from existing systems are incorporated into the UNS numbers. For example, carbon steel, presently identified by AISI 1020, is covered by UNS G11020; free-cutting brass, now identified by CDA 360, is covered by UNS C36000.

CAST IRON

There are several forms of cast iron: white, gray, malleable, and nodular. Cast irons contain more carbon (2 to 4.5 percent) than steels and are easily cast into molds. Cast irons have a melting temperature of 2100°F (1149°C), in contrast to the melting temperature of steel, which is 2500 to 2800°F (1371 to 1538°C). Cast iron may be identified by its brittle fracture and by spark testing (Figure 5.1).

Gray cast irons may be ferritic or pearlitic and may contain alloying elements such as nickel or chromium and varying amounts of silicon. Gray cast irons are classified by the ASTM according to their tensile strength. Table 5.4 compares the ASTM number to its tensile strength in pounds per square inch (psi). Gray cast iron is very brittle, and its tensile strength is much lower than its compressive strength (Figure 5.2). See Chapter 13, "Welding Processes for Iron and Iron Alloys."

FIGURE 5.1. Fractures of *(top)* gray and *(bottom)* white cast iron.

TABLE 5.4. Class of gray iron

ASTM Number	Minimum Tensile Strength (psi)
20	20,000
25	25,000
30	30,000
35	35,000
40	40,000
45	45,000
50	50,000
60	60,000

SHOP TESTS FOR IDENTIFYING STEELS

Steel is usually identified by placing a color code at the end of a shaft. One of the disadvantages of this method is that the marking is often lost. If the marking is obliterated or cut off and the piece is separated from the rack where it is stored, it is very difficult to ascertain its carbon content and alloy group. This points up the necessity of returning stock material to its proper rack. It is also good practice always to leave the identifying mark on one end of the stock material and always to cut off the other end.

Unfortunately, in the shop there are always some short ends and otherwise useful pieces of steel that have lost their identifying marks. In addition, when repairing or replacing parts of old or nonstandard machinery, there is usually no record available for material selection. There are many shop methods a machinist may use

FIGURE 5.2. The thin edges of this cast iron pulley have been broken by hammering. The hub also shows signs of hammering. Cast iron is quite brittle and is easily broken. Wheel pullers placed on the thin edges of a pulley can also cause breakage. The stress should have been applied to the hub when it was being removed.

to identify the basic type of steel in an unknown sample. The machinist can compare the unknown sample with each of the several steels used in the shop. The following are several methods of shop testing that you can use.

FIGURE 5.3. Rolled bars: (top) hot rolled, (center) cold rolled, and (bottom) ground and polished.

Visual Observation

Some metals can be identified by visual observation of their finishes. Heat scale or **black mill scale** is found on all hot rolled (HR) steels (Figure 5.3). These can be either low carbon (0.05 to 0.30 percent), medium carbon (0.30 to 0.60 percent), high carbon (0.60 to 1.75 percent), or alloy steels. Other surface coatings that might be detected are the **sherardized, plated, case-hardened,** or **nitrided surfaces.** Sherardizing is a process in which zinc vapor is inoculated into the surface of iron or steel.

Cold-finish (CF) steel usually has a metallic luster. Ground and polished (G and P) steel has a bright, shiny finish with closer dimensional tolerances than CF. Also cold-drawn ebonized, or black, finishes are sometimes found on alloy and resulfurized shafting.

Chromium nickel stainless steel, which is austenitic and nonmagnetic, usually has a white appearance. Straight 12 to 27 percent chromium is called ferritic and is magnetic with a bluish-white color. Manganese steel is blue when polished but copper colored when oxidized. White cast iron fractures will appear silvery or white. Gray cast iron fractures appear dark gray and will smear a finger with a gray graphite smudge when touched.

Eddy Current Inspection

Eddy current inspection is an electromagnetic induction technique and does not require contact with the material being tested. It is used to measure, identify, or sort dissimilar metals and detect differences in their composition, properties, and microstructure.

Eddy currents create their own magnetic field which can be sensed either through the effects of the field on the primary exciting coil or by means of an independent sensor. Once the baseline characteristics of the eddy current have been established for a given part or raw material (plate, sheet, or bar stock), when placed into the baselined eddy current field, the field becomes disturbed and signals a difference. Identical parts produced from different alloys and/or raw material stock can be easily separated due to chemical, heat treatment, and size differences.

Magnet Test

Most ferrous metals such as iron and steel are magnetic (that is, they are attracted to a magnet), but most nonferrous metals are nonmagnetic. An exception is nickel, which is nonferrous but magnetic. Because U.S. "nickel" coins contain about 25 percent nickel and 75 percent copper, they do not respond to the magnet test, but Canadian "nickel" coins are attracted to a magnet because they contain more nickel. Ferritic and martensitic (400 series) stainless steels are also attracted to a magnet and thus cannot be separated from other steels by this method. (See the section on chemical tests.) Austenitic (300 series) stainless steel are not magnetic unless work hardened.

Hardness Test

Wrought iron is very soft because it contains almost no carbon or any other alloying element. Generally speaking, the *more* carbon (up to 2 percent) and other elements that steel contains, the *harder,* stronger, and less ductile it becomes, even if the steel has been annealed (softened). Thus, the hardness of a sample can help us to separate low carbon steel from an alloy steel or a high carbon steel. Of course, the best way to check for hardness is with a hardness tester. Rockwell, Brinell, and other types of hardness testing were studied in Chapter 3. Not all machine shops have hardness testers available, in which case the following shop methods can prove useful. Hardness testing methods will be discussed in detail in Chapter 26.

***FIGURE 5.4*a.** A piece of keystock (mild steel) is scratched across an unknown metal sample. Since the sample is not scratched, it is harder than the keystock and probably is an alloy or tool steel.

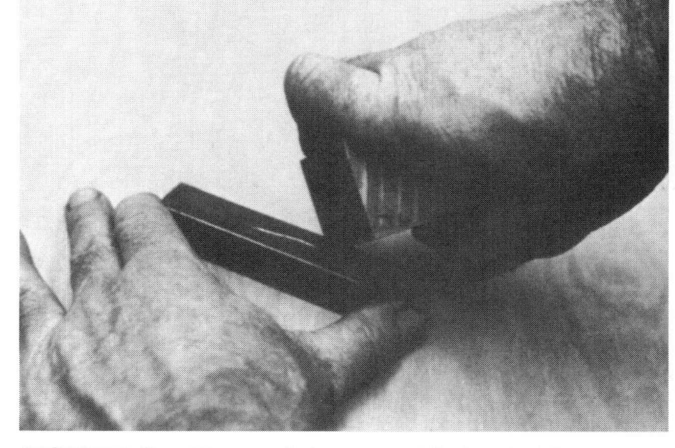

***FIGURE 5.4*b.** The sample is now scratched against the keystock as a further test and it does scratch the keystock.

Scratch Test

Geologists and "rock hounds" scratch rocks against items of known hardness for identification purposes. The same method can be used to check metals for relative hardness. Simply scratch one sample with another and the softer sample will be marked (Figures 5.4*a* and 5.4*b*). Be sure all scale or other surface impurities have been removed before scratch testing. A variation of this method is to strike two similar edges of the two samples together. The sample receiving the deeper indentation is the softer (Figures 5.5*a* and 5.5*b*). For this test all mill scale on the surface to be compared should be removed. Mill scale is quite hard and would give a false reading.

The Moh scale (Table 5.5) can provide a useful means of relating one material to another in terms of relative hardness. The Moh scale hardness of some commonly used materials in the shop is as follows:

Aluminum	2 to 2.9
Brass	3 to 4
Chromium	9
Glass	5
Iron	4 to 5
Lead	1.5
Magnesium	2
Tool steel (hardened)	7.5 to 8.5
Tungsten carbide	9.5
Zinc	2.5

File Tests

Files can be used to establish the relative hardness between two samples, as in the scratch test. The test can be used to determine the approximate hardness of a piece of steel (Figure 5.6). Table 5.6 gives the Rockwell and Brinell hardness numbers for this file test when using new files. This method, however, can only be as accurate as the skill that the user has acquired through practice. Care must be taken not to damage the file, since filing on hard materials may ruin the file. Testing should be done on the tip end or on the edge of the file. File hardness is approximately 50 HRC.

Chemical Tests

Many shop chemical tests are very simple, most of them using chemicals normally found in the shop. More complex chemical testing should be done by a chemist.

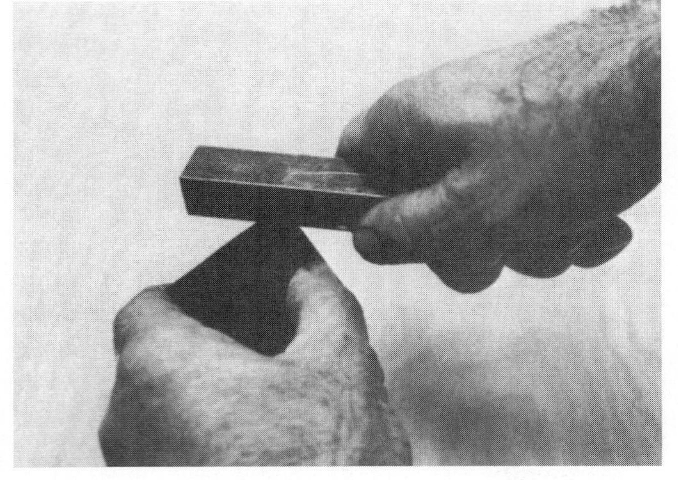

FIGURE 5.5a. An alternate hardness test is to strike the edges of the keystock and the unknown sample together.

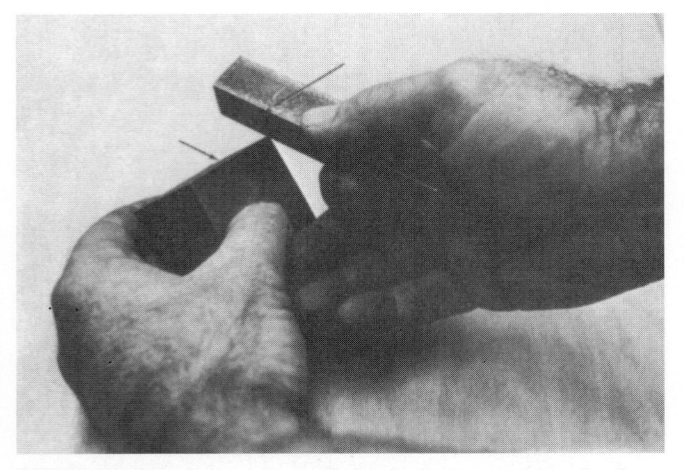

FIGURE 5.5b. It can be seen that only the keystock received an indentation where they were struck together, signifying that the keystock was the softer of the two samples.

TABLE 5.5. Moh scale of hardness

Number	Material
1	Talc
2	Rock salt or gypsum
3	Calcite
4	Fluorite
5	Apatite
6	Feldspar
7	Quartz
8	Topaz
9	Corundum
10	Diamond

The surface to be tested must be very clean and free from all scale or oil. Apply only one drop of the acid with an eye dropper. A solution of 6 percent nitric acid in methanol (wood alcohol) will etch, or darken, carbon steel but will not discolor stainless steels (Figure 5.7). A 10 percent nitric acid solution will etch mild steel almost immediately, whereas the 6 percent solution takes about one minute. A drop of concentrated copper sulfate solution will leave a copper-colored spot on clean iron or steel but not on austenitic stainless steel.

Some stainless steels react to sulfuric and hydrochloric acids. Types 302 and 304 are strongly attacked by sulfuric acid, which leaves a dark surface with green crystals. Types 316 and 317 are attacked more slowly, leaving a tan surface.

FIGURE 5.6. File testing for hardness. (*Lane Community College*)

TABLE 5.6. File test and hardness table

Type of Steel	Rockwell B	Rockwell C	Brinell	File Reaction
Low carbon steel	65		100	File bites easily into metal. (Machines well but makes built-up edge on tool.)
Medium carbon steel		16	212	File bites into metal with pressure. (Easily machined with high-speed tools.)
High alloy steel		31	294	File does not bite into metal except with difficulty.
High carbon steel		42	390	(Readily machinable with carbide tools.)
Tool steel		42	390	Metal can only be filed with extreme pressure. (Difficult to machine even with carbide tools.)
Hardened tool steel		50	481	File will mark metal but metal is nearly as hard as the file, and machining is impractical; should be ground.
Case-hardened parts		64		Metal is as hard as the file; should be ground.

FIGURE 5.7. Identifying stainless steels with an acid test. A 6 percent solution of nitric acid in methanol (nital) has been applied to the stainless steel on the left hand and is being applied to the mild steel on the right. The mild steel is discolored, but the stainless steel is not.

Hydrochloric acid (HCl) reacts with types 304, 321, and 347 very rapidly, releasing gas. Type 302 leaves a pale blue-green solution on the surface, while types 303, 414, and 430F have a spoiled egg odor and leave a heavy black smudge. Steels containing selenium emit a garlic odor when attacked by HCl.

Commercial spot tests are available in kits. Stainless steel and many ferrous and nonferrous alloys can be readily identified using these test kits. See Appendix 2, Table 8.

Spark Testing

Spark testing is a useful way to test for carbon content in many steels. The metal tested, when held against a grinding wheel, will display a particular spark pattern, depending on its content. Spark testing provides a convenient means of distinguishing between tool steel (of medium or high carbon) and low carbon steel. High carbon steel (Figure 5.8) shows many more bursts than low carbon steel (Figure 5.9).

Almost all tool steel contains some alloying elements besides the carbon that affects the carbon burst. Chromium, molybdenum, silicon, aluminum, and tungsten suppress the carbon burst. For this reason spark testing is not always dependable for determining the carbon content of an unknown sample of steel unless it is plain carbon steel. It is useful, however, as a comparison test. Comparing the spark of a known sample to that of an unknown sample can be an effective method of identification for the trained observer. Cast iron may be distinguished from steel by the characteristic spark stream (Figure 5.10). High-speed steel can also be readily identified by spark testing (Figure 5.11).

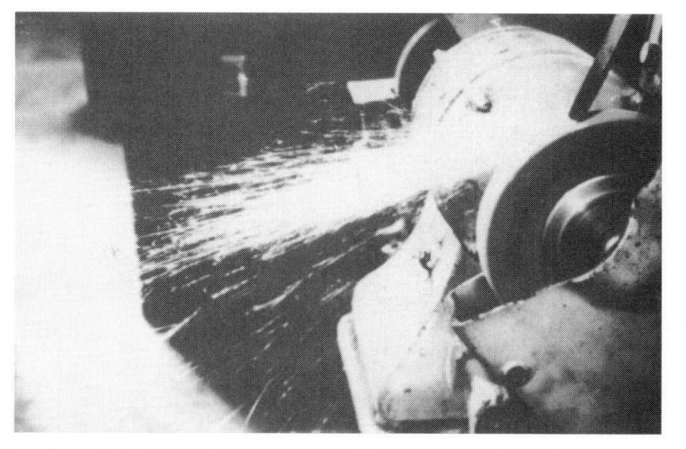

FIGURE 5.8. High carbon steel. Short, very white or light yellow carrier lines with considerable forking, having many starlike bursts. Many of the sparks follow around the wheel.

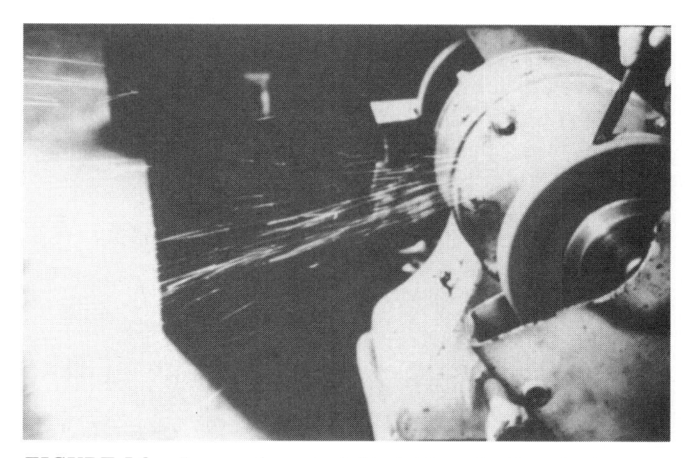

FIGURE 5.9. Low carbon steel. Straight carrier lines having a yellowish color with a very small amount of branching and very few carbon bursts.

The various types of stainless steel may be identified with spark testing. Types 302, 303, 304, and 316 show short reddish carrier lines with few forks. Types 308, 309, and 310 have full red carrier lines with very few forks. Of the martensitic types of stainless steel, types 410, 414, 416, and 431 have long white carriers with few forks. Types 420, 440A, 440B, and 440C have long white-red carriers with bursts. Ferritic types range from white to red carriers with some forks.

When spark testing always wear safety glasses or a face shield. Adjust the wheel guard so that the spark will fly outward and downward and away from you

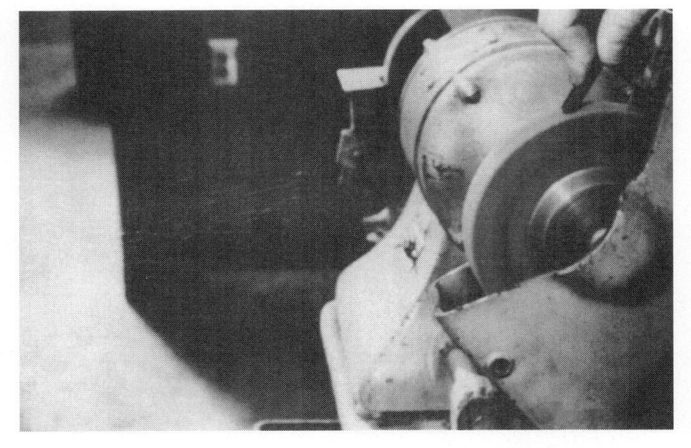

FIGURE 5.10. Cast iron. Short carrier lines with many bursts that are red near the grinder and orange-yellow farther out. Considerable pressure is required on cast iron to produce sparks.

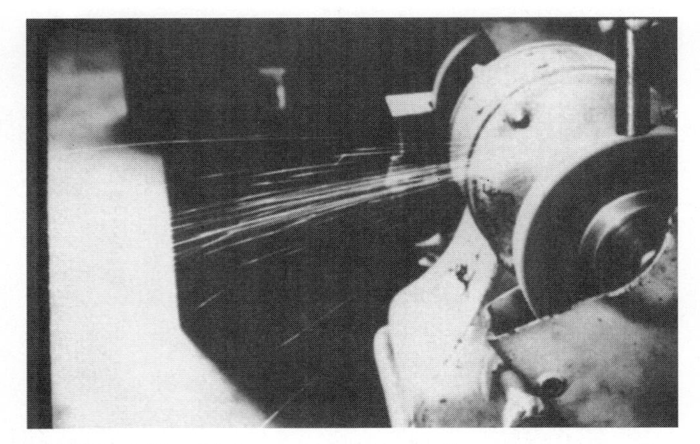

FIGURE 5.11. High-speed steel. Carrier lines are orange, ending in pear-shaped globules with very little branching or carbon sparks. High-speed steel requires moderate pressure to produce sparks.

(Figure 5.12). A coarse grit wheel that has been freshly dressed to remove contaminants should be used.

Machinability Test

As a simple comparison test, machinability can be useful to help determine a specific type of steel. For example, two unknown samples, identical in appearance and size, can be test cut in a machine tool using the same speed and feed for both of them. The ease of cutting should be compared and chips observed for heating, color, and curl (Figure 5.13). Hardness is generally related to ease of machining; harder materials are generally more difficult to machine.

Chemical Analysis for Alloy Identification

Wet chemical analysis is a somewhat outdated method of material identification and is seldom used because of the hazardous chemicals and lengthy processes involved. Wet chemical analysis has been virtually replaced by optical emission spectrometry (OES). OES is a method normally used to determine the base metal composition. The technique employs an electric arc and analyzes the light spectrum for element detection, qualitative and quantitative analysis. It is limited to flat pieces requiring a flat cross section larger than 0.25 inch diameter. Most plain carbon, alloy steels, and tool steels can be identified by OES. Other metals such as titanium

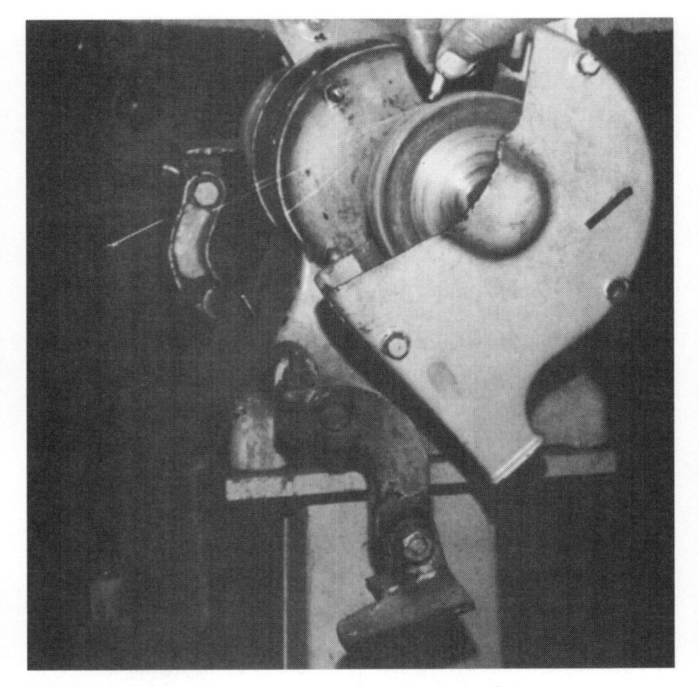

FIGURE 5.12. The proper way to make a spark test on a pedestal grinder. (*Lane Community College*)

FIGURE 5.13. Machinability test. When one of two samples having the same feed and speed shows a darker color (blue) on the chip, it can be assumed that it is a harder, stronger metal. (*Lane Community College*)

alloys, aluminum alloys, copper, nickel, and cobalt base alloys and stainless steels can also be identified by OES.

X-ray fluorescence spectrometry (XRF) is an alternate method used for material identification when determining exotic materials. Carbon, hydrogen, and sulfur determinations are often performed as individual tests in specific testing machines (e.g., Leco testing machines). These and other methods of element identification are discussed in Chapter 27.

SELECTION FOR USES

There are a great variety of carbon, alloy, and tool steels from which to choose when planning a project. Several types of cast irons are also available to the welder, machinist, or machine designer. Some of these are listed according to their carbon content and use in Table 5.7. In later chapters in this book, we present many more principles and concepts regarding the selection of metals for a particular job. For example, many more alloy steels than plain carbon steels are used in the manufacture of tools as the alloy steel gives them special properties.

Several properties should be considered when selecting a piece of steel for a job: **strength, machinability, hardenability, weldability, fatigue resistance,** and **corrosion resistance.**

Manufacturers' catalogs and hand reference books are available for the selection of standard structural shapes, bars, and other steel products (Figure 5.14). Others are available for stainless steels, tool steels, and finished carbon steel and alloy shafting. Many of these steels are known by a trade name.

A machinist is often called on to select a shaft material from which to machine finish a part. Shafting is manufactured with two kinds of surface finish: cold finished (CF), found on low carbon steel, and ground and polished (G and P), found mostly on alloy shafts. Tolerances are kept much closer on G and P shafts. The following are some common alloy and carbon steels.

1. SAE 4140 is a chromium-molybdenum alloy with 0.40 percent carbon. It lends itself readily to heat treating, forging, and welding. It provides a high resistance to torsional and reversing stresses such as in drive shafts.
2. SAE 1140 is a resulfurized, drawn, free-machining bar stock. This material has good resistance to bending stresses because of its fibrous qualities and has a high tensile strength. It is best used on shafts where the revolutions per minute (rpm) is high and the torque moderately low. SAE 1140 is also useful where stiffness is a requirement. It should not be welded.

TABLE 5.7. Uses of ferrous metals by carbon content

Type	Carbon Range (percent)	SAE Number	Typical Uses
Carbon Steels			
Low	0.05 to 0.30	1006	For cold formability
		1008	Wire, nails, rivets, screws
		1010	Sheet stock for drawing
		1015	Fenders, pots, pans, welding rods
		1020	Bars, plates, structural shapes, shafting
		1030	Forgings, carburized parts, keystock
		1111	Free-machining steel
		1113	Free-machining steel
Medium	0.30 to 0.60	1040	Heat-treated parts that require moderate strength and high toughness such as bolts, shafting, axles, spline shafts
		1060	Higher strength, heat-treated parts with moderate toughness such as lock washers, springs, band saw blades, ring gears, valve springs, snap rings
High	0.60 to 2.0	1070	Chisels, center punches
		1080	Music wire, mower blades, leaf springs
		1095	Hay rake tines, leaf springs, knives, woodworking tools, files, reamers
		52100	Ball bearings, punches, dies
Cast Irons			
Gray	2.0 to 4.5		Machinable castings such as engine blocks, pipe, gears, lathe beds
White	2.0 to 3.5		Nonmachinable castings such as cast parts for wear resistance
Malleable castings	2.0 to 3.5		Produced from white cast iron Machinable castings such as axle and differential housings, crankshafts, camshafts
Nodular iron (ductile iron)	2.0 to 4.5		Machinable castings such as pistons, cylinder blocks and heads, wrenches, forming dies

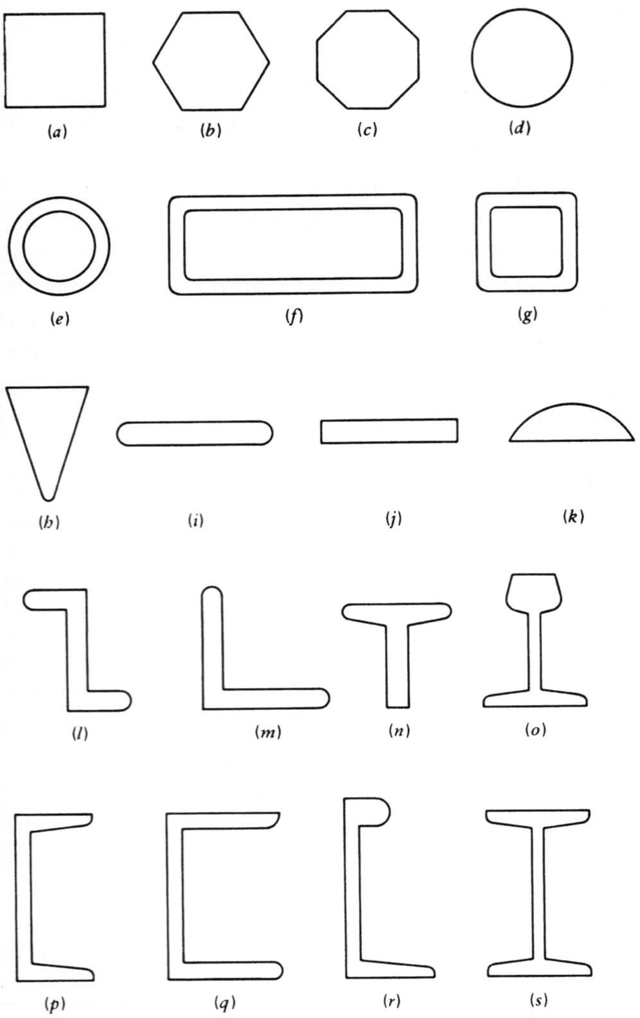

(*a*) Square HR or CR
(*b*) Hexagonal
(*c*) Octagon
(*d*) Round
(*e*) Tubing and pipe round
(*f*) HREW (hot rolled electric welded) rectangular steel tubing
(*g*) HREW square steel tubing
(*h*) Wedge
(*i*) HR flat bar (round edge spring steel flats)
(*j*) Flat bar (CR and HR)
(*k*) Half round
(*l*) Zee
(*m*) Angle
(*n*) Tee
(*o*) Rail
(*p*) Channel
(*q*) Car and ship channel
(*r*) Bulb angle
(*s*) Beams—I, H, and wide flange

FIGURE 5.14. Steel shapes used in manufacturing.

3. Leaded steels have all the free-machining qualities and finishes of resulfurized steels. Leaded alloy steels such as SAE 41L40 have the superior strength of 4140 but are much easier to machine.
4. SAE 1040 is a medium carbon steel with a normalized tensile strength of about 85,000 psi. It can be heat treated, but large cross sections can be hardened only on the surface while the core will be in a normalized condition. Its main advantage is that it is a less expensive way to obtain a higher strength part.
5. SAE 1020 is a low carbon steel that has good machining characteristics. It normally comes as CF shafting. It is very commonly used for shafting in industrial applications. It has a lower tensile strength than the alloy steels or higher carbon steels.

SELF-EVALUATION

1. By what universal coding system are carbon and alloy steels designated?
2. What are three basic types of stainless steels and what is the number series assigned to them? What are their basic differences?
3. If your shop stocked the following steel shafting, how would you determine the content of an unmarked piece of each, using shop tests as given in this chapter?
   a. AISI C1020 (CF)
   b. AISI B1140 (G and P)
   c. AISI C4140 (G and P)
   d. AISI 8620 (HR)
   e. AISI B1140 (Ebony)
   f. AISI C1040
4. A small part has obviously been made by a casting process. How can you determine whether it is a ferrous or a nonferrous metal, stainless steel, or white or gray cast iron?
5. What is the meaning of the symbols O1 and W1 when applied to tool steels?
6. When checking the hardness of a piece of steel with the file test, the file slides over the surface without cutting.
   a. Is the steel piece readily machinable?
   b. What type of steel is it most likely to be?
7. How can you distinguish austenite stainless steel from ferritic and martensitic types by using a magnet?
8. What nonferrous metal is magnetic?
9. List at least four properties of steel that should be kept in mind when you select the material for a job.
10. What alloying element does all stainless steel contain in large amounts that make it corrosion resistant? What other element does stainless steel sometimes contain in fairly large amounts?
11. Name the major characteristic of cast iron as compared to mild steel that limits its use in mechanical parts.

CASE PROBLEM: FIVE UNIDENTIFIED STEEL SHAFTS

A small job shop stocked five kinds of steel shafting in its racks: CF SAE 1020, HR SAE 1040, HR SAE 4140, SAE 4140, G and P, and austenitic stainless steel G and P. Short ends of all these shafts minus their identifying brand were stacked in a box. Not wanting to be wasteful, a machinist decided to use one of these ends for his project. However, the two hot rolled bars looked alike and the two ground and polished bars were similar in appearance. He could identify the CR shaft. How was he able to do that? By what means could he have identified the other four?

WORKSHEET

Objective Correctly identify specimens by comparison by using various tests described in this chapter.

Materials A box of numbered specimens and a similar set of known and marked specimens.

Conclusion Record your results.

Item Number	Test Used	Kind of Metal
1.		
2.		
3.		
4.		
5.		
6.		
7.		
8.		
9.		
10.		

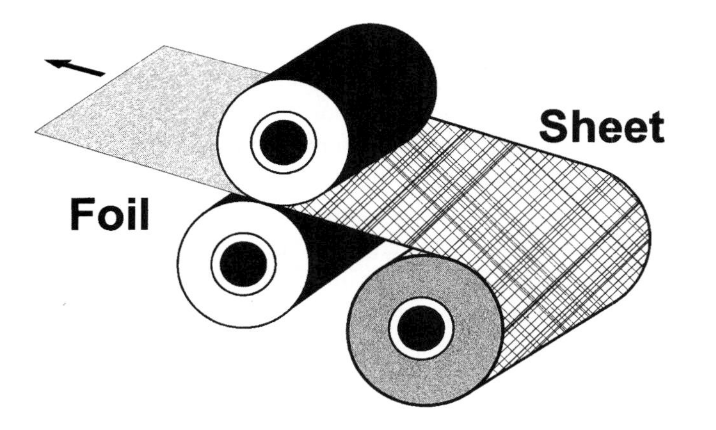

CHAPTER 6

The Manufacturing of Steel Products

After molten steel has been cast into molds of suitable shape, the solidified white-hot metal is ready for processing in a rolling mill. It is there that structural shapes for bridges and buildings and steel rails for railroads are made. Bars made in rolling mills are reprocessed into tools, pipe, and many other products.

This chapter introduces you to the major steel-finishing processes that take these massive forms of metal and shape them into workable materials for manufacture.

OBJECTIVES **After completing this chapter, you should be able to:**

1. Describe how steel is formed into various shapes and products.
2. List the advantages of some processes over others for a given product.

HOT WORKING

Almost all steel that is produced in steel-making furnaces is teemed, or poured, into cast iron ingot molds, where it solidifies as it cools. From these steel ingots finished steel products such as structural shapes, plate, and pipe are made. The ingot mold is removed by "stripping" when the steel has solidified (Figure 6.1). A crane lifts the mold off the ingot.

Before rolling, the ingot must be reheated uniformly throughout as it has become too cold to work. It is placed in a furnace called a soaking pit (Figure 6.2), where it is heated to about 2200°F for 4 to 8 hours. The grain structure of the ingot as in most cast metals is coarse and columnar, making it weak. Hot rolling breaks down the coarse grains and reforms them to make a stronger, finer grained steel (Figure 6.3*a*).

At the higher hot rolling temperatures, the crystal grains are broken down into smaller grains but almost immediately will reform into the former undesirable coarse grain structure. This is called **recrystallization.** However, if hot working is carried out from one operation of plastic deformation to the next on progressively cooling metal, little time is allowed for recrystallization, and the reformed grains tend to remain smaller, resembling to some extent cold-worked steel such as that shown in Figure 6.3*b*. The final grain size is influenced by the original grain size, the rate of cooling, the amount of deformation, and the finishing temperature. Hot working closes up cavities and makes the metal more uniform. Impurities such as slag inclusions and those impurities surrounding the original grains in the ingot become thin, elongated fibers in the rolled metal. The resulting etched microstructure shows plastic deformation which is known as **grain flow** (see Figure 6.4).

Ingots are formed into one of three different shapes, depending on the final products to be made from the steel. These basic forms are **slabs, blooms,** and **billets,** as shown in Figure 6.5. Any piece of steel to be rolled into a plate is first formed into a slab. Slabs are made in a primary mill called a slabbing mill. Blooms are made in a blooming mill, which gives the ingot a cross section of 6 to 12 in. on a side. They are mainly used in the production of rails and other structural shapes. Billets are usually made directly from the bloom and are round or square; their cross section ranges from 2 to 5 in. on a side. Round billets are used in making seamless tubing, while bar stock of various sizes and shapes are also rolled from square billets.

FIGURE 6.1. After steel has solidified in the ingot mold, the mold is pulled off the ingot by a stripper. (*Courtesy of American Iron & Steel Institute*)

FIGURE 6.2. Powerful tongs lift an ingot from the soaking pit, where it was thoroughly heated to the rolling temperature. (*Bethlehem Steel Corporation*)

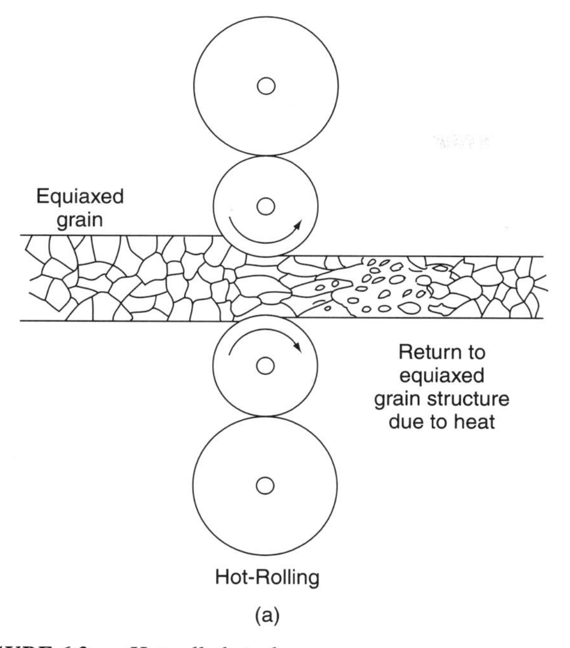

***FIGURE 6.3*a.** Hot rolled steel.

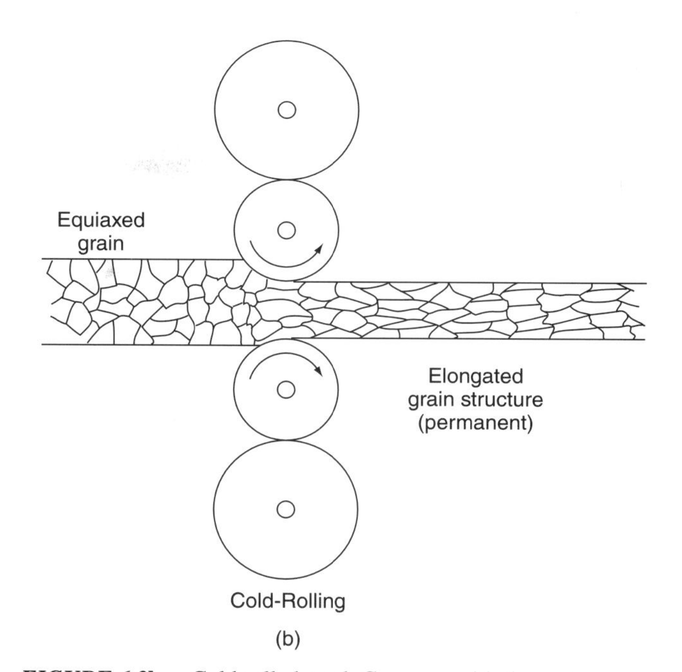

***FIGURE 6.3*b.** Cold rolled steel. Compare with the grain structure in hot rolled steel, Figure 6.3*a*.

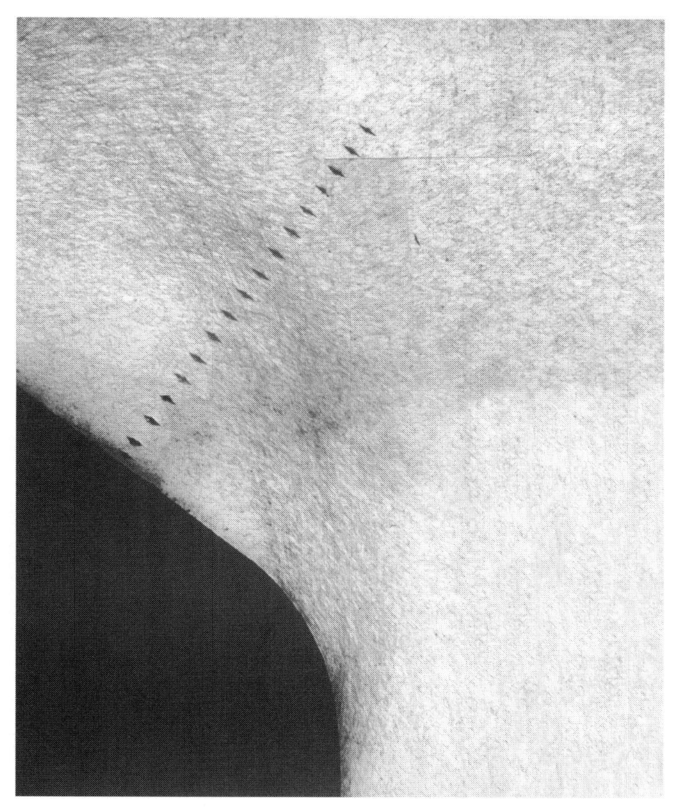

~30× 2% Nital Etch

***FIGURE 6.4*a.**

400× 2% Nital Etch

***FIGURE 6.4*b.** Two examples of grain flow are shown. The fillet radius of a flush head fastener with shear lines opposes the grain flow from head forming. The pattern of evenly spaced black marks are knoop microhardness indentations. The microstructure (right side) exhibits elongated grains with deep grain flow relief from etching a wrought metal.

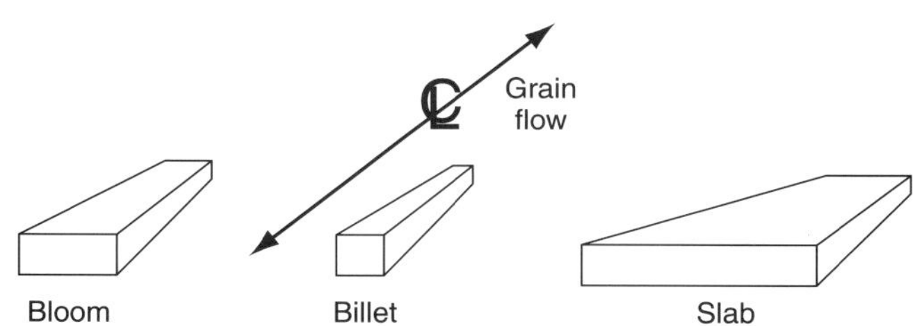

FIGURE 6.5. Ingot formed into blooms, billets, and slabs.

CONTINUOUS CASTING

A relatively new process of steel production that bypasses the ingot stage is called continuous or strand casting (Figure 6.6). The molten steel is converted directly into slabs, blooms, or billets in this process. Slabs can be produced in 45 minutes as compared to 12 hours for the conventional process. The steel is cooled as it passes through a water-cooled mold, and the continuous slab is rolled to size in a series of mill stands. Although some difficulties have been experienced in the use of strand casting for steel, it is becoming more commonly used. Continuous casting of nonferrous metals has been very successful.

ROLLING MILLS

Structural rolling mills (Figure 6.7) produce shapes such as I-beams, channels, angles, and wide-flange beams, as well as some special section such as zees, tees, H-piles, and sheet piling. All these shapes are produced from blooms in a series of mill stands, each mill stand contributing to the final shape. Ordinary structural shapes are made of low carbon steel containing from 0.10 to 0.25 percent carbon. This steel, which has been hot rolled, has a low strength as compared to alloy steels. **Hot rolled** (HR) products have a gray or black mill scale on the surface, which must be removed before further finishing of the steel is done. Hot rolled plate, unlike structural steel, can be formed from a slab of low carbon steel, an alloy steel, or a high carbon steel (Figure 6.8). Hot rolled plate is manufactured as either sheared or universal milled (UM). The sheared plate is cut to the desired size, but the UM plate is formed on the edges by a series of rolls. Plate is designated as shapes from 1/4 in. thick and under 48 in. wide or 3/16 in. thick and over 48 in. wide. Thinner products are called sheets.

Cold rolled (CR) sheet makes up a large part of steel production (Figure 6.9). Hot rolled sheet is cleaned with an acid dip called pickling, followed by a dip in lime water. The sheet is then cold rolled under very heavy pressure, after which it is wound into coils. Some of this sheet steel is used to produce household appliances such as ranges, washers, and dryers, while a vast tonnage of it is used for auto bodies. Very narrow sheets, or strip steel, are usually wound on a roll and used for such manufacturing purposes as press work.

Flat sheet stock is rolled without reheating, a process that permanently deforms and elongates the grain structure of the steel. This process toughens and strengthens the metal and gives it a smooth, bright metallic finish but reduces its ductility—its ability to be deformed or stretched without breaking. The grains are elongated in the direction of the rolling, making the metal more ductile in one axis than the other. This characteristic of cold rolled metals makes them more liable to crack when they are bent in a small radius along the direction of rolling than across the direction of rolling (Figure 6.10).

Annealing, a heat-treating operation, can restore the ductility at the expense of strength. In one such method, the steel is heated to approximately 1600°F (871°C) and very slowly cooled to room temperature. The steel can then be further cold worked without danger of cracking or splitting.

GALVANIZING

When iron is in the presence of oxygen and moisture, it begins to rust. If it is not protected in some way, the iron will eventually revert to its original state of iron oxide. Steel products are dipped in hot zinc to give them a coating resistant to the corrosive effects of the everyday atmosphere. Sheet steel is galvanized to make garbage cans and parts of autos, furnaces, and ventilating systems. Many products such as water pails are galvanized after manufacture. See the flowchart (Figure 1.17) in Chapter 1.

TIN PLATE

Much of the food we eat is preserved in cans made from tin plate. These "tin cans" are actually 99 percent steel and only 1 percent tin. The thin plating of tin protects both the inside and outside of the can from corrosion while giving it a bright, attractive surface. The tin coating is also applied to the steel by a hot dip process. See the flowchart (Figure 1.17) in Chapter 1.

BARS

Hot rolled bars (Figure 6.11) may be round, square, flat, and hexagonal in shape. Bars, like plate, can be low carbon of 0.10 to 0.20 percent, often called mild steel, or they can be alloy or tool steel. Tremendous numbers of steel bars are used to manufacture things such as axles, gears, and connecting rods in automobiles. Tools, such as wrenches and pliers, are also made from steel bars. Tons of steel bars are used every year as reinforcers in concrete construction.

Cold-finished bars are made from hot rolled bar stock and are designated as cold rolled (CR) or cold finished (CF). The hot rolled bar is given an acid bath to re-

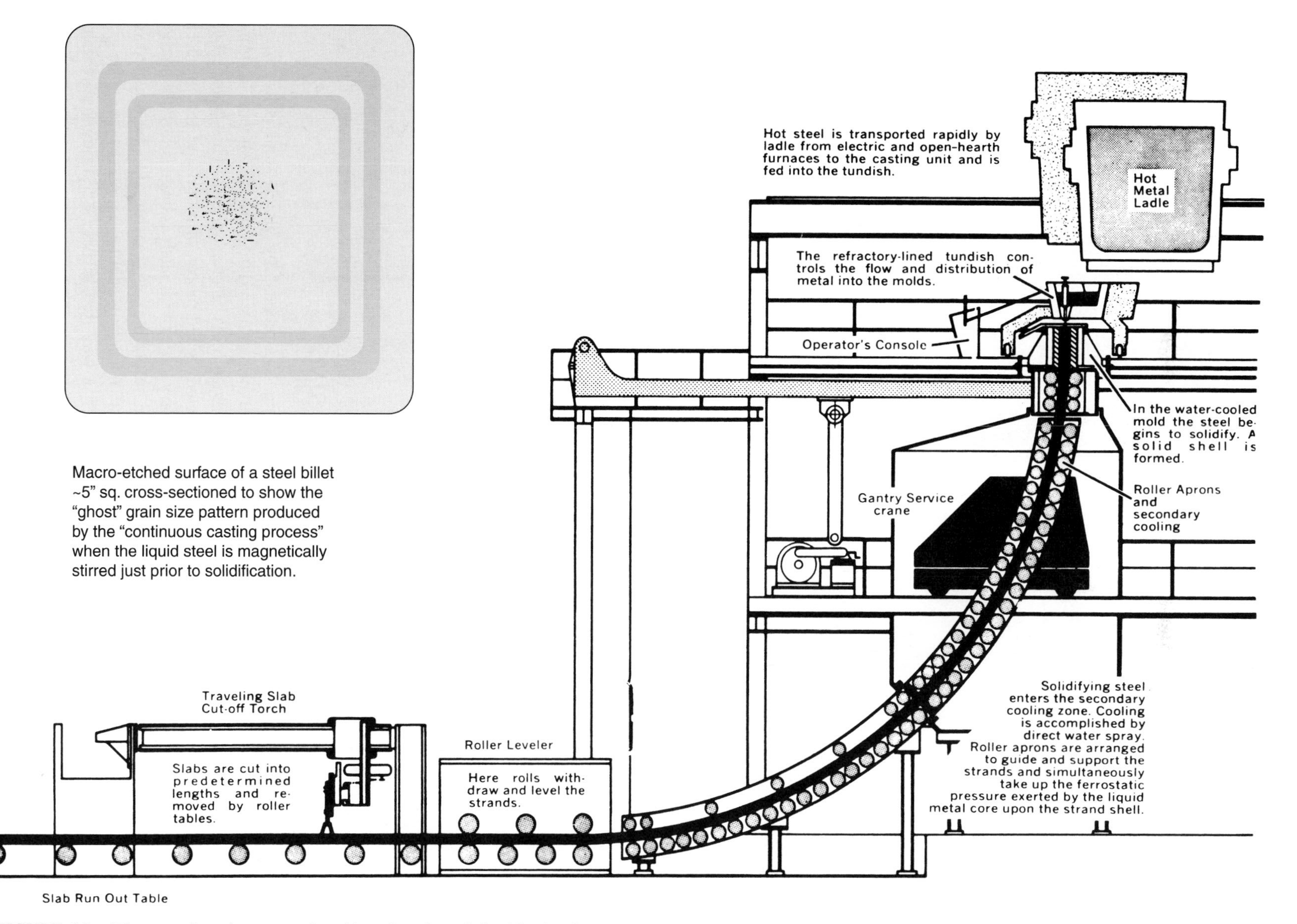

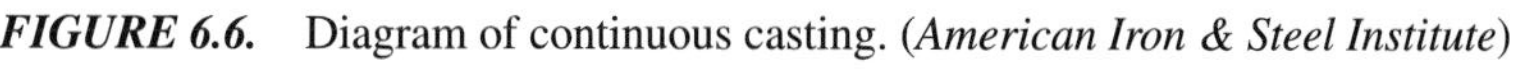

FIGURE 6.6. Diagram of continuous casting. (*American Iron & Steel Institute*)

FIGURE 6.7. Structural steel and rails are rolled from blooms. Standard shapes are produced on mills equipped with grooved rolls. Wide-flange sections are rolled on mills, which have ungrooved horizontal and vertical rolls. (*Bethlehem Steel Corporation*)

FIGURE 6.9. Coils of cold rolled sheet stock. These coils are rolled from descaled hot rolled sheet on high-speed cold reduction mills. (*Bethlehem Steel Corporation*)

FIGURE 6.8. Hot rolled plate emerging from a mill stand. (*Bethlehem Steel Corporation*)

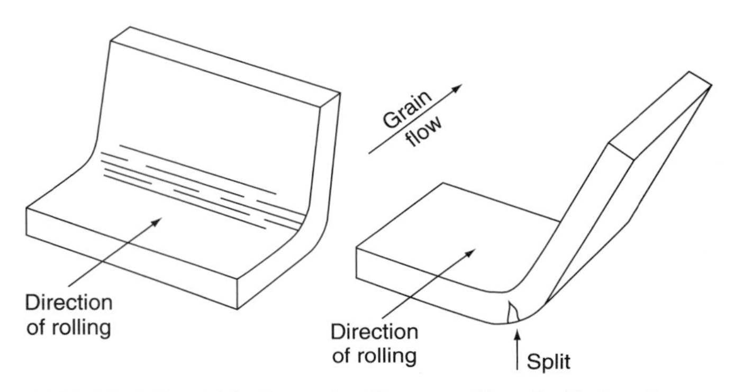

FIGURE 6.10. This shows the fibrous quality of rolled steel called anisotropy. The metal is stronger in one axis than in the other.

move the black scale and is then washed to remove all traces of the acid.

Round bar stock is usually drawn through a die that accurately sizes it and gives a good finish. Cold rolled or cold-drawn steel possesses superior machining qualities as compared to hot rolled steel. Cold-drawn or rolled steel cuts clean, producing a good finish when machined, but hot rolled steel is softer and somewhat gummy. This makes it difficult to get a good finish on these steels.

FORGING

Forging is a process by which steel is heated and then pressed or hammered into specific shapes. Forging is used to produce tools or crank shafts. This is done by heating previously sized pieces of metal and forcing them into a die. Very large parts, such as huge generator shafts, are hot forged before they are machined to the finished size (Figure 6.12). Forging enables a metal to retain the grain flow in such a way that it makes a stronger part than one of equivalent shape that has been cut or machined from bar stock (Figure 6.13). Metals are stronger in the direction in which they have been forged; but if a piece is cut across its grain by machining, it loses some of its strength.

Forged products are full of residual stresses that are often released as the material is being machined. This causes the material to warp during each heavy cut. For

FIGURE 6.11. Steel bars are shaped by rolling between pairs of grooved rails in this high-speed hot mill. (*Courtesy of American Iron & Steel Institute*)

this reason forgings and other hot rolled products should be completely roughed out oversize before any finishing cut is taken. Cold upset forming is similar to hot forging in that a blank is forced into a desired shape in a die (Figure 6.14). This cold-forging operation requires extremely high pressures in comparison to hot forging. However, it is more efficient, making possible high-speed production that converts a blank or slug into a finished part in a single press stroke in some cases. A near net shape (almost to the exact finish size required) is produced, requiring little or no subsequent machining or finishing. Parts made by this process are stronger than those made by hot forging and have a superior metallurgical structure. Due to press size limitations, only relatively small parts are made by cold upset forging processes.

FIGURE 6.12. Between the jaws of a 7500-ton hydraulic forging press, what was once a large glowing ingot is now taking on the desired shape and strength. (*Courtesy of American Iron & Steel Institute*)

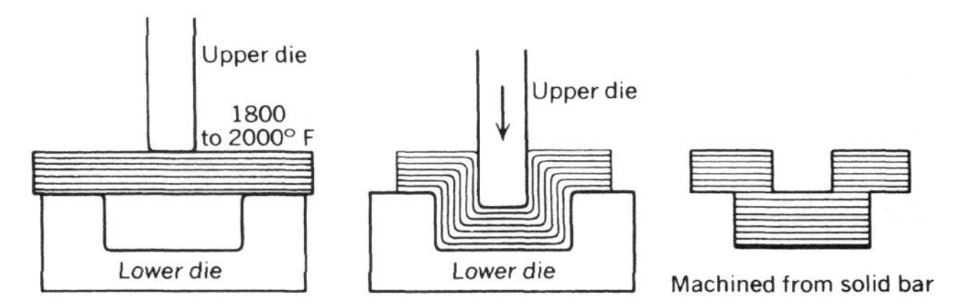

FIGURE 6.13. Grain flow in a solid bar as it is being forged compared to a machined solid bar. (*Machine Tools and Machining Practices*)

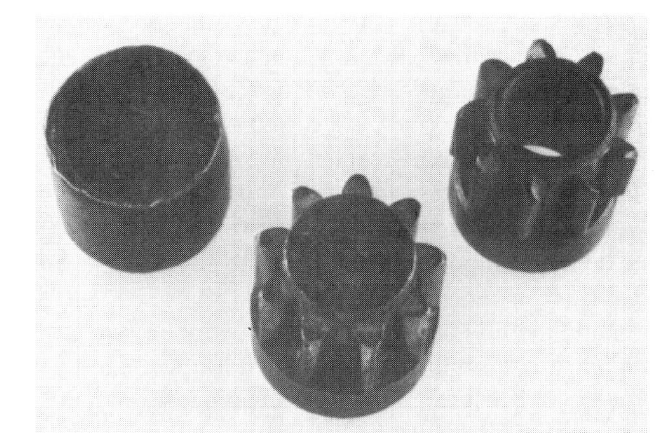

FIGURE 6.14. This gear was cold upset formed in two operations. The blank is at the left, the partially formed gear is at the center, and the finished gear is on the right. An anneal is sometimes needed between operations.

TUBULAR PRODUCTS

A round billet is pierced and rolled over a tube mandrel (Figure 6.15) to form seamless tube. The process produces a rough tube that is finished in a tube mill by passing it through rolls over a ball.

Butt-welded pipe used for water lines in homes is not made from a solid billet but is rolled from flat strip called skelp. The skelp is heated and formed in a series of rolls (Figure 6.16). The hot metal edges are continuously pressed together through a set of pressure rolls and welded securely, thus forming the pipe. Pipes are then cut to length by a flying saw (a metal-cutting circular saw that travels with the moving pipe while cutting it off).

Pipe and tubing are also made by two electric welding processes. Small size tubing is made by the

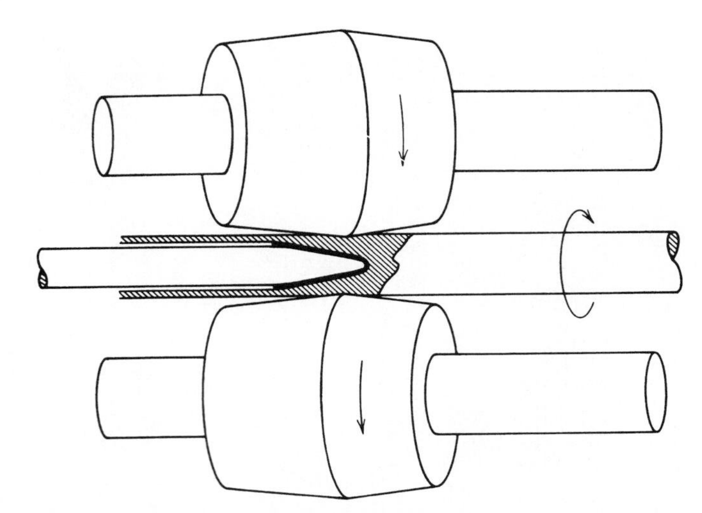

FIGURE 6.15. Piercing of solid billet to make seamless tubing. (*Machine Tools and Machining Practices*)

FIGURE 6.16. Small diameter pipe formed by continuous butt welding. Skelp is welded into pipe as it passes through sets of rolls. A flying saw cuts the pipe into lengths. (*Bethlehem Steel Corporation*)

electric resistance weld (Figure 6.17). This process uses the resistance of the material when an electric current is applied to generate the heat required for welding. The forming of the tube from hot rolled or cold rolled strip is similar to that of butt-welded pipe except that the material is not preheated.

Electric arc welding is used for the manufacture of large diameter pipe (Figure 6.18). The edges are usually joined by the submerged arc method (Figure 6.19). This automatic method of welding uses a spool of wire for an electrode and granulated flux, which is distributed over the weld. This is often a continuous process as the pipe moves along a conveyer. Pipe is usually tested hydrostatically, that is, by capping the ends and pumping in water under pressure. For some uses, pipe is galvanized (zinc coated) to protect it against corrosion.

FIGURE 6.17. In the electric-resistance welding process, skelp from the forming rolls (left) is welded automatically. (*Bethlehem Steel Corporation*)

FIGURE 6.18. Large diameter pipe is made from plate that has been formed in successive shaping processes in the O-ing press. (*Courtesy of American Iron & Steel Institute*)

WIRE AND COLD-DRAWN PRODUCTS

Wire is made by drawing a steel rod through a succession of dies (Figure 6.20), each die being slightly smaller in diameter than the previous one. The wire is pulled by a rotating capstan or drum that is between each pair of dies. The wire is wrapped one or more times around the capstan so that it will not slip. Coolant, which also acts as a lubricant, is used, since this is a cold-forming operation. The finished wire is wound onto a reel (Figure 6.21). The flattened grains produced in wire drawing (and all forms of cold working), as

FIGURE 6.19. The final step in making large diameter steel pipe is welding the curved plates. (*Bethlehem Steel Corporation*)

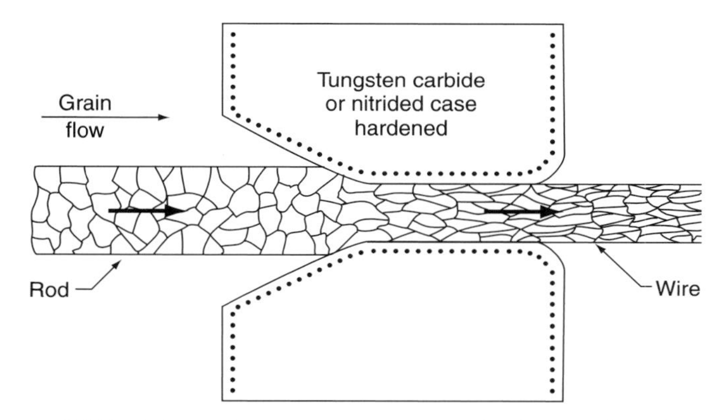

FIGURE 6.20. Enlarged cross section of wire drawing through a die.

shown in Figure 6.20, cause the metal to be harder and stronger than before, when it had larger, regular grains. This is the reason that some kinds of drawn wire are harder to bend than others. For example, wire nails are harder to bend than mechanic's wire. The difference is the amount of cold working.

Steel rods, tubing, and shafting are also produced by a cold-drawing process that is similar to that used in wire production. Steel bars that have been hot rolled are prepared for drawing by pickling to remove scale. Then they are washed in lime. The lime, plus oil or soap, acts as a lubricant for the drawing operation. The die used has a slightly smaller diameter than the hot rolled bar and is mounted on a draw bench, which can be 80 to 100 ft long. Some draw benches can exert as much as 300,000 lb of pulling power and at a rate of 150 feet per

FIGURE 6.21. Battery of modern wire drawing machines. Rod enters at the left, is reduced in size as it passes through successive dies, and is coiled at the right. (*Bethlehem Steel Corporation*)

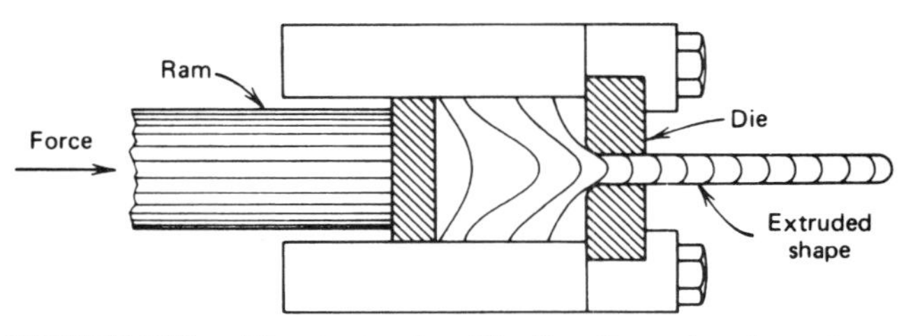

FIGURE 6.22. Direct extrusion. The flow lines show how the metal is forced out of the die by the use of high forces. Softer metals are usually extruded cold, whereas harder metals are brought to a forging heat before being extruded.

minute (fpm). Many different forms can be cold drawn in this way; rounds and tubes, squares, hexagons, flats, and bars are produced.

The cold-drawing process produces a superior product for shafting over the rolling process because it leaves a uniform density throughout the bar. In contrast, a cold rolled product is most highly worked near its surface, while the center portion is less strained. When a machining operation, such as a long keyseat, is performed on one side of the cold rolled shaft, it tends to warp the bar because of the released strain at the machined area on one side. In contrast, a drawn shaft, having the same strain throughout the bar, does not warp when machined unsymmetrically.

EXTRUSION

In the process of extrusion, metal is forced through a die at extremely high pressures, causing the metal to deform plastically and to flow through the die, taking on the shape of the die (Figure 6.22). Soft metals, such as lead and copper alloys, can be extruded cold with pressures

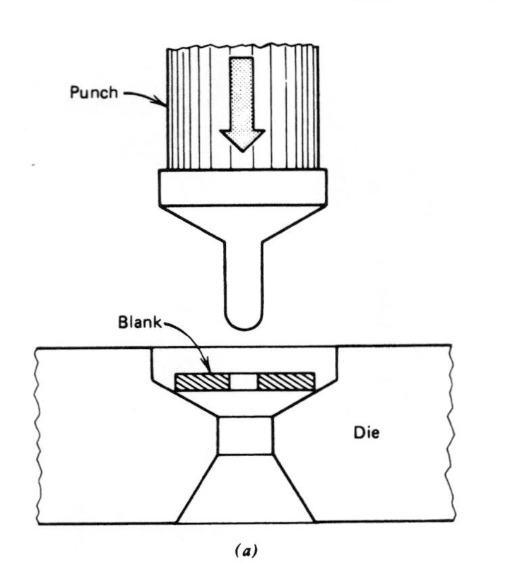

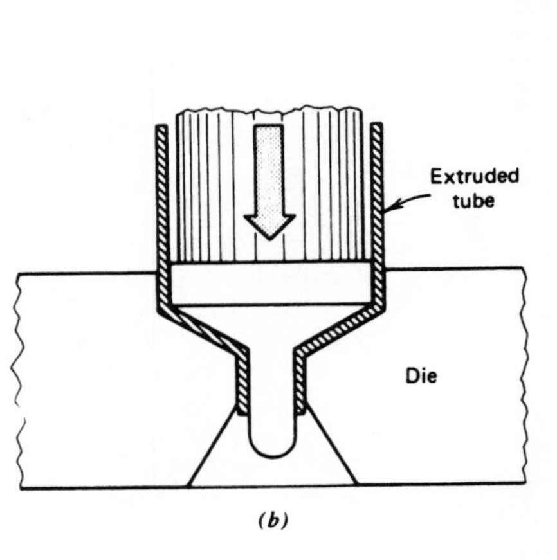

FIGURE 6.23. The method of forming collapsible tubes by impact extrusion. A flat blank is placed in the die (*a*) and the punch is brought down rapidly with a single blow (*b*). The material in the blank then "squirts" upward around the punch. When the punch is withdrawn, a stripper plate removes the tube.

of about 20 to 30 tons/in^2. Aluminum is also extruded cold, but with much higher pressures. Round, square, and hollow forms can be made with this process and with almost any shape.

Impact extrusion (Figure 6.23) is used to produce products from soft metals, usually aluminum, such as collapsible tubes (toothpaste), flashlight cases, and grease gun cases. The process of impact extrusion is simple. A thick slug of metal is placed in the cavity of a die that is held in a horizontal or vertical press. The slug is struck by a punch having the shape and size of the inside of the part. The impact of the punch is sufficient to cause the metal in the disk to flow very rapidly backward through the space between the punch and the die cavity. The part adheres to the punch and is removed by a stationary stripper as the punch is withdrawn. This process is an extremely rapid one, producing pieces at a high rate.

Metals are also formed in presses, either mechanical or punch, and hydraulically powered. Punch presses are used for embossing (making raised numbers or figures as in coining), blanking (making holes or flat shapes by punching), or deep drawing (Figure 6.24). Products such as pots, pans, cartridge cases, and other hollow products are made by deep drawing. This is a cold operation, which is a relatively rapid production process.

Metal spinning (Figure 6.25) is another cold-forming process. A blank is placed in a metal-spinning lathe and forced over a mandrel, which is the shape of the desired product. Missile cones, machine covers, and photographic light reflectors are made by this process.

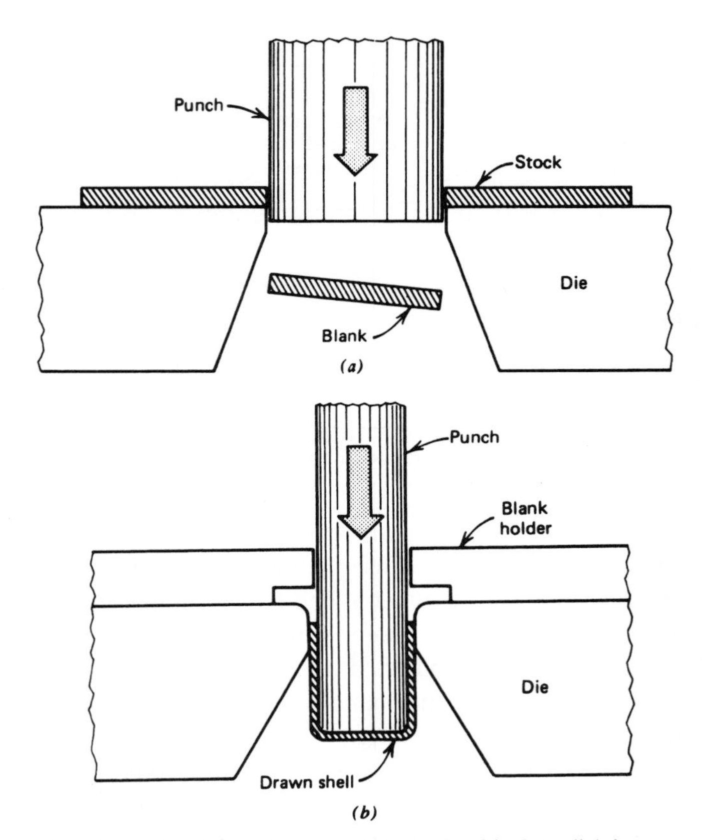

FIGURE 6.24. When deep drawing metal, a blank or disk is punched in a blanking die as at (*a*). The blank is placed in a blank holder in a deep-drawing die as at (*b*), where a punch forces the metal into a cup shape. Since the shape of the blank is totally altered, this is a cold-forming operation.

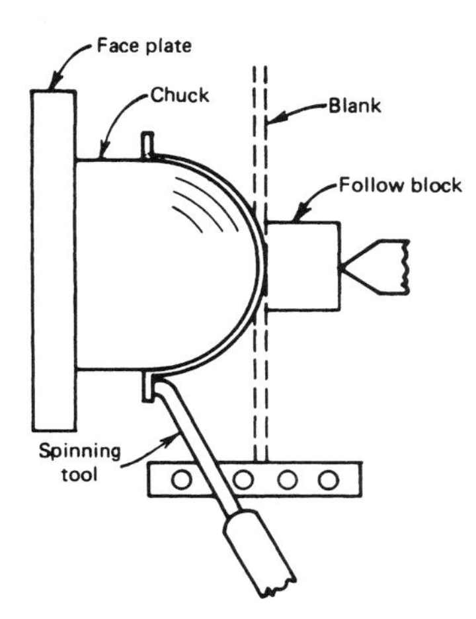

FIGURE 6.25. Flat circular blanks are often formed into hollow shapes such as photographic reflectors. In a spinning lathe, a tool is forced against a rotating disk, gradually forcing the metal over the chuck to conform to its shape. Chucks and follow blocks are usually made of wood for this kind of metal spinning.

CASE PROBLEM: THE MYSTERY OF THE CRACKED BENDS

A sheet metal shop needed to make sharp 90° bends on 12-gauge cold rolled sheet metal using a press brake. A number of these parts were required. During the process of manufacture it was noticed that some of the parts were cracked at the bend while the rest had not failed. A metals specialist was called in and immediately recognized the problem. Can you identify it?

SELF-EVALUATION

1. Why is an ingot heated in a soaking pit before it is rolled into a semifinished steel?
2. How does hot rolling improve the steel in the ingot?
3. What is the approximate carbon content of a low carbon steel bar?
4. Would a cold-finished bar have more or less strength than a hot rolled bar of equal size, shape, and carbon content? Explain.
5. What is the difference in surface texture between hot rolled steel and cold rolled steel?
6. What advantage does forging have over other steel-forming processes?
7. What is the major difference between seamless and butt-welded pipe?
8. What are two methods used to electric weld pipe and tubing?
9. When machining forgings and other hot rolled products, the roughing cuts should all be made before any finish cuts. Why is this?
10. How can a piece of low carbon steel rod be turned into a small diameter wire?
11. What is anisotropy in cold rolled steels?
12. When steel plate, bars, or wire products are cold-worked, what does this do to the metal to make the product more useful?

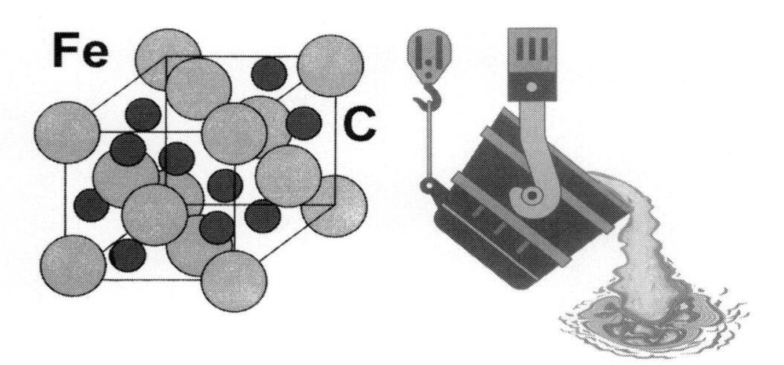

ALLOYING

CHAPTER 7

The Iron-Carbon Phase Diagram

The Iron-Carbon Phase Diagram is a wonderful tool for looking at the effect of carbon in changing the allotropic transformation temperatures of pure iron. As a side benefit and of equal importance, the diagram allows the metallurgist to predict the microstructure resulting from heating and/or cooling various iron-carbon alloys. A basic understanding of the influence carbon has on the behavior of iron is essential in understanding the heat treating of carbon and alloy steels. With each addition of carbon, a microstructure change takes place, identifying it as a distinct alloy which is governed by the rules of solubility, crystal change, and compound precipitation. Cast irons are covered in Chapter 4.

OBJECTIVES **After completing this chapter, you should be able to:**

1. Demonstrate a working understanding of the allotropic forms of iron.
2. Recognize the various microstructures formed by heating and cooling iron-carbon alloys and know their correct names.
3. Identify the areas within the iron-carbon diagram where adverse microstructures are formed.
4. Identify the temperature ranges for the various heat treatments associated with iron-carbon alloys.

PHASE CHANGES OF IRON

Iron, being an allotropic element, can exist in more than one lattice unit structure, depending on temperature. When a substance goes through a phase change while cooling, it releases heat. When a phase change is reached while heating, the substance absorbs heat. This characteristic is used to construct cooling curve graphs. If a continuous record is kept of the temperature of cooling iron, we can construct a graph that will resemble Figure 7.1.

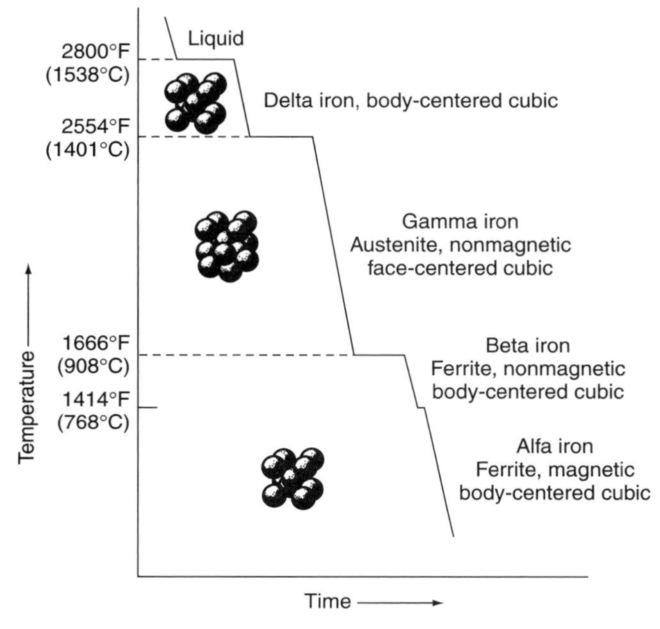

FIGURE 7.1. Allotropic transformation in pure iron.

Each flat segment of the cooling curve represents a phase change. These flat portions are caused by the release of heat during phase changes. At 2800°F, iron changes to a solid, body-centered cubic structure (delta iron). All the remaining changes involve a solid of one lattice structure transforming into a solid of another lattice structure. At 2554°F BCC delta iron changes to FCC austenite (gamma iron). Austenite transforms to BCC ferrite at 1666°F. The next change is not a phase change at all, but it does give off heat. This is the change from nonmagnetic ferrite (beta iron) to magnetic ferrite (alpha iron) at 1414°F.

THE STEEL PORTION

Study the portion of the diagram outlined by dashed lines in Figure 7.2 and you will see how it resembles phase diagrams that you have already studied; however, a little different terminology is used for the lines. Since the metal is now a solid, the terms *liquidus* and *solidus* do not apply. Since *eutectic* means low melting point, this term is not correct, either. The suffix "-oid" means similar but not the same.

The **eutectoid** point appears like a eutectic on the diagram, but it is the lowest temperature transformation point of solid phases, whereas eutectic is the lowest freezing point of a liquid phase. Instead of a liquidus line, we have the line that shows the transformation from austenite to ferrite, called the A_3 line, and the line showing the amount of carbon that is soluble in austenite,

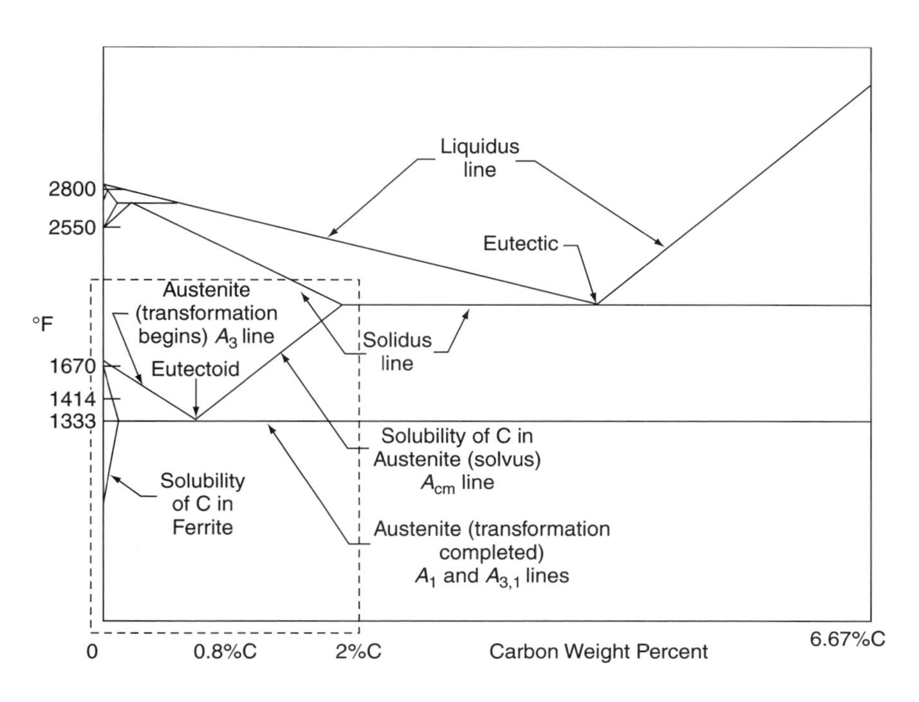

FIGURE 7.2. This simplified *iron-carbon* equilibrium phase diagram shows the key temperature line where allotropic transformation takes place in iron alloyed with increasing additions of carbon. Note the dash box which surrounds the lower left portion of the diagram. Plain carbon steels and tool steels are heat treated in these regions. The far right bottom end of the diagram is the 6.67%C; the percent of carbon in iron-carbide or cementite.

called the A_{cm} line (Figure 7.3). Instead of a solidus line, we have the line that shows where austenite completes its transformation to ferrite and where pearlite is formed. This line is called A_1 to the left of the eutectoid point and $A_{3,1}$ to the right of the eutectoid point.

MICROSTRUCTURES

When the microstructure of iron-carbon alloys (carbon steels) are viewed through the metallurgical microscope, the microstructures may be identified as **pearlite, spheroidized, iron cementite ferrite, delta ferrite, proeutectoid ferrite, martensite, bainite, retained austenite,** and countless combinations of the same. However, there is an easy way to understand the microstructure changes and their relationship with the iron-carbon phase diagram, since the pearlitic formation is distinguishable in low carbon steel at a very low concentration. We begin by viewing the microstructure of a steel with 0.01 percent carbon (1010C) steel (see Figure 7.4). The addition of carbon as an alloying element with iron has a profound effect on microstructure and dramatically increases the overall mechanical properties. As the carbon content is increased, the pearlite formation increased until grains of ferrite are no longer distinguished as individual white appearing grains in the microstructure. Individual ferrite grains disappear into the pearlite formation at about 0.65 percent carbon content. The pearlite grains consist of a lamellar formation of alternating layers of ferrite and iron carbide or cementite. In the microscope, the iron carbide layers appear dark due to incident light not being reflected by the rounded layer edges in contrast to the flat surfaces of the adjacent ferrite layers. Cementite actually appears as a light yellow compound when polished topographically flat.

The formation of a given microstructure is dependent on carbon content and the rate at which cooling takes place or how long the metal is maintained soaking at a specific temperature prior to slow cooling to room temperature. During slow cooling from the austenitizing temperature to room temperature, the crystal structure will change from FCC austenite to a BCC crystal formation. Ferrite and pearlite grains form in hypoeutectoid steels (steels with a carbon content below 0.8 percent) (see Figure 7.5). Hypereutectoid steels (steels with carbon content from 0.8 to 2.0 percent) usually form martensite with grain boundaries of precipitated carbide, due to air hardening characteristics (see Figure 7.6). The specific microstructures formed upon slow cooling plain carbon steels are schematically represented in Figure 7.4. Grain size can vary from very fine or smaller than ASTM 8 to larger than ASTM 1, dependent on alloying elements and thermal cooling conditions causing microstructure transformation.

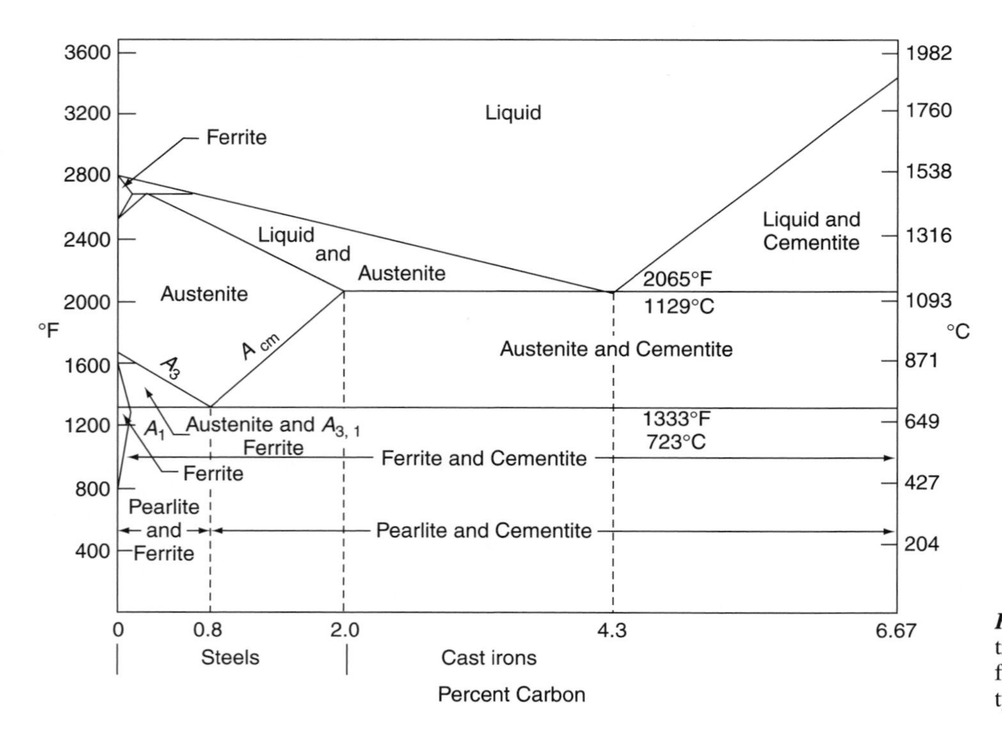

FIGURE 7.3. The principal phase transformations and compounds which form during slow cooling or annealing type heat treatments are shown.

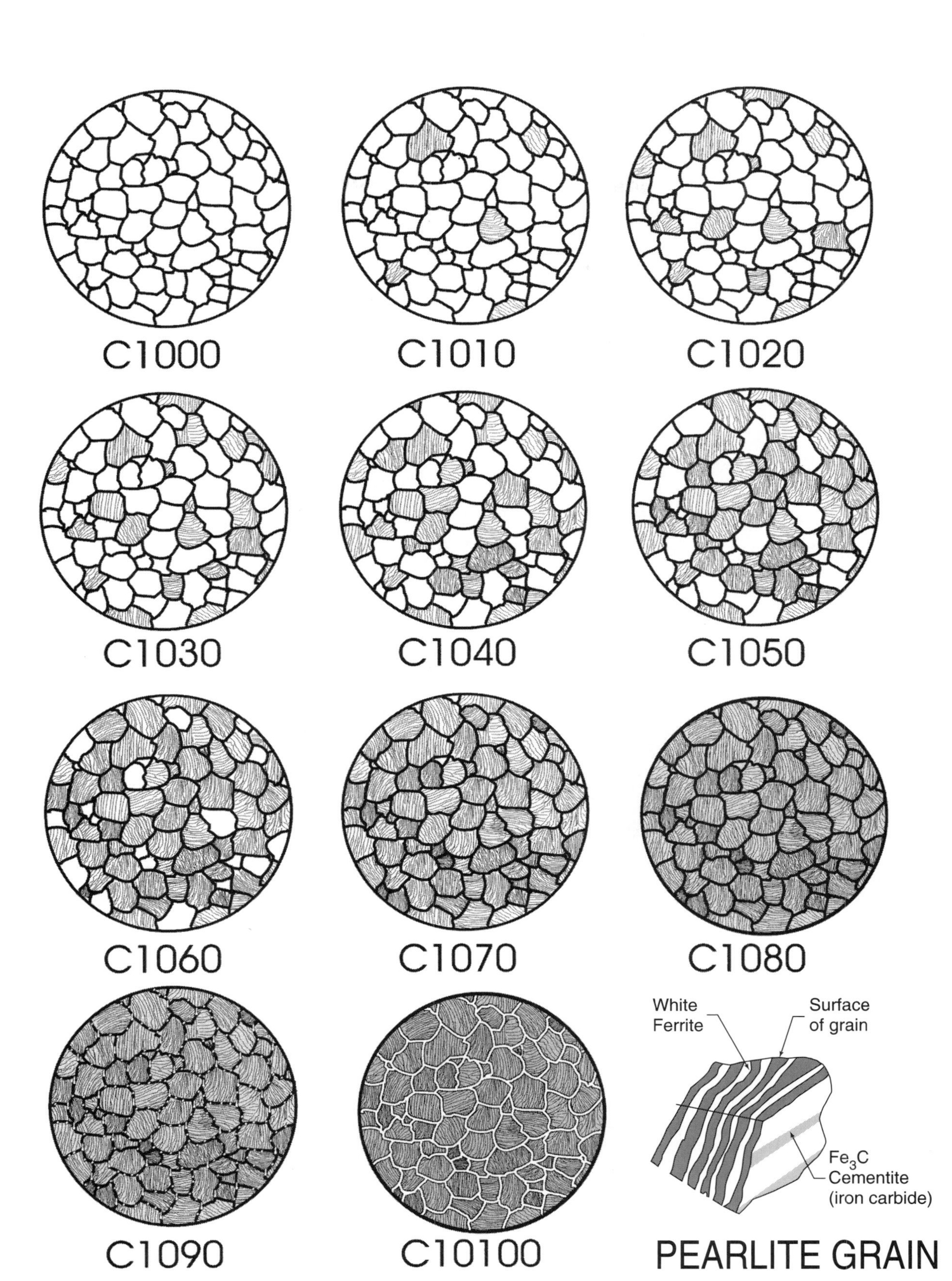

FIGURE 7.4. The microstructure changes which take place in plain carbon steels as a result of additions of carbon. These microstructures form upon slow cooling (air cool) from the austenitizing temperature for each alloy. 100× Magnification, 2% Nital Etch.

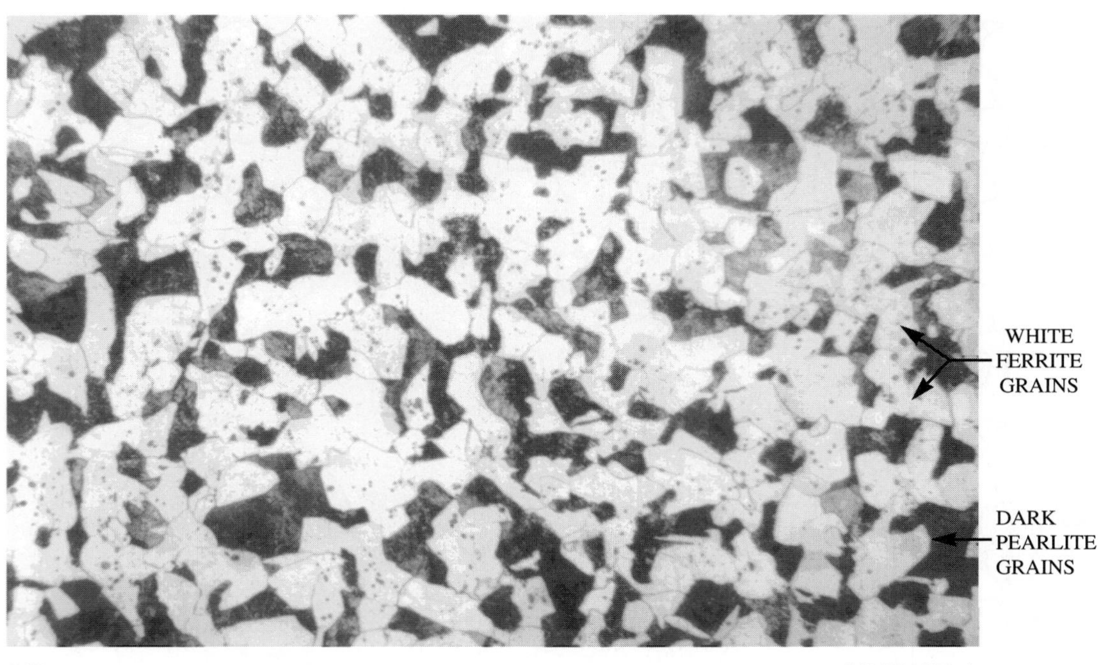

FIGURE 7.5. An 1117 resulfurized steel is shown in this photomicrograph. Gray blotchy areas and spots are sulfur inclusions added for machinability. This steel measured 25 HRC. The grain size is finer than ASTM 8. White grains are ferrite and dark grains are pearlite.

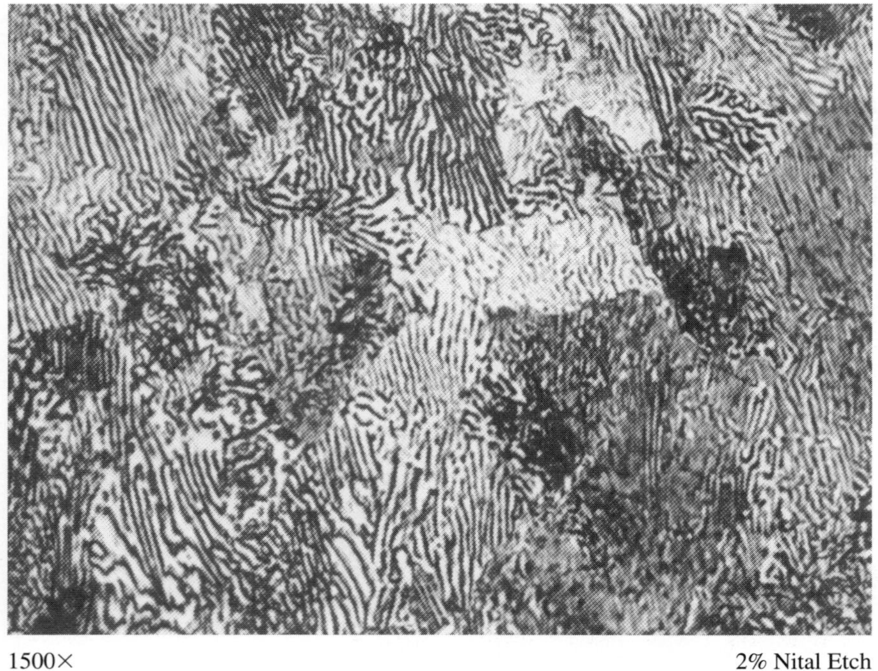

FIGURE 7.6. The microstructure is 100% pearlite in a 1080 carbon steel, slowly cooled from above the eutectoid transformation point.

Rapid cooling or quenching iron-carbon alloys produces a limitless variety of mixed microstructures. In Chapter 11, the heat treatment procedures used to harden steels and produce a martensitic microstructure vary considerably. It is important to understand the primary objective of thorough hardening heat treatment of steels, to produce a consistent homogenized microstructure throughout the metal. Martensite is described as a needlelike microstructure and is most often delineated for optical examination with a 2 percent nitric acid etchant (see Figure 7.7). The rule of thumb when heat treating to harden plain carbon steels is to heat the metal

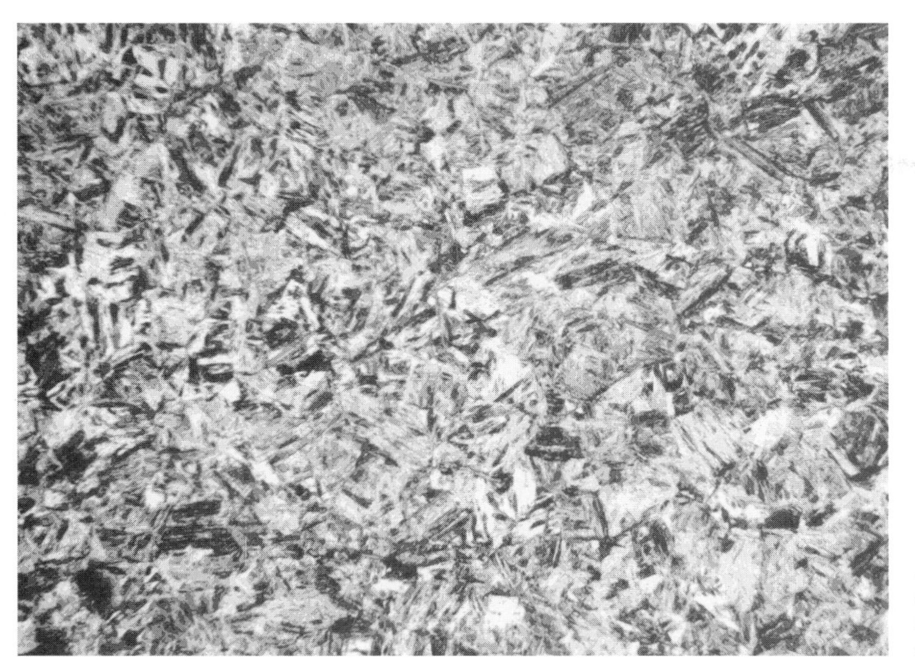

250× 2% Nital Etch

FIGURE 7.7. The needlelike microstructure of tempered martensite delineated with a 2% nital etchant. The steel is AISI 4140 oil quenched and tempered to 40 HRC.

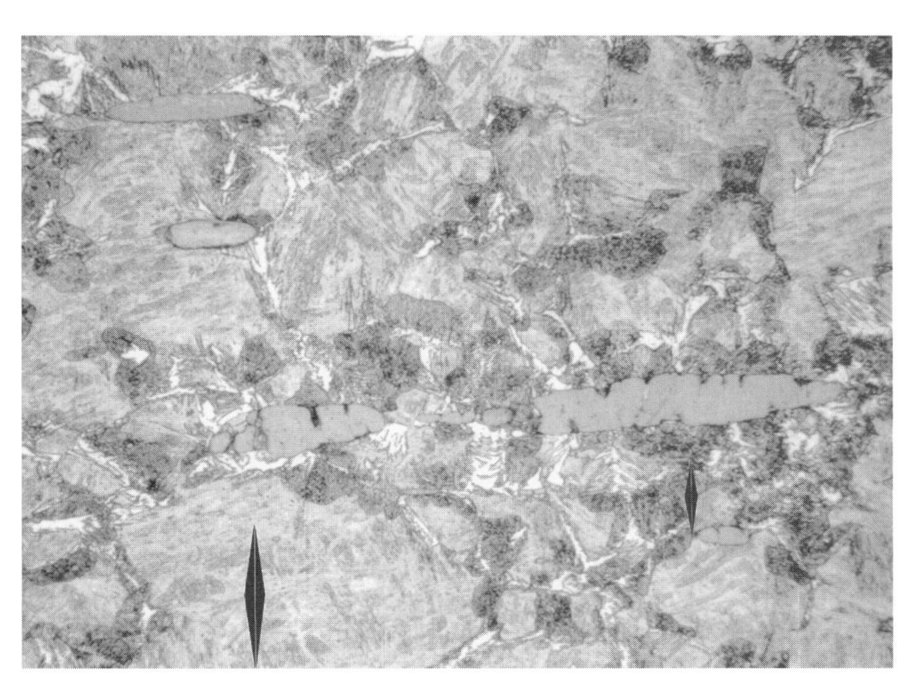

250× 2% Nital Etch

FIGURE 7.8. Mixed microstructures are usually caused by overloading the furnace, which results in insufficient quenching. Note the size difference of the Knoop microhardness indentations. The smaller indentation, lower right side corresponds to a hardness of 45 HRC while the larger indentation is 36 HRC, much softer. The alloy steel is called "Stressproof."

50° to 100° F above the upper critical temperature line, followed by rapid quenching. This rule brings us back to the iron-carbide diagram and the A_3 and A_{cm} (upper critical temperature) lines. The specific carbon content is used to determine the actual austenitizing temperature. This temperature is located on the diagram where the carbon content intersects the upper critical temperature line. Austenitizing is performed at 50° to 100° F above that temperature.

Mixed microstructures of martensite, pearlite, bainite, retained austenite and ferrite occur as a result of failure to quench from an appropriate austenitizing temperature (see Figure 7.8). Mixed microstructures also occur as a result of insufficient heating and too slow of a quench. Once again, examine the triangle areas just above the lower critical temperature line $A_{3,1}$ for hypocutectoid and hypereutectoid steels. These areas of temperature and alloy carbon concentrations indicate

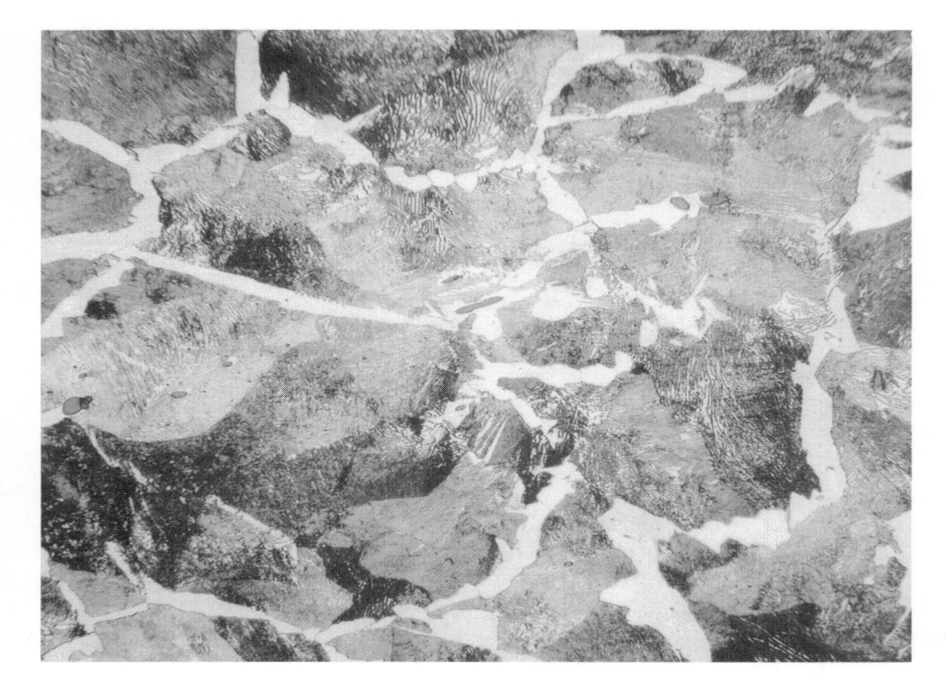

FIGURE 7.9. Proeutectoid ferrite in a matrix of fine pearlite grains.

partial transformation of the crystal structure to austenite upon heating and reversion of the microstructure back to ferrite of and pearlite or pearlite and precipitated carbide upon slow cooling. When reversion to ferrite occurs, the ferritic microstructure is called ***proeutectoid ferrite.*** When reversion of the microstructure to cementite occurs, the carbide forms at the grain boundaries and is called ***grain boundary carbide precipitation.*** The formation of these microstructures does not necessarily mean the microstructure is undesirable, unless the objective of the heat treatment is to produce a fully hardened martensitic microstructure.

It is often difficult to distinguish between proeutectoid ferrite and grain boundary precipitated carbide since both phase transformations form at the grain boundaries. In the optical microscope, proeutectoid ferrite has a jagged appearance (Figure 7.9) while grain boundary precipitated carbide appears as a smooth coating or as droplets in the grain boundaries (Figure 7.10). Physically, the proeutectoid ferrite is very soft in comparison to carbide. Disputes as to which microstructure is present are often easily settled by microhardness testing.

SLOW COOLING HYPOEUTECTOID PLAIN CARBON STEELS

Carbon added as an alloying element to pure iron results in the formation of the compound called **cementite.** Throughout the iron-carbon alloy system, the physical laws of solubility apply, which may be hard to understand since the laws apply to alloyed metals in the solid state. Element diffusion occurs in the solid state and in liquids, only slower in most cases. The diffusion and mixing of alloying elements in iron alloys is assisted by allotropic transformation; however, think of carbon freely mixing with iron in the same context as sugar being mixed into water. Carbon has an affinity for combining with iron to form the compound Fe_3C. In hypoeutectoid plain carbon alloys, when the crystal structure transforms to FCC, the iron carbide breaks down to elemental carbon and iron. The BCC iron becomes FCC austenite and the carbon attempts to evenly diffuse throughout the austenite as the process of allotropic transformation occurs. The homogenization of interstitial carbon in the austenite microstructure is the key which allows carbon entrapment and the formation of martensite to take place when austenite is rapidly quenched.

When cooling, a reverse of allotropic transformation takes place. Ferrite grains, containing almost no carbon are precipitated from the austenite grains during cooling within the A_3 and A_1 lines. During slow cooling, the carbon migrates to the center of the austenite grains causing the carbon-rich austenite to refrain from allotropic transformation. Keep in mind the solubility of carbon in BCC ferrite is 0.0008 percent at 1333°F, the lower critical temperature line and the concentration of the carbon in the austenite will reach a maximum of 0.83 percent. Any remaining austenite grains are forced

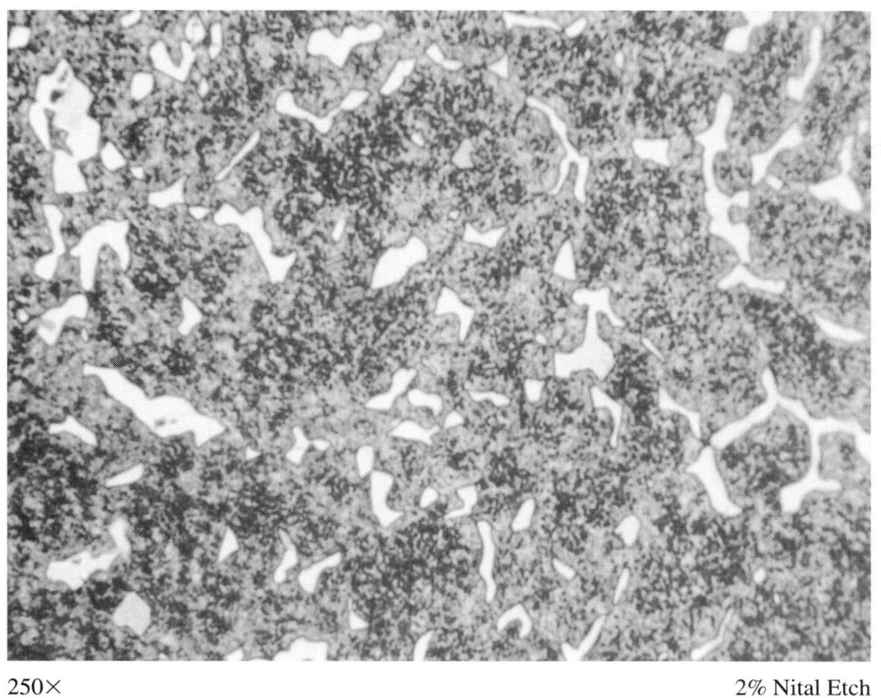
250× 2% Nital Etch

***FIGURE 7.10*a.** Carbide grains (puddling) are formed as a result of failure to remove the excess carbon introduced into the steel during carburizing. This is often seen beginning at the metal surface.

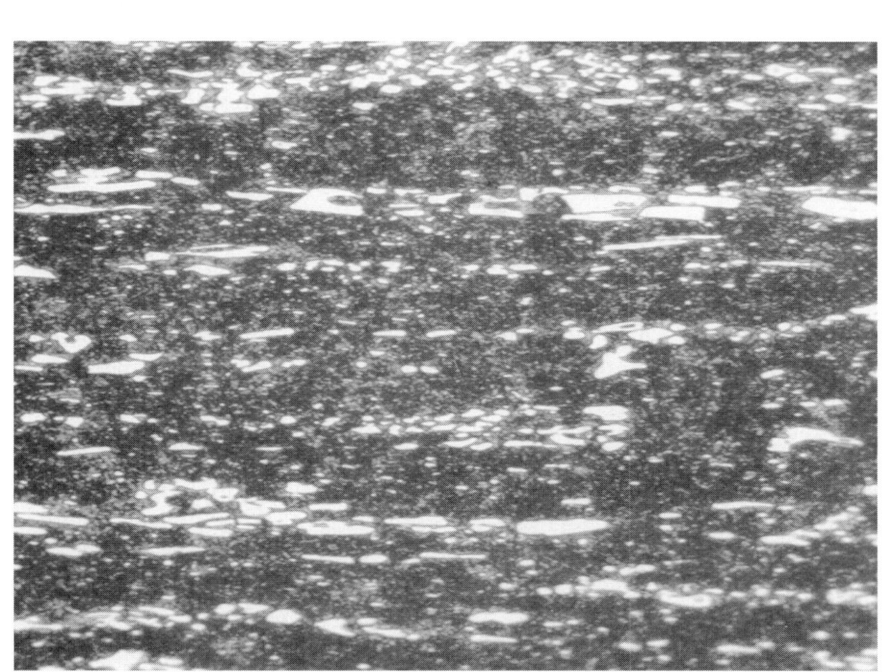
100× 2% Nital Etch

***FIGURE 7.10*b.** Primary carbide (fine white) are distributed evenly throughout this D2 tool steel. Secondary carbides (large white bands) have segregated due to insufficient hot processing as an ingot/bloom.

to undergo allotropic transformation at this temperature. Simultaneously, carbon recombines with the extra five freed atoms of iron to form lamellar iron-carbide and ferrite **(pearlite).** Although the pearlite grain contains alternating layers of 6.67 percent carbon-rich iron-carbide, if plucked from the metal and chemically analyzed, the net carbon content of the pearlite grain will be 0.83 percent due to the pure ferrite layers.

SLOW COOLING OF A HYPEREUTECTOID STEEL

The allotropic transformation of steels with a carbon content above the eutectoid composition of 0.8 percent occurs in nearly the same fashion as hypoeutectoid steels. The major difference is what happens to the excess carbon above 0.8 percent. Excess carbon forms an

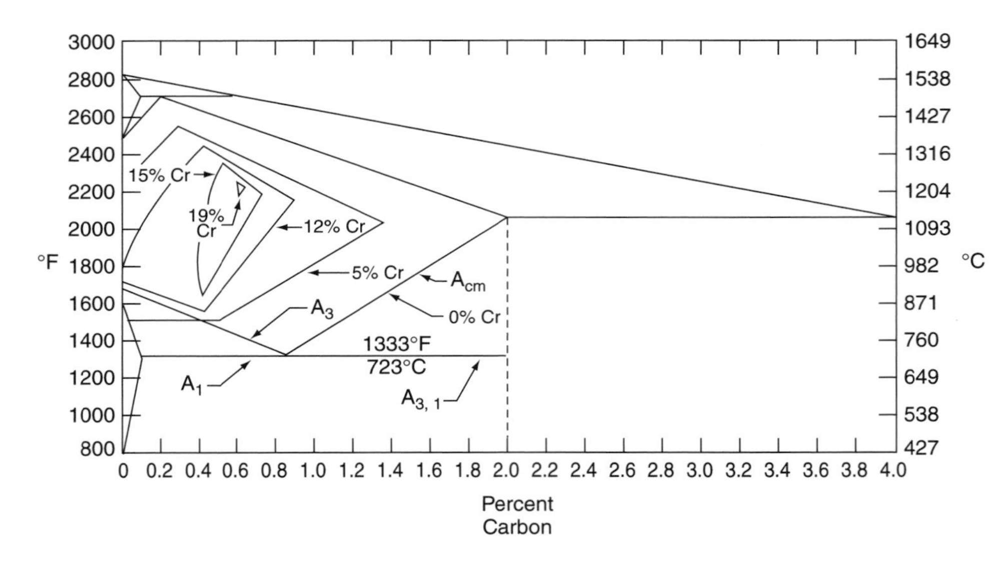

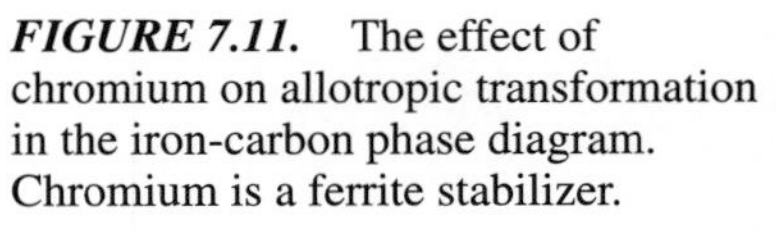
FIGURE 7.11. The effect of chromium on allotropic transformation in the iron-carbon phase diagram. Chromium is a ferrite stabilizer.

iron-carbide or cementite Fe_3C as the metal temperature is slowly reduced to below the $A_{3,1}$ line. The more carbon available in the alloy steel, the more precipitated carbide will form as dispersed droplets throughout the grain structure. The appearance of the cementite differs according to what mechanism introduces the carbide formation. As a precipitated compound, the carbide, called primary carbide, is dissolved during austenitizing or heating the metal above the upper critical temperature line. This carbide is very fine (small droplets) and is precipitated or distributed evenly throughout the steel upon cooling. In addition, larger secondary carbides may be present (see Figure 7.10*b*). These carbides are seldom dissolved completely during austenitizing because the temperature required would overheat the matrix metal and enormously large grains would grow. Cementite puddling occurs during carburizing when the carbon content is at or above 1.2 percent and the metal is quenched directly (see Figure 7.10*a*). This form more closely resembles individual grain formation and is similar in appearance to proeutectoid ferrite; compare with Figure 7.9. Microhardness testing is often used to determine the presence of carbide or proeutectoid ferrite in a microstructure due to the vast hardness difference.

ALLOYING ELEMENTS AND THE IRON-CARBON DIAGRAM

Alloying elements move the transformation lines of the iron-carbon diagram. A common alloying element is chromium. Considering the austenite area of the iron-carbon diagram, it can be seen that the effect of increasing chromium is to decrease the austenite range (Figure 7.11). This will increase the ferrite range. Many other alloying elements are ferrite promoters also, such as molybdenum, silicon, and titanium.

Nickel and manganese tend to enlarge the austenite range and lower the transformation temperature (austenite to ferrite). A large percentage of these metals will cause steels to remain austenitic at room temperature. Examples are 18.8 stainless steel and 14 percent manganese steel. The A_1-$A_{3,1}$ line is also lowered by rapid cooling. Any alloy addition will move the eutectoid point to the left (with a few exceptions such as copper) or, in other words, lower the carbon content of the eutectoid composition.

WORKSHEET 1

Objective Determine carbon content and heat treatment by metallurgical observation.

Materials Metallurgical equipment for encapsulating, polishing, and etching; two unknown samples of carbon steel, one of relatively low carbon and one of very high carbon content; metallurgical microscope (100× to 500× magnification).

Procedure

1. Heat both specimens to above 1650°F and slowly cool them in the furnace (anneal).
2. Prepare the two specimens for microscopic study as explained in Chapter 27 and Appendix 1. Identify the capsule by marking on the bottom or side.
3. Observe each specimen. Estimate the percentage of pearlite and ferrite. Compare your observations with the following:

   100 percent pearlite = 0.8 percent C
   75 percent pearlite = 0.6 percent C
   50 percent pearlite = 0.4 percent C
   25 percent pearlite = 0.2 percent C

Conclusion

1. How much carbon do you estimate for each sample?
2. Draw an iron-carbon diagram and locate each numbered specimen on the diagram for carbon content by drawing a vertical line.
3. Turn in your drawings and conclusions to your instructor.

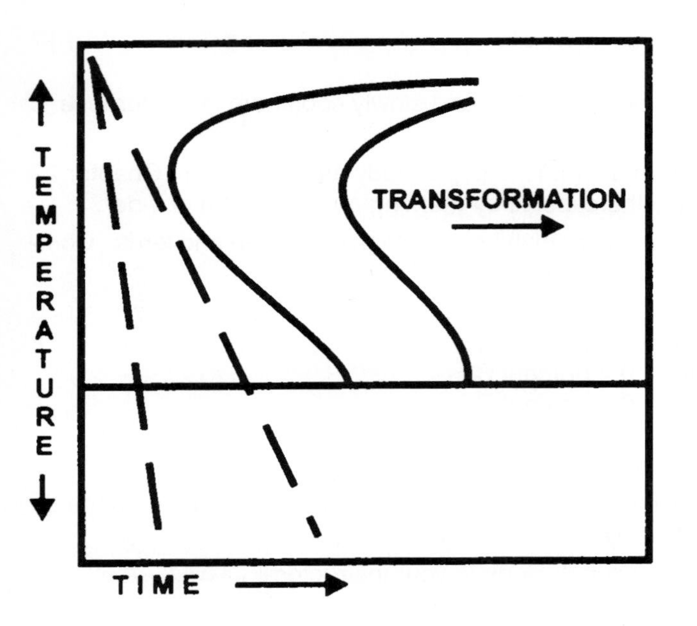

CHAPTER 8

I-T / T-T-T Diagrams and Cooling Curves

When steel is quenched rapidly from the austenitic temperature range, transformation to another crystal structure does not occur instantaneously; it takes time. The isothermal transformation (I-T) diagrams or the time temperature transformation (T-T-T) diagrams are a schematic representation showing the rate of temperature drop versus the microstructure change resulting from the act of quenching a given steel alloy. Keep in mind that the purpose of a through hardening heat treatment is to homogenize a single microstructure transformed from austenite. The I-T/T-T-T diagrams are used by the heat-treat vendor to better understand the various quenching rates and resulting microstructure changes.

OBJECTIVES **After completing this chapter, you should be able to:**

1. Determine the hardenability of steels and their quenching rates by using information gained from the I-T diagrams.
2. Recognize certain microstructures of transformation products produced at various temperatures by the use of the metallurgical microscope.
3. Estimate the hardness of a quenched steel by using the I-T diagram and the microscope.

HARDENING PROCESS

The process of hardening steel is carried out by performing two operations. The first step is to heat the steel to the austenite range (austenitizing), which is heating to a temperature above the upper critical temperature. The second step is that of rapid cooling or quenching near to room temperature or below.

Austenitizing produces the solid solution of carbon in the face-centered cubic structure. The temperature used for austenitization is usually about 50°F above the A_3 or A_{cm} lines (Figure 8.1). Alloys with 0.8 percent carbon or less will become 100 percent austenite at this temperature, whereas steel with more than 0.8 percent carbon will be austenite with some free cementite.

Higher carbon steels that contain carbide-forming elements such as chromium, molybdenum, tungsten, or vanadium require more soaking time at the austenitizing temperature since the complex carbides are relatively slow to dissolve. If the temperature is too low, there may be incomplete solution of carbides and the steel may still contain undissolved ferrite grains that are not beneficial in a hardened tool steel. If the temperature is too high, large grains may form and cause cracking during heat treatment, resulting in failure of the part. Most steel producers publish data sheets containing correct austenitizing temperatures for various alloys.[1]

Quenching undercools the austenite to form a new structure below the Ms temperature. This structure is called **martensite.** The martensite is an extremely hard acicular or needlelike structure.

[1]Reference: *"Heat Treater's Guide,"* 2nd Edition, ASM

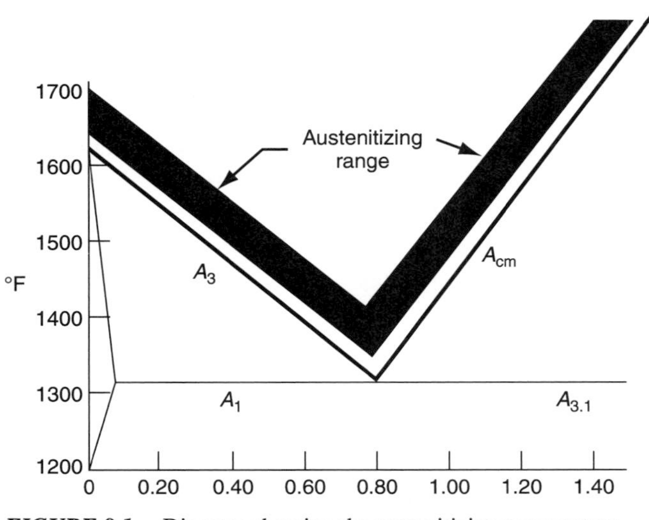

FIGURE 8.1. Diagram showing the austenitizing temperature required before quenching to harden alloy steels and tool steels.

ISOTHERMAL TRANSFORMATION (I-T) DIAGRAMS

Isothermal means the same or constant temperature. I-T diagrams are also called the Bain S-curve or time temperature transformation (T-T-T) diagrams. Once it was discovered that the time and temperature of austenite transformation had a profound influence on the transformation products, a new kind of graph or diagram was needed. The iron-carbon diagram would not do, because it represents equilibrium or slow cooling and austenite transformation takes place under nonequilibrium conditions with various cooling rates. Austenite is unstable below the A_1 line on the iron-carbon diagram and almost immediately begins to transform to some product such as pearlite or bainite. An I-T diagram shows this process very well.

When plotting an I-T diagram, three facts should be kept in mind.

1. When austenite is cooled below the A_1 line to a particular temperature and held at that temperature, it will begin to transform in a given time and will complete the transformation after a given time peculiar to that steel.
2. Martensite is formed only at relatively low temperatures.
3. If austenite transforms at any point in the transformation to a structure that is stable at room temperature, rapid cooling may not change the product already transformed, and the remaining untransformed austenite may not form martensite.

Data for isothermal transformation diagrams are obtained by heating large numbers of small steel specimens of a specific kind of steel to the austenitization temperature (Figure 8.2). They are then abruptly transferred to furnaces or molten salt baths that have been heated to predetermined temperatures below the critical A_1 line.

To study the transformation at 1200°F, a set of specimens is held at 1200°F constant temperature (isothermal). At regular time intervals a specimen is removed and rapidly quenched in iced brine. Microscopic examination will then show martensite if transformation has not yet started, but will show martensite and pearlite (in this case) if transformation has started, and only pearlite if transformation is complete. A mark is placed on the graph, indicating the time and temperature. This procedure is repeated at other temperatures until the entire graph is plotted for that particular steel (Figure 8.3).

The vertical scale on the left represents temperature and the horizontal scale on the bottom represents time. It is plotted on a log scale that corresponds to 1 minute, 1 hour, 1 day, and 1 week. The letters *Ms* can be

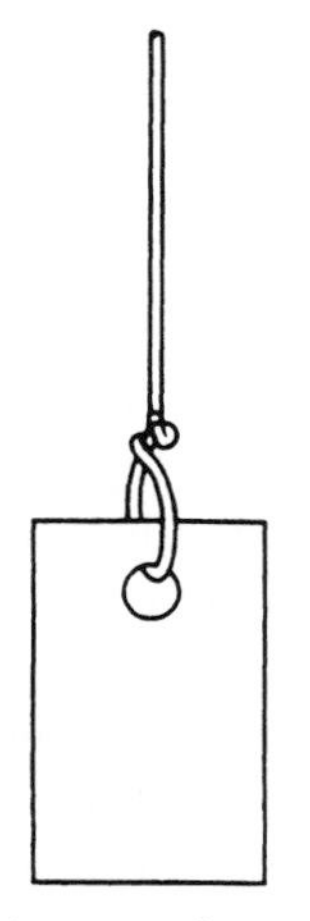

FIGURE 8.2. Isothermal test specimen. (*Machine Tools and Machining Practices*)

found at a specific temperature for each kind of steel. Ms represents the temperature at which austenite begins to transform to martensite during cooling. The Mf temperature is the point at which the transformation of austenite to martensite is completed or near 100 percent during cooling. This is sometimes replaced by a percentage of transformation. Mf is seldom indicated on I-T diagrams due to the fact that complete transformation to martensite is alloy-dependent and may require quenching to a temperature below that of the quench oil in order to fully transform the austenite to martensite. Some alloys require continuous quenching from the austenitizing temperature to a temperature below that of the quench oil. Keep in mind that most quench oils are maintained between 120°–160° F. It is always best to check the *ASM Heat Treater's Guide* to determine the need to continue quenching below the temperature of the quenching oil.

TRANSFORMATION PRODUCTS

Austenite, when cooled below the transformation temperature and held at a constant temperature, decomposes into various transformation products such as pearlite, ferrite, or bainite. Austenite containing 0.89 percent carbon that is cooled quickly and held at 1300°F, for example, does not begin to decompose or transform until after about 3 minutes has elapsed and does not completely decompose until it is at that temperature for more than 1 hour (Figure 8.4). A very coarse pearlite structure has developed at this temperature and the material is very soft. If the austenite is quickly cooled to and held at a lower temperature of 1200°F, decomposition begins in about 5 seconds and is completed after about 30 seconds. The resultant pearlite is coarse grained and slightly harder. At a temperature of about 1000°F, the austenite decomposes extremely rapidly. It takes only about 1 second before transformation begins and 5 seconds to complete it. The resultant pearlite is extremely fine and its hardness is relatively high. This region of the S-curve, when decomposition of austenite occurs, is called the nose of the curve on an isothermal transformation (I-T) diagram.

COOLING CURVES

If the austenite is cooled to temperatures below the nose of the curve (about 600°F) and held at these temperatures for a sufficient length of time, the transformation would produce a product called bainite. If the austenite were cooled quickly to a temperature below the Ms line, the product would then be martensite. As noted, this transformation to martensite is complete at the Mf temperature. These temperatures vary considerably in steels and are a function of carbon content. The Ms and Mf temperatures are lower for high carbon steels than for low carbon steels (Figure 8.5).

If a cooling curve (Figure 8.6) is superimposed on the I-T diagram, it can be seen that it must pass to the left (1) of the nose of the diagram to transform to martensite. If the cooling rate is too slow, however, the cooling curve will cut into the nose of the diagram, showing that a partial or split transformation has taken place at that point (2) and that fine pearlite plus martensite developed in the material instead of the desired martensite. Therefore, the rate of cooling for a hardening quench must be such that the nose of the diagram is to the right of the cooling curve. However, the formation of martensite is temperature dependent, not time dependent.

In some steels there is a problem of retained or untransformed austenite even after correct quenching procedures have been used. Some tool steels will retain austenite even below the Mf temperature. Suitable tempering or subzero treatments will usually fully transform the austenite to martensite. Retained austenite can cause serious problems in hardened tool steels, such as brittleness and cracking, due to a transformation to untempered martensite at a later date, caused by external stresses or thermal cycles.

THE CRITICAL COOLING RATE

Different alloys can affect the shape of the I-T diagrams. An increase in carbon content moves the S-curve to the right (increases the time before transformation takes place). Grain size also has an effect on hardenability

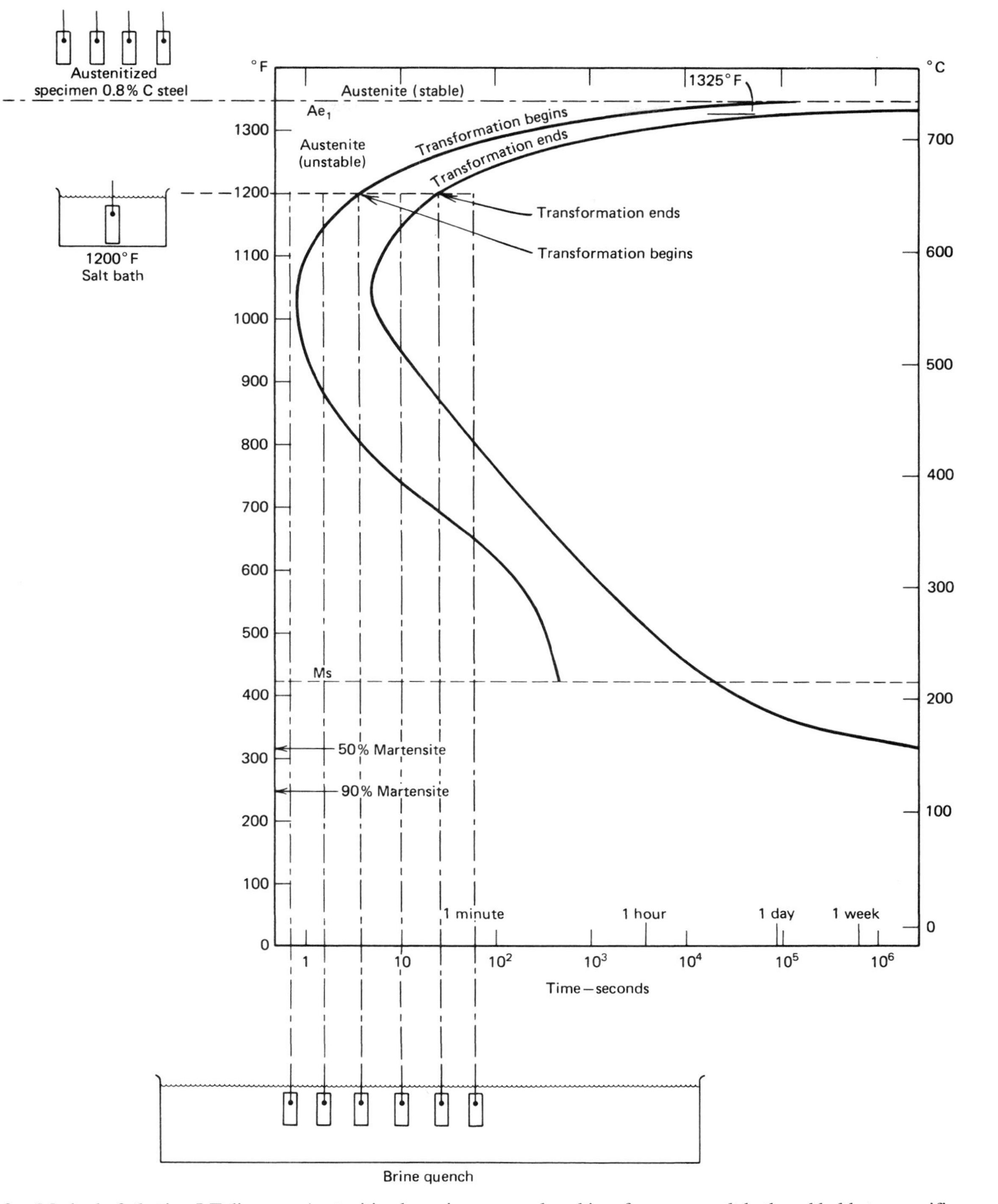

FIGURE 8.3. Method of plotting I-T diagram. Austenitized specimens are placed in a furnace or salt bath and held at a specific temperature. They are removed at specific time intervals and quenched in brine. Each specimen, having been marked for its time period, is then prepared and checked with a microscope for its lack of transformation products (it would be fully martensite in that case) or its percentage of transformation and at what time interval transformation began and ended. Probably many more specimens would be used for a single temperature plot than this diagram shows.

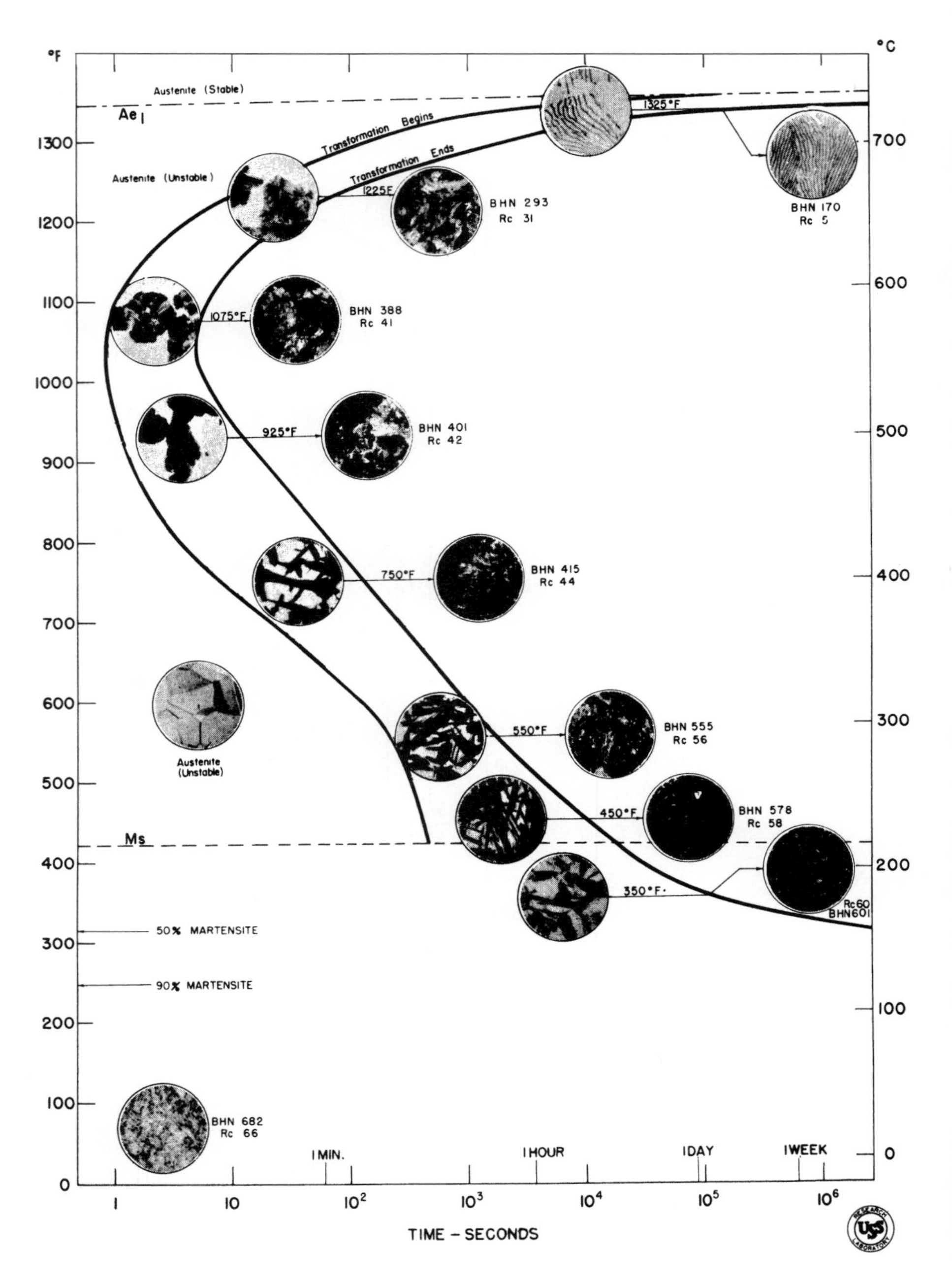

FIGURE 8.4. I-T diagram of 0.89 percent carbon steel. (*Copyright 1951 by United States Steel Corporation*)

(the property that determines the depth and distribution of hardness induced by quenching a ferrous alloy). Larger grain carbon steels also take more time to transform. The addition of alloy to the steel also moves the S-curve to the right.

A plain low carbon steel cannot be hardened for practical purposes because the nose of the diagram is at or falls short of the zero time line, and it would be impossible to avoid cutting into it with the quenching or cooling curve (Figure 8.7). However, with carbon steels at or above 0.30 percent, it is possible to quench rapidly enough to effect a partial transformation to martensite (Figure 8.8). Plain carbon steel of 0.83 percent must be quenched in water to make the quench rapid enough so that it would take place within the 1 or 2 seconds needed to avoid cutting into the nose on the diagram. The critical cooling rate, then, is the rate of cooling that avoids cutting into the nose of the S-curve.

Oil-hardening steels with alloying elements, such as chromium and molybdenum, cause the nose of the diagram to move toward the right, thus increasing the time in which hardening can take place. Also, the shape of

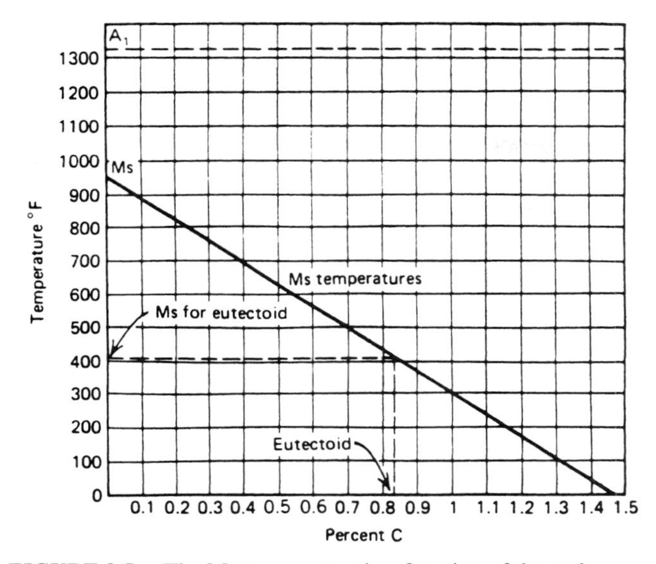

FIGURE 8.5. The Ms temperature is a function of the carbon content and will be further lowered when alloying elements are added. After finding the Ms temperature for steel containing a particular carbon content, subtract the following: 70 times the percentage of chromium, 70 times the percentage of manganese, 50 times the percentage of molybdenum, and 35 times the percentage of nickel (based on Fahrenheit temperatures).

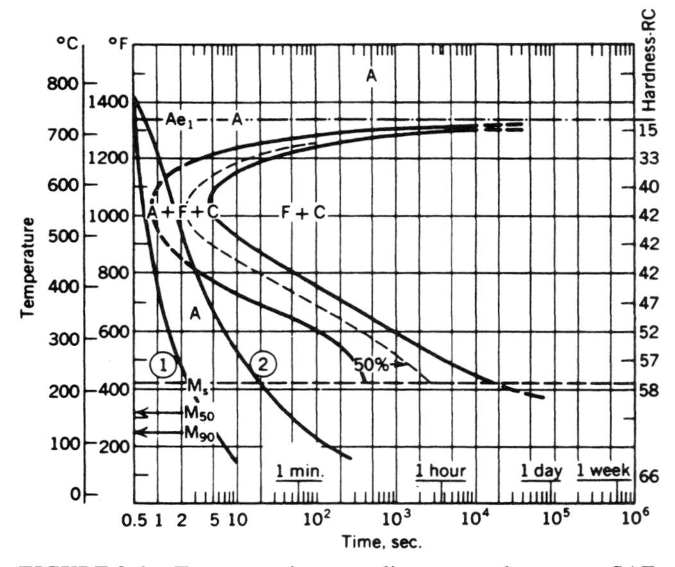

FIGURE 8.6. Two approximate cooling curves shown on a SAE 1095 I-T diagram that illustrate the need for a sufficiently rapid cooling rate to avoid cutting into the nose of the diagram. (*United States Steel Corporation*)

the nose is often changed on the diagram. These changes often allow a great deal of time for the quench to take place. It is easy to see on the I-T diagram how oil-hardening (Figure 8.9) and air-hardening (deep-hardening) steels are affected by cooling rates.

Transformation to the martensite depends on these factors:

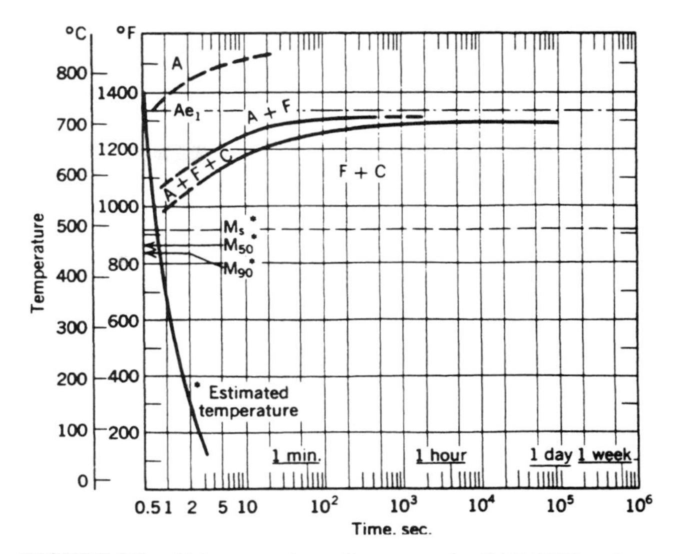

FIGURE 8.7. Brine quench cooling curve for SAE 1008 carbon steel. (*United States Steel Corporation*)

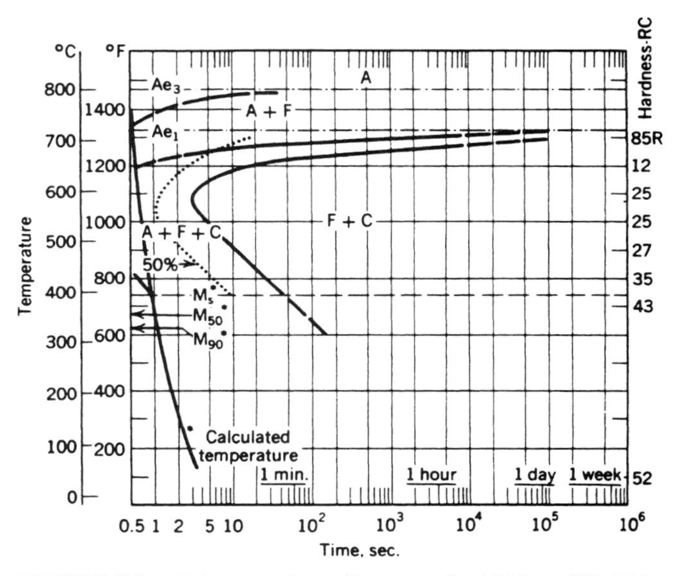

FIGURE 8.8. Brine quench cooling curve for 1034 modified Mn steel. (*United States Steel Corporation*)

1. The chemistry of the steel
2. Grain size
3. The shape of the part
4. The state of heat treatment from the mill
5. The mass of the part
6. The severity of the quench and the quench temperature
7. The hardenability of the material
8. The design and shape of the finished part

The surface area of the part and the thickness (mass of part) have a considerable effect on the cooling rates. A very thin part, such as a razor blade, with a large

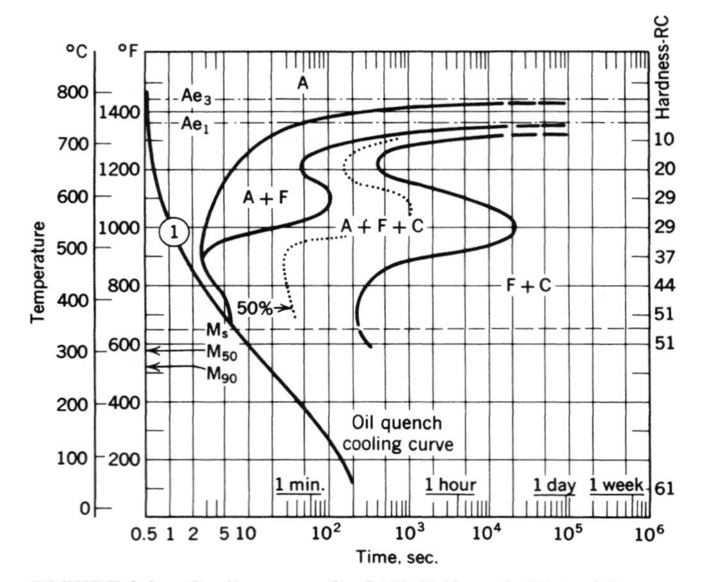

FIGURE 8.9. Cooling curve for SAE 4140 steel. (*United States Steel Corporation*)

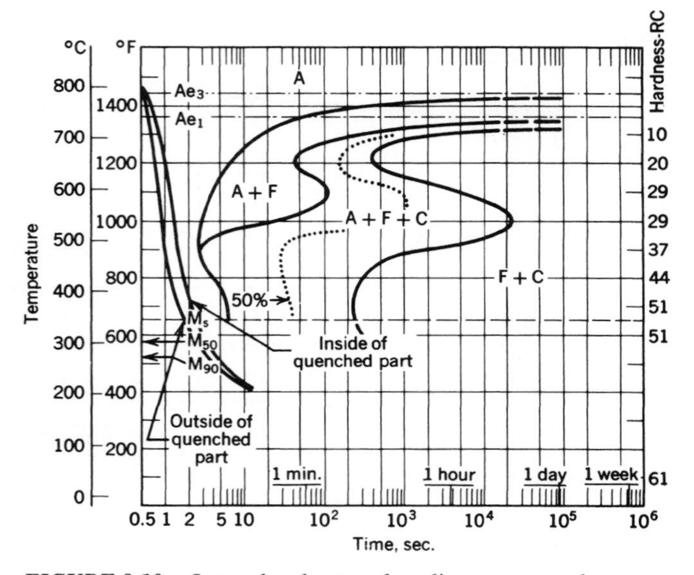

FIGURE 8.10. Internal and external cooling curves on the same part on this I-T diagram show how two different cooling rates can cause the high stresses to build up that sometimes result in quench cracks. (*United States Steel Corporation*)

surface area would have a cooling rate that would be many times greater than a cube of steel 2 or 3 in.^{2} Therefore, a normally water-hardened steel, when extremely small or thin, should be quenched in oil to achieve the proper cooling rate, while the steel block being of the same material would still not achieve critical cooling even in a severe quench of cold salt water.

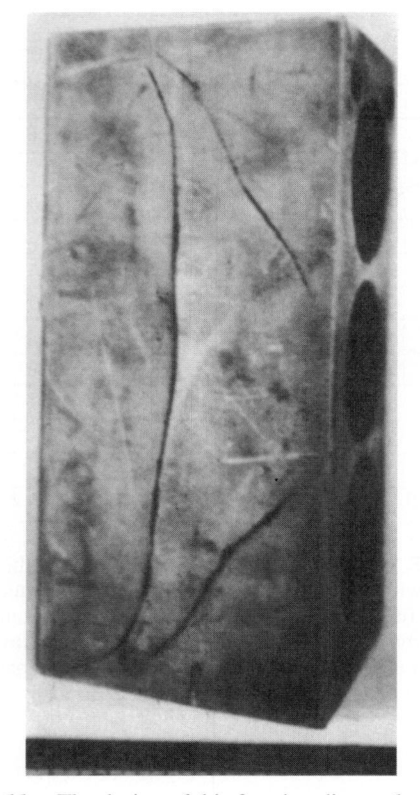

FIGURE 8.11. The design of this forming die, made of Type W1 tool steel, presents an almost impossible problem to the heat treater. Because of the blind holes and the thin section between them, the die cracked during heat treatment. Unless the die can be totally redesigned, the use of an air-hardening steel is imperative. (*Bethlehem Steel Corporation*)

Water-quenched steels normally will harden only to a depth of approximately 1/8 in., while the core is left quite soft. These are termed shallow-hardening steels. A 1 or 2 in. thick, air-cooled steel may harden completely to the core. This is deep-hardening steel. As the time increases in which quenching can take place, the depth of hardening increases.

When quenching rates are drastic, such as in brine or water quench, the stresses in the part caused by the different cooling rates from the interior and exterior of the part can cause warping and cracking (Figure 8.10). Water-quenched steels are particularly prone to this problem. The slower cooling rates of oil and air allow more uniform cooling, and because of this, these steels suffer less cracking and warping. For this reason, when large or heavy sections must be heat treated, an alloy steel that can be oil or air hardened should be selected (Figure 8.11).

SELF-EVALUATION

1. What is the austenitizing temperature for a carbon steel?
2. How is the hard structure, martensite, produced? What is the major consideration when martensite is produced?
3. What are the Ms and Mf temperatures?
4. What is the critical cooling rate?
5. If the cooling curve of a 0.8 percent carbon steel cuts halfway through the nose of the I-T diagram, what would be the resultant microstructure of the metal?
6. How does the carbon content affect the position of the S-curve?
7. What are the two steps needed in hardening steel to produce useful articles?
8. A thick section of W1 steel developed a quench crack when it was hardened. What do you think caused this? What steps can be taken to correct the problem?
9. How can you determine from studying an I-T diagram of very low carbon steel why little or no martensite can be produced when the steel is quenched from an austenitizing temperature?
10. How can you tell by studying an I-T diagram whether an alloy steel has deep-hardening capabilities?
11. Deep-hardening steels have the property of hardening throughout large sections even though the cool rate is slow. How is it possible to produce a slow cooling rate?
12. In question 11, what is the major advantage of a large section that is quenched with a slow cooling rate?

CASE PROBLEM: THE NONHARDENING CHISEL

A shop worker decided to make a specially shaped chisel for a project he was working on. He went to the steel rack and found some hexagonal stock. He cut off a suitable length. He heated it, forged a rough shape, and then used a pedestal grinder to finish the shape. He noticed there were few sparks coming off the grinding wheel, but paid little attention to it. Then he hardened and carefully tempered it by the color method to a yellowish-brown. When he began to use the chisel, instead of cutting, it just flattened out on the cutting edge. Testing it with a hardness tester, it proved to be Rc30. It should have been between 50 and 60 Rockwell. Why do you think the chisel didn't harden properly?

CHAPTER 9

Heat Treating Equipment

The heat treating of metals requires very critical furnace operations to produce the desired changes in the metal. Heat treatment today involves specific protocol for each ferrous and nonferrous alloy. The hardening of metals such as iron alloys or steels will require soaking at temperature in a furnace to homogenize the alloying elements prior to quenching into oil. New vacuum furnaces have the capability of quenching steels by means of inert gas which allows steels to be heat treated in the finish machined condition. Aluminum heat treating furnaces are often constructed with an open bottom to allow the furnace load to be quickly lowered into a quench tank. Study the general equipment discussed in this chapter and follow the specific process directions in the designated specification and you will be able to determine the correct heat treatment equipment for both ferrous and nonferrous alloys. This chapter should be read in conjunction with the processes and procedures described in Chapter 11.

OBJECTIVES **After completing this chapter, you should be able to:**

1. Describe the heat treatment equipment for through hardening alloy steels.
2. Describe the furnace and temperature controls for hardening aluminum.
3. Recognize the physical differences between furnaces and their use.
4. Understand the temperature regulation and override controls used for furnaces.
5. Understand and be able to practice safety procedures in a heat treating facility.

FURNACES

Modern furnaces used in industry for the heat treating of ferrous and nonferrous metals are constructed for a particular application and are usually dedicated to a specific operation, such as hardening or tempering (see color centerfold of industry furnaces). Heat treating furnaces—vacuum, electric, induction, gas, and oil-fueled—can be relatively small to very large, depending on the application. Small tooling furnaces are shown in Figures 9.1*a* and 9.1*b*. Some furnaces are large enough to drive a full-size railroad boxcar into them, holding the part to be heat treated. This type of furnace can be found in shipyards where large component parts, such as propellers for large ships, can be heat treated in support of manufacturing. Vacuum furnaces are used to heat-treat steels and nonferrous metals such as titanium, when the surface of the metal must be protected against contamination and/or chemical change. Induction furnaces are used primarily for surface hardening of steels and for production operations such as soldering electrical contacts. Some furnaces contain sealed chambers or retorts which preclude air from entering the furnace and allow controlled introduction of a specific gas to flow over the workload inside the retort, such as in gas nitriding. Plasma/ion nitriding is a more advanced technique for the surface hardening of alloy steels and tool steels. There are numerous worldwide furnace manufacturers producing an enormous variety of furnaces which are dedicated to particular operations driven by the requirements for the metal or alloy being processed.

Furnaces use various types of controls for temperature adjustment and control. The controls make use of thermocouples which are placed inside the furnace near the workload to control furnace operation by monitoring operating temperature (Figure 9.2*a*). Most furnaces are equipped with two thermocouple temperature controllers; the first is used to regulate the furnace operating temperature ("set" temperature) and the second thermocouple controller is used as a safety override shutoff switch. When the furnace temperature is selected, the furnace will begin heating the chamber. The set temperature thermocouple senses the heat input and regulates the temperature of the furnace by cycling the furnace on and off as needed, to quickly reach the furnace set temperature. The safety controller is designed to monitor furnace operating temperature and is adjusted approximately 5° to 50° above the set temperature (Figure 9.2*b*).

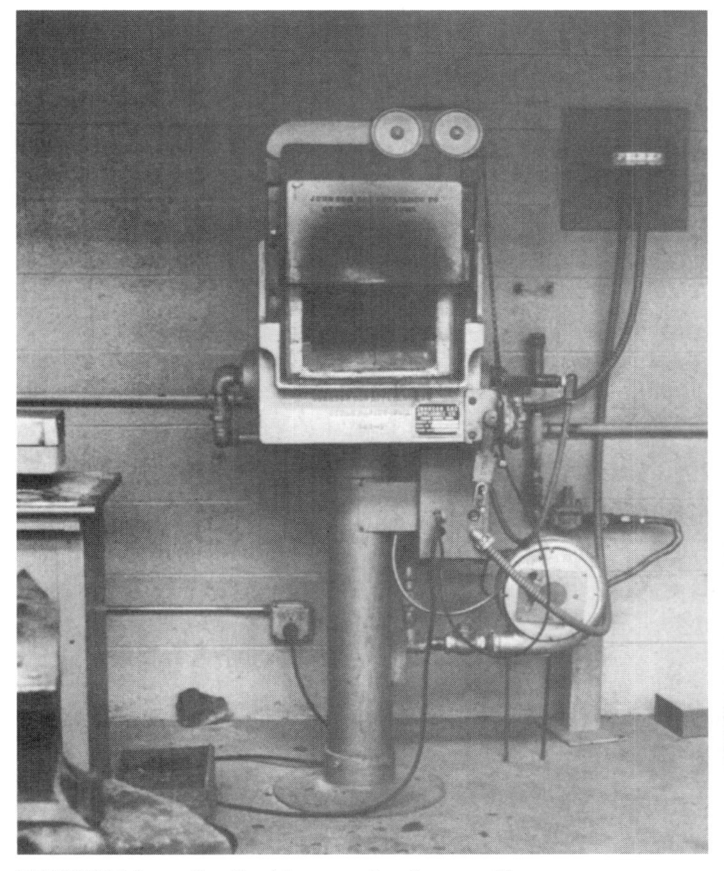

***FIGURE 9.1*a.** Gas-fired heat treating furnace. The gas must *not* be turned on before lighting it. Accumulated gas could cause an explosion. (*Mt. Hood Community College*)

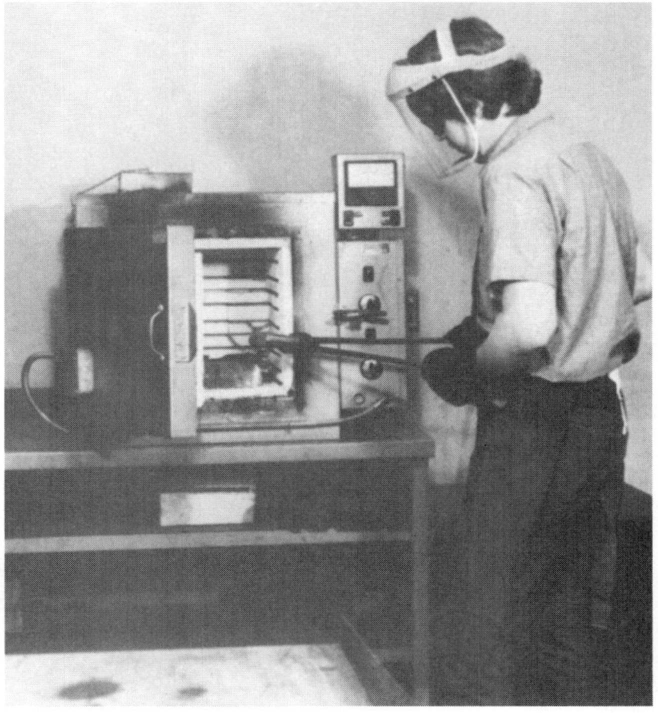

***FIGURE 9.1*b.** Electric heat treating furnace. Part being placed in furnace by heat treater wearing correct attire and using tongs. (*Lane Community College*)

If the furnace set temperature controller fails to function, the safety override temperature controller will shut off the furnace and prevent the furnace from heating above the safety override set temperature.

Successful heat treatment of modern alloys requires strict adherence to furnace temperature control, in addition to performing the required heat treatment according to the governing specification. Furnace load ramp-time or the time it takes a given load of parts to reach the operating temperature of the furnace is extremely important due to the numerous metallurgical changes which take place during heat treatment, such as grain growth, allotropic transformation, element diffusion, and many more. Successful heat treating is predicated upon a thorough understanding of the heating and cooling conditions which create microstructure change. Keep in mind this first principle of heat treatment: Heat treatment changes microstructure and resulting mechanical properties. The true purpose for heat treating a metal is to enhance and/or regulate desired physical properties.

FIGURE 9.2a. Thermocouple. These twisted ends of dissimilar metal wires develop a feeble electric current when heated. The differences in current levels are measured by a control system and registered on a dial face as Fahrenheit or Celsius temperatures. (*Lane Community College*)

FIGURE 9.2b. Temperature control. (*Lane Community College*)

General heat treatment practice can be separated into two categories based upon requirements, aircraft or commercial. Furnaces dedicated to the heat treatment of aircraft parts are usually operated under the requirements of a specific process specification provided by the manufacturer or by one or more government agencies. For example, furnaces in which MIL-H-6875 aircraft steel alloy fasteners are heat treated, hardened, and tempered must be qualified per MIL-H-6875 and the furnace maintained and tested on a monthly basis per MIL-H-6875 requirements. Testing requires coupons to be heat treated in the furnace to be certified and sent to a metallurgical laboratory where the resulting microstructure will be analyzed for compliance to the MIL specification. Tests are performed to certify hardness by microhardness testing and furnace atmosphere compliance by optical microscopic examination of the metal surface for the presence of high temperature oxidation (HTO); or by carbon diffusion, and/or for visual indications of decarburization (loss of carbon content). Qualification of a heat treatment process may also include dissolved gas analysis and ASTM grain size measurement. It is a violation of law to knowingly heat-treat MIL specification parts in a noncertified heat treat furnace and/or to falsely certify compliance.

In contrast, the requirements for commercial heat treatment practices are usually less stringent. Furnaces need not be certified. This of course does not mean the requirements for a proper heat treatment are any less stringent. Parts are parts and to create and control desired changes in mechanical properties, heat treatment must be correctly performed. The major difference is only in the certification process; for commercial heat treatment, compliance is usually verified only by direct hardness measurement of the product following heat treatment.

Ferrous Metal Heat Treatment Equipment

A large variety of furnace designs are used for the heat treating of steels or ferrous alloys. The primary differences in these furnaces are in the temperature capability and in the basic furnace construction. Contrary to what most believe, steels are highly sensitive to chemical reaction with oxygen and must be protected during heat

FIGURE 9.3. Open-air type furnace.

FIGURE 9.4. High temp atmosphere furnace; work exit end.

treatment. Surface protection can be accomplished in a couple of ways: by applying a chemical stop-off or by plating the part with copper when the intention is to heat treat in an open-air furnace. Steels lose alloying elements when exposed to air at temperatures above 450°F, mechanical strength is dependent on the physical strength of the surface metal. The open-air furnaces are usually selected for large batch processing where the condition of the surface metal is of no consequence to the finished product. As an example, annealing of hypoeutectic steels, steels with a carbon content of less than 0.8 percent, is often performed in an open-air type furnace (Figure 9.3). Following annealing or softening, the metal may be machined or cold formed with relative ease.

If a through hardening heat treatment is to follow, the manufacturer may decide to protect the condition of the surface by selecting an atmosphere furnace to heat-treat its parts (Figure 9.4). Atmosphere furnaces come in numerous sizes, shapes, and capabilities, dependent upon the intended function. A continuous belt furnace is shown in Figure 9.5. Parts enter the furnace on a conveyor belt at one end, are heated, and protected by a carbon regulated atmosphere as they pass through to the other end of the furnace. Upon reaching the conveyor end, parts drop off into the quenching oil where rapid cooling takes place. At the bottom of the oil quenching tank, another conveyor belt transports the parts out of the oil and into a basket which is lowered into a washing station where the oil is removed. Once the oil is removed, parts are placed into the tempering furnace where they are heated to temper the as-quenched hardness to the desired hardness.

FIGURE 9.5. Continuous belt furnace.

Many through hardening furnaces are constructed with two or three chambers separated by hydraulic actuated doors. Some have a preheat area, a high temperature soak zone, and a quench area (Figure 9.5). Parts to be heat treated are loaded into an inconel basket and placed into the preheat area where the temperature can be raised between 1000°F to 1250°F. When preheating is complete, an inner door is opened and the basket is automatically transported into the high temperature zone where austenitizing takes place in a controlled carbon potential atmosphere. When ready to quench, the door separating these two chambers opens and a conveyor automatically pushes the basket into the quench chamber where it is immediately lowered into agitated quench oil, maintained at between 120°F and 160°F.

Atmosphere control is the key to nearly scale-free heat treatment of steels. Scale and/or alloy depletion takes place on the surface of steel when the metal is heated above 500°F. In simple terms, the surface of the metal is of one chemical makeup. The atmosphere in contact with the surface is chemically different. As the surface temperature increases, the elements of the atmosphere try to diffuse into and combine with the steel. Likewise, the chemical elements of the steel attempt to diffuse into the atmosphere. Carbon is the most active element in the steel and is easily transported out of the metal to combine with available oxygen. Other alloying elements essential to the strength of the steel also leave the metal at a much slower rate, in an attempt to share atoms by combining with the furnace atmosphere. This propensity to share atoms is minimized by introducing atmosphere into the furnace during heat treatment which is at the same carbon potential as the steel being heat treated. The controlled carbon potential atmosphere displaces and/or replaces the air in the furnace and provides a protective environment for heat treatment. It is impossible to achieve a perfect chemically matched atmosphere protection for alloy steels without being in a vacuum furnace, therefore the surface of steels heat treated in conventional atmosphere controlled furnaces will usually be discolored by the heat treating process. Some steels will form a thin oxide or alloy-rich scale layer on the surface which can be removed by grit blasting or tumble cleaning processes.

Salt pot type furnaces are sometimes used for heat treating tool steels and specialty alloy steels (Figure 9.6). The tool to be heat treated is immersed directly into the salt for the required operation time; heating or cooling is rapid. Salt pot furnaces many years ago used melted sodium cyanide and other dangerous chemicals to quickly transport heat into or out of steels. The handling and liability problems have caused these furnaces to be replaced with fluidized bed furnaces. The fluidized bed furnaces use Al-oxides (the size of beach sand) and gas to provide comparable heating and cooling rates in addition to a protective atmosphere which is less detrimental to the environment than the salt bath.

Batch type furnaces (Figure 9.7) are also used in combination with a controlled atmosphere. The best protection for the surface of a steel against oxidation is provided by a vacuum furnace (Figure 9.8*a-c*). Vacuum furnaces provide moving heat zones in addition to gas fan inert gas quenching capability. Vacuum furnaces are used for heat treating stainless steels, tool steels, and many specialty metals and applications requiring furnace welding and/or brazing capabilities. Baskets made from high heat-resistant RA 330 alloy are commonly used to hold parts during heat treating (Figure 9.9).

QUENCHING MEDIA

Various types of quenching media, their selection based on the type of steel being heat treated and the reaction hot steel has with the quenching media is explained in Chapter 11.

Tempering is a heat treating operation required after performing a hardening heat treatment on alloy and tool steels. The tempering of steels is a means to stress relieve the metal and is performed in special furnaces (Figure 9.10). Tempering heat treatment procedures are covered in Chapter 11.

The flame heats the part to above the austenitizing temperature followed by an immediate water spray quench (Figure 9.11). Individual sections of parts can be hardened by induction heating followed by dunking in an oil or water quench (Figure 9.12). The trend in the induction heat treating industry is to automate the hardening process by automation, thus the quality of the hardened part can be maintained.

HEAT TREATING SAFETY

When heat treating, always wear a face shield, leather gloves, and long sleeves. There is a definite hazard to the face and eyes when cooling the tool steel by oil quenching, that is, submerging it in oil. The oil, hot from the steel, tends to fly upward, so you should stand to one side of the oil tank and not lean over it.

Always work in pairs during heat treatment. One person can open and close the furnace door while the other handles the hot part. The heat treated part should be positioned in the furnace so it can be conveniently removed. This will prevent the heat treater from dropping hot parts and help to ensure successful heat treatment. Atmospheric furnaces should never be opened until the gas supply is turned off. Failure to do so could result in an explosion.

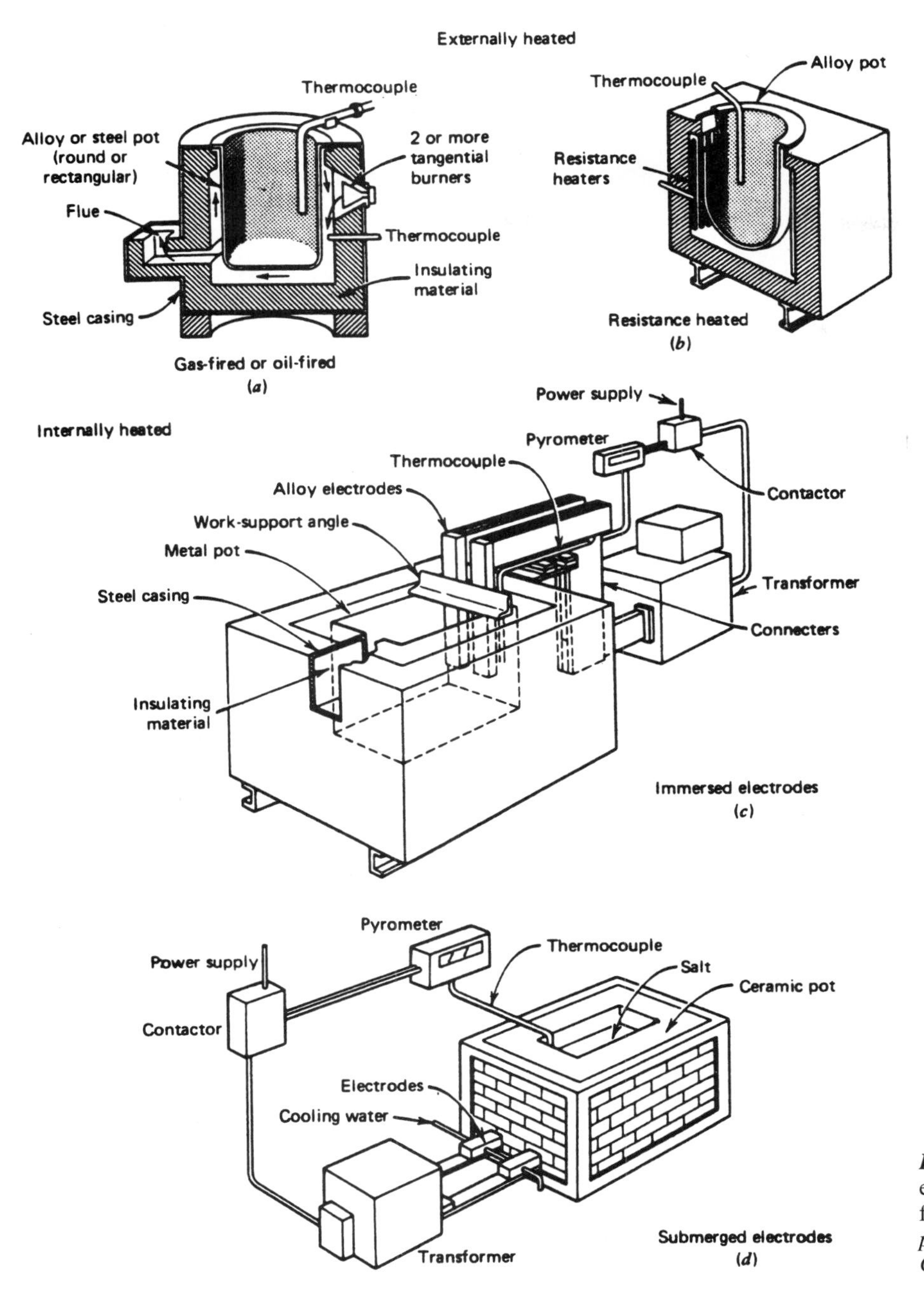

FIGURE 9.6. Pot furnace. Principal types of externally and internally heated salt bath furnaces used for liquid carburizing. (*By permission, from Metals Handbook, Volume 2, Copyright American Society for Metals, 1963*)

*FIGURE 9.7*a.

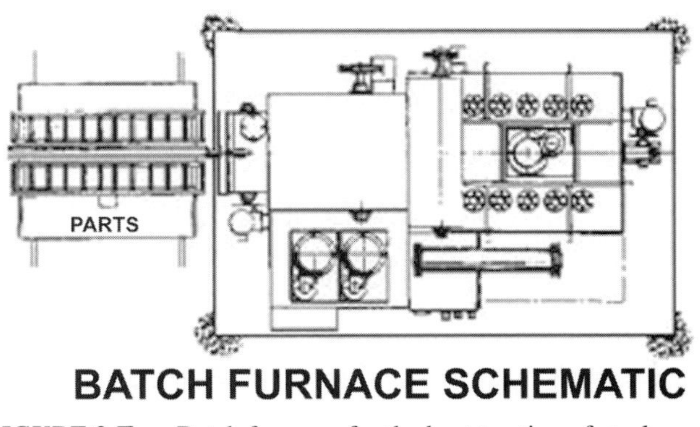

BATCH FURNACE SCHEMATIC

*FIGURE 9.7*b. Batch furnaces for the heat treating of steels.

***FIGURE 9.8*a.** Vacuum furnace.

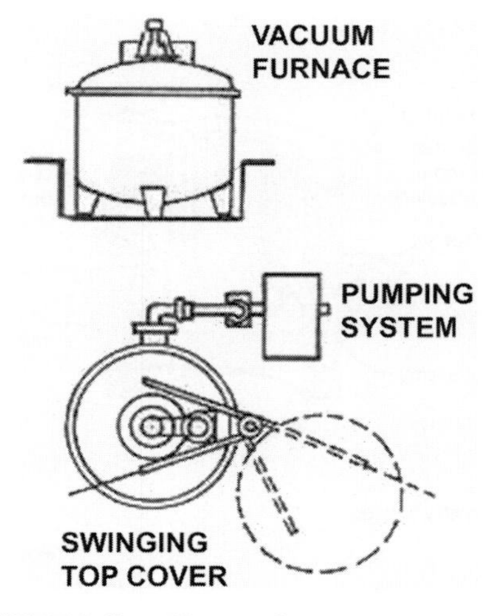

***FIGURE 9.8*b.** Vacuum furnace.

***FIGURE 9.8*c.** Vacuum furnace.

Sheet metal funnel

Carburizing salt

Submerged basket

(*a*)

(*b*)

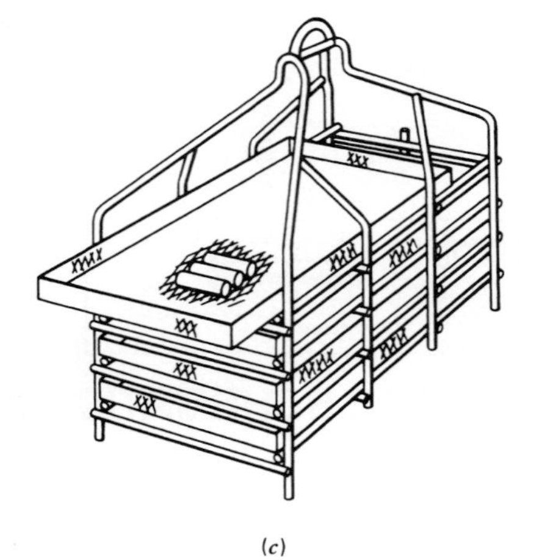

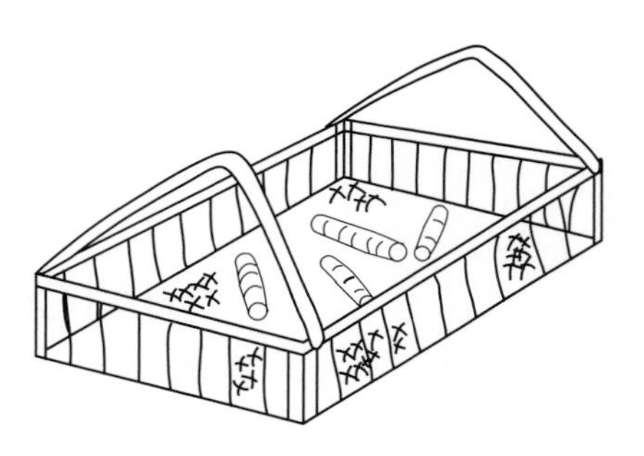

FIGURE 9.9. Baskets for steel heat treating. Baskets for steel heat treating (a–d) are produced in a large variety of shapes and sizes, dependent on the application. The common feature of these heat-resistant baskets is the ability of the metal, usually alloy RA330, to resist thermal erosion and warpage at very high temperatures, sometimes exceeding 1900°F.

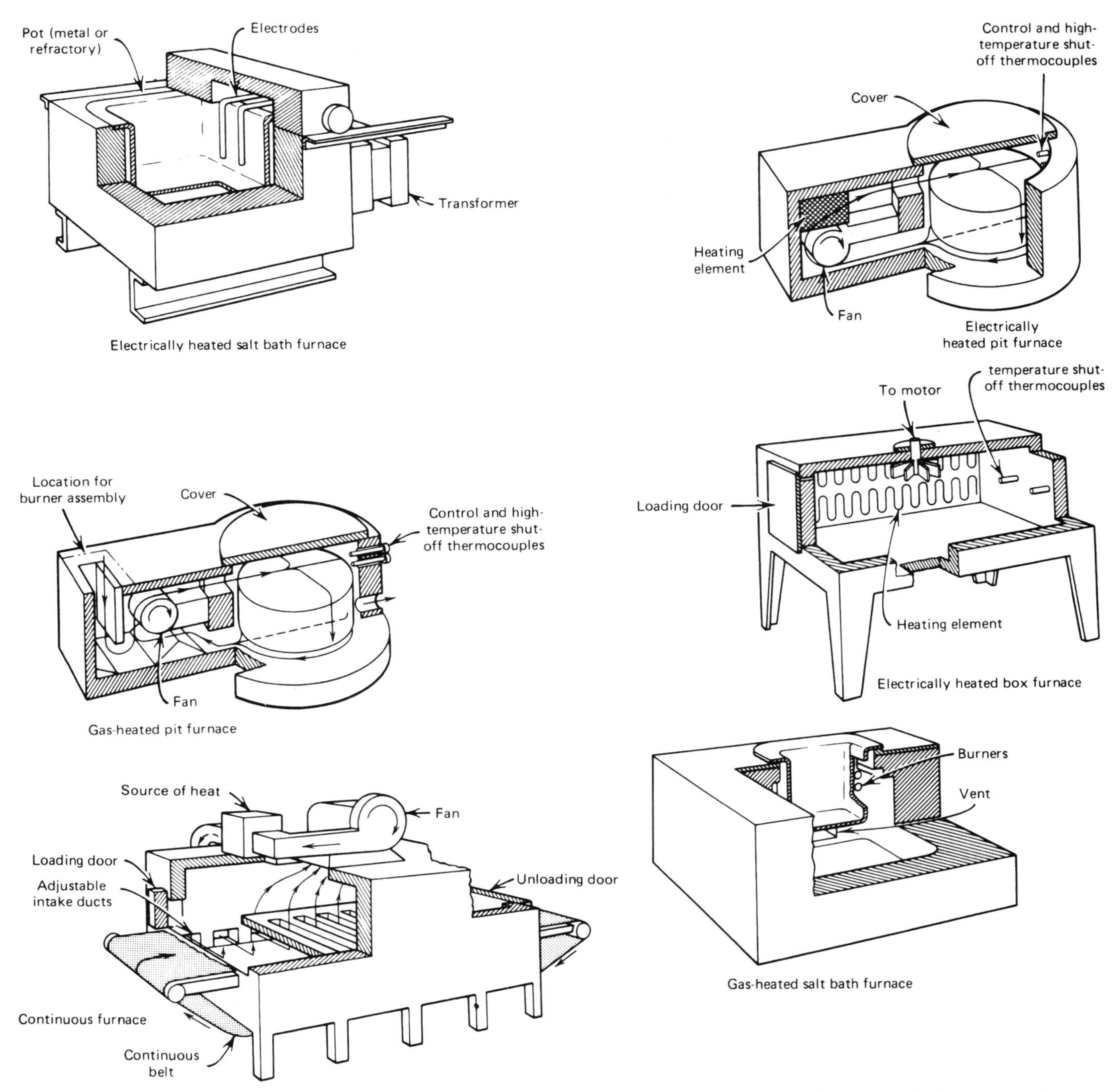

FIGURE 9.10. Types of furnace used for tempering steel. Tempering may also be done in small electric furnaces that have automatic temperature controls. (*By permission, from Metals Handbook, Volume 2, Copyright American Society for Metals, 1963*)

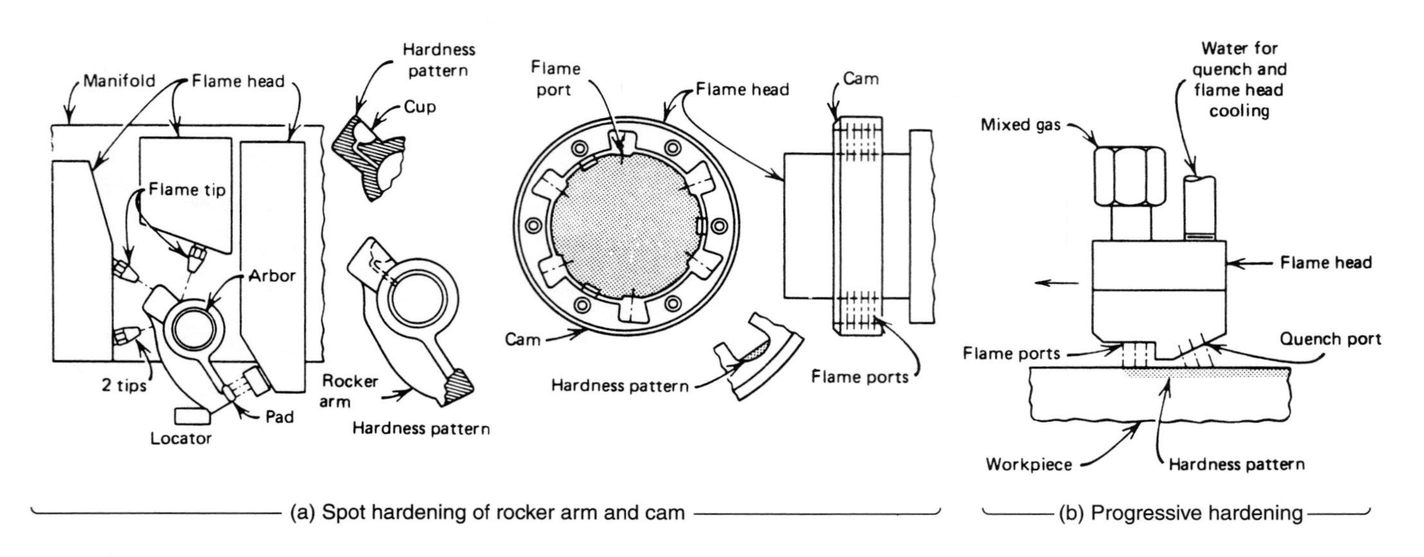

(a) Spot (stationary) method of flame hardening a rocker arm and the internal lobes of a cam; quench not shown. (b) Progressive method

FIGURE 9.11. Flame Hardening. (a) Spot (stationary) and (b) progressive method (continuous heating and quenching) of a moving part. (*By permission, from Metals Handbook, Volume 4, Copyright American Society for Metals, 1981*)

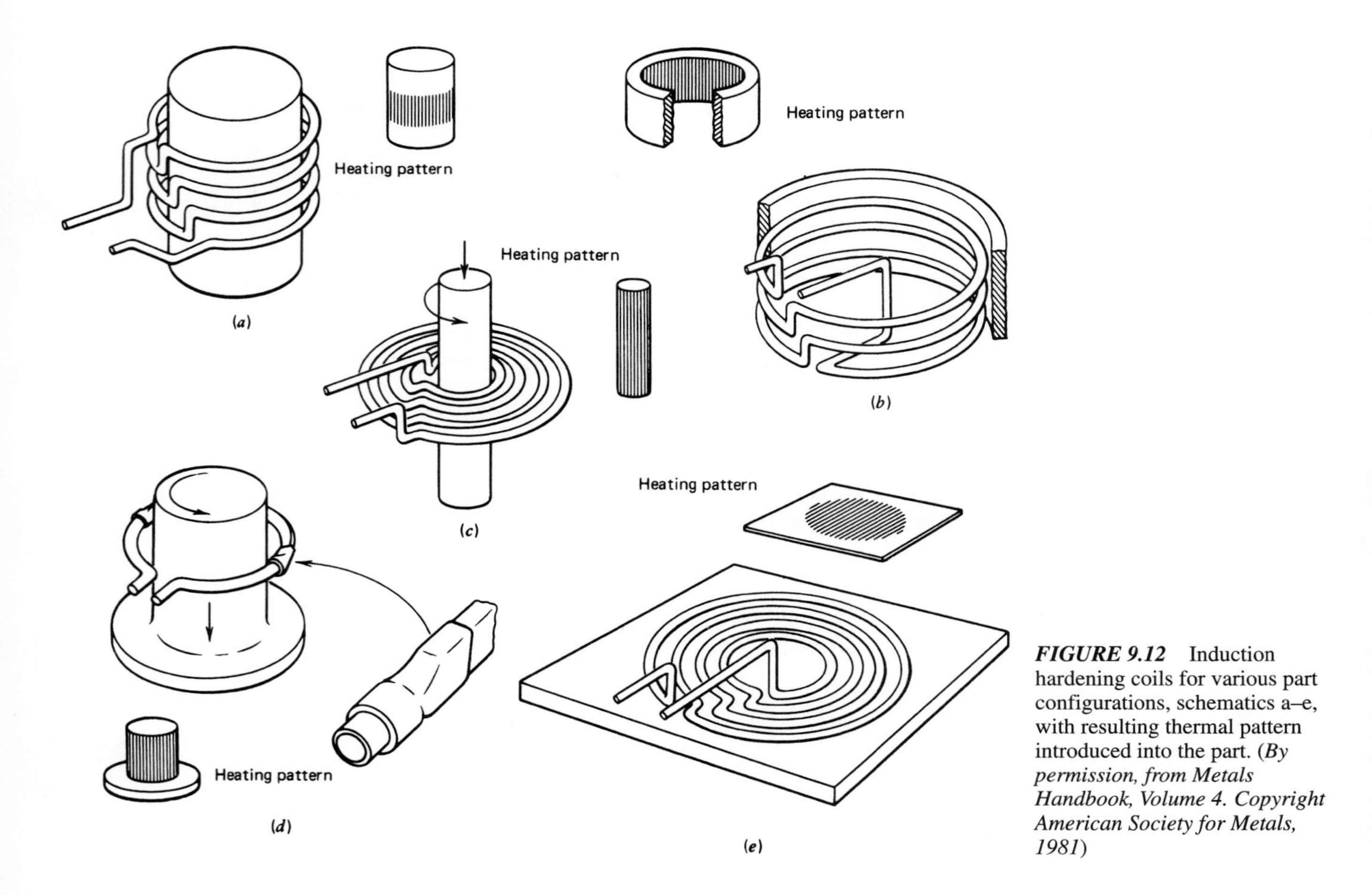

FIGURE 9.12 Induction hardening coils for various part configurations, schematics a–e, with resulting thermal pattern introduced into the part. (*By permission, from Metals Handbook, Volume 4. Copyright American Society for Metals, 1981*)

Safety Note. Very toxic fumes are present when parts are being carburized with compounds containing potassium cyanide. These cyanogen compounds are highly poisonous and every precaution should be taken when using them. Kasenite®, a trade name for a carburizing compound that is not toxic, is often found in school shops and machine shops.

You must read Chapter 11 to complete the Worksheets and Case Problem that follow.

SELF-EVALUATION

1. Name three kinds of furnaces used for heat treating steels.
2. What can happen to a carbon steel when it is heated to high temperatures in the presence of air (oxygen)?
3. Why is it necessary to allow a soaking period for a length of time (which varies according to the kind of steels) before quenching the piece of steel?
4. Why should the part of the quenching medium be agitated when you are hardening steel?
5. Which method of tempering gives the heat treater the greater control of the final product: by color or by furnace?
6. Describe two characteristics of quench cracking that you can recognize.
7. Name four or more causes of quench cracks.
8. In what ways can decarburization of a part be avoided when it is heated in a furnace?
9. Name two types of high-speed tool steels.
10. When distortion must be kept to a minimum, which type of tool steel should be used?
11. What is the advantage of using low carbon steel for parts that are to be case hardened?
12. By which methods of carburizing can a deep case be made?
13. Are parts that are surface hardened always case hardened?
14. Name three methods by which carbon may be diffused into the surface of heated steel.
15. What method of case hardening uses ammonia gas?
16. When a part is overheated to extremes and then quenched, "chicken wire" surface markings can often be seen. What does this indicate about the interior condition of the steel? Can the part be reconditioned by subsequent heat treating?
17. Why should the grasping end of tongs be preheated prior to grasping heated steel in a furnace before quenching?

CASE PROBLEM: THE LOW HARDNESS OF A DIE BLOCK

An air-hardening tool steel was selected in a die shop for a punching die of a moderately light cross section. The steel part was machined from A-10 and correctly preheated at 1200°F. Then it was correctly austenitized at 1475°F. As soon as it was at a red heat throughout, it was removed from the furnace to be quenched in air. It was tested for hardness when it had cooled to room temperature before tempering the part. The hardness should have been between Rc 62 and 65 at that point, but was only Rc 45. What went wrong in the hardening procedure?

WORKSHEET 1

Objectives

1. Correctly harden a steel part.
2. Correctly temper a steel part to RC 48 to 52 (HB 470 to 514).

Materials Two heat-treating furnaces, an oil-quenching bath and accessories, plus a previously machined part of SAE 4140 steel.

Procedure

1. Determine the correct hardening temperature from Table 1. Table 2 provides important mass effect data for SAE 4140 steel. The temperature given in Table 1 may have to be adjusted slightly, depending on the size of the part.
2. Set the furnace thermocouple control to the correct temperature and turn it on.
3. With the tongs, place the part in the furnace.
4. Set the second furnace to the correct tempering temperature and turn it on. Use Table 1 to determine this temperature to obtain a draw temper hardness of RC 48 to 52 (HB 470 to 514).
5. After the part is the same color as the furnace, allow a soaking time of 1 hour per inch of the narrowest cross section.
6. Using face shield and gloves, heat the tongs on the gripping end. Remove the part from the furnace, close the door, and quickly plunge the part into the oil bath, agitating the part until it has cooled. It should still be warm to the touch.
7. Place the warm part immediately into the tempering furnace and hold it at that temperature about 1/2 hour per inch of cross section.
8. Remove the part and allow it to cool in air.
9. Check for hardness. Because of possible decarburization, the readings may be low. Check again after surface grinding.

TABLE 1. Typical heat-treating information for direct-hardening carbon and low alloy steels

			Expected Hardness after Tempering 2 Hr @:								
Grade	**Hardening Temperature**	**Full Hardness**	**400**	**500**	**600**	**700**	**800**	**900**	**1000**	**1100**	**1200**
8620	1650/1750	37/43	40	39	37	36	35	32	27	24	20
4130	1550/1625	49/56	47	45	43	42	38	34	32	26	22
1040	1525/1600	53/60	51	48	46	42	37	30	27	22	(14)
4140/4142	1525/1575	53/62	55	52	50	47	45	41	36	33	29
4340	1475/1550	53/60	55	52	50	48	45	42	39	34	30
1144	1475/1550	55/60	55	50	47	45	39	32	29	25	(19)
1045	1475/1575	55/62	55	52	49	45	41	34	30	26	20
4150	1500/1550	59/65	56	55	53	51	47	45	42	38	34
5160	1475/1550	60/65	58	55	53	51	48	44	40	36	31
1060/1070	1450/1525	58/63	56	55	50	43	39	38	36	35	32
1095	1475/1525	63/66	62	58	55	51	47	44	35	30	26

Note. Temperatures listed in °F and hardness in Rockwell C. Values were obtained from various recognized industrial and technical publications. All values should be considered approximations.

Source. Pacific Machinery & Tool Steel Co.

Conclusion Is your steel part the same hardness that Table 1 indicates it would be at your selected tempering temperature? If not, what reason can you give for the discrepancy?

TABLE 2. Mass effect data for SAE 4140 steel

Single Heat Results									
	C	**Mn**	**P**	**S**	**Si**	**Ni**	**Cr**	**Mo**	
Grade	**0.38/0.43**	**0.75/1.00**	**—**	**—**	**0.20/0.35**	**—**	**0.80/1.10**	**0.15/0.25**	**Grain size**
Ladle	**0.40**	**0.83**	**0.012**	**0.009**	**0.26**	**0.11**	**0.94**	**0.21**	**7–8**

Mass Effect

Size Round (in.)	Tensile Strength (psi)	Yield Point (psi)	Elongation % 2 in.	Reduction of Area (%)	Hardness (HB)
Annealed (heated to 1500°F, furnace-cooled 20°F per hour to 1230°F, cooled in air)					
1	95,000	60,500	25.7	56.9	197
Normalized (heated to 1600°F, cooled in air)					
½	148,500	98,500	17.8	48.0	302
1	148,000	95,000	17.7	46.8	302
2	140,750	91,750	16.5	48.1	285
4	117,500	69,500	22.2	57.4	241
Oil quenched from 1550°F, tempered at 1000°F					
½	171,500	161,000	15.4	55.7	341
1	156,000	143,250	15.5	56.9	311
2	139,750	115,750	17.5	59.8	285
4	137,750	99,250	19.2	60.4	277
Oil quenched from 1550°F, tempered at 1100°F					
½	157,500	148,750	18.1	59.4	321
1	140,250	135,000	19.5	62.3	285
2	127,500	102,750	21.7	65.0	262
4	116,750	87,000	21.5	62.1	235
Oil quenched from 1550°F, tempered at 1200°F					
½	136,500	128,750	19.9	62.3	277
1	132,750	122,500	21.0	65.0	269
2	121,500	98,250	23.2	65.8	241
4	112,500	83,500	23.2	64.9	229

As quenched hardness (oil)

Size Round	Surface	½ Radius	Center
½	HRC 57	HRC 56	HRC 55
1	HRC 55	HRC 55	HRC 50
2	HRC 49	HRC 43	HRC 38
4	HRC 36	HRC 34.5	HRC 34

Note. Some important information about this particular steel (SAE 4140) may be obtained from this table. First, the composition of alloying elements and trace elements is given, plus the grain size (see Chapter 4, Figure 22). Then, strength, ductility, and hardness are given for various round diameters, showing the effect of mass on this steel, and finally, hardness at different depths that shows the hardenability.

Source. Modern Steels and Their Properties, Handbook, 3310, Bethlehem Steel Corporation, 1978.

WORKSHEET 2

Objectives

1. Case harden a piece of mild steel using a furnace or heating torch.
2. Pack carburize a piece of mild steel in a furnace.

Materials A furnace, tongs, carburizing compound, protective clothing (face shield and gloves), and a small piece of low carbon steel.

Procedure for Case Hardening by the Roll Method

1. Heat the part to 1650°F (899°C) and remove from the furnace with tongs.
2. Roll part in carburizing compound.
3. Reheat to 1650°F (899°C).
4. Quench in cool water.

Procedure for Pack Carburizing

1. Place the part in a steel box containing the carburizing compound.
2. Place in furnace set at temperature of 1700°F (927°C). Leave it in the furnace for several hours.
3. Remove the part from furnace and quench in water.

Conclusion

1. Did the piece become hard on the surface? Check with a file.
2. Grind off a small amount and check again. How deep do you think the case is on the part hardened by the roll method? By the pack carburizing method?

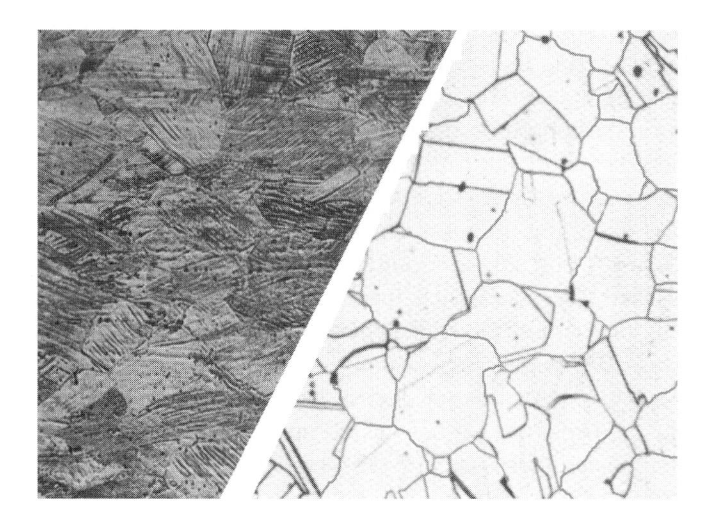

CHAPTER 10

Annealing, Stress Relieving, and Normalizing

Since the machinability and welding of metals are so greatly affected by heat treatments, the processes of annealing, normalizing, and stress relieving are important to the welder and machinist. Also, annealing is important in metals manufacturing processes. Metals that become hard and brittle from cold working need to be softened by the annealing process. You will learn about these processes in this chapter.

OBJECTIVES **After completing this chapter, you should be able to:**

1. Explain the principles of and differences among the various kinds of annealing processes.
2. Test various steels with annealing, normalizing, and stress-relieving heat treatments to determine their effect on machinability and welding.

ANNEALING

The definition of *anneal* is "to soften." Annealing is a heat treatment applied to nearly all metals to remove the effects of work hardening treatments such as forging and or thermal hardening. One of the important properties of metals is their ability to relax the crystal structure. This is accomplished by heating the metal to a prescribed temperature and allowing the metal to cool in a controlled manner. The act of thermal annealing causes the microstructure to reform—A mechanically work hardened elongated grain structure will return to an equiaxed soft grain structure. There are other methods of annealing used in industry such as cryogenic annealing in which the metal is cooled to a very low liquid nitrogen temperature (approximately –300°F). In another method, the metal is vibrated at a natural frequency for the item and as a result of being vibrated, significant residual stresses are reduced.

It is important to note in steels, all thermal heat treatments are specific to the individual metal. Time at temperature plays a critical role in the recovery of a work hardened microstructure. Not all microstructure changes are immediate and specific processes must be followed to the letter to achieve maximum results. Annealing may be performed for a variety of reasons: to make the metal softer and more easily machineable, to remove the effects for mechanical deformation which occurs in operations such as stamping or forging, and to produce a product which can be easily formed into shape.

Preheat Annealing in Preparation for Hardening Steels

Steels are particularly interesting and challenging to anneal, since annealing is performed for a variety of reasons and may yield a wide variety of microstructures and hardness. Most alloy steels which are to be through hardened or are to be heated to the austenitizing temperature range must be annealed prior to allotropic transformation. This is a "given" in the heat treating industry, since cracking upon rapid heating may occur if a preheat annealing operation is not performed. The preheat annealing is performed by heating the metal to a temperature between 1000°F and 1250°F to allow the microstructure to relax and the surface and core of the part to thermally stabilize at the same temperature prior to being placed into the furnace hot zone, usually above 1500°F.

Annealing to Produce a Given Microstructure in Steels

The annealing process, known as "full annealing," involves heating the steel in a furnace to a temperature 50°F above the upper critical temperature recommended for austenitizing the steel (Figure 10.1). Keep in mind when heating a steel above the lower critical temperature line (iron-carbon phase diagram: A_1, $A_{3,1}$ line), the steel undergoes solid-state recrystallization. The upper critical temperature differs for every alloy of steel; a good source for obtaining this information is the *Heat Treater's Guide.*[1] Full anneal cooling is performed slowly at a prescribed rate in a furnace, insulator, or retort. The resulting microstructure is a coarse pearlite/ferrite mixture (Figure 10.2). Machinists will usually appreciate machining the pearlite microstructure due to the chip-breaking properties and relative softness as compared with a tempered martensite microstructure (Figures 10.3*a, b*). The pearlite grain consists of thin alternate layers of iron carbide and nearly pure iron (cementite and ferrite). When the machinist's tool contacts the iron carbide layer, it fractures, allowing the tool to proceed easily through the ferrite layer to the next carbide layer (Figure 10.4). The tool will usually cut through the pearlite microstructure without much difficulty and produce figure-9-shaped chips.

When maximum softness is required, the alloy steel may be spheroidize annealed. This heat treatment causes the carbides to form tiny spheroids in a matrix of ferrite (Figure 10.5*a*) compared with a fully pearlitic matrix (Figure 10.5*b*). Spheroidization of the carbide occurs when the steel is held at 1250°F for a few hours, followed by slow cooling to room temperature. Spheroidization is the preferred annealing heat treatment for martensitic or previously hardened steels. Since allotropic transformation does not occur during spheroidization, high stresses and distortion are minimized. The microstructure of spheroidized hypoeutectoid steels previously martensitic can be distin-

(1) *Heat Treater's Guide: Practices and Procedures for Irons and Steels,* 2nd ed., American Society for Metals International, 1995.

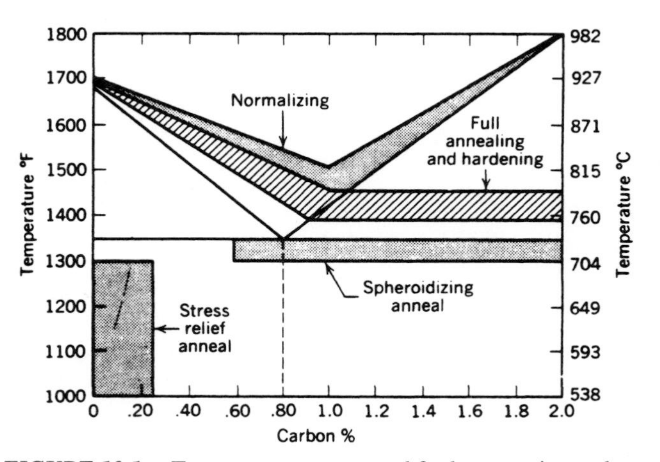

FIGURE 10.1. Temperature ranges used for heat treating carbon steel. (*Machine Tools and Machining Practices*)

FIGURE 10.2. Pearlitic 1040C steel. The dark and light layers in the pearlite grains are iron carbide (cementite) and ferrite. White grains are ferrite or near pure iron.

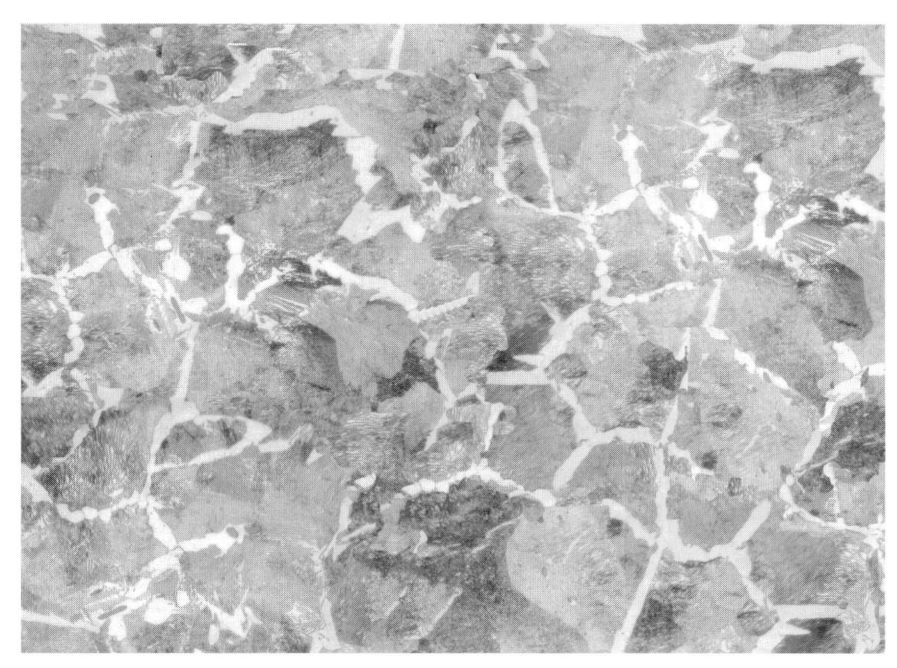

*FIGURE 10.3*a. Pearlitic microstructure in a hypoeutectoid steel.

guished from formerly pearlitic steels by the uneven distribution of spheroidized carbide exhibited by the former pearlitic microstructures. Recrystallization does not take place during spheroidization; the change in microstructure is solely due to phase precipitation and carbide coalescence. The hardness range expected for pearlitic hypoeutectoid steels is between 10 and 35 HRC, which the same steels spheroidized will be 20 HRC maximum.

Full annealing is also used to completely remove the effects of cold work or plastic deformation in the microstructure of steels (Figures 10.6*a, b*). The key to the amount of grain growth and final grain size is dependent upon the time at temperature during annealing, along with the chemical makeup of the steel being heat treated. Steels containing grain refiners, such as boron or vanadium, will exhibit far less tendency to coalesce grains.

***FIGURE 10.3*b.** Martensitic microstructure in a hypoeutectoid steel.

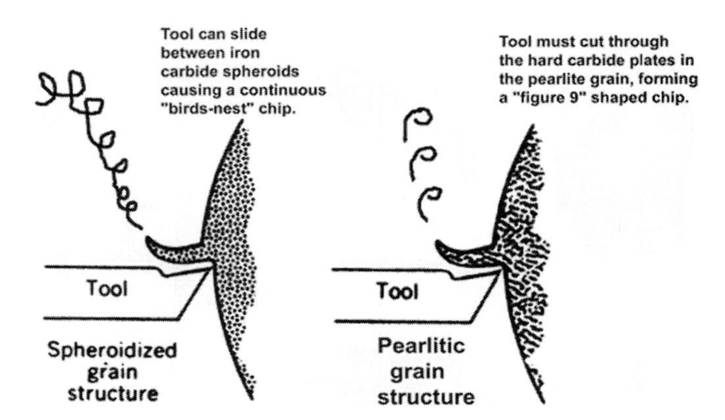

FIGURE 10.4. Comparison of cutting action between spheroidized and normal carbon steels. The most important difference between these two microstructures is that the pearlite grain forms a figure-9 shape when cut and is considered to be the preferred chip-breaking microstructure, while the spheroidize annealed microstructure produces a continuous cutting referred to as a "bird's nest."

Normalizing

Normalizing of the grain structure in steels does not occur unless the steel is specifically alloyed with boron and/or vanadium. These elements are required to nucleate grain formation in the solid state during high temperature soaking. Normalizing is described in most metallurgy and heat treating literature as the heating of the steel to approximately 1700°F followed by cooling in still air or furnace cooling. Hypothetically, a solid-state recrystallization and homogenization of the alloying elements and grain structure take place during normalizing. This thermal treatment is akin to **full annealing.** Severe adverse microstructure conditions such as grain germination, alloy banding, and alloy segregation caused by overheating most alloy steels prior to forging are not removed by normalizing and/or additional heating operations.

Stress Relief and Bright Annealing

What is meant by stress relief annealing? This heat treatment is performed to reduce or minimize the residual stresses and resulting distortion caused by mechanical work hardening processes (hammering, forming, shearing, welding, drilling, and general machining operations). Steels which have been through hardened and require subsequent machining are often stress relief an-

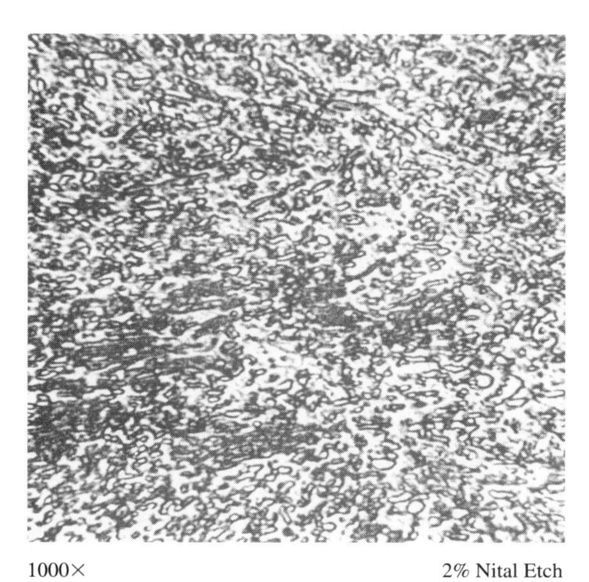

FIGURE 10.5a. 1090C plain carbon steel which has been spheroidize annealed.

FIGURE 10.5b. 1080C plain carbon steel which has been full annealed.

FIGURE 10.6a. Microstructure of flattened grains of 0.10 percent carbon steel, cold rolled (1000×). (*By permission, from Metals Handbook, Volume 7, Copyright American Society for Metals, 1972*)

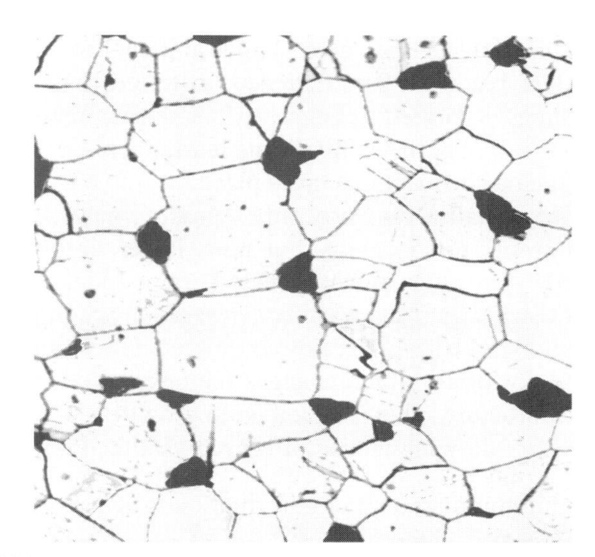

FIGURE 10.6b. The same 0.10 percent carbon steel as in Figure 10.6*a* but annealed at 1625°F. Ferrite grains are reformed to their original state, including the very fine pearlite grains.

nealed. The rule for stress relief annealing hardened steels is *heating should never exceed 50°F below the temperature of the last tempering heat treatment.* Stress relief annealing heat treatment does NOT change the microstructure nor will microstructure changes be visible during optical microstructure examination.

Bright annealing is performed in a sealed retort or a retort filled with inert gas. Bright annealing is selected when the metal surface must be protected from discoloration and/or surface oxidation during annealing. Annealing definitions can be found in the *Metals Handbook* (9th ed., vol. 4).

Annealing of nonferrous metals will be discussed in the appropriate chapters.

RECOVERY, RECRYSTALLIZATION, AND GRAIN GROWTH

When metals are heated to temperatures less than the recrystallization temperature, a reduction in internal stress takes place. This is done by relieving elastic stresses in the lattice planes and not by reforming the distorted grains. Recovery in annealing processes used on cold-worked metals is usually not sufficient stress relief for further extensive cold working (Figure 10.7), yet it is used for some purposes and is called stress relief anneal. Most often, recrystallization is required to reform the distorted grains sufficiently for further cold work.

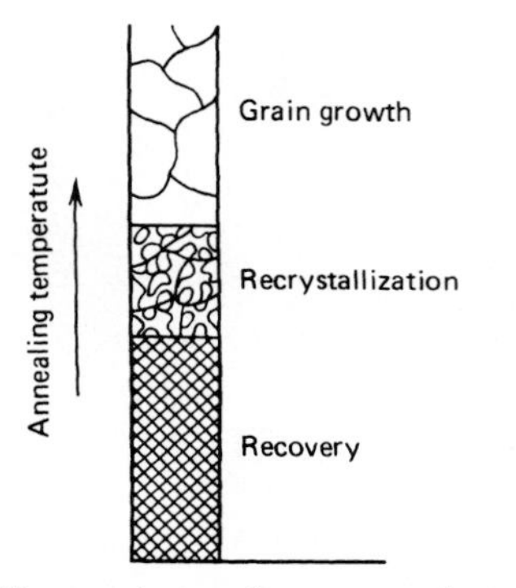

FIGURE 10.7. Changes in metal structures that take place during the annealing process.

TABLE 10.1. Recrystallization temperatures of some metals

Metal	Recrystallization Temperature (°F)
99.999% aluminum	175
Aluminum bronze	660
Beryllium copper	900
Cartridge brass	660
99.999% copper	250
Lead	25
99.999% magnesium	150
Magnesium alloys	350
Monel	100
99.999% nickel	700
Low carbon steel	1000
Tin	25
Zinc	50

Recovery is a low-temperature effect in which there is little or no visible change in the microstructure. Electrical conductivity is increased, and often a decrease in hardness is noted. It is difficult to make a sharp distinction between recovery and recrystallization. Recrystallization releases much larger amounts of energy than does recovery. The flattened, distorted grains are sometimes reformed to some extent during recovery into polygonal grains, while some rearrangement of defects such as dislocations takes place.

Recrystallization not only releases much larger amounts of stored energy but new, larger grains are formed by the nucleation of stressed grains and the joining of several grains to form larger ones. To accomplish this joining of adjacent grains, grain boundaries migrate to new positions, which changes the orientation of the crystal structure. This is called grain growth.

The following factors affect recrystallization.

1. A minimum amount of deformation is necessary for recrystallization to occur.
2. The larger the original grain size is, the greater will be the amount of cold deformation required to give an equal amount of recrystallization with the same temperature and time.
3. Increasing the time of anneal decreases the temperature necessary for recrystallization.
4. The recrystallized grain depends mostly on the degree of deformation and, to some extent, on the annealing temperature.
5. Continued heating, after recrystallization (reformed grains) is complete, increases the grain size.
6. The higher the cold-working temperature is, the greater will be the amount of cold work required to give equivalent deformation.

Metals that are subjected to cold working become hardened, and further cold working cannot be done without danger of splitting or breaking the metal. Various degrees of softening are possible by controlling the recrystallizing temperatures. The recrystallizing temperature, as given in Table 10.1 for low carbon steel, will only affect the stressed or worked ferrite grains, not the carbides (pearlite grains). The higher anneal temperatures are needed to recrystallize pearlite. A practical example of the need for controlled anneal is in the manufacture of nails. Before cold heading the nail, the wire must be drawn to a hardness that will prevent it from bending when struck, but it should not be too hard, since then the cold-heading operation would make the head area brittle. In that case, the head would fall off when struck with the hammer. Perhaps recovery with some recrystallization would be used prior to cold heading the nail.

NONFERROUS METALS

Annealing of most nonferrous metals consists of heating them to the recrystallization temperatures or grain growth range and cooling them to room temperature (Table 10.1). The rate of cooling has no effect on most nonferrous metals such as copper or brass, but quenching in cold water is sometimes beneficial. Annealing temperatures and procedures are very critical with some metals such as stainless steels and precipitation hardening nonferrous metals. See Chapter 15, "Heat Treating of Nonferrous Metals," for further information. The phenomenon of grain growth normally occurs at higher temperatures. When a large amount of deformation is required in one operation, large grains are sometimes preferred, although a surface defect, called orange peel (Figures 10.8*a, b*), is sometimes seen on formed metals having large grains. In this case, a stress relief anneal could be used, that is, recovery without grain growth.

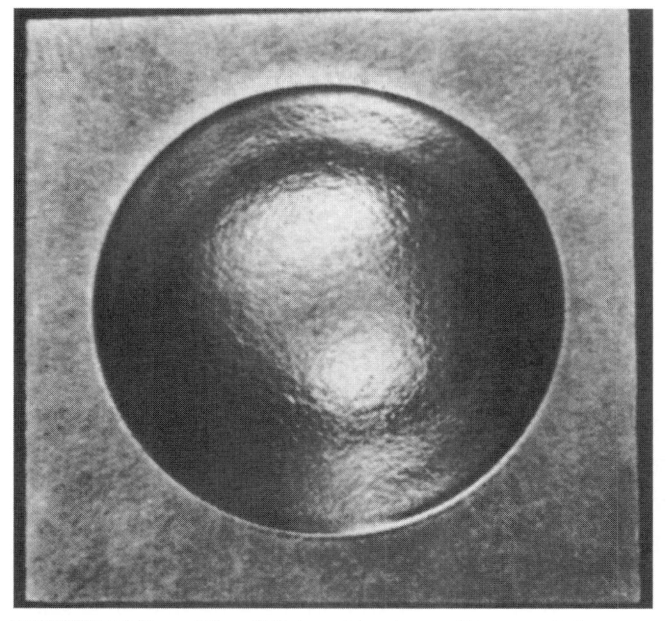

FIGURE 10.8a. Alloy 260 (cartridge brass, 70 percent) drawn cup showing rough surface or "orange peel" (actual size). (*By permission, from Metals Handbook, Volume 7, Copyright American Society for Metals, 1972*)

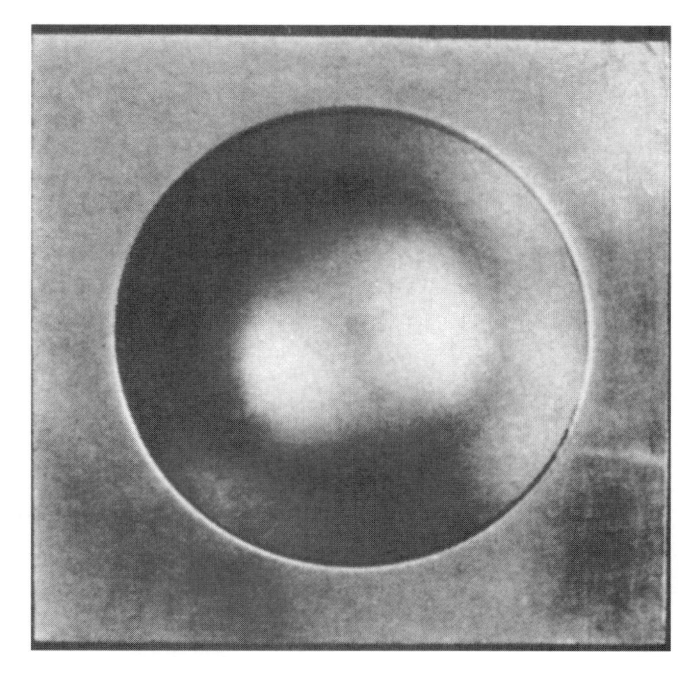

FIGURE 10.8b. Alloy 260 (cartridge brass, 70 percent) drawn cup with a smooth surface, no "orange peel" (actual size). (*By permission, from Metals Handbook, Volume 7, Copyright American Society for Metals, 1972*)

SELF-EVALUATION

1. When might normalizing be necessary?
2. At what approximate temperature should you normalize 0.4 percent carbon steel?
3. What is the spheroidizing temperature of 0.8 percent carbon steel?
4. What is the essential difference between full anneal and stress relieving?
5. When should you use stress relieving?
6. What kind of carbon steels would need to be spheroidized to give them free-machining qualities?
7. Explain process annealing.
8. How should the piece be cooled for a normalizing heat treatment?
9. How should the piece be cooled for the full anneal?
10. What happens to machinability in low carbon steels that are spheroidized?
11. Stress relief anneal or recovery is used only when a slight change is needed in a cold-worked metal. Name the effect that is needed to alter the metal sufficiently for further cold working.
12. If the temperature and time is increased from the process referred to in question 11, what will be the resultant condition of that metal?

CASE PROBLEM: FAILURE OF A TRAILER HITCH

A worker in a large welding and steel fabrication shop asked his employer if he could make a trailer hitch for his pickup truck after hours if he paid the shop for the material used. His employer gave him permission. That evening the worker searched among the many racks of various kinds of steel bars for one suitable for his job. He selected a 1/2 by 2 in. HR bar and sawed off the pieces he needed. He drilled holes, heated areas red hot to bend them, and then quenched them in water. He welded the parts together and assembled the hitch on his truck. When he got home and he lifted the tongue of his trailer over the hitch and dropped it in place, to his great surprise, the hitch broke off and fell to the ground. What had he done wrong?

WORKSHEET 1

Objective Determine the changes in microstructures as a result of annealing heat treatments.

Materials Furnace and metallurgical equipment for microscopic examination, samples of SAE 1010 to 1020 carbon steel and of SAE 1090 carbon steel.

Procedure

1. Take one small specimen of each type of steel and full anneal both.
2. Normalize two specimens, one of each type.
3. Stress relieve two specimens.
4. Spheroidize anneal one of each specimen by holding near 1300°F for several hours in the furnace.
5. Mount, polish, and etch each specimen. Be sure to mark permanently each plastic capsule immediately.
6. Study each specimen with a microscope and record your results.

Conclusion

1. Make circles and draw in each microstructure as you see it. Note the magnification and parts of the structure.
2. Note the changes that have taken place as the result of each heat treatment under each drawing.
3. Turn in your sketches and conclusions to your instructor.

WORKSHEET 2

Objective Determine relative tensile strengths and ductility as a result of heat treatment.

Materials Furnace, tensile testing machine, and four samples of SAE 1040 steel prepared for tensile testing as described in Chapter 3, "The Physical and Mechanical Properties of Metals," Worksheet 1.

Procedure

1. Identify each sample with a different number.
2. Heat the furnace to the quenching (hardening) temperature (about 1650°F) and put three of the samples in the furnace.
3. Quench one sample in water (brine if available) and temper it to 400°F. Record the number and heat treatment.
4. Raise the temperature 50°F and remove one sample to air cool (normalize); record the number.
5. Lower the temperature 50°F and then let the furnace cool slowly to room temperature. Remove the remaining sample and record the number.

Conclusion

1. Draw a stress-strain diagram as seen in Chapter 3 and graph the elastic ranges, yield points, and plastic ranges of the three heat treated samples and the remaining as-rolled sample after testing.
2. Test all four samples in a small tensile tester and record the percentage of elongation and percentage of reduction of area for each on a sheet of paper.
3. Turn in your results to your instructor.

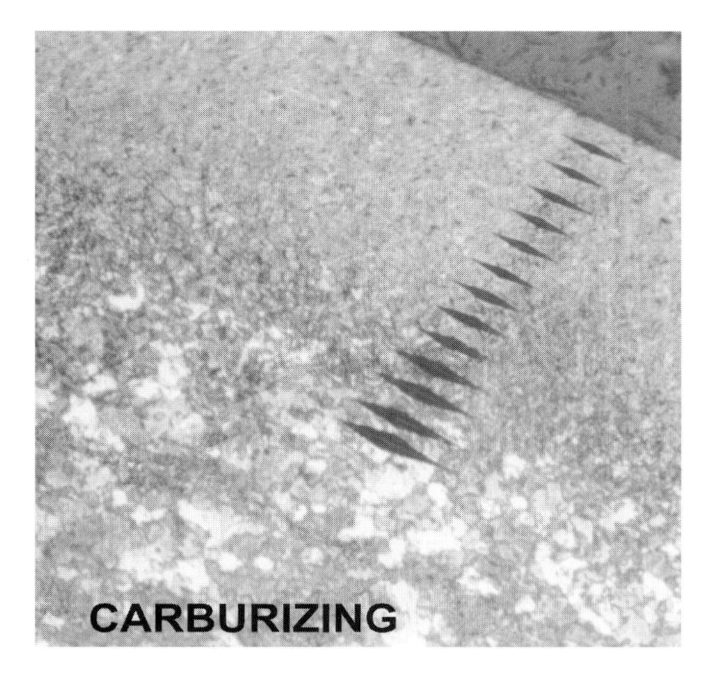

CHAPTER 11

Hardening and Tempering of Steel

Plain carbon steel has been valued from early times because of certain properties. This soft silver-gray metal could be converted into a superhard substance that would cut glass and many other substances, including itself when soft. Furthermore, its hardness could be controlled. This converting of carbon steel into a steel of useful hardness is done with different heat treatments, two of the most important of which are hardening and tempering (drawing), which you will investigate in this chapter.

OBJECTIVES **After completing this chapter, you should be able to:**

1. Correctly harden a piece of tool steel and evaluate your work.
2. Correctly temper the hardened piece of tool steel and evaluate your work.
3. Understand the relationship between tempering temperature and hardness change.

HOW STEELS HARDEN

The mechanism by which alloy steels are hardened is allotropic transformation quenching (ATQ). Iron is allotropic, which means most alloy steels change to another crystal structure, called face-centered cubic (FCC), upon heating above a certain temperature. Upon controlled cooling to below 1333°F, most steels revert to the room temperature crystal structure, called body-centered cubic (BCC). In the absence of alloying elements, iron will always revert to the BCC crystal structure, regardless of the cooling or quenching rate. The addition of allowing elements, in combination with heating, to change crystal structure and homogenization of the alloying elements, followed by a controlled quenching rate, is the secret to hardening steels. To full harden steels, the steel must be quenched when allotropically transformed to FCC or austenite. This chapter will explain the reasons why steels harden, the mechanical properties which develop, and where to find the temperatures and procedures for successful heat treatment of steels.

The Theory

The strength and/or mechanical properties of steels change when they are processed through various hardening heat treatments. Specific mechanical requirements for strength are determined by engineering designers or stress analysts who specify the use of a particular steel alloy. Steels usually come from the mill in soft or annealed condition to facilitate product fabrication and not in the full hardened condition. Steels must be thermally processed in a controlled heat treatment environment to develop the desired mechanical properties.

To harden steels, the given steel must be heated slightly above the temperature where allotropic transformation takes place and all the alloy transforms to a FCC crystal structure. At this elevated temperature, the alloying elements freely mix and homogenize the steel, in a similar manner as sugar dissolves and freely mixes in hot water. In simple terms, alloying elements behave in a predictable manner, following the rules of elemental solubility. That is, as the temperature rises, more alloying elements dissolve and freely mix in the steel; as the temperature drops, elements precipitate out of the steel as free atoms that may form compounds. So, the act of heat treating steels involves heating the steel to dissolve and mix the alloying elements while the steel is in a solid state—a term referred to as *austenitizing.* Soaking at the austenitizing temperature allows the alloying elements sufficient time to migrate and homogenize. Once homogenization is complete, the steel can be rapidly cooled or quenched to room temperature. If slow cooling were to take place rather than a rapid quenching, all the alloying elements would behave according to the rules of solubility and precipitate out or separate from the steel and form complex compounds. The act of rapid quenching traps the homogenized alloying elements within the FCC crystal structure as the steel attempts to form a BCC crystal structure as temperature is reduced to room temperature. Instead, a highly distorted BCC atom shape develops, which is called untempered martensite. Untempered martensite is the hardest condition formed by a through hardening heat treatment. Nearly all the available alloying elements become locked within the untempered martensite crystal structure.

Every AISI designated steel is chemically different due to varying alloying elements. Again, the rules of solubility apply during heat treatment, because the elemental compounds formed originally when the steel is produced by the mill will determine the temperatures at which the steel fully transforms to austenite. Any remaining ferrite or BCC crystal structure in the steel will not be hardened or allotropically changed by the heating process unless it becomes heated to some temperature above the allotropic transformation temperature as noted in Chapter 7. All the microstructure in a steel must become austenite or FCC crystal structure before full hardening can be achieved.

The schematic shown in Figure 11.1 depicts the allotropic transformations of a 0.83 percent carbon steel (eutectoid steel) for both slow heating and cooling. Allotropic transformation to another crystal structure, BCC to FCC and FCC back to BCC, is solely dependent upon the interaction of carbon with iron. Iron has a fixed transformation temperature of 1666°F. The compound known as pearlite is a mixture of 0.83 percent carbon in the form of iron-carbide or cementite layers sandwiched between ferrite layers. Pearlite is really a mixture of layers of 6.67 percent iron-carbide and nearly pure iron. Pluck out and chemically analyze this pearlite grain and the total carbon content will measure 0.83 percent.

The phase diagram shown in Figure 11.2, shows an upper and lower critical temperature line. The temperature of the steel must be raised above the upper critical temperature line to homogenize for a period prior to quenching. Upon heating any steel with a carbon content below 0.083 percent, allotropic transformation does not occur until 1333°F is reached. The carbide layers in the pearlite begin to coalesce and become spheroidal in shape. When raised just above the lower critical temperature line, the iron-carbide compound begins to decompose into carbon and iron. As decomposition takes

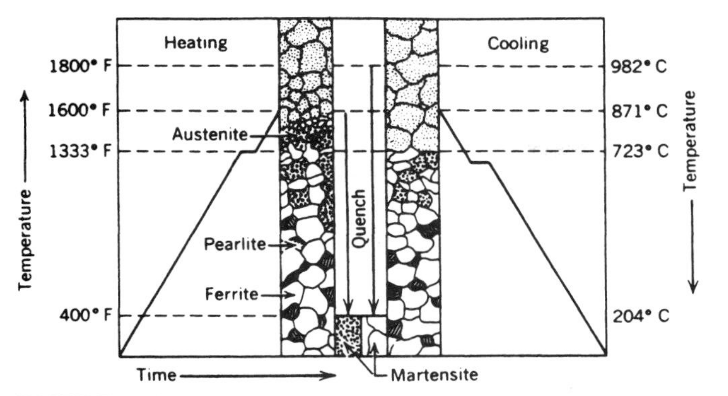

FIGURE 11.1. Critical temperature diagram of 0.83 percent carbon steel showing grain structures in heating and cooling cycles. Center section shows quenching from different temperatures and the resultant grain structure. (*Machine Tools and Machining Practices*)

place, the new mixture becomes a grain of FCC crystal structure. When the carbon content of the steel is 0.83 percent, all the metal will change into grains of FCC crystal structure. The temperature must be raised about 50° to stimulate the atoms into homogenization. Notice how the upper critical temperature line increases in temperature as the carbon content becomes less. Temperature increase is a function of solubility of iron atoms in the newly formed FCC austenitic grains. As the temperature of the steel increases, more BCC iron atoms are dissolved into the austenite grains. Upon absorption, the BCC iron transforms into FCC iron. Carbon is the most critical element alloyed into steel because of the influence on the temperature range in which allotropic transformation takes place.

Tool steels respond to the laws of solubility in the same manner as alloy steels. The only difference is, instead of ferrite, the amount of precipitated primary and secondary carbides present control the temperature at which full transformation to FCC austenite is complete. To understand the chemical reaction, think of the precipitated primary carbides in tool steels above the eutectoid carbon content of 0.83 percent for carbon content as requiring an increase in temperature above 1333°F to dissolve into the FCC austenite crystal structure. The primary carbides are a very fine and, most often, evenly distributed precipitated carbide. The secondary carbides, if present, are very large carbide chunks or grains in the microstructure. The secondary carbides do not completely dissolve in the austenite, only the immediate surface areas partially dissolve. This is why mill processing is very important to the overall strength of the tool steel. A quality tool steel must be free of any secondary carbide network—alloy segregation, necklacing, banding, and agglomeration. The iron-carbon phase diagram now becomes a simple tool which shows the effect carbon con-

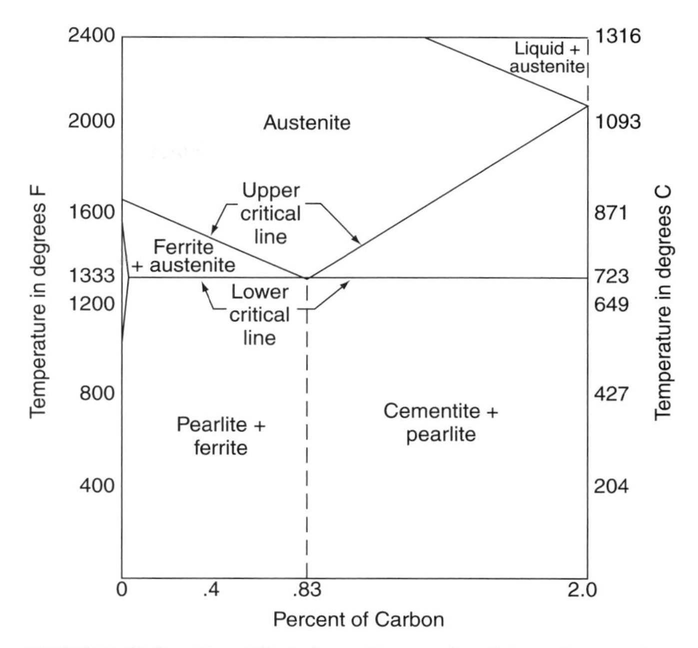

FIGURE 11.2. Simplified phase diagram for plain carbon steels.

tent solubility has on the temperature of allotropic transformation. Every steel is different.

Conditions and Preparations for Hardening Steels

Parts to be heat treated should be clean and free of machining oils, especially when heat treating finished machined parts. Consideration must be given to protecting the surface of finish machined parts from conditions which develop inside furnaces such as decarburization, alloy depletion, carburization, and high temperature oxidation. One of the best methods of surface protection is to plate the steel part with copper (Cu), which acts as a stop-off and is applied to the part by electrolytic plating and in some cases by an electroless plating method. Commercial products are also available for coating parts to protect the surface from exposure to the heat treating atmosphere and/or to mask specific areas of the part.

Reheat treatment or rehardening steels in the martensitic condition first requires a preheating at about 1200°F, followed by slow cooling, to allow the microstructure to fully stress relieve and spheroidize. Although heat treat vendors are familiar with this requirement, when missed, parts often crack during reheating.

Racking parts is very important, especially when performing surface hardening such as carburizing or nitriding. Parts must be positioned in the basket or fixture in such a manner as to allow the furnace atmosphere to surround and flow around the part in an even manner. When parts contact other parts, diffusion does not take place or is minimal, producing a part with an uneven surface hardness. Racking is also very important when the aspect ratio of parts is more than two to one. Steel parts tend to sag or bend when exposed to elevated temperature and must be racked or wired into a vertical position when austenitized. Heat treat vendors usually charge extra for this service.

Engineering Drawing Requirements

"Specificity" is the answer. If there is any confusion as to the requirements for heat treatment, the drawing and the drawing notes and specification requirements have precedence. The terms used in the heat treating industry are often confusing and parts submitted for heat treatment usually do not have the requirements clearly specified. When carburizing, for example, the manufacturer needs to indicate the following heat treat requirements directly on the purchase order.

Requirements:

1. Material Type and Specification.
2. Surface Hardness Requirement.
3. Effective Case Depth.
4. Core Hardness.
5. No Visible Retained Austenite Allowed in Case Microstructure.
6. Do Not Use Ammonia.
7. Mechanical Damage.
8. Method of Cleaning.
9. Method for Preservation / Shipping.

Written on Purchase Order:

1. Material is 93 10 alloy steel. (Specification).
2. 58-62 HRC.
3. 0.030" minimum at or above 50 HRC.
4. 30 HRC minimum.
5. No Visible Retained Austenite Allowed in Case Microstructure.
6. Do Not Use Ammonia.
7. Contact with other parts not allowed.
8. Glass bead clean to remove scale ok.
9. Oil to prevent corrosion / Your Carrier.

The purchase order to the heat treat vendor and the specifications indicated on the drawing should provide sufficient information for successful heat treatment. Any unusual requirements such as fixtures or racking should also be specified. The type of furnace is often a requirement; that is, heat-treat in a vacuum and nitrogen gas fan quench. In recent times, the tendency is to indicate the requirement and leave the method to achieve the requirement up to the vendor. This philosophy has produced an abundance of scrap. Clear and concise information annotated directly on the purchase order will eliminate any confusion about what is expected of heat treatment.

Choosing a Quenching Media

A multitude of quenching media are used in industry; however, the following are nine common quenching media listed in the order of the severity of the quench. The best guide to the selection of a quench media and the recommended procedure for the steel being heat treated can be found in the *Metals Handbook* and the *Heat Treater's Guide,* also published by the American Society for Metals.

1. Fused or liquid salts or metals
2. Brine (water and salt), that is, a solution of water and approximately 10 percent salt concentration
3. Tap water (~60°F)
4. Soluble oil and water mixtures
5. Synthetic liquid
6. Oil
7. Air
8. Furnace cool
9. Cold box refrigeration and/or liquid nitrogen

Selecting the "correct" quenching temperature for alloy steels is difficult at best. In hypoeutectoid steels, the correct quenching temperature is that which produces a fully martensitic core microstructure without evidence of proeutectoid ferrite at the grain boundaries. Recognition of proeutectoid ferrite is easy during optical microscopic examination of a part which has been carburized and the amount of proeutectoid ferrite present can be correlated with the temperature of the part just prior to quenching.

Liquid quenching media go through three stages. The **vapor blanket** stage occurs first because the metal is so hot it vaporizes the liquid. This envelops the metal with vapor, which insulates it from the cold liquid bath. This causes the cooling rate to be relatively slow during this stage. The **vapor transport cooling** stage begins when the vapor blanket collapses, allowing the liquid medium to contact the surface of the metal. The cooling rate is much higher during this stage. The **liquid cooling** stage begins when the metal surface reaches the boiling point of the quenching medium. Since boiling no longer occurs at this stage, heat must be removed by

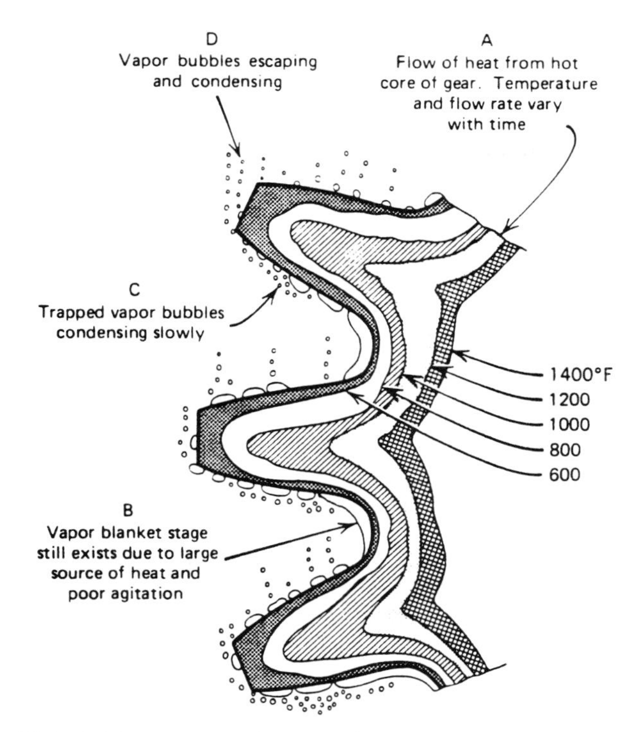

FIGURE 11.3. Stages of cooling. Temperature gradients and other major factors affecting the quenching of a gear. The gear was quenched edgewise in a liquid. (*By permission, from Metals Handbook, Volume 2, Copyright American Society for Metals, 1963*)

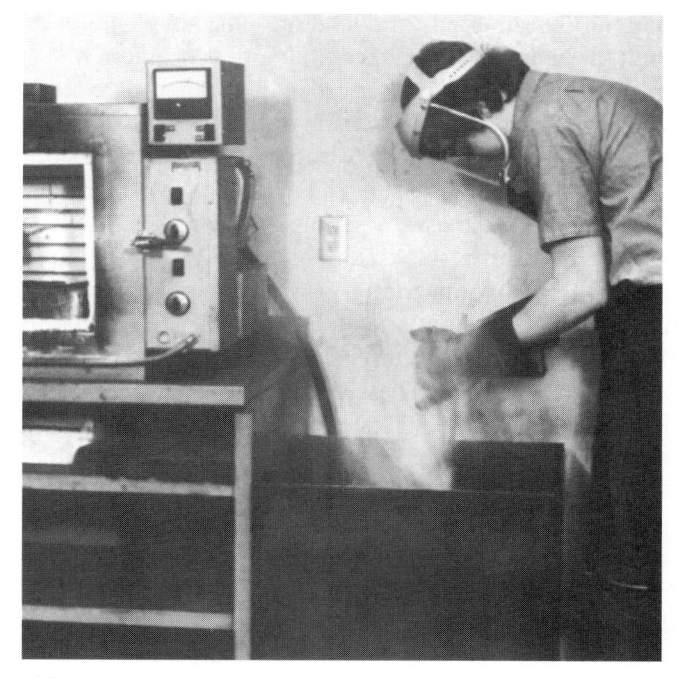

FIGURE 11.4. Beginning of quench. At this stage, heat treater could be burned by hot oil if not adequately protected with gloves and face shield. (*Lane Community College*)

conduction and convection. This is the slowest stage of cooling (Figure 11.3).

It is important in liquid quenching baths that either the quenching medium or the steel being quenched should be agitated. The vapor that forms around the part being quenched acts as an insulator and slows the cooling rate. This can result in incomplete or spotty hardening of the part. Agitating the part breaks up the vapor barrier. An up-and-down motion works best for long, slender parts held vertically in the quench. A figure-8 motion is sometimes used for heavier parts.

Gloves and face protection must be used in this operation for safety (Figure 11.4). Hot oil could splash up and burn the heat treater's face if he or she is not wearing a face shield.

Molten salt or lead is often used for isothermal quenching. The part is quenched at the austenitizing temperature in a molten bath, held about 500° to 600°F, and kept there until it is completely transformed. Austempered parts are superior in strength and quality to those produced by the two-stage process of quenching and tempering. The final austempered part is essentially a fine, lower bainite microstructure. As a rule, only parts that are thin in cross section are austempered; a common example is the power lawn mower blade or a shovel.

In martempering, the part is quenched in a lead or salt bath about 400°F until the outer and inner parts of the material are brought to the same uniform temperature. The part is next quenched below 200°F to transform all the austenite to martensite. Tempering is then carried out in the conventional manner.

Ramp-Time

Ramp-time is the time it takes for a part to achieve temperature equilibrium; meaning, surface and core are at the same temperature. One of the most important rules for successful heat treatment is *time at temperature before transformation takes place.* This rule simply means the part must be at the prescribed temperature for the required time for the complete transformation to occur. Likewise, when quenching, the part must be cooled at the correct rate and to the required temperature before complete transformation of austenite to untempered martensite takes place. The horror in heat treatment is mixed microstructures, because it is impossible to predict the mechanical properties or behavior for steels with them. An example of this condition can be observed when heat treatment baskets are overloaded with parts, and parts at the center of the load never quite reach the temperature where allotropic transformation to austenite takes place. When this basket load of parts is quenched, the parts on the surface become full hardened, while the parts in the middle of the load are not. All of the parts look the same, which makes finding this problem rather difficult until a part fails in service. Another example is when quenching is not complete for the specific steel and the room temperature microstructure becomes a mixture of untempered martensite and retained austenite. Retained austenite is a very serious problem because it is known to be a metastable compound; that is, it will transform without forewarning to untempered martensite upon applied thermal and/or mechanical stress (Figure 11.5). When transformation occurs, the part cracks, usually when in service. These problems can be avoided by following the exact prescription of recommended heat treat practice and by paying strict attention to ramp-time.

It is a common practice, for easy accessibility, to place a brick on the floor of the furnace to raise the part to be heat treated. This practice can quickly ruin the furnace lining if refractory chemistry is not understood. Silicon dioxide (SiO_2) from which ordinary firebrick is made is quite stable at room temperatures, but at hardening temperatures at red heat and above, it is a very active reagent when in the presence of a base material or oxides of metals. Thus, when a firebrick made of silicon dioxide is in contact with a furnace basic lining of dolomite (MgO and CaO) and heated, the lining will dissolve and be "eaten away." The reverse is also true when the furnace that has an "acid" lining of ordinary firebrick comes into contact with a basic substance such as ordinary mill scale (Fe_3O_4 or FeO) or a dolomite brick; the lining of the furnace will dissolve. The terms "acid steel" and "basic steel" in steel manufacture refer to this very thing: the kind of lining of furnaces and its relation to the slag, whether it is acid or basic.

TEMPERING

Tempering, sometimes called drawing, is the process of reheating hardened martensitic steels to some predetermined temperature below the lower critical temperature or Ac_1 line. Proper tempering of a hardened steel requires soaking at the tempering temperature for a certain length of time. The rule of thumb is to temper between 1-1/2 to 3 hours at the tempering temperature followed by cooling to room temperature, which is called "one temper cycle." With any selected full hardened steel in the as-quenched condition, there is a corresponding temper temperature which produces softening to a given hardness (Table 11.1).

Some alloy steels, when cooled slowly after being tempered between 800°F and 1200°F develop a condition known as *temper brittleness,* with a noticeable loss in notch toughness. In simple terms, the steel becomes

TABLE 11.1. Tempering temperature versus hardness HRC for AISI 4340 alloy steel

	C	Mn	P	S	Ni	Cr	Mo
Chemistry	**.38/.43**	**.60/.80**	**.040**	**.040**	**1.65/2.00**	**.70/.90**	**.20/.30**

Temperature (°F)	Hardness (HRC)
400	55
500	52
600	50
700	47–48
800	45
900	42–43
1000	38–39
1100	34

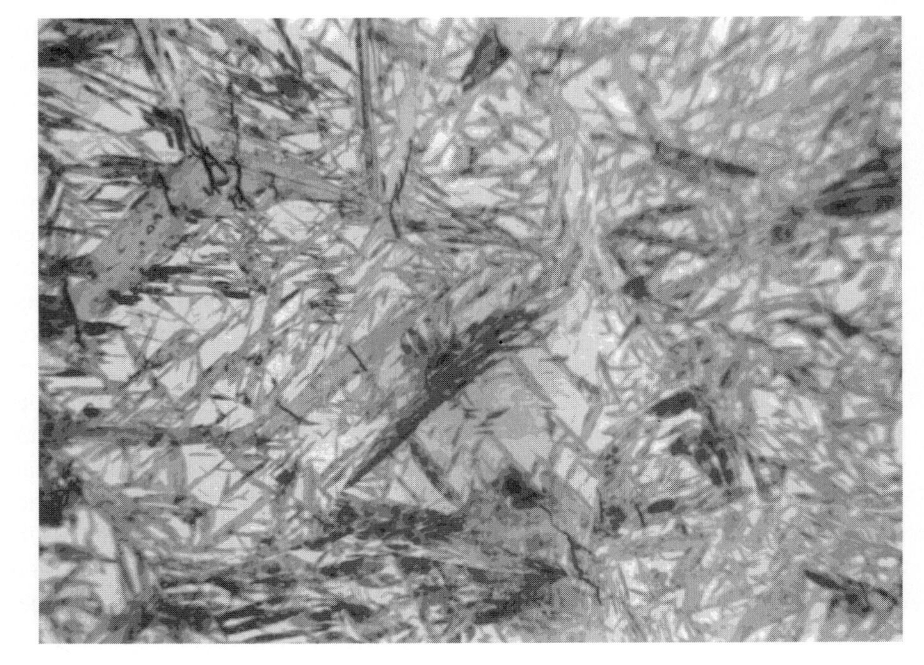

FIGURE 11.5. Retained austenite in a carburized case, white triangles. Microcracking has occurred due to some transformation to untempered martensite.

more fracture sensitive and can crack more easily by a striking blow (impact toughness). Steels that are sensitive to this condition should be quenched from the tempering to eliminate this problem.

Special Tempering

Steels that contain nickel as an alloying element are particularly difficult to temper correctly, because the austenite begins to transform to untempered martensite at a lower temperature Ms, and the Mf is correspondingly lower. In fact, the Mf may be below room temperature; room temperature in a heat treat facility may be as high as 135°F. This means the austenite is never fully or completely transformed to untempered martensite and a percentage remains as retained, mixed with the untempered martensite. Double and even triple tempering in conjunction with cryogenic quenching is a common method used to convert retained austenite to untempered martensite and untempered martensite to tempered martensite. Keep in mind, austenite is a metastable FCC crystal structure; it can be transformed by thermal and/or mechanical stress to BCC untempered martensite. A coarse martensitic grain structure can be seen in Figure 11.6*a,* as compared with a fine martensitic grain structure shown in Figure 11.6*b.*

For a steel to be strong and exhibit predictable mechanical properties, the austenite must be fully transformed to untempered martensite, and the untempered martensite must be tempered to become tempered martensite. Triple tempering is often used to convert retained austenite to martensite. The following describes the transformation

***FIGURE 11.6*a.** The coarse needlelike microstructure of martensite can easily be distinguished from retained austenite. This martensite is typical for low carbon and alloy steels.

***FIGURE 11.6*b.** The very fine needlelike microstructure of martensite as seen in some high alloy steels.

process when the austenite has not transformed completely to untempered martensite. The first tempering only transforms the untempered martensite. The retained austenite is affected by the tempering operation and some of the retained austenite transforms to untempered martensite upon return to room temperature. The steel microstructure, after the first tempering cycle, is composed of tempered martensite, retained austenite, and untempered martensite. The second tempering operation causes the tempered martensite to lose some hardness and gain toughness; the untempered martensite now becomes tempered and is at a slightly higher hardness than the first tempered martensite and a portion of the remaining retained austenite transforms to untempered martensite upon cooling to room temperature. The microstructure, in order of quantity after the second tempering, consists of twice tempered martensite, retained austenite, first tempered martensite, and untempered martensite. Each martensite is a slightly different hardness and mechanical strength, mixed with retained austenite. The third tempering is performed to eliminate all remaining retained austenite; but does it, and what about the different tempered martensites?

The third tempering cycle causes more of the retained austenite to transform to untempered martensite and the martensites are again tempered. Interestingly, some of the steels processed in this manner never get below the detectable limit (greater than 5 percent) to visually see retained austenite in the optical microscope, regardless of magnification. Double or triple tempering should never be used as a means to eliminate retained austenite. The only sure way to eliminate retained austenite from the microstructure of hardened steel is to force all the austenite to transform to untempered martensite during the initial austenitic quench.

Steels with a Mf at or below room temperature should be quenched and tempered in a continuous manner. "Continuous manner" means never to allow the metal temperature to stop dropping and become stabilized. Quenching must include additional cooling capability after oil quenching, such as cold water, refrigeration at −100°F, and/or cooling in liquid nitrogen at −300°F. The secret in transforming all the austenite to untempered martensite on the initial quench follows a basic law of physics: A body in motion tends to stay in motion. It takes little energy to continue moving a body and a great deal of energy to restart the process. This law holds true for transforming microstructure; austenite transformation if stopped for any reason, becomes stabilized as retained austenite. Steels which tend to retain austenite, especially those alloyed with nickel and those which are carburized, need particular attention to performing a continuous quenching to a sufficiently low temperature. It is almost impossible to remove retained austenite from a heat treated and tempered part once the microstructure has become stabilized.

Retained austenite in the microstructure of steels is particularly harmful and often deadly. Without forewarning and usually when the part is in operation, allotropic transformation can occur. Retained austenite is a FCC crystal structure containing 14 atoms in a unit cell. Tempered martensite is a much more complicated crystal structure, and transformation may occur as a body-centered-tetragonal, or a hexagonal, unit cell (see Tables 11.2 and 11.3). Retained austenite is metastable, meaning it can transform to untempered martensite when thermal and/or mechanical stress is applied. Transformation to untempered martensite(s) severely overstresses the microstructure which may crack without forewarning to relieve those stresses.

A part can be tempered in a furnace or oven by bringing it to the required temperature and holding it there for a length of time, then cooling it in air or water. Some tool steels should be cooled rapidly after tempering to avoid temper brittleness. Small parts are often tempered in

TABLE 11.2. Reactions in steel during tempering of martensite

Tempering-temperature range, °F	Reaction	Remarks
77–212	Carbon segregation to dislocations; precipitation clustering.	Clustering predominant in high-carbon steels.
212–482	Epsilon-carbide precipitation. (Sometimes called first stage.)	May be absent in low-carbon, low-alloy steels.
392–572	Retained austenite transforms to bainite. (Sometimes called second stage.)	Occurs only in medium-carbon and high-carbon steels.
482–662	Lathlike Fe_3C precipitation. (Sometimes called third stage.)	Hägg carbide may form in high-carbon steels.
752–1112	Recovery of dislocation substructure. Lathlike Fe_3C agglomerates to form spheroidal Fe_3C.	Acicular fine-grained ferrite structure maintained.
932–1112	Formation of alloy carbides. (Called secondary hardening or fourth stage.)	Occurs only in steels containing Ti, Cr, Mo, V, Cb or W; Fe_3C may dissolve.
1112–1292	Recrystallization and grain growth; coarsening of spheroidal Fe_3C.	Recrystallization inhibited in medium-carbon and high-carbon steels; equiaxed ferrite formed.

TABLE 11.3. Carbides precipitated in steel during tempering of martensite

Carbide	Shape	Crystal structure (and prototype)	Typical solute content (wt %)	Temperature of formation, °F
$Fe_{2-3}C$ (*e*-carbide)	Plates	hcp (*e*-Fe_3N, Ni_3N or Co_2C)	Greater than 0.2C	212–482
Fe_5C_2 (Hägg carbide)	. . .	Monoclinic (Pd_5B_2 or Mn_5C_2)	Greater than 0.8C	392–662
Fe_3C (cementite)	Laths→ spheres	Orthorhombic (M_3C)	All carbon contents	482–1292
CbC	. . .	Cubic (NaCl)	2Cb-0.2C	1022
VC-V_4C_3	Plates	Cubic (NaCl)	2V-0.2C	1022
Mo_2C	. . .	Orthorhombic (ζ-Fe_2N) subcell of hcp	4 Mo-0.2C	1022
W_2C	Needles	Orthorhombic (ζ-Fe_2N) subcell of hcp	6W-0.2C	1112
Cr_7C_3	Spheres	Hexagonal (M_7C_3)	4Cr-0.14C	1022
$Cr_{23}C_6$	Plates	Cubic ($M_{23}C_6$)	10Cr-0.2C	1292
Fe_3Mo_3C	. . .	Cubic (η_1-M_6C)	4Mo.0.2C	1292
Fe_3W_3C	Spheres	Cubic (η_1-M_6C)	6W-0.2C	1292

liquid baths such as oil, salt, or metals. Specially prepared oils that do not ignite easily can be heated to the tempering temperature. Lead and various salts are used for tempering since they have a low melting temperature.

When there are no facilities to harden and temper a tool with controlled temperatures, tempering by color is done. The oxide color used as a guide in such tempering will form correctly on steel only if it is polished to the bare metal and is free from any oil or fingerprints. An oxy-acetylene torch, a steel hot plate, or an electric hot plate can be used. If the part is quite small, a steel plate is heated from the underside, while the part is placed on top. Larger parts such as chisels and punches can be heated on an electric plate until the needed color shows, then cooled in water. See Table 11.4 for oxide colors and temperatures.

When grinding carbon steel tools, if the edge is heated enough to produce a color, you have in effect retempered the edge. If the temperature reached was above that of the original temper, the tool has become softer than it was before you began sharpening it. Table 11.5 on page 157 gives the hardnesses of various tools as related to their oxide colors and the temperature at which they form.

Fracture Toughness and Tempering

Some steels remain brittle and behave in a fracture-sensitive manner (e.g., 4340 alloy steel). Fracture toughness is an important property which can be enhanced by the tempering process. Think of fracture toughness as fracture sensitivity of a product. If a 4340 alloy steel item is subjected to impact pounding, the heat treating process may be modified significantly to obtain enhanced properties. However, for optimum and reproducible results, the process of heat treating must be performed in a continuous manner and without process interruption. The military specification, MIL-H-6875, recommends a double tempering and sometimes a triple tempering to reduce retained austenite and increase fracture toughness. It is best to eliminate retained austenite by completing the transformation process to untempered martensite. Some steels are sluggish to transform and others do not complete transformation until they are cooled to room temperature, meaning approximately 70°F. It is almost impossible to reach this cool temperature in a heat treat facility where furnaces maintain the temperature of the facility at about 120°F or more. Cryogenic quenching is a way to resolve the problem of continuous cooling to fully transform the austenite to 100 percent untempered martensite—so out of the quench oil and into the freezer at about –100°F for CO_2 (dry ice) and/or at –300°F (liquid nitrogen). These cold quenches, when applied as part of the quench cycle, will aid in complete transformation; however, most heat treat vendors will recommend a double or triple tempering. Double and/or triple tempering is the heating of the part to the prescribed tempering temperature for a period of about 2 hours at temperature followed by cooling to room temperature in air. Upon reaching room temperature, the part is immediately returned to the tempering furnace for an additional tempering. This cycle is repeated for triple tempering. The result of these additional treatments is a part with superior wear and fracture toughness characteristics. In addition, as long as the tempering temperature is not exceeded, the steel receiving additional temper cycles will not appreciably change in hardness.

Heat Treating Tool Steels

Tool steels are highly alloyed steels, intended primarily for use as plastic molding, extrusion, forging, die-casting, and roll cutting dies. Tool steels are normally

TABLE 11.4. Temperatures and colors for heating and tempering steel

	Colors	Fahrenheit	Process	
	White	2500°		
		2400°	High-speed steel hardening (2250°–2400°F)	
	Yellow white	2200°		
		2300°		
		2100°		
	Yellow	2000°		
		1900°		
	Orange red	1800°	Alloy steel hardening (1450°–1950° F)	
Heat colors		1700°		
		1600°		
	Light cherry red			
		1500°		
	Cherry red	1400°	Carbon steel hardening (1350°–1550°F)	
		1300°		
	Dark red			
		1200°		
		1100°		
	Very dark red	1000°		
		900°		
	Black red in dull light or darkness	800°	Carbon steel tempering (300°–1050°F)	High-speed steel tempering (350°–1100°F)
		700°		
	Pale blue (590°F)	600°		
Temper colors	Violet (545°F) Purple (525°F)	500°		
	Yellowish brown (490°F) Straw (465°F) Light straw (425°F)	400°		
		300°		
		200°		
		100°		
		0°		

Source: Pacific Quality Steels, "Stock List and Reference Book," No. 75, Pacific Machinery and Tool Steel Company.

processed by the mill in the annealed or soft condition and are subsequently hardened by heat treatment in the finished machined condition. The principle alloying elements in tool steels are chromium (Cr), tungsten (W), molybdenum (Mo), and vanadium (V). These elements readily form carbides which are very hard and wear resistant. Cobalt (Co) and nickel (Ni) are also added to tool steels, but these elements do not form carbides. Cobalt is added to improve red hardness in high-speed steels, and resistance to softening at elevated temperatures. Nickel is added to improve the through hardening properties of the steel.

Hardening alloy steels involves raising the temperature sufficiently above the upper critical temperature line (iron-carbon phase diagram), to allow all the carbides to be dissolved in the newly formed austenite matrix. Once the matrix becomes homogenized with an even distribution of alloying elements, the steel is quenched. The act of rapid cooldown removes the time factor required for the alloying elements to reform into carbides; thus, they are retained and trapped within the crystal structure during allotropic transformation. The result is a highly microstressed and distorted crystal structure called martensite.

TABLE 11.5. Temper color chart

Degrees °C	Degrees °F	Oxide Color	Suggested Uses for Carbon Tool Steels
220	425	Light straw	Steel-cutting tools, files, and paper cutters
240	462	Dark straw	Punches, dies
258	490	Gold	Shear blades, hammer faces, center punches, and cold chisels
260	500	Purple	Axes, wood-cutting tools, and striking faces of tools
282	540	Violet	Springs, screwdrivers
304	580	Pale blue	Springs
327	620	Steel gray	Cannot be used for cutting tools

SOFTER——HARDER

Tool steels have an abundance of carbide-forming compounds which respond to allotropic transformation and thermal homogenization in the same manner as alloy steels. Tool steels differ due to the initial formation and distribution of carbides as produced by the mill. Not all the carbides are dissolved and homogenized during the hardening heat treatment of tool steels; carbides that do not dissolve are called "primary" and those that dissolve are called "secondary" carbides. It is necessary for the tool steel to be sufficiently mill processed by thermal mechanical reduction to physically break up and evenly distribute the primary carbides within the grain structure. When primary carbides are not sufficiently distributed, the grain microstructure will almost always contain a massive grain boundary carbide formation (necklace effect) which robs the surrounding matrix of mechanical property sharing between the matrix and the primary carbide particles. Compare the microstructures shown in Figures 11.7 and 11.8. Carbide necklacing is the result of insufficient hot processing of the ingot. Mechanical strength and overall performance of the tool steel is dependent upon initial mill hot processing and a correct heat treatment.

Tool steels are often stress relieved at 1000°F to 1200°F to remove the effects of rough machining and prior to finish machining. Failure to stress relieve can cause release of internal stresses which result in severe distortion of the heat treated part. Tool steels must be heated slowly to the austenitizing temperature. Slow heating allows the core to heat close to the same rate as the surface metal, minimizing the possibility of cracking during heating. The finish machined tool steels must also be protected from the effects of decarburizing, high temperature oxidation, and surface scaling during heat treatment. Modern vacuum furnaces provide the necessary protective environment and with the additional advantage of a moving heat zone, thermal distortion is minimized (Figure 11.9). Nitrogen and/or argon high-pressure gas fan quenching is employed in the vacuum furnace for a rapid and even quenching of the load. Batch type furnaces with protective atmospheres are also used for hardening tool steels, along with salt bath furnaces and/or fluidized bed furnaces where the tool is hung or immersed directly in the furnace media and removed for quenching by hand.

The hardening temperatures for tool steels is dependent upon the thickness of the part. When the part reaches thermal equilibrium, core and surface reach the temperature required for austenitizing, and the clock begins for the amount of soaking time required—usually about 30 minutes. The exception is in the austenitizing of high-speed tool steels and/or when heating is by

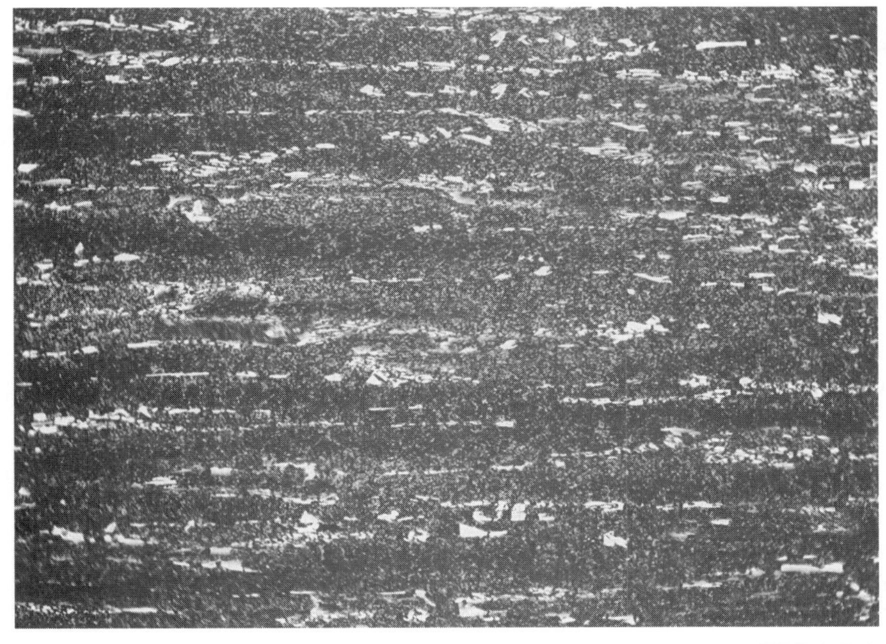

100× Vilella's Etch

FIGURE 11.7. Acceptable carbide distribution in D2 tool steel.

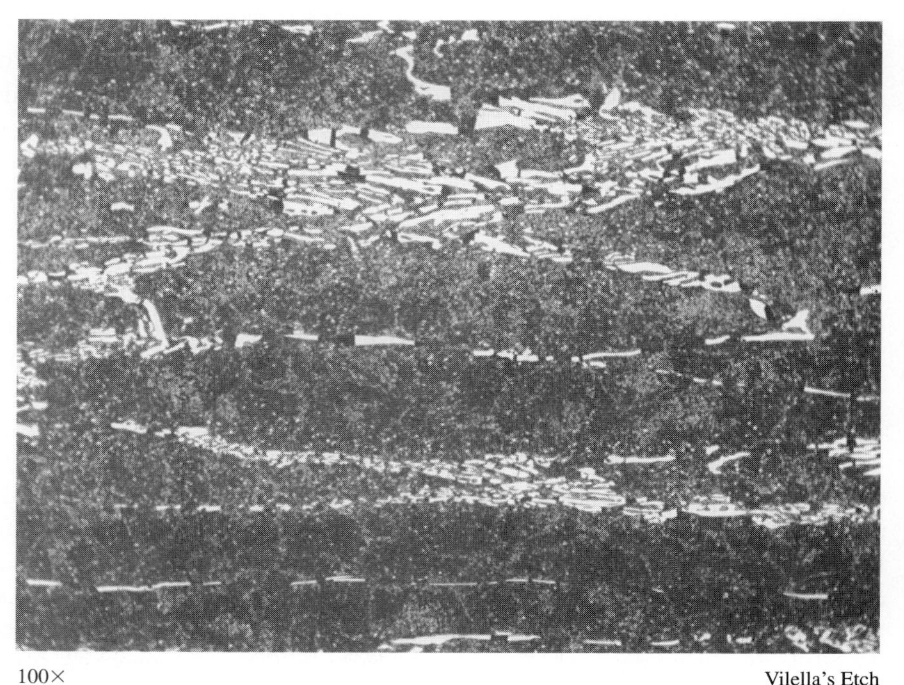

FIGURE 11.8. Unacceptable carbide distribution (necklacing) in D2 tool steel.

FIGURE 11.9. Vacuum furnace

immersion in a salt bath; only a few minutes at temperature are required.

The quenching rate for tool steels must be rapid enough for full transformation to untempered martensite without cracking and severely distorting the part. Air-hardening tool steels are usually placed on refractory brick and allowed to cool in still air. Fan-assisted air quenching is considered a more vigorous method. Oil quenching should be used for low alloyed steels or when water quenching creates problems with cracking and/or distortion. The *Heat Treater's Guide,* by ASM, lists most of the common tool steels and offers suggestions for the choice of quench media.

Tool steels should be tempered immediately after quenching or while the tool is still warm (120° to 160°F). A special "hot cabinet" is sometimes used to prevent the tool from further cooling while awaiting tempering. The tempering treatment reduces hardness while increasing toughness. Two tempers are recommended for most tool steels; however, three are considered necessary for tempering high-speed steels because retained austenite is often sluggish to transform to untempered martensite. The holding time required for tempering tool steels is a minimum of 2 hours at temperature per each temper cycle. Remember, the objective of steel heat treatment is "microstructure homogenization with 100 percent austenite conversion to martensite." Heat treatments for tool steels are shown in Table 11.6.

Many tool steels are used in applications where fracture toughness is a prime concern. Extrusion dies are often made from air hardening H-13 tool steel. The general heat treat practice is to austenitize at between 1825°F and 1905°F followed by quenching in air. Air quenching of H-13 dies is often too slow of a quench rate to produce a fine grain fracture tough martensitic microstructure. The result is a coarse grain microstructure which behaves in a brittle manner. The North American Die Casting Association (NADCA) has published "Acceptance References Annealed H-13 Steel Microstructures 1990," for determining acceptable and not acceptable annealed microstructures. Many of the annealed microstructures appear identical to the quench hardened microstructures of H-13 hot work tool steels

TABLE 11.6. Heat treating temperatures and procedures for various tool steels

Steel	Normalizing (°F)	Preheat (°F)	Austenitize (°F)	Light Sections Holding Time (minutes)	Quenching Medium	Approximate Hardness (RC)
Water-hardening steels W1 through W5	1500 (0.6 to 0.9 C) 1600–1900 (0.9 to 1.5 C)	Stress relief for cold-worked parts, large sections 1200	Uniformly heated through 1450–1525	15	Water (or oil for thin sections)	65–67
Oil-hardening steels						
01 through 07	1600–1650	1200	1475–1575	15	Oil	63–66
Shock-resistant tool steels						
S1	Not recommended	1200	1650–1750	15–45	Oil	60–61
S2, S3, S4	Not recommended	1200–1400	1650–1750	5–20	Brine, water	60–62
S5, S6	Not recommended	1200–1400	1650–1750	5–20	Brine, oil, water	60–62
S7	Not recommended	1200	1650–1750	15–45	Air, oil	60–61
Air-hardening cold-work steels						
A2	Not recommended	1200–1450	1700–1850	20–40	Air	62–65
A4	Not recommended	1200–1450	1500–1600	15–60	Air	62–65
A5	Not recommended	1200–1450	1450–1550	15–60	Air	62–65
A7	Not recommended	1500	1750–1800	30–60	Air	62–65
A8, A9	Not recommended	1500	1800–1875	20–40	Air	62–65
A10	1450	1200	1450–1500	30–60	Air	62–65
High carbon, high chromium cold-work steels						
D1, D2, D4, D5, D6	Not recommended	1500	1700–1850	15–40	Air	61–65
D3	Not recommended	1500	1700–1850	15–40	Oil	61–65
D7	Not recommended	1500	1850–1950	15–40	Air	61–65
Chromium hot-work steels						
H10, H11, H12,	Not recommended	1550–1650	1825–1900	15–40	Air	55–59
H13, H14, H16, H19	Not recommended	1550–1650	2000–2200	5	Air	55–59
Tungsten hot-work steels						
H20 through H25	Not recommended	1500	2000–2300	5	Air, oil	48–57
H26	Not recommended	1500	2000–2300	5	Air, oil	63–64
Molybdenum hot-work steels						
H41 through H43	Not recommended	1350–1550	2000–2200	5	Air, oil, molten salt	54–66
Tungsten high-speed tool steels						
T1 through T9, T15	Not recommended	1500–1600	2200–2300	2–5	Air, oil	63–67
Molybdenum high-speed steels						
M1 through M7, M10, M15, M30, M33, M34, M35, M36, M41, M42, M43, M44	Not recommended	1350–1550	2150–2250	2–5	Air, oil, molten salt	63–66

Note: Preheat and austenitizing temperatures vary slightly for each numbered tool steel. Manufacturers' recommendations should also be consulted prior to heat treating the steels.

published by ASM; *The Metals Handbooks, Volumes 7 and 9; and the Heat Treater's Guide.* The interpretation of these microstructures is quite controversial. The cooling rate of the mill H-13 product will usually produce a martensitic microstructure. The mill will sub-critical anneal the material for many hours, releasing carbon from the martensite. The free carbon combines with iron atoms to form iron-carbide which transforms into a precipitated spheroidal shape during sub-critical annealing. The appearance of the martensite may not change appreciably due to the small size of the spheroidal carbide. The only way to determine if the microstructure has been sub-critical annealed or remains martensitic is by hardness testing.

The quench rate provided for cooling H-13 from the austenitizing temperature is directly related to the property of fracture toughness. When H-13 is cooled too slowly, usually due to part mass, the resulting microstructure becomes a mixture of very coarse and fine grains. This microstructure is very fracture sensitive or brittle. Often, dies never used will crack while awaiting shipment following hardening and tempering heat treatments. An H-13 heat treat study correlating fracture toughness testing (charpy specimens) with various quench rates was conducted for the purpose of establishing standards.(1) The results of this

(1) Mr. Doehler Jarvis, Metallurgical Consulting, Marcellus, New York 13108

study show a direct relationship with quench rate and the formation of a homogeneous tempered martensite microstructure, with a fracture toughness acceptance cutoff established at rates producing "less than 60 percent of maximum fracture toughness."[1] Acceptable microstructures require attention and modification of the cooling rate for H-13 parts of different thickness. The choice of heat treat practice for individual H-13 parts should be based upon the tendency to crack as a result of the heat treat practice, and the end item use requirements of the part. Once determined, the requirements for heat treatment should be formalized by means of a heat treatment processing specification.

Distortion in Heat Treating Steels

Distortion in the heat treatment of steels occurs due to thermal expansion and/or as a result of allotropic transformation. Manufacturers of precision parts are always concerned about controlling distortion and view distortion as a nemesis to part reproducibility. The problem with thermal distortion in hardening steels is complicated by the way parts are "racked" into baskets and how evenly these parts are heated. The shape of the part is very important because irregular-shaped parts tend to distort more readily. Parts heavily machined without being subjected to a stress relief heat treatment prior to austenitizing tend to severely distort during initial heating. Preheating and soaking parts at about 1200°F can help minimize distortion when allotropic transformation begins forming austenite.

Distortion should be expected. Distortion should be determined by heat treating a test sample which has been dimensionally measured prior to heat treatment. Remeasurement of this part in the hardened condition will show the areas where dimensional changes have effected the part. If all the conditions of heat treatment can be reproduced, dimensional changes measured on the test part will reflect actual dimensional changes caused by the heat treat process. Part dimensional adjustments can be made accordingly.

Distortion can be minimized by a simple and symmetrical design; by stress relieving to remove machining stresses prior to hardening; by heating slowly during austenitizing; by quenching as slow as possible, but quick enough to fully transform to the desired microstructure; and by tempering at a suitable temperature as recommended to achieve the temper hardness required for the steel. The *Metals Handbook* (9th ed., vol. 4) is an excellent source for determining generic distortion in steels.

Heat Treatment of Cast Irons

Cast irons may be compared with steels in their reaction to hardening; however, except for gray cast iron, because of their alloy content, cast irons require much higher austenitizing temperatures. Silicon has a major effect on lowering the temperature of graphitizing and retarding the reabsorption of graphite during austenitizing. The response to hardening in cast irons is dependent upon the carbon and alloy content of the matrix. Higher hardness is achieved when higher austenitizing temperatures are used and more carbon is dissolved in the austenite prior to quenching. Cast irons are generally more sensitive than steels to heating and care must be taken to heat cast iron uniformly. Air, controlled atmosphere, and molten salt are used for hardening cast irons.

Gray cast irons are usually subjected to three types of annealing heat treatments. These are ferritizing, full, and graphitizing annealing. Ferritizing annealing is used when it is desired to convert the carbide in the pearlitic matrix into ferrite and graphite. Ferritizing annealing is performed at a temperature range of 1300° to 1400°F. Full annealing is usually performed at a temperature range between 1450°F and 1650°F, causing the carbide to spheroidize rather than to decompose into ferrite and graphite. Graphitizing annealing is used to convert the gray cast iron to graphite and a pearlite matrix. Normalizing is used to restore gray cast irons to the original microstructure and mechanical properties following other heat treatments and after welding. When subjected to hardening and tempering, gray cast irons exhibit wear resistance five times greater than when in the pearlitic microstructure condition. Austempering, martempering, flame, and induction hardening may also be used as hardening heat treatments.

Ductile or nodular cast iron is most often heat treated to produce maximum ductility and good machinability by converting the microstructure to ferrite and spheroidal graphite.

Malleable cast iron pearlitic and ferritic microstructures are both produced by annealing white cast iron. The annealing procedure involves three steps: The first causes neuleation of graphite at a high temperature; the second step or first-stage graphitization consists of holding at a temperature range of 1650° to 1780°F, which causes dissolution of the massive carbides; and lastly, the third stage of the annealing process or second-stage graphitization consists of slow cooling through the allotropic transformation range of iron. Quenching from first-stage graphitization temperature will produce a hardened martensitic matrix.

SELF-EVALUATION

1. If you heated AISI-C1080 steel to 1200°F and quenched it in water, what would be the result?
2. If you heated AISI-C1020 steel to 1500°F and quenched it in water, what would happen?
3. List as many problems encountered with water-hardening steels as you can think of.
4. Name some advantages of using air- and oil-hardening tool steels.
5. What is the correct temperature for quenching AISI-C1095 tool steel? For any carbon steel?
6. Why is steel tempered after it is hardened?
7. What factors should you consider when you choose the tempering temperature for a tool?
8. The approximate temperature for tempering a center punch should be _____. The oxide color would be_____.
9. If a cold chisel became blue when the edge was ground on an abrasive wheel, to approximately what temperature was it raised? How would this temperature affect the tool?
10. How soon after hardening should you temper a part?
11. Name the three major factors on which transformation to martensite depends.

CASE PROBLEM: THE SPLITTING VISE BODY

A student had spent many hours making a precision vise body of SAE 4140 steel. He followed the correct procedures for hardening the part and quenching it in oil. It could have cracked in that operation had he not done it right, but it did not crack. It was near the end of his class period, so he decided to wait until the next day to temper it. Unfortunately, the next day it had a large crack down one side, ruining it. What caused this to happen? How could he have avoided it?

WORKSHEET 1

Objective Harden a piece of tool steel in the form of any tool or part, such as a punch or a chisel that has been previously forged.

Materials An electric or gas furnace, tongs, quenching media (oil or water), and safety equipment (face shield and gloves).

Procedure

1. Assuming the part is a small tool such as a center punch, it should be placed in a furnace that has already been brought up to the correct temperature.
2. Determine the best place to grip the part with the tongs so that it will not be damaged where it is red hot. Use the properly shaped tongs.
3. Make sure you can grasp the part in the furnace to remove it in the proper orientation to enable you to quench it straight in.
4. Heat the end of the tongs so they will not remove heat from the part.

5. When the piece has become the same color as the furnace bricks, remove it and immediately quench it **completely under** in the bath.
6. Agitate it up and down or in a figure-8 motion.
7. Be sure it has cooled below 200°F before you remove it from the quench.

Note If no furnace is available, a torch may be used with a temperature chart. See Table 11.4.

Conclusion Do you think the part got hard? Test with an old file in an inconspicuous place. If a hardness tester is available, a quick test could be taken. For information on hardness testing, see Chapter 26 on Rockwell and Brinell hardness testers.

Is the part as hard as it should be? If not, check with your instructor.

WORKSHEET 2

Objective Temper a part that has just been hardened.

Materials An electric or gas furnace or oxyacetylene torch, fine abrasive cloth, tongs, and safety equipment.

Procedure for Tempering in the Furnace

1. Polish all smooth surfaces of the part with abrasive cloth and remove all oil. A cold furnace should be brought up to the correct temperature.
2. Small parts are then placed in the furnace for about 15 minutes.
3. Remove and cool in air.
4. When this method is used, the striking end of punches, chisels, and other striking tools should be further heated with a torch until they are a blue color on that end. This is done to ensure the safety of the user by preventing the struck end from shattering, since it is softer when tempered to blue.

Procedure for Tempering with a Torch or Hot Plate

1. When using a torch or hot plate, make sure that heat is applied to the body of a punch or a part of the tool that can be softer. Allow the heat to travel slowly out to the cutting edge. This way the colors may be observed.
2. See that the striking end of a tool is blue or gray before the proper color arrives at the cutting end.
3. When the proper color has arrived, quickly cool the piece in water. **Do not delay** or it will be overtempered.

Conclusion

1. Are the colors right according to Table 11.4?
2. If possible, recheck the hardness and compare with Figure 9.9.
3. Leave the temper colors on your project so your instructor can evaluate it. Turn it in for grading.

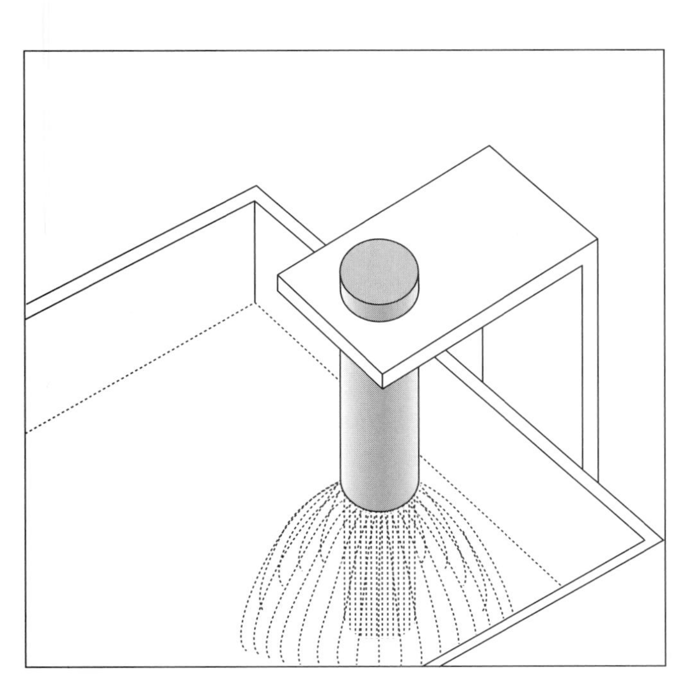

The Jominy Test

CHAPTER 12

Hardenability of Steels

If all hardened parts were less than 1/2 in. thick, then water-hardening plain carbon steels would be sufficient for most purposes. However, many design applications call for the use of large sections of high-strength steels. This requires the designer to specify deep-hardening or high hardenability steels. The difference in hardenability of tool steels is explained and demonstrated in this chapter.

OBJECTIVES **After completing this chapter, you should be able to:**

1. Explain the methods of determining and evaluating the depth of hardening (hardenability) of various steels.
2. Demonstrate and measure the hardenability of a shallow-hardening steel.
3. Demonstrate the use of a mechanical properties chart for predicting the hardness and strength of a hardened and tempered specimen.

HARDENABILITY TESTING AND THE JOMINY END-QUENCH TEST

The Jominy end-quench test is used to determine the depth of hardening or hardenability of various types of steels. In conducting this test, a 1 in. diameter round specimen approximately 4 in. long is heated uniformly to the correct austenitizing temperature for the length of time needed to effect complete austenitization (diffusion of carbon in the austenite). The specimen is quickly removed and placed in a bracket in such a way that a jet of water (or other quenching medium) at room temperature impinges on the bottom face of the hot specimen without wetting the sides (Figure 12.1). It is allowed to remain on the water jet until the entire specimen has cooled. After it has cooled, longitudinal flat surfaces are ground on the side to remove decarburization, and Rockwell C scale readings are taken at 1/16 in. intervals from the quenched end. Since the quenching effect is concentrated on the end surface and the cooling rate diminishes with the distance from the end, the measurement of hardness at each location from the end corresponds to a certain cooling rate and hardness penetration at that depth of the particular type of metal being tested. The data secured by this means are plotted on a graph.

From a study of the curves, it becomes apparent that initial surface hardness is a function largely of the carbon content and that **hardenability** (depth hardness) **depends on the amount of carbon present, the alloy content, and the grain size.** Manganese, boron, chromium, and molybdenum are the chief elements that promote depth hardness, whereas nickel and silicon help to a lesser degree. Although chromium and molybdenum have been most commonly used in the past as an alloy for depth hardening in steels, boron is rapidly supplanting them for this purpose. Boron is the most effective hardenability agent known. In 0.04 percent carbon steels, as little as 0.002 percent boron produces the equivalent of 0.3 percent manganese, 0.35 percent molybdenum, 0.5 percent chromium, or 2 percent nickel. Furthermore, boron is available domestically at a lower cost than the scarcer and imported alloys. However, when the carbon content is over 0.60 percent or if the steel is subjected to low temperatures, boron should not be used.

Figures 12.2 and 12.3 show the different depths of hardening in eutectoid (0.83 percent) plain carbon steel and in SAE 4140. Note that in Figure 12.2, in the upper diagram, the rates of cooling decrease as the distance from the quenched end increases. As you can see, this upper graph is superimposed directly on a drawing of the Jominy end-quench specimen. The vertical line at the end of the specimen represents hardness in Rockwell C-scale increments. The horizontal line represents actual distance along the test specimen. Hardness tests taken along the flat surface are plotted in terms of hardness and distance to form a curve. Point *A* on the curve represents 65 RC hardness and at about 1/16 in. from the

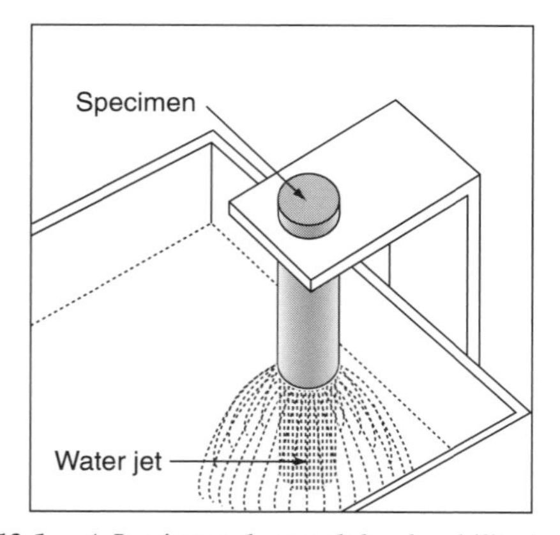

FIGURE 12.1. A Jominy end-quench hardenability test is performed in a tank in which a jet of water comes in contact with only the end of the specimen.

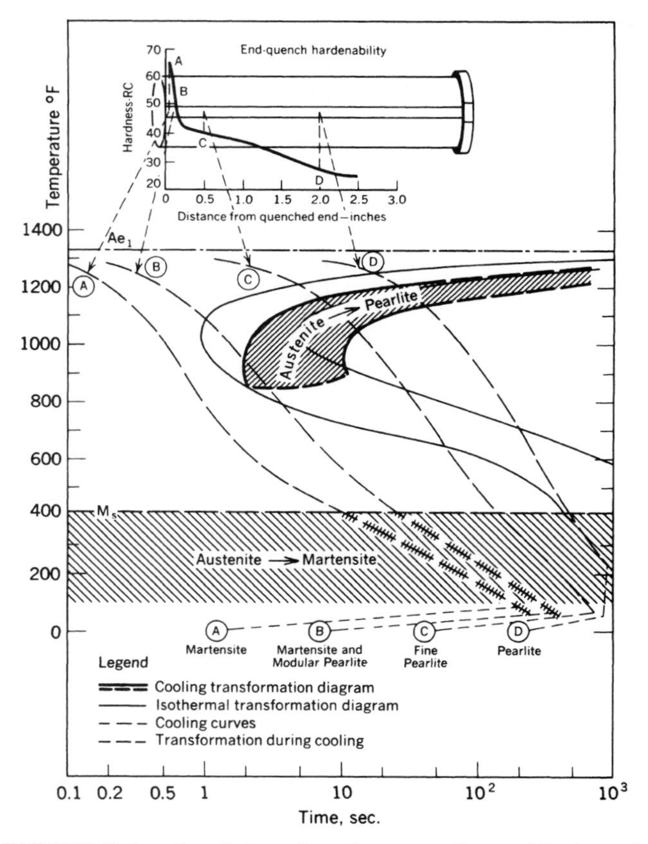

FIGURE 12.2. Correlation of continuous cooling and isothermal transformation diagrams with end-quench hardenability test data for eutectoid carbon steel. (*United States Steel Corporation*)

end. This point is carried down by the dashed arrow to the cooling curve (*A*) on the I-T diagram. Only the pearlite portion of the diagram showing the "knee" is darkened, since that is the area important to this study.

The superimposed cooling curve *A* shows a sufficiently rapid cooling rate to produce 100 percent martensite, but cooling curve *B* with its slightly lower rate of cooling produces martensite and modular pearlite. The hardness as seen in the upper graph at *B* is about 48 RC, and the test was taken at approximately 1/8 in. from the end. In contrast, in Figure 12.3, the Rockwell test at *B* is about 1/2 in. from the end and about 48 RC hardness. Its superimposed cooling curve on the I-T diagram produces martensite, ferrite, and bainite. Ferrite is a possible product because this is a medium carbon steel containing 0.40 percent carbon. Bainite is a relatively hard substance but not so hard as martensite. The chromium and molybdenum in SAE 4140 cause it to be deeper hardening than eutectoid carbon steel. The effect of various types of cooling media on the hardenability or depth of hardening is shown in Figure 12.4.

EFFECT OF MASS ON HEAT TREATED STEEL

It is also true that the mechanical properties of quenched steel depend on the mass of the piece being quenched. If a series of different diameter cylinders of steel were quenched from the same temperature in the same quenching medium and by the same procedure, the mechanical properties will vary depending on the diameters. Figure 12.5 shows the hardness penetration curves of round stock that survey the hardness as it varies from the outside diameter to the center for six different diameters in two steels, C1040 and A4142.

The temperature at which martensite begins to form is called the Ms temperature. This temperature can be lowered considerably by increasing the carbon content. When 0.83 percent carbon steel is quenched to below the Ms temperature, or approximately 400°F, martensite begins to form. Just above 300°F, about 50 percent transformation to martensite (Figure 12.6, line 1) has transpired; just above the Mf temperature, about 90 percent transformation has transpired; and at

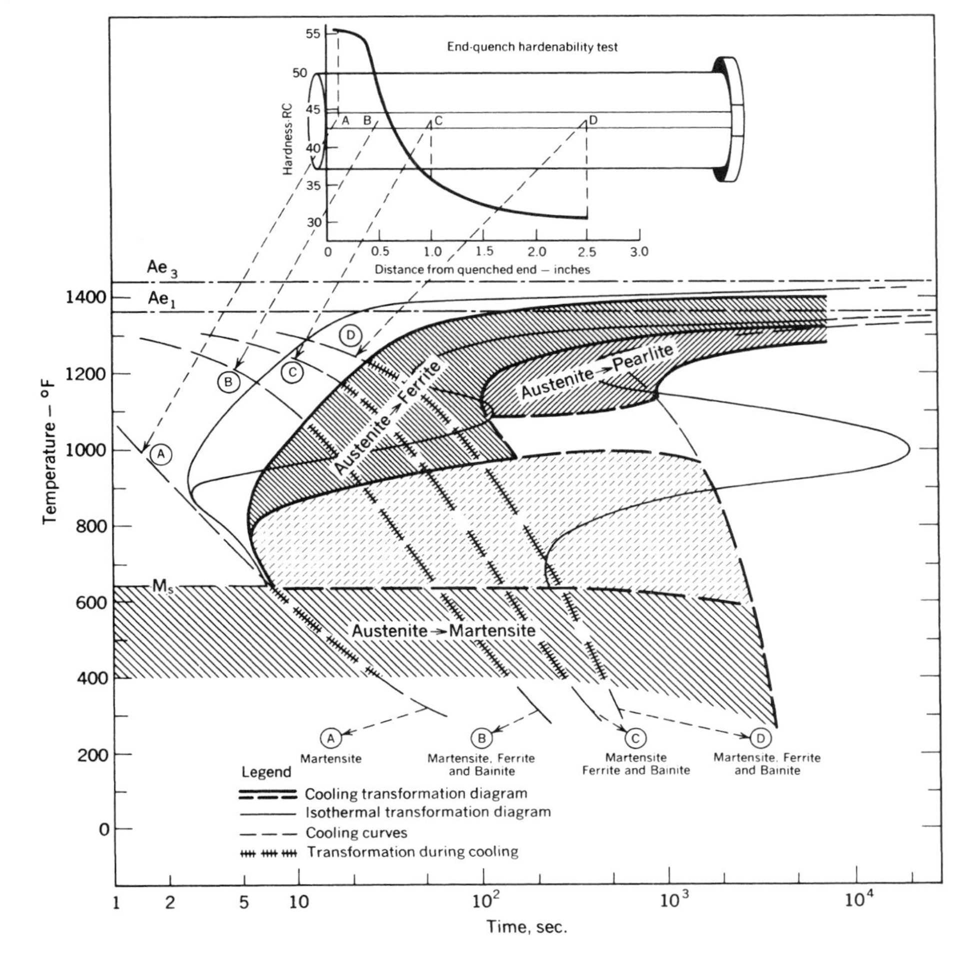

FIGURE 12.3. Correlation for continuous cooling and isothermal transformation diagrams with end-quench hardenability test data for 4140 steel. (*United States Steel Corporation*)

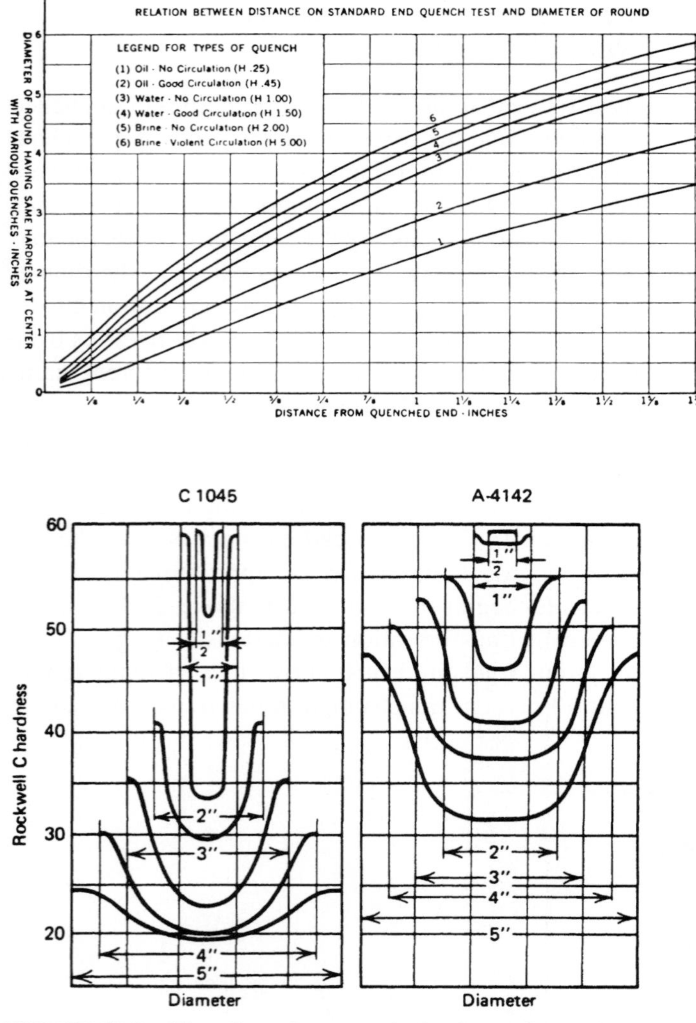

FIGURE 12.4. Cooling curves for various quenching media. (*Courtesy of Pacific Machinery & Tool Steel Company*)

FIGURE 12.5. The effect of mass on the hardness of several cylindrical (round) quenched specimens of different diameters. (*Courtesy of Pacific Machinery & Tool Steel Company*)

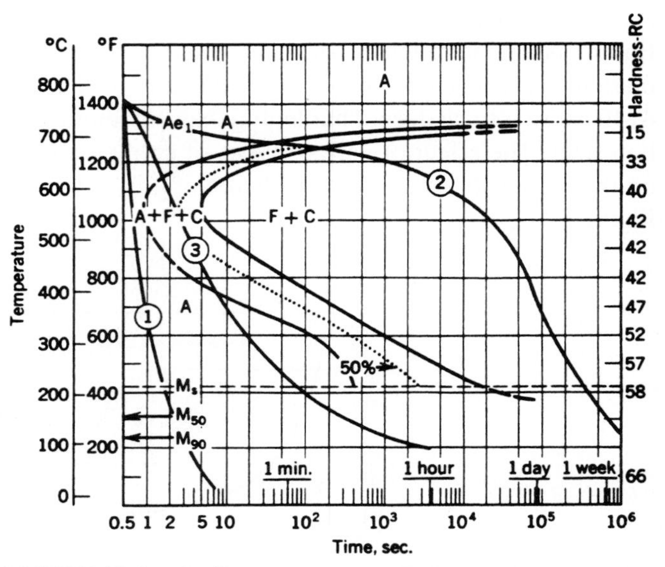

FIGURE 12.6. Cooling curves on an I-T diagram for eutectoid steel. Line 1 shows the quench for undercooling to produce martensite; line 2 is a normalizing curve. Line 3 produces a mixed microstructure of fine pearlite and martensite. (*United States Steel Corporation*)

the Mf temperature, or approximately 200°F, about 100 percent transformation has taken place. When austenitized carbon steel is quenched to below the Mf temperature, it is at its maximum hardness unless there is retained austenite.

ISOTHERMAL HEAT TREATMENTS

When cooling curves are such that they cut across the S-curve on the I-T diagram in various places, certain microstructures are formed. A soft, coarse pearlite develops when a very slow cooling takes place. This would be the case when a part is furnace annealed. When a part is air cooled after heating in the furnace to 100°F above its upper critical temperature, the process is known as normalizing. The cooling curve for normalizing would be approximately through a medium pearlite or upper bainite section (Figure 12.6, line 2) of the S-curve in eutectoid steels, forming smaller, more uniform grains that leave a stronger structure than full anneal will produce. Line 3 represents a cooling rate that results in a split transformation. Fine pearlite forms initially, but some austenite is retained and as the temperature falls and reaches the M_s, it begins to transform to martensite. The fine pearlite that has already formed will remain in the microstructure.

Another method of hardening and tempering is a form of isothermal quenching called **austempering** (Figure 12.7), in which a part is austenitized and quenched into a lead or salt bath held at a temperature of approximately 600°F to produce a desired micro-

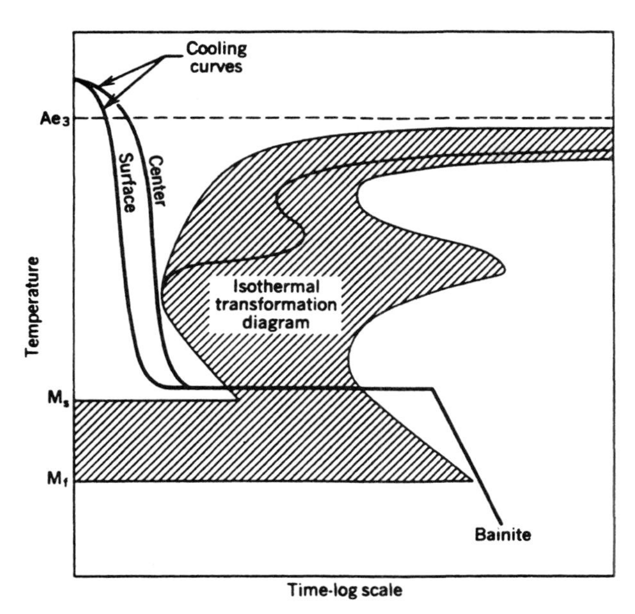

FIGURE 12.7. Austempering. (*Bethlehem Steel Corporation*)

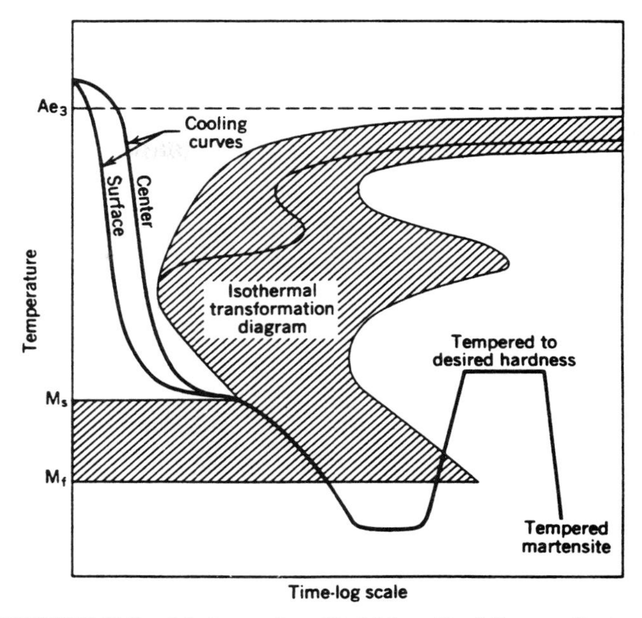

FIGURE 12.8. Martempering. (*Bethlehem Steel Corporation*)

structure of lower bainite. It is held at this temperature for several hours until a complete transformation has taken place. This type of hardening eliminates the need for tempering. Austempering produces a superior product that is much tougher than that developed in the conventional hardening and tempering process. There is one drawback, however: It is confined mostly to small or thin sections. Large, heavy sections of carbon steel cannot be austempered. Austempered ductile iron (ADI) is a rapidly growing application of austempering which is beginning to replace hardened and tempered steels for some applications.

Isothermal quenching is also used for **martempering** (Figure 12.8), in which the austenitized part is brought to slightly over the Ms temperature and held for a few minutes to equalize the interior and exterior temperatures to avoid stresses. Then the quench is continued to the Mf temperature, followed by conventional tempering. **Isothermal annealing** is done by quenching from above the critical range to the desired annealing temperature in the upper portion of the I-T diagram and holding at the anneal temperature for a length of time sufficient to produce complete transformation (Figure 12.9). This method produces a more uniform microstructure than conventional annealing, in which the steel is very slowly cooled.

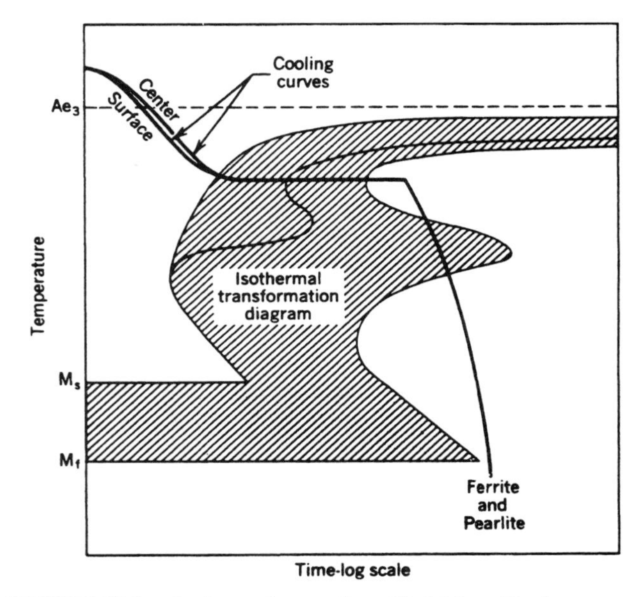

FIGURE 12.9. Isothermal annealing. (*Bethlehem Steel Corporation*)

Tempering

The mechanical properties of a given steel are controlled by the type of heat treatment performed and by the resulting strength developed by the alloying elements. One method by which metallurgists determine the effectiveness of a heat treatment is by measuring the hardness following heat treatment (see Chapter 26). Parts left in the as-quenched condition are highly susceptible to cracking due to serious residual internal stresses which result from allotropic transformation and thermal shock. If the steel is to remain in the full hard condition a "snap tempering" heat treatment must be performed. Snap tempering is used to prevent the tendency for as-quenched to crack prior to being tempered. Following quenching, parts requiring a snap tempering are reheated to approximately 325°F for 1 hour minimum at temperature. Snap tempering will usually cause the loss of about 1 point in Rockwell hardness.

Steels must be tempered to relieve the brittle shock effects and high stresses which develop due to quenching. The act of reheating the full hard steel within the temperature range of 325° to 1250°F causes the untempered martensite crystal structure to expand and release some trapped atoms of alloying elements. The higher the selected tempering temperature, the lower the hardness upon return to room temperature. Steels can be tempered or softened progressively from the full hard condition to the spheroidize annealed condition by subjecting the steel to a predetermined elevated temperature. The usual time at temperature is about 2 to 3 hours followed by air cooling to room temperature (see Chapter 11 for additional details).

SELF-EVALUATION

1. What test is used to determine hardenability?
2. Briefly explain how the test in question 1 is carried out.
3. What relationship does the test referred to in questions 1 and 2 have to the S-curve in the I-T diagram?
4. Refer to the graph in Figure 12.4. What effect does circulation of the quench seem to have on hardenability?
5. Approximately what is the maximum hardness of an austenitized steel of 1.50 percent carbon when quenched to the Mf temperature?
6. What type of microstructure develops in eutectoid steel when it is furnace annealed?
7. What is austempering? Name one advantage.
8. When is the best time to temper? Explain.
9. Explain the difference between the blue brittle tempering range and temper brittleness in some steels.
10. How can you predict the final tempered hardness of a hardened carbon steel that you are preparing to temper?
11. Hardenability is dependent on three things. Name them.
12. Why should a Jominy end-quench specimen be heated for a length of time according to the specifications of that tool steel?

CASE PROBLEM: THE SHEAR BLADE THAT DID NOT HARDEN

A small alligator shear needed a new blade. Instead of ordering it from a supplier, the shop decided to make one since they had the machine shop and heat treating facilities. The blade was rectangular, 2 in. by 4 in. by 12 in. long. The shop had some plow steel (SAE1060) in stock, so the foreman used that to make the new blade. After it was machined, it was placed in a gas-fired furnace for several hours at the correct hardening temperature and then quenched in water. This was followed by tempering at specifications given for plain carbon steel. The required hardness was RC 58, but when it was tested, its hardness was RC 39. Why didn't the blade harden since every step was done properly in the hardening procedure?

WORKSHEET 1

Objectives

1. Learn the effect of tempering on SAE1095 carbon steels.
2. Learn to use the mechanical properties chart to estimate the physical properties resulting from tempering.

Materials Three specimens of 1/2 in. round × 1 in. long SAE1095 steel, Rockwell hardness tester, heat treat furnace, water-quenching tub, and tongs.

Procedure

1. Test the as-rolled specimens for hardness and compare them with the chart (Figure 12.10).
2. Next, place the specimens in a furnace heated to 1550°F. Allow the specimens to reach the same color as the furnace bricks. The steel should be soaked at the austenitizing temperature for a few minutes before quenching.
3. Heat the end of the tongs that will come in contact with the specimens. This will avoid cooling of the specimens before they can be quenched.
4. Remove the austenitized specimens and quench in water. Agitate the specimens for better quenching.
5. Check the hardness of the as-quenched steel. The mechanical properties chart gives the as-quenched hardness of 1095 steel to be 601 Brinell. Use conversion tables if your hardness tests are Rockwell.
6. Place a hardened specimen in a furnace at 400°F for 20 minutes.
7. Remove, cool, and test for hardness. Repeat this procedure with the remaining specimens, one at 800°F and one at 1000°F.
8. Compare your results with the mechanical properties chart (Figure 12.10).

Conclusion

1. Did the results you acquired correspond with the chart? If not, how can you account for the difference?
2. Show your conclusions to your instructor.

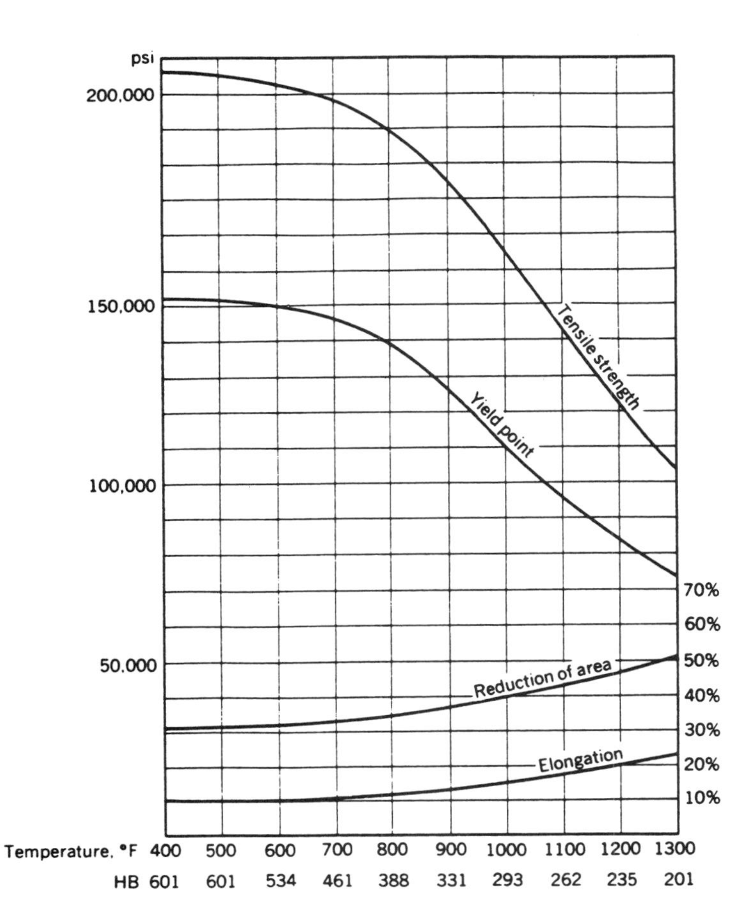

FIGURE 12.10. Water-quenched 1095 steel mechanical properties chart. (*Bethlehem Steel Corporation*)

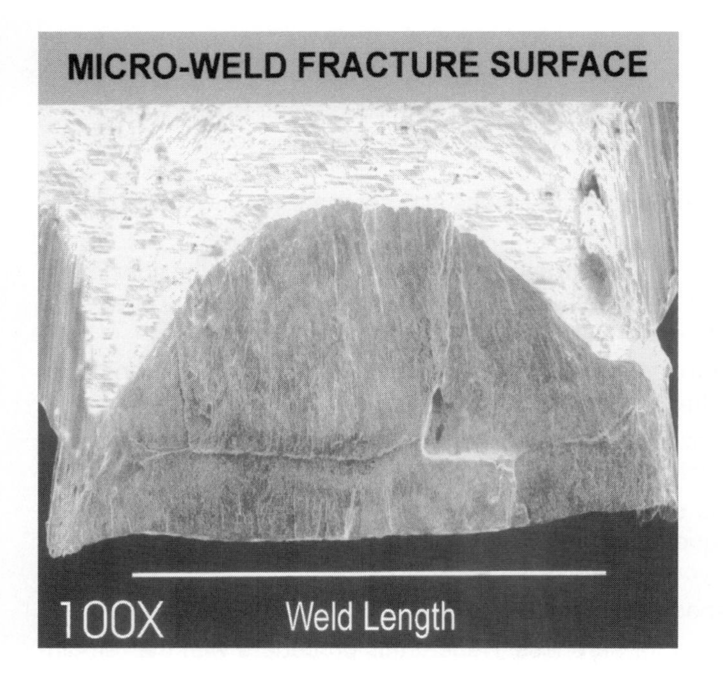

CHAPTER 13

Welding Processes for Iron and Iron Alloys

In the first chapters of this book, you learned the basic principles of metallurgy; that is, process metallurgy involving the reduction of ores and the refining of metals. You also learned about the alloying, casting, and working or shaping of metals to produce finished products and about physical metallurgy, which includes mechanical testing, metallography, and heat treatment. Physical and process metallurgy are both involved in welding processes. In fact, the process of welding is very similar on a small scale to the melting of metals in a furnace and to casting them into the ingot molds.

The welding of metals is quite similar to the manufacture of metals because the elements that go into the making of the metal in the weld or in the furnace and the grain structures that develop as they cool are often the same (Figure 13.1). There is also a similarity in the peening of welds to the hot and cold working of steels. Peening work hardens weld metal just as cold rolling work hardens wrought metal. There are other comparisons, such as thermal cycles that take place in the melting and solidification of metals, that are also related to welding. These cycles cause changes in grain size, strength, and ductility.

There is more included in welding than selecting the electrode or wire and "burning" the rod to make a weld, although much welding is performed in just this way. If the welder is not aware of the type of material in the base metal that is being welded, and does not understand the metallurgical conditions in the weld areas, he or she could experience difficulties such as cracking, porosity, and numerous other problems.

Before modern welding processes were developed, welding of iron and steel was done by a blacksmith with a forge and anvil. The ends of two pieces such as the ends of a wagon tire hoop were scarfed (tapered) and heated to a white heat. Silica sand was tossed into the joint for a flux to remove oxides, and the pieces to be joined were then brought to the anvil. There they were hammered together to cause the joining of the metal by pressure while excess slag was being forced out of the joint. The resultant weld may have been only partially bonded over the weld area un-

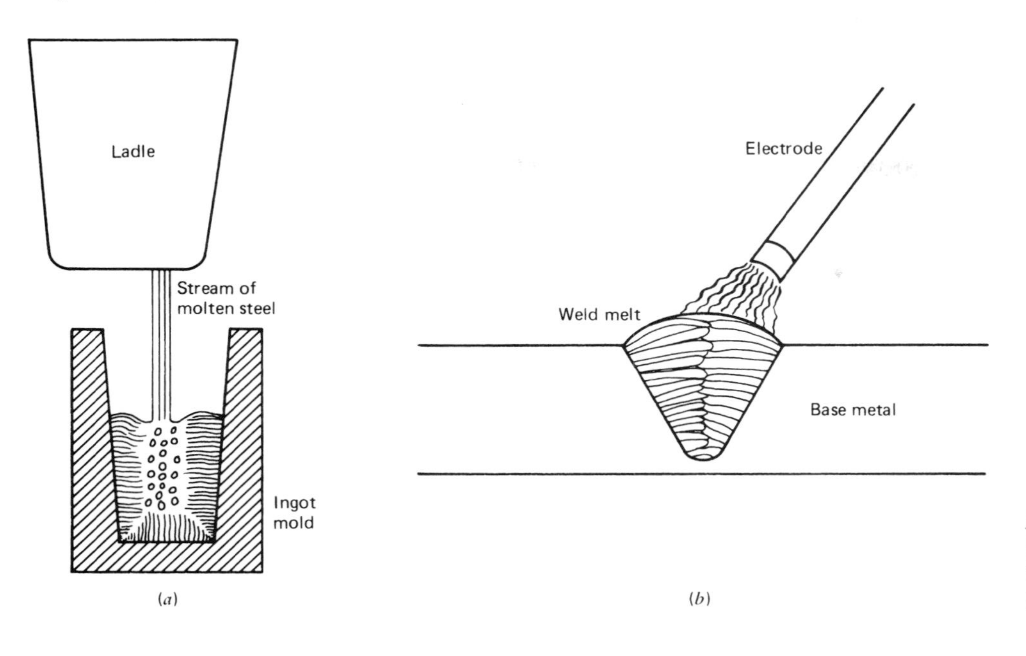

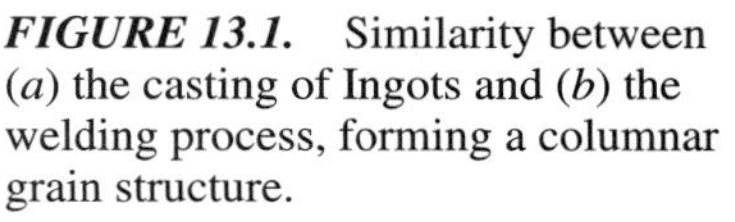
FIGURE 13.1. Similarity between (*a*) the casting of Ingots and (*b*) the welding process, forming a columnar grain structure.

less the correct heat and complete fluxing were used. Modern methods of welding produce more reliable bonding of the metals, but other problems that were not found in forge welding, such as hard brittle zones, have arisen.

An alloy steel has properties that result from some element other than carbon. An alloy steel must contain at least a small percentage of manganese, silicon, and copper; however, all steels contain a small amount of manganese and other trace elements. Tool steels are a special category of hardenable steels. This chapter deals with the weldability and metallurgy of many of the alloy and tool steels.

A great many products used today are made of cast iron and possess varying characteristics. They range from lower-quality gray cast irons used for manhole covers or furnace grates to the more complex high strength alloy irons such as those used in crankshafts and automotive engine blocks. Ductile and malleable irons are used where impact strength is a requirement. This chapter deals with the techniques and problems in the welding of various cast irons.

OBJECTIVES **After completing this chapter, you should be able to:**

1. Describe the effects of welds on the microstructure and properties of several alloy steels.
2. Prepare specimens of cross sections of welds on alloy steels for microscopic study.
3. Describe and analyze the microstructure in a given specimen.
4. Describe the changes in welds and heat-affected zones because of the heat of welding and the effects of these changes upon the welded structure.
5. Select a correct welding process and filler metal for a carbon steel base metal in order to have the optimum metallurgical condition in the weld.
6. Describe the types of welding that are done in industry.
7. Describe the effects of slags and fluxes in welding.
8. Macroetch weld sections for observation of their columnar structure and heat-affected zones.
9. List procedures for making welds on several cast irons.
10. Recognize by microscopic examination weld structures of three cast iron welds.

A great many welding processes such as gas, arc, induction, electron beam, resistance, and pressure welding are used to join steel, but these can all be listed within several classes of welding. Most methods used in welding may be classified as one of the following processes.

1. The application of heat (from electrical resistance, friction, or flame) and pressure (forge welding) without melting of the metal may be classed as **solid phase welding.** The joining is done without filler metals and without melting or changing the base metal. Solid phase welds may also be made by using other energy sources such as ultrasonics or electromagnetic induction.
2. **Fusion joining** (arc, gas, plasma arc, and electron beam welding) requires that the parts be heated until they melt and flow together. Filler metals may also be used.
3. **Liquid-solid phase welding** requires the base metal parts to be heated but not melted. A dissimilar molten metal is used to join the parts together. Brazing and soldering are examples of this kind of welding. Diffusion often occurs but is not necessary for adherence of the filler metal.

In Chapter 5 you learned of a steel classification system that indicated the carbon and metal alloy content of steels. The SAE-AISI systems are used for steels in machinery, tools, products, and bar stock.

Welders in steel construction, pipe lines, and pressure vessels typically use the American Society for Testing and Materials (ASTM) standards to determine material specifications, practices, definitions, and methods of testing (see Table 13.1). These standards for steel all carry the prefix letter *A;* for example, A27-62 denotes a low- to medium-strength carbon steel casting, and A7-61T covers steel for buildings and bridges. The American Welding Society (AWS) also has a system of codes, recommended practices, standards, and procedures. The AWS deals with such areas as welding and testing procedures. AWS also has specifications for welding rods and electrodes.

Most welding on steel is performed on low carbon steel having about 0.08 to 0.2 percent carbon. Let us examine the structures pertaining to these welds. You have already learned that steel undergoes certain changes when it solidifies from the molten state and cools to room temperature. That is, the predominant high-temperature lattice structure is based on a face-centered cube and is called austenite, which cools through a transformation temperature or critical point that marks the change from a face-centered to a body-centered lattice structure called ferrite. The formation of grains takes place during the initial solidification from the molten state, but this grain structure recrystallizes during the phase change from austenite to ferrite. Reheating the metal above the A_1 critical temperature will again recrystallize the grains.

TABLE 13.1. Some ASTM standard numbers for steels

ASTM Number	Type of Steel
A1-58T	Open-hearth carbon steel rails
A27-62	Low- to medium-strength carbon steel castings
A7-61T	Steel for bridges and buildings (tentative)
A8-54	Structural nickel steel
A20-56	Boiler and firebox steel
A36-61T	Structural steel (tentative)
A94-54	Structural silicon steel
A120-61T	Black and hot-dipped zinc-coated welded and seamless pipe for ordinary uses (tentative)
A120-60	Austenitic manganese—steel castings
A216-60T	Carbon steel castings suitable for fusion welding for high-temperature service (tentative)
A240-61T	Corrosion-resisting chromium and chromium-nickel steel plate, sheet, and strip for fusion-welded unfired pressure vessels (tentative)
A415-58T	Hot rolled carbon steel sheets, commercial quality (tentative)
A429-58T	Hot rolled and cold-finished corrosion resisting chromium-nickel-manganese steel bars (tentative)

Note: Complete lists and specifications may be found in ASTM Standards reference books.

The four basic zones in welding are the weld zone, the fusion zone (also called junction zone), and the heat-affected zone and the adjacent zone (Figure 13.2). The weld zone (often called nugget) is the weld melt itself after it solidifies. Melted base metal picked up by the weld usually becomes quite uniformly mixed with the filler metal before the weld solidifies because of turbulence and convection in the weld pool. Although the fusion zone (Figure 13.3) is represented by a fairly narrow area (almost a sharp line), a small amount of diffusion occurs after solidification across the line of fusion (Figure 13.4). When the base metal contains considerable carbon, for example, some of this carbon will diffuse into the melt of the weld zone. This pickup of carbon is the reason that the root pass is often more brittle than other passes; the root pass has more contact with the base metal, hence, more dilution and its cooling rate is also higher. Any composition of high hardenability can promote root cracking. The heat-affected zone is near the weld in the base metal and is so called because it is affected by the heat of welding.

During the time the weld is molten, the atomic structure is not in a lattice arrangement but is amorphous, with a random movement of the iron atoms (Figure 13.5). When solidification occurs, the weld metal begins to form a lattice structure similar to that in the base metal. Nucleation takes place at the boundary line between the molten weld metal and the base metal.

Welds, like ingots, tend to form a columnar structure when they solidify (Figure 13.6). When steel is slowly cooled from the molten state, the grains begin to grow until they join together. The grain size depends on the number of nucleation sites. For example, the casting of metals into ingots generally involves a large mass of material that has a very slow cooling rate. This produces large grains that usually grow in a columnar orientation. Columnar grains in welds are most evident in those having a large weld nugget, high heat input, and a slow cooling rate.

When the weld metal begins to solidify, the crystals grow into the cooling molten steel in lines perpendicular to the direction of maximum heat flow (Figure 13.7). The crystals orient themselves toward the direction of welding and outward from the base metal. A single pass weld generally produces a coarse columnar structure (Figure 13.8) that is somewhat undesirable

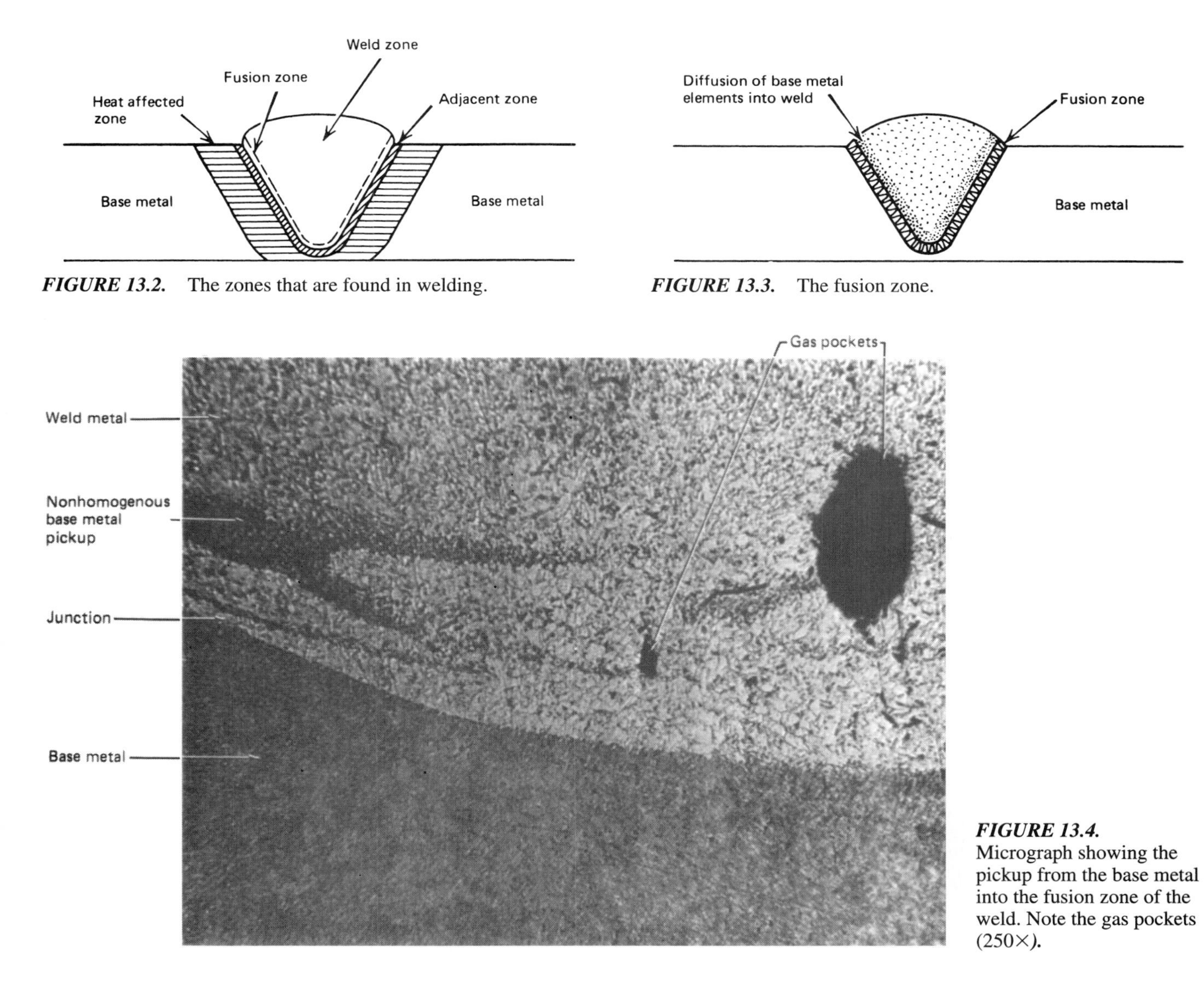

FIGURE 13.2. The zones that are found in welding.

FIGURE 13.3. The fusion zone.

FIGURE 13.4. Micrograph showing the pickup from the base metal into the fusion zone of the weld. Note the gas pockets (250×).

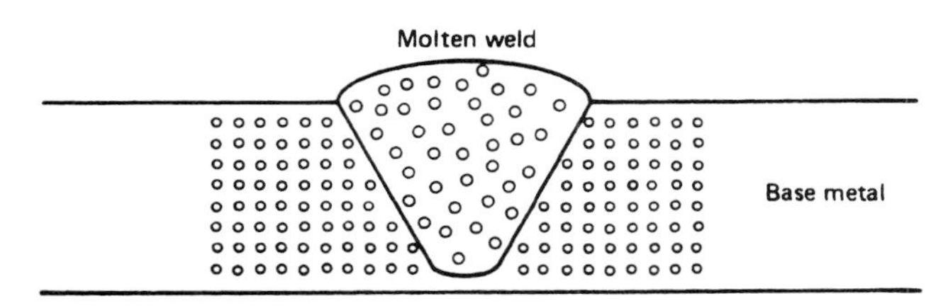

FIGURE 13.5. The lattice structure in the base metal remains uniform while the weld melt shows atoms moving at random.

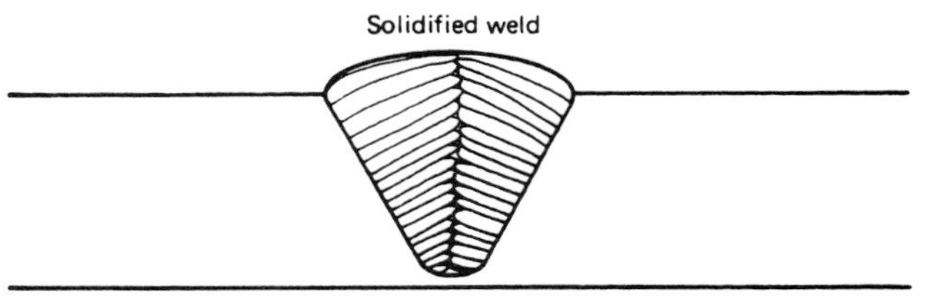

FIGURE 13.6. The typical grain formation of welds. Note the sharp boundary line between the molten weld metal and the base metal.

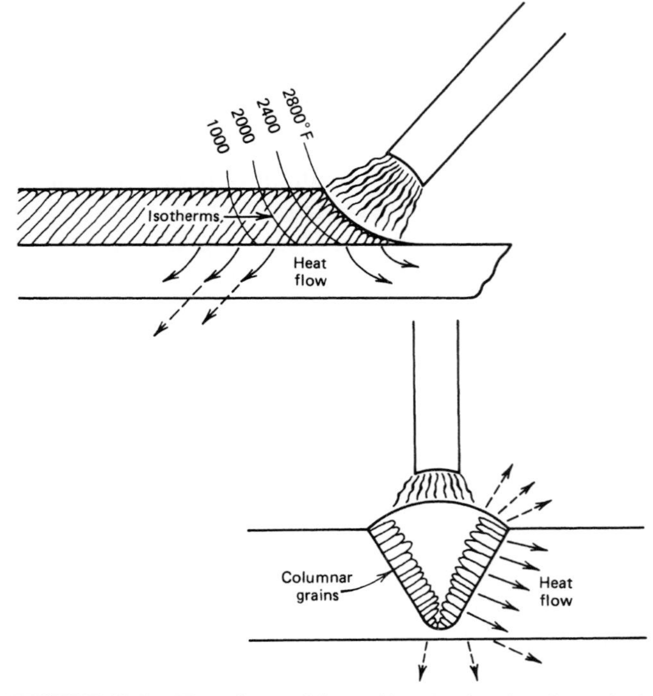

FIGURE 13.7. Two views of the weld melt, showing the typical formation of isotherms.

since it is not as strong as a finer, recrystallized grain structure. A two-pass weld will recrystallize the heat-affected zone of the first pass, thus stress relieving the first-pass weld zones. Because of its finer grain size, the weld zone is usually stronger than the base metal in welded structures (Figure 13.9).

If the cooling rate is slow, the grains in the weld are larger and those in the adjacent heat-affected zone tend to become quite large (Figures 13.10*a* and 13.10*b*). If the cooling rate is rapid, the coarsening effect is minimal. If the weld consists of several passes, the heating effect of later passes will normalize the previously solidified structure, leading to a refinement of the grains since each pass reheats part of the previous weld pass to above the critical temperature.

In the adjacent heat-affected zones near the weld, the base metal grains begin to grow and coarsen when held at higher temperatures for longer periods, usually having little or no hard structures. In the arc welding processes, relatively small quantities of metal are molten and consequently cool rapidly, since the base metal acts as a heat sink (heat absorber) that quickly cools the weld nugget.

The heat-affected zone is most likely to harden in steels containing even small amounts of carbon, since

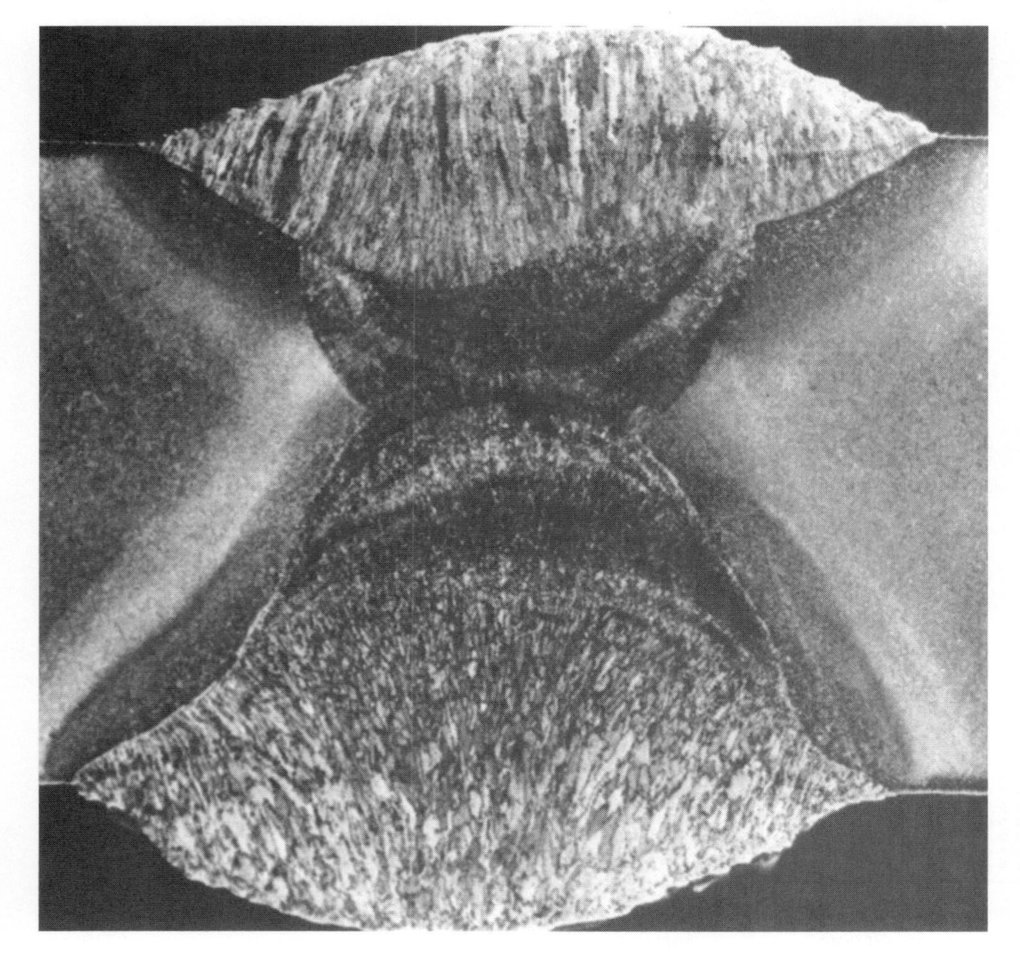

FIGURE 13.8. Macrostructure of an arc weld showing the columnar crystallization in the weld metal and heat-affected zone in the base metal (4×). *(By permission, from Metals Handbook, Volume 7, Copyright American Society for Metals, 1972)*

the cold-base metal quickly quenches the weld and adjacent heated zone as the weld is being made. Multiple passes tend to normalize the previous passes, thus reducing the hardness, as shown in Figure 13.11. Martensitic structures that cause brittleness can often be transformed into softer structures by postheat treatment or prevented by preheating (Figure 13.12).

Various metallurgical structures are also formed in the weld zone as cooling or heating is taking place. Low carbon steel welding rods (0.08 to 0.15 percent C) deposit ferrite grains with some pearlite and do not harden appreciably unless carbon is picked up from the base metal or from carbon-containing contaminants such as oil or grease. When high tensile strength filler metal that contains a higher percentage of carbon is used, a slow cooling produces pearlite (Figure 13.13). Other microstructures may also appear. Bainite, often produced while welding (Figure 13.14), is a dark, acicular (needlelike) microstructure and is harder and tougher than pearlite, depending on the temperature of its formation. Martensite, however, is an extremely hard, brittle structure appearing as a fine, acicular microstructure (Figure 13.15).

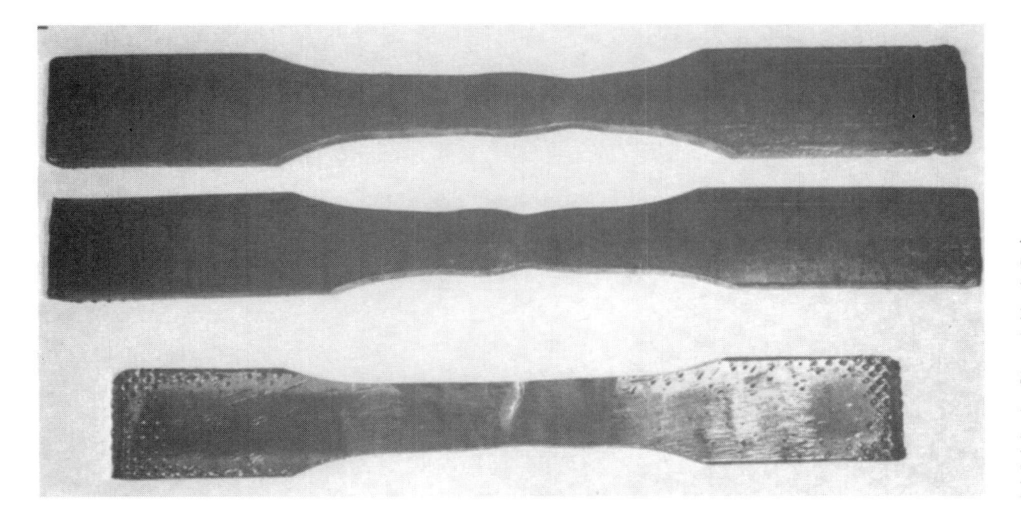

FIGURE 13.9. Tensile test samples of the three different welds. The upper two samples begin to fail (neck down) in the heat-affected zone near the weld, since the larger recrystallized grains in that area are more ductile. The lower test sample began to fail in the weld itself, possibly due to carbon pickup from the base metal and the consequent brittle structure, or to hydrogen embrittlement.

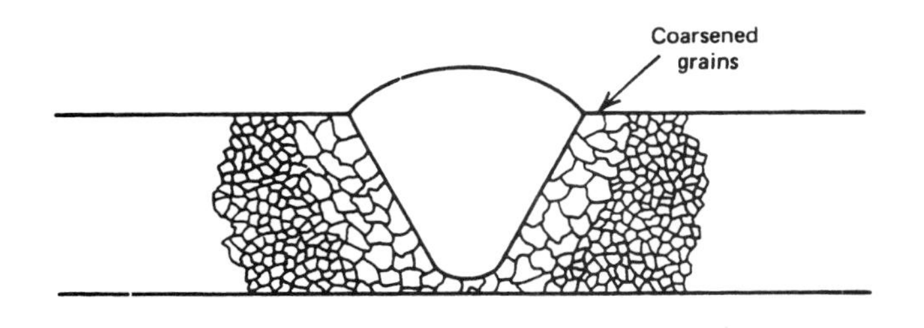

*FIGURE 13.10***a.** The coarsened grains in the base metal in the heat-affected zone are caused by high-temperature grain growth.

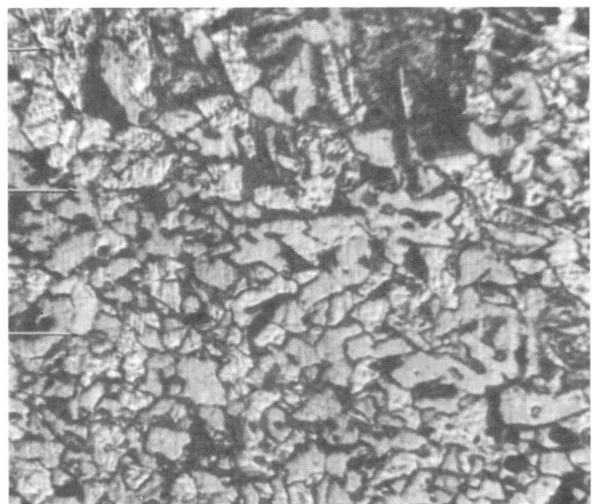

*FIGURE 13.10***b.** This micrograph shows slow cooling in the weld zone (upper right); slow cooling creates irregular grain structures (180×).

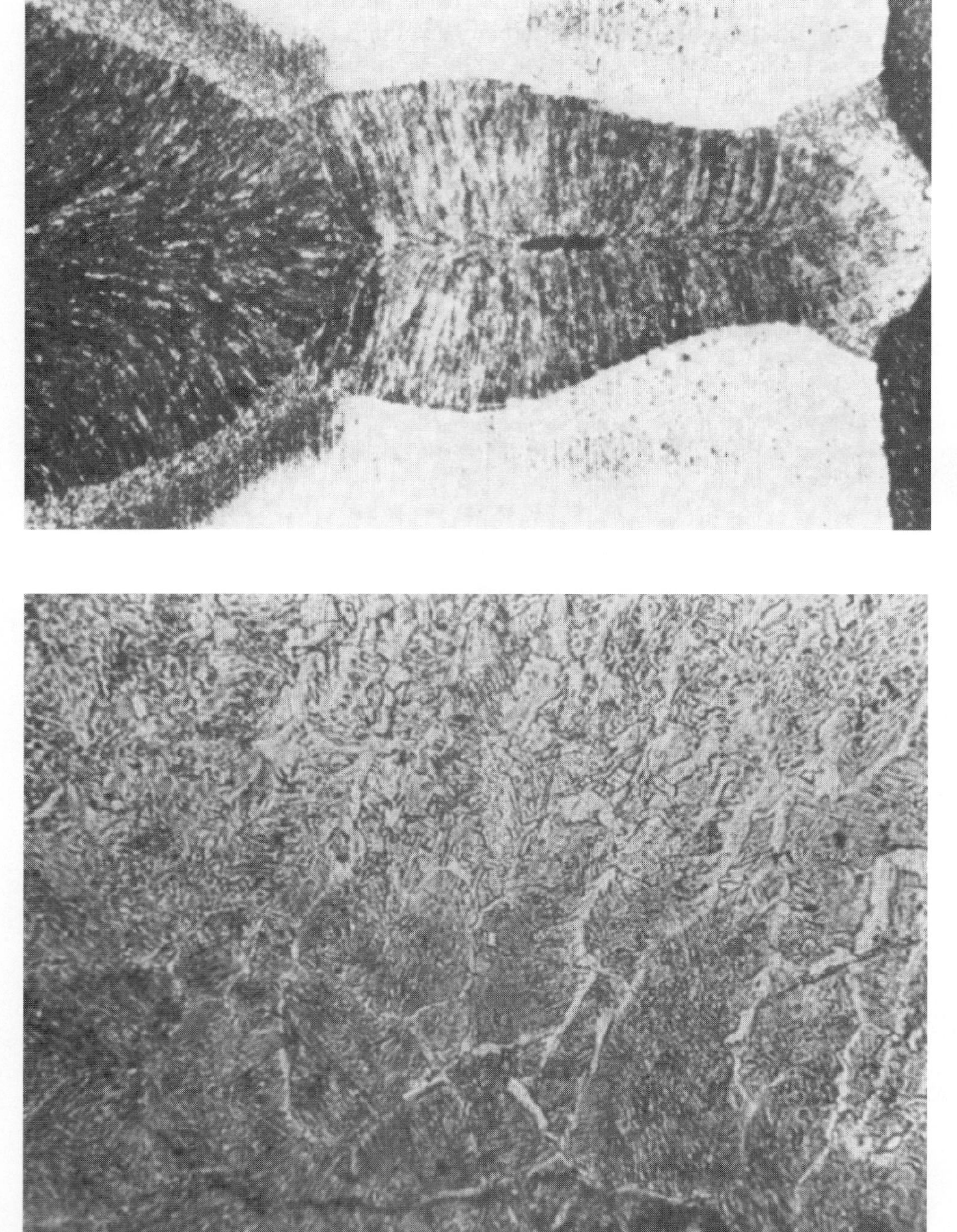

FIGURE 13.11. Section through a two-pass butt weld made in steel pipe by automatic gas metal-arc CO_2 shielding. Successive passes have partially normalized the heat-affected zone. (*By permission, from Metals Handbook, Volume 7, Copyright American Society for Metals, 1972*)

FIGURE 13.12. Micrograph of weld junction zone in SAE 1040 steel with low carbon steel rod. Note the columnar structure in the weld zone (upper half) and the martensite in the base metal (lower half). Also, note the underbead crack in the base metal. A preheat could have prevented the formation of martensite in the base metal and the underbead crack (250×).

The formation of martensite in the heat-affected zone results in many failures in welds because of the brittleness of martensite. Hardenability increases with carbon content and promotes the formation of the hard, brittle martensite structure that may crack after welding.

Preheating and **postheating** minimize the formation of the brittle martensitic structure. Preheating to 200° to 600°F prevents the formation of martensite by inducing slower cooling to the base metal. Postheating by tempering or annealing eliminates brittle martensite that has already formed. Hot cracking can also be reduced by lowering the cooling rate with preheating.

Ferrite structures in weld areas may require a stress relief if the structure is distorted during welding. The stress relief is done by heating to a temperature between 950° and 1200°F for a short period of time. This will recrystallize the ferrite grains. The heating of the structure lowers its yield strength to a value lower than the

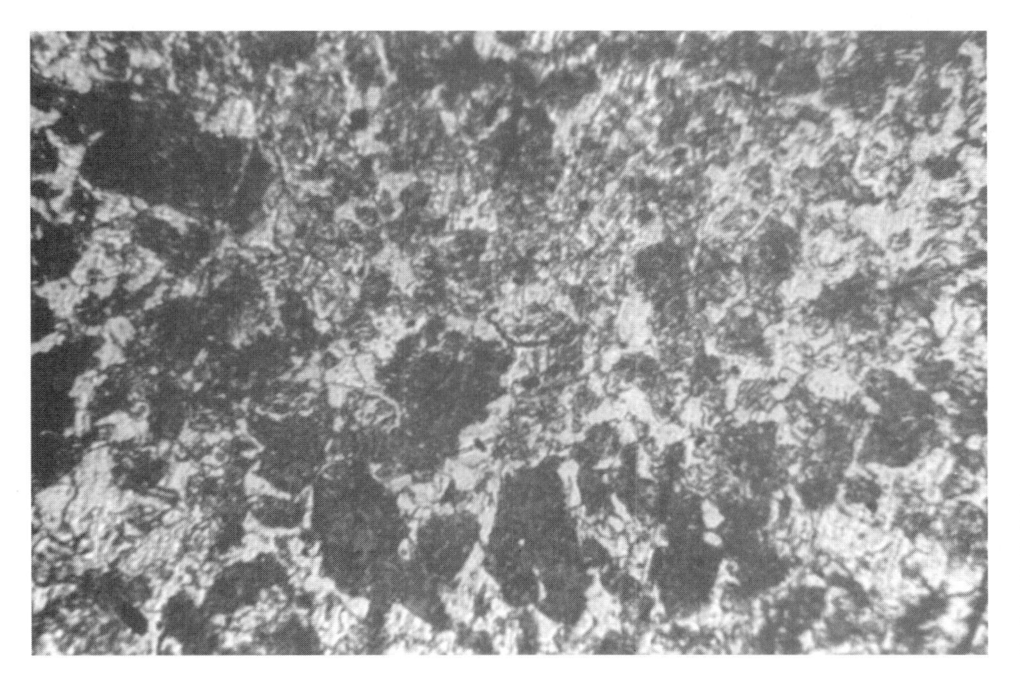

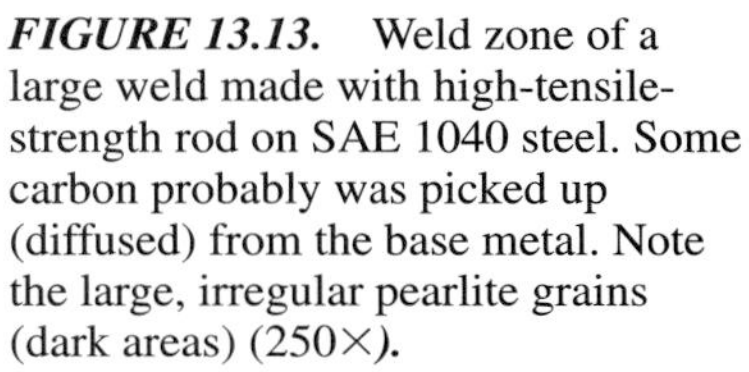

FIGURE 13.13. Weld zone of a large weld made with high-tensile-strength rod on SAE 1040 steel. Some carbon probably was picked up (diffused) from the base metal. Note the large, irregular pearlite grains (dark areas) (250×).

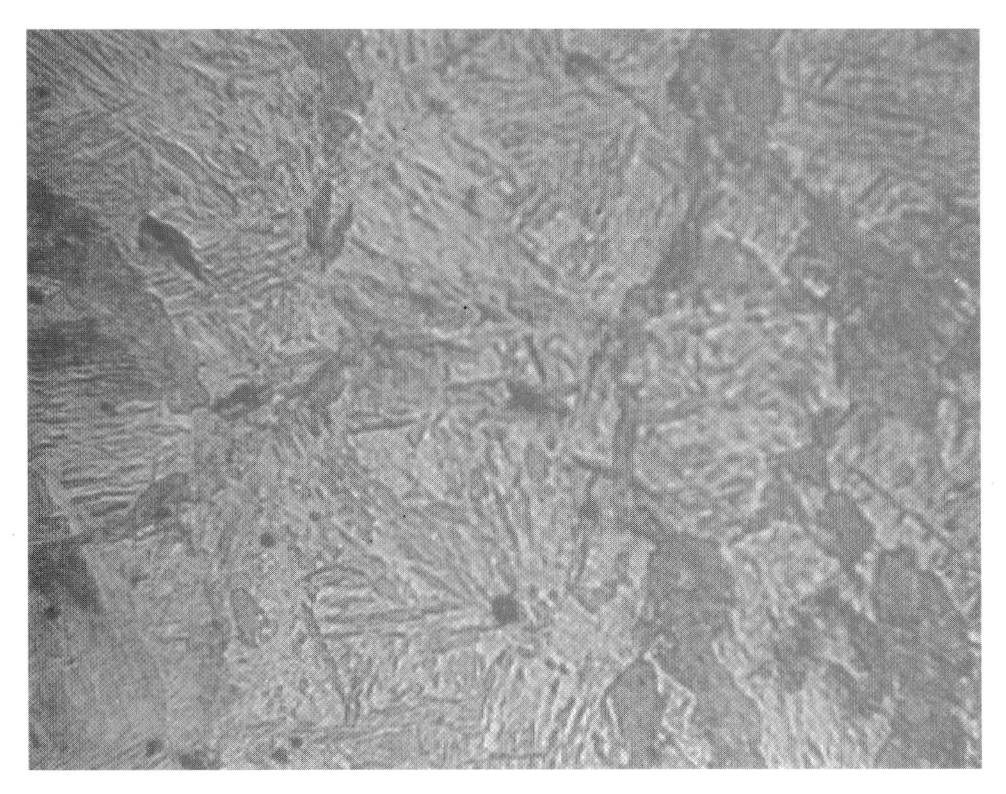

FIGURE 13.14. Bainite, such as this microstructure shows, is often found in welds. The bainite appears as dark acicular areas in a matrix of ferrite. Bainite produces a tougher weld than martensite would (500×).

residual stress value and allows plastic flow to occur, which relieves the stress. Pearlite grains are not affected by the stress relief (Figures 13.16*a* and 13.16*b*). If they are held for several hours at 1200° to 1300°F, however, the lamellar or network formation will spheroidize and ductility will increase. This is not always good, since weld yield strength is reduced considerably. Full anneal at temperatures above A_1 will also reduce weld strength. Therefore, only stress relief anneal is normally applied to steel weldments when anneal is needed to maintain the necessary weld strength. However, much welding on low carbon steel is performed without stress relief anneal.

In arc welding, the various zones in the base metal are confined to a very narrow region extending to $3/16$ in.

500× 2% Nital Etch

FIGURE 13.15. Photomicrograph showing martensite transformation in a 1045 steel weld.

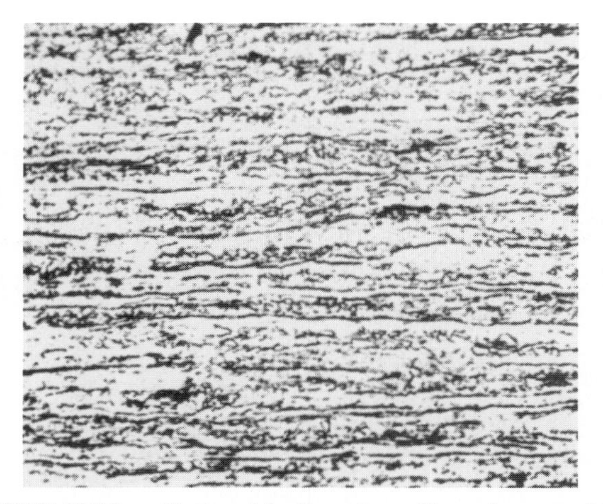

*FIGURE 13.16***a.** Flattened ferrite and pearlite grains caused by cold-working or weld stresses (1000×). *(By permission, from Metals Handbook, Volume 7, Copyright American Society for Metals, 1972)*

*FIGURE 13.16***b.** The flattened ferrite grains are recrystallized when heated to 1025°F, but the pearlite grains are not affected by this stress relief (1000×). *(By permission, from Metals Handbook, Volume 7, Copyright American Society for Metals, 1972)*

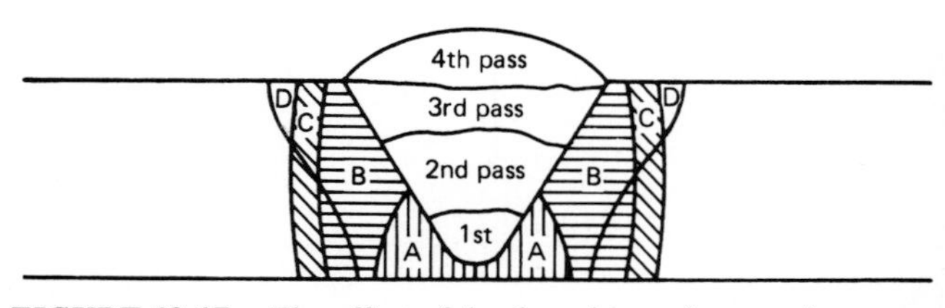

FIGURE 13.17. The effect of the deposition of successive passes on transformations in the base metal.

on either side of the weld edge, depending on the thickness of the material. This is true even when the weld is built up from small passes that tend to normalize (refine) the structure of the earlier passes. The successive passes also affect the structures in the heat-affected zone to a certain extent so that particularly in thick plates you can find alternating layers of heat-affected and partially normalized material.

The tempering effect can be detected by the variations in hardness from point to point in this zone (Figure 13.17). For example, in area *A* the first pass produces a substantial increase in hardness, depending on the carbon content and rate of cooling; subsequently, the hardness at this point falls when tempered by the higher interpass temperatures of later beads. This example is only an approximation, since many factors are involved in the hardening and normalizing of the weld zones. The same occurs in the other areas (*B, C,* and *D*) in the overheated zone. The mechanical properties and structures are eventually averaged out, except perhaps in the small areas in the fourth pass at the corners that tend to retain some hardness, sometimes causing small toe cracks.

Medium or higher carbon and alloy steels tend to "contaminate" the weld metal with carbon or other elements. Thus, the weld zone can also become brittle and develop cracks. Thick mild steel plates contain more carbon than thinner plates of the same ASTM designation, because thick plates cool more slowly while being rolled and therefore have a lower tensile strength and larger grain size. Carbon is added to strengthen the plate and bring it up to specifications.

The strength of the final joint does not altogether depend on the weld metal. If cold-worked steel is heated to about 950°F, which is the temperature of recrystallization, the distorted ferrite grains will recrystallize, and the steel in the heat-affected zone will lose its cold-worked strength. When the base metal is hot-worked steel (soft and low strength), there will not be much change in the strength or hardness of the heat-affected material because it does not recrystallize to any great extent (Figure 13.18).

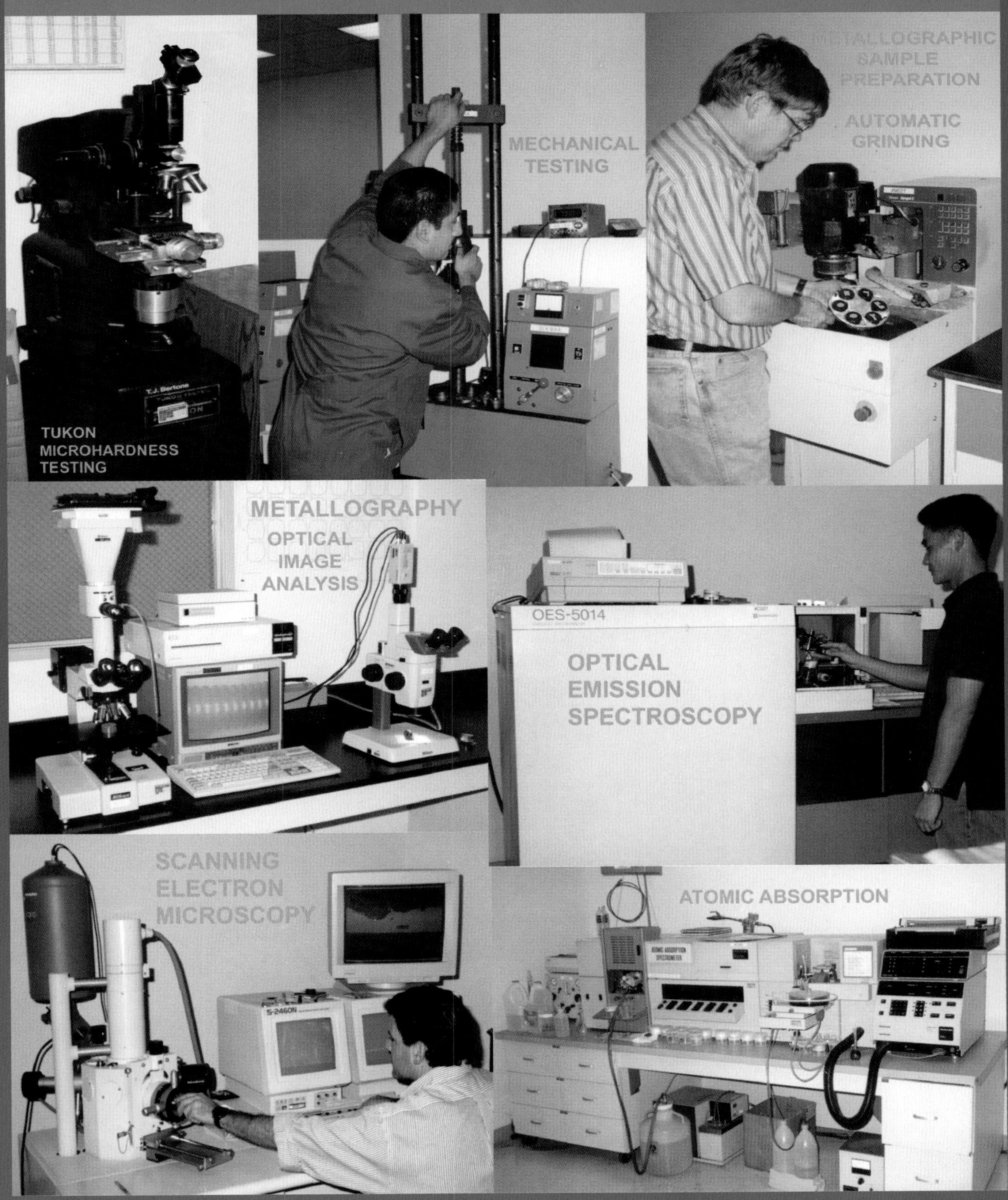

Courtesy of Tom Bertone Consulting, Whittier, CA.

OPTICAL MICROSTRUCTURE ANALYSIS

Index:

1. Proeutectoid Ferrite in a Pearlite Matrix
2. Carburized C1117 with a Retained Austenite in Case
3. Retained Austenite
4. Iron-Carbide in Grain Boundaries
5. Microcracked Lathe Martensite / Retained Austenite in 9310 Carburized Steel
6. Iron-Carbon Diagram
7. High Temperature Oxidation (HTO)
8. Irregular Nitride Case
9. Crack in 4340 Steel (Oxides and Inclusions)

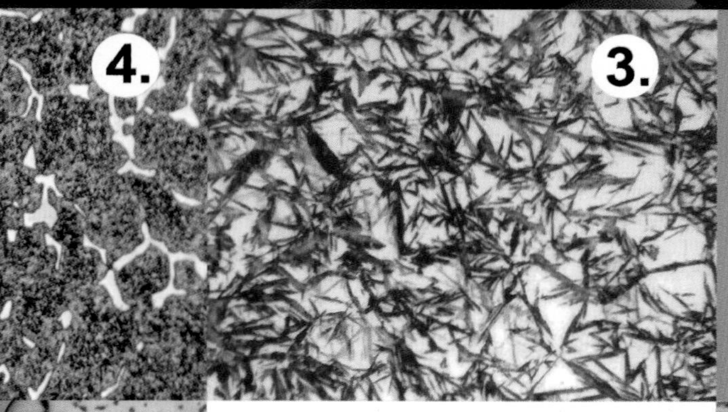

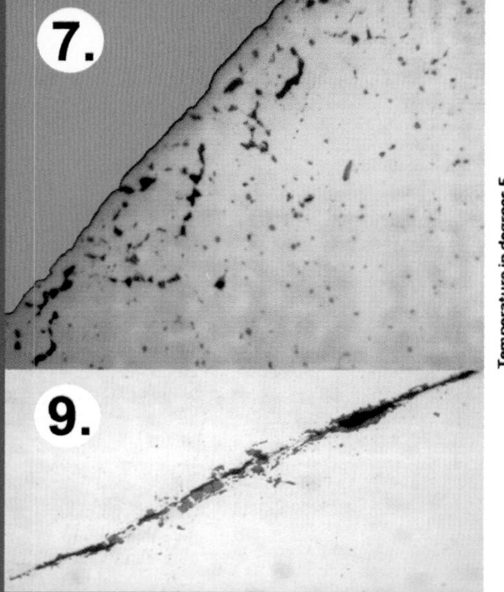

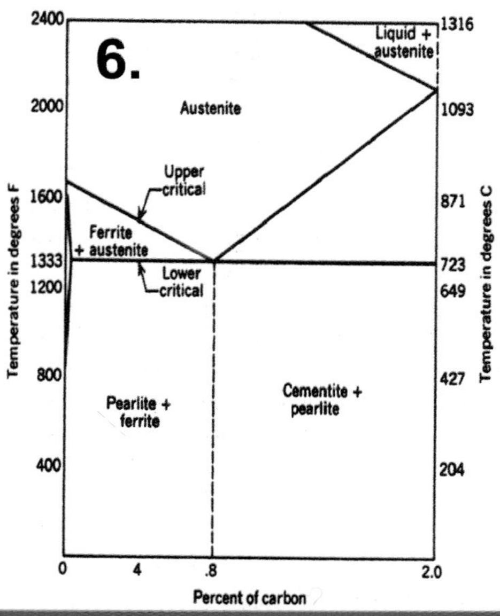

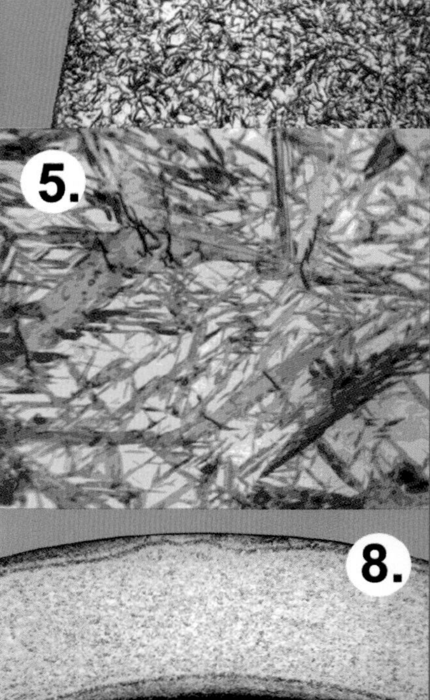

Courtesy of Tom Bertone Consulting, Whittier, CA.

ADVERSE STEEL MICROSTRUCTURES

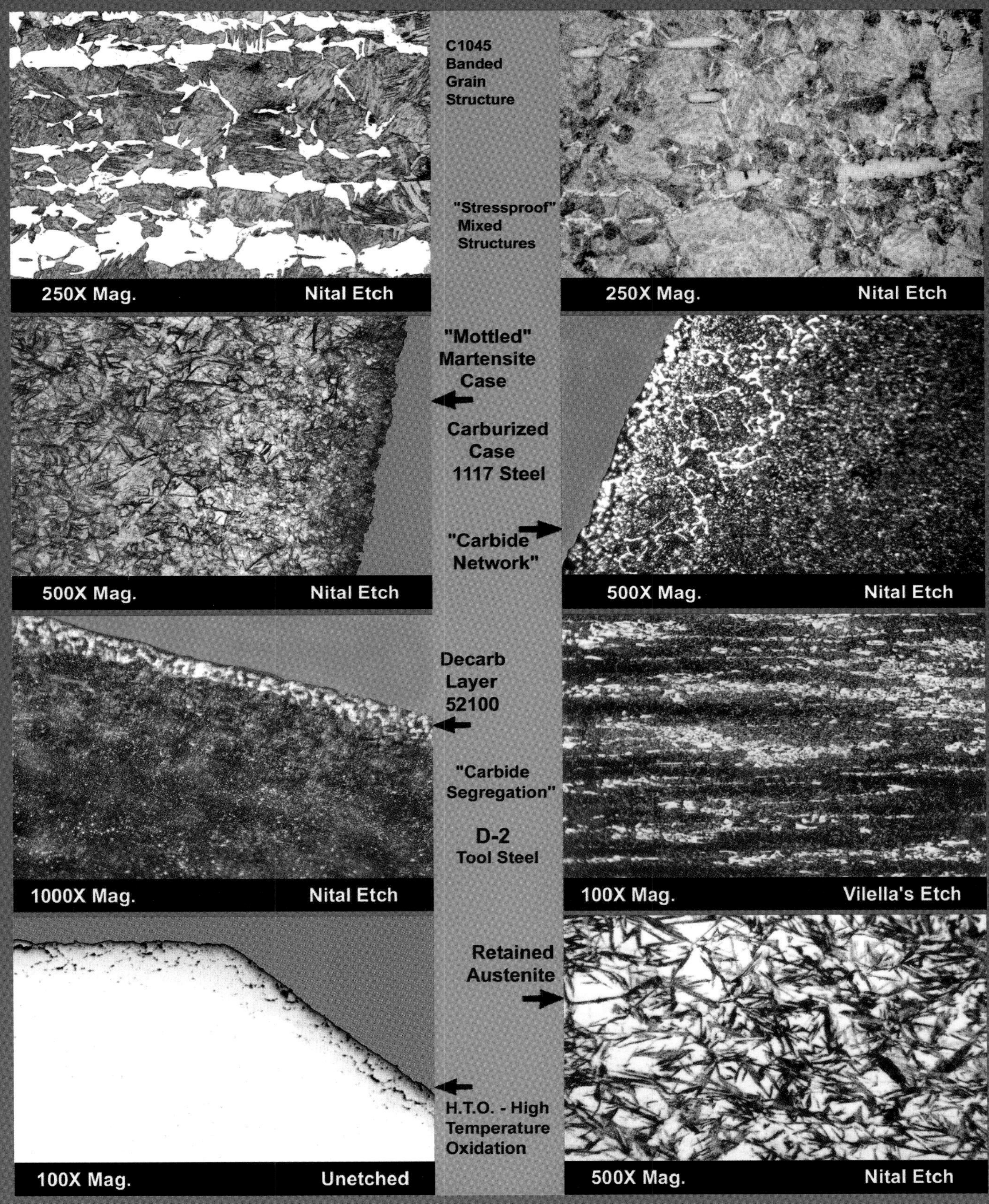

Courtesy of Tom Bertone Consulting, Whittier, CA.

ALUMINUM ALLOY PROBLEMS

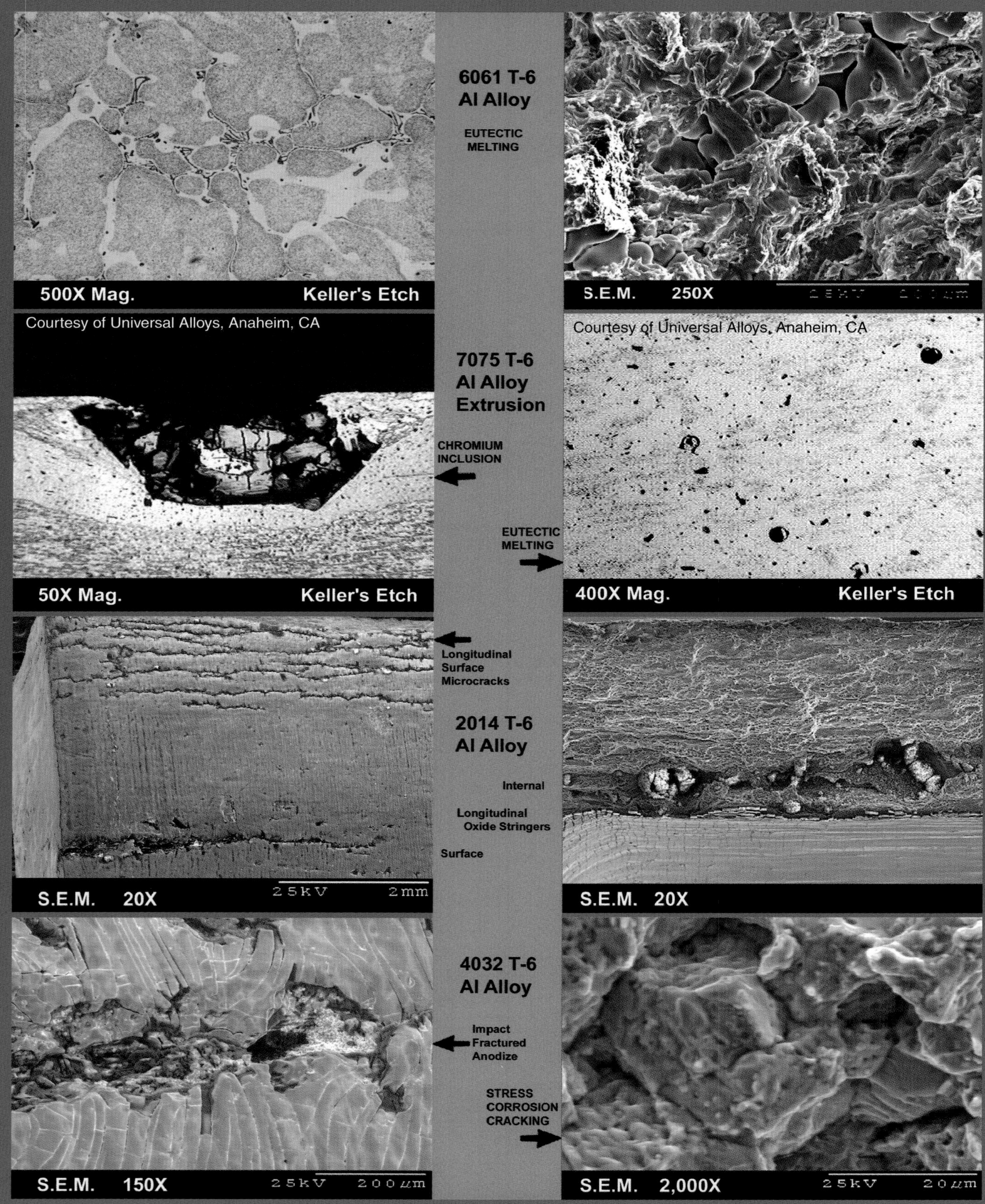

Courtesy of Tom Bertone Consulting, Whittier, CA.

S.E.M. FRACTURE SURFACE ANALYSIS

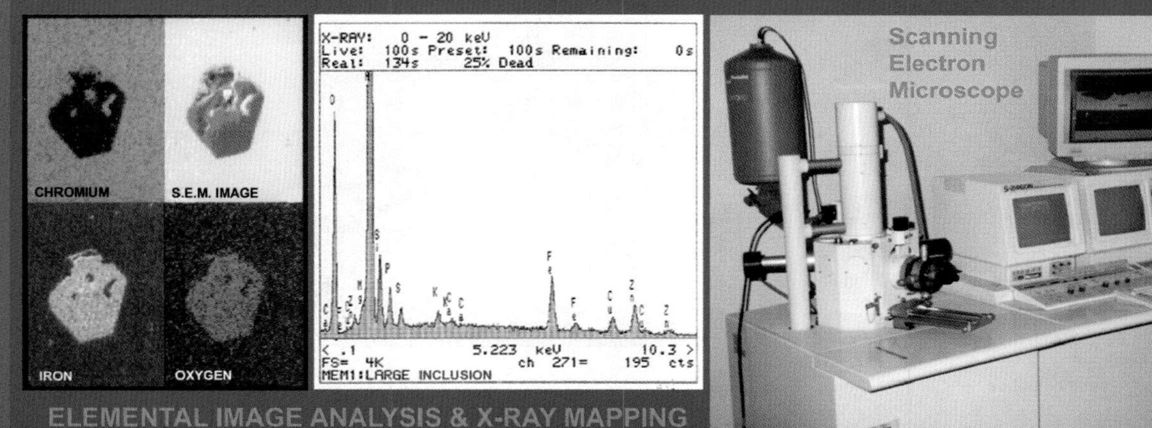

ELEMENTAL IMAGE ANALYSIS & X-RAY MAPPING

Index:

1. Hydrogen Embrittlement in 4340 Steel
2. Quasi-Cleavage & Tensile Overload in 17-7 PH Steel
3. Brittle Fracture in 15-5 PH Steel
4. Tensile Overload by Bending Mode 420 Stainless
5. Stress Corrosion Cracking in 6061 T6 Aluminum
6. Fatigue Cracking GM-7520 Steel Carburized Case
7. Quasi-Cleavage 17-7 PH Stainless

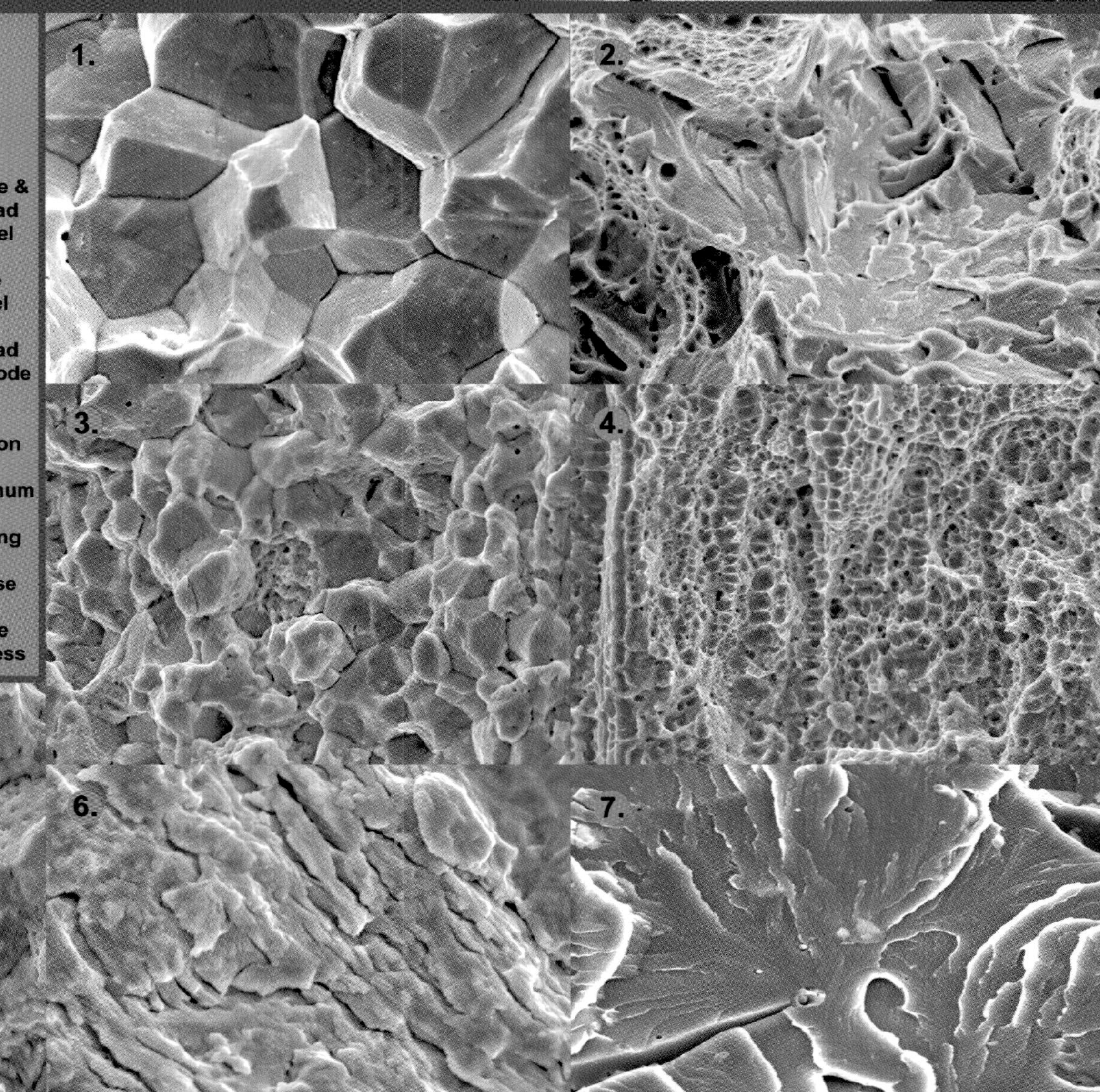

Courtesy of Tom Bertone Consulting, Whittier, CA.

ELEMENTAL X-RAY IMAGE ANALYSIS

(COLOR PIXELS = ELEMENT CONCENTRATION)

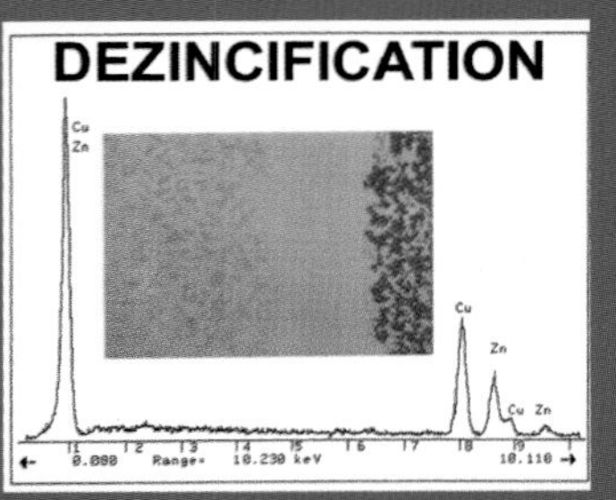

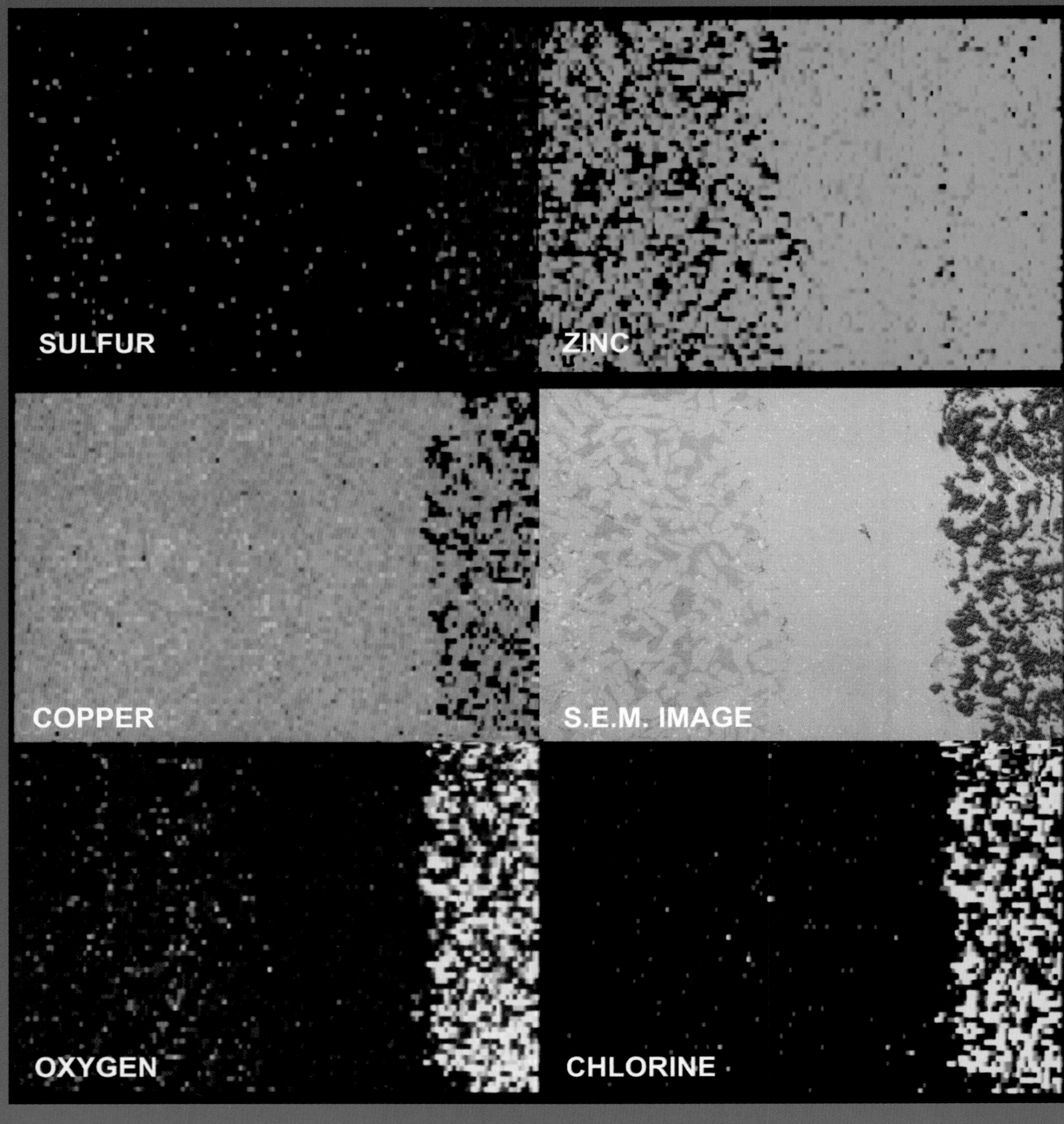

Courtesy of Tom Bertone Consulting, Whittier, CA.

Courtesy of Tom Bertone Consulting, Whittier, CA.

CORROSION ANALYSIS OF METALS

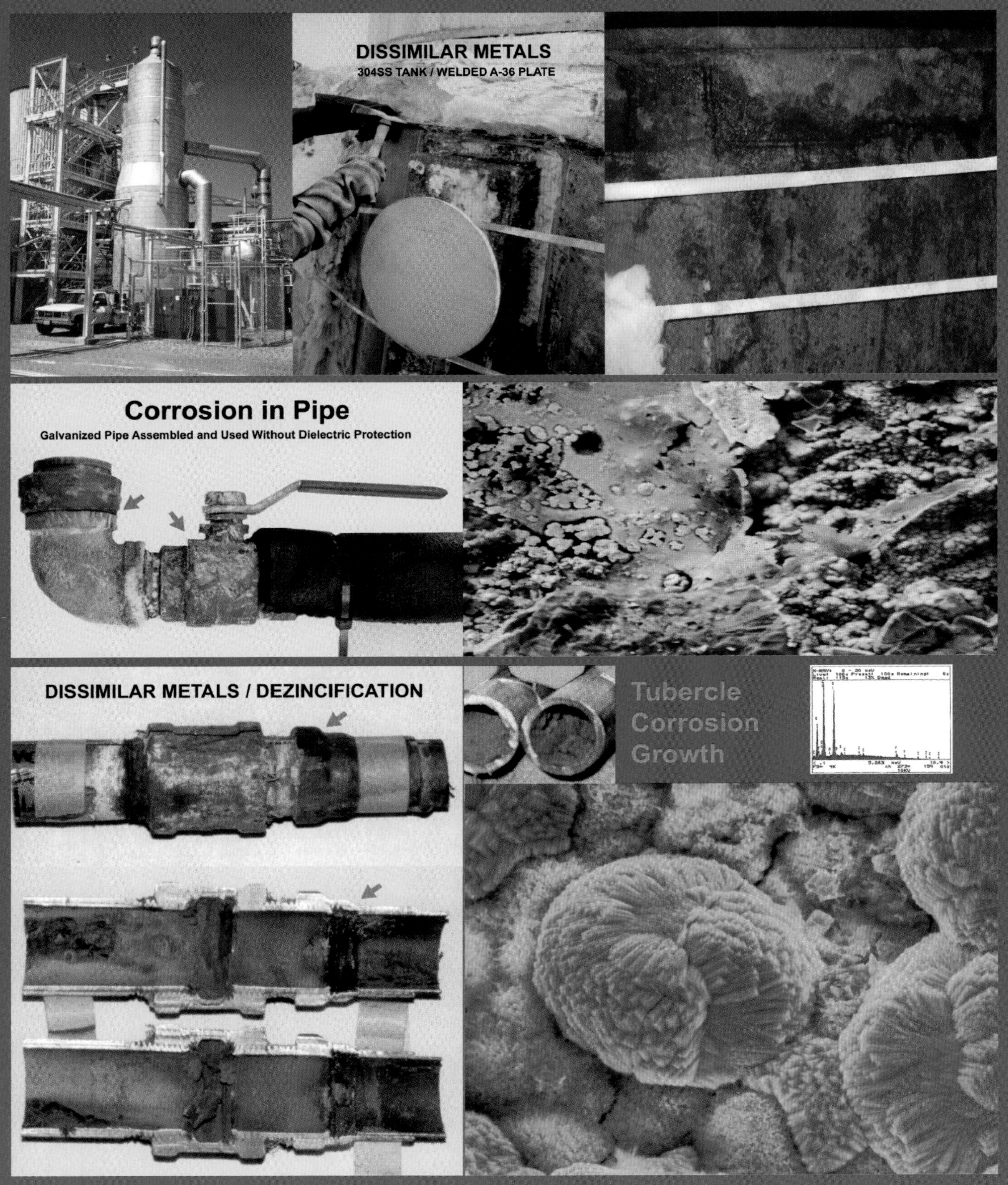

Courtesy of Tom Bertone Consulting, Whittier, CA.

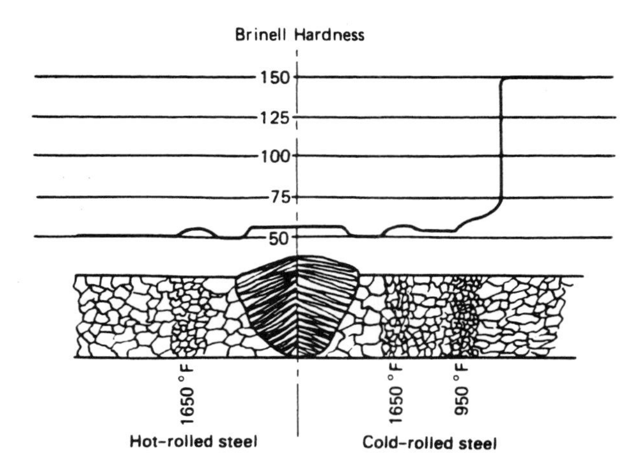

FIGURE 13.18. Single-pass welds in mild steel. The section in the annealed condition shows grain refinement in the heat-affected zone where the temperature reached 1650°F during welding. The graph on the left side shows that welding did not affect hardness. The section on the right shows cold-worked (cold rolled) mild steel before welding. Here grain refinement is seen in the vicinity of the recrystallization zones that reached 950° and 1650°F.

THE EFFECTS OF BASE METAL MASS

The term "heat sink" is often used when referring to the base metal mass. A large mass such as a thick plate will absorb more heat than a small mass. Rapid cooling rates are therefore associated with larger masses; this tends to develop more hardened structures, especially in view of the fact that manufacturers raise the carbon content of thicker plates. Large, heavy plates will not have a heat-affected zone as wide as thinner plates with the same weld size (Figures 13.19*a* and 13.19*b*). It is important to maintain a temperature for the particular steel (Figure 13.20). This is done by preheating, which results in a reduction of the cooling rate, thus preventing the formation of martensite.

PREHEATING

Preheating requires the raising of the base metal to a specific temperature prior to making the weld. Preheat may be localized or involve the entire part, and the temperatures may range from 200° to 600°F. Several methods used to measure the preheat temperature are chalk or crayons that melt at a specific temperature (Figure 13.21), a thermocouple, infrared detector, and/or placing the part in a temperature-controlled furnace. The reduced cooling rate prevents the formation of martensite (Figure 13.22). Even with preheat, it is difficult to effect transformation to the desired pearlite. Preheating is done to prevent cold cracks, reduce distortion and residual stress, and reduce hardness in heat-affected zones. Higher temperature pre-

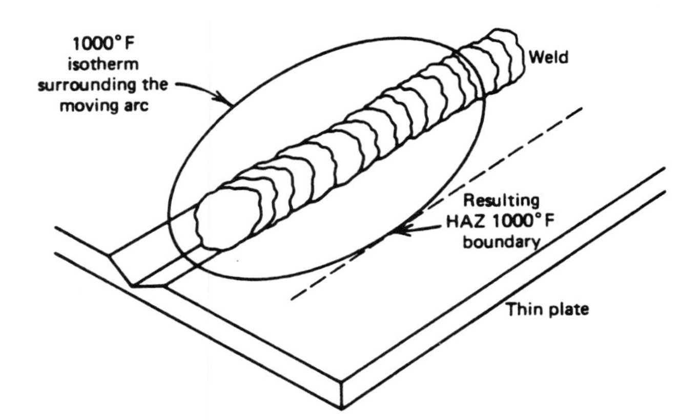

FIGURE 13.19a. A thin plate, when welded, has a wide heat-affected zone (up to 3/16 in.) because of its lower cooling rate.

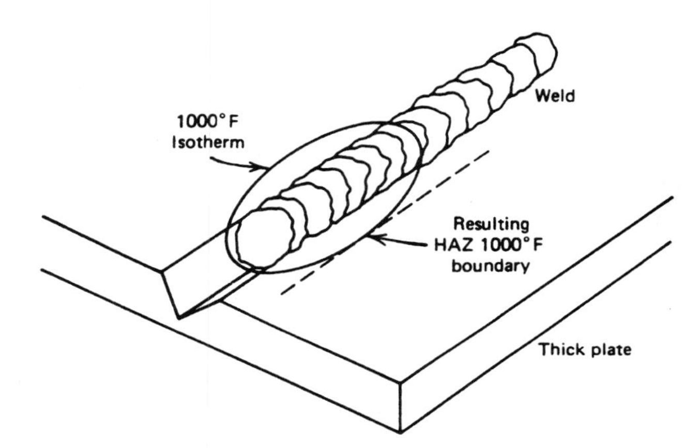

FIGURE 13.19b. A large, heavy plate has a high cooling rate and produces a narrow heat-affected zone.

heat has the disadvantage of widening the heat-affected zone and enlarging the grain structure; multipass welds tend to minimize this condition, however.

POSTHEATING

Postheat treatments for welds are used to relieve stresses, remove the effects of cold work, and increase toughness, strength, and corrosion resistance. Some postheat treatments are stress relief anneal, normalizing, full anneal, hardening, and tempering. These can be carried out in a furnace or by localized heating. Thermal-insulating blankets are sometimes used on very large weldments such as water or fuel tanks, and heat is then applied to the inside.

Hardened structures may be altered and softened by the spheroidizing process as discussed in Chapter 10. Pearlite can also be changed by this process from alternating plates to spheroids of cementite in a matrix of ferrite.

Carbon steels containing more than 0.40 percent C should be both pre- and postheated. Plain carbon tool steel should be welded with an electrode that contains

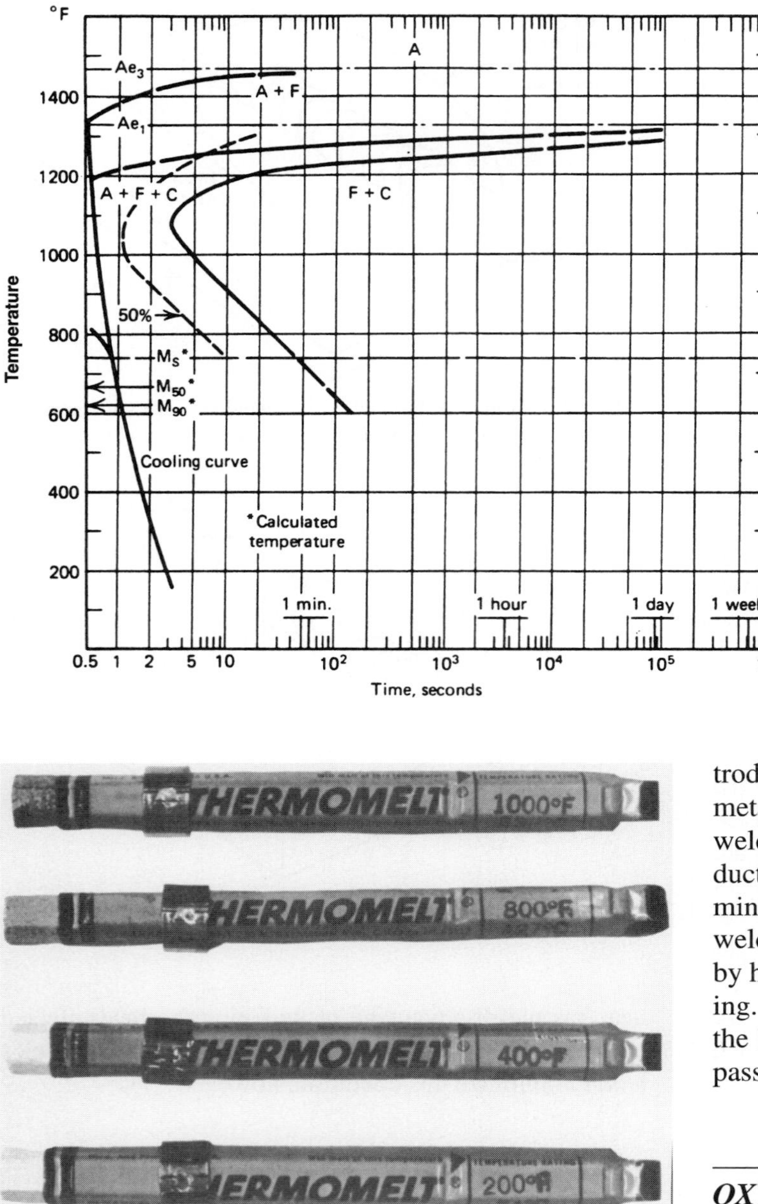

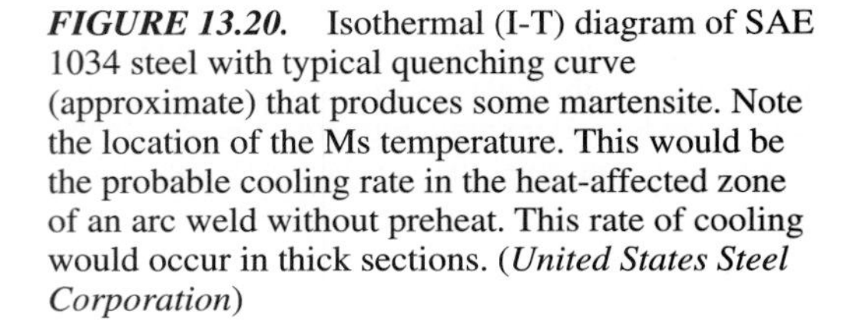

FIGURE 13.20. Isothermal (I-T) diagram of SAE 1034 steel with typical quenching curve (approximate) that produces some martensite. Note the location of the Ms temperature. This would be the probable cooling rate in the heat-affected zone of an arc weld without preheat. This rate of cooling would occur in thick sections. (*United States Steel Corporation*)

FIGURE 13.21. Temperature-measuring crayons are available for checking preheat, postheat, and interpass temperatures of weldments.

approximately the same carbon content. A welded carbon steel tool will lose its hardness due to tempering and/or annealing caused by the heat of the welding process; an additional heat treating process is usually required to restore the tool to its original condition.

PEENING OF WELDS

Welds may be peened when either hot or cold. This can be compared with the hot and cold rolling of metals in the steel mill. Peening on welds when they are cold introduces compressive stresses in the weld and base metal, which helps to relieve tensile stresses caused by welding. This strengthens the weld metal but lowers its ductility. A practice of peening while the weld is hot will minimize the formation of stressed grain structures if weld ductility is a requirement. Peening is either done by hand with a hammer, with air tools, or by shot peening. (*Note:* Do not peen the root pass, as it may displace the base metal or crack the weld. Do not peen the top pass, which will not be stress relieved by any later pass.)

OXY-ACETYLENE WELDING

In oxy-acetylene welding, a fuel gas and oxygen mixture is burned to produce an extremely hot flame. Flame temperatures with oxy-acetylene range from 5800° to 6300°F. When natural gas is used with oxygen, the flame temperature is a little over 5000°F, but when acetylene is burned with compressed air, the flame temperature is 3400°F, too low for welding steel by fusion.

Gas welds are made on steel, brass, aluminum, and copper. Braze welds are usually made by gas welding.

In brazing and braze welding, a metal or alloy of lower melting point than either of the base metals to be joined is used. Some common brazing alloys are bronze, copper, and silver alloys. Brazing is done by heating the base metal to a temperature above the melting point of the brazing rod metal but not high enough to melt the base metal. A flux is usually applied to clean the surfaces, and

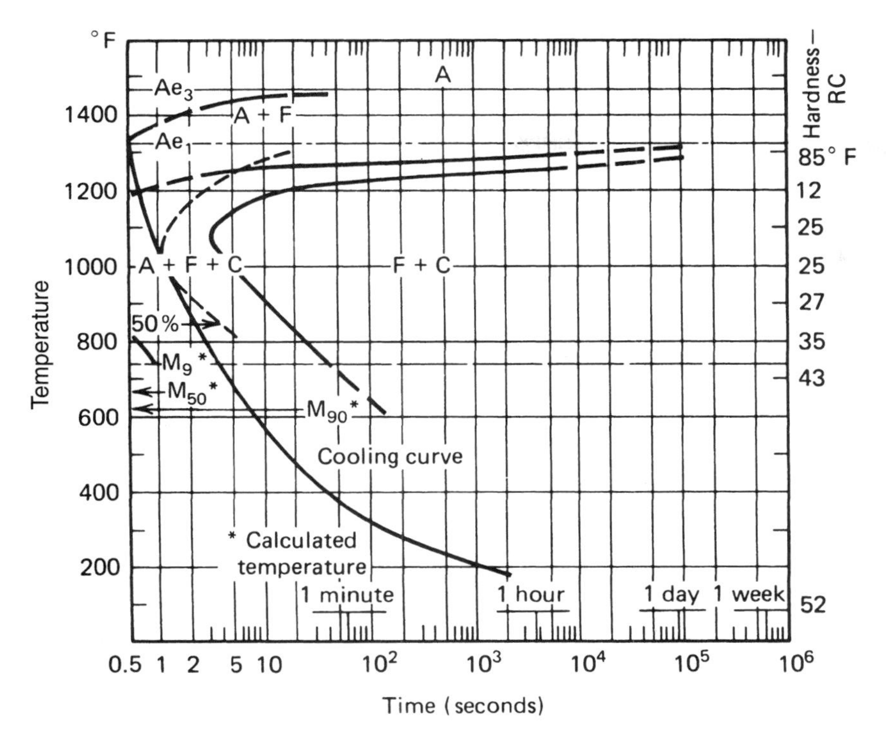

FIGURE 13.22. I-T diagram with cooling curve. Same as Figure 13.20, except that a cooling curve has been added, having a probable cooling rate like that found in most arc welds that have been preheated to 200°F. About 50 percent martensite is produced. (*United States Steel Corporation*)

the melted alloy flows over the surface or between the parts by surface or capillary action. The molten metal must "wet" the alloy by diffusing into the surface to some extent to create a bond that makes a strong joint when the metal solidifies. Figure 13.24*a* shows a braze weld overlay on carbon steel that failed to bond because the surface was not properly cleaned. Brazed welds can also become contaminated by base metal pickup (Figure 13.24*b*) if the base metal is overheated.

Gas welding with steel filler rods is done with a neutral flame (neither excess oxygen nor acetylene) with a low carbon steel filler rod. Welding rods for gas welding are similar to arc welding electrodes of the E-60XX series and also come under AWS specifications.

In the weld zone of gas welds on steel using steel filler rods, the metal is maintained in the molten state for a relatively long time with this welding process, depending on the volume of the molten pool and the thickness of the base metal. Consequently, the grain size increases in the heat-affected zone to a greater distance from the weld zone, depending on the material thickness. The structure of the weld zone in gas welds is very coarse and irregular (Figure 13.25). In the junction zone, which consists of a mixture of weld metal and base metal, the structure is still coarse, but with the slow cooling rate that is typical of gas welding, hard structures such as martensite are not likely to form.

The very coarse grains formed in the base metal, when it is overheated, tend to produce a brittle structure. Failure by cleavage is usually transgranular (through the grain) as contrasted to shear failure, which is often along grain boundaries (intergranular) (Figure 13.26). Cleavage is always a brittle failure occurring suddenly, but shear failures show some ductility. An example of this problem is sometimes seen when low or medium carbon tubing is overheated upon welding or brazing. The failure occurs, not in the weld, but beside it, sometimes as

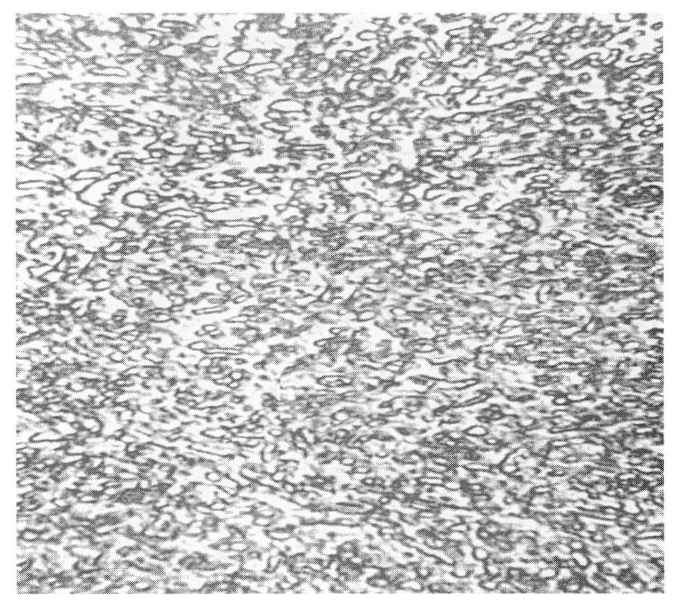

1000× 2% Nital Etch

FIGURE 13.23. This photomicrograph shows a typical spheroidized microstructure of an alloy steel. This micrograph was created by a subcritical heat treatment performed at 1300°F for 22 hours.

FIGURE 13.24a. This bronze overlay on a steel shaft shows a lack of bond in some places. The steel base metal was not adequately preheated and cleaned to ensure "wetting" of the steel with bronze.

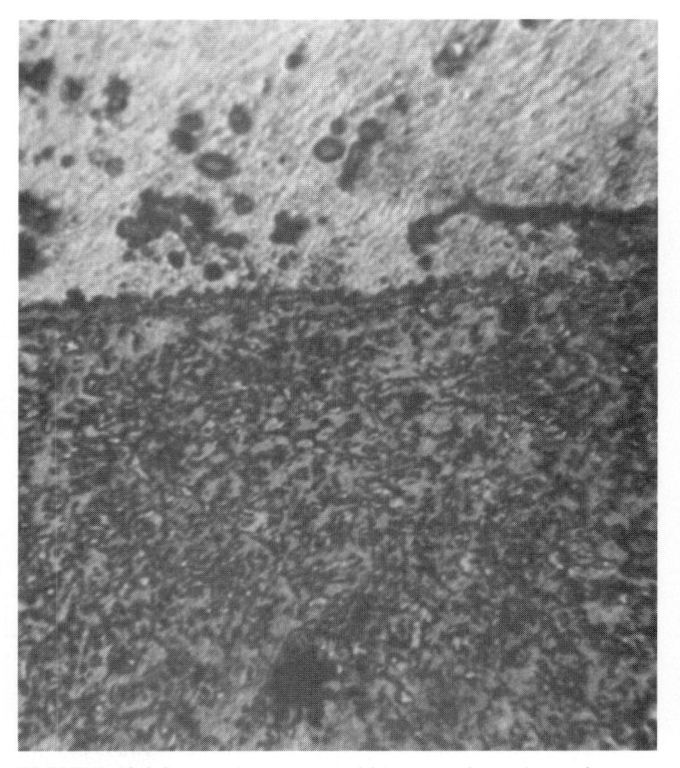

FIGURE 13.24b. This bronze weld was overheated, causing incipient melting of the steel surface and subsequent pickup of the displaced base metal into the molten bronze weld (100×).

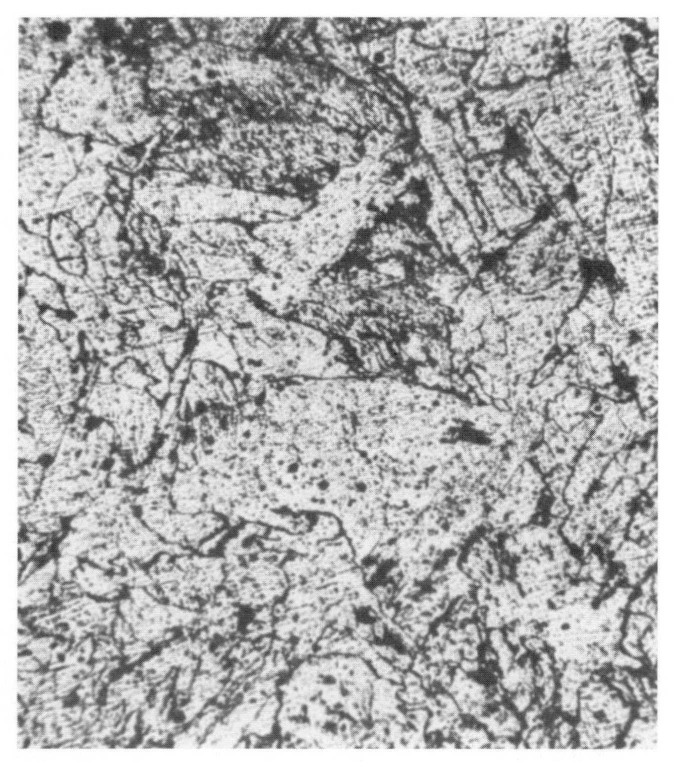

FIGURE 13.25. Structure of the weld metal in an oxy-acetylene weld showing a coarse, irregular structure (500×).

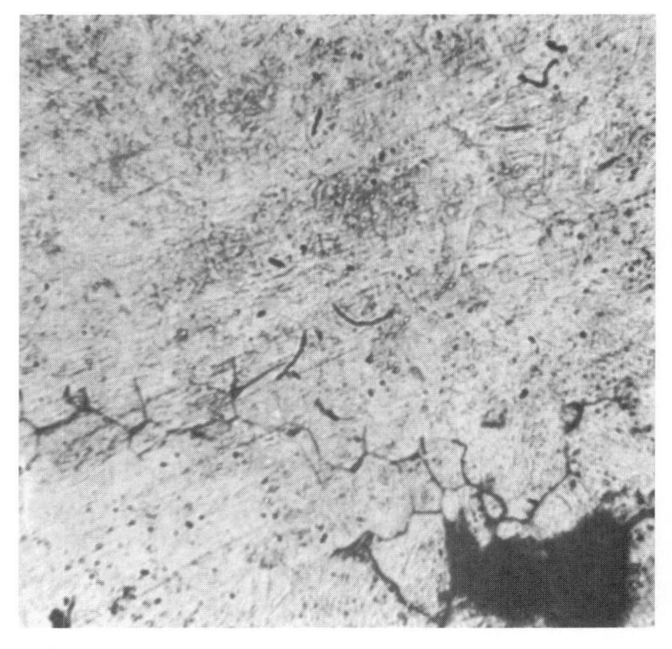

FIGURE 13.26. An intergranular weld failure, possibly developed at high temperature while the weld was solidifying. The black area at the bottom is a gas pocket, probably from hydrogen entrapment (500×).

much as ½ in. away (Figure 13.27). The thicker the base metal is, the more likely the problem will occur. Alloying elements such as manganese, chromium, and molybdenum tend to promote this cleavage pattern in the grains.

Figure 13.28 shows the difference in the distance from the weld axis through the heat-affected grains from the arc-welding and that of the oxy-acetylene-welding process. This grain growth is taking place in the base material that was in the normalized condition prior to welding.

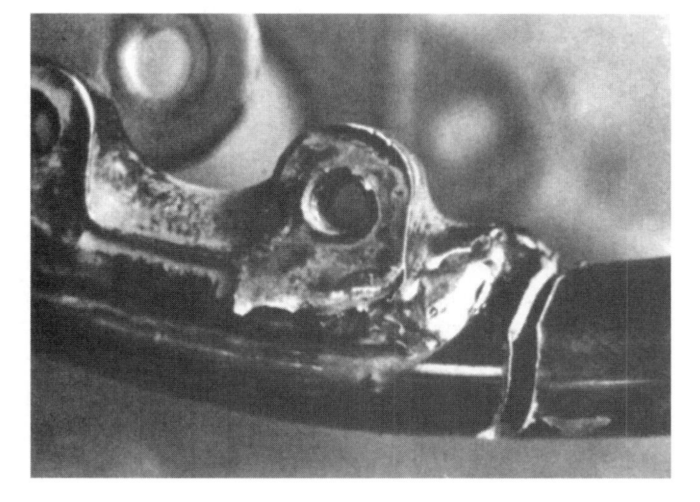

FIGURE 13.27. This tubular motorcycle frame failed in the heat-affected zone of an oxy-acetylene-welded part due to extreme grain growth away from the weld.

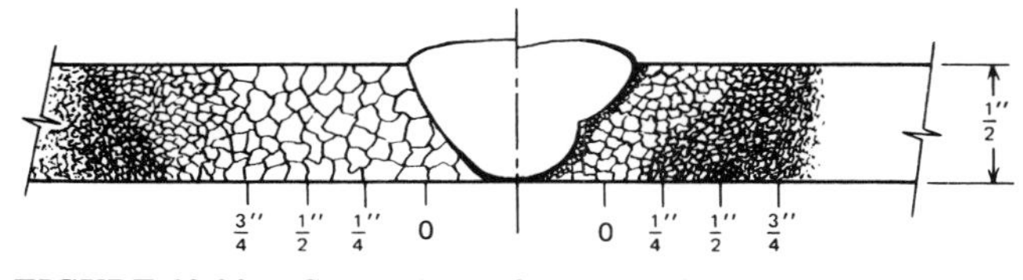

FIGURE 13.28. Comparison of oxy-acetylene welding (left) and arc welding (right).

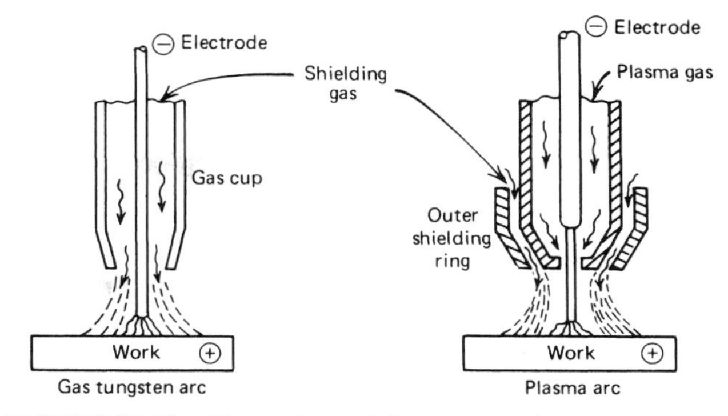

FIGURE 13.29. Comparison of plasma arc and gas tungsten arc-welding torches.

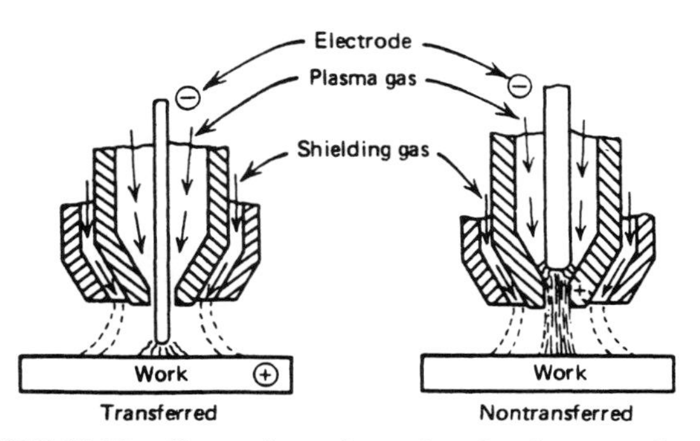

FIGURE 13.30. Comparison of transferred and nontransferred plasma arc torches.

PLASMA ARC WELDING AND CUTTING

In arc welding, the electric current flows between two electrodes through an ionized column of gas called a plasma at a temperature of about 6500°F. A new method of welding called **plasma arc,** or plasma torch, welding is sometimes used in place of the gas tungsten arc (TIG) process (Figure 13.29). The heat in plasma arc welding originates in an arc but not in the same way as in an ordinary arc where it is diffused (spread out). Instead, it is forced through a constricting orifice with a plasma gas that is supplemented with an ordinary shielding gas.

There are two systems: the **transferred arc** and the **nontransferred system.** The workpiece is part of the circuit in the transferred arc system as it is in the ordinary arc-welding system (Figure 13.30). The constricting nozzle surrounding the electrode in the nontransferred system acts as an electrical terminal with the arc forming between it and the electrode end. The plasma gas carries the heat to the workpiece. Gas temperatures are from 10,000° to 60,000°F.

The advantages of plasma arc welding over the gas tungsten arc process are higher welding speeds, narrower welds since there is greater energy concentration, and improved arc stability. The extremely high temperatures generated make possible the welding of high-temperature metals and the melting of refractory materials. Plasma arc heating and welding are most used for the more exotic metals and manufacturing processes and for wear surfacing.

The limitations of this process include the high initial cost of the equipment, the size of the torches, and the fact that the process is limited to horizontal position except for vertical up-welding in the keyhole mode. Plasma welding is limited to approximate thicknesses of 3/8 in. in stainless steel and 1/2 in. on titanium.

Plasma torch cutting has the advantage of making smooth cuts that are relatively free from contamination, with only a shallow melted zone and less metallurgical effects than with oxyacetylene cutting. Unlike oxy-acetylene cutting, a plasma torch can easily cut nonferrous and high-temperature melting alloys. Aluminum, for example, may be cut with a plasma arc in thickness ranging from 1/8 to 5 in. Stainless steel cannot be cut with the oxy-acetylene torch unless a ferrous powder is fed into the cut, but it is easily cut with the plasma arc torch.

FILLER METALS

Many of the welding processes involve a deposition of filler metal. In some cases, a consumable electrode is used. The electrode is covered with a flux. In other cases, a wire is used with an inert gas or a flux core that protects or shields the weld from the surrounding atmosphere. Other types of welding make use of a rod or wire that is melted in the joint by a heat source such as an acetylene torch or by tungsten inert gas (TIG). Brazing and soldering, for instance, make use of the filler metals in this way.

Both submerged arc welding and electroslag welding make use of a flux under which the bare rod or bare wire makes the weld. Different types of wires or electrodes used for welding are classified in the AWS system of classification for filler metals. In this classification system, the initial letters designate the basic process of deposition. The letter *E* stands for electrode, *R* for welding rod, and *B* for brazing filler metal. The combinations of *ER* and *RB* indicate suitability for either of the processes designated. In all the types of welding processes, the following is a list of the standardized filler metals that could be used.

1. Steel with little or no carbon in the composition (0.08 to 0.15 percent carbon)
2. Carbon steel
3. Low alloy steel
4. High alloy or stainless steels and manganese steels
5. Nickel and nickel base alloys
6. Copper base and copper alloys, tin base and tin solders, cobalt base alloys
7. Aluminum
8. Magnesium
9. Titanium

Some electrodes are tubular and contain a granular metal or compound. An example is the surfacing electrode used for depositing tungsten carbide. Table 13.2 gives the designation and composition of some covered electrodes used in carbon and low alloy steel. Table 13.3 shows the mechanical properties of some electrodes.

FLUXES AND SLAGS

When molten iron is exposed to oxygen, the oxygen atoms dissolve in the surface and penetrate by diffusion to some degree throughout the melt. Iron oxide (FeO) begins to form at the surface, and in time the entire melt will oxidize. If the melt is suddenly cooled, the solid metal state retains the oxygen, which causes porosity. Nitrogen and hydrogen also create porosity and cracking in welds by becoming entrapped in the metal as it solidifies.

In the early days of metal arc welding, bare filler wires were used for electrodes, but these produced a poor weld because the arc passed through the atmosphere and entrapped oxygen and nitrogen in the weld, thus producing a very porous and brittle weld. To prevent this contamination from the atmosphere, some form of shielding is necessary such as one of the following methods.

1. Fluxes
2. Slags
3. Gases (controlled atmospheres)
4. Vacuum

Fluxes are primarily used in welding processes to prevent oxidation of the metal by floating as a liquid on the surface of the metal. The flux chemically or physically combines with the surface oxide to remove it.

Shielding slags have a slightly different purpose, as they prevent oxidation of the molten metal and often combine with the melt to produce desired alloys or compounds in the weld metal. Some slags contain deoxidizers such as aluminum to remove trapped oxygen in the melt.

The use of gases for shielding or a vacuum in welding requires no slag to protect the weld surface. Shielding gas such as argon or helium are inert and can surround the arc and the molten pool to protect it from

TABLE 13.2. AWS A5.1-69 and A5.5-69 designations for manual electrodes

a. The prefix "E" designates arc-welding electrode.
b. The first two digits of four-digit numbers and the first three digits of five-digit numbers indicate minimum tensile strength.
    E60XX 60,000 psi minimum tensile strength
    E70XX 70,000 psi minimum tensile strength
    E110XX 110,000 psi minimum tensile strength
c. The next-to-last digit indicates position.
    EXX1X All positions
    EXX2X Flat position and horizontal fillets
d. The suffix (Example: EXXXX-A1) indicates the approximate alloy in the weld deposit.
    —A1 0.5% Mo
    —B1 0.5% Cr, 0.5% Mo
    —B2 1.25% Cr, 0.5% Mo
    —B3 2.25% Cr, 1% Mo
    —B4 2% Cr, 0.5% Mo
    —B5 0.5% Cr, 1% Mo
    —C1 2.5% Ni
    —C2 3.25% Ni
    —C3 1% Ni, 0.35% Mo, 0.15% Cr
    —D1 and D2 0.25–0.45% Mo. 1.75% Mn
    —G 0.5% min. Ni, 0.3% min. Cr, 0.2% min. Mo, 0.1% min. V, 1% min. Mn (only one element required)

Source: The Procedure Handbook of Arc Welding, 12th edition, The Lincoln Electric Company, 1973.

the atmosphere. In the gas metal arc, also termed MIG (metal inert gas), welding process, a wire is used for the electrode, and it is fed continuously while the welding is being performed. An inert gas such as argon or helium and, to a large degree, carbon dioxide is used. Carbon dioxide is not an inert gas and produces a small amount of slag. Some electrodes have a covering that shields the arc. As the covering burns off, it produces a gas that protects the melt from the atmosphere, and a viscous slag forms and solidifies to protect the weld nugget from contamination. Covered electrodes for just about every type of welding are available for the shielded metal arc process. Many different elements or materials are used in various coverings for specific purposes. Cellulose, limestone, asbestos, iron powder, sodium silicate, and many others are used for a rod covering. See Table 13.4 for typical composition of rod coverings.

By contrast, separate fluxes are sometimes used for gas welding. These fluxes are either applied to the joint being welded or applied to the rod by the operator

TABLE 13.3. Typical mechanical properties of mild-steel deposited weld metal

Electrode Classification	Condition: As Welded — Tensile Strength (psi)	Yield Strength (psi)	Elong. in 2 in. (%)	Impact[a] (ft-lb)	Stress relieved at 1150°F — Tensile Strength (psi)	Yield Strength (psi)	Elong. in 2 in. (%)	Impact[a] (ft-lb)
E6010	69,000	60.000	26	55[b]	65,000	51,000	32	75
E6011	70,000	63,000	25	50[b]	65,000	51,000	30	90
E6012	72,000	64,000	21	43	71,000	62,000	23	47
E6013	74,000	62,000	24	55	74,000	58,000	23	
E6020	67,000	57,000	27	50				
E6027	66,000	58,000	28	40[b]	66,000	57,000	30	80
E7014	73,000	67,000	24	55	73,000	65,000	26	48
E7015	75,000	68,000	27	90				
E7016	75,000	68,000	27	90	71,000	60,000	32	120
E7018	74,000	65,000	29	80[b]	72,000	58,000	31	120
E7024	86,000	78,000	23	38	80,000	73,000	27	38
E7028	85,000	78,000	26	26[c]	81,000	73,000	26	85

[a]Charpy V-notch at 70°F except where noted.
[b]Charpy V-notch at −20°.
[c]Charpy V-notch at 0°F.

Source: The Procedure Handbook of Arc Welding, 12th edition, The Lincoln Electric Company, 1973.

TABLE 13.4. Typical functions and composition ranges of constituents of coverings on mild steel arc-welding electrodes

Constituent of Covering	Function of Constituent: Primary	Secondary	Composition Range (%) in Covering on Electrode of Class: E6010, E6011	E6012 E6013	E6020	E6027	E7014	E7016	E7018	E7024	E7028
Cellulose	Shielding gas	—	25 to 40	2 to 12	1 to 5	0 to 5	2 to 6	—	—	1 to 5	—
Calcium carbonate	Shielding gas	Fluxing agent	—	0 to 5	0 to 5	0 to 5	15 to 30	15 to 30	15 to 30	0 to 5	0 to 5
Fluorspar	Slag former	Fluxing agent	—	—	—	—	—	15 to 30	15 to 30	—	5 to 10
Dolomite	Shielding gas	Fluxing agent	—	—	—	—	—	—	—	—	5 to 10
Titanium dioxide (rutile)	Slag former	Arc stabilizer	10 to 20	30 to 55	0 to 5	0 to 5	20 to 35	15 to 30	0 to 5	20 to 35	10 to 20
Potassium titanate	Arc stabilizer	Slag former	[a]	[a]	—	—	—	—	0 to 5	—	0 to 5
Feldspar	Slag former	Stabilizer	00	0 to 20	5 to 20	0 to 5	0 to 5	0 to 5	0 to 5	—	0 to 5
Mica	Extrusion	Stabilizer	—	0 to 15	0 to 10	—	0 to 5	—	—	0 to 5	—
Clay	Extrusion	Slag former	—	0 to 10	0 to 5	0 to 5	0 to 5	—	—	—	—
Silican	Slag former	—	—	—	5 to 20	—	—	—	—	—	—
Asbestos	Slag former	Extrusion	10 to 20	—	—	—	—	—	—	—	—
Manganese oxide	Slag former	Alloying	—	—	0 to 20	0 to 15	—	—	—	—	—
Iron oxide	Slag former	—	—	—	15 to 45	5 to 20	—	—	—	—	—
Iron powder	Deposition rate	Contract welding	—	—	—	40 to 55	25 to 40	—	25 to 40	40 to 55	40 to 55
Ferrosilicon	Deoxidizer	—	—	—	0 to 5	0 to 10	0 to 5	5 to 10	5 to 10	0 to 5	2 to 6
Ferromanganese	Alloying	Deoxidizer	5 to 10	5 to 10	5 to 20	5 to 15	5 to 10	2 to 6	2 to 6	5 to 10	2 to 6
Sodium silicate	Binder	Fluxing agent	20 to 30	5 to 10	5 to 15	5 to 10	0 to 10	0 to 5	0 to 5	0 to 10	0 to 5
Potassium silicate	Arc stabilizer	Binder	(a)	5 to 15[a]	0 to 5	0 to 5	5 to 10	5 to 10	5 to 10	0 to 10	0 to 5

[a]Used (in place of constituent on line above) in E6011 and E6013 electrodes to permit welding with alternating current.

Source: By permission, from *Metals Handbook,* Volume 6, *Welding and Brazing,* 8th edition, copyright American Society for Metals, 1971.

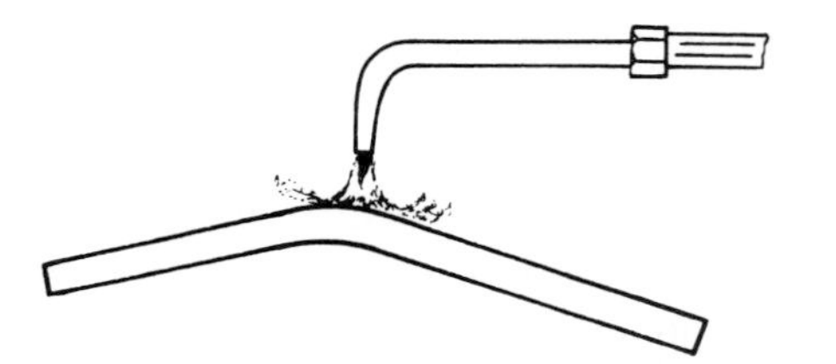

FIGURE 13.31. When heat is applied to one side of a bar of metal, it will bend as shown and return to its original position when cooled.

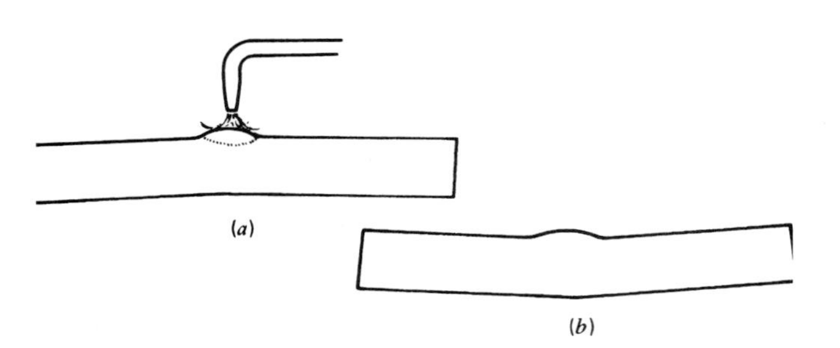

FIGURE 13.32. When heat is applied to a small spot so that it becomes red hot on one side of a bar (*a*), the heated spot will upset because of the restraint of the colder metal surrounding it. When it has cooled, it will be bent in the opposite direction (*b*).

while the rod is in a heated condition. Flux-cored electrodes use a flux in the core of the wire to help the welding process (in some cases with a gas).

SHRINKAGE AND DISTORTION

When heat is applied to steel, it will expand; when it is cooled, it contracts (Figure 13.31). Uneven heating and cooling, or heating and cooling in a localized area, can produce distortions in the base metal (Figure 13.32). Steel expands and contracts at the rate of approximately 0.0000065 in. per degree Fahrenheit change in temperature per inch of length. When steel cools to normal temperatures from a molten condition, this amount of contraction is sufficient to cause a great strain in rigidly held members and in the weld zone, but if they are free to move, considerable distortion can take place in the base metal.

It is also true that iron undergoes shrinkage or a reduction in volume of approximately 1.5 percent when it solidifies. This solidification in the weld zone itself induces a considerable amount of distortion upon the weldment as well as the contraction following solidification while cooling.

Welds on one side of a butt joint, for example, as seen in Figure 13.33, will cause the members to be drawn to that side and tilt toward the direction of welding. If both sides are welded alternately, the stress is equalized so that both members remain in their original position.

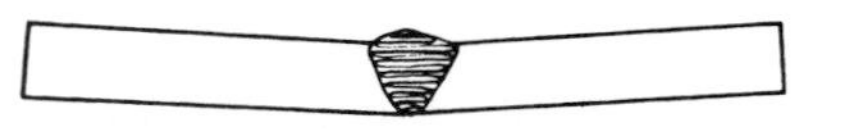

FIGURE 13.33. Welds, as they shrink, tend to move the welded parts out of square. This tendency of welds to shrink and distort weldments should always be kept in mind when making welds.

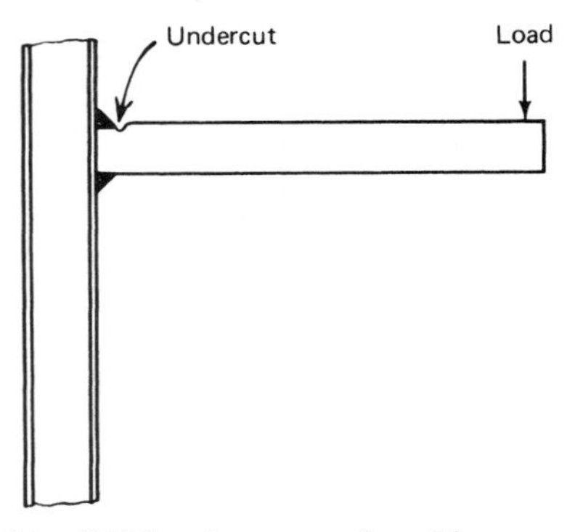

FIGURE 13.34. Weld undercuts, such as this one, create a stress concentration in the notch. Fatigue failures are often initiated by these stress raisers.

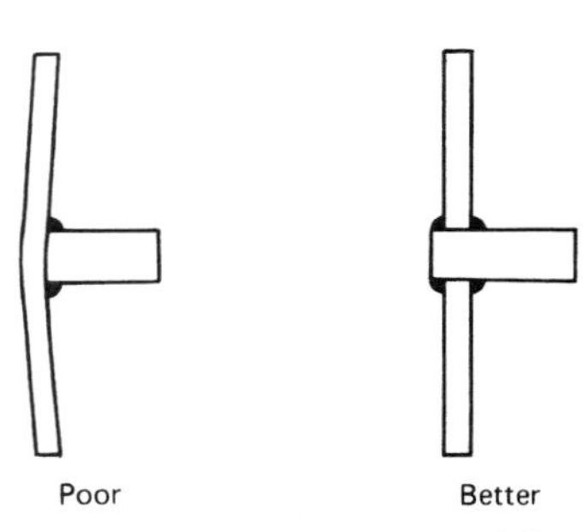

FIGURE 13.35. A piece of solid round stock is welded to a steel plate. The weld causes the steel plate to dish. If the round stock is extended through the part and welded on both sides, the dishing problem is minimized.

If there is an undercut or notch at the edge of the weld, a stress concentration is produced that increases the stress level at that point, since it is an abrupt change in section. A notch such as this, which is called a stress raiser, can initiate failure through cracking or fatigue (Figure 13.34).

Welding that is done on one side of a plate, for example, even if it is relatively light welding, will cause a warping or dishing of the plate as shown in Figure 13.35. In both cases, restraints can be put upon the base metal parts so that they will not distort; but when this is done, considerable internal strain is placed upon the weld and the heat-affected zone of the weld. Sometimes a result of this high stress is failure by cracking. This is a common problem, and it is often overcome by the use of shot peening or peening with air hammers or slag hammers. This is done preferably when the weld is hot so that it is a form of hot working of the steel, although cold peening is more ef-

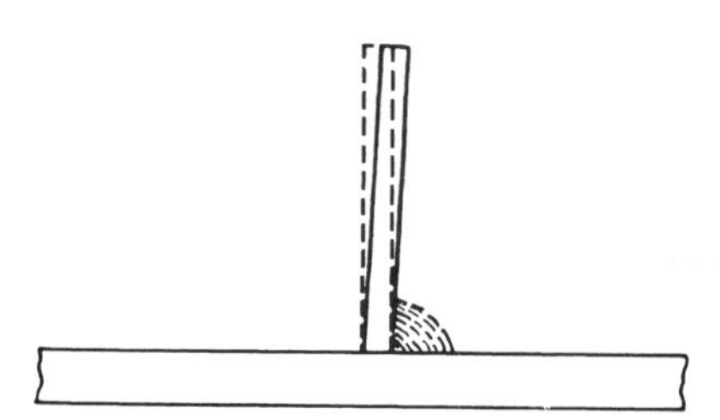

FIGURE 13.36. Biasing a part to be welded to allow for weld shrinkage will help to keep the weldment aligned.

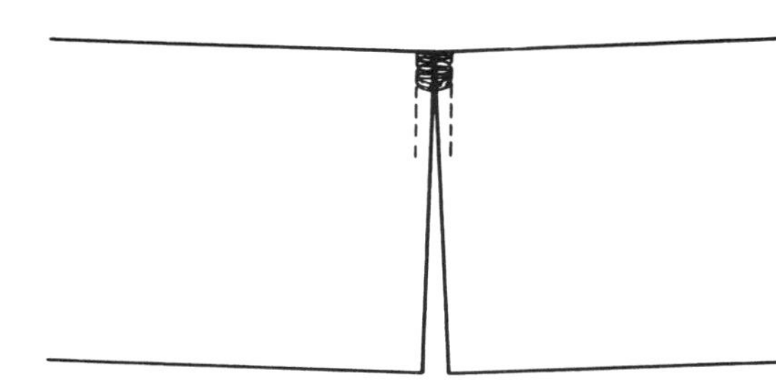

FIGURE 13.37. Plates can sometimes be separated slightly on one end to allow for weld shrinkage and thus avoid restraint.

fective in relieving the primary stress. Of course, stress-relief heat treatment would be even more effective.

Distortion of weldments through weld contractions can be offset to some extent by prepositioning the weldment on a bias that takes into consideration its probable movement or distortion so that when it assumes its final position after being drawn by the weld, it will be in the desired alignment (Figure 13.36). Another example of this is when two plates are butt welded together and the weld is started from one end, they will be drawn together when the weld is finally completed (Figure 13.37). This method eliminates any restraint on the base metal and allows the weld to contract normally with fewer internal stresses. However, it is not always easy to predict the outcome of the final position of the weldment when using this method.

Rolled metals possess a fibrous quality, called **anisotrophy,** in the direction of rolling. Compared with hot rolled steels, cold rolled or drawn steels are even stronger in the direction of elongation or rolling than they are in the transverse (crosswise) axis. Since metals have a tendency to split in the direction of rolling, the welder should be aware of the direction of weld stresses when making a weld so that weld stresses will be lengthwise and not crosswise (Figure 13.38) or through the thickness.

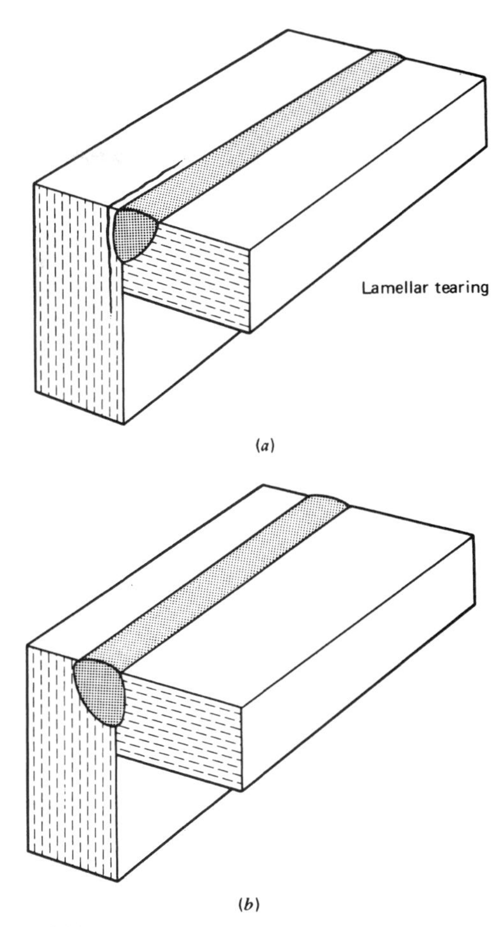

FIGURE 13.38. Lamellar tearing (*a*) and a suggested solution (*b*). (*The Lincoln Electric Company*)

TESTING OF WELDS

The quality assurance of good welds can be promoted by a weld inspection program and by a welder qualification program. The first program is used to assure the company that good welds are being made on its products. The second program is to qualify the welders for specific assignments or projects. Either program involves the following:

1. Visual inspection of each weld
2. Nondestructive testing such as radiographic, magnetic particle, ultrasonic, and liquid penetrant
3. Standard mechanical tests such as bending or tensile tests (These are destructive tests that are made on prototype welds or an occasional sacrificial weld.)
4. Laboratory tests such as hardness, micrographic, and macrographic
5. Field tests; simple methods of determining weldability of base metals with particular electrodes in field conditions where no other equipment is available for testing

Visual Inspection

This requires the good judgment of the welder to interpret the condition of the weld skillfully. Some of the defects

***FIGURE 13.39*a.** Weld having porosity and cracks plus slag inclusions. This is a very poor weld.

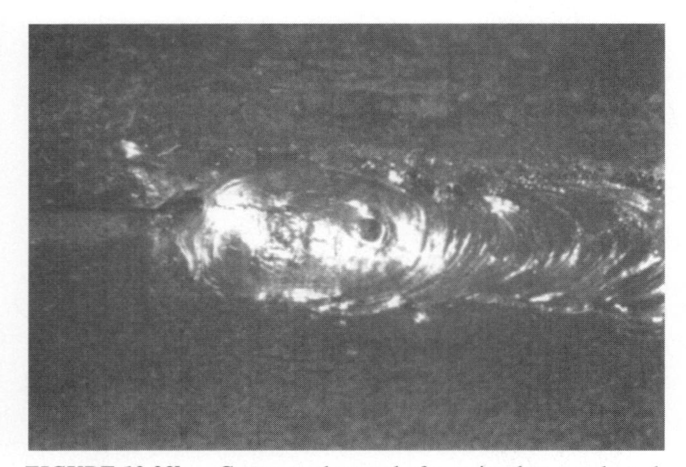

***FIGURE 13.39*b.** Crater crack at end of pass is a hot metal crack caused by removing the electrode too quickly.

he or she might detect in fillet joints and butt joints are an undersize weld, surface porosity, an undercut, and cracks (Figures 13.39*a, b*).

Nondestructive Testing

Internal porosity, surface cracks, flux entrapment, and inclusions can often be detected by X-ray (radiographic techniques), ultrasonic testing, and by fluorescent penetrant or magnetic particle inspection. These nondestructive testing techniques are described in Chapter 20.

Standardized Mechanical Tests

Standardized tests and tensile specimens may be found in publications such as *AWS Standard Qualifications Procedure, ASME Boiler and Pressure Vessel Code,* and *AWS Welding Handbook.* Some of the basic tests that are used are tensile tests to determine proportional limit, yield points, modulus of elasticity, elongation, and reduction of area. Hardness tests are often made on the cross section of the weld (Figure 13.40). Bend tests elongate the outer fibers of the weld to reveal the occurrence of cracks (Figure 13.41). Notch bar, impact tests, or fatigue tests are used to determine endurance limits or the fatigue strength of the weld or base metal. Torsion tests are used to determine yield strength and ultimate shear stress, and creep tests are used on weldments under stress to determine the amount of permanent set in a given time for a given temperature.

Details of the AWS Procedure Qualification Tests may be found in Figures 13.42 to 13.44. Complete specifications for welding qualifications may be found in welding data handbooks.

Laboratory Tests

Hardness testing and microscopic or macroscopic evaluation of welds are often performed on weld sections in the laboratory. Samples of the weld must be taken for the tests, making them destructive types of tests. These

FIGURE 13.40. The hardness of various areas on the weld on Rockwell C-scale numbers.

FIGURE 13.41. In this bend test the weld (bottom) failed along the junction zone of the weld. The top weld began to fail in the center of the weld.

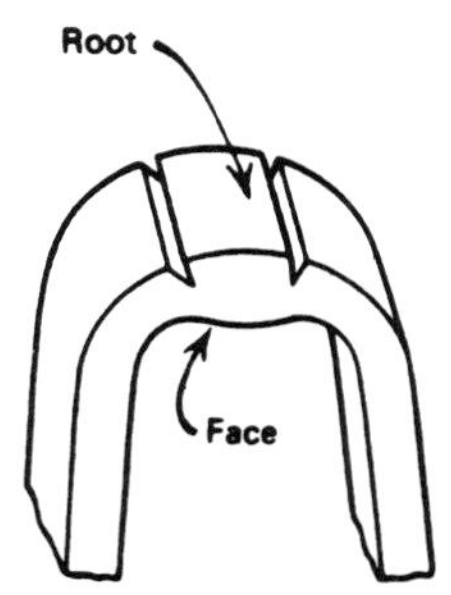

FIGURE 13.42. If weld reinforcement is not removed, stretching is concentrated in two places and failure results. (*The Lincoln Electric Company*)

samples are prepared for microscopic inspection, and the detection of carbon content, inclusions, porosity, and cracking can then be made.

Field Tests

A simple field test, for example, could be the use of spark-testing techniques to determine the carbon content of a base metal since the carbon would greatly affect the weldability of the part. Preheating and postheating may be essential to achieve sound welds. Hardness may be determined to some degree with a file test or a portable hardness tester.

Another very simple field test that welders sometimes use is a fracture test (clip test), in which a small piece of steel is welded at a right angle to the base metal with a fillet weld and then deliberately broken off using a hammer (Figures 13.45*a* through 13.45*d*). If the weld itself fractured, it would indicate that a good bond was maintained on the base metal, but if the weld remained intact and a section of the base metal were torn out, it would indicate that the base metal had too much carbon or alloy content, causing it to harden sufficiently to break in a brittle fracture. Although this very simple test might be useful for some field conditions, it is not satisfactory where many stringent requirements are placed

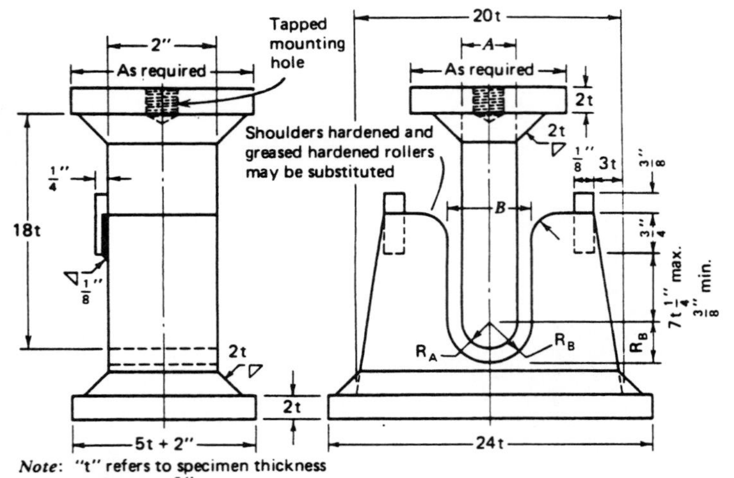

Note: "t" refers to specimen thickness
"t" for AWS test is $\frac{3}{8}$"
"t" for API Std. 1104 is tabulated wall thickness of pipe

(a)

Jig Dimensions	AWS TEST For Mild Steel Mimimum Yield Strength–psi			API Std. 1104 For All Pipe Grades
	50,000 and under	55-90,000	90,000 and over	
Radius of plunger R_A	$\frac{3}{4}$	1	$1\frac{1}{4}$	$1\frac{3}{4}$
Radius of die R_B	$1\frac{3}{16}$	$1\frac{7}{16}$	$1\frac{11}{16}$	$2\frac{5}{16}$

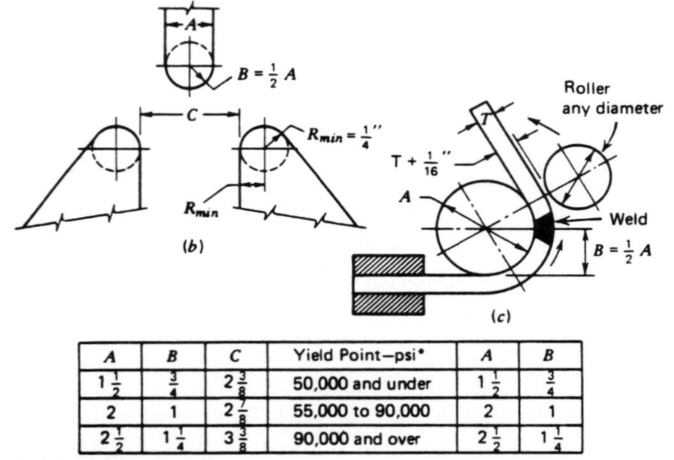

A	B	C	Yield Point–psi*	A	B
$1\frac{1}{2}$	$\frac{3}{4}$	$2\frac{3}{8}$	50,000 and under	$1\frac{1}{2}$	$\frac{3}{4}$
2	1	$2\frac{7}{8}$	55,000 to 90,000	2	1
$2\frac{1}{2}$	$1\frac{1}{4}$	$3\frac{3}{8}$	90,000 and over	$2\frac{1}{2}$	$1\frac{1}{4}$

*Minimum specified

FIGURE 13.43. (*a*) Jig for guided bend test used to qualify operators for work done under AWS and API specifications. (*b*) Alternate roller-equipped test jig for bottom ejection. (*c*) Alternate wraparound test jig. (*The Lincoln Electric Company*)

upon welds such as in building construction and steel pipeline operations. Ultrasonic and radiographic testing are methods that are often used in construction and pipeline welds.

HOT METAL SAFETY

Oxy-acetylene torches are frequently used for cutting shapes, circles, and plates in welding and machine shops. Safety when using them requires proper clothing, gloves, and eye protection. It is also very important that any metal that has been heated by burning or welding be plainly marked, especially if it is left unattended. The common practice is to write the word *HOT* with soapstone on such items. Wherever arc welding is performed in a shop, the arc flash should be shielded from the other workers. **Never** look toward the arc because if the arc light enters your eye, even from the side, the eye can be burned.

When handling and pouring molten metals such as babbitt, aluminum, or bronze, wear a face shield and gloves. Do not pour molten metals where there is a concrete floor unless it is covered with dry sand.

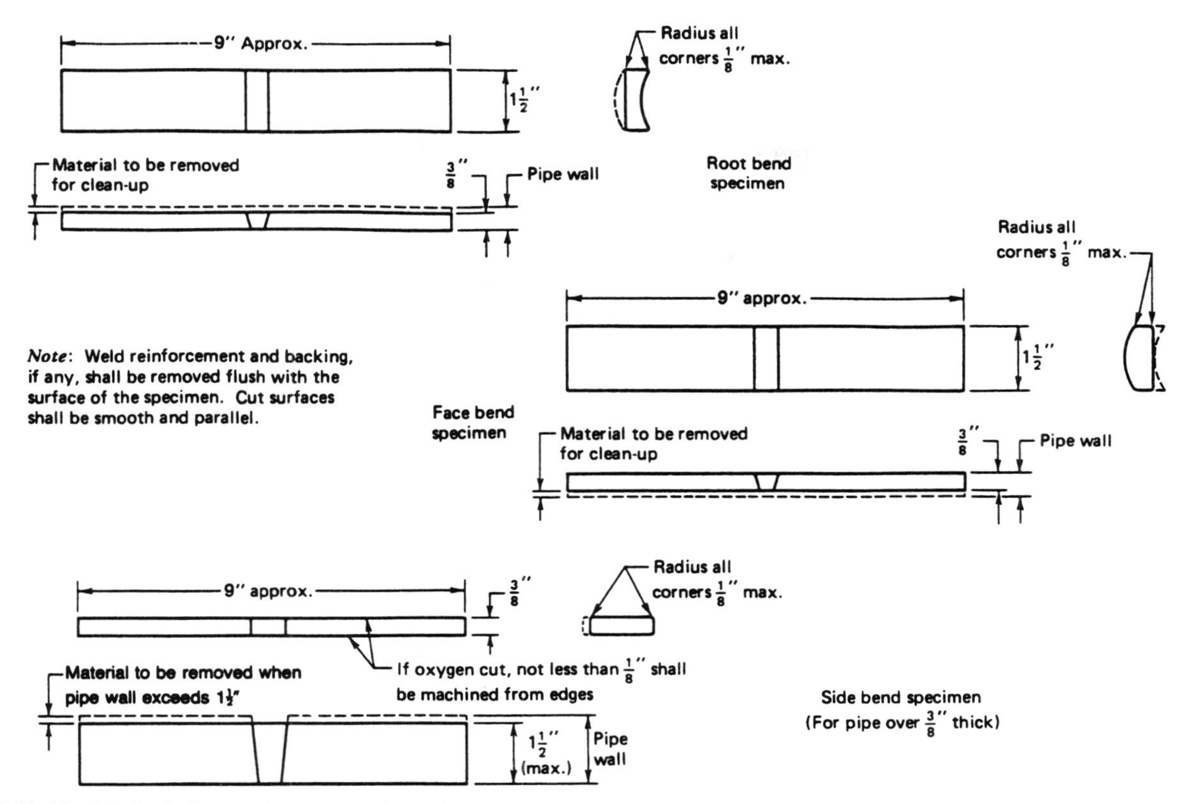

FIGURE 13.44. Method of preparing test specimen from 5 and 8 in. pipe for procedure qualification test. (*The Lincoln Electric Company*)

HAZARDOUS FUMES

Some metals such as zinc and cadmium give off toxic fumes when heated above their boiling point. Some of these fumes when inhaled cause temporary illness, but other fumes can be severe or even fatal. The fumes of mercury and lead are especially dangerous, since their effect is cumulative in your body and can cause irreversible damage. Cadmium and beryllium compounds are also very poisonous. Therefore, when welding, burning, or heat treating these metals, adequate ventilation is an absolute necessity.

Uranium salts are toxic and all radioactive materials are extremely dangerous. When working with any metals with which you are not familiar, it is best to check on toxicity and proper handling by consulting an appropriate reference book or safety representative.

ALLOY STEELS

Among the hundreds of alloy steels, some can be heat treated such as the quenched and tempered alloy steels. Others are high-strength, low alloy steels, stainless steels, abrasion-resisting alloy steels, heat-resisting high alloy steels, and high-temperature service alloy steels. Alloying elements affect hardenability and other properties of steels, all of which greatly affect the weldability of the metal and the condition of the weld.

The purposes of alloying steel with other elements are as follows:

- To increase strength, either at low or high temperatures
- To increase hardenability and improve toughness
- To increase wear resistance
- To improve magnetic properties
- To increase corrosion resistance

The effect of alloying elements on the hardenability of carbon steels was discussed in Chapter 12. There it was noted that the hardenability of carbon steels is increased by most alloying elements and that the element chromium is a ferrite former; that is, chromium dissolves primarily in ferrite, thus reducing the austenite range. Other alloying elements such as molybdenum, silicon, and titanium also tend to reduce the austenitic region. These elements tend to make steels transform into ferrite when cooling.

***FIGURE 13.45*a.** Clip test for determining weldability. (*Lane Community College*)

***FIGURE 13.45*b.** Clip being broken off with hammer. (*Lane Community College*)

CARBIDE FORMERS

Elements that are carbide formers (form hard compounds with carbon) are chromium, tungsten, molybdenum, titanium, and vanadium. These also dissolve to some degree in ferrite and readily combine with any carbon present. All carbides found in steel are hard and brittle, but the carbides formed by chromium, vanadium, and tungsten are extremely hard and have high wear resistance.

***FIGURE 13.45*c.** If the base metal is torn out and the weld remains intact, the base material is not weldable without preheat and postheat operations. (*Lane Community College*)

***FIGURE 13.45*d.** If the weld breaks, but the bond is good in the base metal, it can be considered weldable without preheat or postheat. (*Lane Community College*)

EFFECT OF ALLOYING ELEMENTS ON HARDENABILITY

Figure 13.46 shows how carbon steel is affected by the presence of alloying elements to change the critical range and how the position of the eutectoid is moved to the left. Although the carbon content is reduced at the shifted eutectoid point, the steel will be about as hard as 0.83 percent carbon steel when quenched. It can be seen then that the hardenability is increased in carbon steel by the addition of these alloying elements. Nickel and manganese tend to lower the temperature of the A_1-$A_{3,1}$ line.

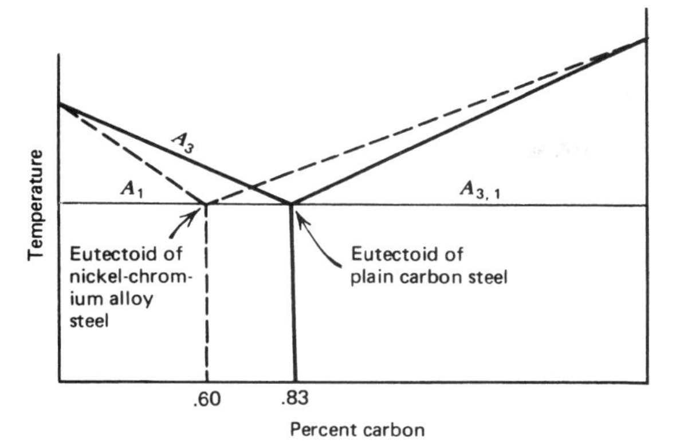

FIGURE 13.46. Iron-carbon diagram showing how eutectoid can move to the left. Most alloying elements tend to shift the eutectoid point to the left, thus increasing the hardenability in the medium carbon range. The nickel-chromium alloy will get as hard as the plain carbon steel.

Rapid cooling rates also tend to lower the A_1-$A_{3,\,1}$ line. When the temperature is lowered sufficiently, austenite cannot transform to ferrite because in the cooler temperature the movement of atoms in the lattice is too sluggish so that the alloy steel remains austenitic at low temperatures. Nickel and manganese also lower the upper critical temperature on heating, whereas molybdenum, aluminum, silicon, tungsten, and vanadium tend to raise it.

TEMPERING EFFECTS OF ALLOYING ELEMENTS

Another important factor in welding is that the complex carbide-forming elements such as chromium, tungsten, molybdenum, and vanadium affect the tempering process in that they raise the tempering temperature for a particular hardness and, in some cases, even increase their hardness when the tempering temperature (or postheat temperature) is raised. This is because of the delayed transformation (decomposition) of retained austenite; that is, not all of the austenite is transformed into martensite when it is quenched. Instead, the resultant precipitation of the remaining austenite into fine carbides takes place during the tempering process. This is known as secondary hardness that would, of course, increase the brittleness of the weld. Some high alloy steels should not be postheated for this reason. These factors are not only important in heat treating alloy steels but are also extremely important to the final condition of the metal when welding it.

SPECIFICATIONS FOR STEELS

Standard ASTM specifications are used to designate carbon and alloy structural steels. Shapes such as wide flange beams, angles, plates, and bars are used for the construction of bridges, buildings, pressure vessels, and other structural purposes. A36, for example, includes carbon steel structural grades of plate, bars, and shapes. Full ASTM specifications may be found in welding data handbooks.

The high-strength, low alloy (HSLA) steels have improved properties over carbon steel such as higher strength, abrasion resistance, and corrosion resistance. These steels are rapidly increasing in importance; millions of tons are now shipped annually. All HSLA steels can be welded, and the filler metals used are different for different grades. The metallurgical condition of base metal and weld metal is very important in these steels. These steels are usually ferritic in structure and are used in the as-rolled condition. Carbon content is usually between 0.07 and 0.22 percent and manganese content is from 0.40 to 1.60 percent. Alloy content includes nickel, chromium, phosphorus, silicon, vanadium, columbium, and copper (Figure 13.47). ASTM specifications for these steels include A242, A441, A572, and A588 SAE steels.

Some high-yield-strength HSLA constructional steels are quenched and tempered at the steel mill for desired properties. For this reason, these construction steels should be fabricated according to the manufacturer's specifications. When strength and weight are a concern, such as in modern fuel-efficient automobiles, some structural parts are being made of HSLA steels. Since parts that are made of much higher strength steel can be smaller or thinner and have the same strength as before, the total weight is reduced. The fact that prolonged heating as in gas welding or heating and bending will weaken these heat-treated steels means that new methods of welding and straightening will be required. Even though the carbon content is usually low, they can have high-yield strengths of 170,000 psi.

C	Mn	P	S	Si	Ni	Cr	Mo	V	Cu	B
0.15	0.92	0.014	0.020	0.26	0.88	0.50	0.46	0.06	0.32	0.0031

FIGURE 13.47. A typical composition of USS "T-1" steel. (*United States Steel Corporation*)

HSLA steels for Arctic pipelines were developed to resist rupture at low temperatures. This required the use of rare metals such as columbium and cerium and also by alloying with molybdenum and silicon. Welding electrodes were designed to be used with these special pipeline steels. Information on these steels may be obtained from the Molybdenum Corporation of America, Pittsburgh, PA.

The constructional alloy T1 Type B has a high-yield strength as shown in Figures 13.48*a* and 13.48*b*. Some wear- or abrasion-resistant alloys are similar but contain higher percentages of chromium and carbon and work harden to 400 BHN or more. A fairly high preheat is used on T1 steels for welding or gas cutting to reduce hard zones. Those containing a higher carbon content require a higher temperature preheat. Preheat temperatures vary with these steels. Low hydrogen electrodes E90XX or E100XX grades are sometimes used for welding these steels.

To avoid thermal cracking, no welding or cutting of these steels should be done on cold metal if the thickness is over 4 in. Manufacturers' data should be consulted before cutting or welding any HSLA steel. Normally HSLA steels up to 2 in. thick are welded with the gas metal arc process; thicker HSLA steels are usually joined by the submerged metal arc, electroslag, or electrogas methods. Ideally, the composition of the welding wire should closely match that of the HSLA steel to be welded. For example, in some cases when the HSLA yield strength is about 60,000 psi, carbon steel welding wires are acceptable. These wires include ER705-2 through ER705-7 and ER705-G. Some of these steels are subject to hydrogen embrittlement and should not be welded when wet. Likewise, low hydrogen stick rod should not be allowed to become damp. The very low carbon high-yield (HY) steels are less susceptible to hydrogen embrittlement and can be welded without preheat. Some of the newer HSLA steels used in shipbuilding are in this category. Complete data on properties, applications, and procedures are available for these steels from the manufacturers.

NICKEL STEELS

The nickel steels contain from 1 to 5 percent Ni, which increases strength, hardenability, and toughness at low temperatures. Nickel (in lower percentages) also reduces the carbon content of the eutectoid and lowers the brittle transition temperatures of steel. When large percentages are added, the steel becomes austenitic. Since higher percentages (10 to 12 percent) of nickel (or manganese) tend to lower the critical temperatures of low carbon steel retarding the decomposition of austenite, rapid cooling rates anneal or soften the steel instead of hardening it, especially if it is fully austenitic. These austenitic steels should not be preheated since complex carbides can form if the steel is held in the 1200° to 1600°F range for any length of time.

CHROMIUM STEELS

Chromium tends to form carbides, which increase the toughness and strength of ferrite. Chromium is also used as an alloying element for steels that are used for magnetic purposes. Chromium steels containing more than 0.30 percent carbon often require solution heat treatment when they are welded in order to inhibit the formation of carbides.

Nickel-chromium steels of the 3XXX series are among the heat-treatable alloy steels and are deep-hardening steels. These steels are used for drive shafts, gears, pins, and axles. Preheat is advised for welding nickel-chromium steels in the carbon ranges above 0.35 percent. No preheat is required in these steels below 0.25 percent C. Low hydrogen rod of similar carbon content should be used for welding these alloys. Choice of filler metal is critical.

MANGANESE STEELS

Manganese is principally used in all steels for a deoxidizer and to control sulfur content, but the steel is classed as an alloy steel when above 1 percent manganese is alloyed with steel. Manganese alloyed with steel promotes deep hardening, strengthens steel, and is a carbide former to some extent. Like nickel, manganese lowers the critical temperature range and decreases the carbon content of the eutectoid.

Austenitic manganese steel containing 12 percent or more manganese has high wear resistance and is used in power shovel teeth and buckets, grinding machinery, rock-crushing machinery, and railway tracks. This alloy has the peculiar property of hardening into large brittle carbides surrounding austenite grains when cooled slowly from 1700°F. On the other hand, if the same alloy, after giving the carbides time to dissolve, is quenched from 1850°F, the structure will be fully austenitic with much higher ductility and strength but softer than before. When this alloy is placed in service and subjected to repeated impact, the hardness increases to approximately

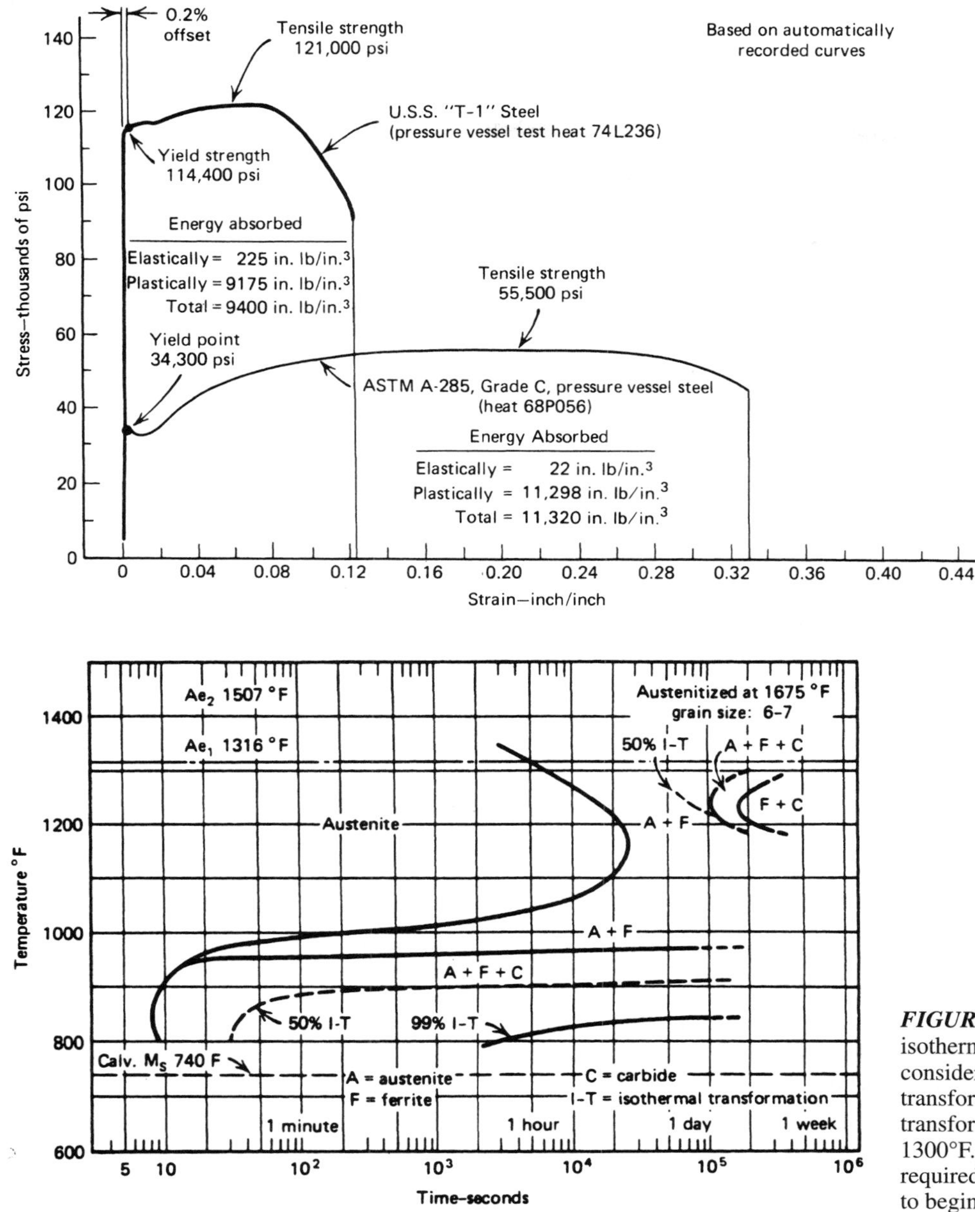

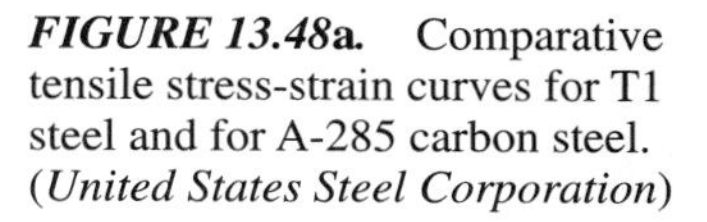

FIGURE 13.48a. Comparative tensile stress-strain curves for T1 steel and for A-285 carbon steel. (*United States Steel Corporation*)

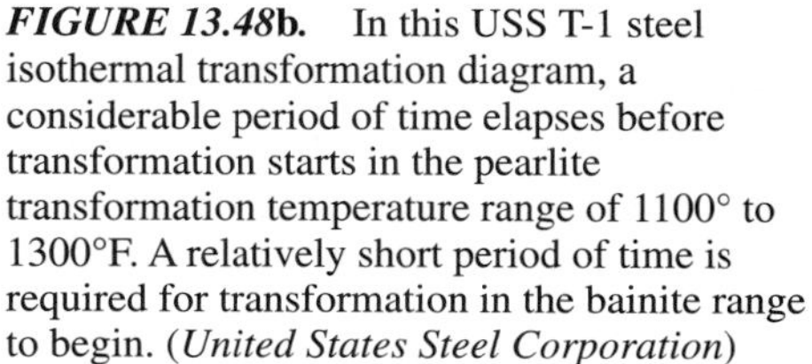

FIGURE 13.48b. In this USS T-1 steel isothermal transformation diagram, a considerable period of time elapses before transformation starts in the pearlite transformation temperature range of 1100° to 1300°F. A relatively short period of time is required for transformation in the bainite range to begin. (*United States Steel Corporation*)

550 BHN. This is because the manganese steels have the property of work hardening to a great extent. Reheating, as in welding, and cooling slowly will embrittle the metal. Low interpass temperatures are therefore needed when welding this metal. Water quench is sometimes used. Peening should not be used. Preheat or postheat is not desirable. An underlayer of austenitic stainless steel is sometimes applied to provide ductility between the work hardened manganese steel and the base metal. This tends to prevent spalling (pieces of weld coming loose on the surface). A disadvantage of this method is that subsequent oxy-acetylene cutting through the stainless steel weld would be difficult.

MOLYBDENUM, TUNGSTEN, VANADIUM, AND SILICON

Molybdenum, like chromium, has a strong effect in promoting hardenability and increases the high temperature hardness of steels. Molybdenum is often alloyed with nickel and chromium to steel.

The effect of tungsten in steel is similar to that of molybdenum in that it has a great effect on hardenability, is a strong carbide former, and causes steel to require a higher tempering temperature because it tends to retard the softening of martensite. Tungsten is mostly used in tool steels.

TABLE 13.5. Full annealing procedure

Stainless Steel	Temperatures °F	Time	Cooling[a]	Hardness (Typical)	
				Brinell	Rockwell
Type 410 Armco 410 Cb Type 403	1550–1650	1–3 h	Slow cool	137–159	B75–83
Armco 416 Type 420 Type 420F	1600–1650	1–2 h	Slow cool	170–201	B86–93
Type 440A	1625–1675	1–2 h	Slow cool	217–241	B94–98
Type 440B	1526–1675	1–2 h	Slow cool	217–241	B95–C20
Type 440C	1625–1675	1–2 h	Slow cool	229–255	B98–C23

[a]25–50°F per hour to 1100°F

Source: Armco Steel Corporation, Middletown, Ohio, *Heat Treating Armco Stainless Steels,* 1966.

Vanadium is also used for tool steels, springs, and bearings where high hardness and wear resistance are required. It is a deoxidizer and a strong carbide former. Vanadium also has the unique capacity to retard grain growth at higher temperatures in steels.

Silicon is used as a deoxidizer also and is much less expensive than vanadium. Above 0.06 percent silicon, it is considered an alloy of steel (silicon steel). Silicon increases strength and toughness. With 3 percent silicon in steel, the steel has useful magnetic properties and is used in electrical machinery.

STAINLESS STEELS

Stainless steels, as you have learned, are alloys of chromium, chromium-nickel, or chromium-nickel-manganese. They are classified in three types: the 400 series martensitic, the 400 series ferritic, and the 300 series austenitic. As you remember, martensite types are hardenable, having sufficient carbon content to harden by quenching; and others, such as ferritic, have no appreciable carbon. The austenitic type, being work hardening, also contains very little carbon. Each of these types presents unique welding characteristics and problems. Proper procedures must be observed and the right filler metals used when welding these metals.

Ferritic chromium grades contain very little carbon so that any brittleness is not caused by hard carbides. Since grain growth is very pronounced in the heat-affected zone in these steels, causing them to be brittle, they should not be used where vibration is a factor. Welded ferritic stainless steel joints possess a high shock resistance at elevated temperatures and may be used for this purpose. Annealing may be done at 1200° to 1550°F, but types 430, 430F, 442, and 446 cannot be normalized.

Martensitic steels contain carbon to 1 percent and may be hardened by heat treatment. Because some types are air hardening, the air cooling of weld zones lowers ductility and raises the hardness considerably. Martensitic steels can be welded, however, in the annealed, hardened and tempered, or hardened condition. Welding will produce a martensitic zone adjacent to the weld. Preheating and control of interpass temperatures are the most effective means of avoiding cracking. Postweld heat treatment is required.

Tables 13.5 and 13.6 show that full anneal and process anneal for these welds are similar to that of plain carbon steels. However, in Table 13.6 the process is done at lower temperatures. Table 13.7 shows annealing procedures for ferritic stainless steel.

Austenitic grades cannot be hardened by heat treatment and contain only small amounts of carbon, but can be hardened by cold working. Heating (welding) of unstabilized grades to temperatures from 800° to 1600°F will cause chromium carbide to form and precipitate out along the grain boundaries (Figure 13.49), reducing corrosion resistance. (Rust will form parallel to the weld on the heat-affected zone.) The chromium carbides can be dissolved by solution heat treatment of the welded part. This is done by heating at 1850° to 2100°F and quenching in the correct medium for the grade. Stabilized grades contain columbium or titanium (Types 321, 347, and 348) and are designed to limit the formation of harmful chromium carbides. The small amount of carbon instead becomes titanium or columbium carbides since these elements have a greater affinity for carbon than does chromium, and the carbides disperse more uniformly in the grains.

Grades having a very low carbon content (304L and 316L) can be heated for short periods without carbide precipitation. Table 13.8 reveals the quite different behavior of austenitic metals in annealing procedures. After heating to the correct temperature, the metal is quenched in water or oil instead of slow cooled to soften it. This restores a ductile grain structure and disperses

TABLE 13.6. Process annealing procedure

Stainless Steel	Temperatures °F	Time	Cooling Method	Hardness (Typical)	
				Brinell	Rockwell
Type 410 Armco 410 Cb Type 403 Armco 416	1350–1450	1–3 h	Any	170–197	B86–92
Type 414	1200–1300	2–6 h	Any	241–255	B99–C24
	1375–1425	2 h	Air	241–255	B99–C30
Type 431	1150–1225	4–8 h	Any		
Type 420 Type 420F	1350–1450	1–4 h	Any	207–223	B94–97
Type 440A	1350–1450	1–4 h	Any	229–248	B97–C22
Type 440B Type 440C	1350–1450	1–4 h	Any	235–255	B98–C23
Type 440F	1350–1450	1–4 h	Any	255–277	B100–C26

Source: Armco Steel Corporation, Middletown, Ohio, *Heating Treating Armco Stainless Steels,* 1966.

TABLE 13.7. Annealing procedures for ferritic stainless steels

Stainless Steel	Annealing Temperature Range, °F	Time[a]	Quench	Typical Hardness (as Annealed)	
				Brinell	Rockwell
Type 405	1200–1500	1–2 h	Air or water	140–163	B77–85
Type 430	1400–1525	1–2 h	Air or water	140–163	B77–85
Type 430F	1200–1450	1–2 h	Air or water	163–192	B85–91
Type 442	1400–1525	1–2 h	Air or water	149–174	B80–88
Type 446	1400–1525	1–2 h	Air or water	159–183	B84–90
Armco 400	1500–1700	3 minutes per 0.100 in. thickness	Air	121–137	B69–75
Armco 409	1500–1700	3 minutes per 0.100 in. thickness	Air	130–150	B72–80
Armco 18 SR	1500–1700	3 minutes per 0.100 in. thickness	Air	163–210	B85–95

[a]One to 2 hours recommended for heavy sections. Hold at temperature 3 minutes for every 0.100 in. (2.54 mm) of thickness.

Source: Armco Steel Corporation, Middletown, Ohio, *Heat Treating Armco Stainless Steels,* 1966.

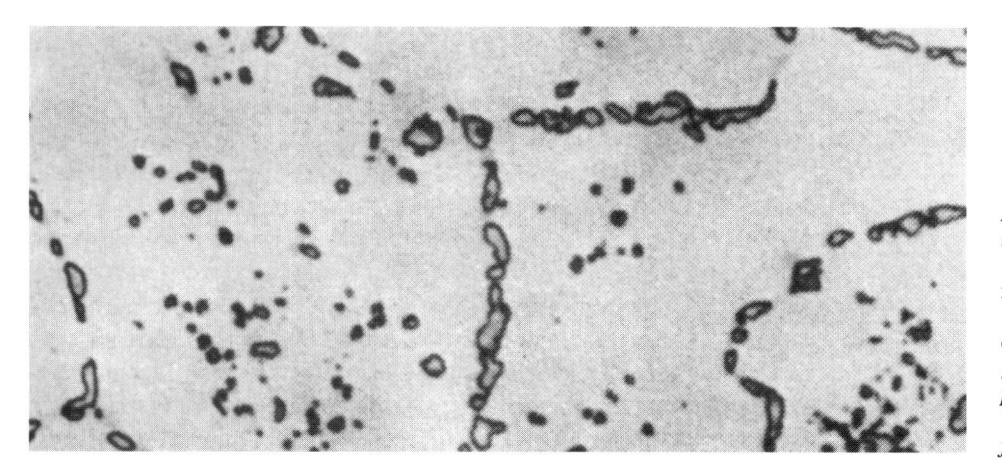

FIGURE 13.49. Type 310 stainless steel hot rolled plate annealed at 1950°F, water quenched in less than 3 minutes, exposed for 27 months at 1400°F, slowly air cooled. Chromium carbides are precipitated at the austenite grain boundaries (250×). *(By permission from Metals Handbook, Volume 7, Copyright American Society for Metals, 1972)*

TABLE 13.8. Solution annealing temperatures for austenitic stainless steels

Stainless Steel	Solution Annealing Temperature Range °F
Type 301	1900–2050
Type 302	1900–2050
Type 302B	1900–2050
Armco 303	1900–2050
Type 303 Se	1900–2050
Type 304	1850–2050
Type 304L	1750–1950
Type 308	1850–2050
Type 309	1900–2050
Type 310	1900–2050
Type 316	1900–2050
Type 316L	1750–1950
Type 317	1950–2050
Type 317L	1750–1950
Type 321	1750–1950
Type 347	1800–1950
Type 348	1800–1950
Armco 18–9 LW	1900–2050
Armco NITRONIC 32	1850–1950
Armco NITRONIC 33	1850–1950
Armco NITRONIC 40	1950–2050
Armco NITRONIC 50	1950–2050
Armco NITRONIC 60	1950

Source: Armco Steel Corporation, Middletown, Ohio, *Heat Treating Stainless Steels,* 1966.

any chromium carbides or prevents their formation at the grain boundaries. The austenitic stainless steels (except free-machining grades) are more weldable than ferritic and martensitic grades. Weld joints are tough.

Precipitation-hardening stainless steel alloys such as 17-4 PH, one of the original grades, can be welded when in a specific heat treat condition. This alloy has low ductility in the short transverse direction when solution heat treated, but resists cracking when heat treated to its minimum strength condition. Cracking in the root pass is a major problem with this alloy. One solution is to use E-308 L electrodes for two root passes and W 17-4 PH electrodes for the balance of the weld passes. Weldments in the newer PH 13-8 MO, PH 14 MO, and 15-5 PH exhibit higher ductility and greater strength regardless of the condition of heat treatment.

Stainless steel may be welded by several processes, such as oxy-acetylene, TIG, GMAW (MIG), or shielded metal arc welding (SMAW). Flux-coated electrodes also provide a fast, economical weld deposit with good penetration.

WELDING TOOL STEELS

Tool steels are similar to alloy steels in that they often contain many of the same alloying elements in varying amounts. Tool steels may be welded if the proper electrode is used and correct preheat and postheat procedures are used. A filler metal must be used that closely approximates the base metal in carbon and alloy content. Preheat is usually necessary to avoid cracking since the HAZ (heat-affected zone) is typically embrittled by welding. Arc-welding procedures are normally used. To restore the tool or part to its original hardness, a hardening and tempering procedure that is correct for that type of tool steel must be followed in most cases. An anneal may be necessary to allow re-machining of the welded part prior to hardening. In some cases, a hardened and tempered tool steel may be welded without anneal by using special electrodes and procedures. For example, austenitic stainless steel electrodes are often used where a ductile joint is required, such as when welding dissimilar metals or broken tools. A buttering technique is used when welds are large, followed by a fill-in of low hydrogen rod.

Welds on high-speed tools may be done either with the tungsten base (T) type tool steel electrode or with the molybdenum high-speed electrode (M). The tungsten type should be used when red hardness is needed and the molybdenum type when full hardness must be maintained as high as 1000°F. Expensive dies and cutting tools are often repaired with these electrodes, but correct procedures must be used for making a successful weld. In most cases, only a preheat is used; the electrode deposit being hard as-welded for use. Heat treating procedures may be used if desired, but distortion or warping may occur in the part and subsequent grinding would then be required. Of course, grinding or sharpening is always required when a cutting tool such as a milling cutter is repaired by a welding process.

Preheating temperature for high-speed steel should be at 1000°F for a length of time that will ensure uniform and thorough heating of the part (this will not affect the hardness of high-speed steel). The temperature should be brought up very slowly. A short arc should be maintained with the proper size electrode to avoid the formation of undercuts and craters. Use a skip-weld procedure with immediate hot peening of the weld to relieve stresses. Do not peen the weld when it is cold.

It is very important to maintain the 1000°F temperature while welding on the part. Sharp changes in temperature will cause cracking in either the weld zone or the heat-affected zone. When the weld has been completed, return it to the furnace and hold it for a few minutes at 1000°F and then allow the furnace to cool slowly to room temperature. Slowly reheat the part to 1000°F to temper and stress relieve it. Again, allow it to cool slowly to room temperature.

High-speed steel tools such as milling machine cutters can be repaired by welding (Figures 13.50*a* to 13.50*f*). Since the weld is usually sufficiently hard as welded, no hardening heat treatments are needed on cutting edges.

FIGURE 13.50a. Broken high-speed milling cutter showing one tooth broken off at the time of fracture.

FIGURE 13.50b. The cutter has been prepared for welding by grinding vee grooves on both sides.

FIGURE 13.50c. The cutter is clamped on a flat surface and the assembly is placed in a cold furnace. The temperature is then brought up slowly to 1000°F.

FIGURE 13.50d. Completed weld on milling cutter showing buildup of broken tooth.

FIGURE 13.50e. Side cutting teeth of cutter being sharpened. Peripheral grinds have already been made.

FIGURE 13.50f. Welded and resharpened cutter now in use.

CRACKING IN THE WELD METAL

Cracking down the center of the weld or microcracking of the weld metal usually takes place at high temperatures (hot cracking) during solidification and is caused by stresses of expansion and contraction in the weld. This is sometimes the result of trying to fill in a gap that is too wide with the weld. High cooling rates tend to increase hot cracking in the weld metal. Slowing the cooling rate will help to prevent hot cracking (Figure 13.51). The end or terminal crater crack is one example of hot cracking. This can often be eliminated by back filling of the crater before breaking the arc.

Preheating is also used to slow the cooling rate in low and high carbon alloy steels. Tables in welding manuals provide correct preheat temperatures for a specific steel and electrode. Some alloy and carbon steels develop coarsened grains if the preheat temperature is too high and the interpass temperatures are consequently above the specified degree (Figure 13.52). If this occurs, a postheat normalizing procedure is necessary to refine the coarse brittle grain in the heat-affected zone (HAZ).

Weld metal cracks may be caused by the following.

1. Wrong choice of filler metal or electrode (For example, a high tensile filler metal on low carbon steel base metal is the wrong application.)
2. Welding conditions; too much gap between base metal parts or pickup of embrittling elements such as carbon, nitrogen, or hydrogen
3. Defects such as porosity, slag or oxide inclusions, and hydrogen gas pockets (hydrogen embrittlement)
4. Too rapid cooling rates (no preheat)
5. Constraint of base metal parts (For example, two heavy plates are clamped rigidly so that no movement can take place while welding: all the stresses involved are concentrated in the weld metal.)

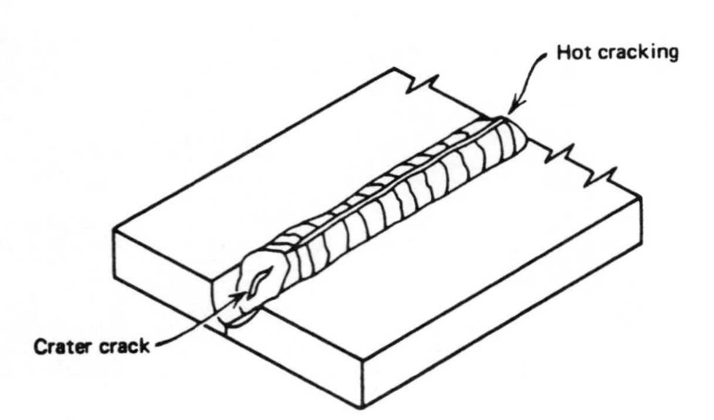

FIGURE 13.51. Hot cracking in the weld is often seen as a separation down the center of the weld.

Cracking in the parent or base metal is a serious problem when welding high carbon or tool steels. These cracks form in the fusion zone or in the heat-affected zone adjacent to the weld bead (Figure 13.53). This is called underbead or hard cracking and cannot normally be detected on the surface of the base metal. These cracks are difficult to detect even with magnetic particle inspection or X ray. Any welding method performed on

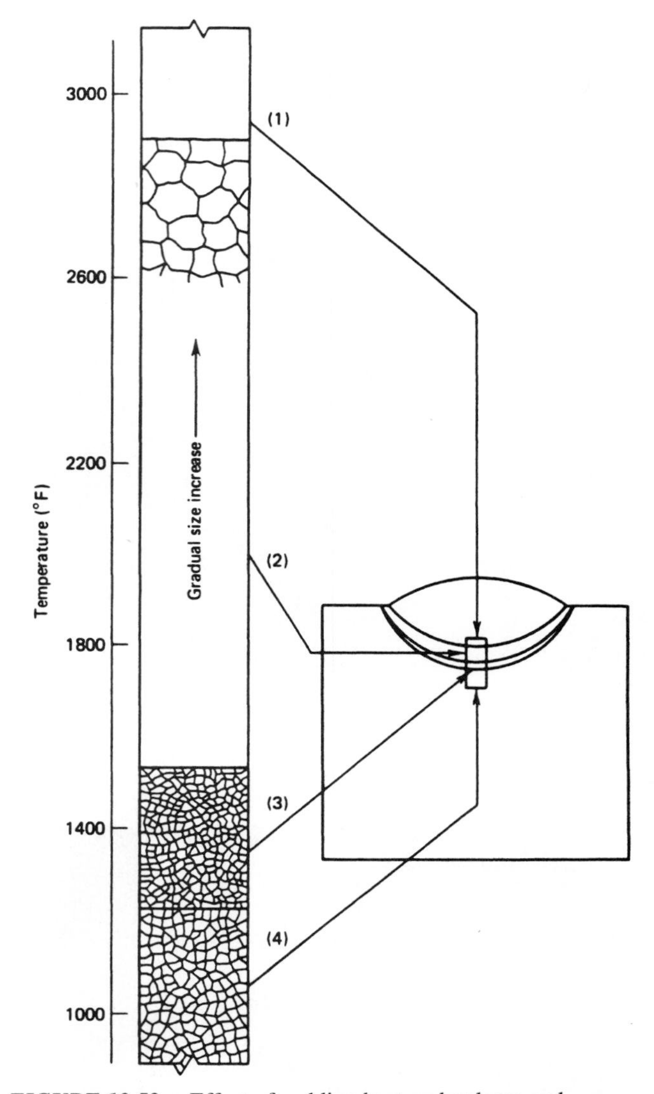

FIGURE 13.52. Effect of welding heat on hardness and microstructure of an arc-welded 0.25 percent carbon steel plate. The schematic diagram represents a strip cut vertically through the weld shown. Significance of the four numbered zones is: (1) Metal that has been melted and resolidified. Grain structure is coarse. (2) Metal that has been heated above the upper critical temperature 1525°F for 0.25 percent carbon steel but has not been melted. This area of large grain growth is where underbead cracking can occur. (3) Metal that has been heated slightly above the lower critical temperature 1333°F but not to the upper critical temperature. Grain refinement has taken place. (4) Metal that has been heated and cooled, but not to a high enough temperature for a structural change to occur. (*The Lincoln Electric Company*)

medium to high carbon steels will result in a hard zone adjacent to the weld if the cooling rate is too fast. The underbead crack follows the contour of the weld bead.

Radial cracks are sometimes a result of welding stress in hard or brittle zones. Hydrogen entrapped in the base metal and weld metal often initiates such underbead cracking. Hydrogen is very soluble in molten steel but is forced out of the metal by diffusion if the cooling rate is slow. Hydrogen-induced cracking is delayed cracking, awaiting diffusion of hydrogen atoms into discontinuities, which rupture under the hydrogen pressure. Hydrogen may be picked up by the use of damp electrodes or by those having a cellulose type of coating. Oil, grease, paint, or water on the base metal will also cause hydrogen embrittlement. When low hydrogen rod that has not picked up any moisture is used on higher carbon steels and preheat is used, underbead cracking is not likely to be a problem (preheating encourages harmless diffusion of hydrogen away from the weld and heat-affected zone).

Toe cracks (Figure 13.53) are caused by stress at the toe of the weld and across the heat-affected zone. Preheating reduces the problem that is usually found in metals with low ductility. A root crack (Figure 13.53) originates in the root bead. This is caused by a rapid cooling rate of the small weld bead; carbon pickup from oily metal intensifies this problem. Preheat should be used.

HARD SURFACING

Machine part surfaces subject to wear and abrasion such as power shovel teeth and rock-crushing machinery are sometimes built up with wear-resistant rod materials. There are scores of hard-facing alloys from which to choose. For a successful weld, you must know the composition of the base metal when making this selection. The proper selection of rod is also important, to have the correct application to the wear problem (impact resistance, abrasive wear, heat, pressure, or corrosion).

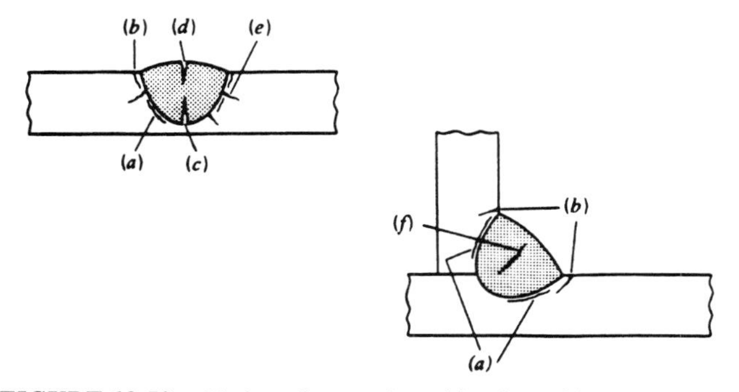

FIGURE 13.53. Various forms of cracking in welds: (*a*) Underbead cracking, (*b*) toe cracks, (*c*) root cracks, (*d*) hot cracking, (*e*) radial cracking, and (*f*) internal cracks.

A tough weld deposit should never be put over a harder, more brittle, hard surfacing weld. These welds will come loose (spall). A hard weld should always be on top of a softer one. Stainless steel is often used as an underlay weld for hard surfacing, as mentioned earlier.

Wear resistance of a hard surfacing weld will decrease when base metal dilution is increased. The oxy-acetylene process produces the least amount of dilution (5 percent), whereas arc deposits vary from 20 to 40 percent (Figure 13.54). Small surface cross checks (cracks) or hairline cracks are desirable in the harder surfacing alloys because they relieve stresses and minimize distortion.

Welding temperatures are very important. A preheat will often eliminate underbead cracking in alloy or carbon steel base metal, whereas high manganese (austenitic) steel should never be preheated, because temperatures above 600°F may remove the tough properties of high manganese steel. High manganese steel derives its wear resistance from its ability to work harden rapidly. As welded, its hardness is about HRC 15 and after being subjected to working or impact, it can increase to over HRC 50. Only arc welding is recommended for manganese steel since oxy-acetylene welding tends to heat the base metal more, thus reducing the cooling rate and causing weld embrittlement.

ADMIXTURE (PICKUP) IN WELDS

Admixture is the combining of the base metal and the molten weld. This natural mixing is of little significance when welding low carbon steels with mild steel electrodes. Alloy and high carbon steels can greatly affect the weld metal through admixture, however. Some combinations of base metal and weld cause segregations or concentrations that weaken the weld. Weld ductility is reduced by carbon pickup and cracking can result. Using welding electrodes that closely approximate the carbon and alloy content of the base metal reduces the effect of the base metal and weld admixture and consequently the difficulties it can cause. In some cases, admixture is an advantage if it strengthens the weld metal.

MICROSCOPIC EXAMINATION OF WELDS

Many of the conditions of weld metal and base metal zones that have been described in this chapter can be identified by using microscopic examination techniques. In

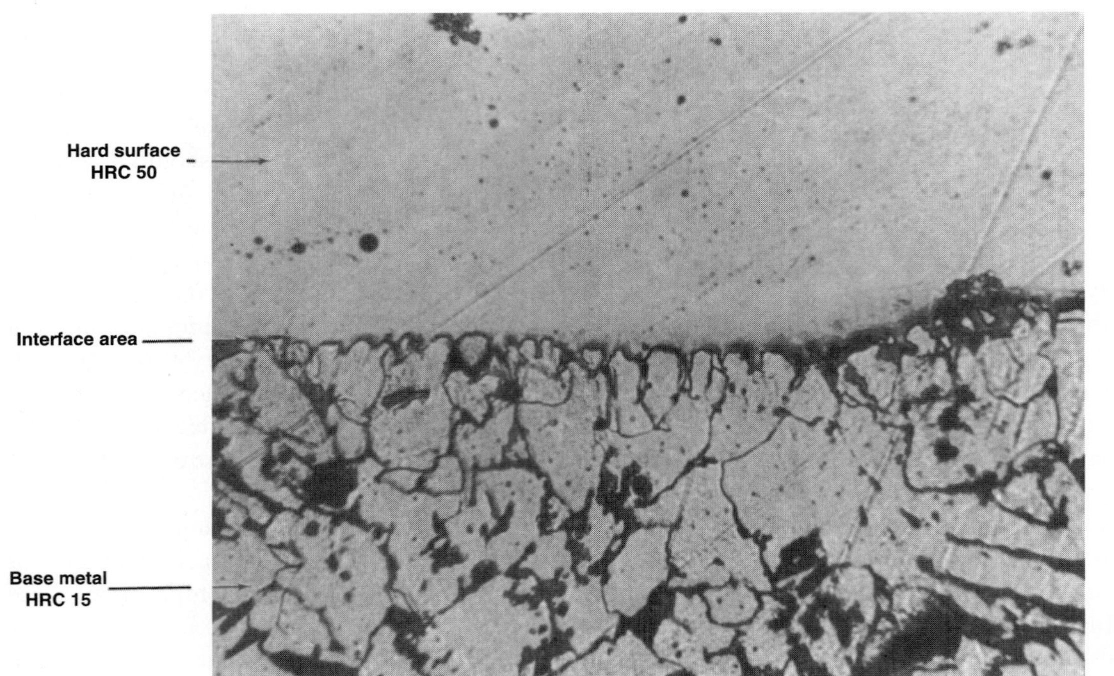

FIGURE 13.54. Hard surface arc weld deposited on low carbon steel.

this course you will purposely make welds with a known electrode on a known base metal (Figure 13.55) to demonstrate this practical means of identifying weld microstructures. Different preheat and postheat temperatures may be used and recorded. The specimen is sectioned by using a metallurgical cutoff abrasive saw, and polishing and etching techniques are followed as explained in Chapter 27.

FIGURE 13.55. Weld passes made on a SAE 1040 steel bar. The piece at left is a section of the bar prior to making the welds. The section at right is polished and etched for microscopic examination.

CAST IRON WELDING

Steel castings and steel weldments are making substantial inroads in the replacement of gray cast irons for the manufacture of machinery. However, ductile iron is fast becoming a good replacement for steel because of its lower cost. The advantages of steel castings are their higher strength. Weldments can often be made at lower cost than steel or cast iron castings. Nevertheless, large quantities of iron castings continue to be made and these sometimes require welding. The greater amount of welding is done on gray cast iron and very little on malleable iron; some ductile iron is welded. The welder should always remember, however, that cast iron is considerably less weldable than low carbon steels. The large amounts of carbon and silicon in cast irons result in a complicated, less ductile, metallurgical weld structure; and under no circumstances should cast iron be welded in areas that will be subjected to high stresses.

Most welding applications on cast iron are found in the repair of casting defects caused by the foundry process. Other applications are in production of welded cast iron assemblies and for the repair of castings that have become damaged or worn in service (Figures 13.56*a* and 13.56*b*).

In every type of cast iron welding, a filler metal is required to accomplish a strong joint. Almost all the common welding processes can be used for welding cast iron. Cast irons are difficult to weld simply because they contain a large amount of carbon and silicon. Gray cast iron can contain anywhere from 2 to 4 percent carbon and 0.5 to 2 percent silicon. The weldability of cast iron depends almost entirely on the type and the distribution of the carbon and the kind of microstructure that develops on cooling. Remember, cast iron can have, in addition to the expected constituents (ferrite, pearlite, and graphitic carbon), all the structures associated with

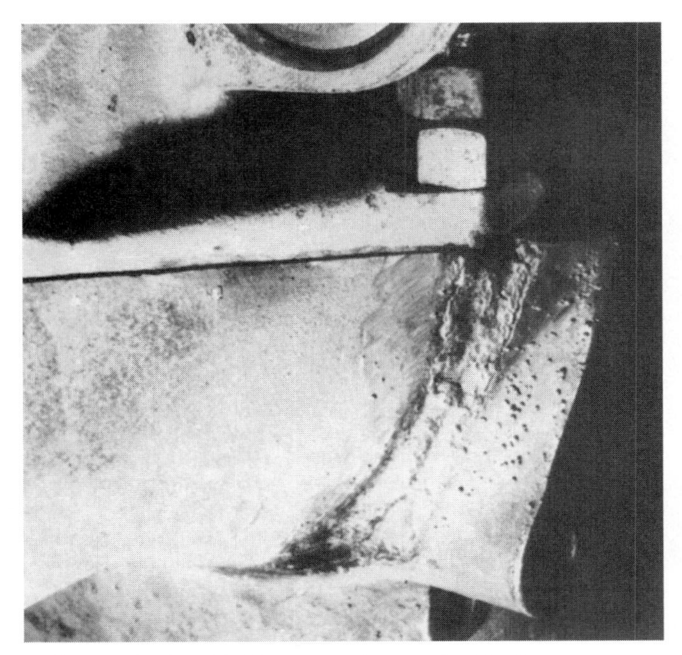

FIGURE 13.56a. Weld on a cast iron engine block.

FIGURE 13.56b. Cracks between cylinders in an engine block are sometimes welded and re-machined before the sleeves are put back in place.

heat-treatable steels, that is, martensite, bainite, free carbides, and so forth.

All cast irons may be welded with the exception of white cast iron, on which welding is not recommended. It is rather difficult to make strong welds on gray cast iron because of its tendency to form carbides near the weld and because of the lack of ductility in the base metal. The brittle behavior of the cast iron is caused by the lack of a plastic thermal region in the base metal, aggravated by low thermal conductivity. On solidification, the weld metal/base metal mixture undergoes rapid thermal contraction causing extreme overstress between the weldment and the heat-affected zone (HAZ). Since the base metal is very brittle and cannot move, cracking often occurs near the weld. Cracking can only be avoided by a very high temperature pre- and postheating of the base metal. This is the reason cast irons are considered very difficult to weld. Class 20 (ferritic) is easier to weld than Class 50 (pearlitic) iron. The microstructure of a cast iron weldment is shown in Figure 13.57. Welding cast irons become even more difficult due to the formation of gas porosity, graphite precipitation, columnar grain structure, and microstructure formations.

The junction zones near the weld tend to form extremely hard and brittle carbides. The weld metal also tends to pick up carbon from the base metal, especially if the filler metal is low carbon steel. This can cause the weld metal to become brittle also. Since gray cast iron has little yield strength and its tensile strength may be only half its compressive strength, the concentrated stress caused by weld shrinkage can cause microcracking in the heat-affected zone and massive cracking in the casting itself.

The method of welding can greatly influence the kind of microstructure that is formed. Brazing or gas welding, for instance, causes less weld stress and weld cracking than arc welding. It is almost always necessary to use some kind of preheating and postheating techniques with the welding process for cast iron.

THE KEEP-IT-COOL METHOD

If preheating is not practical, the welding procedure often used is to keep the casting as cool as possible. Usually only shielded metal arc welds are made when using this procedure and nickel alloy or steel electrodes with special coatings are used. This method is used on a high percentage of repair jobs on large heavy machinery. ENi (austenitic nickel electrodes) or ESt (steel electrodes) are used with this method. Welding can be performed for only a few minutes or seconds at a time, after which the casting must be allowed to cool. Multibead welds are often used on heavy repairs. A second pass of an ENi rod will help to soften the fusion zone of the original pass. Cracks in large castings that cannot be preheated or postheated are welded for a short distance, allowed to cool, and then followed by another short weld. The object is to avoid thermal stresses in the casting by reducing the heat input. Spot preheating is not practical in cast irons because it tends to cause cracking and should not be used. Any localized heating can cause cracking or

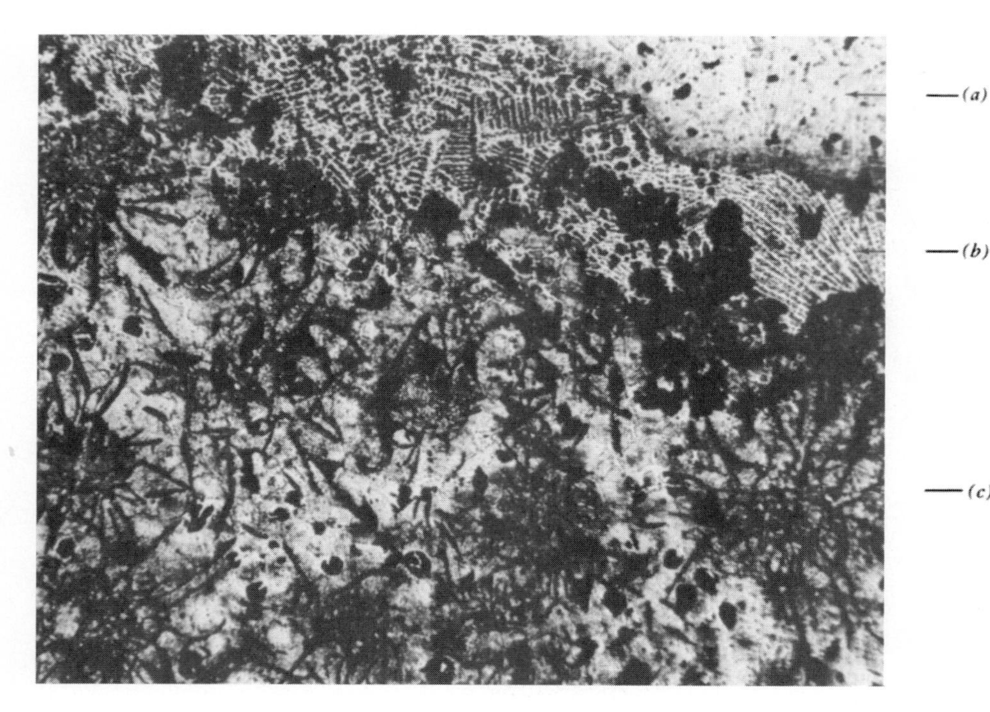

FIGURE 13.57. Microstructure of weld heat-affected showing the formation of carbides. The upper right area (*a*) is the nickel-base deposited weld. The center area (*b*) is a band of brittle dendritic white iron interspersed with small particles of graphite. The lower area (*c*) is a transition from iron carbide to a typical gray cast iron microstructure (100×).

extended cracking in the part that is being welded (Figure 13.58).

One of the difficulties with the keep-it-cool (reduced heating) method is that the porous structure of cast iron may contain foreign materials such as oil, water, and other liquids. These impurities tend to volatilize on heating and produce a weak, porous weld. When preheat is used, many of these foreign materials are driven out prior to welding and consequently the weld is more homogenous without porosity.

The keep-it-cool method has the advantage of providing the narrowest possible heat-affected zone, which can be as small as 0.00015 to 0.00020 in.; but one objection to a very narrow heat-affected zone is that it is more concentrated and contains extremely hard substances that produce tiny microcracks from which larger cracks can later radiate (Figures 13.59*a* and 13.59*b*). Because of the microcracks, the bond between the weld and the base metal is much weaker than a weld that has been preheated. Because of this, when correct procedures are not followed, welds in gray cast iron sometimes fail. If it is at all possible, when using this technique, the use of even a low temperature preheat from 400° to 800°F can make a much stronger weld. The casting after welding should be allowed to cool very slowly and be protected against cold drafts of air.

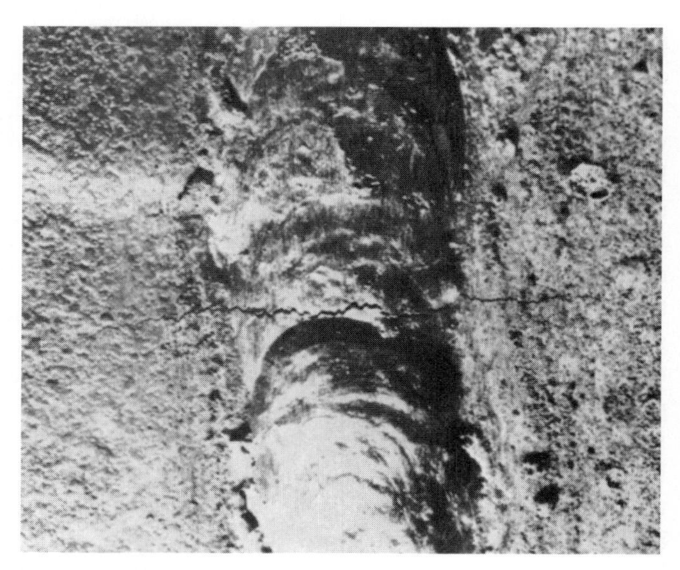

FIGURE 13.58. Transverse cracking in arc weld on cast iron caused by too high heat input on cold-base metal.

PREHEAT AND POSTHEAT

The best way to eliminate hardening in the fusion zone in a cast iron weld is both to preheat and postheat the casting. As with steel, cast iron structures are greatly affected by the rate of cooling of the welds. Large weldments or sections will cause a fast cooling rate in the weld, which will transform the adjacent zone into martensite. On small sections or when preheat is used, the slower cooling rate will prevent the formation of martensite.

If at all possible when making any weld in gray cast iron, it is recommended to make a preheat up to 800°F. The preheat should be held for a period of time before welding (to an hour) to allow the contaminates to vaporize.

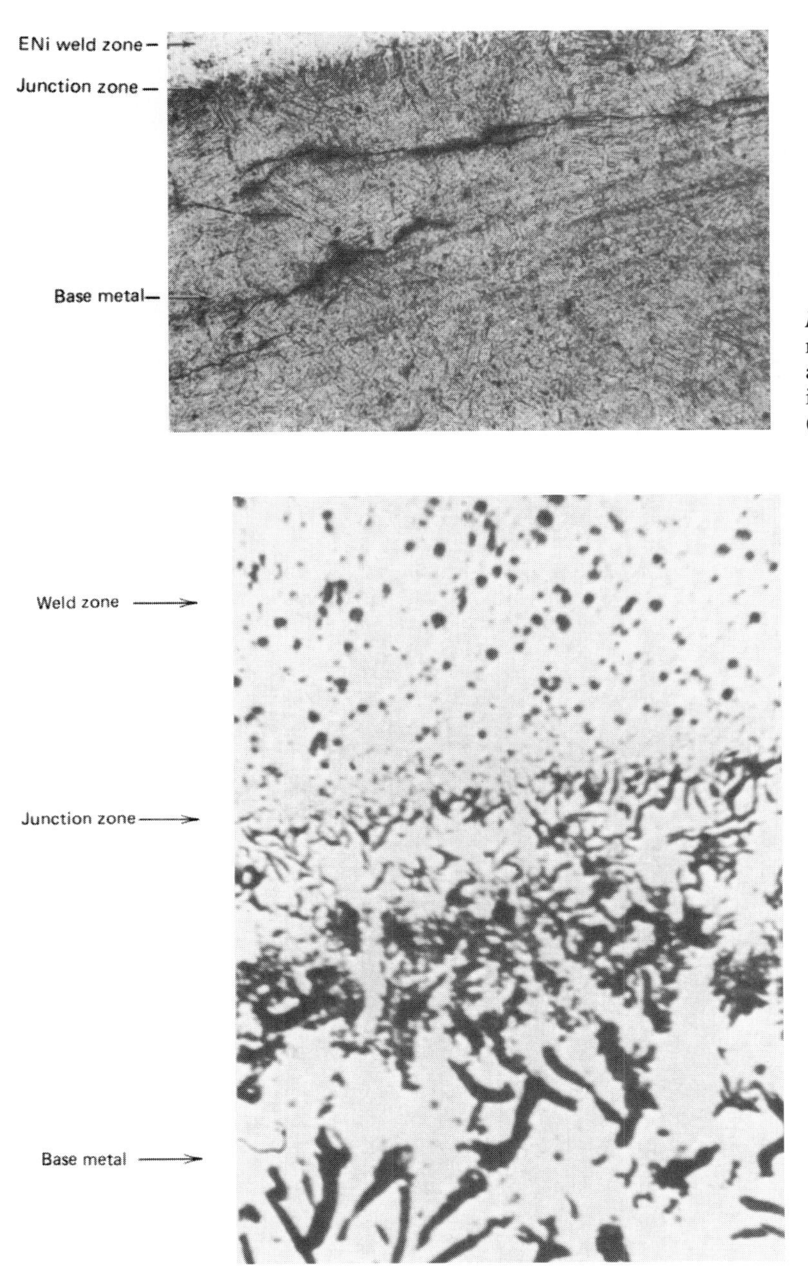

FIGURE 13.59a. The heat-affected zone in this micrograph of a cast iron weld shows a band of martensite adjacent to underbead cracks, which are commonly found in cast iron welds made by the keep-it-cool method (100×).

FIGURE 13.59b. ENi weld on gray cast iron showing good bond with the base metal. This weld was preheated to 800°F. The small black spots in the weld metal are caused by graphite pickup from the base metal. No hard structures or underbead cracking can be seen in this micrograph. Unetched (500×).

Postweld heat treatments may either consist of full annealing or stress relieving. The greatest softening affect may be achieved by annealing from 1650°F. Annealing lowers the as-cast tensile strength of most irons. Stress relief at 1150°F followed by furnace cooling to 700°F usually produces the best results.

Preheat and postheat on large heavy castings should be done very slowly so that localized stresses will not develop and cause cracking in the casting. Also, after preheating or postheating has been done, a slow cooling period should be allowed by burying the casting in an insulating material or by placing a thermal blanket around the casting to keep it warm for a longer period of time (Figure 13.60).

JOINT PREPARATION

For welding cast iron with most methods, the weld groove must be prepared by machining, chipping, or sawing. Grinding should be done only to a limited extent because of its tendency to smear the graphite flakes on the surface to be welded. When this happens, the weld often will not "take" or fuse; that is, it will tend to form into balls and roll off rather than make a bond. Graphite has a higher melting point than iron and will not flux out with most cast iron welding procedures. Steel cast iron electrodes (ESt) may be used where large graphite flakes have been formed from repeated high-temperature heating cycles as in exhaust manifolds and engine castings (Figure 13.61).

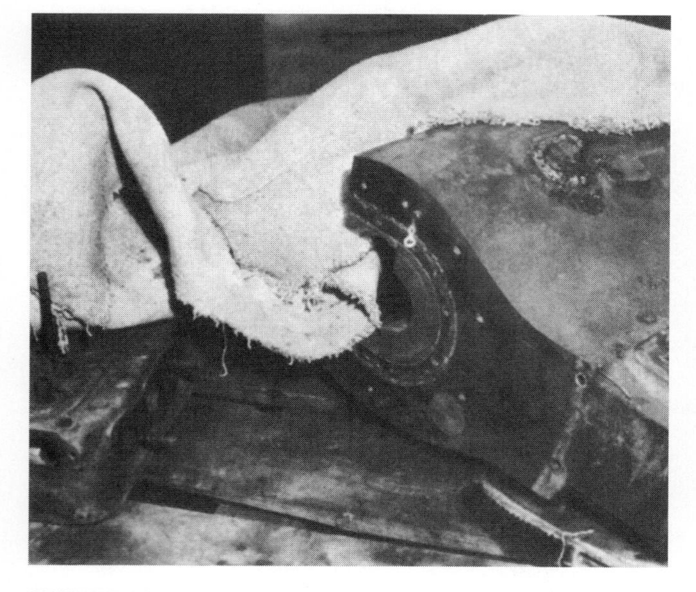

FIGURE 13.60. These castings were wrapped in a thermal (insulated) blanket to prevent them from cooling too rapidly. The blanket has been removed to reveal the castings.

FIGURE 13.61. A difficult cast iron weld was made easier in this casting by using a buttering technique on the sides of the weld surface, using an ESt rod to flux out the large graphite flakes. This was followed by an ENi rod to fill in the weld, giving the weld more ductility.

FIGURE 13.62. Reinforcing a cast iron weld can be done by studding.

If grinding is used for preparation of the V grooves for welding, one way to remove the graphite that is smeared is to heat the surface to be welded with a torch having an oxidizing flame and then allowing it to cool. This causes the surface graphite to form into small, loose spheres that can be removed with a wire brush.

For metal arc welds, the groove can be V shaped with a 60-degree included angle. For braze welding, the groove should be wider, having a 70-degree or more included angle. This provides a wider bond area that is needed for braze-welded joints. A hole should be drilled at the ends of the cracks to be welded in a casting to eliminate more cracking from the welding stresses.

WELDING RODS AND METHODS USED IN WELDING CAST IRON

Steel-type cast iron welding rods (ESt) are very useful when arc welding gray cast irons. These ESt rods have a coating that fluxes out large graphite inclusions making for easier welding; but these welds are usually too hard to machine. Ordinarily steel-welding electrodes will not work well for welding cast iron because they do not have the correct coating. When welding cast iron to steel, the ESt type rod is used, but again the weld will not be machinable. These electrodes are lower in cost than the ENi rods and can be used on small parts where no machining will be needed. A preheat of 400° to 600°F is recommended even on these small parts.

The other type, the nickel rod (ENi), is used when machinability is a factor. Although the ESt rod does not produce a very machinable weld, the ENi rod does, except in the heat-affected zone in some cases. ENi (nickel or nickel-iron) rods are widely used for cast iron welding and are most successful.

Castings can also be gas welded by using one of the RCI types or they can be metal arc welded using an ECI type rod. Since both rod types are made of cast iron, the weld will approach the characteristics of the base metal and be machinable if the casting has cooled slowly.

Studding

One method of reinforcing a weld that is sometimes used is called studding. This method is used on highly stressed parts such as flywheels that are worn or broken and need repair. Studding is also used where narrow thin sections cannot be adequately welded for strength; the reinforcing helps to give the part a longer service life, and is often used to restore gear teeth (Figure 13.62). Where preheat is impractical, such as on large castings and where maximum strength is needed, studding provides the added strength that is often required.

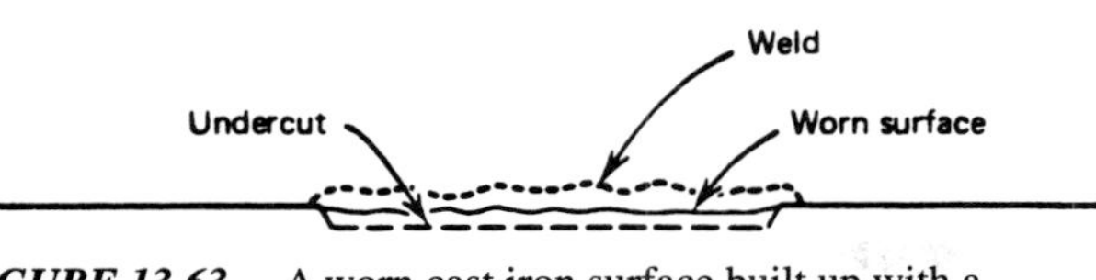

FIGURE 13.63. A worn cast iron surface built up with a machinable rod should be prepared as shown.

Buildup

Nickel rods are often used to build up worn surfaces that are to be re-machined. Sometimes a difficulty with buildup repair on cast iron is that the worn surface is only a few thousandths of an inch deep and the fusion zone of the weld is at the place where the final finish cut will be taken. A hard structure sometimes develops at or under the fusion zone in the base metal when the cooling rate is rapid. Consequently, machinability is sometimes poor even with the so-called machinable ENi rods when it is necessary to make cuts at or under the junction zone. When a situation like this exists, 20 to 30 thousandths of an inch of the surface should be removed by machining prior to welding so that the final finish cut will be taken in the weld zone, which is soft (Figure 13.63).

Gas Welding

If cast iron filler rods are used with gas welding techniques, for example, the postheat should be to a dull red and held for a period of time. But if the arc-welding techniques are used, then a postheat just under a dull red is all that is needed. There should be a soaking period, however, to remove stresses.

Braze Welding

Braze welding cast iron makes a very strong bond and is one of the best ways of welding cast iron, but it does not match the original metal in color and is not acceptable in some cases. The welder must be careful not to overheat the base metal when braze welding. The prepared surfaces must be wider than that type of V groove used for arc welding, and the welder should make certain that the surface is properly wetted with the bronze before he or she fills in the weld. If the part is to be used in high-temperature operations, such as in an exhaust manifold or for stove parts, braze welding should not be used because the weld will fail at high temperatures.

SPECIAL WELDING PROCESSES

A method of making heavy welded joints in large castings at less expense is the use of a buttering technique with an ESt or ENi rod followed by a lower cost rod such as E7018 to fill in the weld. This helps avoid the stresses usually caused by welding. The sides of the groove are first surfaced with the ESt rod and then the center is filled with the lower cost metal. The buttered halves can also be heat treated before the final welding is done to remove stresses in the adjacent and heat-affected zones.

Malleable iron may be welded if the welding preheat temperature is kept fairly low. The recommended welding process is braze welding because welding temperatures are low enough to prevent the malleable iron from reverting to white cast iron. Malleable iron will revert to white iron if it is heated above the transformation temperature.

Metal arc welding on malleable cast iron produces brittle structures just as in other cast irons, and if preheating is not used, the brittle area will be very wide. If cast iron filler rod and gas techniques are used, the malleable iron must go through a postheat treatment that will restore the malleable characteristics of the material. This heat treatment consists of a prolonged heating, annealing, and cooling process.

Nodular cast iron may be welded with nickel ENi rod or buttered and welded with low hydrogen rod. A preheat should be used to 800°F, which limits the formation of martensite and helps to eliminate underbead cracking. Postheat (annealing) also should be used to 1200°F.

SELF-EVALUATION: PART I

1. In what ways are welding and welds similar to the manufacture of metals?
2. Name three classes of welding.
3. If the base metal contains a medium to high carbon percentage (0.40 to 1 percent, for example), what happens in the fusion zone?
4. What are the three basic zones in welds?
5. In which zone is columnar grain structure likely to be found?
6. Is arc welding or oxy-acetylene welding more likely to cause grain growth and large grains in the heat-affected zone?

7. A very rapid cooling rate in a carbon steel base metal can cause the formation of what microstructure?
8. What effect do multiple passes have on the fusion and heat-affected zones?
9. Name three causes of failures in welds.
10. How can preheating help to prevent hard zones in carbon steel?
11. How can postheating eliminate hard zones in welds?
12. Explain why the heat-affected zone in cold rolled steel is not as strong as the unheated base metal?
13. What effect does the mass of the base metal have on cooling rates?
14. What is the purpose for the covering on electrodes or the shielding gases used?
15. Distortion due to shrinkage is mostly caused by (a) the weld or (b) the base metal expansion and contraction?
16. In liquid-solid phase welding, such as brazing, what two things can cause lack of bond between the bronze and the base metal?

CASE PROBLEM: WELD FAILURE IN HEAVY STEEL PLATES: PART I

A massive weldment was being made, similar in shape to Figure 13.37, and care was taken to avoid lamellar tearing. The welder was advised to preheat the 6 in. thick plates with propane torches and use light passes followed by peening with an air hammer. After several hours of steady welding, approximately half the weld was in place when it suddenly split down the center of the weld. All of the ruined weld then had to be removed with an air-arc procedure. Again the weld was made and at about the same place during the welding process, the same thing happened again. Can you offer a solution to the costly problem?

WORKSHEET: PART I

Objectives

1. Macroetch the cross section of a carbon steel weld on low carbon steel using E6013 electrodes to determine the weld zones.
2. Macroetch the cross section of SAE 1040 carbon steel and weld using E11018 electrodes to determine the weld zones.
3. Hardness test a specimen across the various weld zones.

Note Extreme care must be exercised when using acids, to avoid spilling or splashing on your skin. Face masks and rubber gloves should be worn. If spilling on the skin does occur, rinse with **cold** water immediately or apply a bicarbonate of soda solution.

Materials A 10 percent solution of hydrochloric acid in water, nital (5 percent solution of nitric acid in methyl alcohol), Pyrex bowls or beakers, and stainless steel tongs.

Note Carburized or decarburized surface, structure of welds, or depth of hardening may be brought out with nital. Porosity, cracks, and segregation may be seen by using hydrochloric acid. Heating is not necessary for these reagents although higher temperatures accelerate the rate of etching.

Procedure 1

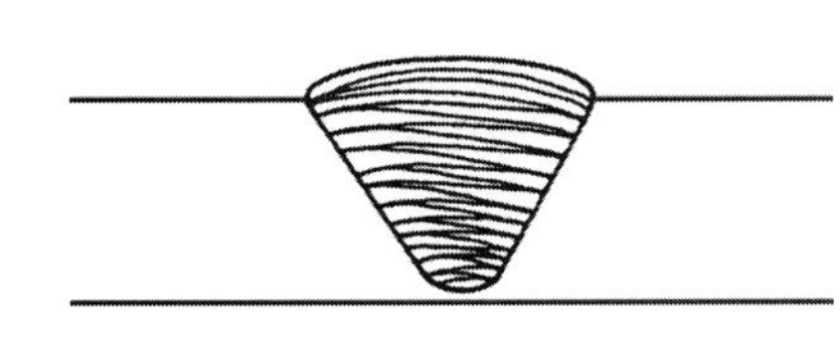

1. Make a butt weld as shown in Figure 13.35 with a mild steel plate and an E6013 rod and a similar one on a 1040 plate with an E11018 rod.
2. Saw a section out of the center of the welded piece.
3. Polish the section on a belt sander.
4. Pour a small amount of nital in a flat Pyrex dish. Pour a small amount of hydrochloric acid (10 percent solution) in another Pyrex dish.
5. Using tongs, place the specimen in the nital for about 5 minutes, then remove and rinse in water.
6. Place the specimen in the hydrochloric acid for about 1 second. Rinse in water.
7. Grains and outline of the weld should now be evident. If not, etch the part again for a longer time.

Conclusion After close visual inspection of the macroetched specimen, can you see the various zones? Describe what you see by writing a short paragraph and make a sketch of your weld section, detailing the parts or zones of the weld.

Procedure 2

1. Check for hardness with a hardness tester across the weld as shown in Figure 13.39.
2. Check at 1/16 in. increments and record the hardness of each test on a sketch of your weld.

Conclusion What did you find out about the hardness of various zones of the weld? Write a paragraph explaining the reason for the variation in hardness.

SELF-EVALUATION: PART II

1. What effect do alloying elements have on hardenability of steels (welds)?
2. Alloying elements such as nickel or rapid cooling rates sometimes inhibit the decomposition of austenite into martensite. What effect does postheating have on this untransformed austenite?
3. Why should high manganese steels (such as those used for hard facing) not be slow cooled or preheated?
4. Which alloying element has the unique ability to retard grain growth at high temperatures?
5. In what way is the brittleness of martensitic stainless steel in the HAZ different from the brittleness in ferritic stainless steels?
6. Name three things that should be remembered when welding on tool steels.
7. SAE-AISI specifications are used for carbon steel and alloy bar stock. What standard specification system is used for carbon and alloy structural steels such as beams, angles, and plates?
8. Describe the location of a hot crack and give two causes of this problem.
9. Welds should never be made where there is oil, grease, paint, or water. Why?
10. Why should a tough but ductile weld not be placed over a brittle, hard surface weld?
11. Should HSLA steels be heated for bending, or gas welded with an oxyacetylene torch? Explain your answer.
12. What is admixture in welds?

CASE PROBLEM: THE RUSTING STAINLESS STEEL TUBING: PART II

A paper mill used stainless steel tubing to carry corrosive liquids in a certain area. There were corrosive fumes that were also in contact with the outside of the tubes. Austenitic stainless steel was used because it effectively resisted this corrosive material. The plant engineer had always specified a certain grade of stainless and never had any trouble with it. However, a section of old pipe had to be replaced and new pipe was ordered by another person when the engineer was away. The pipes were connected by welding as the former ones were. A month later, a mechanic noticed the stainless steel tubes were badly corroded by rust next to all of the welds. The welding rod was called into question, but it was the same as that which was successfully used before. What do you think was wrong with the tubing that caused this problem?

WORKSHEET: PART II

Objective By microscopic examination, identify weld grain structure, admixture, inclusions, carbon content, and condition of the fusion and heat-affected zones.

Materials Three small pieces of SAE 4140 bar or round stock; E6013, E70XX, and E100XX welding rod; metallurgical polishing equipment and microscope.

Procedure

1. Deposit three layers of weld metal on a short length of SAE 4140 steel. Allow it to cool. Prepare three specimens, each representing a different electrode.
2. Make a section cut through the center of each weld with the metallurgical saw or a hacksaw. Do not overheat the weld when cutting it off.
3. The piece should not be more than 1 in. wide to facilitate polishing.
4. Polish and etch the specimen as explained in Chapter 27.
5. Examine the weld metal and heat-affected zones of each sample with the microscope.
6. Make a drawing on a sheet of paper of the welded piece surrounded by several circles in which you may sketch a micrograph representing what you see in the microscope. Draw a line from each circle to that part of the weld it represents (see Figure 13.64).
7. Write a conclusion under each of the three representative specimens on your paper describing the metallurgical condition of the weld.
8. Turn in your specimens and paper to your instructor.

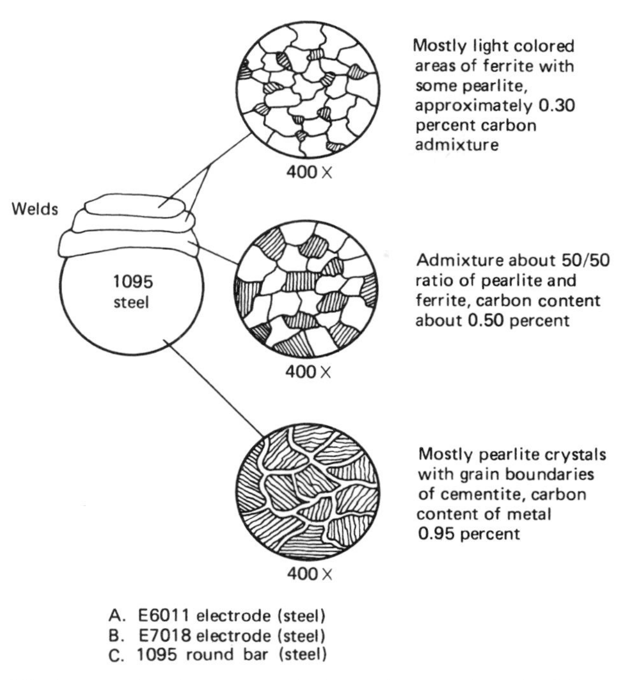

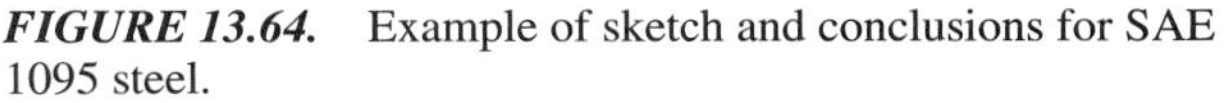

FIGURE 13.64. Example of sketch and conclusions for SAE 1095 steel.

SELF-EVALUATION: PART III

1. Is cast iron normally fusion welded without the use of filler rod? Which cast irons can be welded?
2. What is the most common application of cast iron welding?
3. State the major reasons that cast irons are somewhat difficult to weld.
4. Which kinds of welding methods may be used for welding cast iron?
5. Why does the lack of ductility of the gray cast iron base metal tend to cause cracking near the weld and not usually in the weld?
6. Name one advantage and one disadvantage of the keep-it-cool method of welding gray cast iron.
7. The fusion zone of "machinable" cast iron welds is sometimes hard. What can be done to eliminate or avoid this problem?
8. Why does the weld metal sometimes roll up into balls without fusing to the base metal? How can this be avoided?
9. How can a cast iron weld be reinforced for greater strength?
10. Why must malleable iron go through a prolonged postheat treatment if cast iron rod and gas-welding techniques are used?
11. Name two major advantages of preheating gray cast iron.
12. When a large gray cast iron casting is postheated to reduce internal stresses, how quickly must it be cooled? By a liquid quench or by some other means?

CASE PROBLEM: THE PERSISTENTLY BREAKING SPOKE: PART III

A welder was given the job of welding a cracked spoke on a large gray cast iron pulley. The pulley was 6 feet in diameter and had eight spokes about 2.5 in. in diameter. One of the spokes was cracked at the thinnest diameter. (They were somewhat tapered.) The welder used an air chisel to prepare the spoke for welding. He decided to use the keep-it-cool method of welding, using ENi electrodes. When the weld was about half finished, the spoke broke apart again about 2 in. from the weld with a loud noise. Frustrated, the welder decided to preheat the weld areas, prepare the second break, and weld both of them. The completed welds looked good. Satisfied, the welder went to another part of the shop. Fifteen minutes later, there was an exceedingly loud bang from the locality of the cast iron pulley. The welder thought, what did I do wrong this time? What did he do wrong, and what would you do to make a successful weld on that pulley?

WORKSHEET: PART III

Objective Determine the condition of three cast irons welds by the use of microscopic examination techniques.

Materials Pieces of gray cast iron prepared to make three different butt welds. ENi and ESt electrodes, ECI oxy-acetylene cast iron rod, metallurgical cutting and polishing equipment, and metallurgical microscope (see Chapter 27).

Procedure

1. Make three welds with no preheat using three different rod materials: ENi and ESt by the arc process and the ECI rod by gas weld. Allow them to cool to room temperature.
2. Cut a section of each weld and mark the samples for identification purposes.
3. Polish each specimen, but do not etch.
4. Observe the various zones of the weld with the microscope. Draw a sketch of the specimen and make circles around it. Make micrograph sketches in the circles that approximate what you see. Write a conclusion beneath the circles.
5. Now etch each specimen to bring out other microstructures and repeat the process of making sketches as in step 4. Write a conclusion stating the probable condition of the welds, such as hardness, brittleness, admixture, inclusions, microcracking, and bonding.

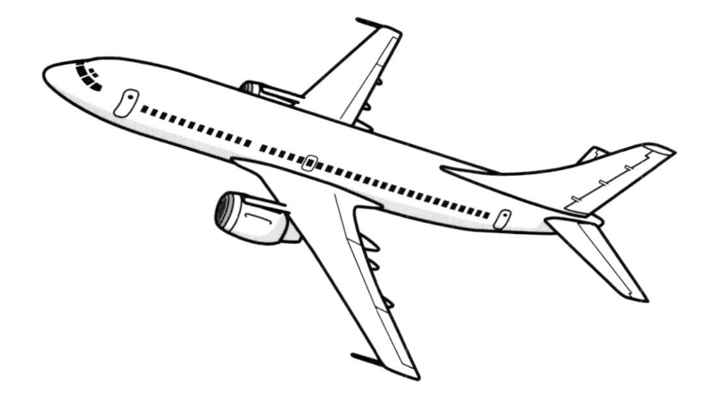

CHAPTER 14

Identification of Nonferrous Metals

Nonferrous metals such as gold, silver, copper, and tin were in use hundreds of years before the smelting of iron; yet some nonferrous metals have appeared relatively recently in common industrial use. For example, aluminum was first commercially extracted from ore in 1886 by the Hall-Heroult process, and titanium is a space-age metal that has been produced in commercial quantities only since World War II.

In general, nonferrous metals are more costly than ferrous metals. It is not always easy to distinguish a nonferrous metal from a ferrous metal, nor to separate one from another. This chapter should help you to identify, select, and properly use many of these metals. Precious metals are discussed in Chapter 18.

OBJECTIVES **After completing this chapter, you should be able to:**

1. Classify some nonferrous metals by a numerical system and identify others by testing methods.
2. List the general appearance and use of the various nonferrous metals.

NONFERROUS METALS

Nonferrous metals are those metals that are not iron base—in simple terms, all metals other than iron. Most nonferrous metals are called by their specific name (i.e., copper, lead titanium). When nonferrous metals are being discussed, however, most expect the subject to be aluminum. In general, metals differ in density and specific gravity, electrical and thermal conductivity, color, and physical properties. Tungsten is a very dense metal as is lead, whereas aluminum and magnesium are low density, and because of their high strength to weight ratio, they are more suitable for use in aircraft. Nonferrous metals, except for nickel and cobalt, are not attracted to a magnet.

One means of identifying nonferrous metals is by chemical spot testing. Titanium alloys, 6A1-4V and 6A1-6V-2Sn (known in industry as 6-4 and 6-6-2), can be separated from other titanium alloys using the author's spot test solution. The chemical formula is as follows: 88 percent water, 2 percent sulfuric acid, 4 percent hydrochloric acid, and 6 percent nitric acid. As with all chemical solutions, safety precautions and guidelines for these chemicals must be strictly followed. When mixed alloys of titanium are submerged into this solution, the 6A1-4V alloy turns a pale green. The 6A1-6V-2Sn alloy turns black and the black residue will remain on the surface after a water neutralizing rinse. Parts can be easily separated by this method. A dip into a dilute hydrochloric acid solution will immediately dissolve the black; then rinse in water and dry. If immersion is not possible, one drop on the surface of the titanium is all that is needed for identification.

A spark test can also be used to separate titanium from aluminum and or stainless steels. Titanium produces a bright unmistakable white deluge of sparks when held against a grinding wheel. Steels produce yellow sparks; the intensity is dependent upon the carbon and alloy content; aluminum does not produce sparks.

The specific gravity of a symmetric or regular shaped solid may be found by measuring the piece to determine its volume and then weighing it on a scale. The density or mass per unit volume thus obtained may then be compared to Table 6, Specific Gravity (Density) and Weights of Some Materials, found in Appendix 1 of this book. Thus, it is possible to identify the metal if it is a pure metal and if the measurements are accurate. The volume and mass of an irregularly shaped solid may be determined by its displacement of a liquid of known density. This experiment is usually carried out in a chemistry or physics laboratory, using special equipment.

ALUMINUM AND AL-ALLOYS

Aluminum is white (or white-gray if the surface is oxidized) and can have any surface finish from dull to shiny and polished. Aluminum has a density (or weight) of 168.5 lb/ft^3 as compared with 487 lb/ft^3 for steel. Aluminum and its alloys are readily machinable and can be manufactured into almost any shape or form (Figure 14.1).

The properties of aluminum and aluminum alloys make this metal one of the most economical and attractive metals for use and acceptance in both industrial and home applications. The working face of the iron used to press clothes was formerly made from a cast iron. Many years ago, when steam irons came into use, rust became a problem. Today, the hot faces of irons are made from aluminum; neither rust nor steam presents a problem for aluminum in this application. Aluminum baseball bats have replaced wood. Internal framework of automobiles were formerly steel; now most are aluminum. Aircraft have certainly come a long way from wooden biplanes to supersonic transports and fighters. New aluminum-lithium alloys and graphite reinforced aluminum composites are now common place in spacecraft applications.

FIGURE 14.1. Structural aluminum shapes used for building trim provide a pleasing appearance.

Aluminum is probably best known for three properties—lightweight, excellent corrosion resistance, and a highly reflective lustrous surface finish. Aluminum weighs about 2.7 Mg/m^3 or about 0.1 pound per cubic inch. It is also nontoxic which makes it safe for use as food product containers and beverages. Aluminum can be any color by anodizing process or texture. Aluminum is the most abundant of all structural metals. Its ore comprises about 8 percent of the earth's crust.

Aluminum is often selected for use in electrical applications due to its high electrical conductivity which is nearly twice that of copper on a weight basis. High-voltage transmission lines are often made with a steel core for mechanical strength surrounded by aluminum wire for conductivity. Electrical buss bars found inside feed-in boxes on homes are normally made from 6061 aluminum alloy. Formerly, aluminum wiring was used to provide electrical service inside homes but was replaced with copper wire due to oxidation problems in making lasting connection to electrical outlets.

Thermal conductivity is a major consideration for use as automotive engines and engine heads because of the rapid heat transfer properties of aluminum. Nearly all of the V-6 engines produced for modern automobiles are made from aluminum alloys. Another advantage is the low melting temperature and ease of casting which aluminum offers over steel and iron alloys. Pure aluminum melts at about 1215°F and most aluminum alloys are poured or cast into molds at 1350°F. Copper melts at 1980°F and iron at 2795°F.

Aluminum has found wide use in structural and decorative applications in modern architecture. Buildings, bridges, storage tanks, towers, highway and city street lighting posts and fixtures made from aluminum alloys have replaced steels due to the light weight and high strength, and when corrosion is a major concern. Aluminum alloys never become brittle at cryogenic temperatures. This property of high fracture toughness allows aluminum alloys extensive use in the aircraft industry. At 38,000 feet altitude, the outside temperature can be as low as –65°F.

Classification of Aluminum

There are several numerical systems used to identify aluminum, such as federal specifications, military specifications, the American Society for Testing and Materials (ASTM) specifications, and the Society of Automotive Engineers (SAE) specifications. The system most used by manufacturers, however, is one that was adopted by the Aluminum Association in 1954.

From Table 14.1 you can see that the first digit of a number in the aluminum alloy series indicates the alloy type. The second digit, represented by an *x* in the table, indicates any modifications that were made to the original alloy. The last two digits identify either the specific alloy or aluminum impurity. An aluminum alloy numbered 5056 is an aluminum-magnesium alloy, where the first 5 represents the alloy magnesium, the second digit represents modifications to the alloy, and the last 5 and 6 are numbers of a similar aluminum alloy of an older marking system. An aluminum numbered 1120 contains no major alloy and has 0.20 percent pure aluminum above 99 percent. New aluminum-lithium alloys that are very lightweight have been developed.

Aluminum and its alloys are produced as **castings** or as **wrought** (cold-worked) shapes such as sheets, bars, and tubing. They can be either cold worked by rolling, drawing, or extruding, or hot worked by forging. Cast aluminum is not generally as strong as wrought aluminum. Aluminum alloys are harder than pure aluminum and will scratch the softer (1100 series) aluminum. A 10 percent solution of sodium hydroxide (caustic soda) will leave a stain on aluminum and aluminum alloy.

Pure aluminum and some of its alloys cannot be hardened by heat treating, but they can be annealed to soften them by heat treatment. This means that hardening them can only be done by cold working (strain hardening). These temper (hardness) designations are made by a letter that follows the four-digit alloy series number.

—F As fabricated. No special control over strain hardening or temper designation is noted.

—O Annealed, recrystallized wrought products only. Softest temper.

—H Strain hardened, wrought products only. Strength is increased by work hardening.

The letter —H is always followed by two or more digits. The first digit, 1, 2, or 3, denotes the final degree of strain hardening.

TABLE 14.1. Aluminum and aluminum alloys

Code Number	Major Alloying Element
1xxx	None
2xxx	Copper
3xxx	Manganese
4xxx	Silicon
5xxx	Magnesium
6xxx	Magnesium and silicon
7xxx	Zinc
8xxx	Other elements
9xxx	Unused (not yet assigned)

—H1 Strain hardened only

—H2 Strain hardened and partially annealed

—H3 Strain hardened and stabilized

The second digit denotes higher strength tempers.

2 ¼ hard

4 ½ hard

6 ¾ hard

8 Full hard

9 Extra hard

For example, 5056-H18 is an aluminum-magnesium alloy, strain hardened to a full hard temper.

Some aluminum alloys can be hardened to a great extent by a process called **solution heat treatment** and **precipitation hardening** or **aging.** This process involves heating the aluminum alloy to a temperature where the alloying element is dissolved into a solid solution. The aluminum alloy is then quenched in water and allowed to age or is artificially aged by heating slightly. The aging produces an internal strain that hardens and strengthens the aluminum. Some other nonferrous metals are also hardened by this process. See Chapter 15, "Heat Treating of Nonferrous Metals."

For these aluminum alloys, the letter —T follows the four-digit series number. Numbers 2 to 10 follow this letter to indicate the sequence of treatment.

—T2 Annealed (cast products only)

—T3 Solution heat treated and cold worked

—T4 Solution heat treated but naturally aged

—T5 Artificially aged only

—T6 Solution heat treated and artificially aged

—T7 Solution heat treated and stabilized

—T8 Solution heat treated, cold worked, and artificially aged

—T9 Solution heat treated, artificially aged, and cold worked

—T10 Artificially aged and then cold worked

For example, 2024-T6 is an aluminum-copper alloy, solution heat treated and artificially aged.

Cast aluminum alloys are often used in applications such as automobile parts, where their low density is important. Sand castings, permanent mold, and die-casting alloys are of this group. They owe their mechanical properties to solution heat treatment and precipitation or to the addition of alloys. A classification system similar to that of wrought aluminum alloys is used (Table 14.2).

TABLE 14.2. Cast aluminum alloy designations

Code Number	Major Alloy Element
1xx.x	None, 99 percent aluminum
2xx.x	Copper
3xx.x	Silicon with Cu and/or Mg
4xx.x	Silicon
5xx.x	Magnesium
6xx.x	Zinc
7xx.x	Tin
8xx.x	Unused series
9xx.x	Other major alloys

The cast aluminum 108F, for example, has an ultimate tensile strength of 24,000 psi in the as-fabricated condition and contains no alloy. The 220.T4 copper aluminum alloy has a tensile strength of 48,000 psi.

CADMIUM

Cadmium has a blue-white color and is commonly used as a protective plating on parts such as screws, bolts, and washers. It is also used as an alloying element to make metal alloys that melt at low temperature, such as bearing metals, solder, type casting metals, and storage batteries. Cadmium compounds such as cadmium oxide are toxic and can cause illness when inhaled. These toxic fumes can be produced by welding, cutting, or machining on cadmium-plated parts. Breathing the fumes should be avoided by using adequate ventilation systems. The melting point of cadmium is 610°F. Its specific gravity is 8.648 and its density is 539.6 lb/ft^3.

CHROMIUM

This hard, slightly grayish metal can take a brilliant polish and is very corrosion resistant. It is reactive to hydrochloric and sulfuric acids but is passivated (made more corrosion resistant) by immersion in nitric acid. Chromium is used in electroplating as a protective and ornamental covering. It is used as an alloying element for electrical resistance units such as nichrome or chromel wire. (See the section on nickel-based alloys later in this chapter.) As an alloying element with steel, an entirely new type of steel is produced, stainless steel, which as the name implies is a very corrosive-resistant steel; this requires about 12 percent chromium in the steel. With about 1 percent chromium, a very tough, strong steel alloy is produced. The specific gravity of chromium is 7.1 and its density is 442.7 lb/ft^3. The melting point is 2964°F.

COBALT

Cobalt is a silver-white metal whose primary use is for alloying with steel to make high-temperature-resistant cutting tools, for example, high-speed steel and "stellite" (a cobalt-chromium alloy) cutting tools. Tungsten carbide particles are combined with cobalt as a binder or matrix material to produce tungsten carbide cutting tools. Heat-resistant jet engine turbine blades are made of alloys containing cobalt such as stellite. Cobalt is harder and stronger than iron or nickel and is not very malleable. Cobalt is resistant to ordinary corrosion in the air, but it reacts with nitric acid and is magnetic below 2128°F. The specific gravity is 8.71 and the density is 543.5 lb/ft^3. Its melting point is 2721.6°F.

COLUMBIUM (NIOBIUM)

This ductile, malleable, slightly bluish metal is chemically related to vanadium and tantalum. It will burn when heated in air. The metals columbium, titanium, hafnium, zirconium, and tantalium alloys are all used in the aerospace industry, in jet aircraft, rockets, missiles, and for nuclear reactors. Columbium is a relatively lightweight metal that can withstand high temperatures and can be used for the skin and structural members of aerospace equipment and missiles. Columbium is used as a carbon stabilizer in stainless steels, but perhaps one of its greatest uses today is as an alloying element in Alaskan pipeline steels. Manganese, columbium, and silicon steel alloys are being used to prevent failure in steels that are subject to extremely low ambient temperatures such as those found in Arctic regions. Ordinary steel will tend to fail at these temperatures by splitting, sometimes for miles, on a pipeline. These fracture speeds can range from 400 to 600 ft/sec. The specific gravity of columbium is 8.4, the density is 523.8 lb/ft^3, and the melting point is 3567.5°F.

COPPER AND COPPER-BASED ALLOYS

Copper is a soft, heavy metal that has a reddish color. It has high electrical and thermal conductivity when pure but loses these properties to a certain extent when alloyed. Copper is very ductile and can easily be drawn into wire or tubular products. It is so soft that it is difficult to machine and has a tendency to adhere to tools. Copper has a high formability and corrosion resistance, and medium strength. To American industry, the term *copper* refers to that element with less than 0.5 percent impurities or alloying elements. Copper-based alloys are those having more than 40 percent copper. The melting point of copper is 1981°F, its specific gravity is 8.89, and its density is 554.7 lb/ft^3.

Tough-Pitch Copper

This copper contains a carefully controlled amount of oxygen, between 0.02 and 0.05 percent. Electrolytic refined tough-pitch copper is the most widely used type for electrical conductors and for building trim, roofing, and gutters. It lends itself to high tonnage production after being cast into wire, bars, and billets for further fabrication. Tough-pitch copper can withstand the ravages of time and weather because it does not harden with age and develop season cracks.

Deoxidized Copper

This type of copper differs from tough-pitch copper chiefly in its lower electrical and thermal conductivity. It also has a somewhat higher ductility and is more readily formed, and for this reason it is the most commonly used copper for the manufacture of tubular products, such as those used in domestic and industrial plumbing.

Oxygen-Free Copper

This type is the purest of commercial coppers (99.92 percent minimum) and therefore is called a high-conductivity copper. It is used in electrical and electronic equipment, radiators, refrigeration coils, and distillers. Industrial copper is classified by a series of SAE numbers; for example, SAE No. 75 is a 99.90 percent deoxidized copper used for tubes.

Low Alloy Copper

Copper is often alloyed with very small percentages of other metals, from a fraction of 1 percent to approximately 2 percent, for the purpose of imparting such qualities as corrosion resistance, higher operating temperature, and increased tensile strength and machinability. These additives include arsenic, silver, chromium, cadmium, tellurium, selenium, and beryllium. Copper cannot be hardened by heat treatments, but with the addition of about 2 percent beryllium it can be sufficiently hardened by precipitation and aging so that it can be used for making springs, flexible bellows, and tools. Because of their nonsparking quality, beryllium-copper tools (Figure 14.2) are used in explosion-hazardous environments such as mines, powder factories, and some chemical plants.

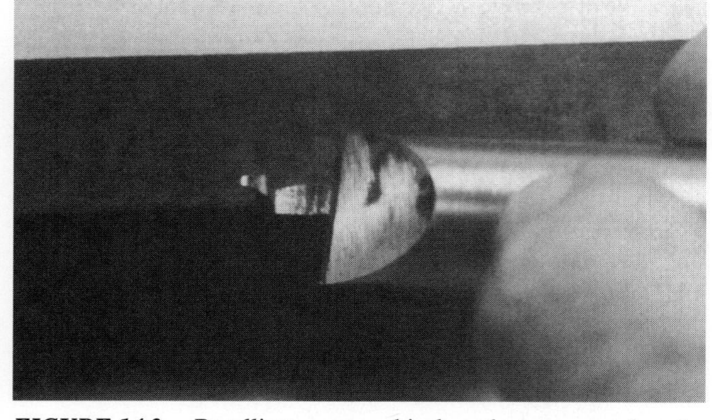

FIGURE 14.2. Beryllium-copper chisel used to remove a burr in low carbon steel.

Since beryllium is quite expensive, an alloy containing only 0.4 percent Be with 2.6 percent Co was developed that is useful for some purposes, but the straight beryllium-copper alloy develops a higher strength and hardness by heat treatment than the cobalt-bearing alloy.

Machining of this metal should be done after solution heat treatment and aging and not when it is in the annealed state. Machining or welding beryllium copper can be very hazardous if safety precautions are not followed. Machining dust or welding fumes should be removed by a heavy coolant flow or by a vacuum exhaust system, and a respirator type of face mask should be worn. The melting point of beryllium is 2345°F, its specific gravity is 1.847, and its density is 115 lb/ft^3.

Brass

Brass is an alloy of zinc and copper. Brass colors usually range from white to yellow, and in some alloys from red to yellow. Brasses range from gilding metal used for jewelry (95 percent copper, 5 percent zinc) to Muntz metal (60 percent copper, 40 percent zinc) used for bronzing rod and sheet stock. A brazing rod has melting temperatures above 800°F. Brasses are easily machined. Brass is usually tougher than bronze and produces a stringy chip when machined. The melting point of most brasses ranges from 1616° to 1820°F, and their densities range from 512 to 536 lb/ft^3.

Bronze

Bronze is found in many combinations of copper and other metals, but copper and tin are the original elements combined to make bronze. Bronze colors usually range from red to yellow. Phosphor bronze contains 92 percent copper, 0.05 percent phosphorus, and 8 percent tin. Aluminum bronze is often used in the shop for making bushings or bearings that support heavy loads (Figure 14.3).

FIGURE 14.3. Flanged bronze bushing.

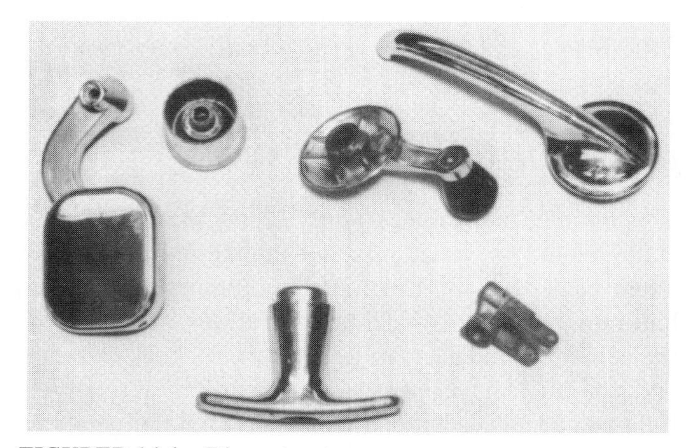

FIGURER 14.4. Die-cast parts.

(Brass is not normally used for making antifriction bushings.) The melting point of bronze is about 1841°F and its density is about 548 lb/ft^3. Bronzes are usually harder than brasses but are easily machined with sharp tools. The chip produced is often granular. Some bronze alloys are used as brazing rods.

DIE-CAST METALS

Finished castings are produced with various metal alloys by the process of die casting. Die casting is a method of casting molten metal by forcing it into a mold. After the metal has solidified, the mold opens and the casting is ejected. Carburetors, door handles, and many small precision parts are manufactured by using this process (Figure 14.4). Die-cast alloys, often called "pot metals," are classified in six groups.

1. Tin-based alloys
2. Lead-based alloys
3. Zinc-based alloys

4. Aluminum-based alloys
5. Copper-, bronze-, or brass-based alloys
6. Magnesium-based alloys

The specific content of the alloying elements in each of the many die-cast alloys may be found in handbooks or other references on die casting. See Chapter 2, "Casting Processes."

INDIUM

Indium, a silver-white metal with a bluish tinge, is malleable, ductile, and softer than lead. It burns in air to form an oxide. A recent development is the use of indium for wear reduction by combining indium and graphite to coat moving parts or by diffusing indium into metallic surfaces.

Small quantities of indium alloyed with aluminum alloys help control the age-hardening properties of these metals. Lead-indium alloys are widely used for solders and bearing alloys. Indium adds the property of greater wettability to brazing alloys. Its specific gravity is 7.28, its density is 453.9 lb/ft^3, and its melting point is 311°F.

LEAD AND LEAD ALLOYS

Lead is a heavy metal that is silvery when newly cut and gray when oxidized. It has a high density, low tensile strength, low ductility (cannot be easily drawn into wire), and high malleability (can be easily compressed into a thin sheet).

Lead has a high corrosion resistance and is alloyed with antimony and tin for various uses. It is used as a shielding material for nuclear and X-ray radiation, for cable sheathing, and for battery plates. Lead is added to steels, brasses, and bronzes to improve machinability. Lead compounds are very toxic and are also cumulative in the body. Small amounts ingested over a long period of time can be fatal. The melting point of lead is 621°F; its specific gravity is 11.342, and its density is 707.7 lb/ft^3.

A **babbitt metal** is a soft, antifriction alloy metal that is often used for bearings and is usually tin or lead based (Figure 14.5). Tin babbitts usually contain from 65 to 90 percent tin with antimony, lead, and a small percentage of copper added. These are the higher grade and generally the more expensive of the two types. Lead babbits contain up to 75 percent lead with antimony, tin, and some arsenic making up the difference.

FIGURE 14.5. Babbitted pillow block bearings. *(Machine Tools and Machining Practices)*

Cadmium-based babbitts resist higher temperatures than other tin and lead base types. These alloys contain from 1 to 15 percent nickel or a small percentage of copper and up to 2 percent silver. The melting point of babbitt is about 480°F.

MAGNESIUM

When pure, magnesium is a soft, silver-white metal that closely resembles aluminum but is much lighter. The density of magnesium is 108.6 lb/ft^3 in contrast with 168.5 lb/ft^3 for aluminum and 487 lb/ft^3 for steel. When alloyed with other metals such as aluminum, zinc, or zirconium, magnesium has high strength-to-weight ratios, making it a useful metal for some aircraft components. Household goods, typewriters, and portable tools are some of the many items made of this very light metal. Magnesium is used as an additive in the ladle in iron foundries to produce nodular iron.

To distinguish between magnesium and aluminum, it is sometimes necessary to make a chemical test. Nitric acid will turn magnesium gray; aluminum will remain unchanged. A zinc chloride solution in water or copper sulfate will blacken magnesium immediately but will not change aluminum (Figures 14.6*a* and 14.6*b*).

Magnesium, although similar to aluminum in density and appearance, presents some quite different machining problems. Magnesium, when finely divided, will burn in air with a brilliant white light. Although magnesium chips can burn in air, applying water will only cause the chips to burn more fiercely. Sand or special compounds should be used to extinguish these fires. Thus, when working with magnesium, a water-based coolant should never be used. Magnesium can be machined dry when light cuts are taken and the heat is dissipated. Compressed air is sometimes used as a coolant. Anhydrous

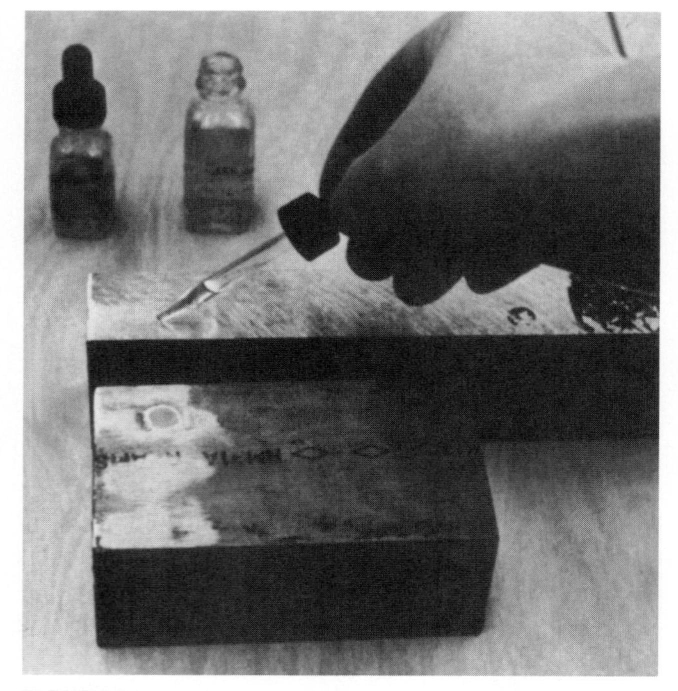

FIGURE 14.6a. Testing to distinguish aluminum from magnesium is done here by applying a weak nitric acid solution. The material at the bottom blackened when the acid was applied, indicating that it is magnesium. When acid was applied to the upper block, the material did not blacken, indicating that it is aluminum.

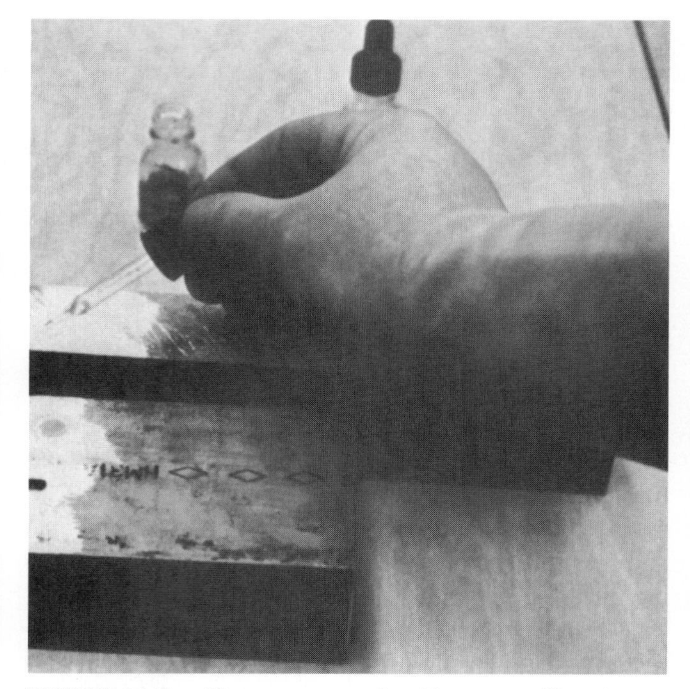

FIGURE 14.6b. The same test made with copper sulfate. A much darker color may be observed on the magnesium, while the aluminum is not affected.

(containing no water) oils having a high flash point and low viscosity are used in most production work. Magnesium is machined with very high surface speeds and with tool angles similar to those used for aluminum.

Cast and wrought magnesium alloys are designated by SAE and ASTM numbers, which may be found in the ASM Metals Reference Handbooks. The melting point of magnesium is 1204°F, and its specific gravity is 1.741.

MANGANESE

This silver-white metal is seldom seen in its pure state for industrial uses but is normally used as an alloying element. In steel production it is utilized as a deoxidizer in the form of spiegeleisen (about 30 percent Mn) and ferromanganese (78 percent Mn). Small percentages of manganese are added to steel to control sulfur. The small amount of sulfur that cannot be removed in the process of steel making combines with iron to produce unwanted iron sulfide, which causes hot shortness (splitting when forged at high temperatures). Manganese, when added to steel, combines with the sulfur to produce manganese sulfide, which does not cause hot shortness and, in fact, increases the machinability of the steel by acting as a chip lubricant. Extra sulfur and manganese are sometimes deliberately added to steel to produce "free-machining steel" or "resulfurized steel." Large percentages of manganese (12 percent or more) cause steel to become austenitic, that is, no longer ferromagnetic.

Thus, high manganese steel can be identified by its nonmagnetic quality. High manganese steel tends to work harden very rapidly and is used in areas where abrasion resistance is needed, as in earthmoving and rock-crushing machinery. Manganese is also an alloying element in some stainless steel. An alloy of 20 percent manganese, 20 percent nickel, and 60 percent copper has high resistance to corrosion and can be heat treated to a greater hardness than any other copper alloy.

The specific gravity of manganese is 7.2 and its density is 488.9 lb/ft^3. Its melting point is 2210°F.

MOLYBDENUM

As a pure metal, molybdenum is used for high-temperature applications and, when machined, it chips like gray cast iron. It is used as an alloying element in steel to promote deep hardening and to increase its tensile strength and toughness. Pure molybdenum is used for filament supports in lamps and in electron tubes. The

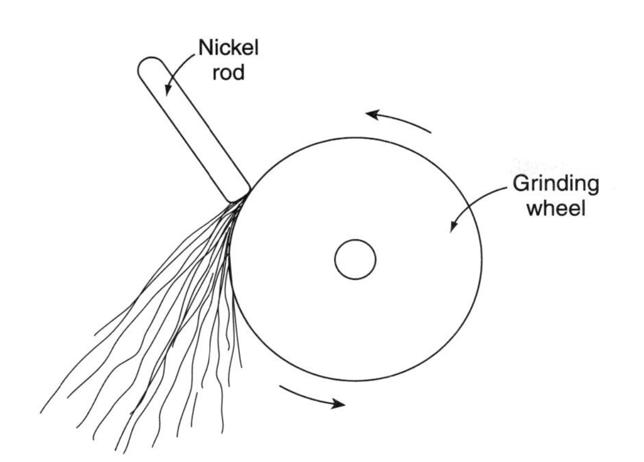

FIGURE 14.7. Nickel. Short circuit lines with no forks or sprigs. Average stream is 10 in. long and has an orange color.

melting point of molybdenum is 4748°F. Its specific gravity is 10.2 and its density is 636.5 lb/ft^3.

NICKEL AND NICKEL-BASED ALLOYS

Nickel is noted for its resistance to corrosion and oxidation. It is a whitish metal used for electroplating and as an alloying element in steel and other metals to increase ductility and corrosion resistance. It resembles pure iron in some ways but has a greater corrosion resistance. Electroplating is the coating or covering of another material with a thin layer of metal, using electricity to deposit the layer.

On a spark test, nickel throws short orange carrier lines with no sparks or sprigs (Figure 14.7). Nickel is attracted to a magnet but becomes nonmagnetic near 680°F. The melting point of nickel is 2646°F, its specific gravity is 8.8, and its density is 549.1 lb/ft^3.

Monel

Monel is an alloy of 67 percent nickel and 28 percent copper, plus impurities such as iron, cobalt, and manganese. It is a tough but machinable ductile and corrosion-resistant alloy. Its tensile strength (resistance of a metal to a force tending to tear it apart) is 70,000 to 85,000 lb/in.^{2} Monel metal is used to make marine equipment such as pumps, steam valves, and turbine blades. On a spark test, monel shoots orange colored, straight carrier lines about 10 in. long, similar to those of nickel. K-monel contains 3 to 5 percent aluminum and can be hardened by heat treatment.

Monel can be distinguished from nickel with the acid test for copper content. Apply one or two drops of nitric acid to the surface of the sample. A blue-green color will be seen in the solution in either case. Next, rub a clean steel rod, such as a nail, in the solution and observe. If the metal is monel, the steel will become copper colored where it was in the solution; if the steel does not become copper colored, the metal may be nickel. This same test can also be used to determine the copper content of ferrous metals. They will show a brown color instead of a copper color if no copper is present.

Chromium, Nichrome, and Inconel

Chromel and nichrome are two nickel-chromium-iron alloys used as resistance elements in electric heaters and toasters. Nickel-silver contains nickel and copper in similar proportions to monel but also contains 17 percent zinc. Other nickel alloys such as inconel are used for parts that are exposed to high temperatures for extended periods.

Inconel, a high-temperature and corrosion-resistant metal consisting of nickel, iron, and chromium, can be distinguished from monel by the acid test. Apply one drop of dilute nitric acid to test samples of inconel and monel. The acid will turn blue-green in about 1 minute on monel but will show no reaction on inconel. The nickel alloys' melting point range is 2425° to 2950°F.

PRECIOUS METALS

In the past, gold has been used mostly for production of jewelry and coinage. Gold used for coinage was often alloyed with copper to give it strength. Silver was also alloyed with copper for the same reasons. Gold and silver have found wide use as commercial metals used for electrical devices, corrosion control coatings, integrated circuitry, mirrors, scanning optics and detection devices, photographic compounds, and electrical equipment. Silver is used in silver solders which have a higher melting point and better corrosion resistance than lead-tin solders. With the onset of the aerospace age, the use of precious metals in electronic devices, computers, and as catalysts for production of new compounds in the polymer industry and in medicine will continue to increase in the future.

Platinum, palladium, and iridium, as well as other rare metals, are even more valuable than gold. These metals are used commercially because of their special properties such as extremely high resistance to corrosion, high melting points, and high hardness. The melting points of some precious metals are gold 1945°F, iridium 4430°F, platinum 3224°F, and silver 1761°F. Gold has a specific gravity of 19.3 and a density of 1204.3 lb/ft^3. The specific gravity for silver is 10.50; its density is about 654 lb/ft^3. Platinum is one of the heaviest of metals, with a specific gravity of 21.37 and a density of 1333.5 lb/ft^3. Iridium is also a heavy metal, with a specific gravity of 22.42 weighing 1397 lb/ft^3.

TANTALUM

Tantalum is a bluish-gray metal that is difficult to machine because it is quite soft and ductile and the chip clings to the tool. It is immune to attack from all corrosive acids except hydrofluoric and fuming sulfuric acids. It is used for high-temperature operations above 2000°F. It is also used for surgical implants and in electronics. Tantalum carbides are combined with tungsten carbides for cutting tools that have extreme wear resistance. The melting point of tantalum is 5162°F. Its specific gravity is 16.6 and it weighs 1035.8 lb/ft^3.

TIN

Tin has a white color with a slightly bluish tinge. It is whiter than silver or zinc. Since tin has a good corrosion resistance, it is used to plate steel, especially for the food-processing industry. Tin is used as an alloying element for solder, babbitt, and pewter. A popular solder is an alloy of 50 percent tin and 50 percent lead. Tin is alloyed with copper to make bronze. The melting point of tin is 449°F. The specific gravity of tin is 7.29 and its density is 454.9 lb/ft^3.

TITANIUM

The strength and light weight of this silver-gray metal make it very useful in the aerospace industries for jet engine components, heat shrouds, and rocket parts. Pure titanium has a tensile strength of 60,000 to 100,000 psi, similar to that of steel; by alloying titanium, its tensile strength can be increased considerably. Titanium weighs about half as much as steel and, like stainless steel, is a relatively difficult metal to machine. Machining can be accomplished with rigid setups, sharp tools, slower surface speed, and the use of proper coolants. When spark tested, titanium throws a brilliant white spark with a single burst on the end of each carrier. The melting point of titanium is 3272°F. Its specific gravity is 4.5 and its density is 280.1 lb/ft^3.

TUNGSTEN

Typically, tungsten has been used for incandescent light filaments. It has the highest known melting point (6098°F) of any metal but is not resistant to oxidation at high temperatures. Tungsten is used for rocket engine nozzles and welding electrodes and as an alloying element with other metals. Machining pure tungsten is very difficult with single-point tools, and grinding is preferred for finishing operations. Tungsten carbide compounds are used to make extremely hard and heat-resistant lathe tools and milling cutters by compressing the tungsten carbide powder into a briquette and sintering it in a furnace. The specific gravity of tungsten is about 18.8, and it weighs about 1180 lb/ft^3.

URANIUM AND THORIUM

Uranium and thorium are both radioactive metals that emit alpha, beta, and gamma radiation. Only the gamma rays have considerable penetrating power; they can be contained, however, with adequate shielding. Gamma radiation is very hazardous to living organisms, and low-level radiation can, over a period of time, cause cancer in humans. However, maximum acceptable dose rates have been established.

Uranium is a silver-white metal in its pure state and can be machined like other metals. It can take a high polish, but tarnishes readily on exposure to the atmosphere and, like other pyrophoric (capable of burning in air) metals, will burn if heated to 338°F; if finely divided, it will ignite and burn in air. Uranium is an important fuel for nuclear reactors. The specific gravity of uranium is 18.7 and its density is 1166.9 lb/ft^3. Its melting point is 3362°F.

Thorium is a dark gray metal that is also used in nuclear reactors and is an important fuel for the breeder type of reactor. Thorium oxide combined with 1 percent cerium oxide is used for gaslight mantles. The specific gravity of thorium is 11.2 and its density is 698 lb/ft^3. Its melting point is 3353°F.

VANADIUM

Vanadium is a silver-white, very hard metal that oxidizes when exposed to air. Most vanadium is used as an alloying element of steels in which a small percentage (usually not over 1 percent) imparts hardness and toughness and increases the resistance to shock or impact. In addition, vanadium steel is used for tools; it is one of the alloying elements in cutting tools such as high-speed steel since it tends to promote a finer grain in steels and to inhibit grain growth at higher temperatures. Vanadium has a specific gravity of 5.69 and a density of 354.8 lb/ft^3. Its melting point is 3110°F.

ZINC

The familiar galvanized steel is actually steel plated with zinc and is used mainly for its high corrosion resistance. Zinc alloys are widely used as die-casting metals. To distinguish zinc from aluminum or magnesium (all three are whitish colored metals), the following spot test is used. Nitric acid does not discolor aluminum but does react with zinc, changing it to a gray color. Hydrochloric acid reacts violently with both zinc and magnesium, leaving a black deposit on both, and it reacts slightly with aluminum, leaving a clean surface.

Zinc and zinc-based die-cast metals conduct heat much more slowly than aluminum. The rate of heat transfer on similar shapes of aluminum and zinc is a means of distinguishing between them. The melting point of zinc is 787°F. Its specific gravity is about 7.10 and it weighs about 440 lb/ft^3.

ZIRCONIUM AND HAFNIUM

The metals zirconium and hafnium are always found together in nature, and all of their properties are similar except one: They have opposite properties in regard to neutron flow in a nuclear reactor. Hafnium has the property of stopping the flow of neutrons by absorbing them, thus stopping the reaction taking place in the reactor core. Zirconium, in contrast, freely allows neutrons to flow through, allowing the fission process to occur. Both of these metals are used in nuclear reactors to maintain precise control.

Zirconium is similar to titanium in both appearance and physical properties, and like magnesium, it will burn in air. It was once used as an explosive primer and as a flashlight powder for photography. Machining zirconium, like titanium, requires rigid setups and slow surface speeds. Zirconium has an extremely high resistance to corrosion from acids and seawater. Zirconium alloys are used in nuclear reactors. When spark tested, it produces a spark that is similar to that of titanium. The melting point of zirconium is 3182°F. Its specific gravity is 6.4 and its density is 399 lb/ft^3.

SELF-EVALUATION

1. What advantages do aluminum and its alloys have over steel alloys? What disadvantages?
2. Describe the meaning of the letter *H* when it follows the four-digit number that designates an aluminum alloy. Describe the meaning of the letter *T* in the designation.
3. Name two ways in which magnesium differs from aluminum.
4. What is the major use of copper? How can copper be hardened?
5. What is the basic difference between brass and bronze?
6. Name two uses for nickel.
7. Lead, tin, and zinc all have one useful property in common. What is it?
8. Molybdenum and tungsten are both used in _____ steels.
9. Babbitt metals, used for bearings, are made in what major basic types?
10. What type of metal can be injected under pressure into a permanent mold?
11. Which is stronger, cast or wrought (worked) aluminum?
12. What can be done to avoid building up an edge on the tool bit when machining aluminum?
13. Should a water-based coolant be used when machining magnesium? Explain.

14. Which type of copper—tough-pitch copper, low alloy copper, or oxygen-free copper—would you choose for manufacturing automobile radiators and electronic and refrigeration equipment?
15. Name one of two nickel-chromium-iron alloys that are used in electric heaters and toasters.

CASE PROBLEM: THE UNIDENTIFIED METALS QUESTION

A shipment of various pieces of nonferrous metals was donated by a large industry to vocational schools. Some of that shipment came to a community college. A metallurgy student was given the task of sorting out the odds and ends of metals and tagging them. He was told that included in the shipment was aluminum and magnesium which have the same appearance; also titanium which looks like aluminum, and zinc which also looks like aluminum. There was also copper and brass which he could readily identify by color, but how could he identify those four look-alike metals? Could you identify them? Explain your methods. There was also an element of risk for the students who would use these metals. What is that danger?

WORKSHEET

Objective Correctly identify specimens by comparison by using various tests that are described in this chapter.

Materials A box of numbered specimens and a similar set of known and marked specimens.

Conclusion Record your results.

Item Number	Test Used	Kind of Metal
1.		
2.		
3.		
4.		
5.		
6.		
7.		
8.		
9.		
10.		

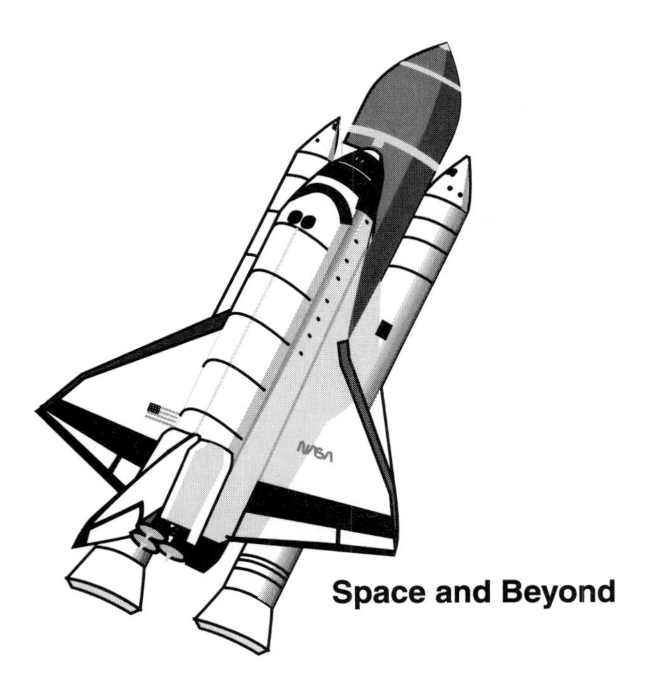

Space and Beyond

CHAPTER 15

Heat Treating of Nonferrous Metals

Prior to 1900 nonferrous metals were not considered to be hardenable in the way that carbon steel could be hardened. Hardening by cold working to a limited extent was the only known way to strengthen nonferrous metals. It was believed that the ancient Egyptians had possessed an art for hardening copper for tools used in building the pyramids. This knowledge was, however, lost to later generations. Today, beryllium copper is used for tools in potentially explosive areas such as powder factories because of its nonsparking quality. It can be hardened to the range of Rockwell C 40 to 50 by precipitation hardening. This may very well be the "lost art" of hardening copper.

A method of hardening aluminum was discovered in Germany by Alfred Wilm in 1906. This new, light, and very strong metal alloy was called duralumin. One of its first uses was for the rigid skeletal structure of the zeppelins of the German navy. These lighter-than-air craft played a part in the air war in World War I. Modern aircraft and space vehicles could not be constructed without these strong, light alloys that can be hardened to exact specifications. This chapter will introduce you to the principles of heat treating nonferrous alloys by solution heat treatment and precipitation hardening.

OBJECTIVES **After completing this chapter, you should be able to:**

1. Explain the reasons underlying the processes of solution heat treatment and precipitation hardening in which hardening takes place.
2. Demonstrate the process of hardening a heat-treatable aluminum alloy.

UNDERSTANDING THE HEAT TREATMENT OF ALUMINUM ALLOYS

Once again, the laws of solubility apply when attempting to understand microstructure changes created when heat treating aluminum alloys. As an example, the major alloying element in 2024 aluminum alloy is copper—alloyed at about 4.3 percent. This quantity of copper dissolves very nicely in aluminum as the temperature approaches the solution heat treating temperature (920°F ±10°F). Any and all compounds formed when alloying with copper, such as the eutectic compound known as copper aluminide ($CuAl_2$), dissolve. Soaking at the solution treating temperature causes homogenization of the grain structure. A rapid quench into water, polymer, or an ethylene glycol and the metal is rapidly cooled to room temperature. Rapid quenching prevents the copper from precipitating out of solid solution and reforming the eutectic compound. With time, natural precipitation and reformation of the $CuAl_2$ will occur. This is called natural aging or "T-4." If the metal is reheated to 375°F, artificial aging "T-6" will occur.

Think of heat treating aluminum alloys in the same manner as dissolving sugar into water. At room temperature, only a small portion of sugar will dissolve into the water. Adding more sugar and stirring furiously will not dissolve the extra sugar. Now, raise the temperature of the water and the extra sugar dissolves. Continue to add sugar while raising the water temperature. When boiling occurs (analogues to reaching the solution treating temperature of the aluminum alloy), all of the sugar is dissolved and if more is added, the extra sugar will not dissolve. This is called a supersaturated solution. Flash freezing (quenching) the supersaturated mixture of sugar and water prevents precipitation and reformation of the sugar crystal. Sugar crystals immediately reform throughout when the mixture is reheated (aging heat treatment).

One of the caveats in the aluminum alloy system is **eutectic melting.** The eutectic compound $CuAl_2$ begins to melt at 935°F. Incipient eutectic melting (the beginning of eutectic melting) is only 5°F above the maximum allowable solution treating temperature. Measurable damage to mechanical properties and corrosion resistance occur immediately if the 2024 aluminum alloy is heated to 935°F or higher (Figure 15.1). Furnace temperature control is critical to successful solution heat treating, as is rapid quenching. Too slow of a quench and $CuAl_2$ phase precipitation will occur during quenching. The metal must be cooled to room temperature within 10 seconds to prevent phase precipitation. The alloy is very soft in the solution heat treated only condition, called "W." To obtain the maximum mechanical properties from the heat treating process, the metal must be reheated or, artificially aged 375°F to accelerate phase precipitation.

Aluminum alloys are hardened by controlled phase precipitation following solution heat treatment. The heat treating temperatures for specific aluminum alloys can be found in the *ASM Metals Handbook* (vol. 2). The temperatures for 2024 aluminum alloy are as follows: liquidus temperature 1180°F; solidus tempera-

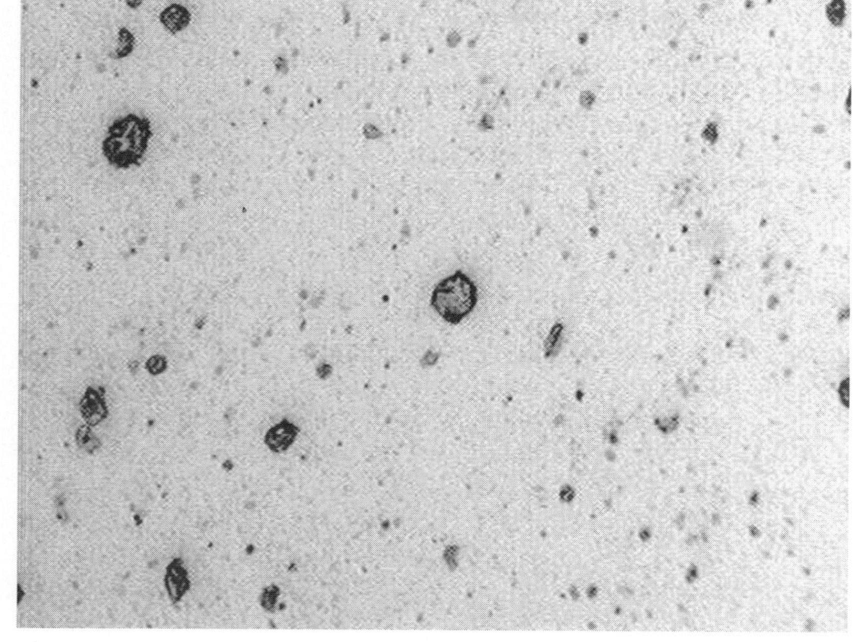

250× Unetched

FIGURE 15.1. Aluminum alloy 2024-T6 extrusion showing rosettes formed by eutectic melting. (*Courtesy of Universal Alloys, Anaheim, CA*)

ture 935°F; and the incipient melting temperature 935°F. Interestingly, the solution heat treating temperature is specific for each alloy and all heat treating temperatures are held within ±10°F. These thermal data are very important to successful heat treatment, since exceeding the solution heat treating temperature, even by just a few degrees, has an immediate deleterious effect on mechanical properties and corrosion resistance of the aluminum.

Annealing of aluminum alloys occurs within the temperature range from about 500° to 800°F. The optimum temperature for annealing aluminum alloys can be obtained from the *ASM Handbook* detailing procedures for the heat treatment of nonferrous metals. Simply, when most precipitation hardening aluminum alloys are exposed to temperatures above 400°F, the rate of eutectic phase precipitation accelerates and particle coalescence occurs with time. The aluminum alloys are softest when the alloying elements are fully precipitated and not in solid solution. Table 15.1 contains solution treating and aging times and temperatures for a few of the more common aluminum alloys.

TEMPER DESIGNATIONS

The temper designation system used in the United States for aluminum and aluminum alloys is a part of the system adopted as an American National Standard (ANSI H35.1). It is used for all product forms, wrought and cast, except ingot. The system is based on the sequences of mechanical or thermal treatments, or both, used to produce the various tempers. The temper designation follows the alloy designation and is separated from it by a hyphen. Basic temper designations consist of individual capital letters. Major subdivisions of basic tempers, where required, are indicated by one or more digits following the letter. These digits designate specific sequences of treatments that produce specific combinations of characteristics in the product. Variations in treatment conditions within major subdivisions are identified by additional digits.

F **As fabricated.** Applies to products shaped by cold working, hot working, or casting processes in which no special control over thermal conditions or strain hardening is employed.

O **Annealed.** Applies to wrought products that are annealed to obtain lowest strength temper, and to cast products that are annealed to improve ductility and dimensional stability. The O may be followed by a digit other than zero.

H **Strain hardened (wrought products only).** Applies to products that have been strengthened by strain hardening, with or without supplementary heat treatment to produce some reduction in strength. The H is always followed by two or more digits, as discussed in the following section.

TABLE 15.1. Solution heat treatment with aging times and temperatures for some commercial hardenable aluminum alloys

Designation	Soaking Temperature (°F)	Soaking Time for Various Thicknesses (Minutes): Up to 0.032 in.	Over 0.032 to 0.125 in.	Over 0.125 to 0.025 in.	Over 0.250 in.
2014-T6	925–950	20	20	30	60
2017	925–950	20	20	30	60
2117	890–950	20	20	30	60
2024	910–930	30	30	40	60
6061-T6	960–1010	20	30	40	60
7075	860–960	25	30	40	60

Note: Soaking time begins after the part has reached temperature.

Designation	Aging Temperature (°F)	Aging Time (Hours)
2014-T6	345–355	2 to 4
	355–375	½ to 1
2017	Room temperature	96
2117	Room temperature	96
2024	Room temperature	96
6061-T6	315–325	50 to 100
	345–355	8 to 10
7075	245–255	24
	315–325	1 to 2

W **Solution heat treated.** An unstable temper applicable only to alloys that naturally age after solution heat treatment. This designation is specific only when the period of natural aging is indicated.

DESIGNATION SYSTEM FOR STRAIN-HARDENED PRODUCTS

Temper designations for wrought products that are strengthened by strain hardening consist of an H followed by two or more digits. The first digit following the H indicates the specific sequence of basic operations, as follows:

H1 **Strain hardened only.** Applies to products that are strain hardened to obtain the desired strength without supplementary thermal treatment. The digit following the H indicates the degree of strain hardening.

H2 **Strain hardened and partially annealed.**

H3 **Strain hardened and stabilized.**

H32 **Strain hardened and stabilized.** (¼ hard)

H34 **Strain hardened and stabilized.** (½ hard)

H36 **Strain hardened and stabilized.** (¾ hard)

H38 **Strain hardened and stabilized.** Full hard = 75 percent

CLADDING OF METALS

Base or less expensive metal plates or sheets and those more subject to the deterioration of corrosion are sometimes covered by a thin sheet of less corrosive metal. These scarcer metals are usually more expensive and the cost would be prohibitive to make the entire plate of the scarcer metal (Figure 15.2). The process of bonding these dissimilar metals together is usually done by applying heat and pressure (Figure 15.3). Some metals such as aluminum and steel are so dissimilar that they cannot be bonded in this way, but they are being bonded by a method known as explosion welding. No heat is needed; an explosive charge provides the extreme pressure needed to make a bond between these metals.

Since pure aluminum is more corrosion resistant than any alloy of aluminum, a thin covering of pure aluminum is sometimes applied to sheets of aluminum alloy that can be hardened. Pure aluminum cannot be heat treated, but clad aluminum alloy can be heat treated in the same way as other heat-treatable aluminum alloys because of the alloy part of the "sandwich." Care must be taken to avoid oversoaking because diffusion of alloying elements can occur in the clad surface. This process would eventually convert the clad surface of pure aluminum into an alloy that has low corrosion resistance (Figure 15.4).

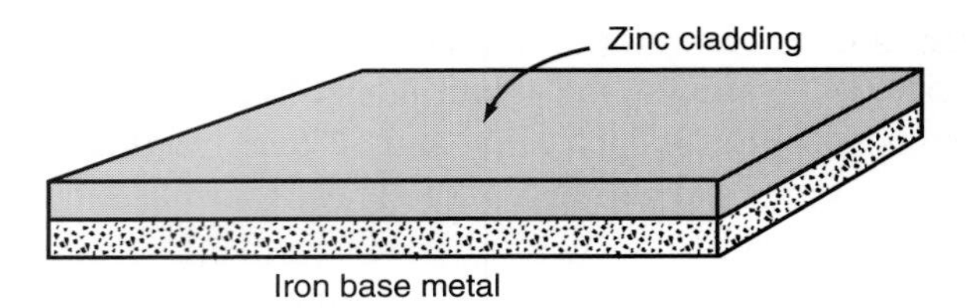

FIGURE 15.2. Metal cladding. By cladding, we use less of the scarce and expensive metals.

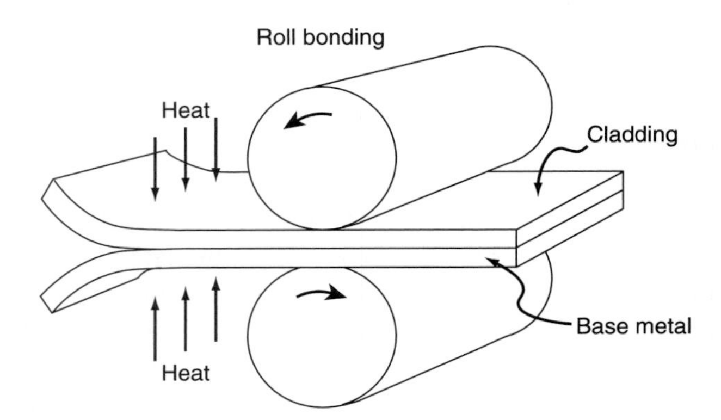

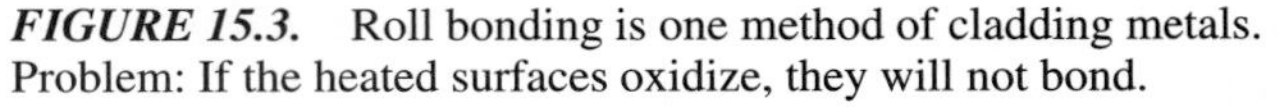

FIGURE 15.3. Roll bonding is one method of cladding metals. Problem: If the heated surfaces oxidize, they will not bond.

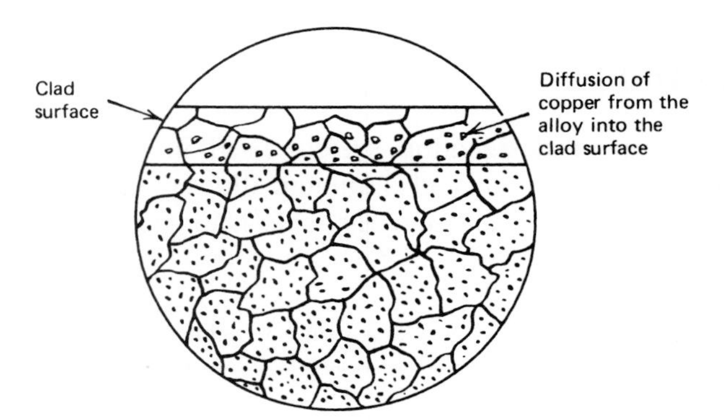

FIGURE 15.4. Diffusion of copper into the pure aluminum-clad surface will reduce its corrosion resistance. Excessive heating time can cause this transition zone of diffusion.

DESIGNATION SYSTEM FOR HEAT TREATABLE ALLOYS

The temper designation system for wrought and cast products that are strengthened by heat treatment employs the W and T designations described in the section on basic temper designations. The W designation denotes an unstable temper, whereas the T designation denotes a stable temper other than F, O, or H. The T is followed by a number from 1 to 10; each number indicates a specific sequence of basic treatments, as follows:

T1 Cooled from an elevated temperature shaping process and naturally aged to a substantially stable condition.

T2 Cooled from an elevated temperature shaping process, cold worked, and naturally aged to a substantially stable condition.

T3 Solution heat treated, cold worked, and naturally aged to a substantially stable condition.

T4 Solution heat treated and naturally aged to a substantially stable condition.

T5 Cooled from an elevated temperature shaping process and artificially aged.

T6 Solution heat treated and artificially aged.

T7 Solution heat treated and stabilized.

T8 Solution heat treated, cold worked, and artificially aged.

T9 Solution heat treated, artificially aged, and cold worked.

T10 Cooled from an elevated temperature shaping process, cold worked, and artificially aged.

Stress-relieved Wrought Products:

T×51 Aluminum products which are stretched to a specific amount following solution heat treatment or after cooling from an elevated temperature shaping process.

T×510 Products that receive no further straightening after stretching.

T×511 Products that may receive minor straightening after stretching to comply with standard tolerances.

T×52 Stress relieved by compressing.

T×54 Stress relieved by combining stretching and compressing.

T42 Solution heat treated from the O or the F temper to demonstrate response to heat treatment, and naturally aged to a substantially stable condition.

T62 Solution heat treated from the O or the F temper to demonstrate response to heat treatment, and artificially aged.

T-35 A T3 with a stretch for stress relief.

T-73 Variation of normal T7 heat treat required to produce T7 condition.

T-651 Cold rolled or cold finished bar stock.

T-6511 Extruded bar stock.

ANNEALING

Aluminum may be stress relieved to remove the stresses of cold working by simply heating to 650°F and allowing to cool. Since this temperature will not dissolve the copper aluminide, the alloy will remain in the annealed condition. Aluminum grain structure can be reformed or recrystallized by this stress relief process. However, for the heating cycle to recrystallize the grains, the piece must first be cold worked or strained to a minimum of 2 percent elongation. Nucleation, or the starting of new grains, begins at points of high stress on the crystal lattice structure. Therefore, a piece that has been highly strained will produce a fine grain structure after recrystallization. Thus, the grain size of the recrystallized metal can be controlled to a certain degree by the amount of cold work performed (Figures 15.5 and 15.6).

Annealing to remove the effects of hardening in hardenable alloys caused by rapid cooling from welding,

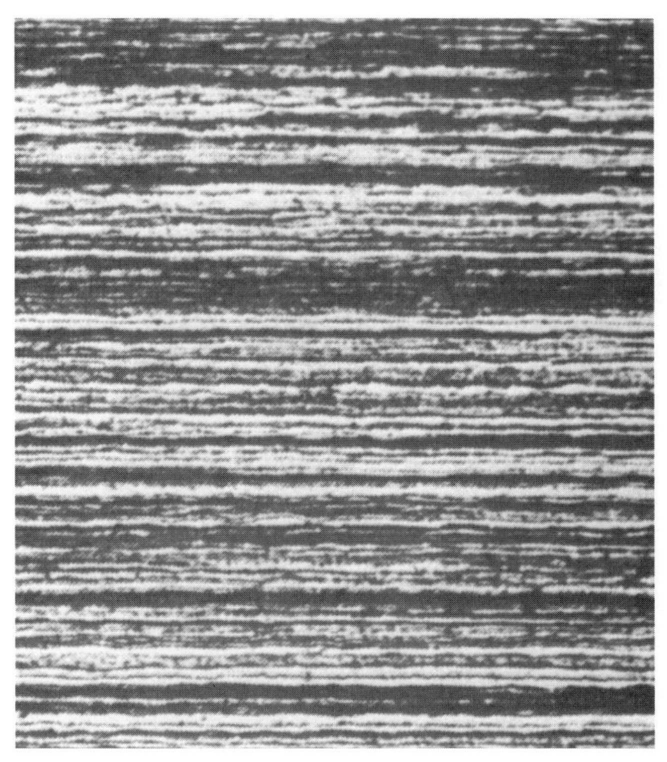

FIGURE 15.5. Micrograph of highly strained cold-worked aluminum (100×). *(By permission, from Metals Handbook, Volume 7, Copyright American Society for Metals, 1972)*

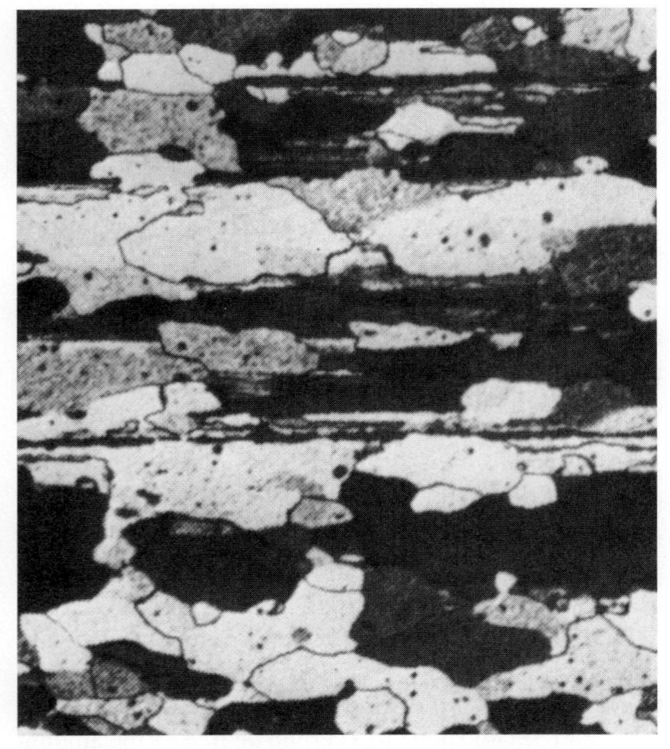

FIGURE 15.6. Recrystallization of the same cold-worked aluminum was produced by stress relief anneal (100×). *(By permission, from Metals Handbook, Volume 7, Copyright American Society for Metals, 1972)*

heat treating, or hot working is done at a higher temperature, about 970°F. Since the copper aluminide is in solution in this operation, rapid cooling would not put the metal in an annealed state. So, in this process, the aluminum is soaked for a period of time at 770° to 825°F. It is then allowed to cool to room temperature. See Table 15.2 for the mechanical properties of some heat-treatable aluminum alloys.

GENERAL NONFERROUS HEAT TREATMENT

Most metals are not allotropic and heat treatment usually means annealing to remove the effects of a work-hardened microstructure. The heat treatment process for strengthening many nonferrous metals involves solution heat treatment followed by precipitation hardening or aging. In simple terms, the alloying elements and/or compounds must be dissolved by heating to a specific temperature at which complete solid-state dissolution takes place. Once the metal has stabilized (held at the solution treating temperature for the required time), it is rapidly cooled to room temperature. For many metals, cooling is accomplished by quenching in a water or polymer solution which rapidly brings down the temperature within a specified time limitation. For aluminum, quenching to room temperature is very rapid, about 10 seconds duration. For other metals, quenching or cooling need not be as rapid. Many stainless steels are solution heat treated in a vacuum furnace and while still inside the furnace, the metal is gas fan quenched to room temperature. Gas fan quenching uses an inert gas—nitrogen or argon—rapidly backfilled into the heating zone in the furnace and directly over the hot parts. Quenching is not as rapid as that achieved with water. The advantage with heating metals in a vacuum is the protec-

TABLE 15.2. Mechanical properties of some heat-treatable wrought aluminum alloys

Alloy and Temper	Tensile Strength (ksi)		Elongation in 2 in. Percent 1/2-in. Dia Specimen	Brinell Hardness Number 10 mm Ball 500 kgf 30 Seconds	Shear Strength (ksi)	Modulus of Elasticity (ksi × 10^3)	Fatigue Limit (ksi)
	Yield	Ultimate					
2014–0	14	27	18	45	18	10.6	13
2014-T3	41	64	—	—	—	10.6	—
2014-T4	40	63	20	105	38	10.6	20
2024–0	11	27	22	47	18	10.6	13
2024-T3	50	70	—	120	41	10.6	13
2024-T4	47	68	19	120	41	10.6	20
6061–0	8	18	30	30	12	10.0	9
6061-T4	21	35	25	95	30	10.0	14
6063–0	7	13	—	25	10	10.0	8
6063-T4	13	24	—	—	16	10.0	10
7075-T6	73	83	11	150	48	10.4	22
7178-T6	78	88	11	160	52	10.4	22

Note 1. Mechanical properties tables for wrought and cast aluminum and other nonferrous metals are available in manufacturers' reference materials.

Note 2. Ksi refers to thousands of pounds per square inch. Only four tempers of wrought aluminum were used for this table as they are particularly relevant to the lab work of this chapter. They are: O—recrystallized, annealed; T3—solution heat treated and then cold worked; T4—solution heat treated and naturally aged to a stable condition; and T6—solution heat treated and artificially aged.

tion from scaling or surface alloy depletion and oxidation for metals requiring heating well above 1000°F.

HEAT TREATING TITANIUM ALLOYS

Although titanium is the fourth most abundant metallic ore (rutile) on the earth's crust, it is a relatively costly metal because of the difficulty of extracting the metal from the ore. Its greatest use is in aerospace, but it is rapidly becoming a metal for more common usage such as for automobile engines, the chemical industry, and superconductors. Titanium generally is grouped into two categories: commercially pure (CP) and alloys. Commercially pure titanium grades are not heat treatable for hardening, but many of the titanium alloys can be hardened by precipitation and aging. However, there is a danger of contamination of the metal during the heating cycle. Oxygen, hydrogen, and nitrogen have a detrimental effect on titanium, but it is particularly sensitive to chlorides such as salt. Even salt from fingerprints can cause stress corrosion when heated to temperatures above 600°F. However, an alpha alloy (Ti-8A1-1MO-IV) has been produced that provides a good resistance to salt stress corrosion.

Titanium alloys are further classified into three major categories: alpha, alpha-beta, and beta. Alpha alloys can be welded and are stable up to 1000°F although they are nonheat treatable. They are strong and tough at cryogenic temperatures. Beta alloys can be heat treated and welded and are stable only to about 600°F. Alpha-beta alloys are heat treatable and stable to 800° to 1000°F. They are not as tough as the alpha alloys and are difficult to weld.

Titanium is an allotropic metal and will change from an alpha hexagonal close-packed (HCP) crystal structure to a beta body-centered cubic (BCC) crystal structure upon heating to or above the allotropic temperature called "beta transus"; about 1650°F. Allotropic transformation for titanium and titanium alloys is a one-way street; exceed the beta transus temperature and almost instant transformation occurs. The beta microstructure is often called a Widmanstatten structure because of the coarse, boxlike interweaving of the grain formation.

Titanium alloys require special protection from oxidation and must be solution heat treated in a vacuum furnace or in a furnace that provides an inert atmosphere such as argon. When exposed to oxygen above ~1000°F, a very hard and brittle titanium oxide layer begins to form on the surface of the metal. This layer, known as "alpha case," is believed to seriously compromise mechanical strength, especially fatigue properties. Titanium is also sensitive to diffusion of nitrogen and hydrogen gases. Following solution heat treatment, titanium alloys must be age hardened, often in an air furnace which causes a bluish oxidizing stain to form on the metal surface.

HEAT TREATING MAGNESIUM ALLOYS

Some magnesium alloys can be solution heat treated and aged, when proper heat treating procedures are used. Extensive study and experience are required to heat treat this material safely. Manufacturers' catalogs and reference books such as the *Metals Handbook* from the American Society for Metals may be consulted for heat treating procedures.

HEAT TREATING NICKEL AND NICKEL ALLOYS

Nickel and its alloys may be annealed, stress relieved, and, in some cases, solution heat treated and aged. Among the hardenable nickel alloys are permanickel 300, duranickel 301, monel 501, inconel 718, and hastelloy R-235. Solution temperatures for these alloys are from 1800° to 2000°F, except for monel 501, which is 1525°F. Aging temperature for permanickel is about 900°F. The other hardenable alloys must be aged for 16 hours at 1100°F and then at 1000°F for 6 hours, followed by 8 hours at 900°F, and air cooled. Annealing is carried out by heating to a predetermined temperature for a period of time and then quenching in water. As with most metals, a scale may be formed at high temperatures. This can be controlled by using a carbon-rich furnace atmosphere or by bright annealing in a closed container with an inert gas.

HEAT TREATING COPPER AND COPPER ALLOYS

Copper and copper alloys may be heat treated for any of the following reasons:

1. Homogenizing: using high temperatures and relatively long times to dissolve and absorb segregation and coring found in castings and hot-worked alloys containing tin and nickel. Homogenizing is usually performed at the mill.
2. Annealing: heating the metal to soften the microstructure by recrystallization. Annealing is done to eliminate the effects of cold work or work hardening.

3. Stress Relieving: a heat treatment used to eliminate the effects of machining or residual stresses which may cause failure in some copper alloys by a mechanism known as "season cracking."
4. Solution Heat Treating and Precipitation Hardening: Solution heat treating of copper alloys is a heat treatment used to remove the effects of work hardening and to soften the microstructure for further fabrication. Solution heat treating is necessary to dissolve and distribute the soluble alloying elements, followed by quenching the metal to prevent phase precipitation. The metal is age hardened at a moderate temperature for about 3 hours to strengthen and harden the alloy by microprecipitation.

Aluminum bronzes containing more than 10 percent aluminum are hardened by quenching from a high temperature of about 1200°F to produce a martensitic type of structure similar to that of hardened steel. This is followed by tempering at a lower temperature.

Beryllium copper and other hardenable alloys are solution heat treated and precipitation hardened. Beryllium copper containing nickel or cobalt is solution treated at 1425° to 1475°F for 1 to 3 hours and quenched in water. Aging time is from 2 to 3 hours at 575° to 650°F. The tensile strength of these hardened alloys ranges from 150,000 to 215,000 psi.

Beryllium copper is usually supplied solution heat treated, aged, and cold worked. It can be machined in this condition with proper tooling. It is, however, sometimes necessary to anneal beryllium copper for further cold working.

SAFETY NOTE

Beryllium is a toxic metal and beryllium compounds are highly toxic. Adequate protection should be used to avoid any fumes caused by any incipient melting or burning of the metal. Beryllium oxide (BeO) solubility in copper is minimal, just more than 1 percent. The primary hazard in machining copper-beryllium alloys is with the beryllium-ozide dust particles created during machining. BeO dust particles are lightweight and can easily become airborne during machining. BeO is highly toxic and known to cause lung problems. Proper containment and handling precautions must be followed.

The annealing temperature for beryllium copper alloy is much higher than that of other copper alloys and the metal is usually held at 1425° to 1900°F for 3 hours and then quenched in water. Other hardenable alloys are annealed at temperatures above 1200°F. Cooling rates are not particularly important for annealing copper and copper alloys that are not hardenable, and they are usually annealed at temperatures below 1400°F.

Stress relieving temperatures for nonhardenable copper alloys range from 400° to 475°F. Stress relief is used when further cold work is necessary where extensive change in mechanical properties is not desirable. The full anneal produces large, coarse grains that leave the metal at a very soft temper and unsuitable for many uses.

Copper alloys may be hardened by solution heat treatment and precipitation (aging) and then can be stress relieved or annealed. The important copper alloys that can be age hardened by precipitation are beryllium copper, aluminum bronze, copper-nickel-silicon, copper-nickel-phosphorus, chromium copper, and zirconium copper.

SELF-EVALUATION

1. Match the correct numbers on the phase diagram (Figure 15.9) with the following.
   - **a.** _____ Solvus lines.
   - **b.** _____ Mushy areas of liquid and solid solutions.
   - **c.** _____ Area of solid solutions.
   - **d.** _____ Line denoting the temperature of highest solubility of the solid solution *a* or *b*.
   - **e.** _____ Mixture of solid solutions *a* and *b*.
2. What is the result when a solution of 4 percent copper in aluminum is quenched so that the cooling curve crosses the solvus line (as shown in Figure 15.7)?
3. What are the two necessary steps needed to harden heat-treatable nonferrous metals?
4. What causes hardening in the aluminum-copper alloy?

5. What is the purpose for using clad aluminum?
6. Aluminum that is not heat treatable may be stress relieved for further cold working by what process?
7. When heat-treatable aluminum alloys are overheated and the surface is blistered, what can be done to correct the situation?
8. Aging of solution heat-treated aluminum can take up to 100 hours before full hardening takes place. How can this process be speeded up in some alloys?
9. Nickel and most other metals form an oxide scale on the surface when they are heated in a furnace in the presence of oxygen. Name two methods mentioned in this chapter to control this problem.
10. When titanium is alloyed with other elements, what effect do they have on the crystal structure?

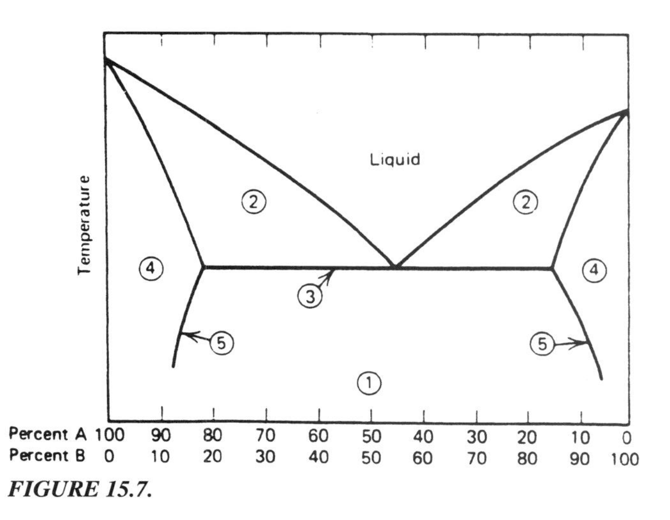

FIGURE 15.7.

CASE PROBLEM: THE BRITTLE DUNE BUGGY PARTS

A mechanic who was building a lightweight dune buggy needed some high-strength aluminum parts. Since he had access to a heat-treating furnace, he decided to do the solution heat-treating and aging process himself. He obtained 2024 aluminum bars and made the parts. Then he heated the parts in a furnace to 1200°F, quenched them, and then allowed them to age naturally. He noticed they had small blisters all over their surfaces and when he began to hammer one in place, it broke in two places. What did he do wrong? Could the parts be salvaged in any way?

WORKSHEET

Objective Demonstrate the process of solution heat treatment and aging.

Materials Furnace and three small strips of 18 to 22 gage 2024-T3 aluminum alloy marked 1, 2, and 3.

Procedure

1. Heat strip No. 1 to 920°F and hold at this temperature for 20 minutes.
2. Quench the strip in cold water.

3. Make a hardness test on both the control (not heat treated) strip No. 2 and the quenched strip. Record the test on a sheet of paper.
4. Heat strip No. 3 to 650°F and allow it to cool. This strip is now in the annealed condition.
5. Note that strip No. 1 and strip No. 3 are both easily bent with the fingers at this stage, but strip No. 2 is quite stiff because it has a hard temper.
6. Make a hardness test each day for 4 days on each specimen and record your results on your paper.
7. On the fourth day, make the bending test with your fingers on all these specimens.

Conclusion

1. Did the hardness change on the solution heat-treated strip each day? How much?
2. Does it bend as easily as the strip that was not heat treated?
3. How does it compare to the annealed strip?
4. Show your results to your instructor.

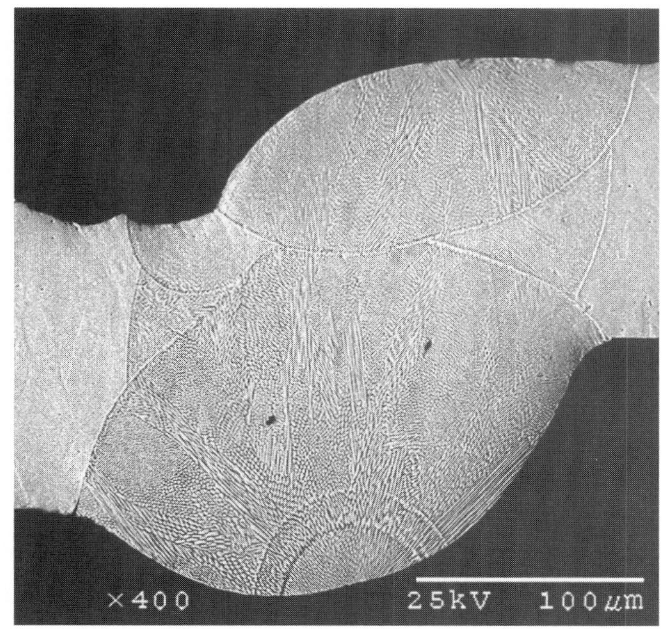

0.004" Weld Bead Produced by Micro-T.I.G.

CHAPTER 16

Metallurgy of Welds: Nonferrous Metals

In the early days of welding, procedures for joining nonferrous metals such as copper and aluminum were either nonexistent or very difficult to use, and consequently very little welding of these metals was done. These difficulties have long since been overcome and nonferrous metals are now easily welded, soldered, or brazed by means of modern technological methods. This chapter deals with welding techniques and methods for such nonferrous metals as aluminum, copper and copper alloys, and monel metal. The emphasis is on the metallurgical condition of the welds and weldments.

OBJECTIVES **After completing this chapter, you should be able to:**

1. List some of the metallurgical conditions and problems related to welding nonferrous metals.
2. Macroetch weld specimens in nonferrous metals to observe and identify the weld structure.
3. Prepare nonferrous welds for microscopic examination to determine the metallurgical condition of the weld.

NONFERROUS WELDING

Although nonferrous metals for most purposes are hardened by cold working, many of them can also be hardened by precipitation heat treatment. Welding on precipitation-hardening alloys presents a somewhat different welding problem than welding on nonhardenable, nonferrous metals. Probably one of the major reasons for past difficulties in welding nonferrous metals has been their great affinity for oxygen, which causes the formation of an oxide film. Many nonferrous metals also have a very high thermal conductivity, requiring a high heat input in the weld area in order to maintain a molten puddle. In some ways the welding of nonferrous metals is similar to the welding of ferrous metals. Grain growth is a problem in both kinds of welding. Porosity, inclusions, cracking, and embrittlement are all common to both types of welding. As you study this chapter, you should refer to Chapter 15, "Heat Treating of Nonferrous Metals," to help clarify some of the metallurgical concepts that are described in this chapter.

JOINING PROCESSES USED FOR ALUMINUM

Of all nonferrous metals, aluminum is probably the most easily welded metal; it can be readily arc welded, brazed, and soldered. All three of these joining processes are used extensively in industrial manufacturing. Aluminum may also be resistance welded, induction welded, ultrasonic welded, gas welded, and pressure welded. In addition, these major groups of welding may be broken down into subgroups that include many other kinds of welding processes.

Probably the most common kind of arc welding done on aluminum in small shops is tungsten inert gas (TIG), where an inert gas is used to shield the aluminum surface from the atmosphere. Gas-metal arc welding (GMAW or MIG) is also used.

The three characteristics of aluminum that make it behave differently from steel are its great affinity for oxygen, its high thermal conductivity, and the fact that it does not change color before reaching its melting point. This persistence of color often causes the operator to have some difficulty in detecting welding temperatures. Aluminum is very active chemically and has such an affinity for oxygen that on clean metal an oxide film forms immediately. This oxide film, which is thin and transparent but very hard, prevents the weld metal from joining with the base metal. If any welding is to be done, this film must be removed. It could be removed mechanically by scraping, by using a wire brush, or by sanding, or it can be dissolved chemically with a flux. A chemical reaction can replace the oxide with another material that does not hinder bonding. Also, the heat of welding can sometimes remove the oxide and float it off into the slag. Aluminum welds are highly subject to contamination. Even the filler metal itself can be contaminated, causing a poor weld. Aluminum sheet or plate to be welded should be carefully cleaned, and preferably the plate should be sheared, sawed, or machined just prior to the welding operation. All lubricants, oils, oxides, and greases should be removed (Figures 16.1*a* and 16.1*b*).

Aluminum does not change color before it becomes molten and only begins to glow a dull red a few hundred degrees above the melting point. This fact causes some difficulties when gas or TIG welding. One way of determining temperatures of the base metal is by marking it with temperature-indicating crayons that melt at a specific temperature. Another method, and the most common one, used to determine the proper welding temperature is to watch the adjacent area near the weld for a "wet" appearance of the surface. This indicates that there is melting at the surface. Because aluminum conducts heat so rapidly, the whole area around the weld tends to become weakened because of the high temperature it develops, especially when gas welding. The base metal, if it is overheated in the vicinity of the melting point, can fail; causing the weldment to collapse near the weld. With arc welding, this problem does not occur since the weld is made so rapidly that there is not sufficient time for the base metal to overheat. In either case, however, mechanical support should be provided for the weldment because it loses strength at higher temperatures.

WELDABILITY OF ALUMINUM ALLOYS

Since most aluminum is strain hardened and most of the aluminum used is in the form of an alloy, the base metal is almost always stronger than the weld metal, which is cast and not strain hardened. The base metal near the weld is softened by recrystallization of the grains; this gives the heat-affected zone more ductility but a lower tensile strength than the rest of the base metal. It is extremely important for the welder to know the kind of aluminum alloy on which the weld is being made. The four-digit system of designation for aluminum alloys may be found in Chapter 14, "Identification of Nonferrous Metals." Of the aluminum alloys, the 1000, 3000, and 5000 series are not heat treatable and can be hardened only by cold working. These and other alloys such as the heat-treatable 6000 series can be gas welded or arc welded with no difficulty. The 2000, 4000, and 7000 series present the greatest difficulty in welding.

Arc welding and resistance welding are preferred when welding is done under controlled conditions. Gas welding is not recommended (see Table 16.1). The 2000, 4000, and 7000 series of aluminum alloys are heat treatable since they contain copper, magnesium, manganese, chromium, zinc, or silicon alloying additions; postweld heat treatments should be used.

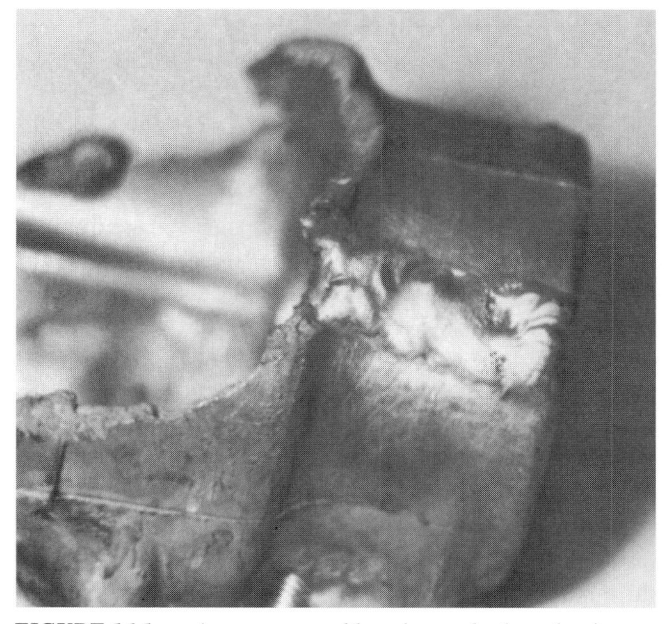

***FIGURE 16.1*a.** A very poor weld made on a broken aluminum casting, showing surface porosity. Detection of surface porosity can usually indicate a weak porous weld (see Figure 16.1*b*).

The heat-treatable alloys such as the 2014, 2017, and 2024 types can be welded only by using special techniques, fixtures, and heat treatments (if necessary).

THE EFFECTS OF WELDING HEAT ON ALUMINUM ALLOYS

The weld zone of almost all aluminum alloys is adversely affected by high temperatures. The weld area becomes weaker than the unaffected base metal because of grain growth. Even on annealed aluminum, the cast structure of the weld itself has a lower strength than the base metal. Sometimes a filler alloy will increase the weld metal strength. When a heat-treatable alloy such as 2024 (which is solution heat treated and aged) is used in a structural application and welded, the heat-affected zone can be adversely affected because copper aluminide precipitates to the grain boundaries. However, there are methods used for welding such alloys.

The heat-treatable alloys of the 2000, 4000, and 7000 series should not be welded unless very stringent controls are used to produce the original heat-treated condition in the weld zones; exceptions are the use of one of the more weldable alloys. Some alloys that can be heat treated by postweld solution heat treatment permit partial or complete restoration of their strength and avoid grain boundary precipitation after welding. Welding of 7075, 7079, and 7178 is not recommended, but 7005 and 7039 were developed for welding.

Aluminum alloys have a tendency to be hot-short; that is, they possess low strength at solidification

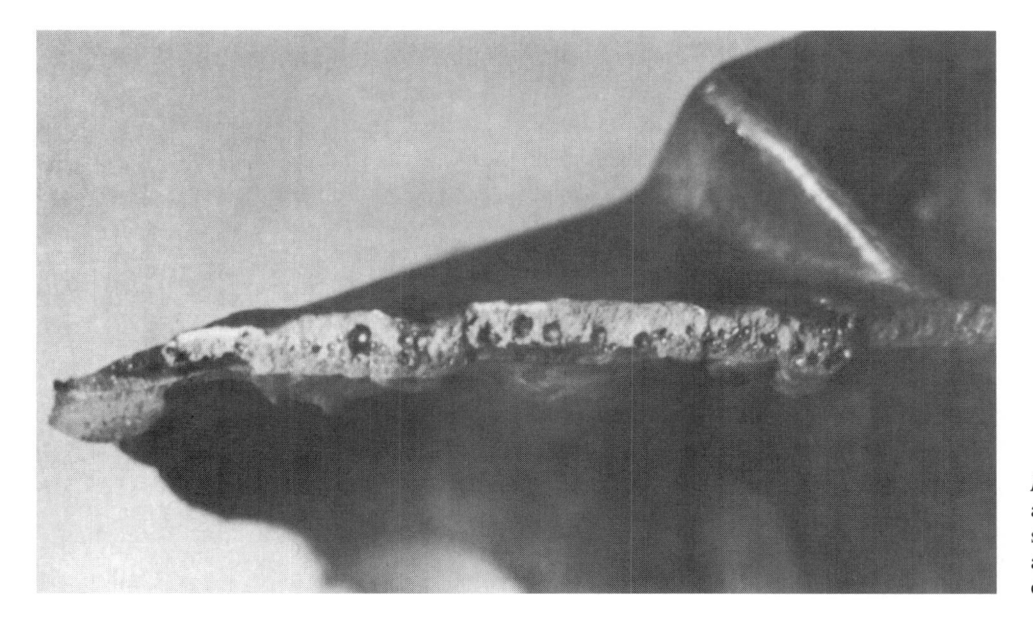

***FIGURE 16.1*b.** Broken section of a weld made on the same casting shown in Figure 16.1*a*. The great amount of visible porosity is due to contaminants on the base metal.

TABLE 16.1. Relative weldability of heat-treatable wrought aluminum alloys and cast aluminum alloys that are not heat treatable

Welding Process	**Wrought Aluminum ASA Alloy Designation**																	**Cast Aluminum ASA Alloy Designation**															
	2014	2017	2024	2218	2219	2618	6061	6063	6070	6101	6201	6951	7005	7039	7075	7079	7118	433.0	A444	208.0	213.0	514.0	B514.0	F514.0	C712.0	A712.0	D712.0	413.0	A514.0	518.0	360.0	380.0
Resistance	A	A	A	A	A	A	A	A	A	A	A	A	A	A	A	A	A	A	C	B	C	B	B	B	A	C	C	C	B	C	C	C
Pressure	C	C	C	C	C	C	B	B	C	A	A	A	B	B	C	C	C	O	O	O	O	O	O	O	O	O	O	O	O	O	O	O
Metal arc	C	C	C	C	C	C	A	A	B	A	A	A	O	O	O	O	O	A	C	B	C	C	C	C	A	C	C	B	C	O	O	O
TIG	C	C	C	C	C	A	A	A	A	A	A	A	A	A	C	C	C	A	B	B	B	A	B	A	A	B	B	B	A	B	B	B
Brazing	O	O	O	O	O	O	A	A	C	A	A	A	B	C	O	O	O	C	A	O	O	O	O	O	B	A	A	O	O	O	O	O
Soldering	C	C	C	C	C	C	B	B	B	A	B	A	B	B	C	C	C	C	B	C	C	O	O	O	C	B	B	O	O	O	O	C
Gas	O	O	O	O	O	O	A	A	C	A	A	A	O	O	O	O	O	A	C	C	C	C	C	C	A	C	C	B	C	O	O	O

Note: Ratings are based on the most weldable temper.
A = readily weldable
B = weldable in most applications; special techniques may be necessary
C = limited weldability
O = not recommended for that joining method

temperatures, and as the weld melt cools, the strain that is placed on the high temperature weld metal sometimes causes hot cracking (or tearing) that is usually evident in the center of a weld. The factors that cause or influence hot cracking are the rate of cooling, the weld size, restraint of the joint, and the solidification temperature of the alloy. For example, pure aluminum melts at about 1200°F, while some aluminum alloys melt at lower temperatures, some as low as 900°F.

The use of preheat helps to minimize the effects of hot cracking by reducing the rate of cooling, but preheating should not be used if the joint is restrained. Also, the mechanical properties of the base metal can be decreased by preheating, especially with the heat-treatable alloys (Figure 16.2). Hot-short cracking can also be controlled by selection of the proper filler metal.

WELDING COPPER

Pure copper presents some welding difficulties because it has an affinity for oxygen and because it contains a small amount of oxygen in the form of oxides that tend to migrate to the grain boundaries, thus weakening the copper. The gas-metal arc process is recommended for welding deoxidized copper, and the weld should be made rapidly to prevent the formation of oxides. Gas welding is not recommended because of its slow rate of deposition and its high heat input. Gas-metal arc welding may be used to join electrolytic and oxygen-free copper, but the welds are generally of poor quality. Shielded-metal arc welding with the proper electrodes is used on some brasses, bronzes, and deoxidized copper.

All the numerous copper alloys are weldable with various types of welding processes. The AWS classification for welding rods and electrodes for copper and copper alloys may be found in welding handbooks.

In copper and copper alloy welding the same problems and defects occur that are found in other types of welding, and usually the same inspection methods can be used. Visual inspection will generally detect porosity and high-temperature cracking. Other defects could be found by ultrasonic, radiograph, and other nondestructive methods of testing. If copper must be welded by oxy-acetylene methods, then it must be deoxidized copper. This is copper that contains a small percentage of

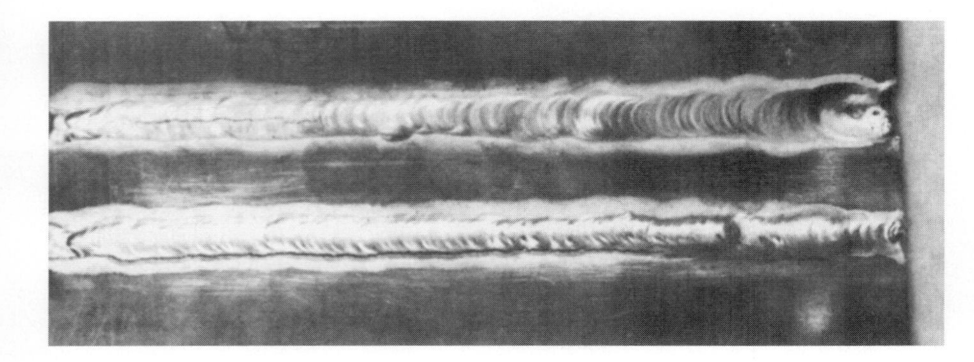

FIGURE 16.2. (Top) Hot cracking in the center of the weld. (Bottom) Toe cracking in aluminum alloy weld caused by overheating.

silicon, phosphorus, or other deoxidizers. Where any weld strength is required, oxygen-free copper should be used. Many of the problems associated with the welding of coppers are also found in aluminum welding, such as high thermal conductivity, which rapidly removes the heat of welding, distributing it throughout the material. Also, the high affinity for oxygen and other gases such as hydrogen and carbon monoxide creates problems. Copper has a high coefficient of expansion that tends to promote cracking in the weld during cooling, especially if there is a restraint in the base metal.

All the hundreds of copper alloys, brasses, bronzes, and beryllium alloys are weldable. The copper alloys containing lead and the high zinc alloys are the most difficult to weld. Lead is volatized (becomes a gas) before the copper melts, and the alloy becomes very weak at welding temperatures; zinc volatizes copiously when attempting to weld Muntz metal.

Some aluminum bronzes and beryllium copper are heat-treatable alloys. A suitable postheat treatment should be used when welding these alloys to restore the original strength and hardness.

WELDING DISSIMILAR METALS

The nickel alloy electrodes are frequently used for joining dissimilar metals. For example, nickel electrodes are used to join nickel to carbon steel by metal arc welding or for welding copper to carbon steel. Austenitic stainless steel or inconel electrodes are also often employed for joining dissimilar metals, such as nickel and stainless steel. Thus, many dissimilar metal combinations are possible; filler metal recommendations should be sought from electrode suppliers or the *AWS Welding Handbook* (7 ed., vol 4).

WELDING NICKEL ALLOYS

Monel, nickel, and inconel are all readily welded. They can be welded with filler metal containing the same analysis as the base metal; the heat effect of welding does not alter the properties of the base metal to any great extent. In nickel, the age-hardening alloys of the K-monel, Z-nickel, and Hastalloy group are hardenable and may be welded by several methods. Shielded-metal arc is usually preferred, but gas welding may be used, although excessive grain growth may take place in the heat-affected zone. These alloys are hardenable simply by heating between 1100° and 1600°F and then cooling slowly. A quenching step is not needed as in aluminum alloys. Therefore, a simple annealing procedure is all that is required after welding to give the weld and the heat-affected zone uniform hardness and higher corrosion resistance.

WELDING TITANIUM ALLOYS

In Chapter 15, "Heat Treating of Nonferrous Metals," you learned about titanium in its various allotropic conditions, such as the alpha and beta transformations resulting from varying alloying elements. For example, the addition of aluminum raises the transformation point of the alpha phase, and chromium, vanadium, iron, or zirconium slow the transformation from beta to alpha when cooling. Alpha alloys can be welded and are stable up to 1000°F. The least weldable group of titanium alloys are those in the mixed or alpha-beta alloys, although some of these have been fusion welded under special conditions. Beta alloys can be welded, but they are somewhat brittle above 1000°F. They are stable to about 600°F. Commercially pure (CP) grades are readily weldable with inert gas shielding. One advantage of CP grades is that the parent metal and the heat-affected zone have equal corrosion resistance. The most important aspect of welding titanium alloys is to avoid contamination by oxygen, nitrogen, and other impurities. This is true of all the reactive metals (those that burn in air) such as zirconium, magnesium, and tantalum. TIG welding is often used for joining titanium where an inert gas is used to prevent contamination. The recommended shielding gases for titanium welding are helium or argon, or mixtures of the two. The weld area must be well shielded to avoid the embrittling effects of dissolved nitrogen, oxygen, and hydrogen. Weld areas should be extremely clean; a very small amount of oil, even from a fingerprint, can cause embrittlement. Titanium is usually welded in the annealed state to prevent embrittlement (Figure 16.3*a*). An example of an unacceptable titanium weld is shown in Figure 16.3*b*.

SOLDERING AND BRAZING

Many nonferrous metals can be joined by soldering (liquid-solid phase). Aluminum, copper, brass, and bronze are among those easily soldered. The solder composition and the flux must be the type designed for the metal to be soldered. Flow solder techniques are often used to join sheet metal parts, whereas some connections, such as for copper plumbing, are assembled by flowing solder into a heated joint by capillary action. Soldering makes a relatively low-strength joint.

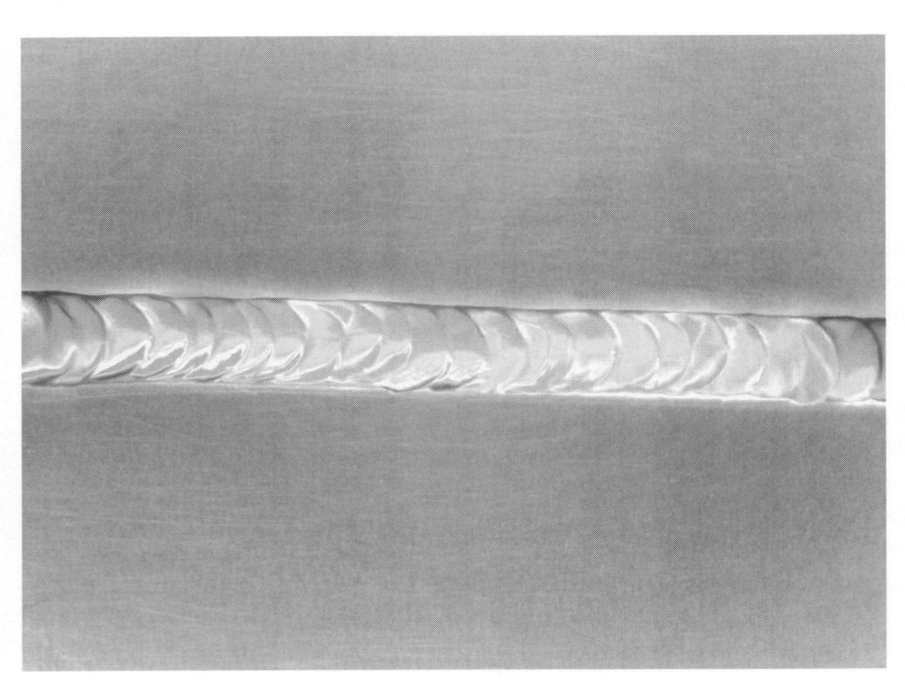

***FIGURE 16.3*a.** Titanium weld done with TIG process. The surface of the weld must be protected from oxidation during welding by an argon purge of surrounding air.

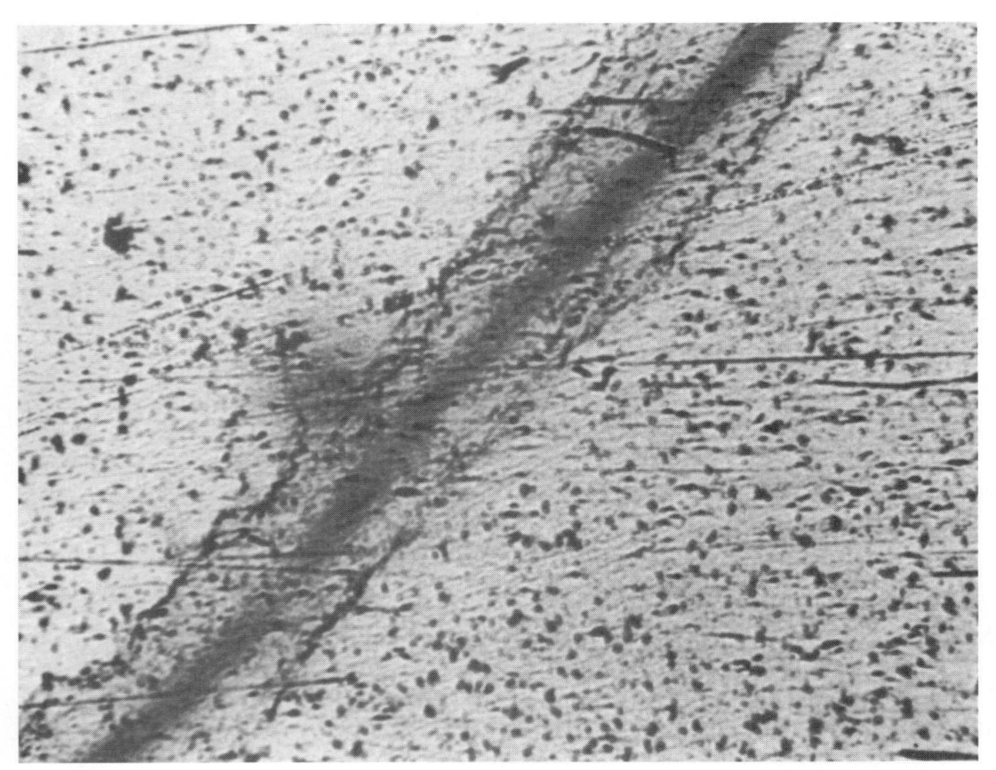

***FIGURE 16.3*b.** Junction zone of a weld on titanium alloy done with TIG process by using argon gas for shielding. This is an incomplete weld, lacking fusion, and would be considered unacceptable.

Torch brazing (liquid-solid phase) is done by flowing a filler metal into a joint. As with soldering, most nonferrous metals can be joined by the brazing process. Welds made by brazing are usually stronger than those made by soldering but are not as strong as any type of fusion joining, because soldering and brazing join metals by adhesion, not by coalescence, as in fusion welding.

Small parts are often furnace brazed for mass production in manufacturing. The preassembled parts with flux and filler metal are passed through a heating chamber, where the brazing metal melts and flows into the joint by capillary action. Furnace brazing makes a fairly strong joint because of the large area usually provided for the joint and the small clearances (0.002 to 0.003 in.) between the parts to be joined.

SELF-EVALUATION

1. What is the single most important reason that nonferrous metals were difficult to weld in the past?
2. Why do some nonferrous metals require a high heat input when they are welded?
3. Why should aluminum be carefully cleaned for welding by brushing or scraping?
4. What takes place in the heat-affected zone of welds in strain- (work-) hardened aluminum? How does this affect strength?
5. Many aluminum alloys that are readily welded of the heat-treatable alloys such as the 2000, 4000, and 7000 series present some difficulties in making strong welds. What is the reason for these difficulties?
6. Oxygen may be present in copper or may be introduced into the weld by the welding process. How can it severely weaken the final weld?
7. When welding heat-treatable nickel alloys, what procedure is needed to restore the original strength?
8. Which kind of filler rods are most useful for welding dissimilar metals?
9. Reactive nonferrous metals (magnesium, tantalum, titanium, and zirconium) are not welded with coated electrodes in the presence of the atmosphere. Why? What forms of welding can be used?
10. How can a welder avoid embrittlement in titanium welds?
11. The most difficult of the copper alloys to weld are those containing lead and zinc. Why is this so?
12. When joining parts by brazing or soldering that have only a few thousandths of an inch clearance between the surfaces, by what means does the molten metal flow into this narrow joint?

CASE PROBLEM: THE COLLAPSING LAWNMOWER FRAME

A welding shop owner brought his lawnmower to his shop to weld a small crack in the side of the frame. It was made of cast aluminum and the blade had thrown a rock into it, causing it to crack. He decided to do a quick job with an arc weld, using an aluminum stick rod. It made such a lumpy, unsightly weld that he ground it all off and prepared the crack for gas welding, using a flux and filler rod. All went well for a few minutes and then suddenly a large section of the casting fell off, leaving a gaping hole where there was formerly only a small crack. What should he have done to avoid this problem? Can this damaged casting still be repaired?

WORKSHEET 1

Objective By macroetching, illustrate changes due to recrystallization in aluminum welds in the heat-affected zone that are accelerated by prior cold working.

Materials Two pieces of ¼ in. thick aluminum, 1000 or 3000 series H0 to H5 (soft) prepared for welding, two more pieces of the same size aluminum H34 to H36 (hard), TIG welder and appropriate aluminum welding rod, a 5 percent solution of sodium hydroxide in methanol, a petri dish, and tongs.

Procedure

1. Preheat to 300° or 400°F and weld each of the specimens together.
2. When cool, section each weld by sawing across them. Make an identifying mark on each specimen.
3. Grind them flat on a belt grinder and finish on 400 grit paper.
4. Pour the sodium hydroxide into the petri dish and, using tongs, place each specimen in the petri dish face up so that you can observe the progress. There must be enough solution to cover the specimens.
5. When the grain structure is visible, remove the specimens and hold them under running water.

Conclusion Which specimen shows the greatest amount of grain growth in the heat-affected zone? Which weld do you think is stronger? If you wish, you may prepare these two specimens for microscopic examination to further study the various weld zones.

WORKSHEET 2

Objective By microscopic examination, determine the effects of welding heat on heat-treatable aluminum alloys.

Materials Two ¼ in. thick specimens of 2024 aluminum alloy preferably in a solution heat-treated, aged condition, a 5 percent sodium hydroxide and methanol enchant, polishing equipment, and a metallurgical microscope (see Chapter 27).

Procedure

1. Make butt welds in two different specimens using a TIG welder.
2. Prepare the welded specimen for polishing and etching.
3. Begin with a light etch. Etch again, if necessary, to bring out the grain structure.
4. Mount in the microscope and set at 400× or 500×.
5. Observe the normal grain in the base metal well away from the weld, then compare them with the grains in the heat-affected zone. You should see some evidence of copper aluminide precipitation at the grain boundaries.
6. Search the adjacent and weld zones for unusual grain structures.

Conclusion Do you think the heat-affected zone has been weakened by welding? How could this be corrected by postheat treatment? Draw a circle and make a sketch of the grain structure that you see in the microscope; identify important areas. Write a conclusion to your paper and submit it to your instructor.

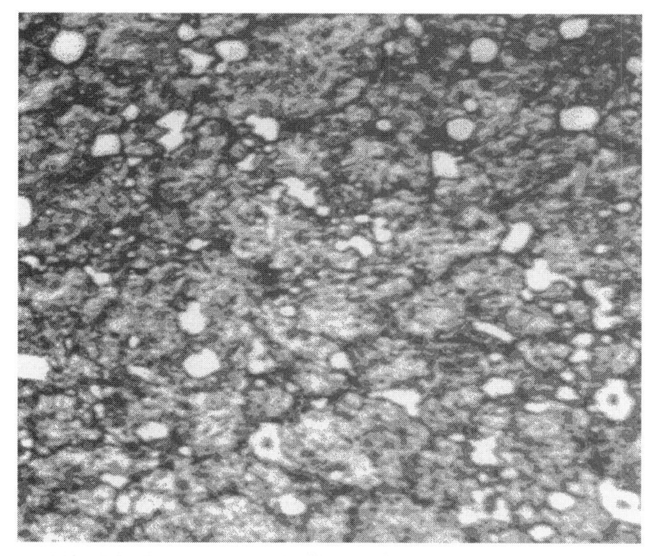

POWDER METAL TOOL STEEL

CHAPTER 17

Powder Metallurgy

Powder metallurgy, commonly referred to as P/M, is one of four major methods of forming metals. The other three are by casting of molten metal, by the plastic deformation of hot or cold metal, and by machining. P/M is the process of producing useful metal shapes from metallic powders.

P/M is essentially the pressing of a blend of metal powders into a useful shape, followed by sintering (heating but not melting) in a furnace. This process is applicable to a wide range of alloy systems offering a broad spectrum of properties and characteristics.

P/M was actually an early development of the automotive industry, which is still the greatest user of P/M products, accounting for more than 60 percent of that market. Yet P/M is used in just about every manufacturing industry for small metallic parts. However, P/M is not always a practical method of metal forming. It is usually not cost-effective and therefore not competitive when a large mass of metal is required for a part. Newer methods of powder forming have also been developed. These methods, such as hot isostatic pressing, powder forging, direct powder rolling of strip, and metal injection molding, vastly increase the possibilities for this method of metal forming. In this chapter, you will learn how P/M parts are made and how they are used.

OBJECTIVES **After completing this chapter, you should be able to:**

1. Describe the methods used to manufacture P/M parts and some of their characteristics.
2. Determine the density of P/M parts by microscopic examination.

ADVANTAGES OF P/M

Very good surface finishes can be obtained, and controlled porosity or permeability for filtration is possible. Powder metallurgy systems are suited to high-volume production of small parts. Almost any combination of alloys can be used to produce high-temperature components and very hard or tough products such as tungsten carbide tools. Close dimensional tolerances can be maintained. Machining is eliminated or reduced. Scrap losses as found in conventional methods of manufacture are eliminated by the powder metallurgy method.

Small, intricate parts can be produced at a rate that is as much as three times greater than conventional means. Tooling costs are relatively low. Certain unique products, which in some cases cannot be produced by any other method, can be made with a great variety of combinations of metals and nonmetals. The self-damping nature of P/M parts helps to quiet operations where a ringing noise is a problem. Wrought steel gears are usually noisy and are often used in business machines, air conditioners, and similar products. P/M gears greatly reduce the noise level in such applications. P/M parts may be heat treated, machined, plated, or impregnated with lubricants and other materials.

DISADVANTAGES OF P/M

P/M has some limitations, but many of these are being overcome by new methods. In general, P/M parts have a lower resistance to corrosion than solid metals, especially if they are quite porous. Also, parts made by the P/M process generally have poorer plastic properties (impact strength, ductility, and elongation) than conventional metals.

MANUFACTURE OF P/M

Some of the products that are made by P/M are cutting tools such as tungsten carbide inserts and cermets (ceramic tools). Precision parts such as cams, gears, and links are also made by P/M. Such techniques make possible the manufacture of antifriction materials such as self-lubricated porous bearings, filters such as those used in gasoline lines in automobiles, and high-strength magnets such as the well-known alnico magnet. High-strength metal parts used at high temperatures, such as turbine blades in jet engines, which cannot easily be fabricated by conventional methods, are made possible by P/M (Figure 17.1).

FIGURE 17.1. Some of the hundreds of metal parts made by powder metallurgy.

There are three basic steps in the manufacture of P/M parts.

1. *Blending.* Metal powders are mixed together with alloy additions and lubricants until they are thoroughly blended.
2. *Compacting.* The blended metal powders are fed into a precision die and pressed, in most cases, at room temperature with pressures from 10 to 60 tons/in.^{2} The part, called a green compact or briquette, is ejected from the die.
3. *Sintering.* The green compact is heated in a controlled atmosphere furnace to just under the melting point of the metal to bond the compressed powders into a strong structure.

POWDER METALLURGY PRESSES

Several different methods are used to form metal powders into briquettes prior to sintering. By far the most common is **simple die compacting,** in which presses are used. In advanced processes, such as isostatic pressing, the briquette is formed without a die or the need of a press.

Presses used for powder metallurgy are similar to those used for press-working operations. They can be either mechanically or hydraulically actuated or a combination of both. Some specific requirements for P/M presses follow.

1. The length of stroke must be sufficient to compress the powder.
2. The length and speed of the pressure and ejection strokes must be adjustable.
3. The presses must be able to apply sufficient pressure in both directions of pressing to provide as nearly as possible uniform density of pressing.
4. The press stroke, in production presses having many operations, must be synchronized with other automatic operations such as powder transfer.

The part shape, size, and density, plus the rate of production, are the most important factors in selecting a press for a given powder metallurgy application. Mechanical presses have the advantage of a high production rate. As many as 2000 compacted parts per hour can be produced on eccentric presses, but they are limited to less than a 30 ton pressure; however, toggle type presses may have a 500 ton pressure.

SIMPLE DIE COMPACTING

Mechanical (eccentric) presses are most often used in simple die compacting of small parts. Compaction pressure of 20 to 30 tons/in.^{2} is usually required. This means that parts made at high speed on a mechanical press should not have more than 1 in.^{2} of surface area on the compacted ends unless a toggle or hydraulic press is used. As a rule of thumb, a press cavity in a die should be deep enough to hold powder about 2 1/2 times the final briquette size to allow sufficient volume for the metal powder. This stroke length would eliminate the use of some mechanical presses if the finished part is quite long.

Powder metallurgy presses range from small (Figure 17.2) to very large (Figure 17.3). Hydraulic presses are used when long compacting strokes are needed and for large powder metallurgy parts. Hydraulic presses can have a force up to 5000 tons. In general, hydraulic presses have slower operating speeds than mechanical

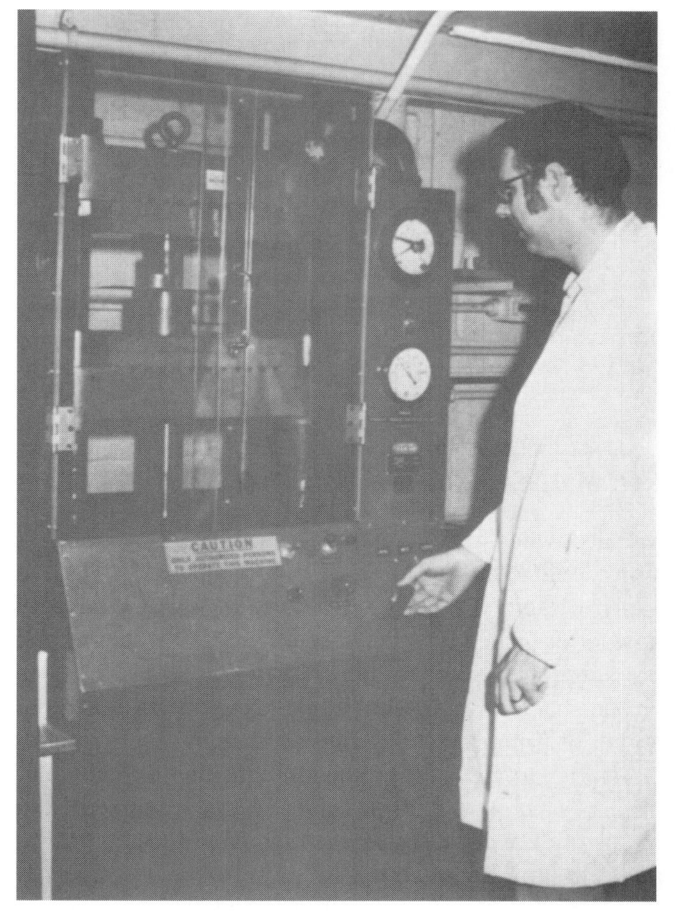

FIGURE 17.2. Small hydraulic powder metallurgy press used for experimental purposes in a research laboratory. (*Photo courtesy of the Bureau of Mines*)

FIGURE 17.3. Large hydraulic press used for briquetting metal powders or sponge for experimental purposes. (*Photo courtesy of the Bureau of Mines*)

presses, but for short-stroke work such as repressing or coining, relatively high speeds can be produced.

COLD AND HOT COMPACTION

Metal powder forming can be separated into two divisions: **cold compaction** and **hot compaction** (Figure 17.4). Simple die compacting can be done with cold powder or heated powder. This affects the density of the pressing. The **density** of the pressed article also depends on the shape of the part, particle size, pressure, and the length of time in the sintering furnace. Hot pressing, in which the article is pressed while at a high temperature, produces a density approaching that of rolled metal, and has a higher ductility than cold-pressed and sintered parts. In simple die compacting, pressure is applied to the article in only one axis, often with a punch (plunger) above and one below to help maintain an even density. However, the highest densities occur nearest the punches and lowest in the center, farthest from the punches. This tends to limit the length of pressing. In contrast, some P/M parts require very low densities. Porous filters and prelubricated sleeve bearings are two examples in which the metal powders are combined with sawdust or a volatile substance that is removed by the heat of sintering, leaving voids or pores in the finished article. Some very porous filters are sometimes made by "loose sintering" in a mold with little or no pressure and then forged or cold worked to the required density and shape.

Simple die compacting has several advantages, such as speed of production, simplicity, economy, and repeatability; but it does have its limitations. Die-compacting methods cannot produce parts with thin sections or undercuts. Holes can be produced only if they are perfectly parallel to the movement of the punches. Only relatively small length-to-diameter ratios are practical. To overcome some of these deficiencies, several methods are used. Split-die techniques are used when cross holes or undercuts are necessary, but other methods besides die compacting have been developed. Some complex shapes such as those in Figure 17.5 cannot be produced by simple die-compacting processes.

ISOPRESSING

Some shapes, such as hollow hemispheres or hollow cones, do not lend themselves to direct pressing in conventional presses. Isostatic presses are a relatively new development in which the powder is contained in a plastic bag in a metal can of the shape required (Figure 17.6). The chamber is pressurized at 50,000 to 60,000 psi with a liquid or gas, which compresses the powder from all directions. Isopressing is not a rapid production method but is presently used where conventional methods do not apply or where it can improve a product with more uniform densification.

In cold isostatic pressing (CIP), the powder is loaded into molds made of rubber or other elastomeric material and subjected to high pressures at room temperature inside a pressure chamber. Pressure is transmitted to the flexible container by water or oil. The parts are removed and sintered, followed by a secondary operation if needed. In hot isostatic pressing (HIP), an inert gas such as argon or helium is used to provide pressure. The gas is reclaimed after each batch of pressings. Preforms are often used to help maintain the required shape. HIP provides more density than CIP. Both processes are useful only for particular applications that cannot be done by more rapid production methods. A high rate of production for CIP would be 120 parts per hour. However, tooling costs are very low compared with die compacting.

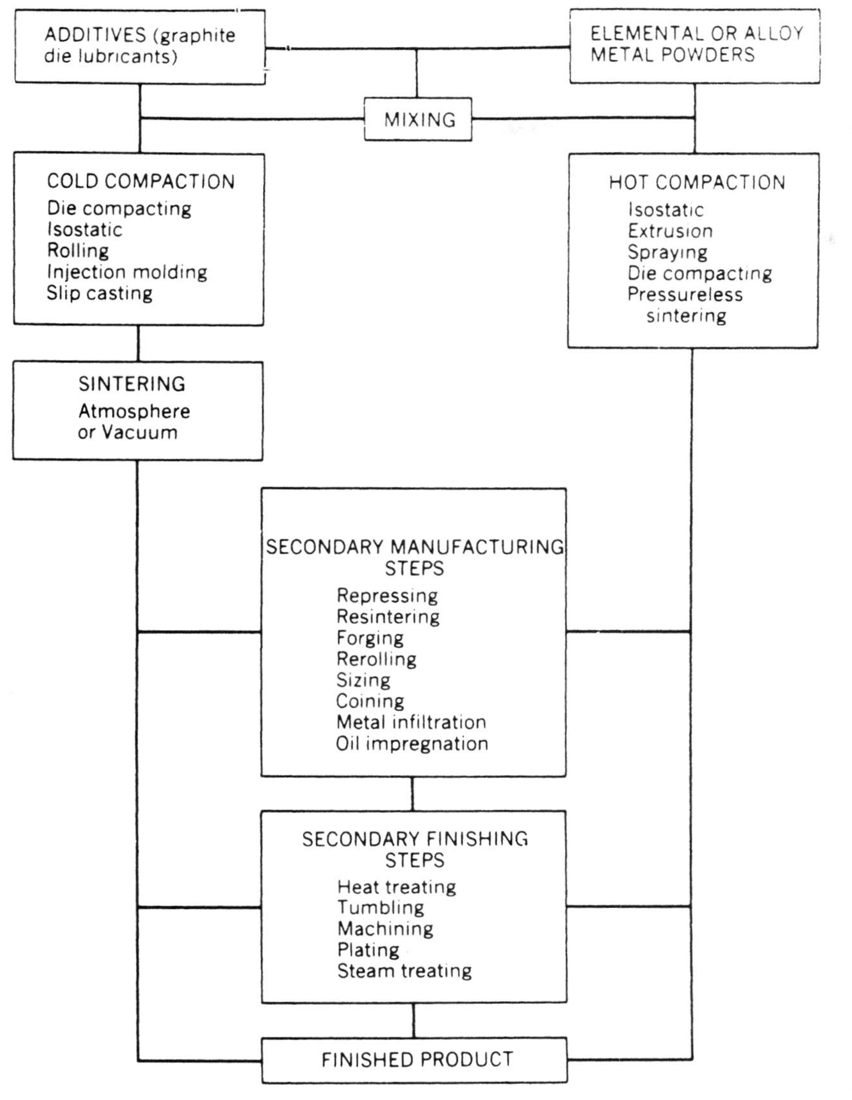

FIGURE 17.4. The P/M process.

FIGURE 17.5. Complex shapes such as these require special molds and techniques to avoid incomplete filling and variations in density.

METAL POWDERS

Metal powders used for powder metallurgy range in size from microscopic, less than 0.0001 in., to about 0.002 in. in diameter. Powders are available for almost any of the metals and alloys. Metals and other elements such as graphite, which are insoluble in the solid state, are easily combined as mixed powders and compacted to produce a strong metal part. Powders are classified according to particle size and shape in addition to other considerations, such as the presence of impurities and the metallurgical condition of the grains.

Metal powders are produced by several methods, some of which are listed here.

1. *Mechanical.* Milling, using crushers, rollers, and ball mills (Figure 17.7) where metals are disintegrated to

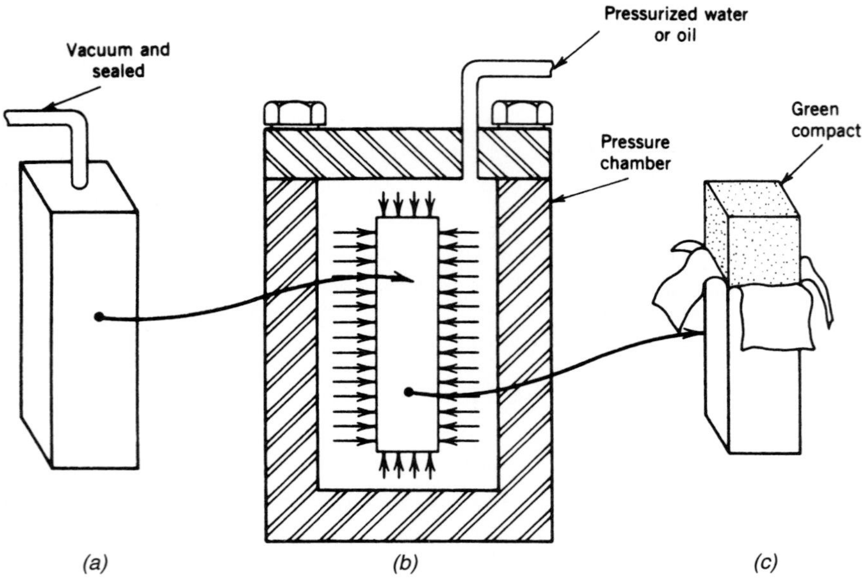

FIGURE 17.6. Isostatic pressing. (*a*) Prepared powder is placed inside a flexible container or mold. A vacuum is drawn in the mold and it is then sealed. (*b*) The powder and mold are then placed in a pressure chamber, into which water or oil is pumped under pressures of 15,000 psi or more. (*c*) The green compact is removed from the pressure chamber and the container is stripped off. (*John Neely and Richard Kibbe, Modern Materials and Manufacturing Processes, copyright 1987, John Wiley & Sons, Inc.*)

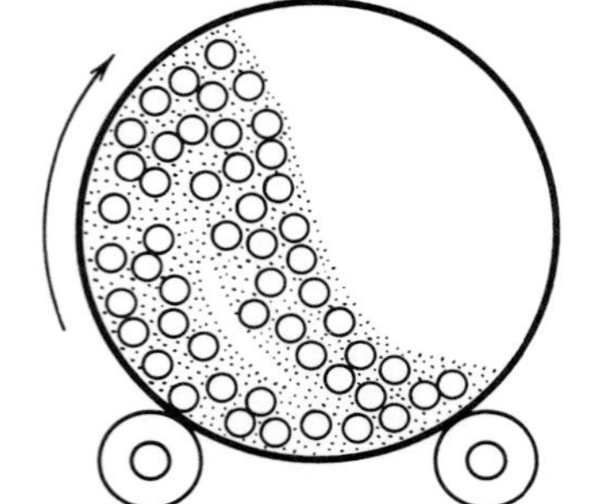

FIGURE 17.7. The action of the ball mill is shown as a continuous grinding as the drum rotates.

produce the size of powder required. The final grinding is usually done in a rotating ball mill that has many steel balls that impinge upon the powder and grind it to the needed size.

2. *Chemical.* Reduction, which converts metals from the oxides (ores) directly to metal powders at a temperature below their melting point. For example, iron powder is produced by direct reduction from either iron ore or mill scale.
3. *Shotting.* The method of making powders (usually rather coarse) by passing molten metal through a sieve and dropping the particles into water.
4. *Atomization.* The spraying of molten metal to produce powders. This process is generally only used with low-temperature metals such as tin, lead, and zinc.

When the powders are produced by any of these methods, they must be classified and blended to be ready for use. Vibratory screens or sieves are used to separate or classify the granules according to particle size. Other vibrating mills aid in the blending and mixing of various components such as would be found in sintered tungsten carbide that has a cobalt matrix powder plus the tungsten carbide granules. Two or more metallic powders may thus be blended together to impart special properties to the final product. The mixed and blended powders are pressed into a die in a hydraulic or crank type press to form the correct shape. The powder is fed through a chute into an opened die cavity; space in the die is allowed so that when the punch exerts the pressure on the powder, cold welding the particles together, the part is formed into what is known as a green briquette. The part is then ejected from the die and is hard enough at this stage to be handled, but it can be damaged if dropped or mishandled (Figure 17.8).

The depth and shape of the die cavity, the amount of powder, and the amount of pressure all determine the density that is required. The briquette is then moved to a furnace where it is sintered, that is, held at a predetermined temperature for a given length of time. Here the powders begin to bond together (Figure 17.9). The steps needed to produce a powder metal part are as shown in Figure 17.10.

SECONDARY OPERATIONS

Unless more precision is needed, the parts are now ready for use. For greater precision, the parts can be sized in a coining die that brings them to the needed dimension. Other secondary operations include impregnating the parts with antifriction material such as graphite or oil to make self-lubricating bearings. The density and strength of the part can be increased by infiltrating the pores with a lower melting point alloy. Sintered products can also be modified by machining, plating, and heat treatment (Figures 17.11*a* and 17.11*b*).

SINTERING

In **solid phase sintering** (the most common method), the green compact part, or briquette, must be sintered in a carefully controlled atmosphere furnace from 1/2 to 2 hours at 60 to 80 percent of the melting point of the lowest melting constituent. **Liquid phase sintering** is carried out above the melting point of one of the alloy constituents or above the melting point of an alloy formed during sintering. Sintering, a solid-state process, develops metallurgical bonds. The important changes that take place during this process are as follows:

1. *Diffusion.* This takes place especially on the surface of the particles as the temperature rises.
2. *Densification.* Particle contact point areas increase considerably. When this happens, there is a loss in porosity that reduces voids. Also, there is an overall decrease in the size and some distortion of the part during the sintering process. A coarse, green briquette must be made larger to allow for this shrinkage.
3. *Recrystallization and grain growth.* This occurs between particles in a contact area, causing the grain and lattice structures such as those in a solid metal to join, but still leaving voids or holes between the particles, depending on the amount of compression in the green briquette and the time of sintering.

ADVANCED PROCESSES

Some newer processes involving metal powders have been developed for special application and difficult-to-work metals. **Powder forging (P/F)** involves a compaction step first, where the green compact is formed. After sintering, the preform is hot forged in a second set of dies to provide a full density and near-net shape, often requiring no subsequent machining or grinding. Connecting rods for engines and other high-strength parts are being made by this process.

Metal powder injection molding is a newer process that was borrowed from the plastics injection-molding process. In fact, in some variations of this

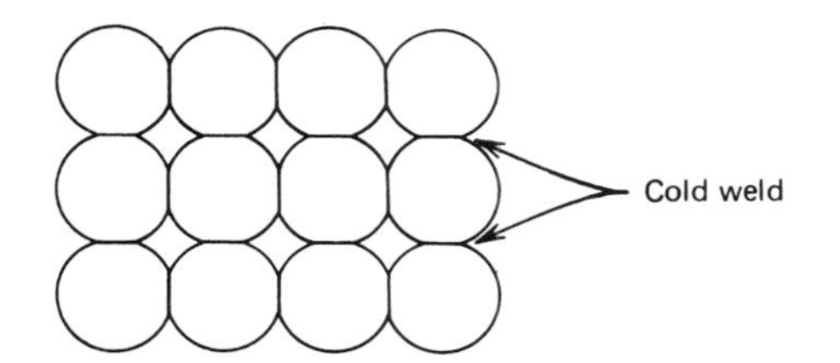

FIGURE 17.8. The compressed particles in the briquette are cold welded together at this stage with very weak bonding.

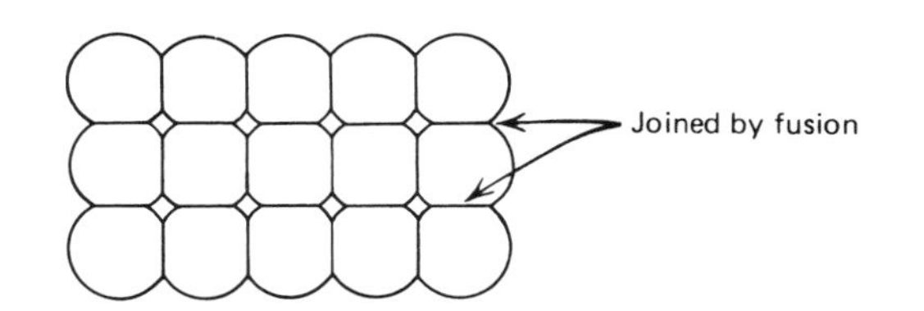

FIGURE 17.9. After sintering, the particles in the briquette become fused together.

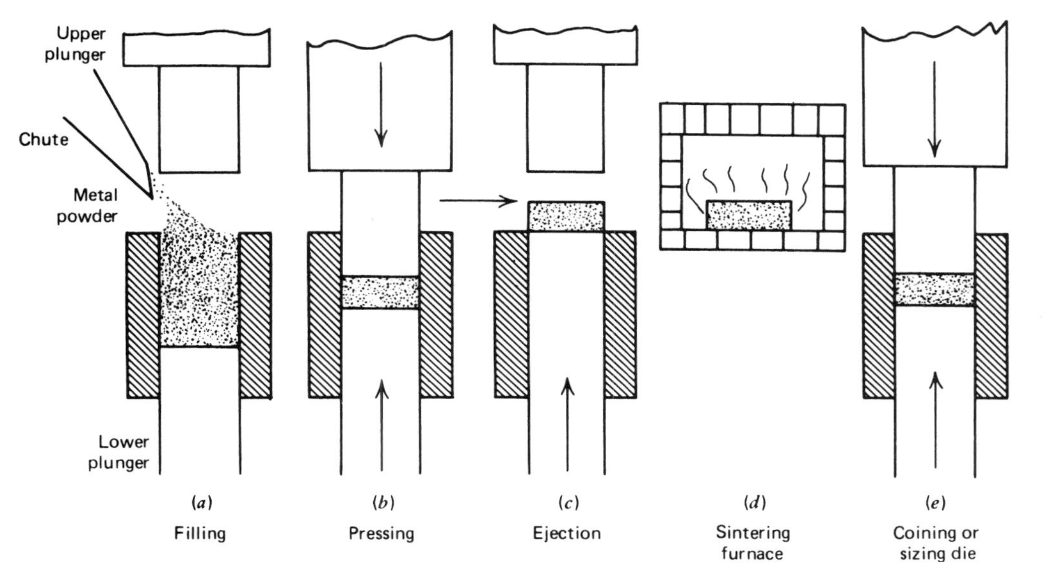

FIGURE 17.10. The depth of die cavity and length of plunger stroke is determined according to the density required. (*a*) A measured amount of metal powder is placed in the die cavity through the chute. (*b*) Pressure is applied. (*c*) The briquette is ejected from the die cavity. (*d*) The parts are sintered at a specified temperature for a given length of time. The parts are now ready for use. (*e*) If more precision is needed, they can be sized in a coining die.

process, plastic injection-molding machines can be used. To inject powder into molds, the particle size must be much finer than that used in conventional processes. This "dust" is combined with a thermoplastic binder, which is later removed by heating in an oven before sintering. Thin walls and odd shapes can be made with this process, but injection molding is not a high-speed process compared with simple die compacting.

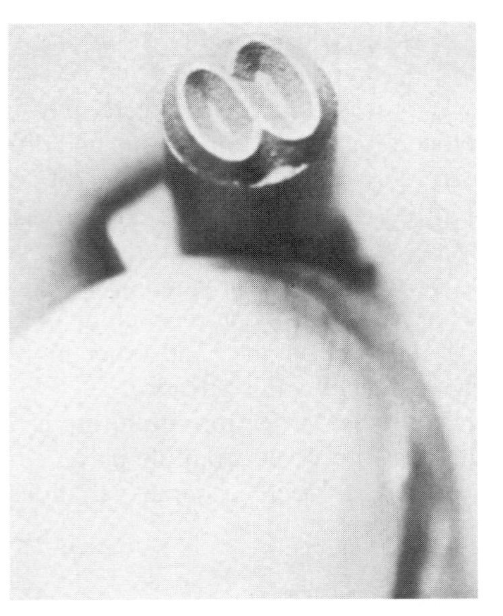

*FIGURE 17.11***a.** P/M number stamp that has been heat-treated to RC 58.

Metal powder-to-strip technology involves the direct rolling of a powder slurry with a cellulose binder into a thin metal strip. After drying, the strip is compacted between rolls and then sintered, first to remove the binder and then to bind the particles. It is rolled again to provide more density. Difficult-to-work metals, such as titanium and zirconium, and bimetal strip can be produced by this method.

Powder extrusion is similar to other extrusion processes except that metal powders are used. These must be placed within a metal can, which is evacuated and sealed. This assembly is then heated and extruded as a single unit. Metal bars, tubing, and other shapes can be made from metal powders by this process.

P/M PRODUCTS AND THEIR USES

Porous, prelubricated bearings are used in cases when normal lubrication would not be practical or when extra lubrication could contaminate a product, as in the food and textile industries. Small motors can be lubricated for life by using porous sintered bronze bearings that are prelubricated, sometimes known as Oilite® bearings.*

Machining of wear surfaces in porous bearings is not recommended, because the voids become partially closed during cutting operations such as reaming. This seals the

*Oilite® is a trademark of the Chrysler Corporation.

*FIGURE 17.11***b.** Micrograph of hardened P/M number stamp showing the dense structure. The particles are surrounded by a secondary alloy that increases strength and density (500×).

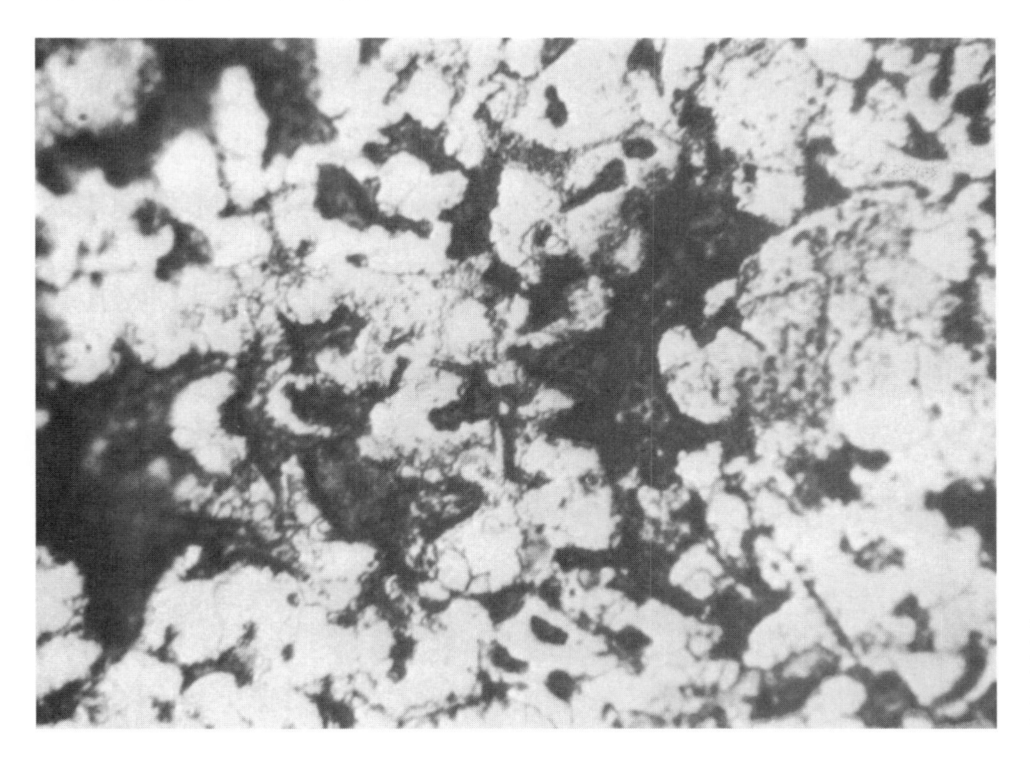

FIGURE 17.12. Micrograph of a porous bronze sintered bearing (500×) showing large voids (black areas) that are impregnated with oil or graphite.

lubricant in the bearing so that it cannot circulate on the wear surface. It is the capillary action of the bearing that allows the impregnated oil to flow onto the bearing area, especially when the temperature rises (Figure 17.12).

Porous iron-graphite bearings are extensively used in machines since they have a higher wear resistance than babbitt or bronze and are economical to manufacture. Aluminum-base antifriction materials are also made by the P/M process. These also contain graphite, iron, and copper in small percentages. Copper-nickel powders are often formed as a layer on a steel strip and sintered. The sintered strip is impregnated with babbitt metal and is then formed into insert bearings for automobile and aircraft engines.

Creep-resistant alloys are made from tungsten, molybdenum, tantalum, and other powders such as carbides, nitrides, borides, and silicides. These articles are used in high-temperature applications such as jet engine impeller blades, heating elements, and electronic equipment.

Very hard alloys for cutting tools are made by the P/M process. Tungsten carbide, tantalum carbide, and titanium carbide are all used to give various properties to the "carbide" family of cutting tools. These powders are usually bonded in a matrix of powdered cobalt. Ceramic cutting tools are also made by the P/M process (Figure 17.13). High-speed steel end mills are now being produced by P/M process. These are said to provide an increase in tool life of three to five times that of conventional high-speed steel end mills used in the same applications.

FIGURE 17.13. Ceramic cutting tools such as this insert are produced from aluminum oxide powders by the P/M process.

Friction materials such as brake shoes for automobiles are often made by P/M. Such friction materials are usually in the form of bimetal powder materials that are bonded to a steel base. Strips, discs, and other forms are made this way.

Mechanical parts are extensively made by the P/M process. Gears, cams, levers, and link mechanisms are but a few of the many types of small precision P/M parts produced by this process (Figures 17.14*a* to 17.14*c*).

P/M is widely used with many applications besides those presented in this chapter. Continuing development in this field, which has and will continue to have a great economic importance for the metals industry, is to be expected.

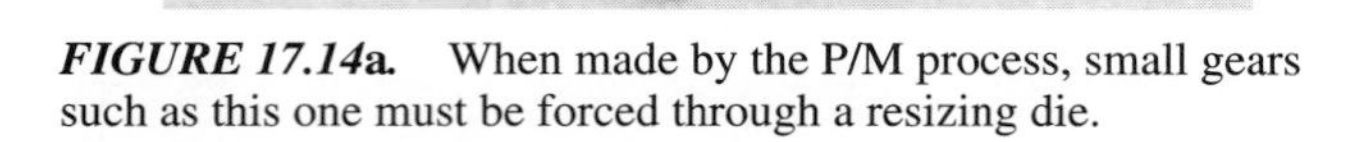

***FIGURE 17.14*a.** When made by the P/M process, small gears such as this one must be forced through a resizing die.

***FIGURE 17.14*c.** These gear pump parts that are made by the P/M process are examples of the very high precision attainable with this method.

500× 2% Nital Etch

***FIGURE 17.14*b.** Micrograph of the gear in Figure 17.14*a* shows a high density with some porosity (500×).

SELF-EVALUATION

1. Although P/M was used in ancient times to some extent, modern powder metallurgy as we know it was an outgrowth of what industry?
2. Name some kinds of unique parts that can be made by P/M.
3. State the three usual steps for making P/M parts, assuming that the mixed and blended powder is already prepared.
4. What holds the green briquette together before it is sintered and what bonds it together after it is sintered?
5. Name some advantages and disadvantages of P/M.
6. In what way does hot pressing improve sintered parts?
7. The most rapid method of making small P/M parts is done with mechanical presses. What is it called?
8. What P/M parts are made by mixing a volatile substance such as sawdust with the metal powders?
9. A high production using CIP would be 120 parts per hour. How many parts per hour may be possible using an eccentric press?
10. Which products are more cost-effective when using P/M, small or large parts?

CASE PROBLEM: THE SLEEVE BEARING PROBLEM

A machinist had pressed a new prelubricated, porous sintered bronze bearing in an electric motor. The rotor shaft wouldn't fit into the bearing. Measuring the bore, the machinist discovered that it was a few thousandths of an inch undersize. He removed it, put it in a lathe, and bored out the excess material. The bearing was pressed in again, the motor assembled, and put into use. A few days later the motor was back in the shop with the bearing burned out. What had happened to this prelubricated, sintered bearing to cause this failure?

WORKSHEET

Objective Determine the relative porosity of several sintered parts by microscopic examination.

Materials Several discarded, sintered metal parts with widely diverse uses, such as a porous bearing or gas line filter, and parts with more density, such as gears or levers. Specimen mounting and polishing equipment. A metallurgical microscope and magnifying glass.

Procedure

1. Fracture each specimen in a vise with a hammer. Cover it with a shop cloth before striking it and wear safety glasses.
2. Observe the fractures with a magnifying glass.
3. Encapsulate the parts for polishing. Put an identifying mark on them.

4. Polish the samples with successively finer grits of paper and finish with 600 grit. Do not polish on a motorized polishing wheel and do not etch. This process must be done with extreme care since the surface metal tends to slide over and cover the pores if excessive pressure is applied while polishing.
5. The specimen is now ready for microscopic study. Use 100× power at first. Compare the various specimens at this magnification.

Conclusion How do the structures of these samples compare with conventional metals? How did each sample compare with the others? Make a sketch of each sample as you see it in the microscope. Write your conclusion and give it to your instructor.

CHAPTER 18

Precious Metal Processing

"There's ***GOLD*** in them there hills," and plenty of other precious metals such as silver, and platinum group metals. Their functions in jewelry, coins, and bullion, and as catalysts in catalytic converters to help control automobile exhaust emissions, is well understood. The use of precious metals in the electronics, communications, chemical manufacturing, and pharmaceutical industries has exceeded all expectations. Foodstuffs depend more and more on the use of fertilizers produced with the aid of a platinum-rhodium catalyst which is woven in the form of a gauze. Certain organometallic compounds containing precious metals are used in chemotherapy for the treatment of cancer.

As you cross a mountain stream along a seldom used trail in California, you look down through the clear water and see something sparkle. You quickly scoop the glitter in your pan and slosh the water to rinse away the gravel and expose a nugget about the size of a dime. It feels heavy as you bite down to check hardness. Gold, a soft yellow metal, is panned in this same way every day by those with the "fever."

OBJECTIVES **After completing this chapter, you should be able to:**

1. Describe some of the methods used to find and mine precious metals.
2. Understand the different uses and properties due to alloying gold.
3. Explain what is meant by "karat" rating of gold.
4. Describe uses for platinum group metals.

PRECIOUS METALS TRADE PRACTICES

Everyone wants to know what is meant by the term *karat.* My ring is 14 karat gold; how much gold is actually used? Well, there are 24 karats in pure unalloyed gold. So, since gold is found in the native state as pure gold, it is naturally 24 karats. In precious metals, a karat designates purity. Twenty-four karats equal 100 percent pure gold. Twelve karat metal indicates it is 50 percent pure and the remaining half is made up of alloying elements. Thus, 10 karat gold has 41.7 percent gold and the remainder is alloying elements; 14 karat gold is composed of 58.3 percent gold; and 18 karat gold has 75 percent gold. Depending on the alloying constituents (silver, copper, zinc, nickel, platinum, and palladium), yellow, red, green, or white karat gold may be formed.

By weight, such as in a precious stone, a carat weighs 200 milligrams or .2 grams. Carats apply only to precious stones and not to metals. Note the difference in spelling; a "carat" versus a "karat." Karat for metals refers to purity and not to weight. Weights and measures can cause confusion. Just remember a karat in metals refers to the concentration or percent of precious metal in the alloy.

Avoirdupois weight is based on, the pound equaling 16 ounces and the ounce equaling 16 drams. *Apothecary* weight is based on, the pound equaling 12 ounces and the ounce equaling 8 drams. The weight measure for precious metals is by the troy ounce or the apothecary ounce. One troy ounce is also equal to 1.097 avoirdupois ounces. Precious metals are sold and traded by the troy ounce or in markets where the metric system is used.

Fineness in precious metals refers to the weight portion of either metal in the alloy, expressed in parts per thousand. For example, 1000 fine gold is 100 percent pure gold. Gold metal which is traded is at least 995 fine or higher. Sterling silver is 925 fine or 92.5 percent pure silver and 75 parts copper or 7.5 percent copper. The U.S. practice, set by the Federal Trade Commission, requires any article which is labeled gold to contain at least 10K, ± 1/2K. Regulations are also set for gold plate, gold filled, and clad gold. Trading in the platinum group metals is conducted on the New York Mercantile Exchange. Commercial grade platinum is 99.9 percent pure. Higher purity platinum for use in thermocouples and resistance thermometers is designated Pt 67 and is 99.999 percent pure.

SILVER

Silver is a bright white metal which is very soft and malleable. Silver is not readily oxidized at room temperature but is attacked by sulfur in the air and becomes tarnished. In commercial applications, silver is noted for superior thermal and electrical conductivity, softness and ability to be easily formed, and corrosion resistance. Silver is used in photography, brazing, medicines, dentistry, coinage, as catalysts, and for nuclear control rods. Silver solders are widely used in the plumbing industry and have replaced toxic lead solders for assembly of copper water pipes. Silver used in electrical contacts is often alloyed with cadmium oxide (CdO) up to about 20 percent. The introduction of cadmium helps prevent sticking, welding, and resistance to arc erosion.

GOLD

Gold is a very soft, malleable, bright yellow metal. Gold is highly corrosion resistant and is easy to alloy with other metals. Gold, because of its esthetic beauty and enduring physical properties, remains unchallenged for use in jewelry and in the arts. Approximately 55 percent of the gold mined today is used in the jewelry industry. Gold also finds use in the electronics industry and dentistry, and is a good infrared reflector. Gold use in dentistry has decreased over the years due to the development of stronger, less expensive, chemical-resistant epoxies to fill cavities. Thin films of gold are used as thermal barrier coatings on windows in high-rise buildings. Gold has also found use as a means of controlling temperature by preventing excessive heating of astronaut spacesuits due to solar radiation. Gold is used to decorate porcelain, glassware, and exotic furniture as gold flake or foil.

GOLD ALLOYING

Alloying elements are commonly alloyed with gold to increase hardness and to vary the color and mechanical properties of the alloyed gold. Yellow, green and red gold alloys are based on the gold-silver-copper alloy system which is often modified by additions of zinc or nickel (see Figure 18.1). The spectrum was produced using a scanning electron microscope X-ray analyzer. Keep in mind, 14K gold contains about 42 percent alloying elements. The alloying of gold offers increase in mechanical strength, wear resistance, hardness, color variation, and resistance to fire cracking or heat checks.

White gold was first introduced as a substitute for platinum jewelry. The primary alloying elements are gold, nickel, and copper. Zinc from 5 to 12 percent is also added to modern white gold.

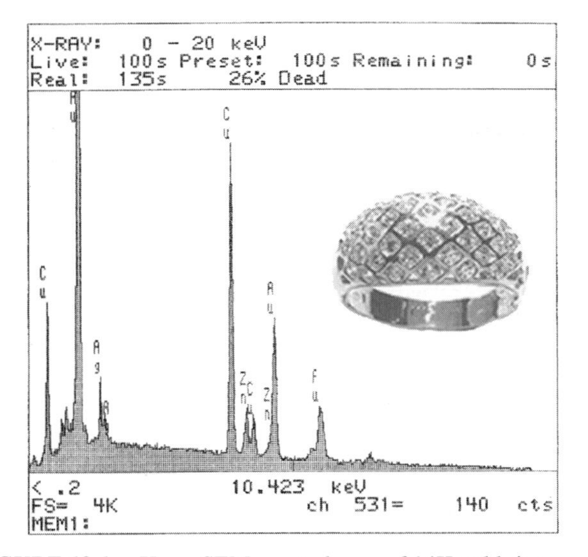

FIGURE 18.1. X-ray SEM spectral scan of 14K gold ring. Elements present are copper (Cu), silver (Ag), zinc (Zn), and gold (Au).

Dental gold alloys consist of varying amounts of precious metals dependent upon their use. Wrought gold alloys and castings are used in dental restorations such as cover inlays, crowns, multiple unit bridges, and partials. The lost wax casting process is universally used since castings are small, ranging in weight from 1.5 to 40 grains. Gold foil, mat gold, and powdered gold are used for fillings. Solders for dental appliances consist of gold, silver, copper, and zinc. Copper and silver percent are varied to control color and working characteristics.

GOLD MINING

Gold has been mined around the world for thousands of years successfully, because it is most often found in the pure metallic or native state as nuggets. Probably the best documentation of the numerous methods of mining gold is during the history of the California gold rush at Sutter Creek in 1849. There are numerous types of gold mines found in California and throughout the world. For example, *placer mines* may be either deep or shallow in which gold is imbedded in clay, sand, or with gravel. A *quartz mine* has gold bound together or encased in rock or quartz. Quartz mines are usually deep tunnel type mines. *Open pit mines* are those that dig into the surface earth, removing ore which may be processed or refined in a different location. Many large open pit mines are in operation in Brazil and New Guinea.

When the gold rush in California began, gold was found in shallow mines such as sand bars in rivers, beds of ravines or gulches, and even in flat areas. Because gold is a heavy metal, far heavier than the sand and surrounding rock, as water erosion occurred, chunks of gold were broken into smaller parts, nuggets, or flakes and deposited on the bottom of stream beds, beneath rocks, and at the base of sand bars in rivers. Upon finding gold, miners would stake a "claim" on the area and file the property claim at the local county seat.

Panning for gold at a river or stream claim was a popular means of recovering gold and the "black sand" which was rich in elemental gold. A popular but dangerous method of extracting gold from the ore or capturing very fine particles is by washing in mercury. Gold readily dissolves in mercury, the same highly toxic mercury used by the dentist as an amalgam to fill tooth cavities. The mercury amalgam would be poured off into a retort. The retort or iron container was capped with a funnel connected to a spiral cooling tube, somewhat similar to an alcohol distiller. Heat would be used to drive off the mercury, leaving very fine gold dust at the bottom of the pan. The gold dust would be fired or melted into solid gold. The mercury vapor would be cooled by the cooling tube and recondensed into liquid for reuse. This method of extraction is still in heavy use today, especially in countries such as Brazil and New Guinea where people have not been educated about the hazards of mercury.

Another less aggressive means of gold ore mining is with the *sluice box*. The sluice box is from 4 to 6 feet wide and about 3 feet deep with a thick, one-piece wooden flat bottom. The paving of the sluice box is sometimes done with wood planks or hard flat rocks standing on edge about a foot thick and so placed as to be least affected by water flow. Lateral boards are placed across the rock bed about 6 to 8 feet apart so as to hold them in place. Mercury (quicksilver) is added to the sluice in a spray so as many gold particles as possible can come in contact with the mercury.

The sluice box is the most commonly used tool in mining aside from the shovel and pan. Sluice boxes are used in dredges, long toms (a longer version of the sluice), rocker boxes, and even on large-scale open pit mines. Today, many sluice boxes are made of steel or lightweight aluminum. In the nineteenth century, sluice boxes were homemade out of wood and burlap sack.

Many modern miners take pride in their homemade sluice boxes. Some miners spend hours perfecting the riffles in their sluice boxes so they are the right angle, or sanding down the bottom surface so the boxes have a perfect smoothness. The gold is separated from the mercury by distilling; the residual is "gold dust."

Hydraulic mining was used in 1852 to obtain large quantities of precious metal ore by washing away sides of hills and banks of streams and rivers with pressurized water and blasting powder. Water under tremendous force was sprayed through a tapered metal nozzle, in a similar manner as modern firemen might direct water when fighting a fire. This method of man-made water erosion was very destructive to the environment and caused bitter arguments between miners and agriculturists in early California. Although hydraulic erosion produced a great amount of gold, it did irreparable damage to the landscape and, by washing tons of tailings downstream, damaging fertile valleys and farmland.

THE PLATINUM GROUP METALS

The platinum group metals (PGM) consists of six metals: platinum (Pt), palladium (Pd), rhodium (Rh), ruthenium (Ru), iridium (Ir), and osmium (Os). Ore containing the PGM is mined, concentrated, and smelted in Montana and the resulting matte was sent to Belgium for refining and separation of the individual metals. Refined PGM metals are also recovered as a by-product of the copper-refining process in Texas and Utah. PGM are widely used in the automotive, electronic, electrical, chemical and petroleum-refining, and dental and medical industries. The automotive, chemical, and petroleum-refining industries use PGM primarily as catalysts, whereas the others use PGM because of their chemical inertness and refractory properties.

Platinum group metals are processed worldwide, with resources estimated to total 100 million kilograms; U.S. sources are estimated at 9 million kilograms. Domestic mine production of platinum and palladium remains unchanged yearly, despite slight improvements in the average price of both metals—approximately $400 per troy ounce. The consumption of these metals is directly related to automotive and truck sales, and catalytic converter use. Some automotive companies have substituted palladium for the higher priced platinum in catalytic converters. Platinum is a more resistant metal to sulfur and lead poisoning in gasoline engine emissions. Palladium is more effective as a catalyst for controlling emissions from diesel-powered vehicles.

SELF-EVALUATION

1. Explain how precious metals are weighed.
2. Name the metals in the platinum group.
3. How are precious metals such as gold and silver hardened?
4. Explain how gold is extracted from ore by using mercury.
5. Explain some of the hazards of using mercury.
6. Explain the use of a sluice box.
7. What percent of pure gold is contained in a 10K ring?
8. What is meant by *fineness* in the precious metal industry?
9. What metal(s) would you alloy with gold to change it to white gold?
10. What is a troy ounce?

CASE PROBLEM: IS IT REALLY GOLD?

A rider on horseback was on a mountain trail in King's Canyon in northern California, when he noticed sparkling and glitter in the creek bed as he crossed. Immediately he jumped off his horse, and began panning for gold. "I'm rich! I'm rich!" he exclaimed. After about an hour, he had managed to collect about 6 ounces. How can he determine if indeed what he found was really gold? Can you name a material that resembles the appearance of gold and has fooled many inexperienced miners for years?

WORKSHEET

Objective Determine the amount of gold in a piece of jewelry and the actual value of the gold based on current gold market pricing.

Materials A gram measuring scale; a simple calculator; and a piece of karat stamped solid gold jewelry.

Procedure

1. Make sure the gold jewelry selected is truly solid gold and not plated gold. Wash and dry the jewelry before weighing.
2. Observe the karat stamp and record the K value.
3. Carefully zero the scale or balance you will use to weigh the jewelry. Weigh the jewelry and record the amount.
4. Determine the current price of gold per troy ounce, and calculate the value of the pure gold present in your jewelry.

Conclusion Is the value of the actual gold close to what you expected? Can you determine what the price of an ounce of gold would be, based on what you paid for the jewelry? This value is called “markup.”

CHAPTER 19

Corrosion of Metals

With the exception of some noble metals, all metals are subject to the deterioration caused by ordinary corrosion. Iron, for example, tends to revert to its natural state of iron oxide. Other metals revert to sulfides and oxides or carbonates. Buildings, ships, machines, and automobiles are all subject to attack by the environment. The corrosion that results often renders them useless and they have to be scrapped. Billions of dollars a year are lost as a result of corrosion. Corrosion can also cause dangerous conditions to prevail, such as on bridges where the supporting structures have been eaten away, or on aircraft in which an insidious corrosion called intergranular corrosion can weaken the structural members of the aircraft and cause a sudden failure.

Those who work with metals must have a knowledge of the principles involved in metal corrosion to be able to use the correct preventive measures. It is the purpose of this chapter to help you understand these principles of corrosion and the methods by which corrosion can be offset.

OBJECTIVES **After completing this chapter, you should be able to:**

1. Demonstrate how oxygen in water affects the rate of corrosion of iron.
2. Demonstrate how an oxygen concentration cell works to corrode stainless steel.
3. Demonstrate how active metals can displace ions of a less active metal.

TYPES OF CORROSION

Many recognized modes of corrosion destroy metal at a cost of billions of dollars each year. The first grouping of corrosion modes is **concentration cell corrosion** in which three failure modes are recognized: **crevice corrosion, tuberculation,** and **underdeposit corrosion** (Figures 19.1*a,* 19.1*b,* and 19.1*c*). Each of these forms of corrosion are common in closed cooling systems such as in a house or apartment. Concentration cell is corrosion attack associated with the nonuniformity of the aqueous environments at a metal surface. Corrosion occurs when the environment near the metal surface differs from area to area, creating anodes and cathodes (regions of differing electrical potential). Anode areas lose metal to cathode areas by corrosion at a cell. Areas in a pipe may be shielded and, as a result, often act as anodes. An example of shielding may occur in the threaded areas where two sections of threaded pipe are coupled together. Gaskets, flanges, bolt holes, and riveted areas may be a natural source for crevice corrosion. Metals are often attacked by decomposing biological materials.

Virtually all metals are attacked to some degree if the environment is sufficiently harsh. The microstructure of a metal or alloy can also influence the rate of or susceptibility to corrosion. In austenitic or 300 series stainless steels, when quenching from the annealing temperature is too slow, a chromium-rich phase depletes from the grains and forms as a phase deposit in the grain boundaries. This is known as sigma phase precipitation and is the primary cause of intergranular corrosion in this group of stainless steels (Figure 19.2). Metals may be attacked by chlorine, sulfates, hydrogen, and other potentially corrosive ions which weaken protective surface oxides (passivated surfaces) and accelerate corrosive attack. Coupling dissimilar metals such as copper to galvanized pipe is a sure way to cause massive and rapid decomposition of both metals. Likewise, introduction of contaminated or dirty well water into a plumbing system for a pressure test and allowing stagnant well water to remain in the pipes without flushing is a sure way to generate localized tuberculation and corrosion pits (Figure 19.3).

Crevice Corrosion

The crevice must be filled with water such as in a water pipe and adjacent surfaces must also be in contact with water. Liquid must fill the crevice, sometimes only a few thousands of an inch wide, or corrosion will not occur. Crevice corrosion is a type of concentration cell corrosion or corrosion caused by the concentration or depletion of metal ions, dissolved salts, oxygen, and other gases. The buildup of cells results in deep pitting which often penetrates to the outside wall, causing the pipe to leak.

Tuberculation

Tubercles are mounds or clumps of corrosion product and deposit that cap localized corrosion activity such as pits (Figure 19.1*a.*). Tubercles are often seen blocking the flow of liquid through water pipes and are formed when surfaces are exposed to oxygenated waters. High concentrations of alkalinity, sulfur, and chlorine can stimulate tubercle formation. Tubercle growths are highly structured and often crystalline in appearance (Figures 19.4 and 19.5).

Underdeposit Corrosion

Underdeposit corrosion is a term applied to localized corrosive activity that occurs beneath a deposit. The result is usually a severely pitted surface. Underdeposit corrosion also occurs between tubercles and base metal, forming spectacular crystal formations (Figure 19.6).

Oxygen Corrosion

The rate of corrosion in iron alloys and steels is nearly proportional to the amount of dissolved oxygen in the water. Corrosion does not occur when dissolved oxygen is not present. When the water pH is between 4 and 10, the corrosion rate of iron is nearly constant. When iron is exposed directly to more acidic solutions, rapid attack occurs and hydrogen is evolved. When the pH rises above 10, the corrosion rate of iron falls off sharply.

Dissolved oxygen in water will attack normally passive metals such as copper. Aluminum and aluminum alloys are essentially unaffected by dissolved oxygen in water due to the presence of an adherent oxide film. The passive surface layer of austenitic stainless steels provides protection against attack by dissolved oxygen in water as long as the alloy contains at least 11 percent chromium. Heavily oxygenated waters aggressively attack zinc, the protective plating on galvanized pipes.

Biologically Influenced Corrosion

Active biological corrosion directly accelerates or establishes new electrochemical corrosion reactions or the interaction of microorganisms with materials to produce new corrosion chemistries. Sulfate-reducing bacteria cause the most localized cooling water corrosion associated with bacteria. Many bacteria produce acids which attack metals. Other bacteria are metal depositing and cause oxidation. Some form slime which contributes to corrosion by consuming oxygen and stimulating the

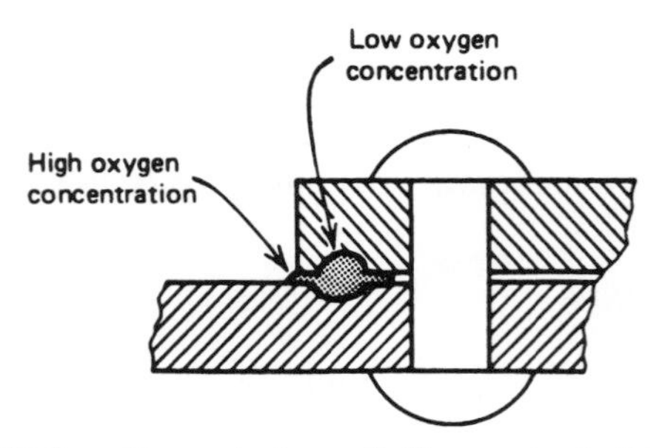

*FIGURE 19.1*a. Concentration cell. The crevice in this concentration cell hinders the diffusion of oxygen and causes high- and low-oxygen areas. The low-oxygen area is anodic. When the solution surrounding a metal contains more metal ions at one point than another, metal goes into solution where the ion concentration is low.

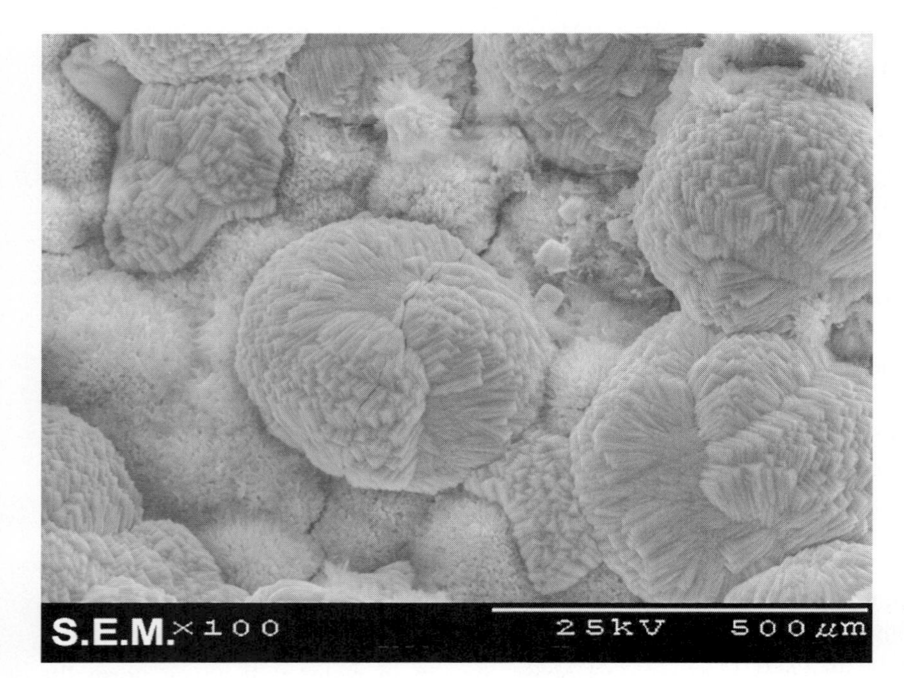

*FIGURE 19.1*b. Huge tubercles such as these were found growing on the inside surface of hot galvanized steel water pipes.

*FIGURE 19.1*c. Underdeposit corrosion. Corrosion growth/crystal formations developing beneath a large tubercle.

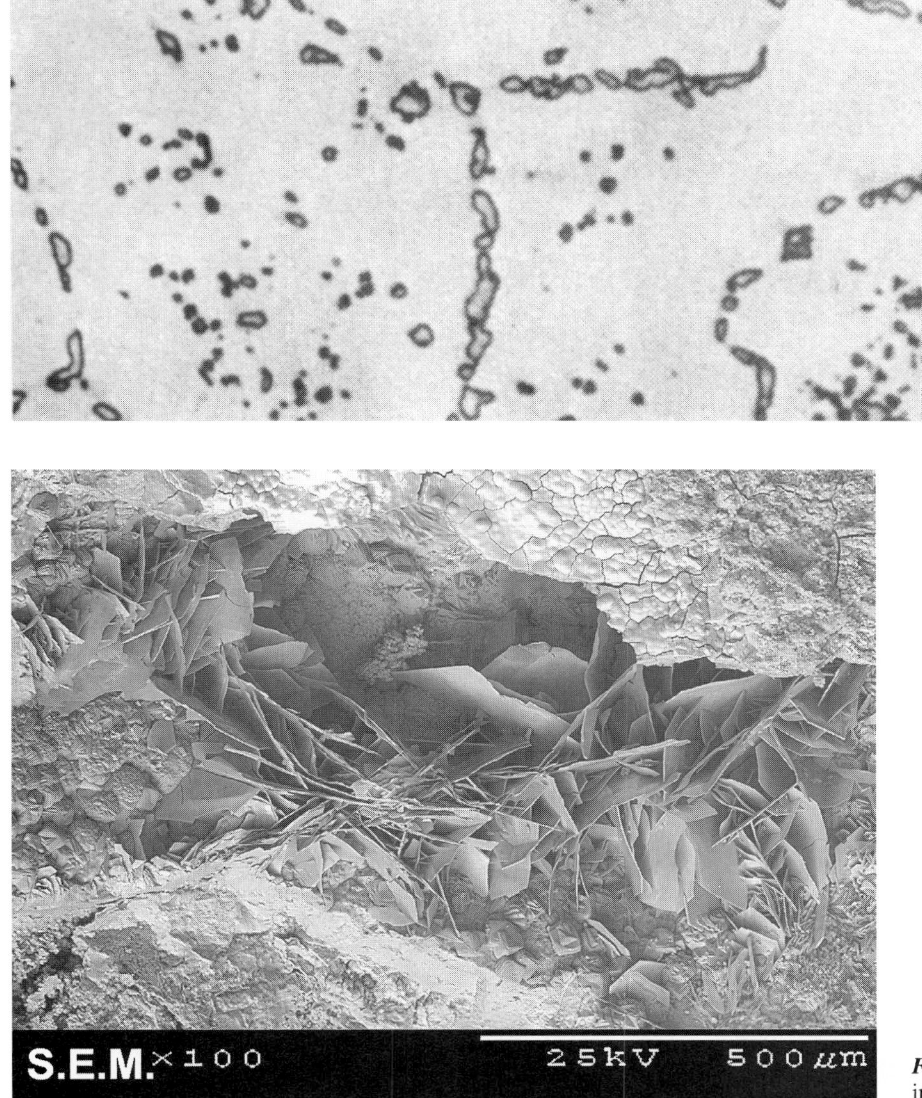

FIGURE 19.2. Sigma phase precipitation (chromium carbide phase formation) at the grain boundaries of a 310 stainless steel. (250×) (*By permission from Metals Handbook, Volume 7, Copyright American Society for Metals, 1972*)

FIGURE 19.3. Corrosion product growing inside a larger pit in the wall of a water pipe.

FIGURE 19.4. The outer shell of this tubercle was opened to view the highly crystalline structure inside.

FIGURE 19.5. Cubic crystalline formation at 500× magnification, found inside the tubercle in Figure 19.4.

FIGURE 19.6. Underdeposit crystalline formation growing between a heavy corrosion deposit inside a copper pipe used for hot water.

formation of oxygen cells. Other bacteria form nitrate by oxidizing ammonia to nitrate.

Stress Corrosion Cracking

Stress corrosion cracking (SCC) occurs in austenitic stainless steels due to interaction between tensile stress and a particular corrodant. In cross section, stress corrosion cracks form tight, branched cracks which propagate as intergranular grain boundary cracks (see Figure 19.7). Austenitic stainless steels are particularly sensitive to SCC in the presence of chlorine. SCC can also occur if the microstructure has been compromised by chromium (sigma) phase precipitation during heat treating. SCC can occur in copper-based alloys due to exposure to ammonia, sulfur dioxide, and nitrates. Some titanium alloys develop SCC due to exposure to ethanol, methanol, ocean salt water, and hydrochloric acid. Carbon steels can develop SCC when stressed in the presence of a concentrated caustic, ammonia, nitrate solu-

FIGURE 19.7. Stress corrosion cracking in A-286 steel. (*Photo courtesy of Stork-MMA Laboratories, Huntington Beach, CA)*

tions, and carbonate/bicarbonate solutions. Solutions can become concentrated in the following ways: by localized stagnation, by evaporation, and within thin films of condensate known as vapor spaces.

Corrosion Fatigue

Corrosion fatigue describes the generation of cracks resulting from the combined effects of cyclic stress and corrosion. Corrosion fatigue cracking in metals is distinguished from SCC by producing a transgranular type fracture with little or no branching.

Erosion Corrosion

Erosion corrosion is the accelerated degradation of a metal due to the action of corrosion in combination with erosion caused by a rapidly moving fluid. Metal is removed as solid particles or as ions. Erosion corrosion often takes place at a right angle bend in pipes containing a fast moving and normally nonabrasive fluid. This failure mode is distinguished from other modes by comet tail and horseshoe formations, in addition to irregular flow patterns formed on the metal surface.

Cavitation Damage

Cavitation damage is often experienced on aluminum boat propellers. The mechanism occurs as a result of instantaneous formation and collapse of vapor bubbles in a liquid subject to rapid and intense localized pressure changes (e.g., a high-speed boat propeller). Cavitation produces sharp, jagged, and spongelike metal loss without the formation of corrosion products. The best way to protect against this mechanism is to upgrade to a more resistant alloy or material.

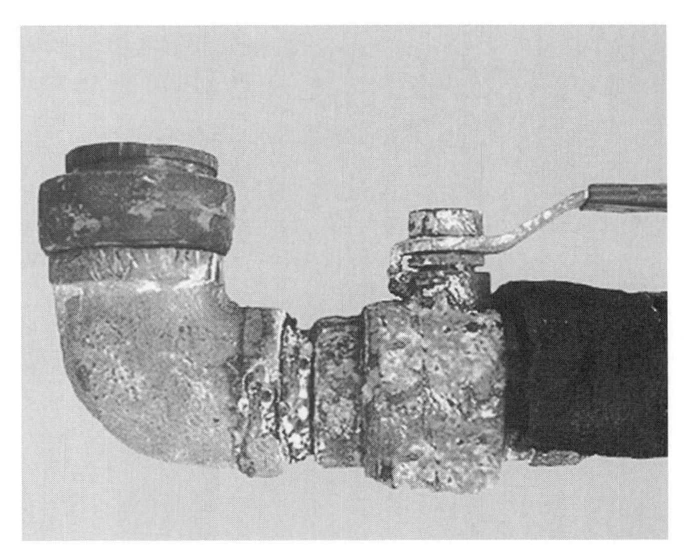

FIGURE 19.8. The corroded valve shown in this photograph is brass. It failed due to dealloying of the copper and zinc.

DEALLOYING

Dealloying occurs when one or more of the alloying elements separate to form individual grains and are

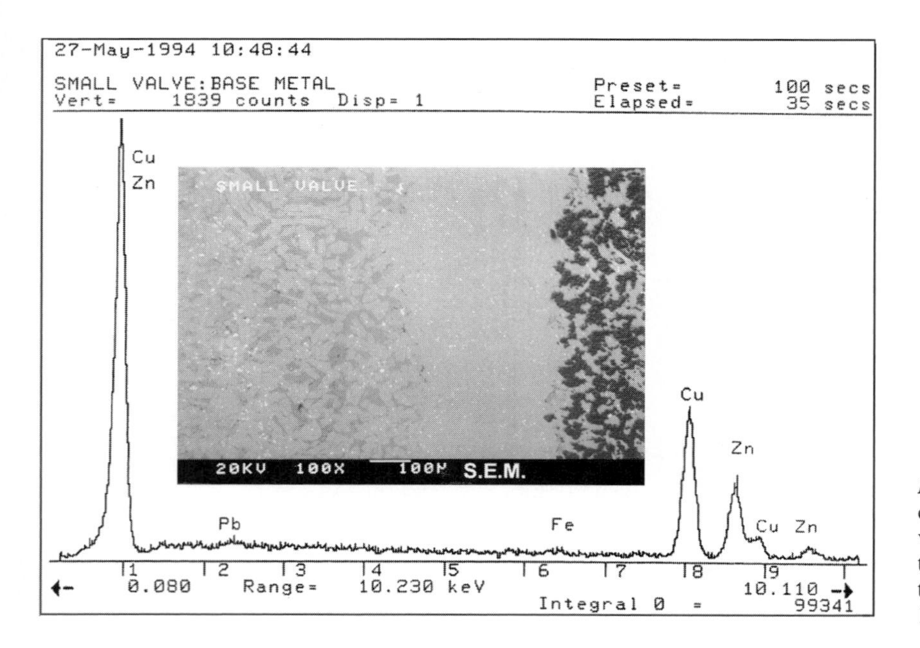

*FIGURE 19.9***a.** This photo montage shows the cross-section microstructure of the corroded valve in Figure 19.8. The S.E.M. X-ray scan of the elements in the valve cross-section indicates the correct element ratio for brass (see Figure 19.9*b*).

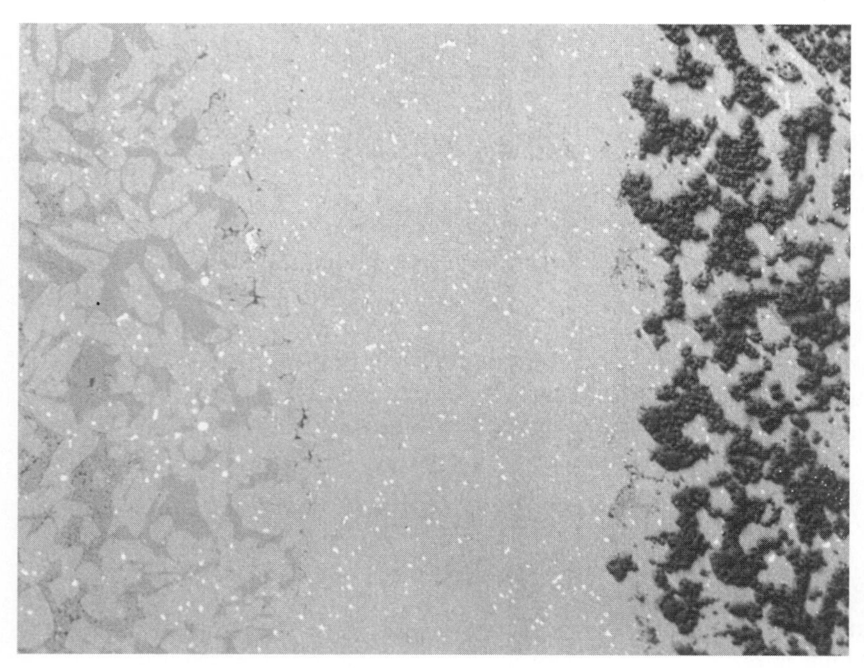

150× Unetched

*FIGURE 19.9***b.** The microstructure on the left side contains dealloyed copper. The center area is normal appearing brass. The right side shows decomposing zinc grains covered with oxidation.

selectively leached out of the metal (see Figure 19.8). Dezincification is the term applied to the dealloying of brass. The zinc separates from the copper and is selectively removed from the metal (see Figures 19.9*a* and 19.9*b*). When nickel dealloys, the process is called denickelification. Corrosion of the dealloyed element causes a porous condition which is often accompanied by weeping leaks. In copper pipe systems, dezincification is a common failure mode and is aggravated by the use of bronze fittings and valves coupled directly to the copper pipe by silver-tin soldering.

INTERGRANULAR CORROSION

Another form of corrosion is intergranular corrosion. This takes place internally. Often the grain boundaries form anodes and the grains themselves form cathodes, causing a complete deterioration of the metal in which it simply crumbles when it fails (Figure 19.10*a*). This often occurs in stainless steels in which chromium carbides precipitate at the grain boundaries. This lowers the chromium content adjacent to the grain boundaries, thus creating a galvanic cell. Most types of austenitic stainless steel when held at

***FIGURE 19.10*a.** The very large grains are outlined in this zinc-based die casting because of intergranular corrosion. The metal has deteriorated at the grain boundaries, which have become anodes.

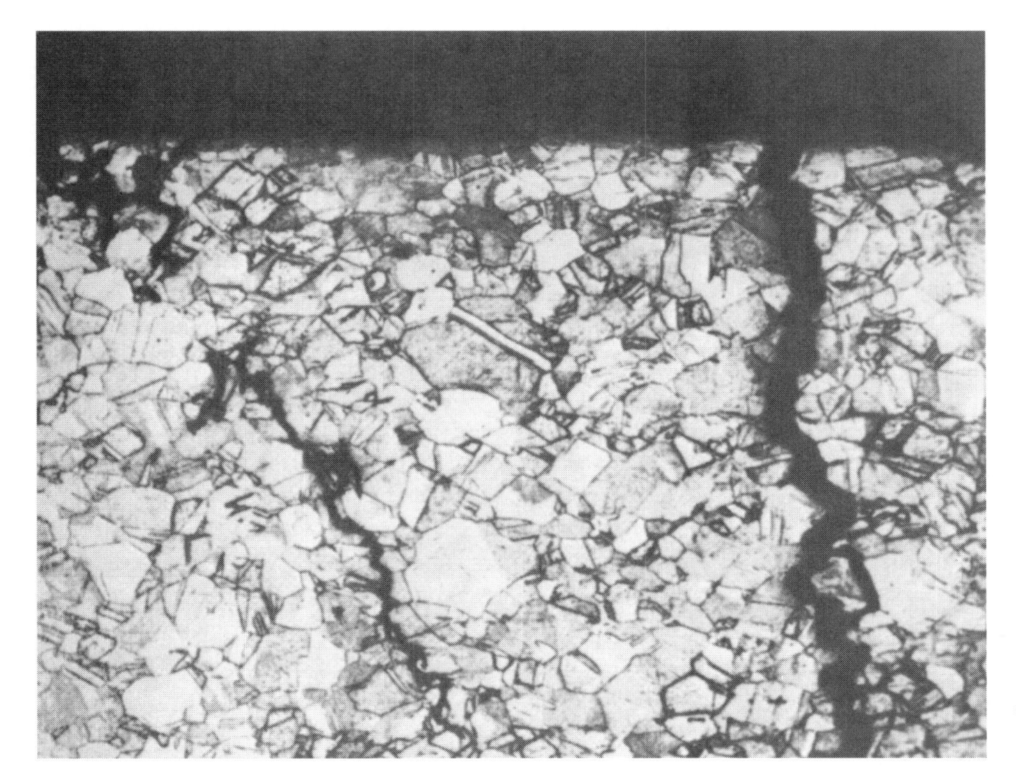

***FIGURE 19.10*b.** Typical intergranular stress-corrosion cracks in alloy 260 (cartridge brass 70 percent) tube that was drawn, annealed, and cold reduced 5 percent. The cracks show some branching. (*By permission, from Metals Handbook, Volume 7, Copyright American Society for Metals, 1972*)

temperatures of 1500°F and cooled slowly are subject to this kind of intergranular corrosion (Figure 19.10*b*).

MATERIAL DEFECTS AS A SOURCE FOR CORROSION

Material defects such as laps, folds, gouges, seams, laminations, and weld-seam defects contribute to the problem of corrosion. Surface defects may cause the metal to fail but more often they act as triggering mechanisms for other failure modes. The severity of the defect is the critical factor governing any interaction which may produce failure. Laps, folds, laminations, and gouges may lead to crevice corrosion or act to increase sensitivity to other failure modes such as corrosion fatigue or SCC. Incorrect welding practice when welding stainless steel can leave the weldment sensitized and susceptible to

SCC. The use of the wrong filler metal in welding can compromise corrosion resistance of the weld in addition to mechanical strength.

The most common form of corrosion is a deterioration of metals by an electrochemical action. It is generally a slow and continuous action. High-temperature scaling and the formation of oxides on metals is oxidation corrosion. The oxide of iron formed at high temperatures is black; it is often called mill scale. Corrosion in metals is the result of their desire to unite with oxygen in the atmosphere or in other environments to return to a more stable compound, usually called ore. Iron ore, for example, is in some cases simply iron rust. Corrosion may be classified by the two different processes by which it can take place: **direct oxidation corrosion,** which usually happens at high temperatures, and **galvanic corrosion,** which takes place at normal temperatures in the presence of moisture or an electrolyte.

DIRECT OXIDATION CORROSION

Oxidation at high temperatures is often seen in the scaling that takes place when a piece of metal is left in a furnace for a length of time. The black scale is actually a form of iron oxide, called magnetite (Fe_3O_4) (Figure 19.11). This oxide coating is also called mill scale because it is formed on heated ingots or slabs that are rolled in steel mills. The red-hot steel is constantly scaling, since it is in contact with the oxygen in the atmosphere.

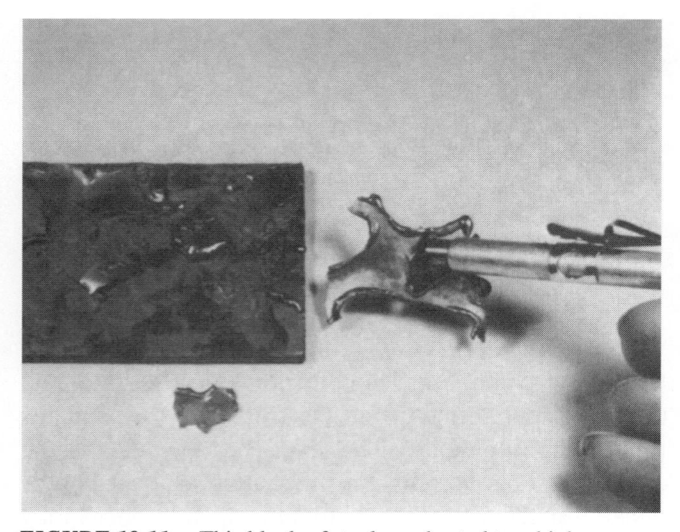

FIGURE 19.11. This block of steel was heated to a high temperature in a furnace in the presence of air. It is covered with a loose black scale, which is magnetite-iron oxide (Fe_3O_4). Magnetite, as the name implies, is magnetic.

GALVANIC CORROSION

Galvanic corrosion is essentially an electrochemical process that causes a deterioration of metals by a very slow but persistent action. In this process, part or all of the metal becomes transformed from the metallic state to the ionic state and often forms a chemical compound in the electrolyte. On the surface of some metals such as copper or aluminum, the corrosion product sometimes exists as a thin film that resists further corrosion. In other metals such as iron, the film of oxide that forms is so porous that it does not resist further corrosive action, and corrosion continues until the whole piece has been converted to the oxide (Figure 19.12).

POSITIVE AND NEGATIVE IONS

Certain elements have common properties and are arranged into groups in the Periodic Table (see Appendix 1, Table 5). Eight groups in vertical columns contain elements that are able to form compounds, and one group (0) on the right is inactive (inert). The horizontal rows are periods of elements and are arranged in steps in increasing order of atomic numbers. The periodic law states that the properties of elements are periodic functions of their atomic numbers. The atomic number is equal to the number of protons within the nucleus of the atom or the sum of all orbiting electrons.

Valence is the combining power of an atom and refers to the bonding force of atoms. Ions (atoms having a positive or negative charge) of some elements always have a fixed charge such as $+1$ or $+2$ (that is, they are missing electrons in the outer shell); others such as iron vary in valence. When iron has a valence of $+2$, it is called ferrous; when it has a $+3$, it is called ferric. Metallic atoms form positive ions, and nonmetals form negative ions. Ions having opposite charges can often combine to form compounds with various bonding arrangements. The charge of the ion is written with a positive or negative sign at the upper right of the symbol; Fe^{2+} for iron, Zn^{2+} for zinc, and Cl^- for chloride are examples. When a group of atoms in a chemical relation have an electrical charge, they are called radicals. An example is the hydroxyl ion composed of one oxygen atom and one hydrogen atom with a charge of -1, which would be written OH^-.

ELECTROLYTES

An electrolyte is any solution that conducts electric current and contains negative or positive ions. Corrosion

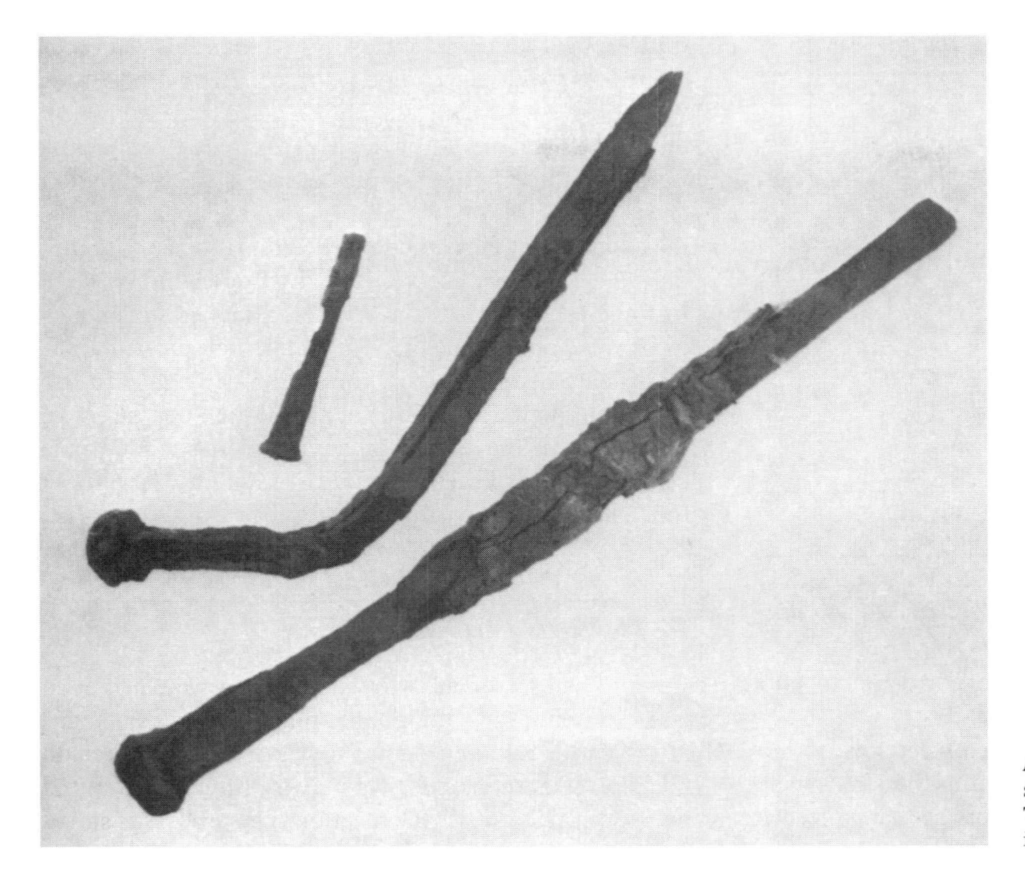

FIGURE 19.12. These large iron spikes were found on an ocean beach. They are almost completely changed into iron oxide rust.

requires the presence of an electrolyte to allow metal ions to go into solution. The electrolyte may be fresh or salt water and acid or alkaline solutions of any concentration. Even a fingerprint on clean metal can form an electrolyte and produce corrosion. There must be a completed electric circuit and a flow of direct current before any galvanic action can take place. There must also be two electrodes—an anode and a cathode—which must be electrically connected, (see note at end of chapter.) The anode and cathode may be of two different kinds of metals or located on two different areas of the same piece of metal. The connection between the anode and the cathode may be made by the metal itself or by a metallic connection such as a bolt or a rivet.

In the galvanic or two-metal form of corrosion with dissimilar metals, as shown in Figure 19.13, a metal that is lower or nearest the more active end of the galvanic series (see Table 19.1) is the anode, and the cathode is a metal toward the more passive end or the end of least corrosion. This is called coupling. However, if a single piece of metal—iron, for example—is immersed in slightly acid water that contains oxygen, corrosion will take place. In this case, there are anodes and cathodes, and the deterioration of the metal occurs at the anodes. Very tiny, well-defined cathode and anode areas are formed all over the piece of metal. However, they

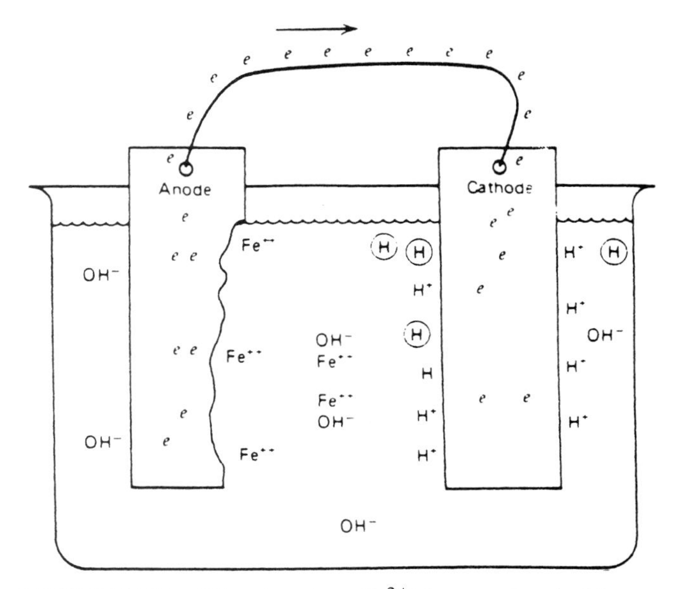

FIGURE 19.13. When iron ions (Fe^{2+}) become detached from the anode, electrons flow to the cathode and neutralize positively charged hydrogen ions, releasing them from the cathode surface.

TABLE 19.1. Galvanic series of metals and alloys in seawater

Anode End (Highest Corrosion)	
Magnesium	Bronze
Beryllium	Copper-nickel alloys
Aluminum	Nickel
Cadmium	Tin
Aluminum alloys	Lead
Uranium	Copper
Manganese	Inconel
Zinc	Silver
Plain carbon steel	Stainless steel (passive)
Alloy steels	Monel
Cast iron	Titanium
Cobalt	Platinum
Stainless steel (active)	Gold
Brass	Cathode End (Least Corrosion)

may shift, causing a very uniform corrosion to take place.

Various metals react differently to electrolytes. For example, aluminum surfaces exposed to the atmosphere quickly form a thin, invisible oxide coating that protects it from corrosion from ordinary weathering. However, if it is exposed to certain substances or conditions, it can deteriorate very quickly. Many acids have no effect on aluminum but alkalis attack the surface skin and, therefore, are corrosive to aluminum. Certain soaps or detergents that contain alkaline substances, when used to clean aluminum, actually can cause corrosion unless the excess is rinsed off. In contrast, nickel is almost completely resistant to corrosion by alkalis, especially caustic soda, and is resistant to attack by most other reagents with the exception of sulfur dioxide or ammonia when mixed with water.

HOW CORROSION TAKES PLACE

When corrosion of a metal occurs, positively charged atoms are released or detached from the solid surface and enter into solution as metallic ions while the corresponding negative charges in the form of electrons are left behind in the metal. The detached positive ions bear one or more positive charges. In the corrosion of iron, each ion atom releases two electrons and then becomes a ferrous iron carrying two positive charges. Two electrons must then pass through a conductor to the cathode area (Figure 19.13). (Without this electron flow, no metal ions can be detached from the anode.) The electrons reach the surface of the cathode material and neutralize positively charged hydrogen ions that have become attached to the cathode surface. Two of these ions will now become neutral atoms and are released generally in the form of hydrogen gas. This release of the positively charged hydrogen ions leaves an accumulation and a concentration of OH negative ions that increases the alkalinity at the cathode. When this process is taking place, it can be observed that hydrogen bubbles are forming at the cathode only. When cathodes and anodes are formed on a single piece of metal, their particular locations are determined by, for example, the lack of homogeneity in the metal, surface imperfections, stresses, inclusions in the metal, or anything that can form a crevice, such as a washer, a pile of sand, or a lapping of the material.

CATHODIC POLARIZATION

In more neutral electrolytes such as pure water or even a sodium chloride solution, the release of hydrogen gas at the cathode is very slow; thus, very little corrosion takes place at the anode. This layer of hydrogen on the cathode surface slows the reaction. The process is called cathodic polarization (Figure 19.14*a*). If, however, oxygen is dissolved in the electrolyte, a reaction can take place. Oxygen reacts with the accumulated hydrogen to form OH^- ions or water (H_2O). This process permits corrosion to proceed, oxygen acting as a cathodic depolarizer (Figure 19.14*b*). Often a further combining of the corrosion products and the hydroxyl (OH^-) ions from the cathodic reaction takes place; when ferrous ions, for example, are released from the anode, they combine with the hydroxyl ions at the cathode to form ferrous hydroxide. This then becomes oxidized by the oxygen present to form ferric hydroxide $[Fe(OH)_2]$, which precipitates as ordinary iron rust (Figure 19.15).

The action of oxygen in accelerating corrosion in water or in an electrolyte can be demonstrated by placing iron turnings into glass containers of water. Oxygen can be bubbled through one container for a period of time. After several hours, it can be seen that in the container saturated with oxygen, the metal has begun to rust, and in the one without oxygen, the metal still has not corroded. Placed in still water like a stagnant pond, iron will not corrode for years because the water is oxygen free. In contrast, iron placed in running water containing oxygen often rusts very quickly. Also, when water spray or raindrops impinge on the surface of iron, the iron is constantly in the presence of oxygen from the atmosphere, which accelerates corrosion.

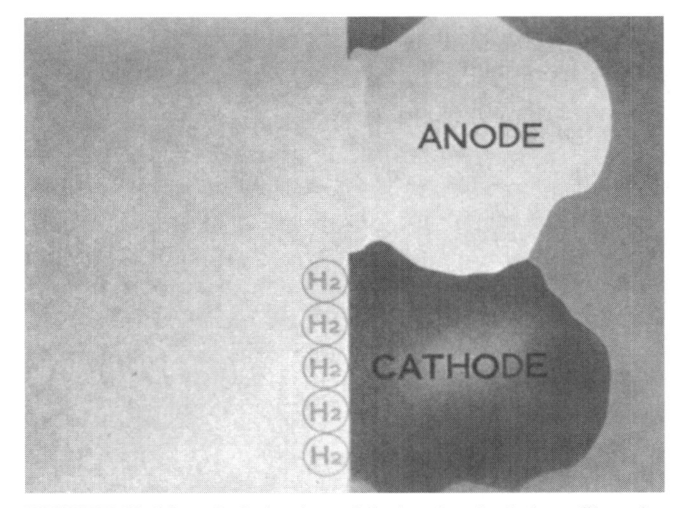

FIGURE 19.14a. Polarization of the local cathode by a film of hydrogen. (*Courtesy The International Nickel Company, Inc.*)

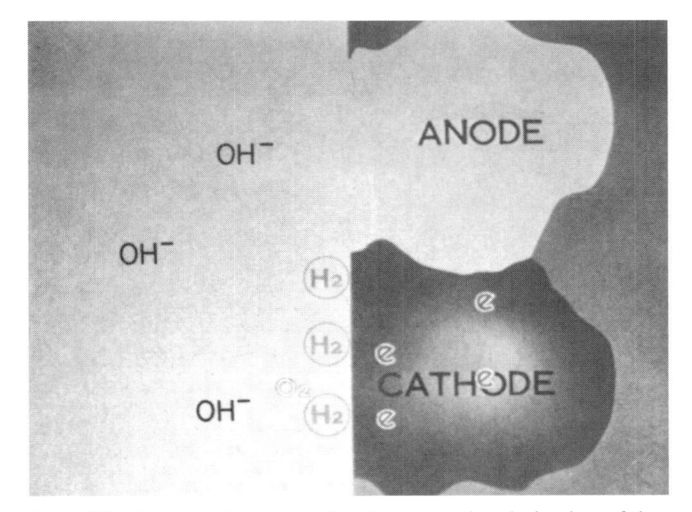

FIGURE 19.14b. Removal of hydrogen or depolarization of the cathode caused by oxygen in water. (*Courtesy The International Nickel Company, Inc.*)

THE GALVANIC SERIES

Anodes and cathodes can form at various places on the surface of a metal, but on the surface of dissimilar metals the corrosion rates produced are often greater. Some metals have a greater tendency to corrode than others because they are more active chemically and tend to become anodes. Others are less active and are cathodic. Gold, for example, is very inactive and is cathodic and has about the lowest possible corrosion potential. The **galvanic series** of metals and alloys shows this difference in the activity of metals in seawater (Table 19.1). The galvanic series (also called the electromotive series) is so named because of the direct electric current

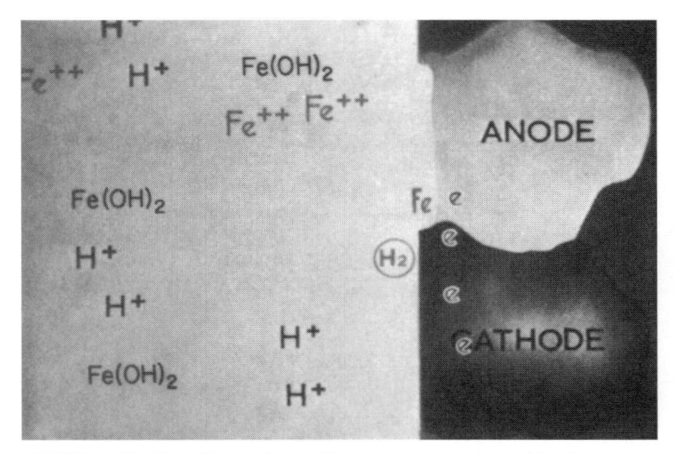

FIGURE 19.15. Formation of ions at the anode and hydrogen at the cathode. This is the process of rusting on iron. (*Courtesy The International Nickel Company, Inc.*)

produced (galvanism) when two dissimilar metals are immersed in an electrolyte. At the more active end, we find that the metals that readily become anodic to iron are magnesium and zinc. Except for those metals at the extreme opposite ends (gold and magnesium), any given metal may be either an anode or a cathode when it is coupled with another metal in the series.

In the galvanic series, the metals become increasingly active toward the anode end. For example, if potassium is placed in water, a violent reaction takes place, but gold or platinum in strong nitric acid will show no evidence of attack. In the electromotive series, a metal may displace any metal below it from chemical combination. Iron can displace copper from a copper sulfate solution, with the copper ions acting as receivers for electrons in the corrosion of the iron. The displaced copper forms a coating on the corroded iron surface. This takes place when a solution of copper sulfate is placed on a clean surface of iron; iron atoms are detached (corroded), leaving the surface copper plated. In this way hydrogen ions act as electron accepters or receivers for any metal that is more active than hydrogen in the galvanic series. Another example is the displacement of silver from a silver nitrate solution by copper, with the silver ions becoming the electron acceptors and the copper being corroded; a silver coating is formed on the surface of the copper. This can be demonstrated by placing a strip of copper in a silver nitrate solution; a film of silver will then be deposited on the copper by ion transfer. The rate at which galvanic action and corrosion take place depends on the degree of difference in electrical potential and the resultant current flow. If zinc is coupled with copper, a large current will flow, but if brass is coupled with copper, a very small current will

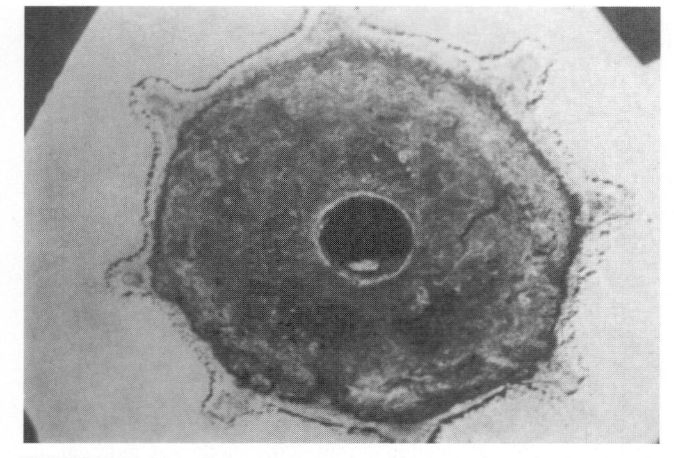

FIGURE 19.16. Galvanic corrosion of magnesium that surrounds a steel core. (*Courtesy The International Nickel Company, Inc.*)

flow and less reaction will be seen. A galvanic corrosion of magnesium that is in contact with a steel core can be seen in Figure 19.16.

RATE OF CORROSION

The speed at which corrosion proceeds depends on several factors: the type of corrosion product, the kind of electrolyte, galvanic current density, environmental condition (crevices, rate of electrolyte flow past the cathode, concentration cells), and the availability of oxygen as a depolarizer. The current density can be greater on small electrodes than on large ones, if a large cathode is coupled with a small anode. The current density for its size can be very high, since the cathode provides a relatively larger area for depolarization of hydrogen by any available oxygen, thus allowing more current flow from the small anode and causing a large number of metal ions to leave the anode. If a small cathode is coupled to a large anode, however, the reverse is true; the cathode provides only a small surface for depolarization. This limits the current density, and very little corrosion occurs at the anode (Figures 19.17*a* and 19.17*b*).

There is also a difference in rate of corrosion in steel having different alloying elements, such as copper, nickel, or chromium steels, which have less tendency to corrode than most other alloy steels. However, as a general rule, all metals having almost no alloying elements or any other kinds of contaminating materials, such as pure iron or pure aluminum, tend to have very low corrosion rates. Wrought iron containing almost no carbon will tend to resist corrosion or rust more than carbon steel. Ancient wrought iron artifacts more than

*FIGURE 19.17*a. Influence of area relationship between cathode and anode that are illustrated by copper-steel couples after immersion in seawater. Copper rivets with a small area in steel plates of a large area have caused only a slight increase in corrosion of steel. (*Courtesy of The International Nickel Company, Inc.*)

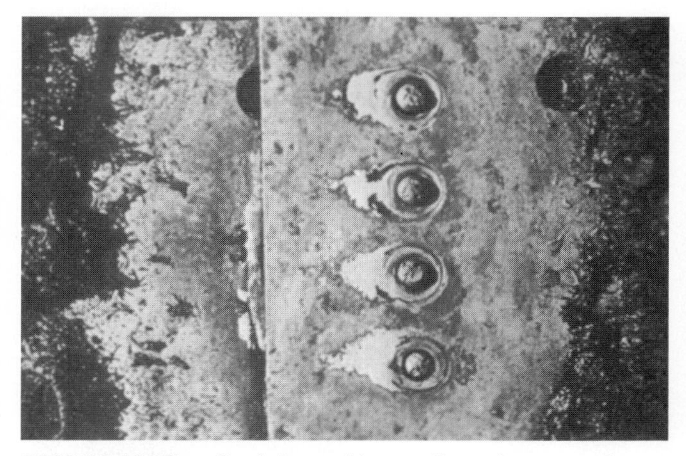

*FIGURE 19.17*b. Steel rivets with a small area in copper plates of a large area have caused severe corrosion of steel rivets. (*Courtesy The International Nickel Company, Inc.*)

a thousand years old have resisted corrosion over the years. The iron tower at Delhi, India, and the wrought iron nails used by the Vikings in their ships are examples of this resistance to rusting. Other forms of protection used to inhibit the process of corrosion are organic and inorganic coatings, the use of cathode protection, and inhibitors. Inhibitors are usually liquids that are placed in the electrolyte or other corrosive medium to render it inert. They are sometimes used as a coating on metals.

Rates of corrosion are determined by practical tests in certain atmospheres (Figures 19.18*a* to 19.18*c*). The progress of corrosion is plotted on a graph showing loss of weight in relation to the amount of time the specimen was subjected to a given environment (Figure 19.19).

FIGURE 19.18a. Corrosion of steels in marine atmosphere. (*Left*) Low copper steel; (*center*) ordinary steel; and (*right*) nickel-copper-chromium steel. (*Courtesy The International Nickel Company, Inc.*)

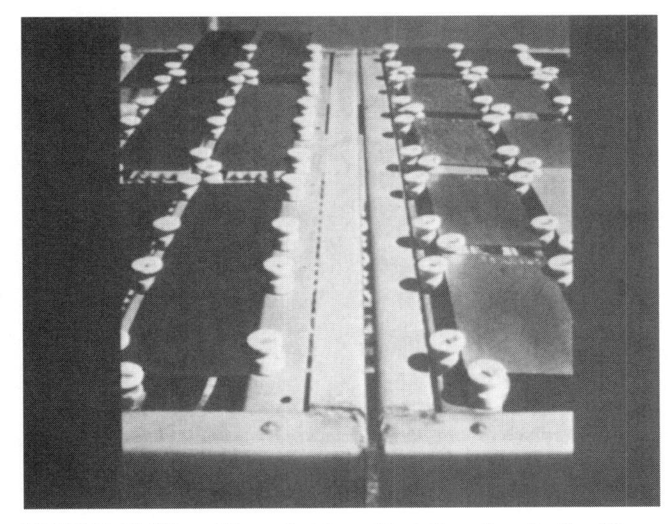

FIGURE 19.18b. Mirror finish on Hastelloy alloy C after 10 years of exposure in marine atmosphere. (*Courtesy The International Nickel Company, Inc.*)

PROTECTION AGAINST CORROSION

Cathodic protection is often used to protect steel ships' hulls and buried steel pipelines. This is done by using zinc and magnesium sacrificial anodes that are bolted to the ship's hull or buried in the ground at intervals and electrically connected to the metal to be protected. In the case of the ship, the bronze propeller acts as a cathode, the steel hull as an anode, and the seawater as an electrolyte. Severe corrosion can occur on the hull as a result of galvanic action. The sacrificial anodes are very near the anodic end of the galvanic series and have a large potential difference between both the steel hull of

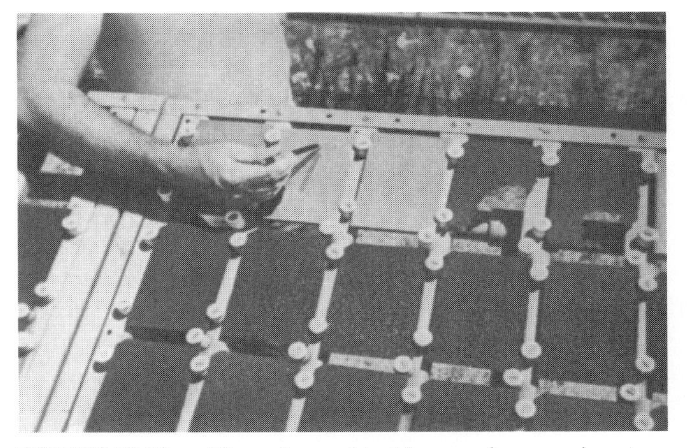

FIGURE 19.18c. The racks used in this corrosion experiment, made of monel-nickel-copper alloy, show only a superficial tarnish after years of exposure in marine atmosphere. (*Courtesy The International Nickel Company, Inc.*)

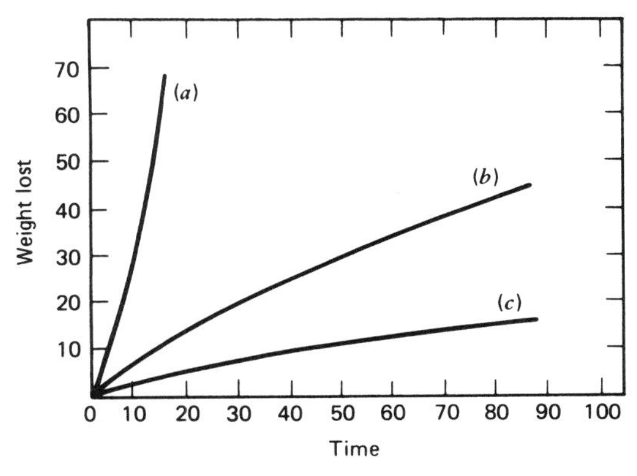

FIGURE 19.19. Progress of corrosion on steels with different resistance to atmospheric corrosion. (*a*) Steel containing very small amounts of copper. (*b*) Plain carbon steel. (*c*) Steel containing nickel, chromium, and copper.

the ship and the bronze propeller. Both the hull and propeller become cathodic and consequently do not deteriorate. The zinc or magnesium anodes are replaced from time to time (Figures 19.20*a* and 19.20*b*).

Stray electric currents are often a problem found on boats, pumps, and improperly grounded machinery. It can be produced by external electric current flow or by the potential difference between two dissimilar metals in an electrolyte. More corrosion damage is caused by stray direct currents than by stray alternating currents. A very valuable method of providing cathodic protection to ship's hulls, underground lines, and steel structures involves the application of a direct current to the structure requiring protection. The structure serves as the cathode side of the circuit, thus protecting it. When the current is applied from an external source, the local action current density is reduced, ideally to zero. This method has no value in protecting structures or

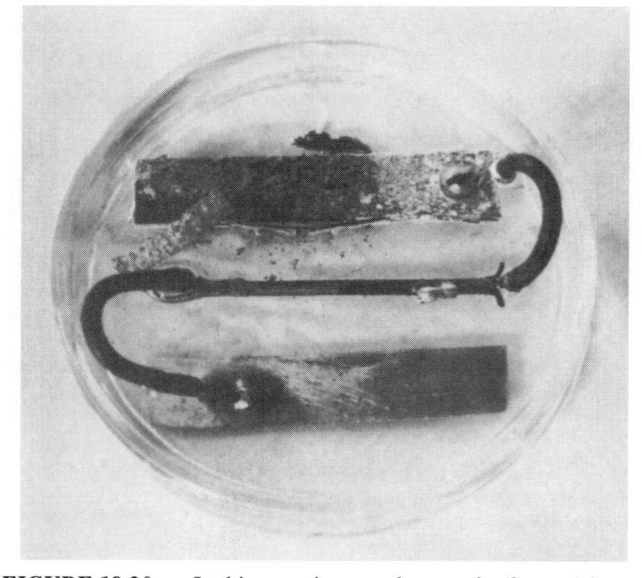

*FIGURE 19.20*a. In this experiment, a brass strip (*bottom*) is soldered to a nail (*center*) and a magnesium strip (*top*) bolted to the nail. The magnesium has started to deteriorate in the sodium chloride electrolyte.

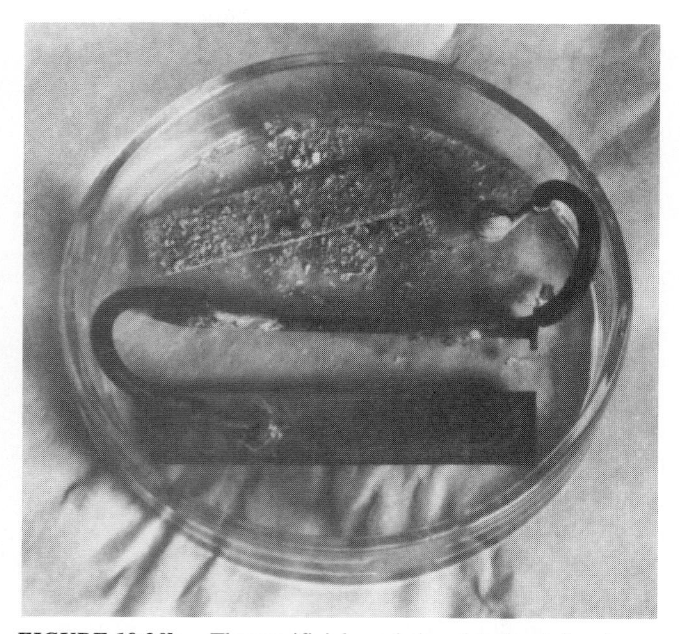

*FIGURE 19.20*b. The sacrificial anode has deteriorated and separated from its connection. The iron nail has not corroded since it was cathodic.

surfaces that are continuously in the atmosphere or in nonconducting liquids. The structures must be buried in soils or immersed in liquids that will conduct electric currents.

The selection of materials is of foremost importance. Even though a material may be normally resistant to corrosion, it may fail in a particular environment or if coupled with a more cathodic metal.

Zinc has been used on iron and steel for years for cathodic protection, but now a somewhat different kind of cathodic protection is being used on concrete. This cathodic protection was developed about 9 years ago by the California Department of Transportation. They discovered that running an electric current through a zinc coating on concrete draws away chloride ions that tend to penetrate the concrete and corrode the steel reinforcement inside. As the interior steel corrodes, it swells. That cracks the concrete, allows more moisture inside, and accelerates the process. Bridges near salt water are especially vulnerable to such deterioration. The zinc coating is applied with a special applicator that melts the zinc and sprays it against the concrete.

FIGURE 19.21. The concrete bridge situated near the Pacific Ocean has been protected from salt water deterioration with a unique method of cathodic protection. The entire concrete structure has been coated with zinc to which a constant electric current is applied. (*Reprinted with permission from The Eugene Register-Guard, January 23, 1992*)

Many Oregon coast bridges are of historical concrete arched construction. Figure 19.21 is one such 61-year-old bridge at Cape Creek. The renovation involved repairing broken concrete, sandblasting, grounding exposed metal, running a maze of wires across the structure, and spraying the bridge with 125,000 pounds of zinc coating. It is believed the bridge will now stand indefinitely. However, the zinc coating may have to be replaced in 20 years. New concrete bridges in Oregon must have an epoxy coating on the reinforcing bars at a minimum of 0.015 inch. This coating is so tough that a bar can be bent to shape without breaking the coating.

Coatings are extensively used to prevent corrosion. They are classified as follows.

1. Anodic coatings (anodizing)
2. Cathodic coatings (electroplating, chrome, copper, nickel)
3. Organic and inorganic coatings
4. Inhibitive coatings (red lead, zinc chromate)
5. Inhibitors (placed in the electrolyte)

Metal coatings may be listed as follows.

1. Hot dip process (galvanizing)
2. Metal spraying
3. Metal cementation (diffusion of atoms into the surface of a metal part)
   a. Sherardizing (metal parts heated with zinc powder to 600° to 850°F)
   b. Chromizing
4. Metal cladding (rolling a "sandwich" of metal with the outer layers of a corrosion resistant metal). For example, clad aluminum sheet on which a thin outer layer of pure aluminum covers the strong aluminum alloy that is not so resistant to corrosion.

Anodizing is the process of thickening and toughening the oxide film. On metals such as aluminum, the thicker layer of oxides increases resistance to further corrosion. Anodized films not only provide a hard wear surface but also provide a somewhat porous base for paint or other coatings.

The familiar electroplating process is similar to galvanic corrosion in that metal ions are detached from the anode and are deposited on the cathode to provide protection from corrosion. In most cases, the noble metals that are cathodic in the galvanic series are used for plating materials. For example, tin plate is used extensively for food containers. If there is a break in the plating material, corrosion will begin at that point if the plating is cathodic to the base metal. If a break occurs in tin plate on iron, the tin is more cathodic than iron and therefore accelerates the corrosion of the iron at the break (Figure 19.22). In air-free conditions, however, as when food is sealed in "tin" cans, the tin plate is generally anodic to steel.

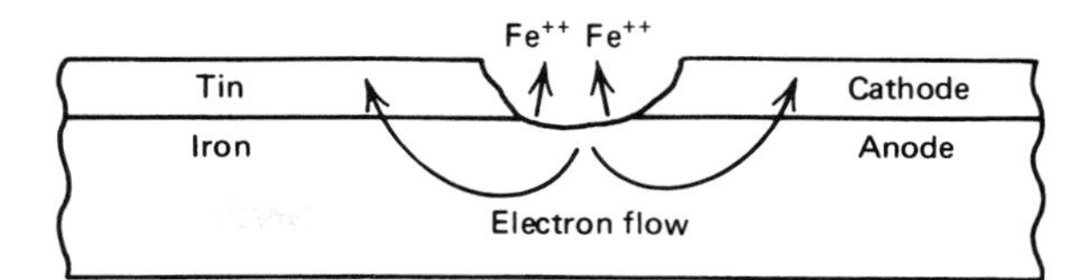

FIGURE 19.22. The action of corrosion on tin plate.

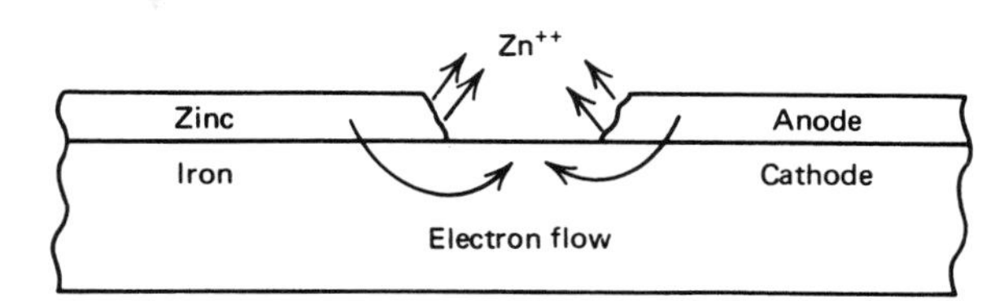

FIGURE 19.23. The action of corrosion on zinc plate (galvanized iron).

Zinc-plated steel is used for roofing materials and many other products. Zinc is more anodic than iron, and if there is a break in the zinc plate on iron, the zinc will be a sacrificial metal in this case and will corrode instead of the iron (Figure 19.23). Large patches of zinc may corrode away on galvanized sheet iron before denuded areas become sufficiently wide to allow anodic areas to form on the steel. Only then will it begin to rust.

Chemical inhibitors used in the corrosive medium make it inert, that is, unable to transfer metal ions from the anode to the cathode. Some types are used as a protective coating on the surface of the metal. Sodium phosphate in water can be used to produce a passive (inactive) ferrous oxide (Fe_2O_3) film on steel.

Organic coatings such as paint, tar, grease, and varnishes prevent corrosion by keeping the corrosive atmosphere from contacting the surface. Inorganic coatings such as vitreous enamels or even mill scale also create a barrier, but one drawback is their brittleness. Corrosive attack can take place where the coating is chipped or broken.

Metal cementation or diffusion of a material into the surface of metal is done by applying heat. Carburizing of low carbon steel is one example of metal cementation. Zinc powder can be diffused into the surface of steel by heating both powder and steel together to a few hundred degrees Fahrenheit. This is contrasted to the hot-dip zinc process or electroplating in that here the powder penetrates to some extent into the surface of the steel. Silicon, aluminum, chromium, and many other elements are diffused or cemented into steel and other metals.

Ion implantation can be used on a wide range of materials: tool steels, alloy steels, copper, aluminum, and phospher bronze. This process has been used to increase the hardness and wear resistance of tools, dies, and ball bearings by implanting nitrogen ions into the material's crystal lattice, which tends to lock lattice defects into place.

Ion implantation has developed into a widely used method of protecting steel from rusting. In the past, cadmium has been extensively used as a plating for protecting steel and aluminum, but there were many problems. Cadmium electroplating on high-strength steel often caused hydrogen embrittlement which led to premature failure of the part. Also, its toxicity is harmful to workers and to the environment.

Aluminum, because of its nearness to cadmium in the galvanic series and because it does not cause the problems stated above, is the ideal replacement for cadmium. However, aluminum is not easy to apply to steel by the usual methods. Ion implantation of aluminum is considered the best method for applying aluminum coatings on steel. The basic equipment required for ion vapor deposition (IVD) is a steel chamber, a vacuum pumping system, an evaporation source, and a high-current, high-voltage supply. The chamber is evacuated and backfilled with an inert gas. A high negative potential is applied between the parts being coated and the evaporation source. The gas becomes ionized and creates a glow discharge around the parts to be coated. This cleans the steel part. An evaporated aluminum wire (1100 alloy) is fed into heated crucibles and the resultant aluminum vapor passes through the glow discharge, becomes ionized, and is impinged on the part surface, thoroughly plating it. A typical plating cycle takes about 90 minutes.

Note. The study of galvanic corrosion as presented in this chapter is based on the current flow theory in which the flow of current is considered to be opposite to the electron flow and in which the **anode** deteriorates because of corrosion as the electrons and positive ions leave it. In contrast, the electron theory as used in the study of electronics shows the **cathode** with a negative charge as the emitter of electrons and it deteriorates (as in a vacuum tube) while the anode with its positive charge (as the plate in a vacuum tube) does not.

Positive and negative signs have not been used to avoid confusion about polarity. The terms *anode* and *cathode* as related to the current theory are in common use in industrial circles and probably will not be changed to the electron theory terminology for many years, if at all. The terms *anodizing* and *cathodic protection* are examples of common usage of current flow theory terminology.

SELF-EVALUATION

1. Corrosion may be classified in what two ways?
2. What is the process by which galvanic corrosion takes place?
3. Atoms may exist as positive or negative ions. For the most part, do metals become negative or positive ions?
4. Name three or more necessary conditions that are required for any galvanic corrosion to occur.
5. What role does hydrogen play in cathode polarization? How does this affect the rate of corrosion?
6. What happens to cathodic polarization when oxygen is dissolved in the electrolyte?
7. In the galvanic series, which would be more active (anodic), iron or copper?
8. Why will a galvanic couple having a large cathode and a small anode corrode much more rapidly than one having a large anode and a small cathode?
9. Some alloying elements such as copper in steel tend to reduce corrosion but, in general, in what conditions are metals least likely to corrode?
10. How can some metals such as stainless steel, which are normally resistant to corrosion, rust under a washer, a crevice, or a pile of sand?
11. A bronze propeller on a ship is normally cathodic and the ship's steel hull normally anodic, causing it to corrode. How can this problem be overcome?
12. Ion implantation is widely used today on the surface of metals. What are two major advantages in its use on steel?

CASE PROBLEM: THE CORRODED TRAILER FRAME

A family owned a relatively new travel trailer that was built on an aluminum frame. The plywood flooring was covered with carpeting. They took very good care of it, frequently shampooing the carpet. A mechanic who serviced the trailer told them the aluminum frame was pitted and almost eaten through in some places from corrosion, and that expensive repairs would have to be made. The corroded frame members were all around the outside edge and were just beneath the place where the outside wall and floor met. They had never traveled near the ocean or on salted roads. Can you explain this? How could a new aluminum frame be so badly corroded? How could this have been avoided?

WORKSHEET 1

Objective Demonstrate the importance of oxygen in the corrosion of iron in water.

Materials Two 500 milliliter (mL) Erlenmeyer flasks or equivalent, rubber tubing, glass tubing, oxygen tank, distilled water (deaerated by boiling), iron (low carbon steel) turnings, benzene, and acetone.

Procedure Pour 250 mL of distilled water in each of the Erlenmeyer flasks. Degrease the iron turnings by rinsing first with the benzene and then with the acetone. Allow to dry and weigh two 50 g samples and put one sample in each flask. Using a short length of glass tubing as a bubbler in one flask, connect it with the rubber tubing to the oxygen tank. Allow oxygen to bubble through one flask for a period of several hours.

Conclusion Observe the development of rust on the iron turnings. Which flask, the one with the still water or the one with the oxygenated water, shows rust forming on the turnings? Explain why you think this happened.

WORKSHEET 2

Objective Demonstrate the development of corrosion in crevices because of oxygen concentration differential.

Materials One strip of AISI 410 chromium steel 2×6 in., two small pieces of wood 1×1 in., a strong rubberband, one 400 mL beaker or equivalent, and a 3 percent solution of sodium chloride, containing 3 or 4 mL of a 5 percent ferric chloride solution.

Procedure Clamp the two pieces of wood on opposite sides of the stainless steel strip with the rubberband and insert it into the 400 mL beaker. Fill the beaker with the sodium chloride solution and set it aside for a period of 6 weeks or more. It may be necessary to replace the solution about once or twice a week after corrosion has begun.

Conclusion After the test period, remove the sample of stainless steel and inspect under the crevice provided by the blocks of wood. Has any corrosion begun? If so, how can a normally corrosion-resistant metal such as stainless steel begin to rust in a crevice area?

WORKSHEET 3

Objective Illustrate ion displacement in a solution by a more active metal in the galvanic series.

Materials Four small beakers or other glass containers, 5 percent solution of copper sulfate, 5 percent solution of silver nitrate, 10 percent solution of hydrochloric acid, and small strips of zinc, copper, and low carbon steel.

Procedure

1. Immerse a strip of copper in the 5 percent silver nitrate solution and note the deposition of silver on the copper surface. Copper displaces silver from solution because copper is more active (anodic) than silver in the galvanic series and suffers from corrosion or loss of copper ions in the process.
2. Immerse a strip of copper in the 10 percent hydrochloric acid and note the absence of any reaction. Hydrogen (in the acid) is more anodic than copper in the galvanic series and is not displaced by copper.

CHAPTER 20

Nondestructive Testing

Nondestructive testing methods are among the most useful tools of modern industry. Quality assurance and control go hand in hand with nondestructive testing. The reliability of manufactured parts, bridge structures, pipelines, and critical aircraft parts depends on nondestructive testing systems. Ultrasonic inspection, X rays, and particle and penetrant inspection methods have been used for decades. Other processes have been developed, but what is rapidly changing is the way the information gathered by the testing methods is processed by modern electronics into more useful information. This often eliminates the need for skilled operators to interpret information. Many of these testing systems are discussed in this chapter.

The emphasis is on metals, but with the increasing use of plastics and composites, nondestructive testing is also applicable for those materials. Since aircraft or any other mechanism in use will sooner or later experience service failures, a damage tolerant design philosophy has emerged. The assumption is that when a structure is put into service, there will always be flaws such as discontinuities in the lattice structure or places where cracks can be formed and propagated. NDT is, therefore, a necessary function of this philosophy in order to assess any possible damage from time to time during the lifetime of the mechanism.

Fracture mechanics has a close relationship with nondestructive testing, but is used primarily to evaluate failures after they have taken place. See Chapter 27 "Failure Analysis."

OBJECTIVES **After completing this chapter, you should be able to:**

1. Name the several methods of nondestructive testing and explain the specific uses and operation of each.
2. Use testing equipment or inspecting test pieces.

Nondestructive testing, as the name implies, in no way impairs for further use the specimen that it tests. Usually these tests do not directly measure mechanical properties, such as tensile strength or hardness, but are intended to locate defects or flaws. When machines or parts have large safety factors built into them, there is little need for nondestructive testing. However, many products used in aircraft, space technology, and other industries require a high level of reliability. This is achieved by inspection at the time of manufacture and, in some cases, continued testing during the service life of the part. The most common types of nondestructive tests are **magnetic-particle inspection, fluorescent-penetrant inspection, ultrasonic inspection, radiography (X ray and gamma ray), eddy current inspection,** and **visual inspection.**

MAGNETIC-PARTICLE INSPECTION

There are several methods of magnetic-particle inspection (Magnaflux) that are used to detect various kinds of flaws in ferromagnetic metals such as iron and steel (Figure 20.1). A magnetized workpiece is sprinkled with dry iron powder or submerged in a liquid in which the particles are suspended. The Magnaflux Corporation has developed a method called Magnaglo in which fluorescent-magnetic particles are suspended in solution. The solution is flowed over the magnetized workpiece, which is then viewed under a black (fluorescent) light. Surface cracks that are only a few millionths of an inch wide can be found by this method (Figure 20.2).

The wet method is useful for inspection in the manufacture of parts and for maintenance inspection. The dry method is generally used for inspection of welds, large forgings, castings, and other parts having rough surfaces. The basic principle of magnetic testing is shown in Figure 20.3. A magnetic pole is formed at a crack, which causes the magnetic powder to concentrate at that point. When a part is magnetized lengthwise, as shown in Figure 20.4, transverse (crosswise) cracks can be detected. This is done by energizing a coil around the bar. If an electric current is passed through the bar, however, a circular magnetic field results and the defects can be found that are lengthwise to the part (Figure 20.5). Because the part must often be demagnetized after testing, this feature is built into the machine. This system of inspection is limited to the magnetic metals such as iron and steel.

Magnetic-particle inspection is also applied to automated product lines. Steel billets are tested for inclusions and a scanning system marks them for later sorting and surface grinding. Automotive parts are similarly tested on a production line and then indexed past inspection and demagnetization stations.

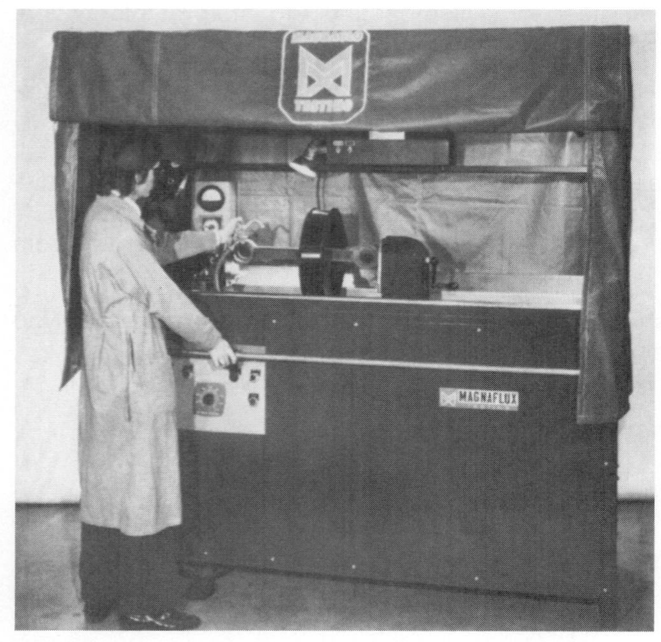

FIGURE 20.1. Magnaflux type "H-600" series machine. A versatile shop unit that will handle the requirements of all magnetic particle testing up to the capacity of the machine. (*Courtesy of Magnaflux Corporation*)

FIGURE 20.2. Front axle king pin for a truck as it appeared (left) under visual inspection, apparently safe for service; (center) with dangerous cracks revealed by Magnaflux; and (right) with the same cracks revealed by fluorescent Magnaglo, under black light. (*Courtesy of Magnaflux Corporation*)

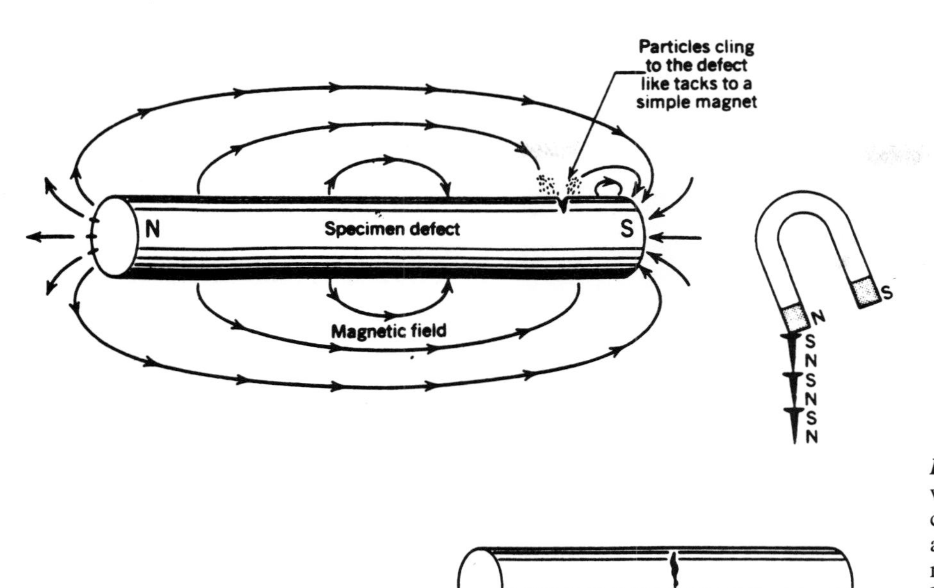

FIGURE 20.3. By inducing a magnetic field within the part to be tested, and by applying a coating of magnetic particles, surface cracks are made visible; in effect the cracks form new magnetic poles. Particles cling to the defect like tacks to a simple magnet. (*Courtesy of Magnaflux Corporation*)

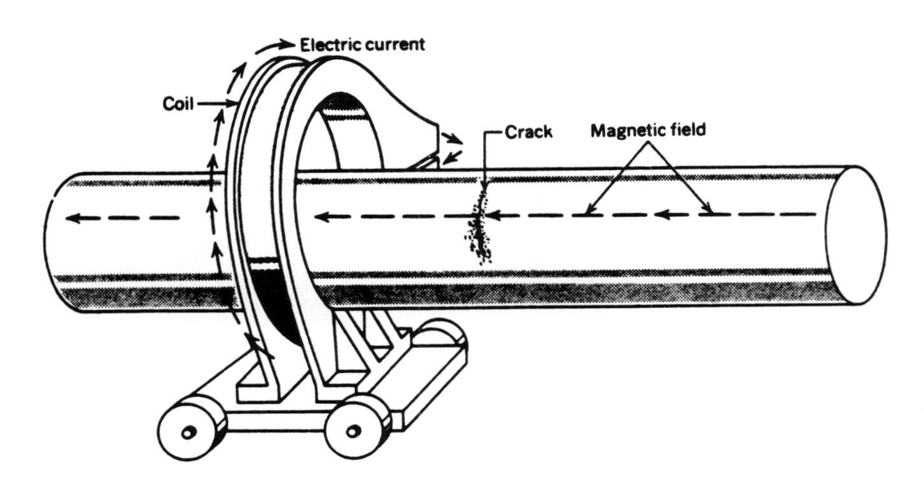

FIGURE 20.4. Longitudinal method of magnetization. (*Courtesy of Magnaflux Corporation*)

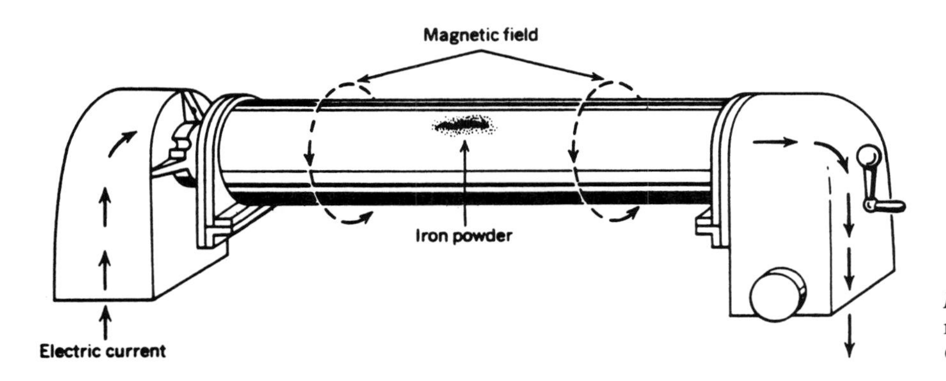

FIGURE 20.5. Circular method of magnetization. (*Courtesy of Magnaflux Corporation*)

FIGURE 20.6. Freshly reground carbide tipped tools as they appear; (left) to normal visual inspection, perfectly good; (center) test results with Zyglo penetrant; and (right) with supersensitive Zyglo Pentrex. All cracks, however small, are found. (*Courtesy of Magnaflux Corporation*)

FLUORESCENT-PENETRANT INSPECTION

Invisible surface cracks, porosity, and other surface defects are also found by this method in metals such as iron, steel, aluminum, bronze, tungsten carbide, and nonmetals such as glass, plastics, and ceramics. Fluorescent-penetrant inspection is widely used for testing and inspection on these materials. Zyglo, the Magnaflux Corporation copyrighted name for this method, is similar to Magnaglo in the use of black light to make the defects glow with fluorescence (Figure 20.6). Once applied, Zyglo penetrant is drawn into every defect no matter how fine or deep, but it must be given the time to do so. The length of time depends on the type of defect. The surface film is rinsed off with water, developer is applied to draw out the penetrant, and the part is inspected for defects under black light where cracks and other flaws will fluoresce brilliantly. Zyglo systems are available in many sizes, ranging from hand-portable test kits to huge automated systems.

FIGURE 20.7. Spotcheck is used to locate fatigue crack in punch press frame. (*Courtesy of Magnaflux Corporation*)

DYE PENETRANTS

A similar method of nondestructive testing using dye penetrants is visual inspection, but without the black light and fluorescent penetrant. As with Zyglo, it may be used on almost any dense material. This method, called Spotcheck by the Magnaflux Corporation, works as follows:

1. Surfaces must be clean and dry prior to soaking.
2. Dye penetrant is applied to the defect and allowed to soak. Soaking time should be sufficient to get the penetrant into fine cracks.
3. Remove excess dye penetrant, but do not rinse out cracks.
4. The developer is applied. Allow enough time for the developer to find very small cracks.
5. Inspection shows a bright colored indication marking the defect.

Spotcheck (Figure 20.7) is available in sealed pressure spray cans or for brush or spray gun application. Its advantages are portability for remote uses (Figure 20.8) or for rapid inspection of small sections in the shop, low cost, and ease of application.

ULTRASONIC INSPECTION

The pulse-echo system and the through-transmission system are two methods of ultrasonic inspection (Figure 20.9) that are used to check for flaws in metal parts. Both systems use ultrasonic sound waves (millions of cycles per second) in their testing.

In the pulse-echo system, a pulse generator produces short electrical bursts that activate a transducer/transceiver (a device that converts mechanical energy

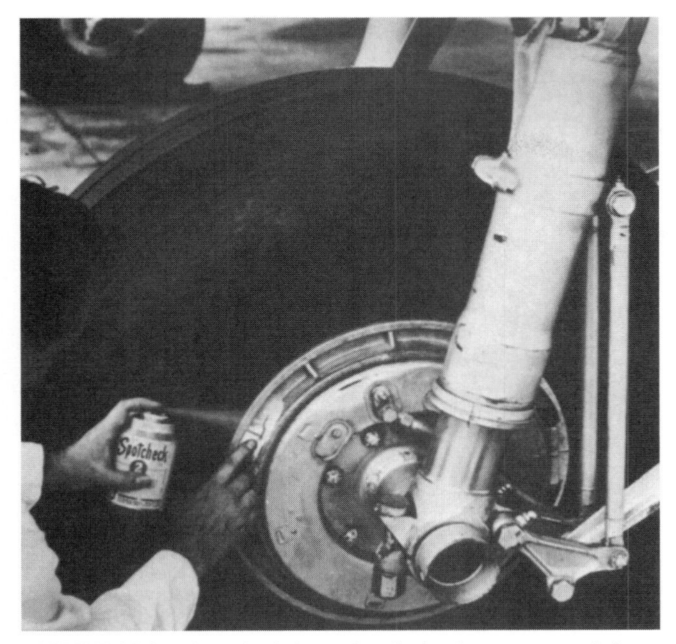

FIGURE 20.8. Spotcheck is used to find a dangerous crack in an aircraft wheel. (*Courtesy of Magnaflux Corporation*)

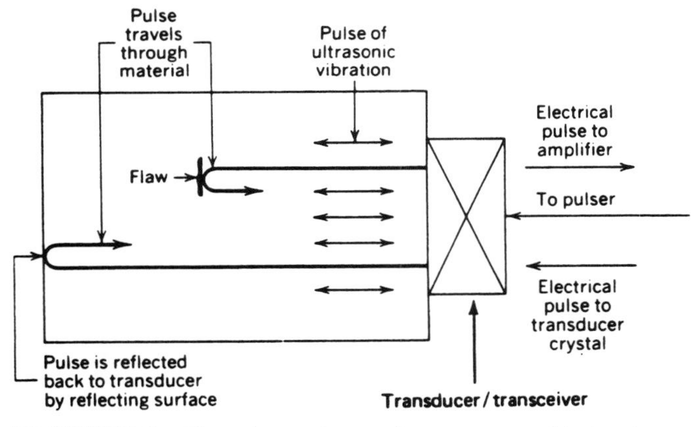

FIGURE 20.9. How ultrasonic sound waves are used to locate flaws in material. (*Machine Tools and Machining Practices*)

into electrical energy and electrical energy into mechanical energy) that is fastened to the metal being tested. In the case of immersion testing, no transducer contact is necessary, since the water acts as a connector. The signal pulse is seen as a pattern or "pip" on an oscilloscope screen when the sound wave enters the test piece (Figure 20.10). Part of the sound wave is reflected back when it reaches the other side of the material and shows on the oscilloscope as a second pip. If there is a flaw in the specimen, a smaller pip will be seen between the other two. Since the distance between the pips on the oscilloscope screen represents elapsed time of the reflected pulse, the distance to a flaw can be accurately measured. A trained operator is required to interpret the results.

The through-transmission method uses a transducer on each side or end of the test piece (Figure

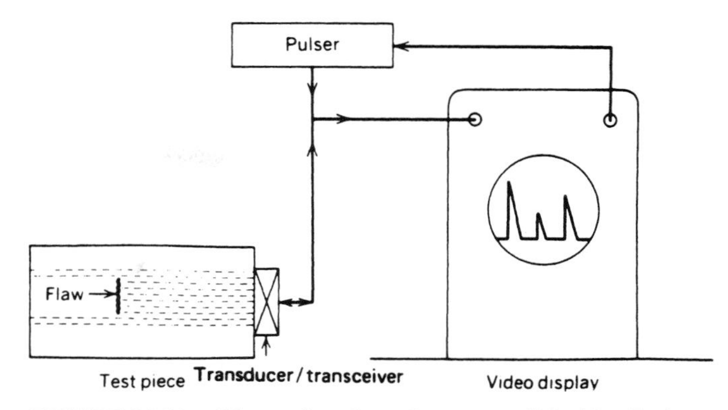

FIGURE 20.10. Ultrasonic pulse-echo system. (*Machine Tools and Machining Practices*)

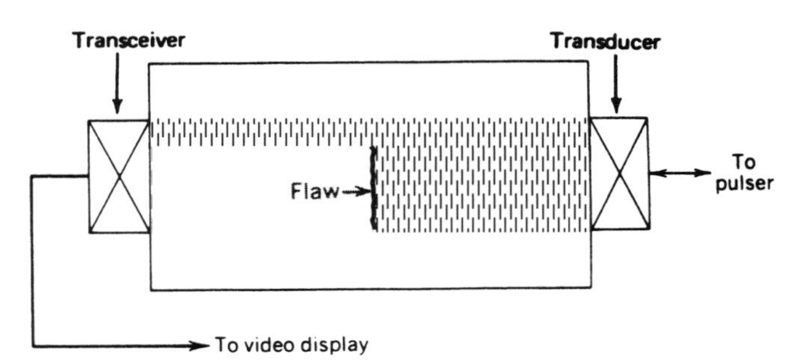

FIGURE 20.11. Through-ultrasonic transmission. (*Machine Tools and Machining Practices*)

20.11). The signal pulse enters the material at the transducer, travels through the material to be received by the transceiver, and is translated into another signal shown on an oscilloscope. If no flaws are present, a clear, strong signal will be seen on the oscilloscope, but if the material contains any flaws, a weaker or distorted signal will be seen. The reduced signal can be set up to actuate a bell or light to alert the operator when similar parts are being checked.

Ultrasonic inspection is used in production for inspection of manufactured parts and for testing structures such as pipelines and bridges. Internal defects, cracks, porosity, laminations, thickness, and weld bonds are tested in metal or nonporous nonmetallic structures (Figure 20.12). Some testing equipment is relatively light and portable (Figure 20.13), and some requires the part to be immersed in a liquid. Test results are immediate and accurate with this method. Tests can be made through long bars of steel or through thin sections that require testing on one side only. The reflecting surface must be parallel to the testing surface, however.

One of the most promising developments in ultrasonic inspection techniques is a hybrid system developed by researchers at Battelle's Pacific Northwest Laboratories in Richland, Washington. This system consists of linear holographic imaging and acoustic emission testing which is designed to monitor the growth of cracks

FIGURE 20.12. Portable ultrasonic machine instruments can be used for routine manual inspection under field conditions. This is very useful for structural weld testing and corrosion surveys. (*Courtesy of Magnaflux Corporation*)

FIGURE 20.13. The ultrasonic method results in swift inspection of critical weldments and dependability. (*Courtesy of Magnaflux Corporation*)

in remote locations such as offshore drilling platforms and nuclear reactors. Another possible use could be remote monitoring in aircraft, ships, or submarines. Also, ultrasonic testing at high temperatures is being developed to test hot-rolled ingots of steel to determine if any defects are so bad as to require scrapping or cutting out.

RADIOGRAPHIC TESTING

Radiographic testing utilizes the ability of X rays (Figures 20.14 and 20.15) or gamma rays (Figures 20.16 and 20.17) that are emitted from radioactive materials, such as radioactive radium or cobalt 60, to pass through solids. The test results are determined from a radiograph, which is a film exposed to radiation that has gone through the test materials. Shadows on the radiograph reveal defects, since the radiation will pass through a void, crack, or area of lower density at a greater intensity and will appear darker on the negative (Figures 20.18*a* and 20.18*b*). This testing method is widely used for forgings, castings, welded vessels including welds on pipelines, and corrosion analysis on pipelines and structures. Portable equipment is used for field inspection. Since there is a radiation hazard, only trained technicians should use the equipment. The test is quite sensitive and provides a permanent record. However, a new, low-intensity, portable X-ray device has been developed that

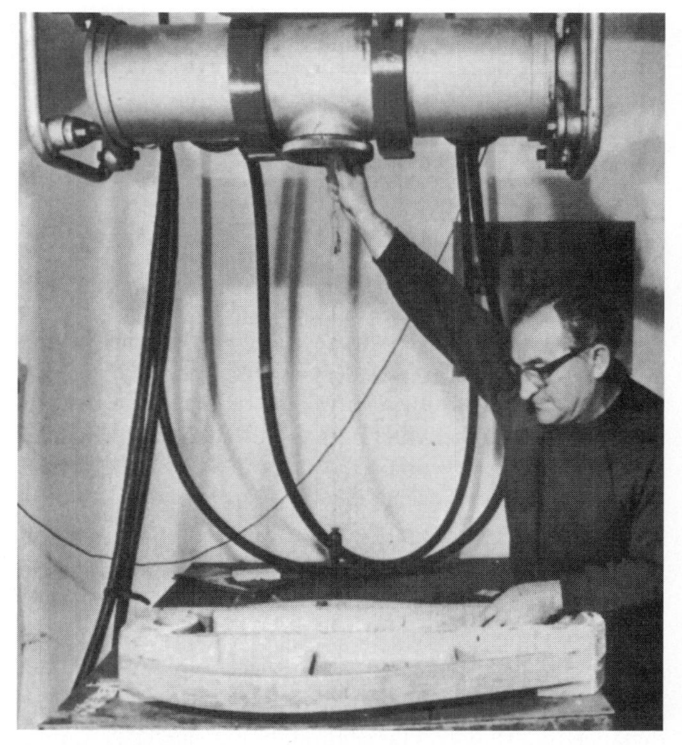

FIGURE 20.14. X-ray method of testing. A specialist makes a test of pilot run parts to evaluate manufacturing techniques and procedures. (*Courtesy of Magnaflux Corporation*)

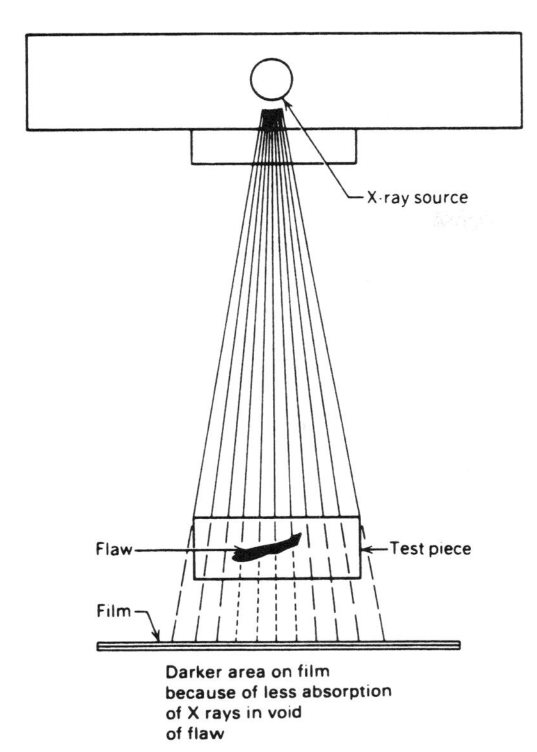

FIGURE 20.15. How X-ray inspection works.

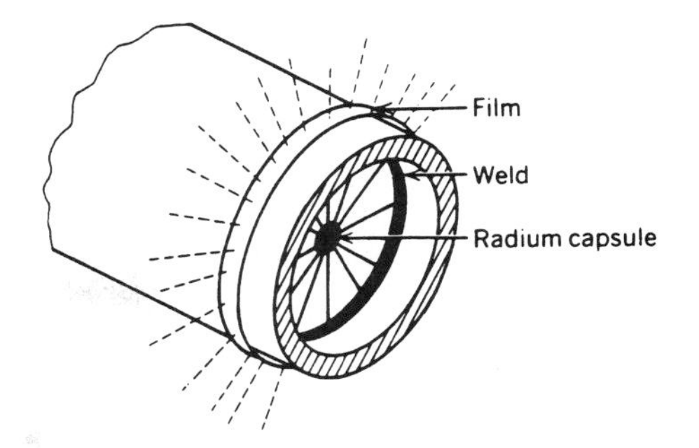

FIGURE 20.17. How gamma-ray inspection works.

*FIGURE 20.18***a.** This gear looks perfectly sound on the outside surface. (*Radiograph taken with a Hewlett-Packard Faxitron® Model 43804N Cabinet Radiographic Inspection System*)

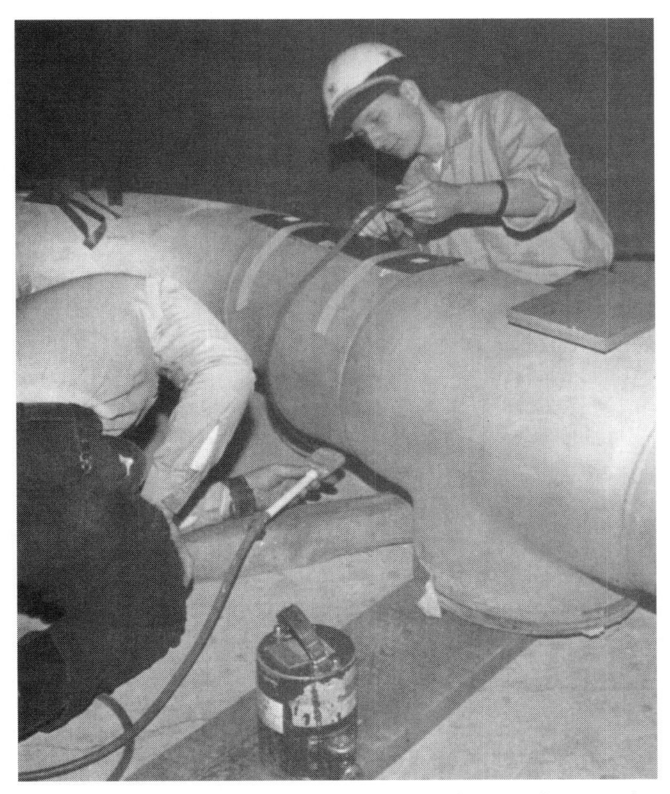

FIGURE 20.16. An experienced technician is preparing to make gamma-ray inspection of a section of pipe. (*Courtesy of Magnaflux Corporation*)

*FIGURE 20.18***b.** A radiograph of the gear in Figure 20.18*a* was made, and it revealed considerable porosity. The black spots in the darkened region (near the hub) are holes in the metal. Of course, this would weaken the casting; the part should be discarded. (*Radiograph taken with a Hewlett-Packard Faxitron® Model 43805N Cabinet Radiographic Inspection System*)

uses low-level radioactive isotopes. A person can operate this device for several years and incur no more radiation exposure than an annual X ray , according to the manufacturers. Objects can be viewed on a screen in real time (instantly) and can also be recorded on a Polaroid film. Portable X-ray analyzers are also used for verifying and sorting metals for alloy content and impurities. These devices are ideal for sorting scrap metals for producing high-integrity steel and nonferrous metals.

A radiation detector may be used to determine material thickness, since the radiation that passes through the test material decreases as the thickness increases. A moving, continuous strip of material can then be constantly monitored for thickness without any physical contact with the radiation source or detector.

EDDY CURRENT TESTING

Eddy current inspection can test only electrically conducting materials. An alternating current in a coil produces a corresponding magnetic field. Eddy currents, which flow opposite to the main current, are produced in the test material if the coil is placed in, near, or around it. The eddy currents then produce a change in impedance in the magnetic field, which is converted into voltage that is read on a meter or oscilloscope (Figure 20.19). This method is used for detecting seams, surface cracks, or variations in thickness and for sorting alloys in various compositions and heat treatments. Physical differences in mass, dimension, and shape can also be detected.

Eddy current testing has the advantage of being very rapid; it need not touch the specimen and can be used for automatic inspection. One limitation of this inspection method is the fact that eddy currents are present only at the surface of metallic specimens; defects at or near the center of parts or rods will not be detected.

The magnatest system used either 50 or 10 Hertz (Hz), which are the most effective frequencies for magnetic testing (Figure 20.20). In the absence of a test object, the secondary signal pickup coil delivers a zero voltage output signal to the instrumentation because the no-load voltage-compensation transformer removes the empty-coil voltage signal from the output to amplifier. When a test object is placed in the coil assembly, an induction signal due to the presence of the test object is superimposed on the empty-coil voltage signal that is picked up by the secondary coil. Thus, an indication is read on the instrumentation.

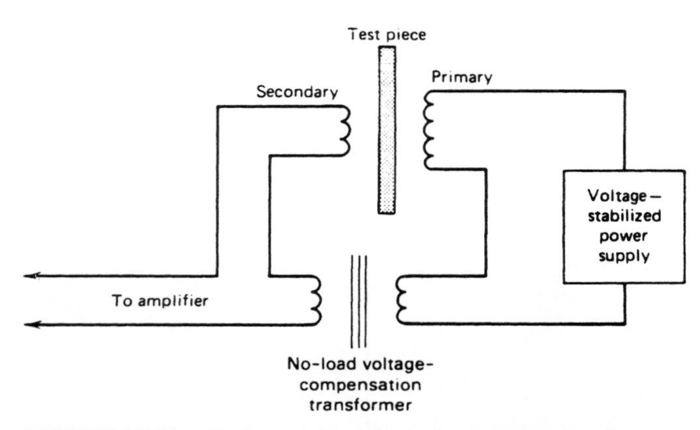

FIGURE 20.20. Basic circuitry of a no-load-compensated single-cell electromagnetic test system.

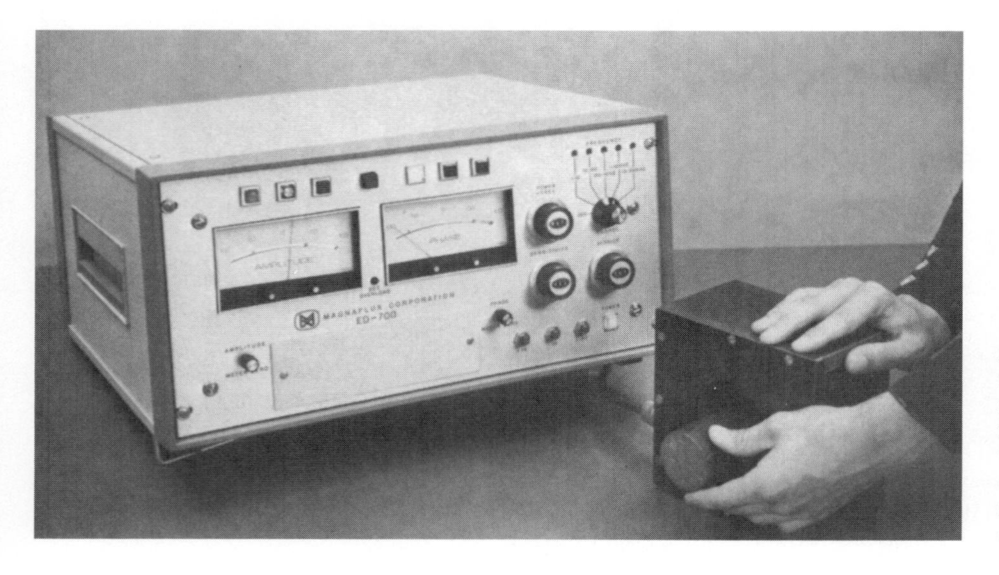

FIGURE 20.19. Magnatest ED-700 for nondestructive comparison of magnetic materials. Used to separate metallurgical properties as alloy variations, heat-treat condition, hardness differences, internal stress, and tensile strength. (*Courtesy of Magnaflux Corporation*)

SELF-EVALUATION

1. What is the purpose of nondestructive testing?
2. Name a limiting factor of magnetic particle inspection.
3. How must the part be magnetized to form new magnetic poles on a crosswise crack? On a lengthwise crack?
4. Describe how fluorescent penetrant works to reveal defects. On what materials can this method be used?
5. What are some of the advantages of using a dye penetrant method (without black light)?
6. Name the two basic systems of ultrasonic testing. Briefly explain their differences.
7. What can be determined in a specimen by the use of ultrasonic inspection systems?
8. How can X rays help us to detect flaws in solid steel or other metals?
9. In what way does gamma-ray inspection differ from X-ray inspection?
10. Explain some of the uses of eddy current inspection. To what metals is it limited?
11. Is nondestructive testing used exclusively on metals?
12. Are some mechanisms not subject to failure and therefore do not need to be tested? Would the effects of a catastrophic failure be a factor in making a decision whether to make periodic NDT tests?

CASE PROBLEM: THE CRACKED CRANKSHAFT

A small aircraft was in a hanger for routine inspection. The engine was dismantled and the parts checked for wear and fatigue cracking. Somehow, the crankshaft was overlooked and it was installed. Later the engine failed but the pilot was able to land safely. Metal fatigue had been found in the corner of one of the throws of the crankshaft and it had finally failed in service. What would have been the best method of detecting this potentially disastrous crack in the crankshaft had it not been overlooked?

WORKSHEET

Objective Inspect for cracks or other flaws by using any nondestructive testing equipment or method available.

Materials Three different mechanical parts that have been in rough service, but appear sound otherwise.

Procedure Check to see if your shop has a fluorescent penetrant, dye penetrant set, magnetic inspection equipment, or any other type of nondestructive testing equipment. Choose the type of inspection available to you and obtain the operation instruction manual. Follow the step-by-step instructions for testing the three specimens.

Conclusion List your results.

	Specimen 1	Specimen 2	Specimen 3
Type of material			
Kind of test			
Type of flaw			
Location of flaw			
Service condition (Safe or unsafe for use)			

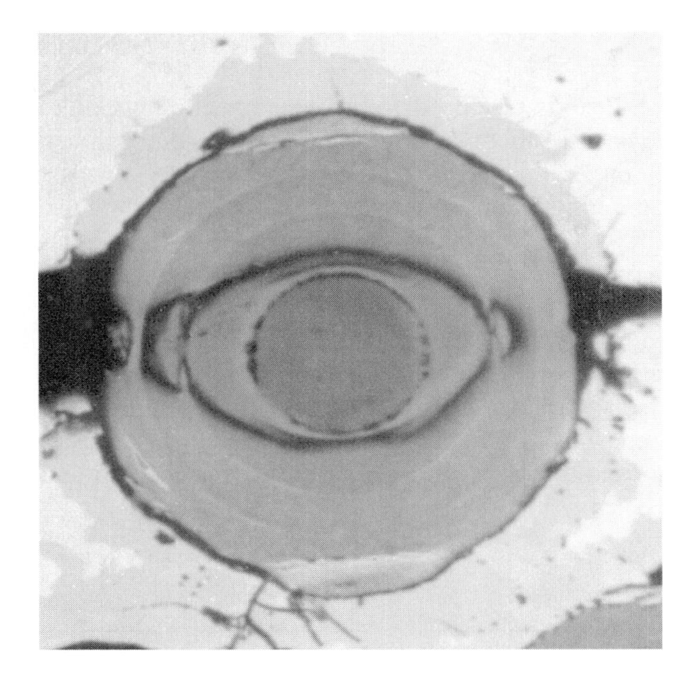

CHAPTER 21

Composite Materials

The development of composite materials, both metal matrix and graphitic or carbon-carbon composites, during the past 30 years has opened a vast new world of material combinations, exhibiting dynamic and unusual physical and mechanical properties not possible to obtain with ordinary metals and alloys.

OBJECTIVES **After completing this chapter, you should be able to:**

1. Describe the characteristics of metal matrix composites.
2. Prepare metallographic samples of both whisker shown in F21.3, and fiber-reinforced composites.

METAL MATRIX COMPOSITES

Composite materials have been in use for a long time as golf club handles, tennis rackets, fishing rods, electrical contacts, and fiberglass (a composite mixture of fine glass fibers held together by resin). The purpose of combining a high-strength fiber with a mechanically weak matrix material is to force the combination to share mechanical properties. The intimate contact between the high tensile strength glass fiber in fiberglass and the soft ductile resin causes a sharing of mechanical properties (see Figure 21.1). The combination produces a man-made product known as fiberglass which resists impact, holds form, and is structurally superior in mechanical strength when compared with the stand-alone mechanical properties of the fiber or the resin matrix.

In the early 1960s, the aerospace industry became interested in the development of composite materials because of the demanding conditions of supersonic flight and rocket propulsion. The combination of extremely high strength filaments such as boron-tungsten core, boron-silicon core, and boron-carbon core filaments added a new dimension in rigid fiber-reinforced composite manufacturing (Figures 21.2*a,* and 21.2*b*). At about the same time, the automotive industry found a need for highly abrasive-resistant materials for use as cylinder seals in production engines (Figure 21.3).

Metal matrix composites find great use in the abrasive cutting industry where silicon carbide and diamond particles held by a soft metal are deposited on a steel blade and used to cut everything from ceramic tiles and building bricks to concrete and asphalt (see Figure 21.4).

Pistons produced from hypereutectic aluminum alloys with silicon content at or above 18 percent also fall under the guise of a composite material. Cast aluminum-silicon hypereutectic alloys are finding use as brake rotors and as cylinder liners or sleeves in automotive engines. In these applications, heat is eliminated by the excellent thermal properties of the aluminum matrix; and wear, in contact with special piston sealing rings, is prevented by the abrasive-resistant silicon carbide particles. Aluminum-lithium alloy composites exhibit extremely high strength and are being used to cover the outer surfaces of spacecraft and missiles.

GRAPHITE FIBER METAL MATRIX COMPOSITES

When lubricity and electrical conductivity are of prime importance, graphite fiber metal matrix composites are often used. Extremely fine graphite fibers are metallic coated by a vapor deposition process or by immersion in the liquid metal, followed by extruding and hot pressing into bundles. Bundles can be individually shaped and/or thermally combined to form a block or billet of composite material. The result is a fiber-reinforced metal composite which may be machined into product or further processed by hot rolling as a wrought product. The optical microstructure of some graphite fiber-reinforced metal matrix composites are shown in Figures 21.4*a* through 21.4*c*.

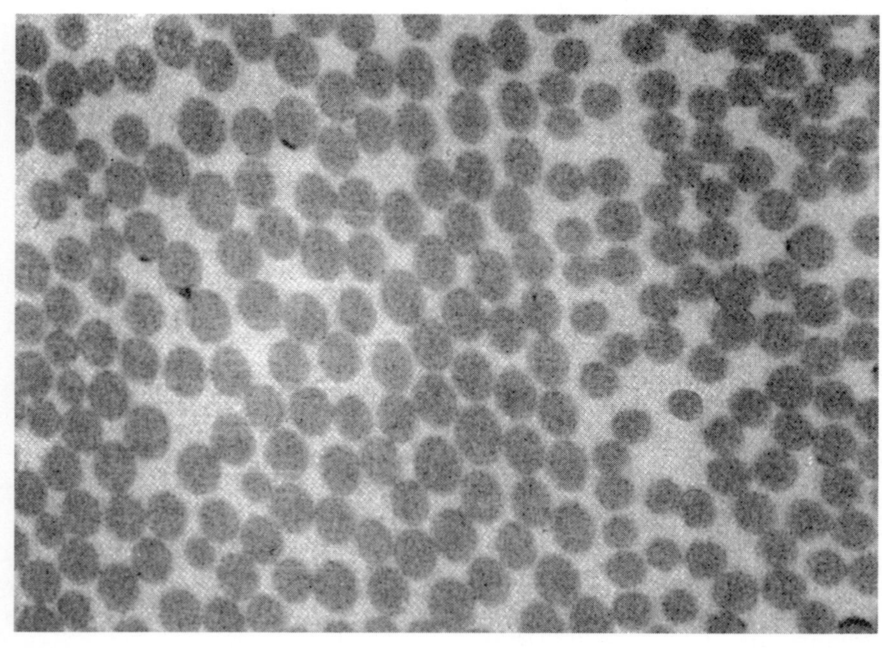

1000× As-polished

FIGURE 21.1. Glass fibers in a resin matrix—fiberglass.

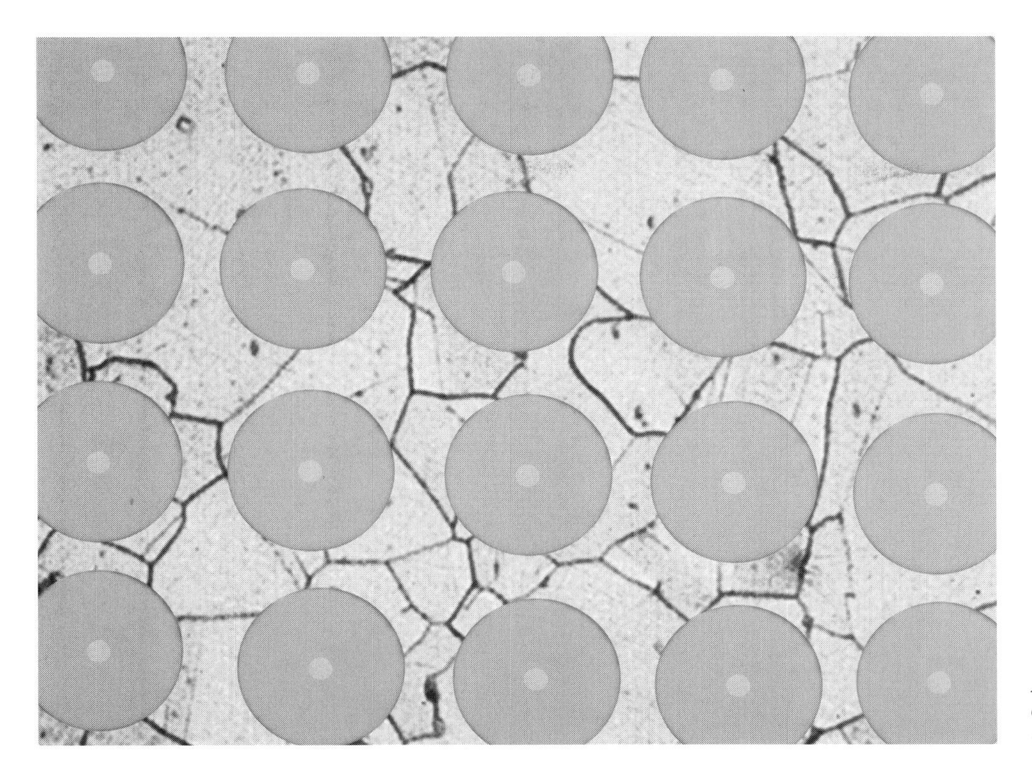

FIGURE 21.2a. Boron-tungsten core in nickel matrix at ~500× (etched).

FIGURE 21.2b. Boron carbon core filaments in an aluminum matrix.

500×

METALLOGRAPHY OF COMPOSITES

The metallography of hard fiber metal matrix composites presents an unusual challenge for the metallographer. The abrasive resistance of the hard fibers is extremely high in comparison with graphite fibers. Sectioning must be performed using a diamond-impregnated saw with minimal contact pressure against the composite being cut. Also, the composite must be cooled by water or a light oil during cutting. Any bending or impact damage caused by cutting will not be easily removed during rough grinding of the mounted composite, so proceed slowly. Once cut, wash and dry the composite. Examine the cut surface using a low-power macroscope and carefully remove any burrs that remain from cutting. Most composites will be mounted in a

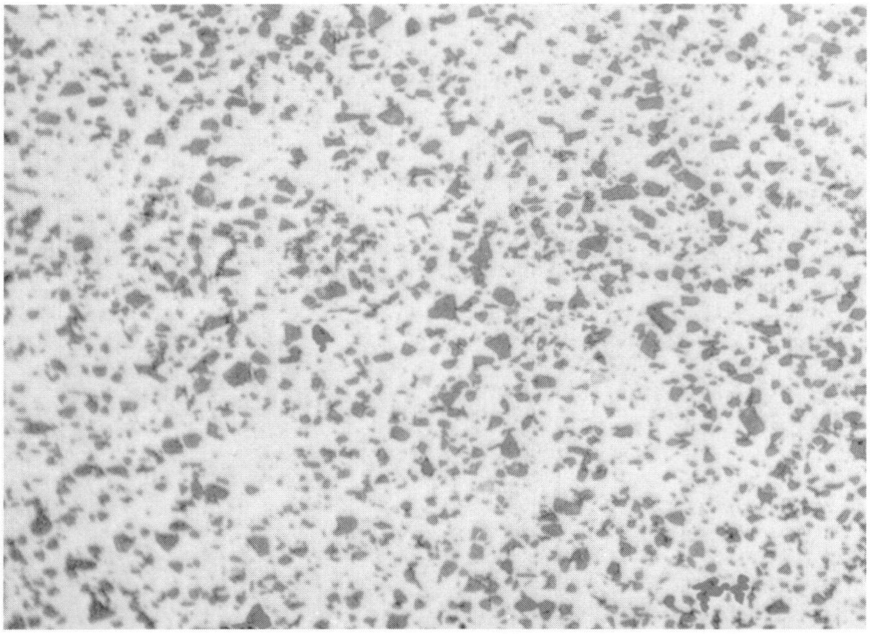

FIGURE 21.3. The tiny silicon carbide particles are distributed throughout this aluminum matrix composite. ~500×

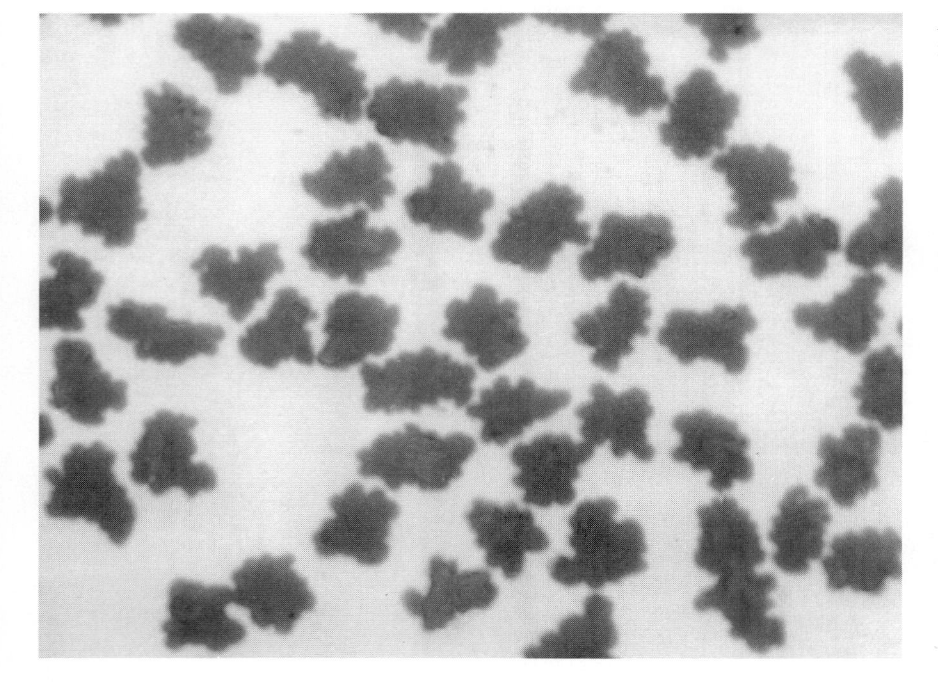

FIGURE 21.4a. Leaf-shaped graphite fibers in copper matrix. ~2000×

clear resin which takes a few hours to cure to a clear, hard consistency.

Rough grinding of graphite fiber-reinforced composites is performed wet, using water or WD-40 as a lubricant, on 600 grit paper until the level of interest is reached. The 600 grit grinding is done by hand, sliding the specimen back and forth. Hand grinding minimizes bending strain and/or movement of the fibers during grinding. A one-step final polish to a scratch-free finish suitable for examination by optical or scanning electron microscopy is easily obtained by rotary polishing with medium pressure on a soft cloth using artificial 1 micron diamond paste with WD-40 as a lubricant. Polishing only takes about 20 seconds. This metallographic sample preparation technique is also used for carbon-carbon, metallic particle impregnated graphitized composites, and nearly all other metals (see Chapter 26).

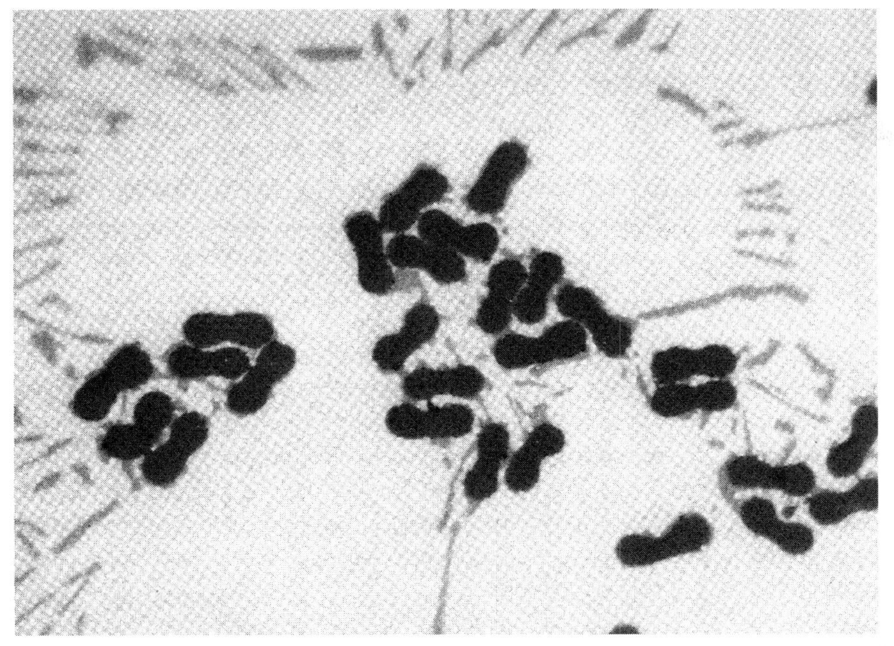

FIGURE 21.4b. Dog-bone shape graphite fibers in an aluminum alloy matrix.

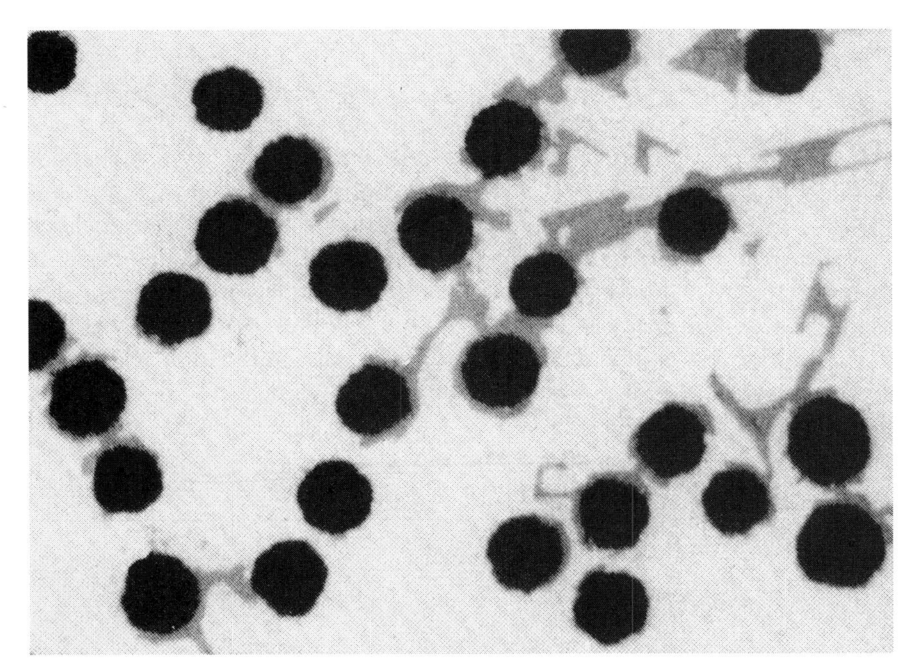

FIGURE 21.4c. Round graphite fibers in an aluminum alloy. The gray phase is a silicon compound.

MECHANICAL PROPERTIES

Certain mechanical properties for matrix metals and alloys, such as aluminum, copper, nickel, and titanium, are relatively low when compared with their composite counterparts. Fibers and filaments used in composites have extremely high tensile strength and a modulus of elasticity approaching 500 million psi. Steel has a modulus of about 30 million psi. A high modulus of elasticity produces the effect of stiffness and springlike behavior. However, common metallic properties—ductility, fracture toughness, shear strength etc.—are relatively low in fibers and filaments. When combined with metals, the chemical-mechanical interface between fiber and metal matrix allows sharing of mechanical properties. Thus, the best of both worlds is achieved from a strength standpoint.

During the early stages of fiber composite development, total chemically and mechanical joining of the fiber-matrix interface was difficult to achieve. The silicon-carbon-nickel matrix composite on the first page of this chapter suffers from several process-related problems. The filaments, although very strong, chemically reacted to form an intermetallic compound (gray irregular formation appearing to flow from the filament into the matrix nickel). Although time at temperature during the process of metallic coating was kept to a minimum, extensive filament dissolution occurred. During the second stage of processing, metallic layers containing individual filaments are sandwiched together and thermally bonded. Once again, bonding was incomplete, probably due to insufficient cleaning. Fiber-reinforced composite fabrication techniques are among the most guarded industrial secrets due to the numerous problems encountered. The objective is to assemble the composite fibers in a manner in which there is both mechanical and chemical bonding between fiber and matrix metal. When too much intermetallic is formed, the mechanical strength of the interface is compromised and is often brittle rather than elastic.

COMPOSITES FOR ELECTRICAL CONTACTS

Most electrical composites are produced by powder metallurgy processing. They are classified into three groups: carbide base and refractory, silver base, and copper base. The first of the manufacturing processes is called **infiltration,** which is used for carbide base and refractory composites. Metal or carbide powder is first blended into the desired composition, compacted into a form called a skeleton, and furnace sintered. The porous skeleton is infiltrated with copper or silver, producing the most densified composites, approaching 97 percent density. The contact is sometimes chemically etched to improve performance and corrosion resistance following impregnation, so that only pure silver appears on the surface.

Composite electrical contacts are also produced by a press-sinter process, a press-sinter and re-press process; by a press-sinter-extrude process, by a preoxidize-press-sinter-extrude process, and by a process known as coprecipitation. Each of these processes requires blending or mechanical mixing of powders, compacting them into a skeleton form, sintering or heating to a temperature very close to the melting point of the lowest melting point material, performing impregnation, and processing the composite through extrusion dies as a compacting process, or by chemical precipitation to densify the composite.

In general, composites for electrical use have unique requirements and properties. It must have (1) high electrical conductivity to minimize the heat generated during passage of current, (2) high thermal conductivity to dissipate the resistive and arc heat developed, (3) high reaction resistance to chemical changes resulting from use in harsh environments, and (4) immunity to arcing damage caused by the making and breaking of electrical contact. The melting point of the composite must be high enough to minimize arc erosion, metal transfer, cold welding, or thermal welding and/or sticking. Hardness must be high to provide good wear resistance and extended contact life.

SELF-EVALUATION

1. Explain what is meant by a "composite material."
2. Name some composite materials you might find at home.
3. What mechanical properties are important when evaluating fiber-reinforced composite materials?
4. How do fibers in composite materials interact with the base metal?
5. What are some of the problems in preparing metallographic cross sections of hard fiber reinforced composite materials?
6. What is a whisker (see Figure 21.3) enforced composite metal?

CASE PROBLEM: CAN I FIX THAT CRACK?

An angry golfer threw his expensive composite driver in the air after the ball missed the green. The club landed on a concrete walkway. When it was retrieved, the golfer found the shaft had a longitudinal hairline crack near the middle of the shaft. He decided to repair the shaft at home and used a super-glue to cover the crack. Do you believe this method of repair will work? Explain your answer.

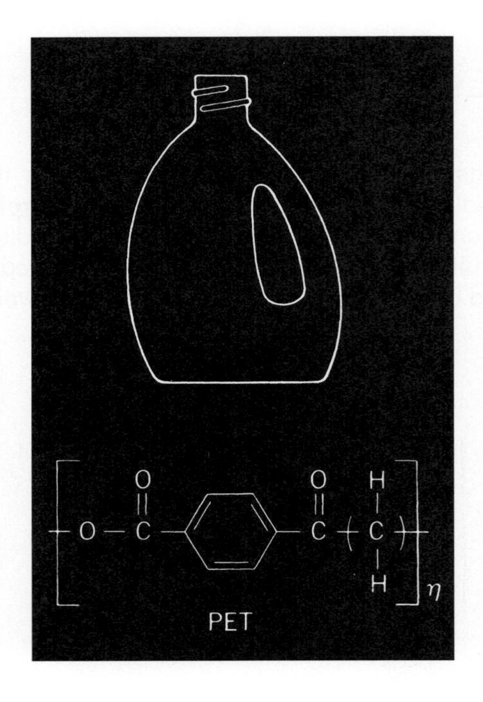

CHAPTER 22

Plastics and Elastomers

In Chapter 4 the atomic and crystalline structure of metals was discussed. You learned there about bonding of atoms in solids and in metals. Some different structures, such as crystalloids and colloids (not found in metals), may be found in nonmetals. Some organic materials, called polymers, can exhibit several different kinds of microstructures, and these are the basic structure of plastics. Elastomers, that is, rubbers and rubbery materials, are also made of polymers or long-chain molecules, which for the most part are derived from petroleum. Raw materials other than petrochemicals used in plastics manufacture are coal, water, limestone, sulfur, and salt. All of these are quite plentiful and inexpensive. As petroleum becomes scarcer and more expensive, it may be replaced to some extent in the future for plastics manufacture by certain latex and oil-producing plants, some of which are found in the desert in the southwestern corner of the United States.

From the first piece of bone or stone used for a tool or lump of clay for a container, materials have shaped the future. Today it is plastics that will shape the future in the manufacture of high-technology and general consumer products. Plastics have lost their "cheap" image and are now formulated and tailored for the precise requirement of a given product. The fastest growing market for new plastics is in building construction because in that field plastics have the highest potential for the replacement of conventional materials. Many of these plastic replacements are superior to the ones they replace. Some of these products are siding, pipe and conduit, plumbing fixtures, windows and doors, and floor materials.

Automobiles will be much lighter in the future as manufacturers gradually replace metal and glass with plastics. Of course, steel will still be necessary in many areas because of its greater strength. So-called plastic engines have been designed, but pistons, crankshafts, and other high-stress and high-temperature parts are still made of metal.

Many of these high-strength plastics used in aircraft and automobile manufacture are composites, that is, fiber-reinforced plastics. These are bonded with high-strength adhesives. One disadvantage of adhesives in the past for automobile manufacture is that adhesives take time to cure. Cars must be built on the assembly line at the rate of about one per minute, and welds have no cure time. New, fast bonding methods have been developed so that plastic auto bodies can be assembled competitively.

One example is a single-part epoxy that can be used to fasten sheet metal by using induction heating to speed the adhesive cure. One new method is claimed to have a 4-second bond. Complete cure then takes place in the paint oven.

In this chapter, many of these new materials will be presented. We will cover their chemical structures, terminology, and uses.

OBJECTIVES **After completing this chapter, you should be able to:**

1. Explain the chemical structures of several plastic materials and the reason for their particular behavioral characteristics.
2. Describe the process by which a sticky substance, such as latex, can be made elastic and resilient.
3. Identify kinds of plastics and rubbers and some of their uses.

PETROCHEMICALS

Many of today's plastic and rubber materials are derived from petroleum distillates, often called petrochemicals. Before 1920, gasoline and kerosene were obtained by the distillation of crude oil in a retort. The lighter elements, called naphthas, were considered useless and were burned off to get rid of them. Later, it was discovered that many useful chemicals could be derived from these lighter elements. This was closely followed in the 1930s by the discovery of the process called catalytic cracking. This process uses heat to induce a chemical reaction with the help of catalysts, usually composed of refractory oxides of aluminum, silicon, and magnesium. High-temperature boiling components of petroleum are broken down into lower boiling point components that are suitable for gasoline. The remaining gaseous material is made up of compounds having from one to four atoms. Among these are ethylene, propylene, and the butylenes. Natural gas, from which the liquefied gases butane and propane are produced, is also a source of propylene, ethylene, hydrogen, and methane through the process of thermal cracking.

These petrochemicals can be formed into strong, tough plastics or rubbery materials through the amazing process of chemical synthesis.

Acetylene gas, made from the hydration of calcium carbide, is also a source of plastic and rubber materials. Calcium carbide is a synthetic substance, produced in an electric furnace from carbon (in the form of coke from coal or petroleum) and calcium oxide. Acetylene is also synthesized from natural gas by the high temperature cracking of natural gas. Probably the greater amount of acetylene gas used today is produced from methane rather than from calcium carbide. From acetylene many products have been derived, but the most important of them are chloroprene and polychloroprene (or neoprene). Neoprene was the first synthetic rubbery material to become commercially successful.

From ethylene is derived ethyl alcohol, tetraethyl lead (a gasoline additive), ethyl benzene, styrene and polystyrene, ethylene glycol (antifreeze), and polyethylene (from which plastic squeeze bottles are made). Polyesters are also a derivative of ethylene. Ethylene glycol forms a methyl ester from which polyester is subsequently made. It is a polyester from which fibers are made for clothing (Dacron®), magnetic tape for recording (Mylar®), and many other products.

Polystyrene is one of the most important industrial plastics today. It is a clear, glasslike product that dissolves in many solvents and is rather brittle. This thermoplastic is used where its clear, transparent qualities are useful, as in window panes and moldings. Polystyrene, although useful for many products, is also made into a foam usually with carbon dioxide gas, making a strong, lightweight material known as styrofoam, which is used for insulation and packaging.

Cellulose is found in all plants but is in its purest form in cotton. It is a polymer and is very useful in the form of cellulosic textiles such as rayon. When cellulose is treated with nitric acid, cellulose nitrate is formed. This product is well known as guncotton or nitrocellulose. It can be used as an explosive or as a lacquer. The plastic called celluloid is made by compounding nitrocellulose with camphor and is used for lacquers in auto painting. Plasticizers are added to give it more resiliency.

Epoxies are very strong but brittle thermosetting plastics. They may be cured with heat or at room temperature. Epoxy resins may be hardened by combining with a catalyst or they can be single-component epoxies. These versatile plastics are used as adhesives or are fiber reinforced for added strength. They may be cast in a mold for tooling, jigs or dies, or for potting electrical hardware. Epoxies bond to almost any material and are used to make a durable surface on concrete floors or to bond materials to concrete. They are also used extensively as household adhesives. Table 22.1 lists some of the products that are derived from the petrochemical monomers.

TABLE 22.1. Products from petrochemicals

Products	**Petrochemical Monomer (Raw Materials)**
Antifreeze fluid, Mylar®	Ethylene glycol
Butyl rubber	Isobutene
Epoxy resins, polycarbonates (Lexan®), acetone, phenol, bisphenol-A	Isopropylbenzine
Ethylene glycol antifreeze, Mylar®, Dacron®, ethyl alcohol	Ethylene
Mylar poly-para-xylene (Parylene®)	Xylenes
Neoprene rubber	Chloroprene
Nylon synthesis, neoprene, polybutadiene rubber (source)	Butadiene
Orlon: chloroprene and neoprene rubber	Acetylene
Phenol, phenolic resins, styrene, polystyrene	Benzene
Phenolic plastics	Toluene
Phenolic resins, epoxy resins	Phenol
Polypropylene, isopropylbenzine, isopropyl alcohol	Propylene
Polystyrene	Styrene
Polystyrene (nonsoluble)	Divinylbenzene
Production of acetylene	Methane
Synthetic rubber	Isoprene

ORGANIC MATERIALS

Organic materials such as cotton, wood, plastics, rubber, and resins are typically **polymers.** Polymers are made from small molecules called **monomers,** and the monomer molecules are the building blocks from which the polymer chain is built (Figure 22.1). These building blocks of materials derive their names (monomer and polymer) from the Greek *mono* (one), *poly* (many), and *meros* (parts). Polymer chains are monomer cells that are combined to produce these strong, versatile materials (Figure 22.2). **Polymerization** is a chemical process in which many molecules are linked together to form one large molecule.

It is these long-chain molecules (often called high polymers) that impart the properties of resiliency and elasticity to the organic substances we call plastics. Early discoveries in organic chemistry that produced substitutes for natural materials were not widely accepted and were considered poor replacements. For example, in 1866 John Wesley Hyatt first produced a synthetic material he called **celluloid** by combining camphor and nitrocellulose (guncotton). This new product was used as a substitute for ivory, which comes from the tusks of animals. Soon products such as combs and billiard balls were made of this new synthetic material. Celluloid had two bad qualities: It had a tendency to burn rapidly when ignited and it cracked and became discolored with age. Celluloid was not the first polymerized plastic, since the basic component of cellulose nitrate is cellulose, which in cotton or wood is already a high polymer.

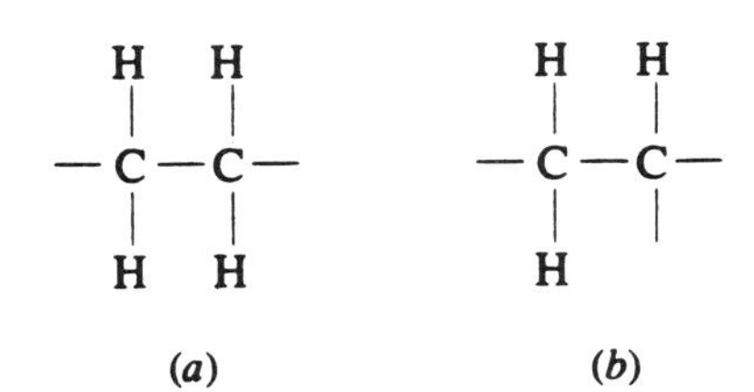

FIGURE 22.1. A monomer unit (*a*) can become a branch source unit (*b*) if it loses one or more atoms.

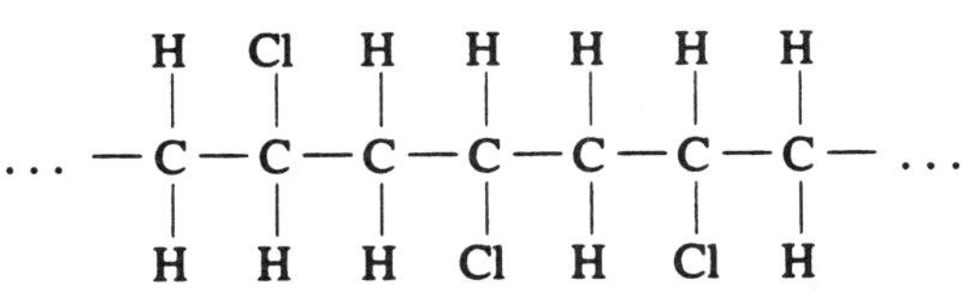

FIGURE 22.2. Giant-chain molecules may be formed from monomers as in Figure 22.1*b* by combining other atoms such as chlorine (Cl) to the carbon and hydrogen atoms to produce rigid polyvinyl chloride (PVC) as shown here.

Soluble substances can be divided into two classes, the **crystalloids** and the **colloids.** The crystalloids, such as table salt and sugar, when dissolved in water easily pass through a semipermeable membrane, but colloids are thick, viscous substances, like table jelly or gelatin, which will not pass through a semipermeable membrane. This colloid behavior is seen in the synthetic plastics and rubber. A knowledge of colloid chemistry is necessary for any in-depth study of plastic materials.

STRUCTURE OF PLASTICS

Synthetic materials, commonly known as plastics, have become so widely used as to replace other materials in manufacturing and construction. The only materials used in greater quantities are steel, concrete, and paper. The plastics are organic materials of which the basic element is carbon; the other elements that comprise the chemical structures of various plastic materials are oxygen, hydrogen, chlorine, and fluorine. Sodium is used in inomers and silicon in silicones.

The first high polymer to be made from small molecules of raw materials was the strong, hard plastic called bakelite, which was developed by the Belgian chemist Leo Baekland in 1908. He was able to join the small molecules of phenol and formaldehyde to produce large, long-chain molecules (Figure 22.3). Bakelite, a thermosetting plastic, is still widely used, especially for electrical components.

Types of Plastics

The three basic types of polymerization are **addition, copolymerization,** and **condensation.** Addition polymerization (also called linear polymerization) occurs when similar polymers join to form a chain (Figure 22.4). Polyethylene is one example of a linear polymer. These linear polymers are called **thermoplastics,** meaning that they can be softened or melted by heating. The other class of plastic is called **thermosetting,** because the plastic is set or hardened by heat during the molding operation and cannot be resoftened. Oxygen is usually used in thermosets as a cross-linking agent, which hardens the material and greatly reduces the mobility of the long polymer chains. Therefore, thermosets are more brittle than the thermoplastics. The combining of two or more different kinds of monomers is called copolymerization; this is analogous to the alloying of two or more metals. Many engineering (high-strength) plastics are copolymers. ABS is a copolymer of acrylonitrite, butadiene, and styrene. It is a very

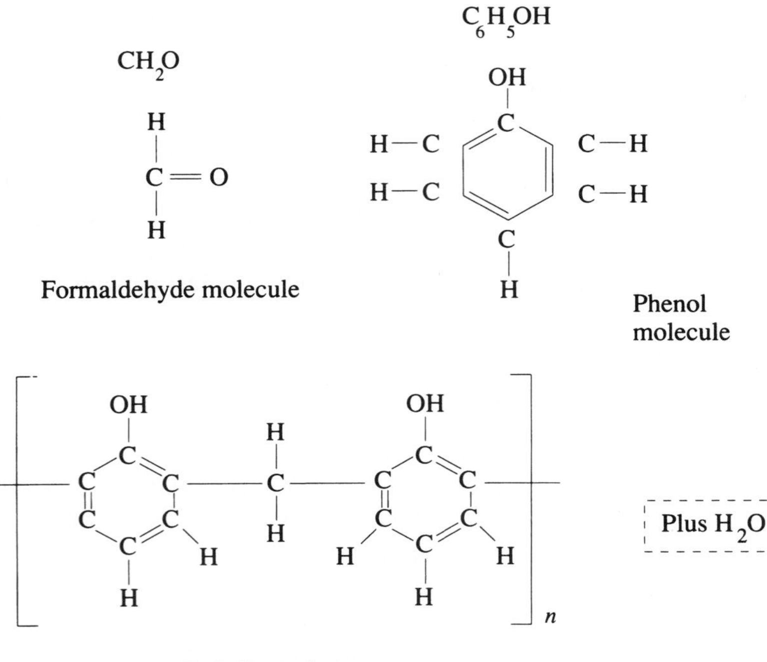

FIGURE 22.3. The basic chemical structure of bakelite as it is produced from formaldehyde and phenol monomers. (*John Neely and Richard Kibbe, Modern Materials and Manufacturing Processes, Copyright 1987, John Wiley & Sons, Inc.*)

tough plastic. Engineering plastics incorporate additional elements besides carbon into the polymer chain to improve their properties. One of the first of these was nylon, a polyamide. Condensation polymerization is a process of combining long-chain molecules to form more complicated chains. These chains when coiled exhibit plasticity, but when they are cross-linked, they lose plasticity and become elastic. Rubber is an example. Some plastics such as polyethylene exhibit definite crystalline microstructures, whereas others such as polystyrene, with their long chains, are amorphous like glass, having no apparent order.

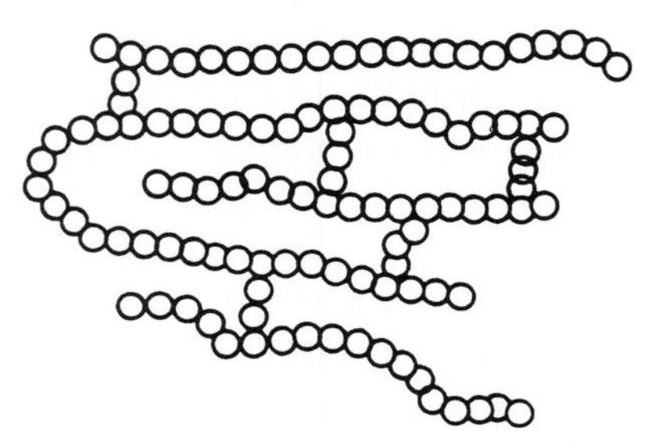

FIGURE 22.4. The spheres in this long-chain molecule represent atoms or monomers. This cross-linking is only one of several ways that polymers can be strengthened. Some chains are rigid, as found in heat-resisting materials, whereas others are elastic as in polyethylene, nylon, and cured rubber.

Characteristics of Plastics

Plastics have several general characteristics in common. Of all the engineering materials, they are the lightest in weight. Since they have a relatively low thermal conductivity, they are all relatively good heat-insulating materials. Nearly all plastics are good electrical insulators, but a recently developed polyacetylene plastic is a relatively good conductor of electricity and shows great promise for plastic wire, batteries, and motors. Most plastics can be obtained in a wide variety of colors, transparent or opaque. Plastic products can usually be produced at a fairly high production rate, depending on the process used.

Each year, hundreds of new plastic blends and alloys are developed and patented, most of them having difficult-to-pronounce names. Simply keeping track of these changes could be a full-time job for a plastics design engineer. However, plastic materials manufacturers provide technical information and technicians to keep the product manufacturer updated. Chemical names of plastics are normally used because trade names give no information concerning the type of plastic involved.

One problem associated with our widespread use of plastics is that of waste disposal and recycling. Unlike metals, a very small percentage of plastic is recycled at the present time. Most plastics are not biodegradable and they will last almost forever. New, short-life plastics have been developed that deteriorate when exposed to sunlight. They are now being used in packaging for food items.

Plastics are relatively corrosion resistant but they can be used only within a limited temperature range that is less than 550°F. If they are burned, they often produce toxic compounds, which is a concern in their use as a construction material.

FABRICATION OF PLASTICS

Several different methods are used to form plastic materials into desired shapes. Several methods of forming plastics are shown in Figures 22.5 to 22.14.

Hand layup (Figure 22.5) is the simplest method of forming plastic resins into desired shapes. Often, a glass fiber mat is impregnated with a resin such as epoxy and then draped over a form or placed in a mold and allowed to harden.

Thermoforming (Figure 22.6) is a method of making a sheet of plastic conform to the shape of a mold by softening it with heat.

Injection molding (Figure 22.7) is a process in which molten plastic is heated and forced into a mold by injection under pressure. The molded part quickly solidifies, and the mold is opened to eject the part. The mold is then closed and the cycle continues. A very large percentage of the plastic items we use every day are made in this manner.

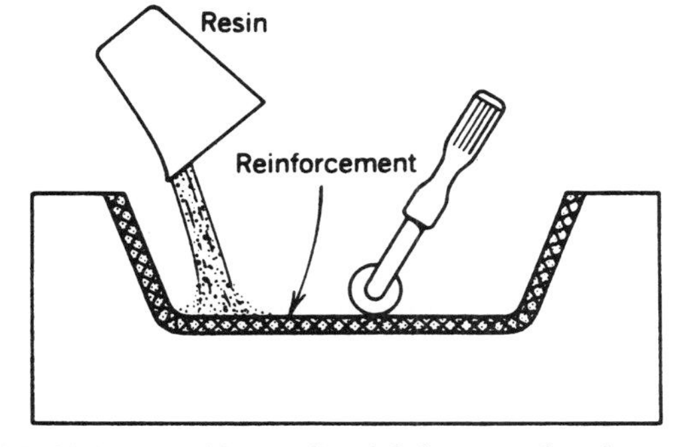

FIGURE 22.5. Hand layup. Special shapes such as boats, tanks, and aircraft parts are formed over a mold with fiber mats impregnated with a plastic resin. These are placed by hand and then allowed to harden.

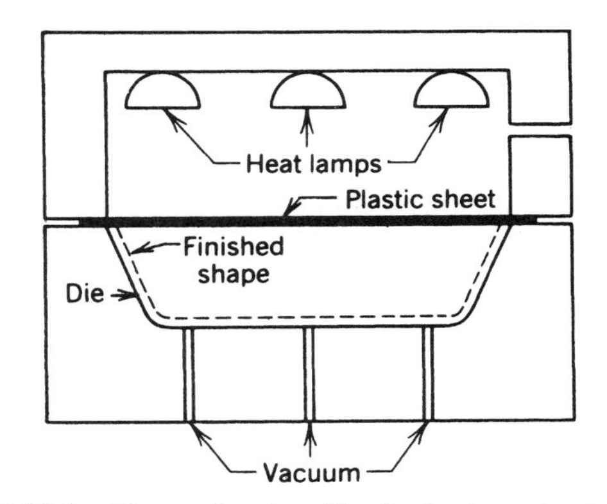

FIGURE 22.6. Thermoforming. Plastic sheets are heated to soften them and are then forced into the mold by the use of a vacuum.

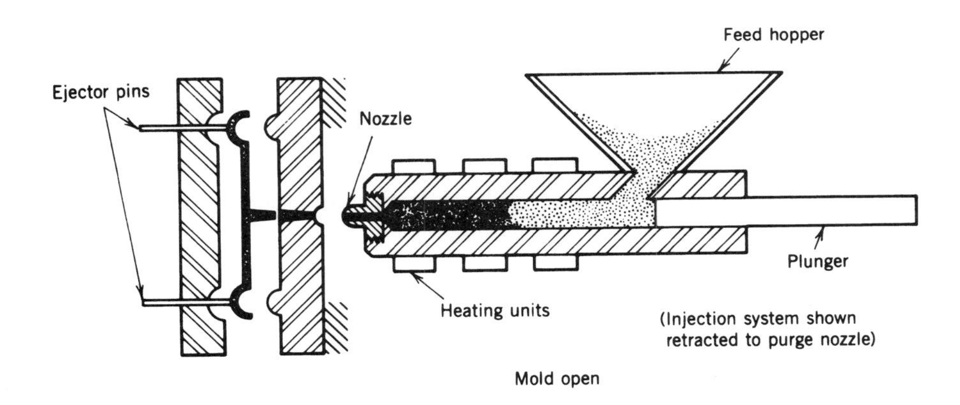

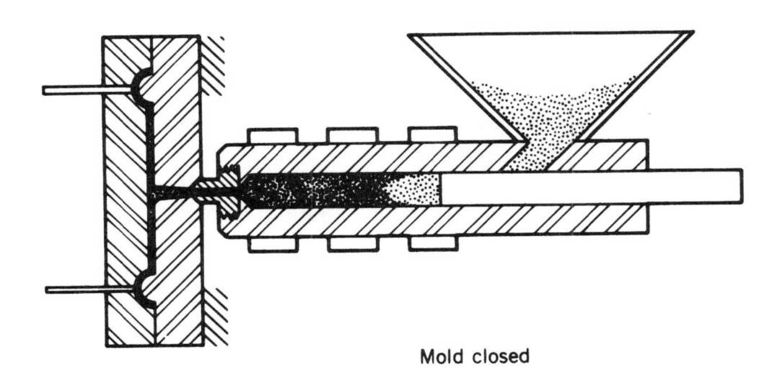

FIGURE 22.7. Principle of injection molding. Granulated thermoplastic moves forward by the ram and passes through a series of heaters that melt the plastic. It is then injected with pressure into a closed mold, where the molten plastic quickly solidifies. The mold is then opened and the part with the sprue is ejected. This cycle is then automatically repeated. (*John Neely and Richard Kibbe, Modern Materials and Manufacturing Processes, Copyright 1987, John Wiley & Sons, Inc.*)

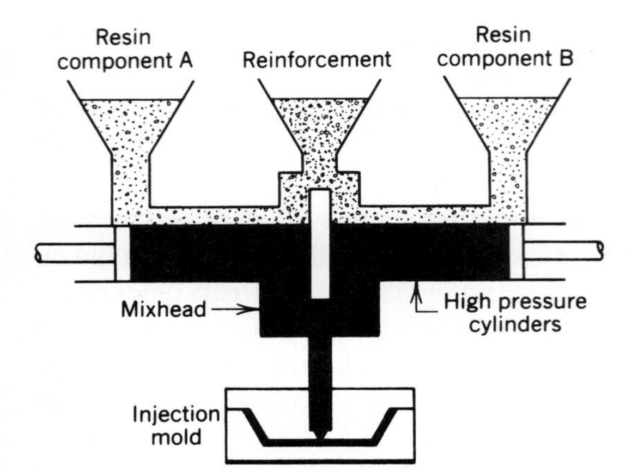

FIGURE 22.8. Reaction injection molding (RIM). Certain plastics and elastomers can be blended for cross-linking or polymerization at the time of injection without heat when RIM is used.

Reaction injection molding (RIM) (Figure 22.8) is a newer development of injection molding. In this process two basic resins (monomers) are combined just as they enter the mold, a chemical reaction takes place at low heat, and the polymer is created at that moment. The RIM process does not require the plastic material to be heated before molding.

Compression molding (Figure 22.9) is a common method of forming thermosetting plastics, since most of them cannot be injected into a mold. In this process, granulated plastic is placed in a heated mold and compressed with a plunger having the required shape until the object is formed. It is then removed from the mold with an ejector pin.

Transfer molding (Figure 22.10) is similar to compression molding, but the plastic is not melted in the mold as in compression molding. Instead it is preheated in an upper chamber and forced into the mold by a separate plunger. This method is also used mostly for thermosets.

Casting is another method of molding polymers. It is not possible to heat plastics to be sufficiently fluid for casting purposes, as is done with metals. A liquid polymer or monomer, or one dissolved in the other, is poured into the mold. Solidification is achieved by either the completion of polymerization or the cross-linking of the polymer resins. One of the more utilitarian uses of the casting process is in the making of dentures. Casting is often used for art forms and decorative molding and for encapsulating (potting) electrical components.

Blow molding (Figure 22.11) is a method borrowed from the glass-making industry and is used to make bottles and other containers. An external split mold is closed around a thick-walled heated tube of plastic. Compressed air is blown inside the plastic tube, which expands to fit the configuration of the mold. The mold is then opened, the bottle injected, and the cycle continued.

Filament winding (Figure 22.12) is a process in which a continuous strand of fiber is impregnated with resin and drawn onto a rotating mandrel of the required shape. It is usually spun back and forth, creating diagonal plies for great strength. This method is used to make shock-resistant tanks and other cylindrical products.

Extrusion (Figure 22.13) in plastics is the same process as that used in metals. A granular plastic is heated and forced through a die with a screw.

Pultrusion (Figure 22.14) is a method of forming a composite by pulling it through a die instead of pushing it. A long, continuous fiber mat cannot be forced through a die, but it can easily be pulled. Long, slender, reinforced parts can thus be formed by pultrusion.

Spinning of polymer filaments to produce fibers is the source of such artificial clothing materials as

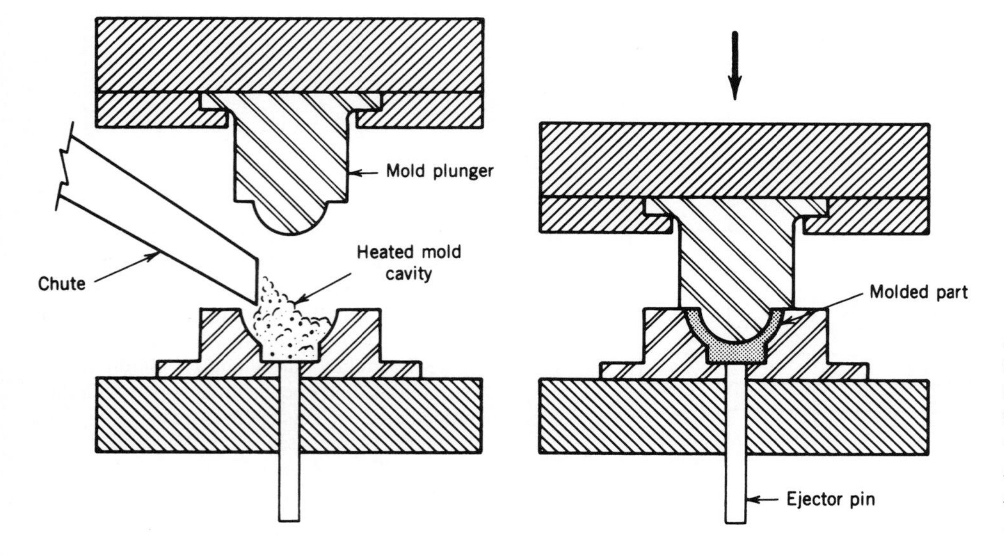

FIGURE 22.9. Compression molding. Granulated plastic is placed in a heated mold (*left*) and compressed (*right*) until the object is formed. It is then removed with an injector pin. (*John Neely and Richard Kibbe, Modern Materials and Manufacturing Processes, Copyright 1987, John Wiley & Sons, Inc.*)

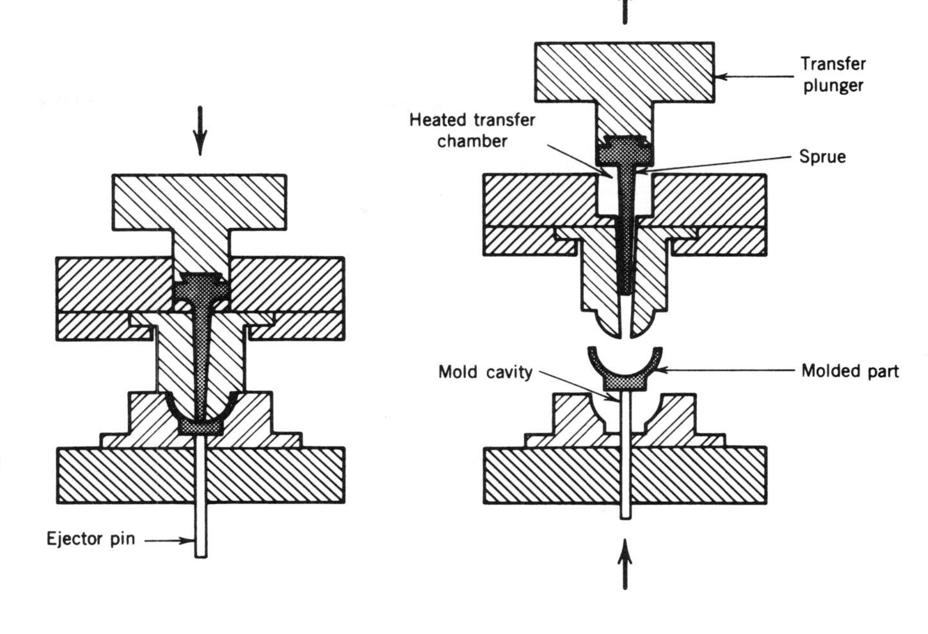

FIGURE 22.10. Transfer molding. Plastic is heated in an upper chamber and forced into the mold (*left*) by a second plunger. The object is formed and ejected (*right*). (*John Neely and Richard Kibbe, Modern Materials and Manufacturing Processes, Copyright 1987, John Wiley & Sons, Inc.*)

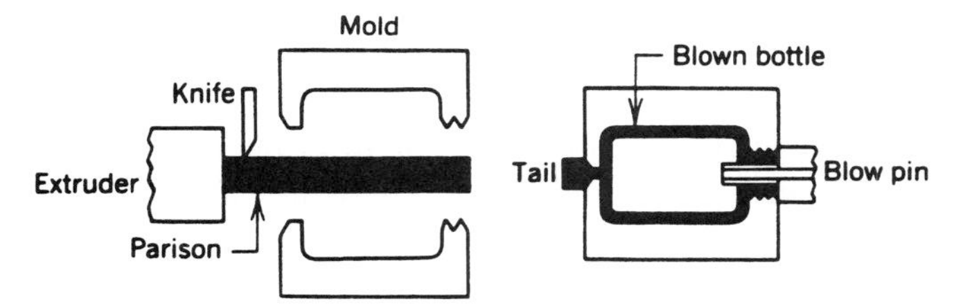

FIGURE 22.11. Blow molding. In extrusion blow molding, a hollow tube called a parison is extruded into a mold and cut off. The mold closes, a blow pin is inserted, and the tube is pressurized with air, expanding the tube (which is still heat softened by the extrusion process) to conform to the cold mold. The blown bottle hardens instantly, the blow pin is removed, the mold opened, and the bottle ejected. In injection blow molding, the parison is formed with an injection-molding machine.

Dacron®, nylon, and rayon. This process is actually a form of extrusion in which the liquid polymer is forced through a spinneret, a metal plate containing many small holes. Rayon is spun from cellulose acetate in an acetone solution, the filaments are dried in a heated chamber, and the solvent is driven off. Nylon and Dacron®, however, can be sufficiently liquefied by heating to be forced through a spinneret.

The following groups of thermoplastics and thermosets are the most common types in use today, but there are hundreds of blends and newer developments not covered here.

THERMOSETTING PLASTICS

As the name of this process implies, all plastics can be shaped or formed, but only when heated; hence, the prefix *thermo*. After once being shaped by the cure process (cooling or chemical), thermosets cannot be reheated

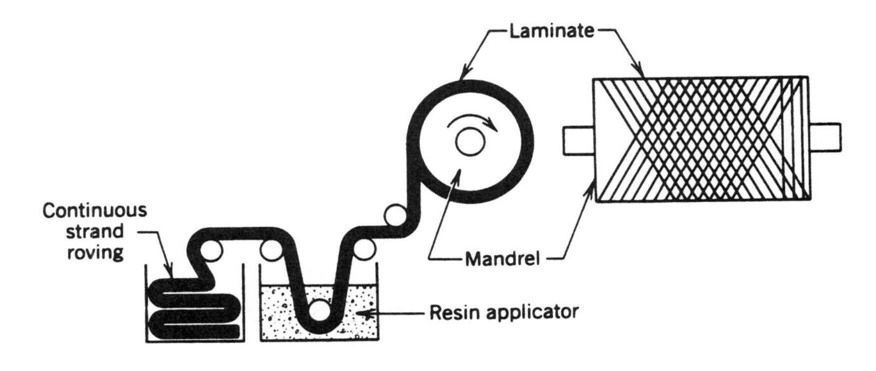

FIGURE 22.12. Filament winding. A continuous strand or mat of fiber (usually glass) is drawn through a resin applicator and onto a rotating mandrel or form. The filament is moved back and forth, creating a very strong laminate.

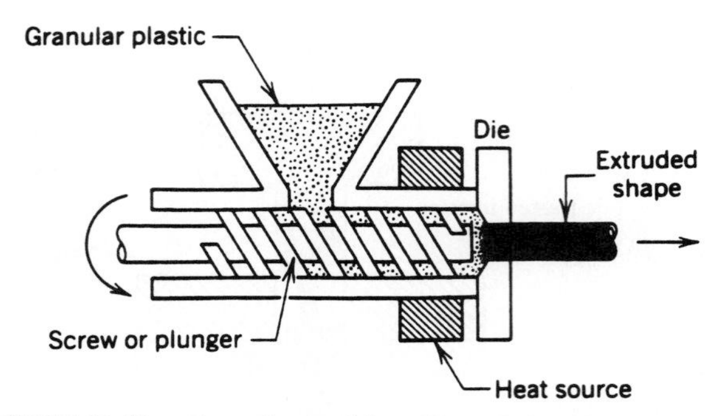

FIGURE 22.13. Extrusion molding. Heated plastic granules are forced through a die in extrusion molding to form various shapes such as rods, tubes, and angles.

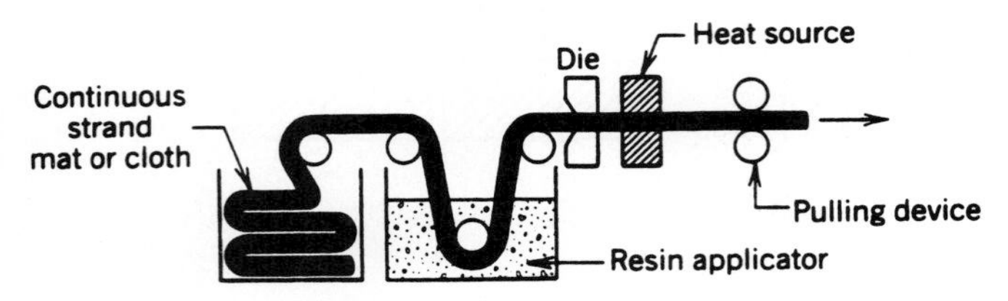

FIGURE 22.14. Pultrusion molding. When a glass fiber mat is added to a long shape to reinforce it, ordinary extrusion will not work. Such composite shapes are made by pulling the resin-impregnated glass fiber through the die instead of pushing it.

and reshaped. Additional heating over that which they are designed to withstand will destroy them by breaking the cross-links in the polymer chain. Some types of thermosets do not require heat to harden but are cross-linked by combining liquids in a definite proportion. This mixture then hardens in a preset time. Some examples are epoxy, silicone, polyurethane, and polyester. Unlike thermoplastic scrap, thermoset scrap cannot be reheated for recycling, but it can be ground up and combined with a suitable binder to make a kind of polymer concrete.

As a general rule, thermosets are harder and more brittle than thermoplastics and must be combined with fillers and reinforcing agents to give them the mechanical properties needed. For the most part, thermosets cannot be injection molded, the most rapid production method, but some types can be injection molded using special techniques. Thermosets in general must be heated to a certain temperature for the molecules to bond by cross-linking. After curing, they are chemically insoluble and will not return to their original state by reheating.

Allyl Plastics

Produced from an alcohol, the allyl plastics are clear and scratch resistant, and are used for prisms, lenses, and window glass. They resist most solvents such as benzine, gasoline, and acetone. Allyls have high strength but are quite expensive.

Aminos

These urea-formaldehyde thermosets are produced by reacting urea with formaldehyde in the presence of a catalyst and mixing in a filler. These plastics are used for a few products, but their major use is as adhesives for bonding furniture, plywood, and foundry sand cores. More familiar uses of the aminos are seen in the melamines (melamine-formaldehyde), which are similar to but more complex than the ureas and have better resistance to acid and alkali attack. They are also more shock, heat, and water resistant, making them useful for bonding exterior plywood. Melamines are used for dinnerware, scratch-resistant furniture surfaces, and electrical parts such as terminal blocks and circuit breakers. The aminos have an unlimited range of colorability.

Epoxies

Epoxide chemicals are generally any compound containing the three-membered ring epoxide group. Epoxy resin may be any number of thermosetting materials made of polyethers and dihydric phenol reacted in the presence of an alkali. However, the epoxy principally used, because of its low cost and availability, is epichlorohydrin and Bisphenol-A. Epon, Epi-rez, and Araldite are just a few of the many trade names.

Some epoxies are poured at atmospheric pressure to encapsulate electronic parts. These are called potting compounds. As with other thermosets, most applications require fillers and reinforcing agents to impart needed mechanical properties. Epoxies have better heat and chemical resistance than the other thermosets, which is one of the reasons they are used extensively as composites in aircraft construction. When they are damaged, a patch is applied and heat cured with a repair pad or blanket or with a flow of hot air. Epoxies are often hand laid up with glass fiber blankets or by filament windings. Epoxy-glass laminates are also used extensively for electronic circuit boards. Epoxies are the most widely used family of plastics in all of industry, commercial and military.

Phenolics

Phenolic plastic is a heat-reacted, cross-linked product of phenol and formaldehyde. The phenolic-formaldehyde plastics, such as bakelite, have been around for a long time, but newer phenolics with superior qualities have been compounded. Most glass-reinforced phenolics used today for electrical components have a UL temperature index of 311°F. They are often used in hotter environments, such as on electrode holders for arc welding. A newer short-glass phenolic has been developed that will withstand temperatures ranging from 500° to 700°F in soldering applications.

Phenolic laminates consist of (1) glass, carbon fabric, or (2) other reinforcements to stabilize canvas, paper, or linen to reinforce the plastic, and are formed under pressure. Wood flour is often used as a filler.

Polyesters

Polyesters are a class of plastics related to the alkyd type. Alkyds are produced by the reaction of an alcohol and an acid. The name comes from the words ALC*ohol* and *ac*ID. Their most important use is in paint and lacquers. There are two different types of polyesters: *saturated* and *unsaturated.* The saturated types are produced as film and fibers known as Dacron® and Mylar®. The unsaturated types are used in molding and casting automobile parts and for such applications as boats, tanks, building panels, and automobile bodies with the use of glass fiber.

Some polyesters are thermoplastic molding compounds that can be injection molded and are chemically listed as polybutylene-terephthalate (PBT) and polyethylene terephthalate (PET). Like all polyesters, fiber reinforcing, usually glass (15 to 30 percent), is added to improve strength and heat resistance. There is an exceedingly large market for this thermoplastic polyester PET. Over 1 billion pounds a year are used in the bottle and film industries alone. Because of its high strength (up to 23,000 psi when glass reinforced) and good heat resistance, it has many other applications, such as pump and power tool housings and sporting goods.

Silicones

The silicone family of semiorganic polymers consists of chains of alternating silicon and oxygen atoms, modified by various organic groups attached to the silicon atoms. Silicone plastics include elastomers, adhesives, potting compounds, and oils. One application of silicone is room-temperature-vulcanized (RTV) rubber, which is used to make flexible molds for casting prototypes. Silicone rubber can withstand temperatures up to 500°F without breaking down. Because of their excellent electrical properties, silicones are extensively used in the electrical industry for encapsulation and components. Silicones are also used as sealants and for caulking because they do not lose their rubbery qualities or crack with age and have good bonding characteristics.

THERMOPLASTICS

Probably the greatest advantage thermoplastic materials have over the thermosets is the cost-saving advantage of higher production rates of molded parts. Injection-molding machines (Figures 22.15*a* and 22.15*b*) have a faster cycling time than compression- or transfer-molding processes. Also, deflashing and other secondary operations are not needed as they often are when molding thermosets.

Thermoplastics are generally less brittle and more flexible than thermosets, and so parts can be designed with snap-fit parts and "living hinges" that can flex thousands of times without breaking. Thermoplastics can be bonded by the process of ultrasonic welding as well as with adhesives.

Thermoplastics are utilized in a myriad of applications such as automotive, construction, medical, and household.

Acronitrile Butadiene Styrene (ABS)

This copolymer consists of monomers of styrene plastic and acrylonitrile and butadiene rubbers. By varying the amount of both of these rubbers, a wide range of specific materials having widely different properties is made available. Styrene was first introduced in 1938 and was called *polystyrene.* It has a very low strength, but it is one of the least expensive plastics and for this reason it is still used for disposable products. Styrene acrylonitrile (SAN) is a styrene copolymer that has a better chemical resistance and higher strength than polystyrene. However, ABS is by far the most outstanding of this family of plastics. Its high impact resistance and moderate cost enables ABS to be used for automobile grilles, safety helmets, luggage, and some furniture parts. It can be blended with polyvinyl chloride (PVC) to become flame resistant.

Acetals

Acetals are crystalline polymers that have excellent engineering properties such as high strength, high stiffness, high tensile strength, fatigue endurance, impact strength, good dimensional stability, and resistance to creep under load. For this reason they are classed among the *engineering resins.*

There are two types of acetals: copolymer and homopolymer. Depending on its structural makeup, an acetal copolymer silicon fluid concentrate provides good wear resistance where plastic parts slide against plastic parts. Mineral-filled acetal copolymer resins have applications

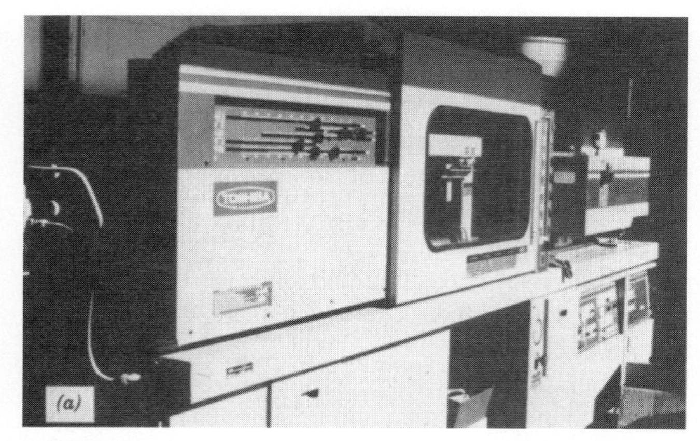

FIGURE 22.15a. Modern injection-molding machine with microprocessor that automatically controls cycling times. (*Hergert's Industries, Inc.*)

for such products as toys, zippers, and stereo and video cassette cases. Acetal homopolymer resins are used to make plumbing fixtures, replacing brass and zinc parts, and other hardware items. The DuPont acetal homopolymer Delrin® is used for gears and other high-stress products. These resins resist solvents and most alkalis, but contact with most acids should be avoided.

Acrylics

Acrylics are noncrystalline polymers known for their clarity and light transmission qualities. Acrylics can be made in brilliant, transparent colors and are used for the manufacture of automobile taillights, jewelry, and novelty items. Lucite® and Plexiglas® are acrylics that are made in the form of sheets, tubes, and rods. Acrylics are not very shock or scratch resistant, and they will soften at temperatures over 200°F.

Cellulosics

Unlike other thermoplastics, cellulosics do not melt, a requirement for injection molding, but with proper chemical treatment they can be injection molded. Cellulosics tend to creep or cold flow when under load. There are five basic groups: cellulose nitrate (not much used today), acetate, butyrate, propionate, and ethyl cellulose. Cellulose plastics are made into thin sheets and sold under the name *cellophane.* The ethyl type has high shock resistance and is used for products that may be accidentally dropped, such as flashlight cases and toys. In general, cellulosics are the toughest of all plastics.

Fluoroplastics

These fluoropolymer plastics are paraffinic hydrocarbon polymers in which a fluorine atom replaces some

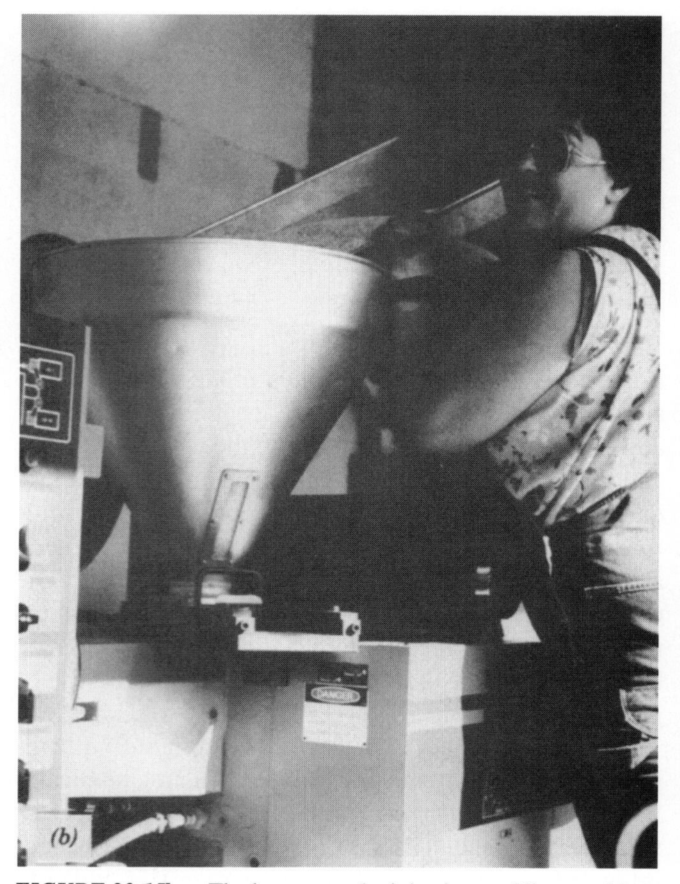

FIGURE 22.15b. The hopper on the injection-molding machine must be filled frequently while the machine is in operation. (*Hergert's Industries, Inc.*)

or all of the hydrogen atoms in the molecule. Fluoropolymer resins are characterized by their high temperature stability and the fact that they have the highest chemical resistance of all plastics. They also have the lowest coefficient of friction of any plastic and for that reason are widely used for nonstick baking or frying pans that do not need to be oiled. The fluorocarbon family includes polytetrafluoroethylene (TFE), which has the greatest chemical resistance but is difficult to mold; chlorotrifluoroethylene (CTFE); fluorinated ethylene propylene (FEP), which is more easily molded but is more expensive than the other types; and polyvinylidene fluoride (PVF_2), which is a crystalline polymer used in the building industry. Surface coatings made of these materials have exceptional resistance to sunlight and weather.

Polyamides

Nylon is a generic name given by DuPont to numerous variations of polyamides. Most nylons are semicrystalline and share the properties of toughness, low friction, fatigue and impact resistance, and inertness to aromatic hydrocarbons. However, all nylons absorb some moisture. The most popular nylons used in the United States are type 6, type 6/6, type 11, and type 12. Types 6 and 6/6 are the most commonly used and the least expensive, but type 6/10 is more flexible and has low moisture absorption and so can be used in water mixing values for washing machines and similar applications because it does not swell or change dimensions as some plastics do when in contact with water.

Nylons possess some incredible qualities when reinforced with glass fibers. They can replace metals in many applications such as bicycle wheels, automobile speedometer gears, and lawn mower carburetors. Nylons 11 and 12 are commonly used in film and tubing.

Polycarbonate

Polycarbonate based on bisphenol A, which is catalytically combined with carbonyl chloride, is one of the most important as far as variety of uses. Polycarbonate resins have properties that are similar to those of polyamides (nylon), and they can be used in many of the same applications. They are characterized by high stiffness combined with high-impact strength. For this reason they are used for unbreakable bottles and as a replacement for window glass. The General Electric polycarbonate Lexan® is an excellent glazing product that is made with a special surface coating that resists the damaging effects of ultraviolet light. Polycarbonates are used for lenses, goggles, safety helmets, and other applications where optically clear material is required or where colored formulations are needed. It has good resistance to dilute acids, but is attacked by alkalis and aromatic hydrocarbons.

Polyethylene

Several methods of polymerizing ethylene lead to the formation of polyethylene. Polyethylene resins are noted for their good chemical resistance, low-temperature impact resistance, low coefficient of friction, good dielectric strength, and flexibility with high flex life, which makes them excellent materials for "living hinges." Low-density polyethylene (LDPE) is flexible and transparent to translucent. Uses include squeeze bottles, boil-in packaging, soft gaskets, and ice cube trays. High-density polyethylene (HDPE) is stiffer and more opaque and is used for pipelines, rigid detergent bottles, wire insulation, garbage pails, and dishpans. Polyethylene terephthalate (PET) is widely used for beverage bottles and is a material that is highly recyclable. A process called *blow molding* is used to form plastic containers.

Polyurethanes

Manufacture of polyurethanes generally consists of combining urethane in linkages formed by the reaction of toluene diisocyanate and diphenyl methane diisocyanate.

Polyurethane resins may be either thermosetting or thermoplastic and can be rigid or flexible, depending on formulation. They have properties of both rubbers and plastics. Foamed flexible polyurethane is used in seat cushions for cars and furniture. Rigid foamed urethane is used as an insulating material. Urethane rubbery plastics are used for industrial truck solid tires, caster wheels, drive belts, floor coverings, and roofing materials. Metal objects such as furniture are sometimes covered with a tough, resilient skin which provides a colorful, damage-resistant surface, while the steel core provides the high strength. Urethanes may be cast, injection molded, or coated. Synthetic leathers for shoes and handbags are often made of reinforced polyurethanes.

Polyvinyl Chloride

The most common of the vinyl polymers is polyvinyl chloride (PVC). Vinyl chloride (vc) monomer is the basic precursor of PVC. This vc is transformed into PVC by any number of free radical addition polymerization mechanisms. There are two basic types: flexible and rigid. The flexible type is used where resilient, tough plastic is required. Uses include garden hoses, flooring, gaskets, floor mats, and toys. It is often used as a replacement for rubber in many applications. Rigid PVC compounds are extensively used in pipe and pipe fittings, rain gutters, siding, and other building materials. PVC can be foamed and fabricated by most processes.

Alloys

Alloys are blends of two or more thermoplastic polymers that have intermediate properties between those of each polymer. For example, ABS and rigid PVC are alloyed to produce a stiffer, fire-resistant, self-extinguishing material, having a higher heat distortion temperature than is normal for rigid PVC. Thousands of these blends have been developed for specific requirements for use in aircraft and automobile manufacture, building construction, and for consumer products.

Composites

In the plastics industry, a **composite** is a term used as a general reference to the various kinds of reinforced plastics that use one or more types of reinforcing materials. The five general types or kinds of composite materials are laminar, fiber, flake, separate particles, and

FIGURE 22.16. A piece of this fiberglass sheet has been split off, showing the interior with the glass fibers. Note that they lay randomly thus giving the sheet equal strength in all directions.

filled. Laminars are layers of different materials bonded together. Fibers are either long, as for filament winding, or chopped into short lengths. Flakes, such as aluminum or copper, are used for decorative effects. Separate particles are added for decorative purposes, such as stone pebbles for storefronts. A filled composite is an open structure like a sponge with a liquid resin forced into it by pressure or vacuum and then polymerized.

Both thermoplastic and thermosetting plastic resins are reinforced with various fibers to increase strength and other properties. Reinforced thermosets are imbedded with fibrous materials, usually glass fiber in the form of mats, fabrics, or chopped fibers. Laminated plastics are piles of sheet material usually impregnated with a thermosetting resin and bonded together with heat and pressure. The principal resins used in laminated products are phenolic, polyester, epoxy, and silicone.

Most thermoplastics can be reinforced with glass fiber or specialty fibers such as ceramic, boron, or graphite. The strength of any thermoplastic resin can be increased by adding glass fiber (Figure 22.16), but there are some undesirable effects in its use. These include an increase in opaqueness, occasional fiber appearance on the surface, and some difficulty in electroplating.

Manufacturing Composite Materials

Composite products may be produced by virtually all methods of plastic fabrication. Figures 22.5 to 22.12 show some of these processes.

These products possess greatly superior mechanical properties compared with simple plastics. They have a much greater strength-to-weight characteristic, a high resistance to impact, and a higher tensile strength.

Composites are widely used in automotive, aerospace applications, and pleasure boat manufacture. Some of these lightweight materials are as strong as some metals and have more resistance to fatigue and corrosion. Because of this, many more applications of these materials will be seen in the future.

ADDITIVES

Antioxidants

Additional ingredients are often used in formulating plastics so as to tailor the product with desired properties. These materials provide resistance to damage by ultraviolet for such resins as polystyrene, polyethylene, polypropylene, and ABS. They also impart melt-flow retention, making them easier to mold.

Fillers

Fillers used in resins reduce the cost of plastic products, lower the part weight, and often increase physical properties such as tensile and compressive strength. Common organic fillers are wood flour, ground walnut and pecan shells, rice hulls, and cellulose. Mineral fillers include calcium sulfate, calcium carbonate, talc mica, clay, and silica. Glass and ceramic microspheres are sometimes used, although they are much more expensive than organic or mineral fillers.

Stabilizers

Some plastic materials such as styrene and vinyl are subject to breakdown in the presence of oxygen or heat. Phenols can be added to styrene to inhibit degradation, and PVC polymers are stabilized by barium-cadmium-zinc compounds. Other stabilizers may be used to mitigate degradation of many plastics.

Plasticizers

Some plastic materials such as vinyls are normally hard, brittle materials, but by adding a plasticizer they can be made soft and flexible. Plasticizers make the difference between rigid and flexible PVC pipe and tubing.

Colorants

Pigments and dyes for plastics give them their brilliant colors, which are important for providing sales appeal. Some colorants are designed to give a solid, opaque color, whereas others tint the clear resin to impart a translucent coloration. Colorants must be able to completely disperse evenly throughout the molten plastic and must have heat stability; that is, they must not change color from molding heat.

Blowing Agents

Foamed plastics have been around for some time, but structural foamed (SF) plastics having an outer skin of hard, tough material are showing great promise for building, aircraft, and automotive construction. These plastics have extremely high strength-to-weight ratios. A plastic resin such as polystyrene is foamed by injecting an inert gas such as nitrogen into the molten plastic as it is forced into the mold. This creates tiny bubbles, and where the material touches the mold wall, a skin is formed. This creates a porous interior sandwiched between solid plastic skins. There are many other blowing agents besides nitrogen. The familiar disposable white coffee cup is foamed in a mold from expandable beads of polystyrene. Plastic foams can be either rigid or flexible, depending on the kind of plastic used and the formulation.

Finishes

Plastics may be given surface finishes for sales appeal or wear, scratch, and chemical resistance. Plastics can be plated, metallized, or enameled. Texturing in the mold provides raised or depressed designs in geometric patterns such as basket weave, pebble, or leather texture. These often help to mask manufacturing flaws such as flow lines, pin holes, sink marks, and other flaws, thus lowering part rejection. In high-technology applications such as aerospace uses, many finishes are integral to the design (e.g., for thermal control).

Corrosion Resistance of Plastics

Corrosion of plastics can be caused by certain organic solvents, acids, alkalis, and sunlight. Deterioration from these causes varies widely; however, certain cellulose esters used for a basis of lacquer formation are completely insoluble to any known solvent, as is also the case with polytetrafluorethelyne. Some are highly flammable, some slow burning, and still others self-extinguishing. For example, cellulosic molding compounds will decompose in the presence of strong acids and strong alkalis and are soluble in ketone (especially acetone) and esters, but burn slowly. Nylon is resistant to common solvents and alkalis but is attacked by strong acids, and epoxys are generally resistant to corrosion from any of these except for a few strong acids. The many plastics and their varied resistance to corrosion may be found in engineering materials handbooks or manufacturers specifications.

MACHINABILITY OF PLASTICS

Nonmetallic materials are very poor conductors of heat; therefore, heat dissipation is concentrated in the tool and cutting edge. Surprisingly, conventional tool materials tend to break down on many soft materials because the temperature rise is sufficient to cause tool breakdown. When machining thermoplastics and other materials having low heat conductivity at high speeds, inadequate heat distribution may result in the melting of the workpiece and a dulled or damaged tool. However, many thermoplastics such as nylon and acrylic (Figure 22.17) have good machining qualities when the proper tools and cutting fluids are used.

Thermosetting materials do not soften when exposed to heat, but heat dissipation is still low. Filler materials such as glass fiber, paper, or cotton used for reinforcement can be very abrasive and shorten tool life considerably.

Drills for plastics should have an included angle of 60 to 90 degrees and zero rake (Figures 22.18*a* and 22.18*b*). Special drills for plastics are made with large, polished flutes. Some types are carbide tipped.

Tools for plastics vary considerably according to their heat and abrasive resistance and the required rake and clearance angles. On some materials such as the

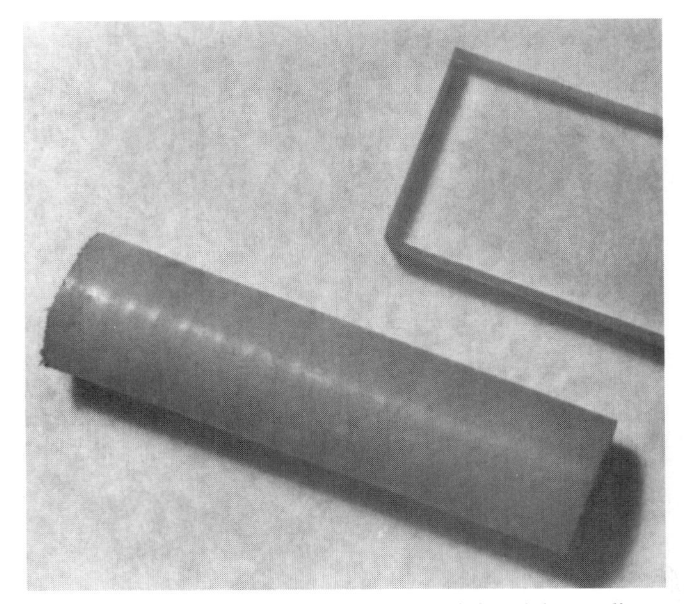

FIGURE 22.17. Both the nylon rod on the left and the acrylic plastic sheet on the right have good machining qualities.

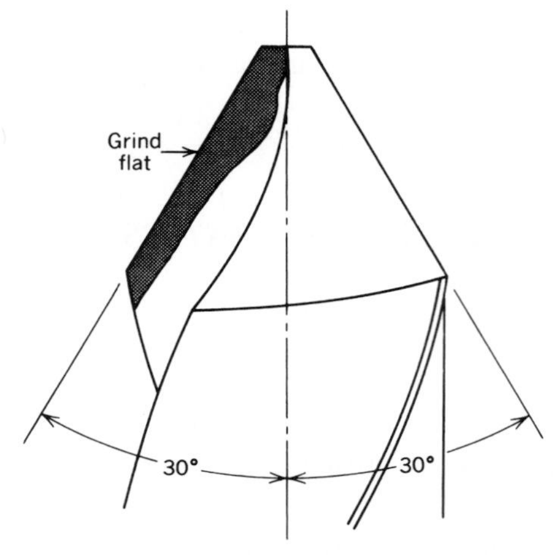

FIGURE 22.18a. A drill used for most plastics such as acrylics, fiber, and hard rubber should be sharpened as illustrated. The rake is ground with a zero angle. (*Machine Tools and Machining Practices*)

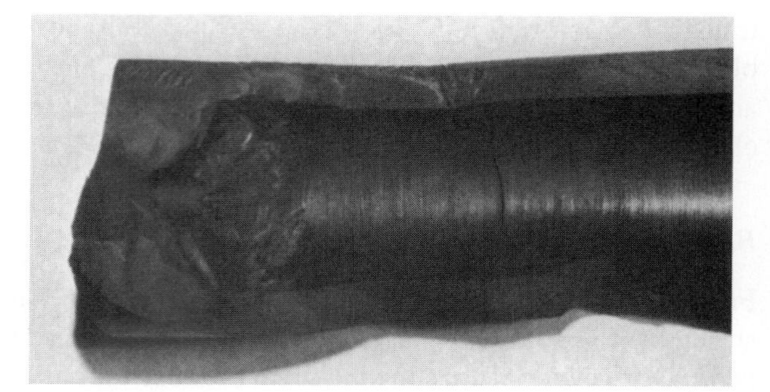

FIGURE 22.18b. This round nylon rod was being drilled with an improperly shaped drill when the drill suddenly grabbed and split the part into several pieces.

acrylics, cutting tools tend to gouge and dig into the surface if a positive rake is used; therefore, a zero rake is needed in these cases. High-speed, carbide, ceramic and diamond-tipped tools are all used for machining plastics. These tools must always be kept very sharp. Diamond-tipped tools can, in many cases, increase production 10 to 50 times over that of carbides when applied to the machining of nonmetallic materials. Table 22.2 gives the machining characteristics of some plastics and other nonmetallic materials.

Standard taps and dies may be used for threading in nonmetals, but they must be sharp. Threading, reaming, milling, and other machining operations on plastics are performed dry or with coolants such as a blast of air, soluble oil, a water-soap solution, or plain water.

Ultrasonic machining is also performed on plastic materials. With this method, taps or other shapes are inserted into the work to form the desired shape with little machining stress.

Laser machining is used for cutting and machining all plastics, especially the highly abrasive composites. Its major use has been for cutting shapes out of sheet materials and for making holes. Although finishes can be smooth when low cutting rates are used, edges are often rough when cutting speeds are rapid. Yet, because of the absence of tool contact during laser machining, thermoplastics, thermosets, and composites can be machined at high speeds without crazing, cracking, or mechanical degradation of the edge.

JOINING PLASTICS

Plastics are often joined by the use of adhesives, which must be of the correct type for each variety of plastic. Thermoplastics can also be joined with heat. The familiar heat sealing of plastic bags is an example. The hot-air welding of thermoplastics is done in a way similar to the torch welding of metal. A hot-air gun using heated compressed air is used and a filler rod is melted into the joint.

Ultrasonic welding is also used to bond thermoplastic parts. Butt, scarf, and shear joints are bonded

TABLE 22.2. Machining characteristics of plastics and elastomers

Material	Drill Point Angle (Degrees)	Cutting Tool	Single Point Rake Angles (Degrees)	Cutting Speeds	Feed Rate (per Revolution)	Coolant/Lubricant Required
Nylon	118	High speed	0–5 positive	250–500	0.002–0.005	Dry or soluble oil
Rubber	118	High speed	0–hard positive	200–700	0.010–0.025	Dry or water
Phenolics	118	Carbide	10–15	700–900	0.003–0.006	Dry or water
Vinyls	60	High speed	0	250–500	0.003–0.005	Dry or water
Acrylics	60	High speed	0–3 negative	200–500	0.002–0.006	Dry or water
Teflon	90	High speed	0	200–500	0.004–0.006	Dry, air, or soluble oil
Epoxies	60	High speed or carbides	0–3 negative	300–700	0.004–0.006	Dry, air, or soluble oil
Acetals	118	High speed	0	700–850	0.002–0.005	Dry or soluble oil

Source: Machine Tools and Machining Practices, 1977.

with high-frequency vibrations that cause the material to melt between the surfaces of the joints.

TESTING

Many of the same tests that are used for metal are used for plastics, such as tensile, hardness, and shock resistance. The Izod-Charpy test is used for the latter (Figures 22.19*a* and 22.19*b*). Other tests are electrical resistivity (dielectric strength), cold flow, softening temperature (heat resistance), flammability, dimensional and color change on aging, water resistance, acid, alkali and solvent resistance, specific gravity, and peel tests.

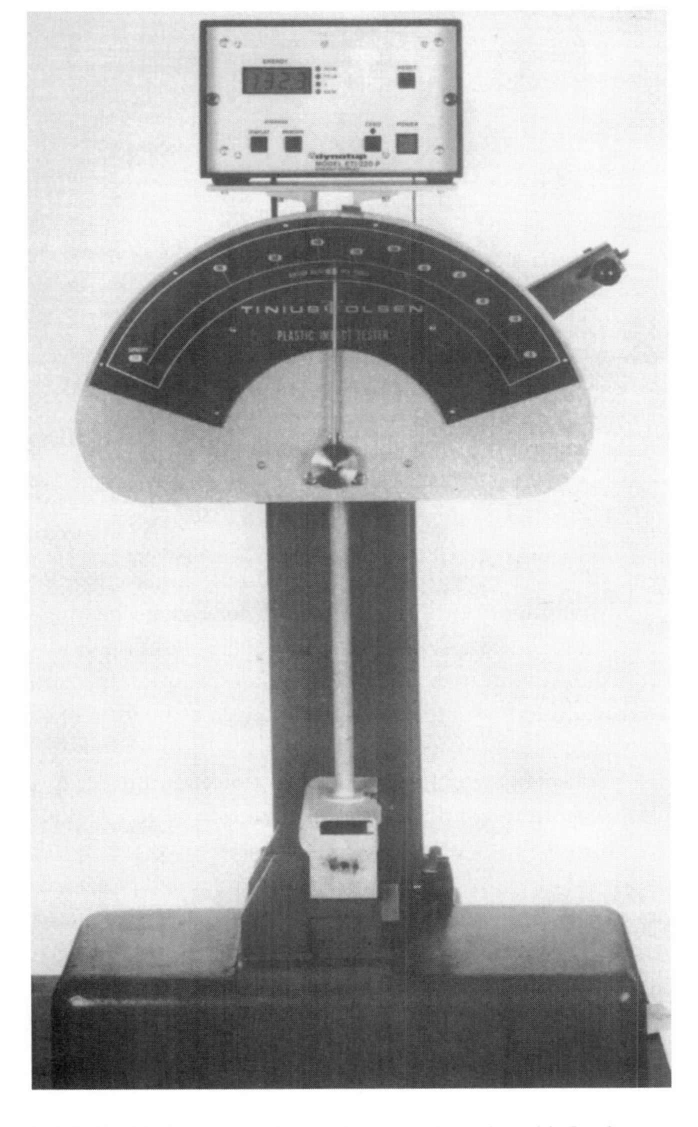

***FIGURE 22.19*a.** Plastic specimens are tested on this Izod-Charpy impact testing machine. (*Tinius Olsen Testing Machine Company, Inc.*)

RUBBER

Natural rubber latex is a gummy, sticky substance that has little value in its natural state; it can be stretched, but it will not return to its original shape, having very little elasticity: The long-chain molecules simply stretch out. Natural rubber is obtained from a thick, milky fluid (latex) that oozes from certain plants when they are cut. Most of this natural latex comes from the Para rubber tree (*Hevea brasiliensis*) of South America, southeast Asia, and Sri Lanka. Natural rubber must be broken down, by kneading or mastication between hot rolls, and vulcanized. The hot vulcanization process was discovered by Charles Goodyear in 1839. He combined natural latex with sulfur and heated the mixture to the melting point of sulfur. The result was the cross-linked polymer that has great elasticity and little plasticity that we know of as rubber.

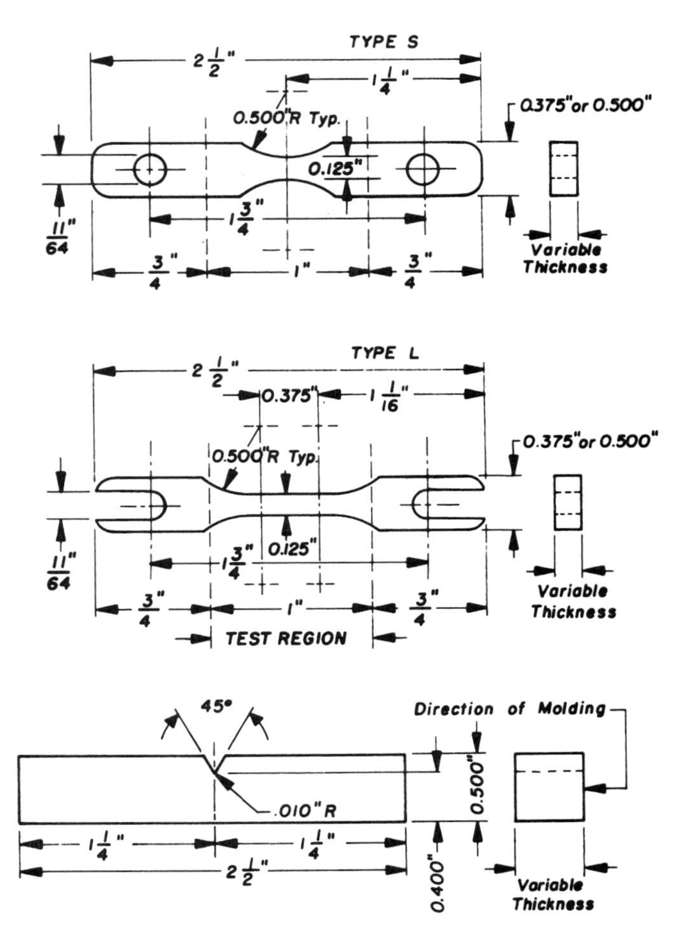

***FIGURE 22.19*b.** Plastic specimens for tension impact tests (top two diagrams) should be made according to these specifications. The Izod test specimen for impact testing is shown in the bottom diagram. (*Tinius Olsen Testing Machine Company, Inc.*)

FIGURE 22.20. The severe cracking on this rubber tire is due to long exposure to ultraviolet in sunlight.

Elastomers

Rubbers and elastomers may be either natural or synthetic. The term *elastomer* has been generally accepted as a broad term for those materials possessing rubbery qualities such as high resilience and extensibility. These include the synthetic rubbers such as neoprene and rubbery plastics. The term *rubber* is also now used for many of these synthetic products. With the exception of the "hard" rubbers, these materials all have the property of high deformability and elasticity. Some products have an extensibility of as much as 1000 percent without rupture and can then recover to nearly their original shape. However, working deformations do not exceed 100 percent. Unlike metals, rubber does not yield at higher stresses; instead it stiffens as the stress is increased.

Additives

Rubber can be made hard (as in ebonite), soft, tacky, or resistant to oxidizers by using certain additives. Accelerators are added to speed the vulcanization process. Synthetic rubbers have been developed to have certain properties not found in natural rubber. However, the first such synthetic rubber was developed in the early 1930s as a substitute for natural rubber. It was first derived from acetylene gas and was a long-chain molecule called **polychloroprene,** better known as neoprene. Acetylene gas was derived from calcium carbide, but most of this gas is now derived from the petroleum distillate methane. Natural rubber is somewhat more resilient than neoprene, but it tends to deteriorate in sunlight (ultraviolet light) (Figure 22.20), and it swells when in contact with petroleum oil. Neoprene has better resistance to these factors. However, chemical additives such as antioxidants, plasticizers, and stabilizers are now added to natural rubber and other elastomers to overcome these difficulties.

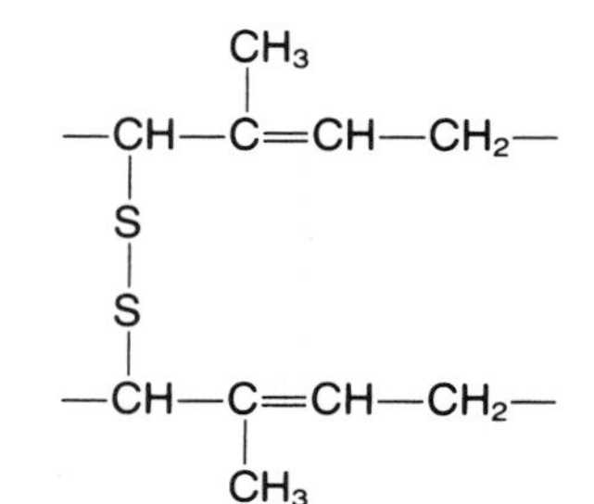

FIGURE 22.21. General formula for vulcanized rubber. (*John Neely and Richard Kibbe, Modern Materials and Manufacturing Processes, Copyright 1987, John Wiley & Sons, Inc.*)

Chemically, natural rubber is polyisoprene, a hydrocarbon that has the empirical formula C_5H_8. A general formula for vulcanized rubber is shown in Figure 22.21. Many synthetic rubbers have been developed, most of which are derived from petroleum distillates (petrochemicals) such as acetylene, butadiene, isoprene, and chloroprene (Figure 22.22).

Rubber is an extremely useful material. Modern society could scarcely do without such products as automobile tires, shock absorbers for machines, belting, wire covering, O-rings, and seals (Figure 22.23). Hard rubber (ebonite) has been used for many years as an industrial plastic material. Products such as battery boxes and combs are made of ebonite.

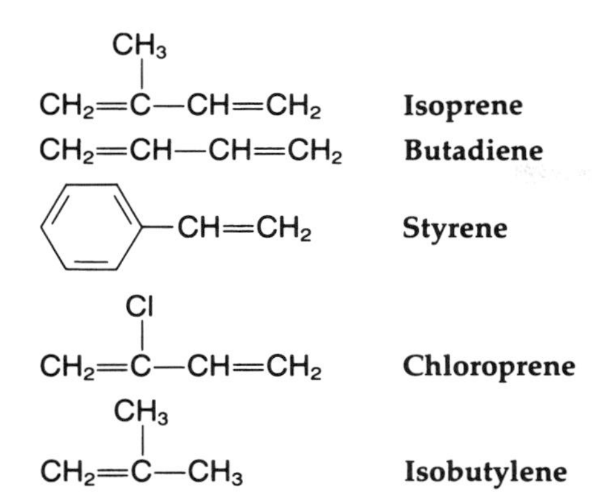

FIGURE 22.22. Rubbers and elastomers. (*John Neely and Richard Kibbe, Modern Materials and Manufacturing Processes, Copyright 1987, John Wiley & Sons, Inc.*)

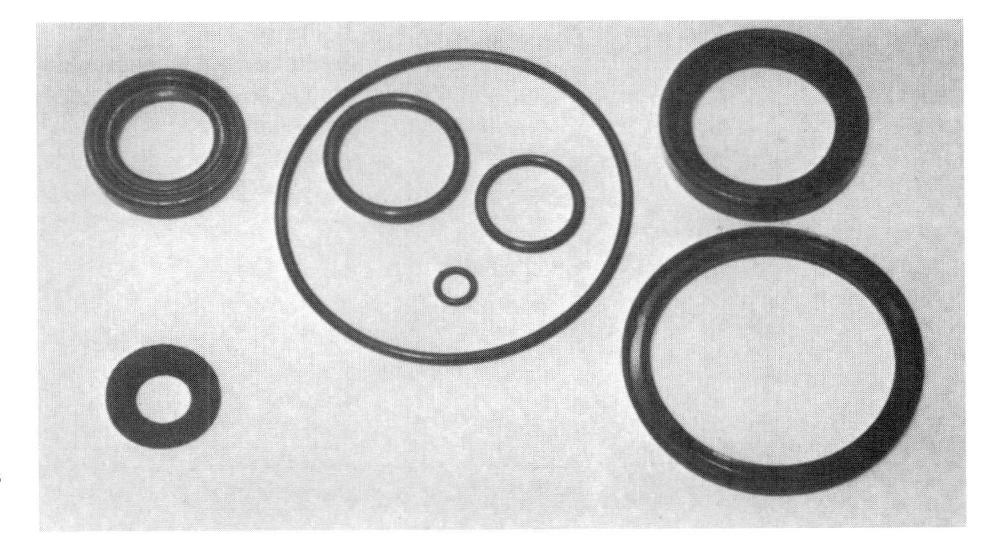

FIGURE 22.23. Many rubber products such as these O-rings and seals are used in hydraulic and other machinery.

SELF-EVALUATION

1. Name two conditions that cause the breakdown of organic materials.
2. Plastics may be divided into two major groups. Name them.
3. Which of the two groups of plastics is the kind that can be resoftened by heat after the part is made?
4. What is the source of natural rubber?
5. What is the source of synthetic rubbers such as neoprene, isoprene, butadiene, and chloroprene?
6. The first synthetic rubbery material used commercially was derived from acetylene gas. What was it called?
7. Name three derivatives of ethylene.
8. Why do conventional tool materials often break down when machining plastics?
9. Name two methods of joining plastics.
10. Define a polymer.
11. Why are nylons and acetals called engineering resins?

12. Would polystyrene be a good plastic from which to make gears? Explain.
13. What is a plastic composite?
14. What does a blowing agent do?
15. What is blow molding?
16. Which rubber material is more likely to deteriorate and show surface cracking, natural or synthetic?

CASE PROBLEM: THE EXPENSIVE CLEANING JOB

A complex electronic device was malfunctioning intermittently. The operator took off a cover plate and could see dust and lint inside on the plastic parts. Assuming this to be the problem, he obtained a spray can of cleaner that was supposedly designed for electrical equipment and sprayed the electronic component. Suddenly, springs, pieces of plastic, and wire coatings flew all over. The electronic component was destroyed and had to be replaced. The operator looked on the can label and found it contained acetone. What do you think the molded plastic parts in the electronic device were probably made of? What, if anything, did this operator learn about plastics?

CHAPTER 23

Ceramic Materials

Ceramic materials such as clay, glass, and stone have always been useful materials. Mortar (sand and hydrated lime) has been used for centuries to bind stone and brick in building structures. The Romans discovered a way to make fluid stone that would solidify into a hard mass. It was called Pozzolan cement because it was first made from the volcanic ash at Pozzuoli, Italy. After this ash was combined with hydrated lime, it began to harden. Modern portland cement, from which concrete foundations, sidewalks, and dams are made, is similar in some ways to that ancient Roman cement. In recent times, the addition of reinforcers of steel, glass fiber, and polymers to portland cement has produced composites with new properties that are useful for many products.

A new and rapidly growing technology is that of advanced ceramics, also called structural ceramics. These were first introduced in 1971 for high-temperature use in gas turbines that operated at 2506°F. Silicon nitride and silicon carbide were used for these parts. New automotive engine parts that withstand high temperatures and wear better than their metal counterparts are being developed. Indeed, an entire ceramic engine, one that needs no lubrication or cooling system, may be possible in the future.

Among these new high-technology ceramics are "whiskers" of alumina or boron to reinforce metals such as aluminum. These metal-ceramic composites give a new dimension to high-strength metals.

Refractories, glasses, and abrasives are also types of ceramics. This chapter will introduce you to the structure and uses of these inorganic materials.

OBJECTIVES **After completing this chapter, you should be able to:**

1. Describe the composition, characteristics, and uses of several inorganic materials.
2. Explain the characteristics of, and some uses for, advanced ceramic materials.
3. Explain the principle of hydration in portland cement.
4. Show how a very plastic material, common clay, can be hardened by firing.
5. Describe the structure of glass.

CERAMIC MATERIALS

Ceramic materials include clay, glass, silica, graphite, asbestos, limestone, and portland cement (from which concrete is made). Fired earthenware is formed on a potter's wheel (Figures 23.1*a* through 23.1*d*) and fine china is made from porcelain (a pure form of kaolin) clays. All these inorganic materials are held together by ionic and covalent bonds that are more rigid than the metallic bond. However, because these bonds are very strong, the ceramic materials have greater heat and chemical resistance than organic materials. Ceramics are also usually good electrical and thermal insulators.

Local stress concentrations may exceed the bond strength, however, and cause a brittle failure. These materials, unlike metals, have few slip planes to absorb local stresses, which is the reason that ceramics are brittle materials. Ceramic materials have a high compression strength, but low tensile strength. This feature makes them useful for load-bearing and supporting structures in building construction.

Ceramic products have been machined or ground for many years. Marble and slate are examples. Many ceramics cannot be machined by traditional methods. Ultrasonic, abrasive jet, electron beam, and laser beam are applicable machining processes for ceramic materials.

***FIGURE 23.1*a.** The process of forming (throwing) a clay pot on a potter's wheel. Clay is being prepared by wedging (kneading).

CLAYS

Several types of clays are mined. Ordinary clays are used in making brick and in building construction and for firebrick. These ordinary clays are composed of alumina and silica in various proportions, with other impurities present, such as iron oxide (which gives it a red color), manganese oxide, potash, magnesium, and lime. Kaolin (a white clay that is mostly composed of alumina and silica) is used in the manufacture of earthenware, fine china, porcelain, paper products, and firebrick.

Fire clay has less than 10 percent impurities and is utilized for the manufacture of clay products and for joining firebrick in furnace construction. Ordinary fireclay may be used for temperatures up to 3000°F.

***FIGURE 23.1*b.** Lump of clay is being molded into a round shape.

After clays are mined, they are crushed and ground to fine particles and passed through a screen; the larger particles are recrushed. Water is mixed with the clay in a *pug mill* in a process called *tempering*.

The bonding of silica and alumina atoms forms a chain that develops tiny, flat plates of clay that tend to flow over one another like a deck of cards when they are wet; this gives clay its plastic character. When dried, the plates become linked together solidly as in unfired clay objects, and when they are fired, they fuse together into a hard, glasslike substance. The glassy appearance is due to both a liquid and solid phase that forms between the chains when they are fired. This is also true of the porcelains that are made from almost pure kaolin.

ENAMELS

Porcelain enamels that are composed of quartz, feldspar, borax, soot ash, and other substances are used to coat articles such as stoves, refrigerators, pots, and pans. This material can be applied to iron, steel, or aluminum for wear and decorative appearance.

BRICK

Bricks are a building material made from clay and sand or shale. The temperature of firing and the proportions of various materials that are used determine the characteristics

FIGURE 23.1c. Hole is formed into the center and the sides are being raised.

FIGURE 23.1d. Pot is now in final stages of finishing and will be dried and later fired.

and kinds of brick produced. As the temperature of firing increases, the strength and porosity increases and the bricks become darker. These darker bricks are most suited for face brick on the outside of buildings where they must withstand all weather conditions. Common brick is more porous and is used for the underlying or basic construction of buildings. Mortar, composed of sand, lime, and cement, is utilized for joining the bricks into a solid construction mass such as a wall or foundation.

REFRACTORIES

A refractory material can withstand high temperatures without breaking down (spalling or melting). Refractory brick that is used in furnaces is a very common example, and without refractories, modern steel making would not be possible (Figure 23.2). Refractory materials can be cast or built of brick, which is made when fireclay or other material is used as mortar to bond the firebrick together.

The silicate structures commonly found in most rocks and clays (when fired) give them their characteristic properties of hardness and stability. Silica (silicon dioxide, SiO_2) combines in the unit structure of a tetrahedron (Figure 23.3). When linked together as a chain, the units become Si_4O_7 (Figure 23.4); this is the form in which it is usually found in minerals and rocks. Combined with water and alumina (aluminum oxide, Al_2O_3), the resultant aluminum silicate is the basic component of the clays, which are complex silicates containing attached water molecules.

Graphite is also an excellent refractory material since it cannot be spalled (that is, pieces cannot be broken off by thermal shock), because of its high thermal conductivity. Most refractories such as firebrick can withstand temperatures slightly higher than 3000°F before they break down. Graphite tends to oxidize in the presence of air and can be used up to 6000°F. Amorphous carbon and graphite are both forms of the element carbon. Usually, carbon is mixed with a binder such as clay and is formed by molding and pressure. Both carbon and graphite exhibit low tensile to high compressive strength ratios. They also have the unusual property of gaining in strength as the temperature rises. After being baked in an oven to harden and strengthen it, the carbon is either ground or machined. Graphite is relatively easy to machine, but carbon requires carbide, ceramic, or diamond tools because it is very abrasive.

The abrasive silicon carbide can also be used as a high-temperature refractory, but it is quite expensive to utilize in this way. Refractory bricks containing large quantities of chromium oxide, referred to as chromite refractories, are especially well suited for high temperature use in steel melting furnaces. Magnesite brick, composed largely of magnesium oxide, is also used for this purpose. Insulating firebricks are made from ordinary fireclay, but to provide porosity, the clay is combined with sawdust or coke, which is burned out when the brick is fired.

FIGURE 23.2. Firebricks are to be used for lining a furnace.

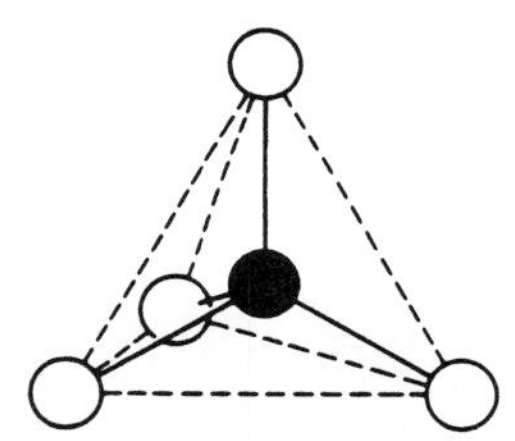

FIGURE 23.3. Unit structure of silicon dioxide (SiO_2). The darkened circle is silicon and the plain circles represent oxygen atoms.

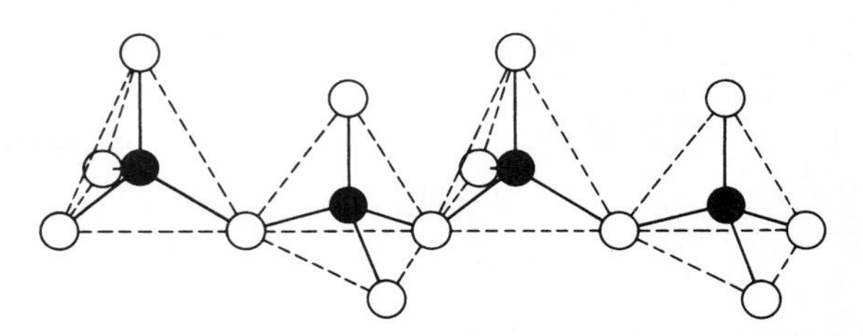

FIGURE 23.4. Silicate units can link through any of the oxygen atoms as shown. This is one way in which a chain of silicate tetrahedra can be joined. The dark circles represent silicon atoms and the plain circles are oxygen atoms.

Abrasives, such as aluminum oxide and silicon carbide, are also ceramic materials. Most abrasives in use today are manufactured synthetics rather than the natural abrasives such as emery that were used in the past. Aluminum oxide abrasives are produced from bauxite ore that is calcined and placed in an electric furnace with steel turnings and coke. A fused ingot of aluminum oxide is formed. The ingot is allowed to cool; it is then crushed and screened for grade size.

Silicon carbide is also produced in an electric furnace. Silica, coke, sawdust (to produce porosity), and salt are charged in the furnace. A silicon carbide ingot is formed, which is allowed to cool and is then crushed, cleaned, screened, and graded for grit size. Grit (grain) size refers to the particle size that will pass through a screen or grid with a given number of holes; that is, a number 100 grit size will pass through a 100 mesh screen that has 100 meshes per linear inch; the size of this particle is 0.010 in. in diameter. Particle sizes range from very coarse (numbers 6 to 12), coarse (numbers 14 to 24), medium (numbers 30 to 60), fine (numbers 70 to 120), very fine (numbers 150 to 240), and flour size (numbers 280 to 600).

Some manufactured refractory materials such as aluminum oxide are machined with diamond tools before firing. Silicon, a metalloid (which is a nonmetal that has some characteristics of a metal or forms an alloy with one), is very hard and abrasive and can be successfully machined with a polycrystalline diamond tool (Figure 23.5).

FIGURE 23.5. Megadiamond insert turns rebonded fused silica cylinder. (*Courtesy Megadiamond Industries*)

ADVANCED CERAMICS

Electronic ceramics used for solar cells, lasers, optical fibers, piezoelectric devices, and integrated circuits account for most of today's market and will continue to dominate the advanced ceramic market for some time to come. However, the structural ceramics promise to become replacements for metals and plastics in many areas of manufacture.

Polycrystalline aluminum oxide fiber is being used to strengthen aluminum piston rods since it is an ideal material for reinforcing nonferrous metals. This not only increases the stiffness and fatigue strength three or four times, it increases the high temperature capability of the metal as well. Silicon carbide and silicon nitride have been leading ceramic candidates for gas turbine research. Silicon nitride is produced in hot pressed, hot isostatically formed, and sintered shapes. Its properties are high strength (100,000 psi in bending up to 1858°F), good resistance to thermal shock, low thermal expansion, low thermal conductivity, high hardness, and good corrosion resistance. Also, silicon nitride cutting tools increase productivity in machining cast iron. Missile randomes are made of this material, as are turbocharger rotors. Zirconia and partially stabilized zirconium oxide (PSZ) show great potential for automotive engine parts, such as ceramic pistons, sleeves, valves, and seats.

Combinations of these ceramics as well as metals and ceramics are being developed for special purposes. Many other spin-off technologies and applications, such as in tool and die work, wire dies, ball and roller bearings, and for wear applications will be seen in the near future.

There are at present some drawbacks that must be overcome before these ceramic materials will be as widely accepted as metals and plastics. One major problem is brittleness and imperfections that could lead to failure. Ceramics tend to develop microscopic cracks and flaws during production that can later increase in size and lead to breakage. Such production steps as joining ceramics to other materials and grinding or machining pose greater difficulties than they do with most metals, thus increasing the cost of ceramics. No doubt these problems and challenges will be resolved in the future.

CONCRETE AND MORTARS

Silica, when combined with sodium and water, becomes sodium silicate or water glass, which, when dried, forms a silica gel. This material was once used for bonding grinding wheels. Similarly, portland cement, used for pavements, foundations, and other construction, contains a large proportion of calcium silicate; it also forms a silica gel when combined with water and allowed to harden. The difference is that portland cement is a "hydraulic" cement that hardens in the presence of water and does not need air to dry. In fact, just the opposite is true; if it dries out before it is "cured," the cement will not get hard and will crumble away. Concrete made with portland cement is often poured under water with forms separating it from the water for structures such as bridge supports. This process of hardening is called *hydration.* During hydration, the cement particles begin to form a gel on their surfaces when in contact with water. Whiskers of silica then begin to grow outward from each particle, intermeshing with fibers of adjacent particles to form a hardened, interlocking mass.

Portland cement is basically composed of clay and limestone (calcium carbonate, $CaCO_3$), with small amounts of iron oxide (Fe_2O_3) and dolomite ($MgCO_3$). These materials are ground to a powder and fired in a large rotary kiln. The resulting "clinker" is again ground to a fine powder as a final operation (Figures 23.6*a* and 23.6*b*). Cements are ground to about 44 microns particle size, or 325 mesh. The cement powder is shipped in bulk or in sacks that weigh about 100 pounds each.

Mortars for brick construction are composed of sand, hydrated lime, and portland cement. Concrete is a mixture of portland cement, water, and a suitable aggregate such as gravel and sand. Expanded, lightweight aggregates such as pearlite or vermiculite are sometimes used in concrete when weight is a factor. Concrete aggregate is composed of graded sizes of clean (washed) gravel and coarse sand to fill the voids (spaces between the gravel). Sand may have a particle of grain size that varies from 100 mesh to 3/16 in. Gravel is measured by the average aggregate diameter, which ranges from 3/16 to 1½ in. and occasionally up to 2½ to 3 in.

Several proportioning methods are used to determine the concrete mixture by either volume or weight. An approximate method used for small batches of concrete is to refer to the cement content as a unit followed by fine aggregate (sand) and coarse aggregate (gravel). A 1:2:3 mixture would contain one part cement, two parts sand, and three parts gravel. A proportioning method employed, when a large quantity of concrete is involved, uses the 94 lb net weight (about 1 ft^3) bag of cement as a measure for the cement. The number of bags or sacks of cement used per cubic yard of aggregate serves as a measure. A 5-sack mixture is considered to be a standard mix for most purposes.

Proportioning is also done according to the weight of the aggregate. With this method, the bulk density of the gravel and sand must be known in order to determine the correct proportion of each. The gravel forms, more or less, a group of spheres with voids or holes between them (Figure 23.7). Sand also consists of smaller spheres with voids. Ideally the cement fills the voids in

FIGURE 23.6a. Powdered cement materials are fired in this large rotating kiln, producing a clinker at the Oregon Portland Cement Company.

FIGURE 23.6b. Large ball mills, such as this one, are used for raw grinding basic materials and clinker in the production of cement at the Oregon Portland Cement Company.

the sand and the sand-cement mixture fills the voids in the gravel.

A method of expressing various mixtures that are used for premixed concrete is based on its expected ultimate compression strength when it is completely cured. Since a standard mix, for example, should reach at least 2000 psi, a 2000, 2400, or 3000 psi mix might be requested for a particular job. Compression tests are used to determine concrete strength after the curing period. Standard mixes for concrete sidewalks, foundations, and driveways should have a breaking strength of 2400 to 3000 psi and a high strength concrete up to 7000 psi. Additives, such as chemicals to speed the hardening or setting rate or to entrain or homogenize the entrapped air, are sometimes added when they are needed for a specific job.

The strength of concrete is determined largely by the proportion of cement used per a cubic yard of aggregate; a standard 5-sack mix is more economical than a 6-sack mix and is employed for most purposes; however, a 6- or 7-sack mix results in a higher-strength concrete because of the extra cement that is added. Strength is also dependent to some extent on the amount of mixing water used; too much water produces a weak concrete. A slump test is sometimes used to determine water content (Figures 23.8*a* to 23.8*c*). Most concrete, for testing purposes, reaches its highest strength in 28 days, but for all practical purposes, it continues to grow in strength for 25 years after pouring if it is in a damp environment.

The sharpness of the sand aggregate also affects the final strength of the concrete to a considerable degree. Sharp glacial sand, found in the Midwest where ancient glaciers formed sand, produces a very high strength concrete, but rounded coastal or river sands produce a weaker concrete if all other factors are the same. Beach sand is a very poor aggregate for concrete because it is not only rounded but also contains salt, which slows the curing of concrete and prevents the concrete from attaining its full hardness. Any humus or dirt of any kind in the aggregate will also lower the ultimate compression strength of concrete. If concrete is poured in freezing temperatures without covering it or protecting it from the low temperatures,

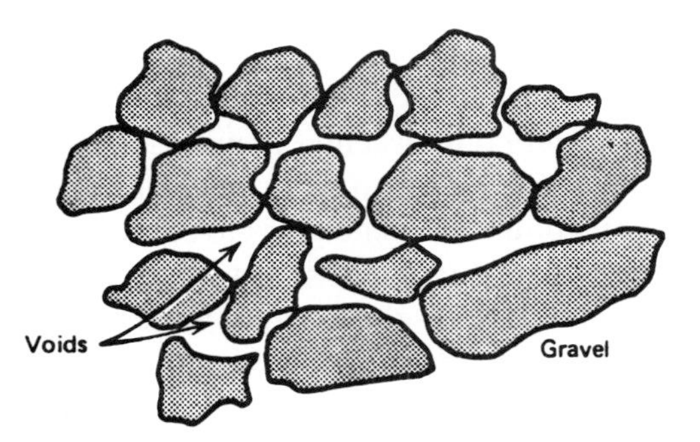

FIGURE 23.7. Gravel showing voids. Sand and cement fill the voids to make concrete.

*FIGURE 23.8***a.** Mold used to make slump test.

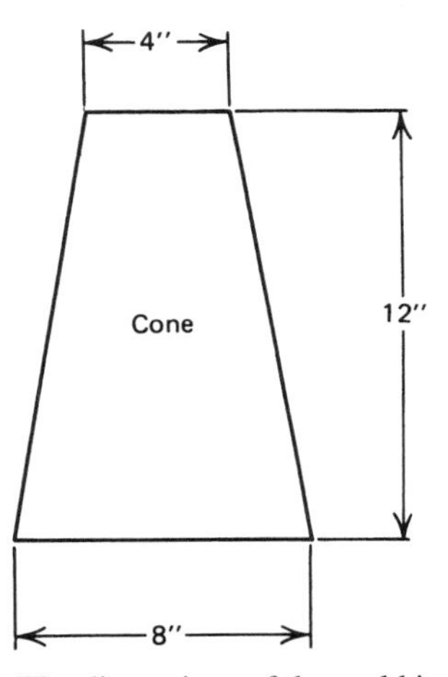

*FIGURE 23.8***b.** The dimensions of the mold in Figure 23.8*a* are given here. The concrete mixture to be tested is placed in the mold and rammed with a steel rod. The excess at the top is struck off and the mold removed vertically.

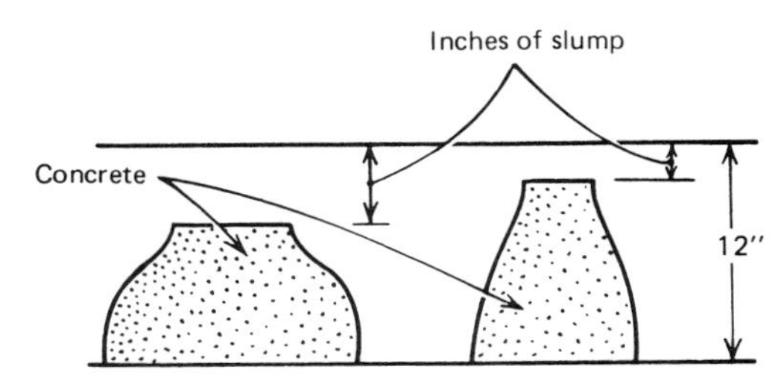

*FIGURE 23.8***c.** After the mold has been removed, the concrete will slump (flatten out) an amount that depends on its consistency, which is controlled mostly by mixing water.

it can be damaged or ruined altogether. If just the surface freezes, it will later flake off and be unsightly. However, if it freezes completely through, it will not harden at all and will have to be removed and replaced.

Concrete is used extensively in the construction of building foundations, as pavements for streets and highways, bridges, and building construction (Figure 23.9). Since concrete has a low tensile strength as compared to its high compression strength, concrete load support beams must be of massive size. Prestressing concrete beams provides the tensile strength of steel wires. Long, slender, more graceful beams and concrete structures are made possible with this technique (Figure 23.10). The steel wires do not corrode because they are buried in the concrete where oxygen cannot penetrate. Steel wires and reinforcing rods embedded in concrete can rust if sodium chloride is in contact with the concrete. It can slowly seep through tiny pores and cracks in the concrete reaching the steel. The rust that forms enlarges the steel rod and causes increased cracking, thus accelerating the process. Some concrete bridges have collapsed because of the proximity of salt to the concrete. Wires and rods are often covered with epoxy in new construction.

Another form of concrete construction is sometimes called gunite, a method of using compressed air to blow a specially prepared concrete mixture onto a surface. It is used to form a protective covering on steel construction and to line drainage ditches and the inside of tunnels. Tunnels are lined by blowing the concrete mix onto the inside surface that is covered with a steel mesh and reinforcing bars. The concrete is mixed with a high slump and a consistency that allows it to "stick" when blown onto a surface. Probably the most common use of this method is for building backyard swimming pools (Figure 23.11). Some advantages of this way of placing concrete are that it produces a very high strength, dense concrete and facilitates the construction of artistic and complex shapes without the use of expensive forms.

Sewer and water pipes are made of concrete and a useful cement-asbestos concrete is used to make "transite" pipe. Transite pipe is sometimes machined on lathes to form fitted joints. The dust from machining this material can cause lung irritation and should be removed with an exhaust fan. The dust is also very abrasive on tools and machine ways must be protected to prevent damage.

GLASS

Quartz is composed of silicon dioxide (silica), from which the amorphous structure known as glass is made by fusing the silica with an oxide such as soda or lime. This combination causes the glass to solidify at a lower temperature than quartz, and instead of forming the crystalline structure of quartz, it solidifies directly into an amorphous or random lattice structure.

Glass is actually a supercooled, rigid liquid. It is in a metastable state, that is, it can go to a stable lower energy state only by passing through a higher energy state. The molten glass is cooled slowly to prevent crystallization.

FIGURE 23.9. Concrete foundations are being poured for a large building.

FIGURE 23.10. Graceful structures such as this curved concrete bridge are made possible by the technique of prestressing concrete beams.

FIGURE 23.11. Placing concrete in a backyard swimming pool with the gunite process.

Additives can alter the glass, giving it special properties. For example, the common soda-lime glass softens at lower temperatures and is used for windows, glassware, and lightbulbs. Lead-alkali-silicate glass has an even lower softening temperature; it is known as flint glass, and is used for optical purposes and crystal glass for tableware. Borosilicate glass, with its low thermal expansion, does not tend to crack when unevenly heated. It is therefore used for household cookware, laboratory glassware, and large telescope mirrors.

Glass has an excellent chemical resistance, except to hydrofluoric acid, which is often used to etch glass objects. Glass also has excellent colorability; either translucent or opaque coloring can be imparted to glass products.

Silicon (in alumino-silicate structures, sand, and many other compounds) is the second most plentiful element in the earth. This is why glass is a relatively inexpensive material.

Glass is a very strong substance in some ways. A glass plate, for example, will support a heavy load without taking a permanent set (deforming). However, since glass is also a very brittle and notch-sensitive material with low tensile strength, a single scratch could cause a sudden failure when loaded in tension.

Glass is often *annealed* to remove internal stresses. Annealing consists of reheating the glass to a temperature below the melting point and slowly cooling to room temperature. To strengthen glass, it is often *tempered.* In this process, the glass is also heated to just below its melting point (about 1000°F) and quickly cooled in an air blast or oil bath. This causes the surface of the glass to be in compression while the inside of the glass object is in tension. In this condition, glass is less likely to break when placed in tension. Safety glass is often tempered glass. Glass that is spun into fiber, known as fiberglass, is used with other components such as plastic resins to produce very tough materials that are not brittle or notch sensitive.

SELF-EVALUATION

1. Why are ceramic materials such as glass, fired clays, and stones rigid and brittle?
2. In what way do ceramic materials have high strength?
3. What is a refractory material and what is it used for?
4. Name the two principal ingredients in clay.

5. Abrasives such as aluminum oxide and silicon carbide are both produced in an open-hearth furnace. True or false?
6. Portland cement contains sharp sand and gravel or vermiculite. True or false?
7. The strength of concrete is determined mostly by the properties of cement and mixing water plus the sharpness and cleanliness of the aggregate. True or false?
8. How can slender beams be used in constructions that are made of a substance such as concrete that has a low tensile strength?
9. What is glass made of?
10. Define a composite material.
11. Portland cement hardens and bonds together in the presence of water. What is this process called and how does it progress to form a solid mass?
12. In the area of advanced ceramics, how are structural ceramics used to strengthen metals such as aluminum? Name three uses for silicon carbide and silicon nitride besides their uses for abrasives.

CASE PROBLEM: THE CASE OF THE CRUMBLING CONCRETE

A concrete foundation for a house was poured one day. The aggregate was clean and sharp and the correct proportions of cement and water were used. That night the temperature fell to −15°F. Several days later the forms were removed and one piece of the foundation stuck to the forms and came off. The entire foundation was weak and crumbling off. What happened? Did the workmen remove the forms in the wrong way? What do you think caused this problem? What could they have done to prevent this?

FRAMING LUMBER

CHAPTER 24

Wood and Paper Products

Trees, or timber as known by the lumber industry, are one of our most valuable renewable resources. The homes we live in, the paper we write on, furniture, and many other products depend on our northern, southern, and western forests. You will learn in this chapter about the many operations required to process the raw timber into useful products.

OBJECTIVES **After completing this chapter, you should be able to:**

1. Describe the structure of wood and how it is processed to make lumber and plywood.
2. Explain the process of making paper.

Timber, especially standing timber, refers to a stand of commercially valuable growing trees. Forest products consist of anything made of the woody parts of trees such as boards, planks, timbers (wood beams), plywood, and paper.

In North America, there are basically two kinds of forests that are harvested for commercial purposes: softwoods and hardwoods. Both hardwoods and softwoods are used in our civilization in a variety of applications. Hardwoods are typically used for furniture making. Conifers (softwoods) are used extensively for lumber products. Large stands of forests in the southern and western United States are harvested for lumber, plywood, and particle board (Figure 24.1).

HARDWOODS

Deciduous trees shed their leaves each year and are classed as hardwoods. Oaks, maples, and birch are in this group and grow in the northern forests from Minnesota east and south to Virginia. Sweet gum, ash, and tupelo are found in Texas and the South. Hickory, oak, black walnut, and basswood grow in the prairie states. Balsa, although it is very light in weight, soft, and porous, is classed as a hardwood because it comes from a deciduous tree.

FIGURE 24.1. A stand of Douglas fir trees that will be harvested for lumber and other products. (*Weyerhaeuser Company*)

SOFTWOODS

Coniferous trees have cones and needles instead of flat leaves and are sometimes called "evergreen" trees; that is, they do not lose their needles in the fall but keep them throughout the year. The coniferous trees are softwoods such as the southern and northern pines; western softwoods include the Douglas fir, white fir, pines, hemlocks, and spruces.

Softwoods such as fir, pine, and spruce often have greater hardness and strength than some hardwoods. The sitka spruce, for example, is classed as a softwood because it has needles, but is considered to have the highest strength-to-weight ratio of any wood. It is noteworthy that sitka spruce was once used for structural frames in aircraft. It is also used to build musical instruments because of its resonant qualities.

STRUCTURE OF WOOD

Growth takes place in trees in the cambium layer, which is the only living part of the tree. Nutrients and moisture are carried to the leaves through this outer layer. As the tree grows, rings of inner cambium cells become inactive in winter, lose their moisture, and become sapwood. The sapwood eventually becomes part of the lifeless core or heartwood, which is the useful part of the tree (Figure 24.2). These growth rings, called annular rings, indicate the age of a tree as a new growth ring is formed every spring. The heartwood is composed of a cellular structure made of fibrous cellulose. These cells are held together by a natural adhesive called lignin (Figure 24.3).

Each growth ring consists of two distinct layers. The inner part of the growth ring, which is produced early in the growing season, is called earlywood or *springwood.* The outer layer is called latewood or *summerwood.* Differences in springwood and summerwood occur in different species. In oaks, Douglas fir, and southern yellow pine, the springwood is less dense, coarser in texture, and softer than the summerwood. In Figure 24.3 this difference can easily be seen in the large boxy cells of springwood and the flat-

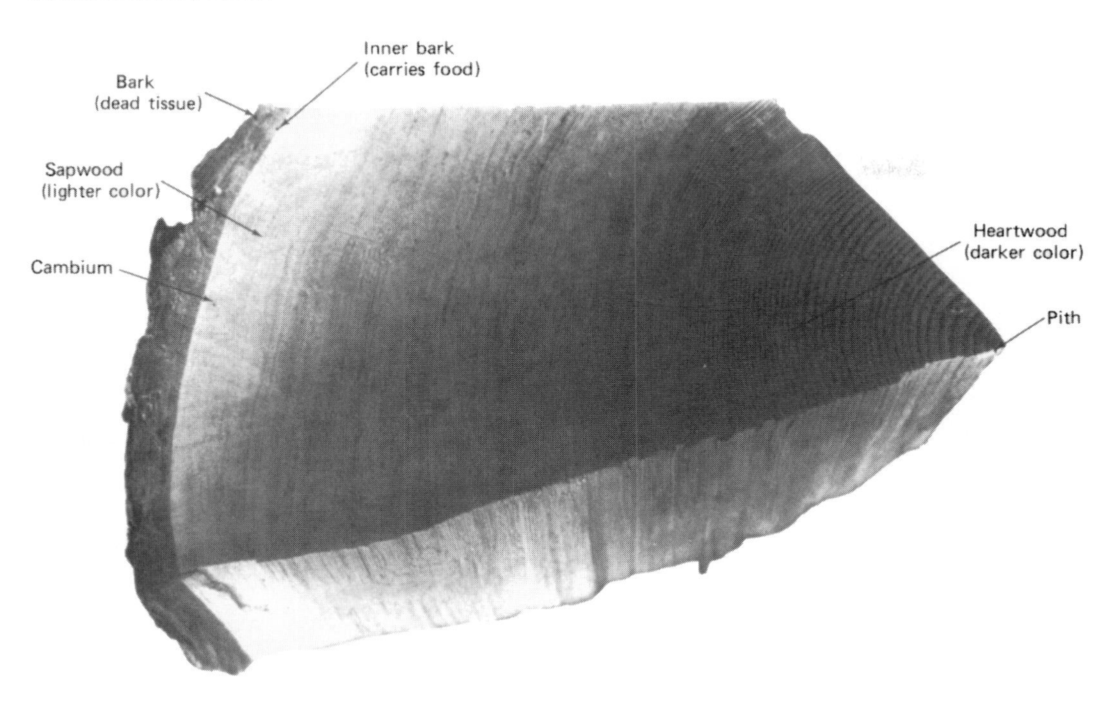

FIGURE 24.2. The cambium layer, which is the living part of the tree trunk, carries nutrients from the roots to the leaves, buds, and flowers. The bark is dead tissue that protects the tree. The lifeless core (the heartwood) is the principal source of lumber and pulp, and the bark and cambium supply wood distillates, chemicals, and fuel.

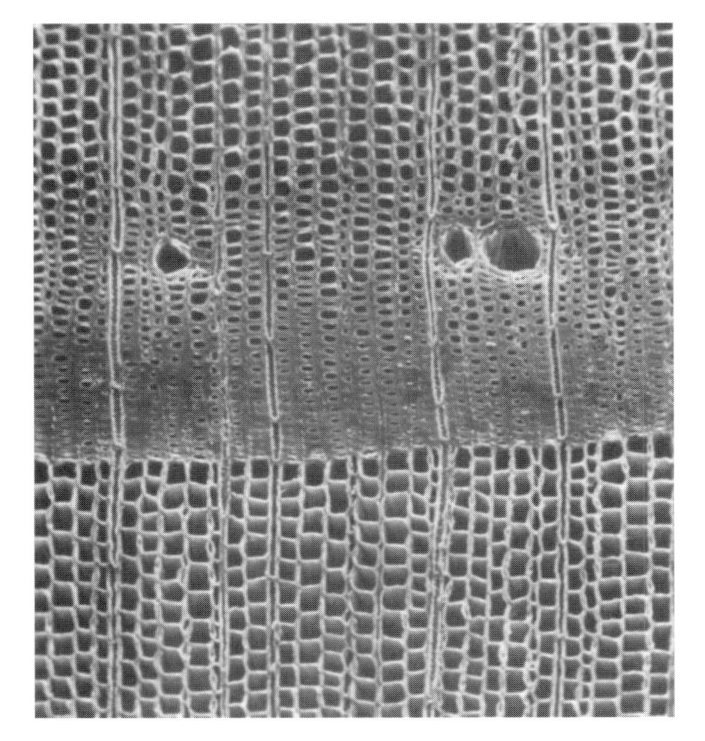

FIGURE 24.3. Transverse surface of a conifer or softwood (Douglas fir) illustrating the differences in cell shape and wall thickness between earlywood and latewood fibers. The latter are thicker walled and flattened radially. Three vertical resin ducts are also seen here in cross section (100×). (*Scanning electron micrograph courtesy R. A. Parham of The Institute of Paper Chemistry, Appleton, Wisconsin*)

tened cells of summerwood. However, for other species such as the maples, birches, and gums, the growth ring is uniform throughout its width and no importance is attached to the springwood and latewood. The percentage of summerwood, therefore, is an indicator of wood quality.

LUMBER PRODUCTION

When these trees are ready for harvesting, they are felled (Figure 24.4) and trucked to the mill (Figure 24.5). The logs, while awaiting processing, are not allowed to dry out as this would cause cracking and splitting. The logs are stored in either a mill pond (Figure 24.6) or a cold deck where they are kept wet with sprinklers (Figure 24.7). The logs are first sawed into large slabs on the head rig (Figure 24.8) from which various lumber products (boards, planks, or timbers) are made. Figures 24.9*a* to 24.9*c* show the various cuts that can be made to utilize the log in the best ways. The lumber is further cut to usable sizes on other sawing machines. It is then carried along on a green chain (a set of moving chains that carries newly sawed lumber slowly along a long table). Here it is graded or selected for different uses.

Wood can have defects such as knots, dry rot, cracks, or pitch pockets, which explains why lumber

FIGURE 24.4. A tree being cut down (felled) in a Douglas fir forest. (*Weyerhaeuser Company*)

FIGURE 24.5. Logs being trucked to the mill. (*Weyerhaeuser Company*)

FIGURE 24.6. Logs stored in a mill pond. (*Weyerhaeuser Company*)

must be graded (Figure 24.10). Softwood lumber is graded into three major categories: yard lumber, shop lumber, and timbers. Yard lumber is divided into select, common, and dimension lumber. The select (finish) lumber is graded A, B, C, or D according to the appearance of the board. Select grades are used for window frames, doors, and trim because they can have no defects and must be clear of knots and other blemishes.

The common grades of yard lumber are divided into five numbered grades. Number 1 grade common is free from structural defects that have no loose knots. Number 2 has some structural defects. Number 3 can have loose or open knots and be twisted or warped. These grades are used for construction. Numbers 4 and 5 are not acceptable for construction specifications because they may have pitch or dry rot.

FIGURE 24.7. Pile of logs, called a cold deck, is waiting to be processed in the mill. (*Weyerhaeuser Company*)

FIGURE 24.8. Logs are first sawed on the head rig. Other machines cut these slabs into smaller boards. (*Weyerhaeuser Company*)

Shop lumber is used for such purposes as furniture parts, cabinets, and window sashes. Timbers are used for structural purposes. A 4 in. or larger dimension is considered to be a timber.

The lumber, after it has been graded, is either dried in the open air in stacks or is kiln dried by being subjected to heat in an enclosed area. This operation tends to shrink the wood to some extent. This shrinkage can cause warping, depending on how the board was sawed from the log (Figures 24.11*a* and 24.11*b*). Rough-sawed lumber is then finished by planing in a machine with whirling knives. By-products of sawing and planing in lumber mills are sawdust and chips. These can be processed into particle board or pressed logs for burning (Figure 24.12). Wood products are sometimes immersed in chemicals

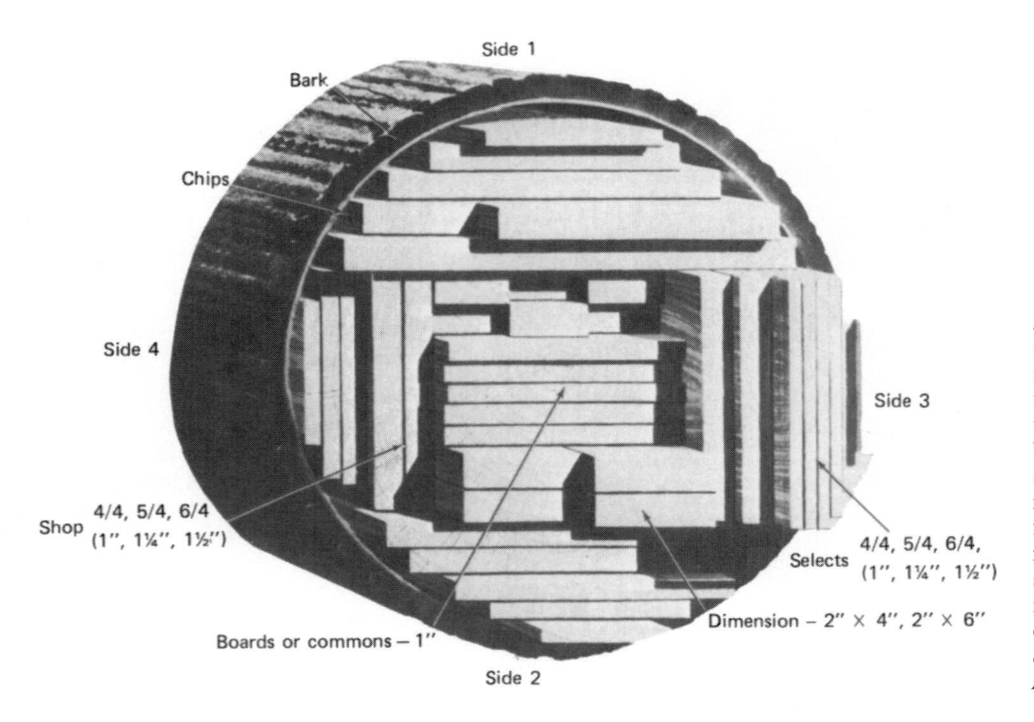

***FIGURE 24.9*a.** The sawyer opened the log marked Side 1 and sawed off five thicknesses before turning the log and sawing Side 2. He then removed five pieces of lumber before turning the log and sawing Side 3. After removing eight cuts, he rotated the log and took five cuts from Side 4. By this time, he had reduced the log to a timber 12 in. square and had the choice of reducing it further or transferring the timber to other saws in the mill. (*Photo courtesy Western Wood Products Association*)

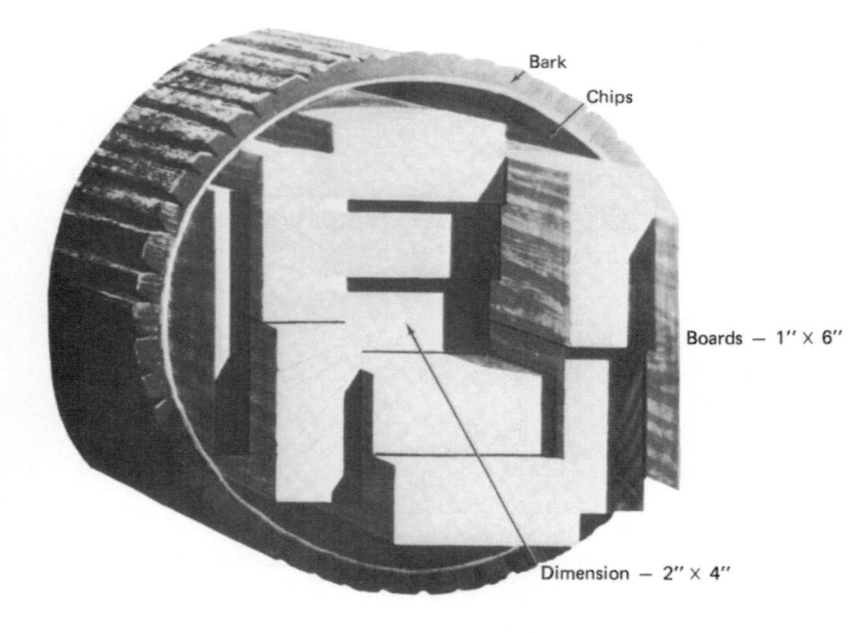

***FIGURE 24.9*b.** This method is used by mills specializing in sawing small logs. The whole log passes in one motion through a series of circular or band saws and wood chippers, which reduce it to 2- or 4-in. thick pieces. These pieces are then turned flat and transported through a series of saws, which cut them in one operation into 2 × 2's, 2 × 3's, or 2 × 4's. Such pieces, when cut 10 ft or shorter, may be specially graded and called studs. (*Photo courtesy Western Wood Products Association*)

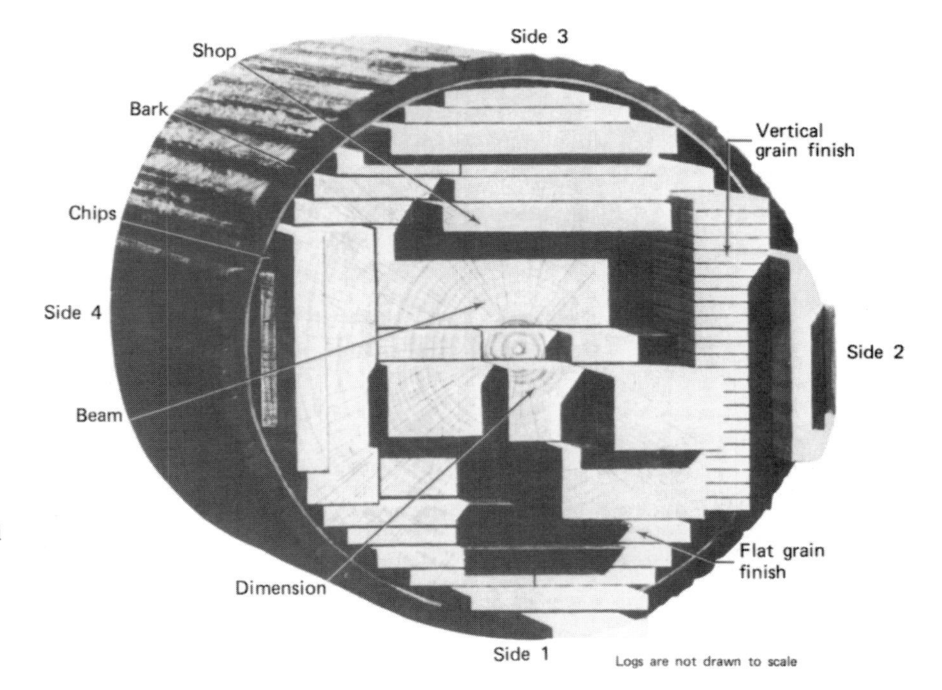

*FIGURE 24.9***c.** This method is called "sawing around the log." The sawyer took six cuts from Side 1, then turned the log to begin sawing Side 2. The same procedure is followed with Sides 3 and 4 until a 16-in. square timber remained. The same choices are available as with Figure 24.9*a* to reduce it as shown or market it as a timber 16 in. square. (*Photo courtesy Western Wood Products Association*)

FIGURE 24.10. Lumber being graded. (*Weyerhaeuser Company*)

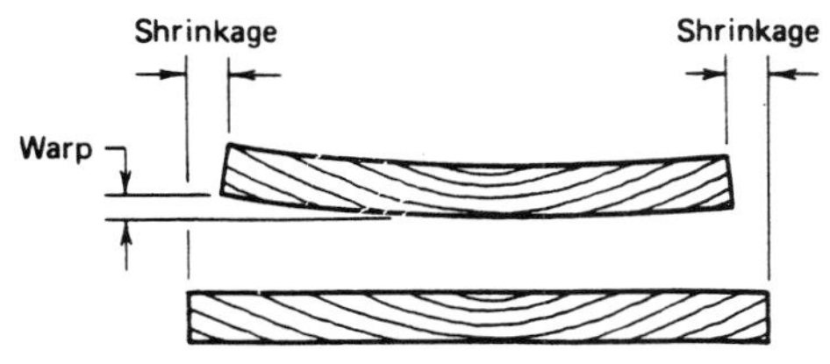

*FIGURE 24.11***a.** Warping is a disadvantage with plain sawed lumber. Wood products always shrink when dried.

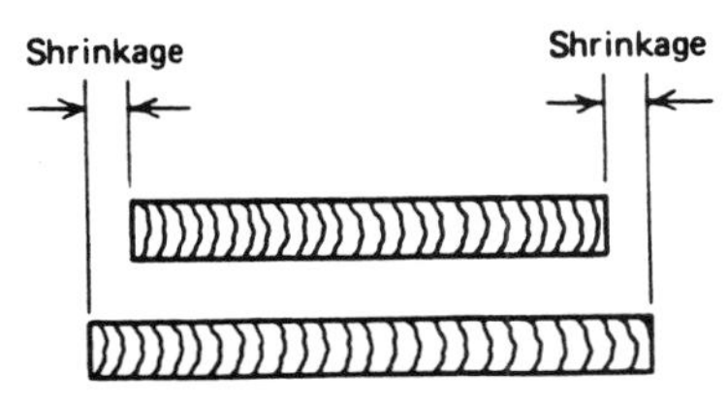

*FIGURE 24.11***b.** Shrinking without warping is the major advantage of quartersawing (sawing perpendicular to the annular rings).

FIGURE 24.12. Sawdust is pressed to make logs for fireplace use. (*Weyerhaeuser Company*)

FIGURE 24.13. The graceful arched beams in this construction project are laminated and glued. The trusses are still covered with paper to protect them from the elements while construction is in process.

such as wood preservatives to prevent dry rot and fire retardants.

A rather unique recent development of the wood products industry is the laminated construction of beams that make possible graceful shapes (Figure 24.13) of strong, lightweight trusses that are used to support roofs in building construction. Flat boards are laid together and clamped in fixtures that form the shape of the beam. Strong adhesives are used to bond the boards together. Many of the weaknesses of sawed beams (cut from a single log) are eliminated with the lamination method. Knots, cracks, and rot can severely weaken a sawed beam, but in laminated beams the best boards, free from defects, are used on the top and bottom sides of the beam where the stress is the greatest.

PLYWOOD

Plywood is made by gluing veneer (thin slices of wood) together in alternate layers in which the direction of the grain is placed at right angles to the previous layer. Since wood is stronger in one direction than in the other, crossing the grains greatly increases its strength. This crisscross stack of veneer and adhesive is then placed in a hot press that has a series of heated plates or platens. Pressure is applied until the glue has hardened (Figure 24.14). The plywood sheets are then removed from the hot press, trimmed to size, and sanded to give them a smooth, finished surface. The thickness of the finished plywood sheets depends on the number of plies or layers of veneer. Veneers are sliced from logs by either rotating them against a knife (Figure 24.15) or slicing them with a reciprocating motion (flat cut) across the log (Figure 24.16).

Various hardwood veneers are used in the manufacture of furniture, and special cuts are needed for this purpose to provide a satisfactory wood grain, which can be either the flat cut or rotary cut. Softwoods, such as fir or pine, are usually cut into veneer by the rotary method for making plywood. Plywood is mostly used in building construction (Figure 24.17). Like lumber, plywood is also graded by a classification system.

HARDBOARD AND PARTICLE BOARD

Hardboard and particle board are made from wood fibers such as sawdust or wood chips. Hardboard is bonded together by its own natural adhesive (lignin) while under heat and pressure. Adhesives are added to particle board to bond the fibers together when they are compressed. Both processes are somewhat similar, but hardboard is stronger and more dense than particle board.

In 1924, William H. Mason discovered the process of making hardboard, which was then produced under the name "masonite." Standard hardboard is a dense and stable material that is used for making furniture parts,

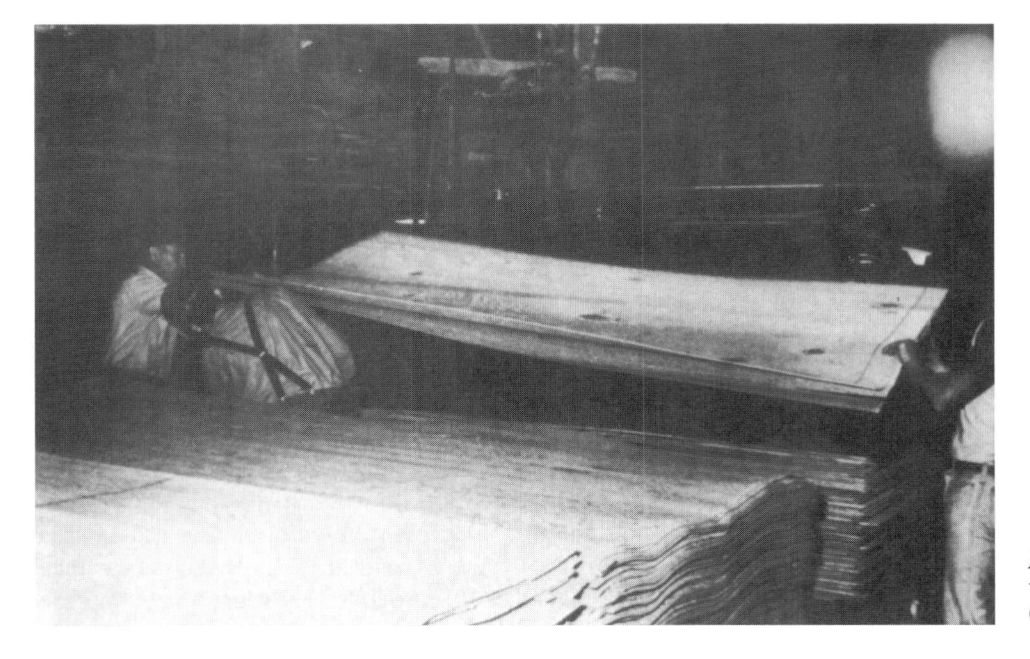

FIGURE 24.14. Stacks of glued veneer being put into a hot press. (*Weyerhaeuser Company*)

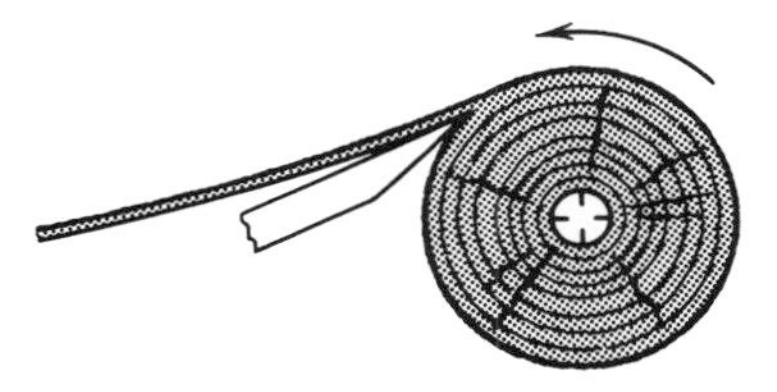

FIGURE 24.15. Rotary cut method of making veneer.

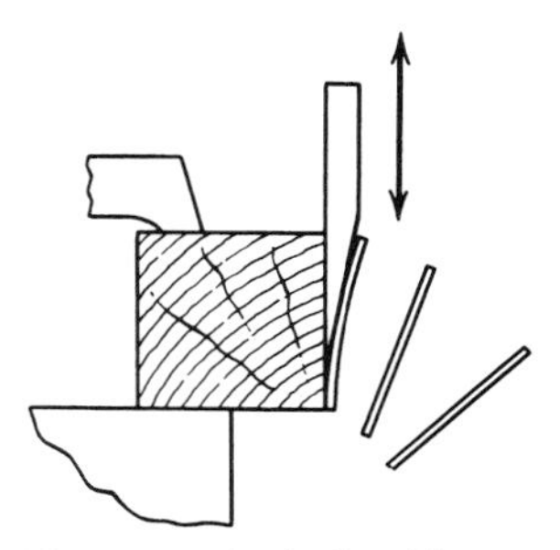

FIGURE 24.16. Flat cut method of making veneer.

FIGURE 24.17. Lumber and plywood are used extensively for building construction.

drawer bottoms, pegboard, and many other products. Tempered hardboard is treated with chemicals to make it more dense and water resistant. It is often made with a colored shiny surface for cabinets and bathroom walls.

The manufacture of particle board is a continuous process in which raw materials flow into one end of a mill and the finished board is prepared for shipping on the other end. Figure 24.18 is a flowchart showing the progression of these automatic processes.

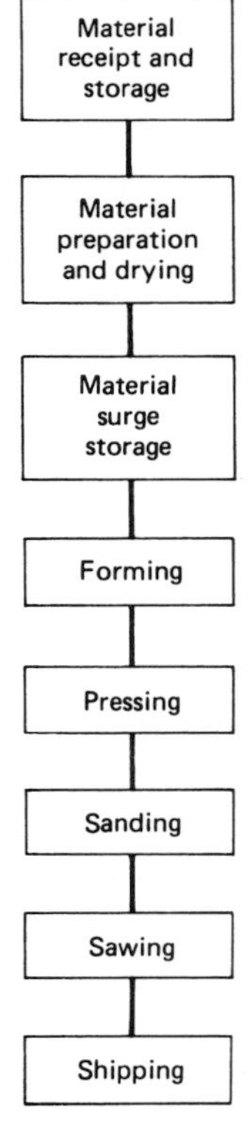

FIGURE 24.18. Flowchart for particle board manufacture. (*Weyerhaeuser Company*)

Particle board is made of wood chips or sawdust that is mixed with an adhesive and formed into a mat (Figure 24.19). Ten to 20 mats are then placed in a hot press between platens and put under pressure for 20 to 30 minutes (Figures 24.20*a* and 24.20*b*). When the boards are removed, a somewhat dense, hard material is formed (although it is not as dense as hardboard). In Figure 24.20*a,* the press with its heated platens (where steam lines are located) is charged or loaded with cauls (metal plates that hold the mat of mixed sawdust particles) from the charging machine (on the right), which raises and lowers, allowing it to be loaded with cauls from the conveyor. In Figure 24.20*b,* a machine similar to the charging machine removes the pressed boards from the press and unloads them on a conveyor in a spaced sequence (Figure 24.21). From there, they are transferred to machines that saw them to size and sand them for a finish (Figure 24.22). Particle board is used extensively for furniture making and building construction. Thin plastic, simulated wood grain surface and other textures are sometimes cemented to the surface of particle board for furniture and wall panels.

PAPER

Our modern civilization could scarcely get along without paper. Papermaking is one of the world's oldest industries. Ancient Egyptians made a kind of writing material from a native grass called papyrus. The art of papermaking seems to have been discovered by the Chinese and from there spread to Europe. Many materials such as linen or cotton rags, wood, and other vegetable fibers have been used to make paper. Perhaps the idea of

FIGURE 24.19. A mat of sawdust and adhesive mixture is formed on a large metal plate called a caul. The cauls move down a conveyor and are placed in a charging machine ready to be inserted in the hot press. (*Bohemia Inc.*)

FIGURE 24.20a. Particle board hot press. A number of heated platens are used to press the mat into a dense board. Large hydraulic cylinders provide the necessary force. (*Weyerhaeuser Company*)

papermaking derived from observing the wasps that build their nests of wood fiber and secretions from their bodies. Paper is used for the printing and publishing of books, magazines, newspapers, building materials, many kinds of containers, and other products.

Softwoods such as spruce, balsam fir, hemlock, and southern pine are usually used as pulpwood for papermaking, although hardwoods are sometimes used. The logs are cleaned and the bark, pitch, and foreign materials are removed. They are then cut into suitable lengths. The clean wood is conveyed to a chipper or a grinder, and the particles are placed in a digester to remove the lignin.

There are four major processes for pulping wood for paper manufacture: the groundwood process, the sulfite process, the soda process, and the sulfate process (also called the Kraft process).

Groundwood Process. During the groundwood or the mechanical process, the wood is ground on a rough grindstone to produce a fine fiber that is usually steamed to soften it. Although this does not produce a chemical change, the paper made from it is not stable, causing it to discolor and often to disintegrate in use.

The other three processes are chemical in nature.

Sulfite Process. In the sulfite process, the chips are cooked in a digester in an acid liquor under pressure. The cooking process in the bisulfite of lime liquor removes most of the noncellulose material in the wood.

FIGURE 24.20b. This is the unloading end of the hot press where the compressed boards are removed and sent on their way to finishing machines. (*Weyerhaeuser Company*)

FIGURE 24.21. Here the particle boards, stripped from the metal cauls, are transferred to a conveyor, where they move at a high rate of speed to the next operation. (*Weyerhaeuser Company*)

FIGURE 24.22. This board is being turned 90 degrees on a transfer conveyor for a finishing operation. The board turner in the background is used to turn the board over for sanding. (*Weyerhaeuser Company*)

The chief advantage of the sulfite process is that the pulp can be easily bleached to produce white writing and printing paper.

Soda Process. Caustic soda is used as a solution in the digesters to remove the unwanted lignin from the wood pulp. This process is used for short fiber woods such as poplar and other deciduous trees.

Sulfate Process. This process is similar in some ways to the soda process in that an alkali (sodium sulfide produced from sodium sulfate) is used in place of part of the caustic soda in the cooking liquor. The sulfate process, or **Kraft process,** makes possible the use of fibrous substances that are difficult to break down by other processes (Figure 24.23). The Kraft process takes less time than the other methods and produces a pulp of exceptional

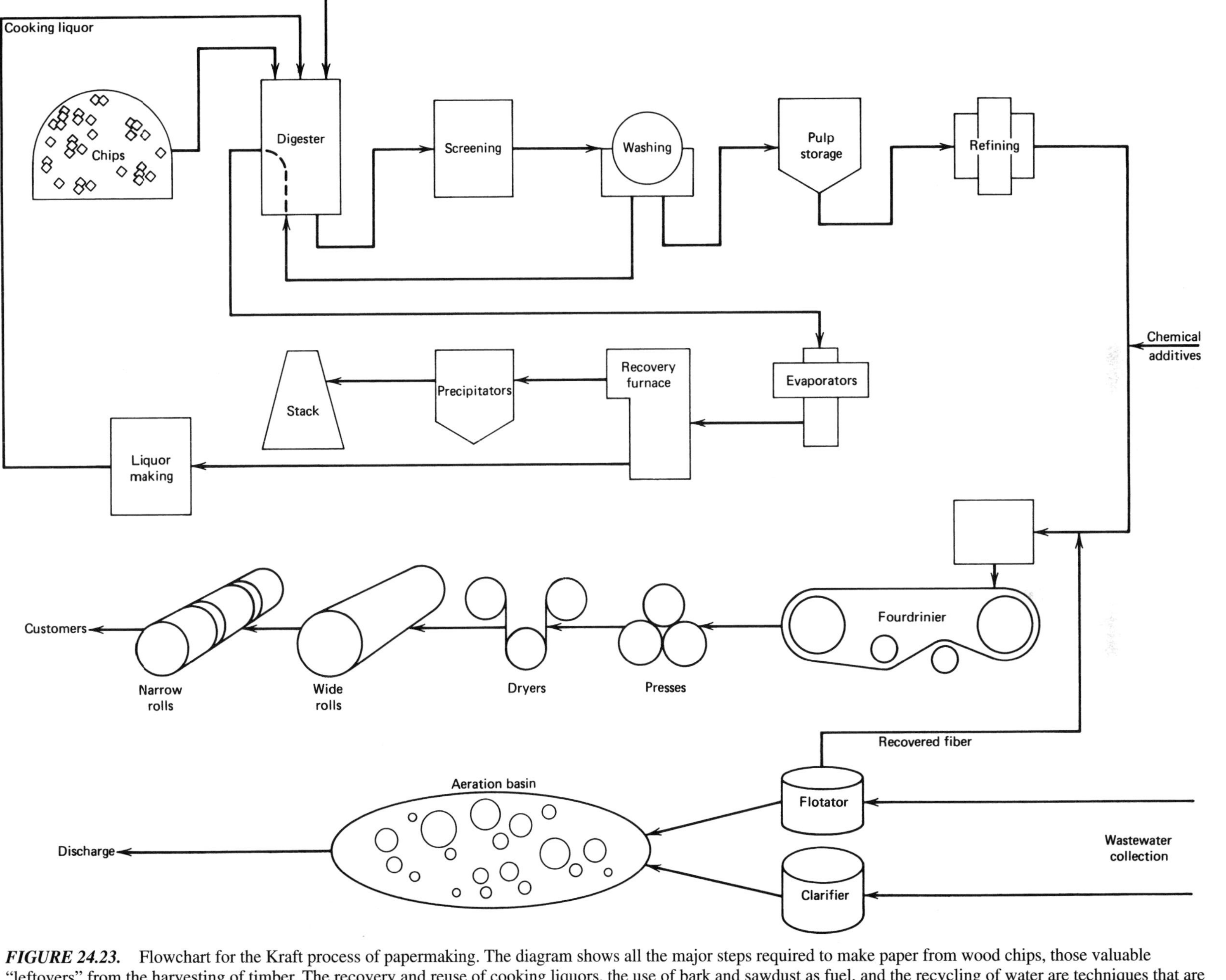

FIGURE 24.23. Flowchart for the Kraft process of papermaking. The diagram shows all the major steps required to make paper from wood chips, those valuable "leftovers" from the harvesting of timber. The recovery and reuse of cooking liquors, the use of bark and sawdust as fuel, and the recycling of water are techniques that are designed to reduce wastes. (*Weyerhaeuser Company*)

strength (Figure 24.24). Sulfate pulp is manufactured from pine in the southern United States, spruce and eastern hemlock in the northeastern United States, and western hemlock in the Pacific Coast states. Other materials used are straw, jute, rags, and waste paper. These materials are often used to make cheap, brown paper for corrugated cartons and other containers. The Kraft process is mostly used in the South and the Far West.

The Paper Machine

In the Fourdrinier machine (named for its inventor), a fiber mat is continuously formed, which is dried, then rolled between a series of heavy steel rolls, and finally made into a thin, hard sheet (Figure 24.25). The Fourdrinier machine is the type of paper machine that is most used to form the pulp into a continuous ribbon on an endless belt of bronze or plastic screen. As the pulp moves forward, water drains from it. The pulp is then squeezed in a series of rolls and is further dried on steam-heated rolls. The finishing rolls further press the paper into a dense product of controlled thickness. The finished paper is then wound on large rolls for shipment (Figures 24.26*a* and 24.26*b*).

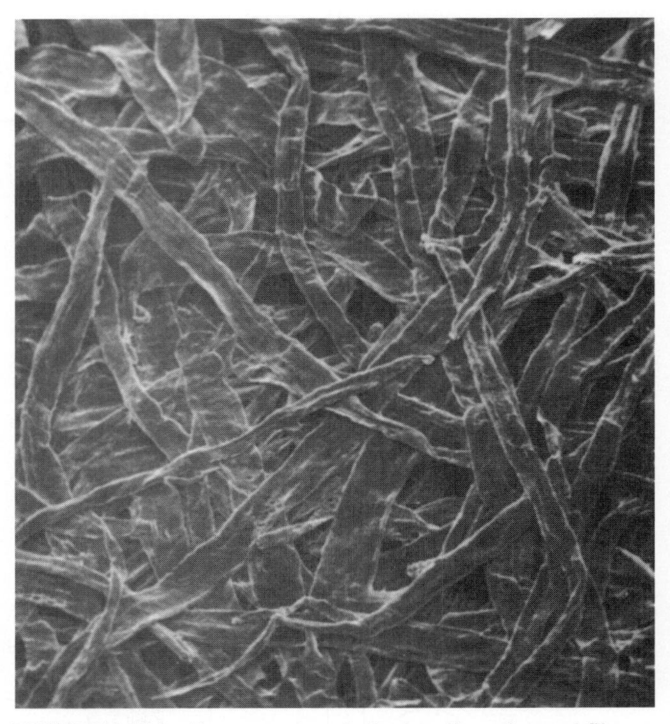

FIGURE 24.24. Paper composed of primarily earlywood fibers of southern yellow pine. They are thin walled and collapse readily into ribbons to form a dense sheet (120×). (*Scanning electron micrograph courtesy R. A. Parham of The Institute of Paper Chemistry, Appleton, Wisconsin*)

FIGURE 24.25. Fiber mat in the Fourdrinier machine. (*Weyerhaeuser Company*)

*FIGURE 24.26*a. At the end of the machine, the paper is wound up on huge rolls. (*Weyerhaeuser Company*)

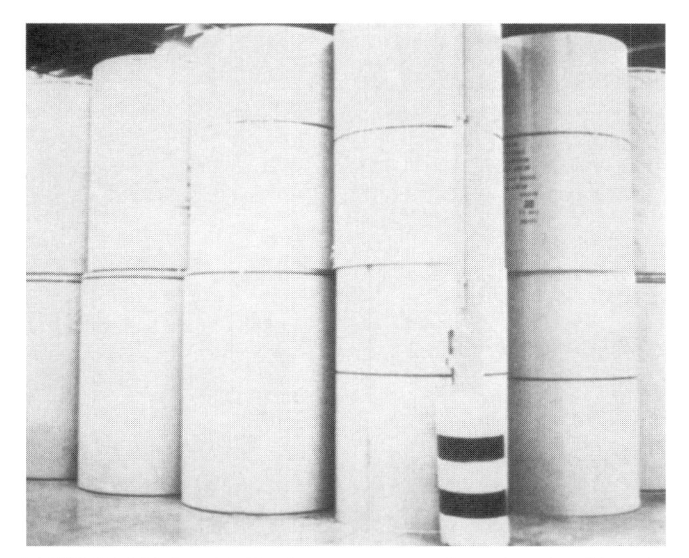

*FIGURE 24.26*b. Rolls of paper are stored, ready to be shipped. (*Weyerhaeuser Company*)

SELF-EVALUATION

1. What are the two classes of wood used for lumber products?
2. Name several coniferous trees.
3. In which part of the tree does the growth take place?
4. How can the age of a tree be determined?
5. Describe the major processes in a sawmill that are required to make finished lumber.
6. Name two by-products of the sawmill and two uses.

7. Why is plywood stronger than ordinary sawed lumber?
8. What is the reason that lumber is graded and a classification system used?
9. What are hardboards and particle boards made of and used for?
10. Briefly describe how paper is made.
11. Why are laminated wood beams much stronger than the same size sawed beams?
12. Is the heartwood (inside the cambium layer and sapwood) a living part of the tree?

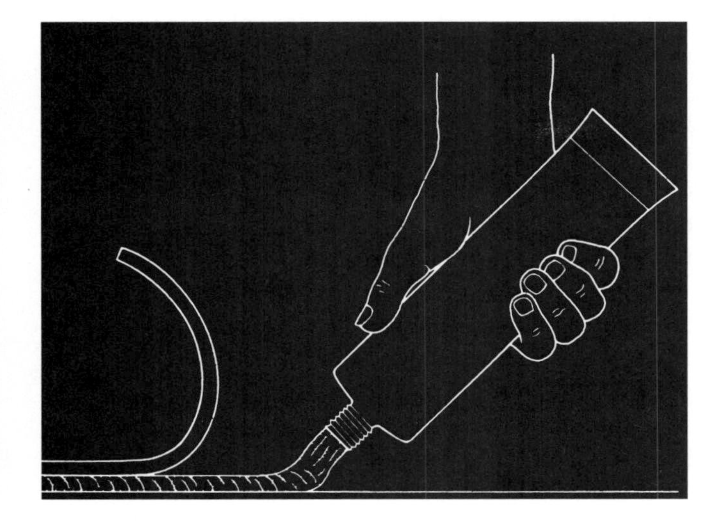

CHAPTER 25

Adhesives and Industrial Lubricants and Gases

Prior to the twentieth century, the only adhesives available were made from the hides and hooves of animals or parts of fish. Vegetable materials such as tree gums, starch, and casein were also used. None of these products provides a very strong bond, and they are easily loosened by the action of moisture. They were used mostly to bond wood for furniture making. In recent times there has been almost a revolution in the development of adhesives, especially in the area of synthetics. Many of the modern synthetic plastics are also used as adhesives.

There are many of these synthetic adhesives, each designed for a specific use. For example, the epoxies are used to bond metal and composites on aircraft and automobiles, some types offering shear strengths up to 10,000 psi. Also, many adhesives double as sealants.

Because of this rapidly expanding use of adhesives in manufacturing, many conventional fasteners such as nails, screws, bolts, and rivets will be replaced. Of course, where parts need to be assembled or disassembled for maintenance purposes, cap screws and bolts will still be needed. Structures that are bonded with adhesives resist damage from vibration better than riveted or bolted assemblies. The bond is made over the entire surface instead of in a few places, as in a riveted or nailed joint where each connector must resist high stress concentrations that can result in cracking. Metal connectors such as bolts or rivets can be a cause of corrosion that requires the flow of electricity, but adhesives prevent electrical current flow because they normally have high electrical resistance, thus reducing corrosion.

Modern industry could not operate without the use of certain gases, oils, greases, and asphalts. All these products are discussed in this chapter.

OBJECTIVES **After completing this chapter, you should be able to:**

1. Explain the basic principles of adhesive bonding.
2. Evaluate a variety of glues and adhesives in terms of their correct applications.
3. State the major advantages of adhesive bonding over other joining methods.
4. Identify several of the more common industrial oils, greases, and gases used in manufacturing and other mechanical applications.

ADHESIVES

Bonding

In all materials, including metals, adhesion requires molecular contact. Adhesion is the result of attractive forces that exist between atoms or molecules; these are secondary valence forces called *van der Waals forces.* Adhesion requires that atoms be no more than a few tens of angstroms from each other. An angstrom is one ten millionth of a millimeter. If it were not for the fact that nothing is truly smooth in the real world, everything would stick together; even glass is not smooth enough to allow its atoms to come close enough to effect adhesion. To overcome this obstacle, a liquid is used that bridges the gap—a liquid that will harden to provide sufficient bond strength. Surface energies such as surface tension in liquids, trapped air, and contaminants can weaken the adhesive bond. For that reason, to realize maximum bonding strength, absolute cleanliness of the surface to be bonded is necessary. This is the reason that aircraft panels are assembled with adhesive in a "clean room" called a layup room, where technicians wear white gloves, socks, smocks, and caps. The air is filtered and slightly pressurized in the layup room. This is where reinforcing plies of aluminum are bonded to the fuselage.

Metals must be cleaned with degreasers or with etching baths such as sulfuric acid, sand blasted, or anodized (as in the case of aluminum). Water is almost always a contaminant and should be removed just before bonding. The correct adhesive should be used for a particular substrate; for example, an anaerobic adhesive such as cyanoacrylate should never be used on rough or porous surfaces. For that use, it simply would not harden.

Adhesives harden in various ways. Some types harden through the evaporation of a solvent or water. Some harden through chemical reaction, as in the case of mixing a hardener with resin when epoxy is used. Other adhesives harden by freezing after being melted, as with the hot-melts used in a glue gun.

Selection of Adhesives

Table 25.1 lists a number of adhesives and their uses. There are many other types of adhesives, some for general use and some for special purposes. Many of these adhesives have trade names or names that describe their use, such as wood glue, particleboard adhesive, tub and tile sealer, or rubber cement (not for rubber but for paper). With the vast array of available adhesives having varying applications, it would be difficult to choose the best adhesive to bond a broken glass object together or mend a loose wooden rung on an antique chair.

Anaerobic cyanoacrylates, the so-called super glues that "stick in seconds," really do just that. This type would work very well to bond the broken glass object just mentioned. They will also bond metal, plastic, and rubber. An anaerobic adhesive is one that will not harden in the presence of air; therefore it does not work well for porous materials. This adhesive should be used sparingly; a drop is usually enough. If excess adhesive gets on your skin, it can stick your fingers together. A solvent or acetone (nail polish remover) is available to remove the glue in that case. Cyanoacrylates resist water to a certain extent after curing (a full cure takes 48 hours), but where moisture resistance is needed, epoxy would be a better choice. These adhesives have a rather short shelf life. Age and weather slowly deteriorate bonds.

Acrylic is like cyanoacrylate and epoxy; it is a high-strength adhesive. It is almost as fast setting as the "instant" glues, but it will bond to a wider variety of substrates: metals, woods, glass, ceramics, cork, concrete, and some plastics such as ABS, vinyl, acrylic, nylon, and phenolic. Like epoxy, it has an ability to fill gaps (cyanoacrylates do not). Unlike epoxy, it is moisture sensitive.

Acrylic is a two-part adhesive like epoxy, but the chemistry is different. The components are not mixed together. The activator is brushed on one surface, and the adhesive resin is applied to the other with a syringe. Curing takes place when the parts are mated. It has one unusual quality: It will bond on oily surfaces as long as the surface is free of loose dirt. The oil is adsorbed by the adhesive.

Casein is a natural glue made from milk protein. It is a beige powder that you mix with water. It is mostly used as a wood glue for furniture and is moisture resistant but not waterproof. It bonds well even on resinous or oily woods such as yew, teak, and lemonwood. Parts should be clamped together for several hours. Another wood glue, often referred to as white glue, is a polyvinyl acetate solution—a water-based synthetic adhesive. It works only on porous and semiporous materials such as wood, leather, cork, cardboard, porous ceramics, and polystyrene foam. The reason for this is that in order to bond, the water must be removed, and so it could not bond materials such as glass. A variety of white glue used for schools is especially formulated to set up more slowly, and it will wash out of clothing after it is completely dry.

Hide glues are also classed as wood glues and are usually sold as ready-to-use liquids. These traditional glues have poor moisture resistance.

TABLE 25.1. Adhesives and their uses

Name	Advantages	Disadvantages	Uses
Anaerobic cyanoacrylates (popularly called super glue)	High strengths (2000 psi) in a few minutes	High cost, thicknesses greater than 0.002 in. cure slowly or not at all, does not cure in presence of air	Production speed in bonding small parts, especially metal to metal and metal to plastics
Acrylic	Clear with good optical properties, ultraviolet stability	Moisture sensitive, high cost	Bonding plastics
Cellulose esters and ethers	Very soluble	Moisture sensitive	Bonding of organic substances such as paper, wood, and some fabrics
Epoxies: modified, nitrile, nylon, phenolic, polyamide, vinyl	Low-temperature curing, adheres by contact pressure, good optical qualities for transmitting light, high strength, long shelf life	Cost, low peel strength in some types	Bonding microelectronic parts, aircraft structural bonding, bonding cutting tools and abrasives
Phenolics (Bakelite)	Heat resistance, low cost	Hard, brittle	Abrasive wheels, electrical parts
Polyimides	Resistance to high temperature (600°F)	Cost, high cure temperature (500° to 700° F)	Aircraft honeycomb sandwich assemblies
Polyester (anaerobic)	Hardens without the presence of air	Limited adhesion, cost	Locking threaded joints
Polysulfide	Bonds well to various surfaces found in construction	Low strength	Caulking sealant
Polyurethane (rubbery adhesive)	Strong at very low temperatures, good abrasion resistance	Low resistance to high temperatures, sensitive to moisture	Flexible and rigid foams, protective coatings, vibration damping mounts, coatings for metal rollers and wheels
Rubber adhesives: butadienes, butyl, natural (reclaim), neoprene, nitrile	High-strength bonds with neoprene (4000 psi), oil resistance (nitrile)	Low tensile and shear strength, poor solvent resistance	Adhesives for brake linings, structural metal applications, pressure-sensitive tapes, household adhesives
Silicones	High resistance to moisture, can resist some temperature extremes	Low strength	Castable rubbers, vulcanized rubbers, release agents, water repellents, adhesion promoters, varnishes and solvent type adhesives, contact cements
Urea-formaldehyde resins	Low cost, cures at room temperatures	Sensitive to moisture	Wood glues for furniture and plywood
Vinyl: polyvinyl-butyral-phenolic vinyl plastisols, polyvinyl acetate and acetals, polyvinyl chloride	High room-temperature peel, ability to bond to oily steel surfaces (vinyl plastisols)	Sensitive to moisture	Bonds to steel (brake linings), safety glass laminate, household glue, water and drain pipe (PVC), household wrap

The most common wood glues used today in construction and by cabinetmakers are hot-melt adhesives. These are applied with a glue gun, and are polyamide or polyethylene-based thermoplastics that usually melt at 200° to 300°F, and harden in 10 to 40 seconds, achieving half of their strength in about 1 minute and 100 percent in 24 hours. Hot-melts can be used on any porous surface, although some types can be used for plastics or metals. These adhesives make a quick but low-strength bond having a shear strength of only a few hundred pounds per square inch.

Resorcinol is another adhesive used primarily for wood joining, but it can be used on other porous materials. Its greatest advantage is its superior strength and moisture resistance. It is more expensive than the other wood glues.

An adhesive **phenolic** formulation similar to the phenolic-formaldehyde plastics is used in the manufacture of exterior plywood and in bonding of laminates such as formica. Not only do phenolics have good heat resistance, they are not moisture sensitive as most wood glues are.

The **urea-formaldehyde** wood glues have been widely used for years in bonding wood in furniture and interior plywood where moisture is not likely to be a factor. These adhesives are low cost and cure at room temperature but are quite sensitive to moisture.

Cellulose cements have been used for many years as household cement, a catchall name for many different formulations today, such as cellulose nitrate or vinyl or styrene resin. These can be cleaned with acetone. Almost any material can be joined with these glues, but they have low strength, and their solvents will attack some plastics. They harden through solvent evaporation in about 10 minutes, but full strength takes 24 hours. They are flammable, and their fumes can be toxic if breathed for any length of time. They are somewhat water resistant but not waterproof.

Epoxies are high-strength adhesives that harden by chemical reaction. Usually a resin and a hardener are mixed together just before use. There are several different epoxy formulations for varying uses. Nitrile epoxies are often used on aircraft for bonding honeycomb structures (Figure 25.1). Epoxies usually come in two containers (tubes, cans, or syringes), or in two-part bars or ribbons to make a putty for filling holes. Some epoxies have a steel powder filler, which gives them more strength. Epoxies have good water resistance, but to get a fast-hardening epoxy formulation, there is a trade-off in a loss of moisture resistance and strength. Epoxies will bond to almost any surface including most metals, woods, glass, ceramics, concrete, and some plastics. Some plastics such as polyethylene, polypropylene, and Teflon are unbondable. Most epoxies are generally not recommended for flexible surfaces because they become rigid when hardened, and they do not resist peeling pressure very well even though they have high shear and tensile strengths.

Silicone rubber adhesives are actually room-temperature-vulcanizing synthetic liquid rubbers. Since they keep their rubbery qualities after hardening, they should also serve as sealants. They have a very high resistance to temperature extremes from –60° to 450°F, and they are water resistant. In fact, the glass in frameless aquariums is joined with silicone adhesives. They are excellent gap fillers. Silicones do not have the strength of epoxies, but they will hold a seal for years. Silicone caulks have fewer adhesive qualities than the adhesives and sealants.

Polyurethane adhesives possess high strength compared with epoxies and are a one-component product. They will do the same job as epoxy with the added advantage of flexibility, so they can be used on leather, rubber, vinyl, and fabrics. Moisture in the air acts as a catalyst and triggers the cure. Polyurethane adhesive actually swells when hardening, and the parts should be clamped together for 2 hours. This characteristic allows it to fill gaps and voids. It makes an extremely good bond with almost anything, including your hands. Rubber gloves should be used when working with this adhesive. If it dries on your hands, you cannot remove it with any solvent; it must be abraded off with sandpaper or pumice stone.

Contact cements are also made of synthetic rubber, usually neoprene, dispersed in a solvent or water. Usually a coat is applied to both surfaces to be joined and allowed to dry. These cements stick instantly and permanently. They are used only for non-load-bearing applications such as countertop laminates and furniture veneer.

Adhesive Testing

Certain destructive tests are used to determine joint strength and efficiency for various adhesives and substrates. Among these tests are shear, compression, tensile, impact, and peel tests. Cleavage (a variation of the

FIGURE 25.1. Cutaway views of sandwich honeycomb structures used for aircraft.

peel test), tack (adhesion), creep, and fatigue are also common. Ultrasonic testing is used to some extent on bonded assemblies. Two of several specifications that apply to adhesive bonding are the Military, Federal Adhesive Specifications and those of the American Society for Testing and Materials (ASTM).

INDUSTRIAL OILS AND LUBRICANTS

Without lubricants such as oils and greases, machines would all cease to operate within a very short time. Lubricants are used in engineering and manufacturing practice for two reasons: (1) to diminish friction between two moving surfaces of machine parts, such as shafts and sleeve bearings, or machine slides and (2) to decrease the friction between a cutting tool and the material to be cut, as in a machining operation. However, when the primary function of a lubricant is to dissipate heat, it is referred to as a coolant.

Lubricants may be classified as liquids, solids, or semisolids and from their sources as animal, mineral, or vegetable. Animal lubricants are obtained from the fat of common animals. Vegetable oils include linseed oil, soybean oil, cottonseed oil, castor oil, and rape seed oil. Mineral oils are obtained from crude oil by the refining process. Some crude oils, such as those found in Pennsylvania, contain less gum-forming and asphaltic components than others, such as those found in western states, notably Texas. For this reason, the refining process varies considerably to remove unwanted components and add others such as antigumming chemicals and antioxidants. The hydrocarbon molecules found in lubricating oils may be grouped into classes known as paraffins, napthalenes, and aromatics. The paraffins and napthalenes are the most stable and are the principal constituents of lubricating oils.

Synthetic oils are obtained by chemical processing of various organic and inorganic materials. They are most valuable where widely contrasting and higher temperatures will break down ordinary petroleum oils. For example, the bearings in a jet engine may be subjected to temperatures as high as 450°F (232°C) at speeds up to 10,000 rpm.

Solid lubricants include graphite in colloidal form and molybdenum disulfide as a fine powder. Dry lubricants can maintain an effective lubrication even at temperatures as high as 700°F (371°C). Dry lubricants are often bonded as a coating on bearing surfaces. When contained in oil, they provide a residual coating of lubricant so that when starting from a position of rest, a bearing surface will always have some lubrication.

Greases are valuable for their stay-put property and adhesiveness to metallic surfaces. Heavy greases are used for large external gears that are not encased in an oil bath and for wheel bearings and steering linkages on automobiles and in many other ball and roller bearing applications.

Grease is composed of a lubricating oil held by absorption and capillary attraction in the fibrous or granular structure of a thickening agent, usually a clay type soap or thickener. Synthetic greases also use a clay type thickener and are particularly useful where high temperatures are encountered.

Friction takes place as the result of contacting moving surfaces, even when they appear to be smooth. Abrasive wear is caused by contact between two metals that move against each other. Sometimes a foreign nonmetallic material such as sand or grit gets between the surfaces and causes rapid wear. Erosion (by contact with moving liquids or gases) is a form of wear that is often seen in pipe fittings and valve faces.

Wear is usually the result of the constant rubbing of the contact areas that may either break off small particles or displace the metal to flatten and make it smoother. This is often what happens to new bearings when they are "run in." A pressurized film (hydrodynamic lubrication) completely eliminates metallic contact when in motion (Figure 25.2) and wear is kept at a minimum. Boundary layer or adhesive film lubrication (Figure 25.3) is frequently used when it is not possible to maintain hydrodynamic lubrication. The addition of graphite, molybdenum disulfide, phosphides, or similar high-pressure lubricants to the oil or grease will provide the boundary layer that reduces friction. Hypoid oils and greases were developed for the automotive industry to provide boundary lubrication for the intense pressures generated in automobile differentials.

Industrial process oils include those used for quenching tool steel, making shoe and furniture polish, concrete form oil, tanning oil, waterproof coatings, fly spray for cattle, cutting fluids, and for many other applications.

Oils for hydraulic machinery are specially prepared with antifoaming agents, controlled viscosity, corrosion inhibitors, and a high flash point so that they will not burn at lower temperatures. Both petroleum and synthetic oils are used in industrial hydraulic machines.

ASPHALTS

Asphalt is a black, sticky material that is also separated from crude petroleum by various refinery processes.

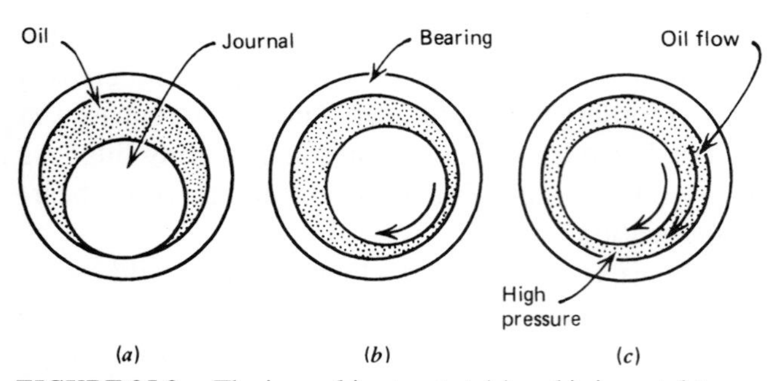

FIGURE 25.2. The journal is at rest at (*a*) and is in metal-to-metal contact. At (*b*), as the journal begins to turn, it rolls upward to the right, but after adherent oil clinging to its surface begins to build up pressure and a separating oil film between the journal and the bearing, the journal moves to the left as shown at (*c*). Bearing friction is largely dependent on the force that is necessary to shear the oil film.

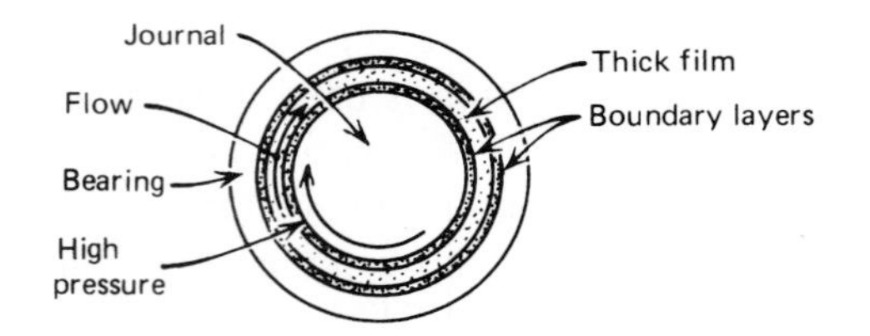

FIGURE 25.3. Thick film or boundary lubrication almost always provides a lubricating surface even when the parts are not moving. The boundary layers do not move to any great extent because of adhesion, and the shear is confined to the center area of the thick film. The oil flow is always slower than the movement of the journal, depending on the amount of shear.

Some asphalts occur naturally in pits or lakes that are residues of crude oil in which the lighter fractions have evaporated over a period of many thousands of years. There are a number of products in the family of asphalt materials. Asphalts can generally be divided into two categories: paving asphalts (asphalt cements) and liquid asphaltic materials. Asphalt cements that are thermoplastic solids can be heated to a liquid condition for such applications as roofing tar and paving. Liquid asphalts are not heated but are diluted with solvents, such as propane or oil to provide the desired consistency. Some are liquefied by emulsifying them with water. These asphalts include road oils, kerosene cutback asphalts, and gasoline (naphtha) cutbacks. Both solid (heated) and liquefied (cold application) asphalts are used for paving. The liquefied asphalts harden over a period of time, but the molten (heated) asphalt hardens when cool.

The penetration test is used to measure the consistency of hardness of asphalt cements. The ASTM standardized test uses a needle similar to a sewing machine needle. It is loaded with a given weight and the point is allowed to penetrate the asphalt for a given length of time. The depth of penetration, measured in 0.10 mm increments, determines the relative hardness.

Asphalts are used in the manufacture of battery cases, shoes, waterproof paper bags, and electrical insulation. They are also used in the field of irrigation and water conservation for waterproof linings of reservoirs and canals and for waterproofing foundations of buildings. However, plastics are beginning to replace asphalt products for many of these applications.

INDUSTRIAL GASES

Compressed air may be considered an industrial gas since it is used everywhere to operate air cylinders and pneumatic rock drills, to clean and ventilate, and to supply oxygen for combustion. Air is composed of oxygen, nitrogen, and some rare gases and has a pressure of 14.7 psi at sea level. Air contains moisture that condenses when it is compressed and then collects in air lines. This must be removed with water traps for most industrial air uses such as spray painting or operating air equipment.

Ammonia (NH_3) gas is used for nitriding, a type of case hardening of steel, and for refrigeration. **Nitrogen** is a gas that is used for charging hydraulic accumulators.

Oxygen is used where heating and burning is required, as in furnaces, cutting torches, and rockets. Oxygen can be dangerous since oils, greases, and other hydrocarbons will explode in the presence of pressurized oxygen.

Helium is also an inert gas that is used as a refrigerant, in balloons and blimps, and for welding. **Argon** is an inert gas that is used in electric lamps and for welding. **Carbon dioxide** (CO_2), when frozen, is the familiar "dry ice." It is used in fire extinguishers, in soft drinks, and as a shield gas for welding. Carbon dioxide is not an inert gas and disassociates into oxygen and carbon monoxide when heated by welding. It is therefore used for welding metals such as steel and iron that are not so affected by the presence of oxygen. It is not suitable for welding aluminum, titanium, and copper.

Fuel gases such as hydrogen, acetylene, natural gas, propane, and butane are burned with oxygen (O_2) or in air for heating purposes. The highest temperature is obtained by burning acetylene in oxygen (6200°F). For information on the many (more than 100) industrial gases, their uses, and toxicity, obtain a good reference book such as the *Matheson Gas Data Book,* The Matheson Company, Inc.

SELF-EVALUATION

1. Before the age of synthetics, what materials were used for adhesives, pastes, and glues?
2. Name two advantages of adhesives over riveted or bolted joints.
3. How should a bonded surface be prepared?
4. The so-called super glues, the cyanoacrylates, can join two smooth metal surfaces with a single drop for each square inch and almost immediately gain a bond strength of 2000 psi. However, they will never harden in the presence of air. Why is this?
5. Sealants are usually rubbery, flexible materials. Name one or more synthetic materials that are used as sealants.
6. Name three fuel gases and two inert gases that are used for industrial purposes and explain their uses.
7. Hydrodynamic lubrication reduces wear because the thick film lubricant keeps the moving parts separated. When the operating pressure is so intense that the thick film cannot be maintained, wear is the result. What kind of lubrication would eliminate rapid wear in this case?
8. Paving asphalts (asphalt cements) are liquefied by heating. How are liquid asphalts liquefied?
9. Why would model airplane glue (cellulosics) be a poor choice to cement fuselage panels on a jet aircraft?
10. If a liquid adhesive cannot get any closer than 1000 angstroms to the substrate atoms because of surface impurities, what is the underlying reason for the lack of bonding?
11. Metal connectors such as bolts and rivets are being replaced in some applications by adhesives. Name two advantages of adhesives over metal connectors on metal assemblies.
12. Which would be the better adhesive to use in a moist environment, cyanoacrylate (super glue) or epoxy?

CASE PROBLEM: THE PEELING STORAGE SHED

A homeowner needed a tool shed in his backyard so he decided to build it himself. He went to a cut-rate lumber yard where he spotted a pile of plywood panelling which was priced at a very low cost. He built his tool shed one summer and by the next spring all of the outer layers of the plywood veneer had begun to peel off. What was the reason for this problem? What could he have done to prevent it?

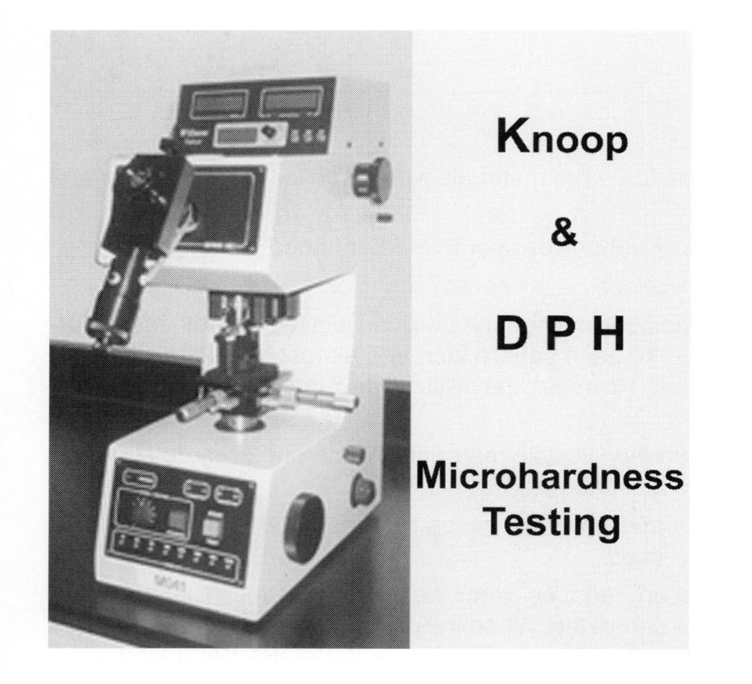

CHAPTER 26

Hardness Testing

Hardness by definition is the resistance of a material to penetration. We assume hardness to mean brittleness but this is not necessarily true. Many metals can be very hard and still maintain ductility. Hardness is also associated with a material being brittle: The harder the material the more brittle it becomes. This statement is also misleading. The new aluminum-lithium alloys used in aerospace applications are extremely strong, impact resistant, and very hard. Bubble gum is very soft and chewy; but cool it to below 32°F and it becomes surprisingly hard with almost no impact strength.

This chapter explains the use of various hardness testing machines and their calibration. Brinell, Rockwell, and microhardness testers are the most commonly used hardness testers for industrial and metallurgical purposes. Heat treaters, quality control inspectors, and many others in industry use these machines on a daily basis. This chapter will lead you to a correct understanding of the calibration of these testers and proper procedures for hardness testing.

OBJECTIVES **After completing this chapter, you should be able to:**

1. Understand the operation of common industrial hardness testers.
2. Be able to calibrate a hardness tester.
3. Understand the differences between loads and indenters.
4. Make hardness tests using a Brinell, Rockwell, and/or microhardness tester.

HISTORY

Hardness is defined as (1) the resistance of a material to penetration; (2) the ability of a material to resist being permanently deformed; (3) elastic hardness; and (4) the resistance to abrasion. Historically, we have recognized the properties of hardness and, more recently, have associated hardness primarily with brittle metal behavior. During the last 75 years, numerous methods to quantify hardness have been developed: The Moh hardness test, a scratch test kit containing 10 materials of varying hardness (chalk to diamond), was used mostly by geologists in the field as a method of separating and identifying ores. It allowed comparison and crude identification based on the principle that *the softer material will be scratched by the harder.* The Shore scleroscope, used before and during World War I, operated on the principle of *elastic rebound* (Figure 26.1). This instrument dropped a hardened ball bearing from a fixed height to the perpendicular surface of the material (usually rubber) being tested, and the operator measured the return height of the ball. The higher the return, the harder the material being tested. Testing, at best, was subjective, as were many other methods documented in metals history books. The Brinell hardness tester, discussed in detail in this chapter, was also developed and used to measure hardness at the beginning of the twentieth century and remains in wide use today. The file test was another method which was employed in industry, comparing a metal's abrasion resistance to that of a hardened steel file. The file test is used today as a quick and crude method to determine a metals response to a file-hard carburizing heat treatment.

The Rockwell hardness tester was invented in 1919 by Stanley P. Rockwell, a metallurgist who was working in a ball bearing manufacturing plant. He developed a satisfactory method for testing and controlling hardness of bearing components and many other parts. The Rockwell hardness tester and test blocks have long been registered as trademarks in the United States and throughout the world as a standard method of measuring hardness.

FIGURE 26.1. Model D scleroscope. Small parts may be tested with this model. Scleroscopes are simple to operate and do not mar finished surfaces. (*Shore U.S.A. Trademark #757760, Scleroscope U.S.A. Trademark #723850*)

INSTRUMENT CALIBRATION

The objective of hardness testing is to repeatedly measure and record hardness accurately and within ± 0.2 hardness points, regardless of testing range. To perform hardness testing at this level of accuracy, an instrument calibration procedure is required for the Rockwell tester. Instrument calibration is required by government contractors, ISO 9001 and ANSI 9004 Quality Systems, to be performed daily for several reasons: (1) when a new operator uses the machine; (2) when the indenter, test fixtures, and/or load range is changed; and (3) when testing produces a significant variance in hardness greater than 10 points above or below the hardness value indicated on the calibration block used to calibrate the machine. Machine hardness is compared with the hardness of a *calibration block* which is traceable to a certified standard. The importance of performing calibration cannot be underemphasized, because the measure of a correct response to heat treatment is usually based on a hardness test. In today's aerospace manufacturing environment, compliance with ISO 9001 and ANSI 9004 Quality Systems requirements for hardness testing also demands the maintenance of a log in which actual calibration hardness values of the testing machine are permanently recorded.

The property of hardness tested by the Rockwell and Brinell hardness testing machines is defined as "the resistance of a metal to penetration." Ferrous and nonferrous metals and alloys, and nonmetallic materials and composites, vary in hardness from nearly butter soft to almost as hard as diamond. For example, iron-carbide cutting tools used by machinists are much harder than alloy steels hardened by heat treatment. Alloy steels can be as hard as 65 HRC or tempered and/or annealed to a soft condition less than 20 HRC. A hardness of 80 C scale is considered to be the hardness of diamond. Full spheroidized, annealed, low alloy steels are approximately 20 HRC or at the bottom of the C scale. Nonferrous metals, such as aluminum and copper, are much softer than annealed steels and require a larger indenter and a different testing range to measure hardness. Plastics and many other nonmetallic materials exhibit hardness often softer than metals.

When considering a corresponding tensile strength based on a hardness value, remember that the values listed for tensile strength are based on testing *plain carbon steel.* Alloy steels measuring the same hardness as a plain carbon steel will usually exhibit higher mechanical properties due to the influence of the additional alloying elements. The tensile values are listed only as a reference value (Figure 26.2 and Table 26.1) for Rockwell and Brinell testing.

TABLE 26.1. Hardness comparison chart with indentations

Brinnell 10 mm-Ball 3000 kgm		Rockwell			Tensile	Brinnell 10 mm-Ball 3000 kgm		Rockwell			Tensile
Dia.	**No.**	**C**	**A**	**15-N**	**MPSI**	**Dia.**	**No.**	**C**	**A**	**B**	**MPSI**
2.20	780	70	86.5		384	3.75	262	26	63.5		123
2.25	745	68	85.5	93.0	368	3.80	255	25	63.0		120
2.30	712	66	84.5	92.5	352	3.85	248	24	62.5		118
2.35	682	64	83.5	92.0	337	3.90	241	23	62.0	100	115
2.40	653	62	82.5	91.0	324	3.95	235	22	61.5	99	112
2.45	627	60	81.0	90.0	311	4.00	229	21	61.0	98	110
2.50	601	58	80.0	89.5	298	4.05	223	20	60.5	97	107
2.55	578	57	79.5	89.0	293	4.10	217	18	59.5	96	103
2.60	555	55	78.5	88.0	287	4.15	212	17	59.0	96	102
2.65	534	53	77.5	87.0	283	4.20	207	16	58.0	95	100
2.70	514	52	77.0	86.5	273	4.25	202	15	57.5	94	99
2.75	495	50	76.0	58.5	256	4.30	197	13	57.0	93	95
2.80	477	49	75.5	85.0	246	4.35	192	12	56.5	92	93
2.83		48	74.5	84.5	237	4.40	187	10	56.0	91	90
2.85	461	47	74.0	84.0	231	4.45	183	9	55.5	90	89
2.90	444	46	73.5	83.5	221	4.50	179	8	55.0	89	88
2.95	428	45	73.0	83.0	215	4.55	174	7	54.0	88	87
3.00	415	44	72.5	82.5	208	4.60	170	6	53.5	87	85
3.03		43	72.0	82.0	201	4.65	166	4	53.0	86	83
3.05	401	42	71.5	81.5	194	4.70	163	3	52.5	85	82
3.10	388	41	71.0	81.0	188	4.75	159	2	52.0	84	81
3.15	375	40	70.5	80.5	181	4.80	156	1	51.0	83	80
3.20	363	38	69.5	79.5	170	4.85	153		50.5	82	74
3.25	352	37	69.0	79.0	165	4.90	149		50.0	81	72
3.30	341	36	68.5	78.5	160	4.95	146		49.5	80	70
3.35	331	35	68.0	78.0	155	5.00	143		49.0	79	68
3.40	321	34	67.5	77.0	150	5.05	140		48.5	78	68
3.45	311	33	67.0	76.5	147	5.10	137		48.0	77	68
3.50	302	32	66.5	76.0	142	5.15	134		47.0	76	66
3.55	283	31	66.0	75.5	139	5.20	131		46.0	74	64
3.60	285	30	65.5	75.0	136	5.25	128		45.5	73	62
3.65	277	29	65.0	74.5	132	5.30	126		45.0	72	62
3.70	269	28	64.5	74.0	129	5.35	124		44.5	71	60
*3.73		27	64.0	73.5	126						

*Brinell gloss diameter marked thus, *, have no corresponding number, but are included since the equivalent Rockwell represents the limits for a commonly used aircraft specification.

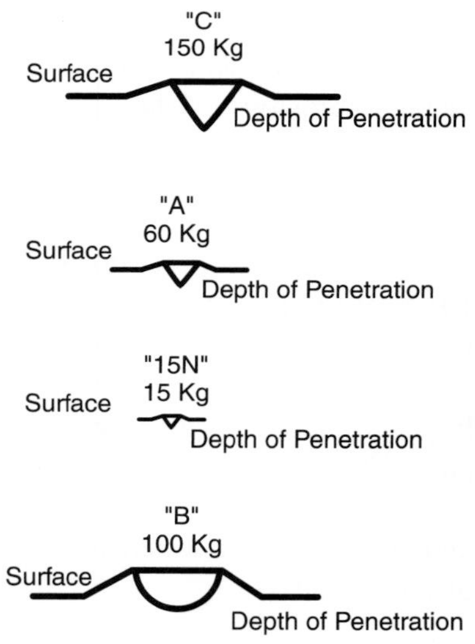

"C"
150 Kg
Surface
Depth of Penetration
"A"
60 Kg
Surface
Depth of Penetration
"15N"
15 Kg
Surface
Depth of Penetration
"B"
100 Kg
Surface
Depth of Penetration

FIGURE 26.2. A common hardness comparison chart showing the correlation in load and indenter with the more commonly used hardness scales. Note the size and depth of indentation in comparison with the applied load.

TABLE 26.2. Hardness and tensile strength comparison table

Hardness Conversion Table

Brinell		Rockwell			Brinell		Rockwell		
Indentation Diameter (mm)	**No.[a]**	**B**	**C**	**Tensile Strength, 1000 psi Approximately**	**Indentation Diameter (mm)**	**No.[a]**	**B**	**C**	**Tensile Strength, 1000 psi Approximately**
2.25	745		65.3		3.75	262	(103.0)	26.6	127
2.30	712		—		3.80	255	(102.0)	25.4	123
2.35	682		61.7		3.85	248	(101.0)	24.2	120
2.40	653		60.0		3.90	241	100.0	22.8	116
2.45	627		58.7		3.95	235	99.0	21.7	114
2.50	601		57.3		4.00	229	98.2	20.5	111
2.55	578		56.0		4.05	223	97.3	(18.8)	—
2.60	555		54.7	298	4.10	217	96.4	(17.5)	105
2.65	534		53.5	288	4.15	212	95.5	(16.0)	102
2.70	514		52.1	274	4.20	207	94.6	(15.2)	100
2.75	495		51.6	269	4.25	201	93.8	(13.8)	98
2.80	477		50.3	258	4.30	197	92.8	(12.7)	95
2.85	461		48.8	244	4.35	192	91.9	(11.5)	93
2.90	444		47.2	231	4.40	187	90.7	(10.0)	90
2.95	429		45.7	219	4.45	183	90.0	(9.0)	89
3.00	415		44.5	212	4.50	179	89.0	(8.0)	87
3.05	401		43.1	202	4.55	174	87.8	(6.4)	85
3.10	388		41.8	193	4.60	170	86.8	(5.4)	83
3.15	375		40.4	184	4.65	167	86.0	(4.4)	81
3.20	363		39.1	177	4.70	163	85.0	(3.3)	79
3.25	352	(110.0)	37.9	171	4.80	156	82.9	(0.9)	76
3.30	341	(109.0)	36.6	164	4.90	149	80.8		73
3.35	331	(108.5)	35.5	159	5.00	143	78.7		71
3.40	321	(108.0)	34.3	154	5.10	137	76.4		67
3.45	311	(107.5)	33.1	149	5.20	131	74.0		65
3.50	302	(107.0)	32.1	146	5.30	126	72.0		63
3.55	293	(106.0)	30.9	141	5.40	121	69.8		60
3.60	285	(105.5)	29.9	138	5.50	116	67.6		58
3.65	277	(104.5)	28.8	134	5.60	111	65.7		56
3.70	269	(104.0)	27.6	130					

Note 1. This is a condensation of Table 2, Report J417b. SAE 1971 Handbook. Values in ( ) are beyond normal range and are presented for information only.

Note 2. The following is a formula to approximate tensile strength when the Brinell hardness is known.

$$\text{Tensile strength} = \text{BHN} \times 500$$

[a]Values above 500 are for tungsten carbide ball; below 500 for standard ball.

Source: Bethlehem Steel Corporation, *Modern Steels and Their Properties,* seventh edition, Handbook 2757, 1972.

To satisfy the principle of resistance to penetration and be considered a universal hardness testing machine, the hardness machine must be designed to use a variety of different size indenters and load ranges. In the Rockwell machine, a steel ball indenter 1/2 in. in diameter may be used to apply a 15 kg standard load to the material being tested. Although this load may seem quite heavy, when applied over a large area such as the contact area of the 1/2 in. ball indenter, the overall applied stress is quite low. The machine is designed to apply load on and into a material using an indenter which is a standard size. Therefore, anyone testing the same material, anywhere in the world, if using a calibrated hardness testing machine, can measure the hardness of the material and record exactly the same value. Methods for the calibration of the Rockwell, Brinell, and Tukon type hardness testing machines are included in this chapter.

The recording of Rockwell hardness may be 50 HRC, Rc 50, or any combination that indicates the measured hardness is 50 on the Rockwell C scale of hardness. These data were obtained using a 150 kg load and a brale indenter by Rockwell method. The Rockwell C scale offers the capability to measure the hardness of very soft annealed steels to very hard carburized or through hardened heat treated steels. A file, for example, is usually heat treated to 50 HRC; a spring steel may achieve a hardness of 60 to 65 HRC and a good knife blade may measure 48 to 52 HRC. Since the C scale

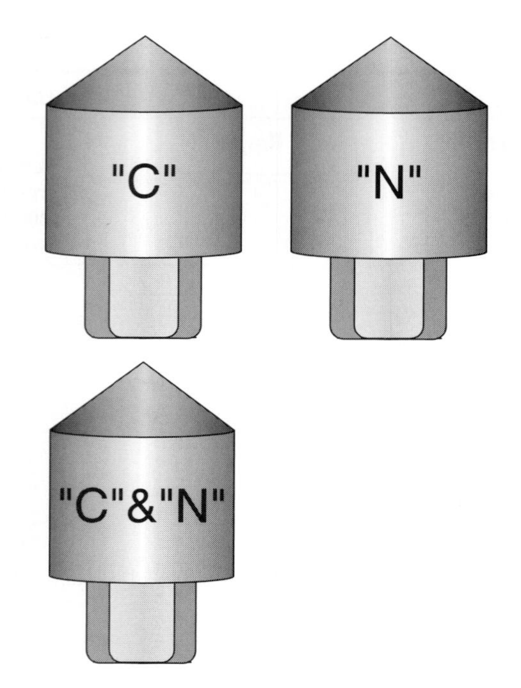

FIGURE 26.3. The 1/16 in. ball indenter is shown with the standard brale diamond indenter. Lettering on the face of the brale indenter indicates use in a standard Rockwell as a superficial Rockwell tester.

below 20 is somewhat unreliable, the B range should be used for softer materials. Each hardness range is overlapping, which provides the opportunity and convenience of scale choice and conversion of a hardness to another hardness scale if desired: 50 HRC = 75.9 HRA. Oftentimes it is necessary to reduce the load to a lower range when testing brittle metals and metallic compounds, such as carbides, to avoid cracking.

There are two basic types of penetrators or indenters used in the Rockwell tests (Figure 26.3). The 1/16 in. diameter steel ball is used for the "B" hardness range and the testing of relatively soft metals such as aluminum, copper and copper alloys, and babbit metals. Although the ball is a 62 HRC alloy steel bearing, testing with the ball indenter is limited to materials softer than 100 HRB (top of the B scale) to prevent damage to the indenter. Ball indenters for use on very soft metals and plastics are available in sizes 1/16 in., 1/8 in., 1/4 in., and 1/2 in. (Table 26.3). The brale indenter has a spheroconical, ground diamond tip and may either be a "C" brale, "N" brale, or "C and N" brale, dependent upon the type of Rockwell tester. The "C" brale indenter is used on a standard Rockwell tester with load ranges limited to 60 kg, 100 kg, and 150 kg. This indenter is not interchangeable with the Rockwell superficial tester which uses an "N" indenter. The superficial tester is limited to testing loads of 15 kg, 30 kg, and 45 kg. The "N" brale indenter tip is more precisely ground to a tighter tolerance. The combination Rockwell tester has the capability of using all Rockwell load ranges and the indenter is marked "C and N." This indenter provides accuracy for all ranges of brale Rockwell testing.

THE MEANING OF CALIBRATION

Instrument calibration is performed to ensure that a hardness testing machine is working properly and within operating parameters. The procedure often includes calibration of the operator. The advantages of instrument calibration are many; most importantly, calibration provides a picture of data variance and is a measure of the accuracy of a hardness test.

When calibration is correctly performed, the data gathered are said to be "actual or true hardness values." Thus, when calibrated hardness data are measured and recorded by another facility, identical results will be obtained.

Hardness calibration test blocks are usually traceable to the National Bureau of Standards. The C scale represents the hardness range most often used to measure steels. Three test blocks are required to certify instrument linearity through the full range of measurable hardness, that is, from full hardened steel at 65 HRC to very soft steel at about 20 HRC. The ideal hardness value of these test blocks should be 58 HRC, 40 HRC, and 28 HRC. The rule for selecting a test block is that the block should be within 10 HRC points of the part

TABLE 26.3. Penetrator and load selection

Scale Symbol	Major Load Penetrator	Dial kg	Figures	Typical Applications of Scales
B	1/16 in. ball	100	Red	Copper alloys, soft steels, aluminum alloys, malleable iron, etc.
C	Brale	150	Black	Steel, hard cast irons, pearlitic malleable iron, titanium, deep case-hardened steel, and other materials harder than B-100
A	Brale	60	Black	Cemented carbides, thin steel, and shallow case-hardened steel
D	Brale	100	Black	Thin steel medium case-hardened steel, and pearlite malleable iron
E	1/8 in. ball	100	Red	Cast iron, aluminum and magnesium alloys, and bearing metals
F	1/16 in. ball	60	Red	Annealed copper alloys, thin, soft sheet metals
G	1/16 in. ball	150	Red	Phosphor bronze, beryllium copper, malleable irons. Upper limit G-92 to avoid possible flattening of ball
H	1/8 in. ball	60	Red	Aluminum, zinc, lead
K	1/8 in. ball	150	Red	
L	1/4 in. ball	60	Red	Bearing metals and other very soft or thin materials. Use the smallest ball and heaviest load that does not give an anvil effect.
M	1/4 in. ball	100	Red	
P	1/4 in. ball	150	Red	
R	1/2 in. ball	60	Red	
S	1/2 in. ball	100	Red	
V	1/2 in. ball	150	Red	

Source: Wilson Instruction Manual, "Rockwell Hardness Tester Models OUR-a and OUS-a," American Chain & Cable Company, Inc., 1973.

(*Note. Rockwell* is a registered trademark of American Chain & Cable Company, Inc., for hardness testers and test blocks.)

to be tested. Therefore, when measuring the hardness of a 38 to 40 HRC part, a test block within 10 hardness points should be used to verify testing machine accuracy.

The property of hardness tested by the Rockwell and Brinell hardness testing machines is defined as "the resistance of a metal to penetration." Ferrous and nonferrous metals and alloys, along with nonmetallic materials and composites, vary in hardness from nearly butter soft to almost as hard as diamond. For example, iron-carbide cutting tools used by machinists are much harder than alloy steels hardened by heat treatment. Alloy steels can be as hard as 65 HRC or tempered and/or annealed to a soft condition less than 20 HRC. Rockwell 80 HRC is considered to be the measure of hardness of diamond; soft steels easily shaped and formed are often within the 20 HRC range. Nonferrous metals, such as aluminum and copper, are usually much softer than annealed steels—usually far below 20 HRC—and for testing purposes, require a larger indenter and a different testing range to measure hardness. Plastics and many other nonmetallic materials exhibit hardness often softer than metals.

To satisfy the principle of resistance to penetration and perform as a universal hardness testing machine, hardness machines are designed to use a variety of different size indenters and load ranges. In the Rockwell machine, a steel ball indenter 1/2 in. in diameter may be used to apply a 60 kg standard load to the material being tested. More commonly used hardness tests are listed in Table 26.2. Although this load may seem quite heavy, when applied over a large area such as the contact area of the 1/2 in. ball indenter, the overall applied stress is quite low. The machine is designed to apply load on and into a material using an indenter which is a standard size. Therefore, anyone testing the same material, anywhere in the world, if using a calibrated hardness testing machine, can measure the hardness of the material and record exactly the same value.

Typical table model Rockwell hardness testers are shown in Figures 26.5*a* and 26.5*b*. The automatic readout digital Rockwell tester (Figure 26.5*a*) offers direct electronic readout of hardness and can be connected directly to a computer for data recording. This tester provides the operator with a complete range of loads and indenter selection and automatically performs the test. The standard superficial Rockwell tester (Figure 26.5*b*) requires manual selection of the correct weight for the desired load range and the crank handle must be released and at the completion of movement, returned gently to the rest position when performing a hardness test.

USING THE ROCKWELL HARDNESS TESTING

The Rockwell hardness test is made by applying two loads to a specimen and measuring the difference in depth of penetration in the specimen between the **minor load** and the **major load.** The minor load is used on the

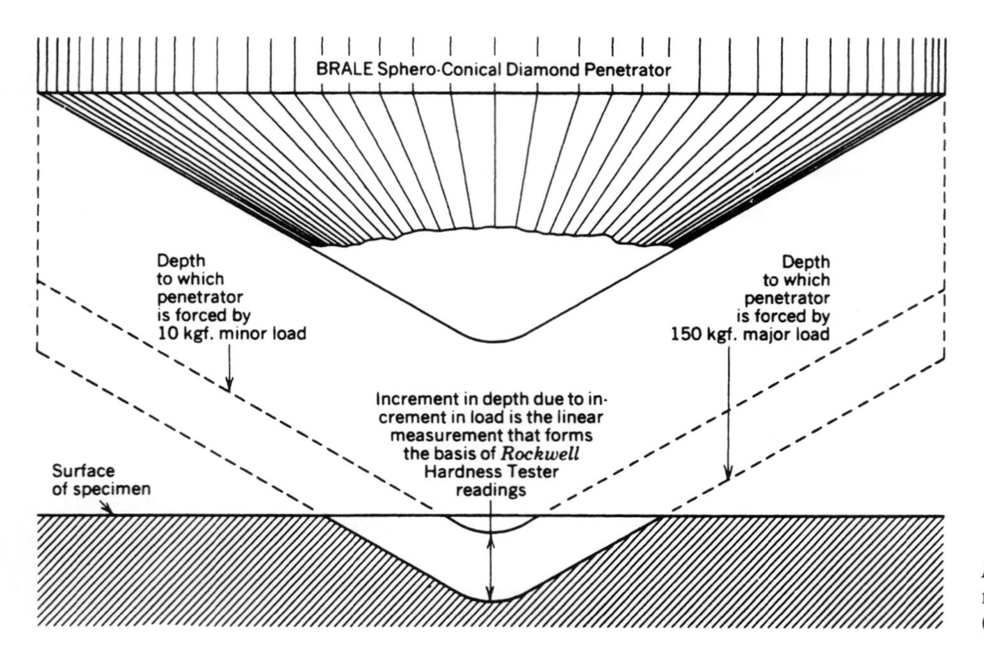

FIGURE 26.4. Schematic showing minor and major loads being applied. (*Wilson Instrument Division of Acco*)

*FIGURE 26.5***a.** The automatic readout digital Rockwell tester is one of many new automatic hardness testers in use today. This tester is a universal tester capable of performing tests using all load ranges and indenters. *(Micromet® Rockwell-type hardness tester, digital model. Photograph courtesy of Buehler Ltd.)*

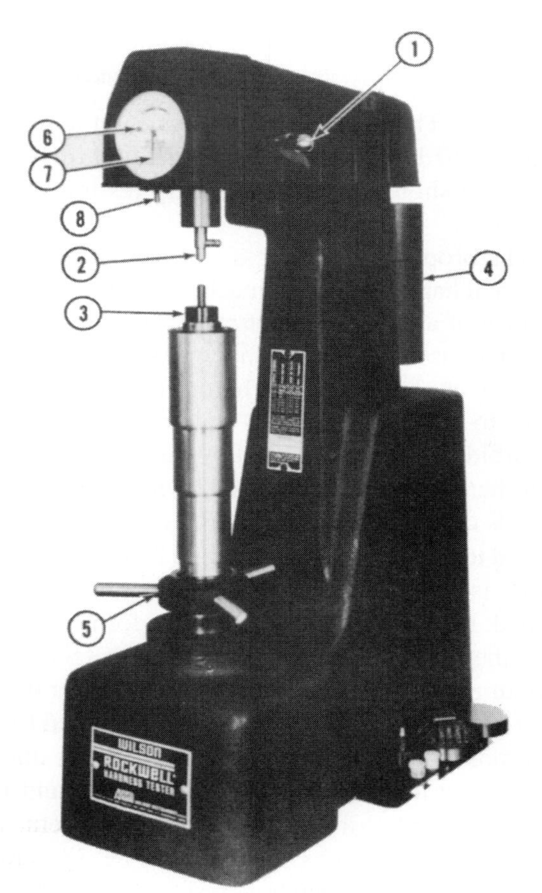

*FIGURE 26.5***b.** The standard superficial Rockwell hardness tester provides a full range of hardness testing capabilities as indicated in Table 26.2.

1. CRANK HANDLE.
2. PENETRATOR.
3. ANVIL.
4. WEIGHTS.
5. CAPSTAN HANDWHEEL.
6. SMALL POINTER.
7. LARGER POINTER.
8. LEVER FOR SETTING THE BEZEL.

standard Rockwell tester to eliminate errors that could be caused by specimen surface irregularities. The minor load is 10 kilograms of force (kg) when used with 60, 100, or 150 kg major load and 3 kg on the superficial tests where the major loads are 15, 30, and 45 kg.

The major load is applied after the minor load has seated the indenter firmly in the work. The Rockwell hardness reading is based on the additional depth to which the penetrator is forced by the major load (Figure 26.4). The depth of penetration is indicated on the dial when the major load is removed. The amount of penetration decreases as the hardness of the specimen increases. Generally, the harder the material is, the greater its tensile strength will be, that is, its ability to resist deformation and rupture when a load is applied.

ROCKWELL HARDNESS PROCEDURE FOR AUTOMATIC TESTER

1. Check log book for daily calibration.
2. Clear all data from the hardness tester, and push "clear".
3. Select the range desired, and push "regular or superficial" and push "test".
4. Select the correct calibration test block (within 10 hardness points of that expected of the part being tested).
5. Perform five hardness indentations on the calibration block and determine machine variance or correction factor (CF).
6. Test the hardness of the part; discard the first test. This is "measured" hardness.
7. Recheck machine variance CF and convert "measured" hardness to "actual" hardness. Record actual hardness data.

SURFACE PREPARATION AND PROPER USE

When testing hardness, a surface condition is important for accuracy. A rough or ridged surface caused by coarse grinding will not produce results that are as reliable as a smoother surface. Any rough scale caused by hardening must be removed before testing. Likewise, if the workpiece has been decarburized by heat treatment, the remaining softer "skin" should be ground off the test area.

Error can also result from testing curved surfaces. This effect may be eliminated by grinding a small flat spot on the specimen. Cylindrical workpieces must always be supported in a V-type centering anvil, and the surface to be tested should not deviate from the horizontal by more than 5 degrees. Tubing is often so thin that it will deform when tested. It should be supported on the inside by a mandrel or gooseneck anvil to avoid this problem.

Several devices are made available for the Rockwell hardness tester to support overhanging or large specimens. One type, called a jack rest (Figure 26.6*a–f*), is used for supporting long, heavy parts such as shafts. It consists of a separate elevating screw and anvil support similar to that on the tester. Without adequate support, overhanging work can damage the penetrator rod and cause inaccurate readings.

No test should be made near an edge of a specimen. Keep the penetrator at least 1/8 in. away from the edge. The test block, as shown in Figure 26.7 should be used to check the calibration of the tester every time the indenter and/or load range is changed.

CALIBRATING THE ROCKWELL HARDNESS TESTER AND MAKING A TEST

1. Using Table 26.3 and/or Table 26.4, select the proper weight and indenter (Figure 26.8). Make sure the load trip lever is pulled forward.
2. Select the calibration block with a hardness within ±10 hardness points of the part you intend to measure.
3. Without changing indenter and/or load range, place the part to be tested on the anvil in the same way as the calibration block in Figures 26.7 and 26.9.
4. By turning the handwheel, bring the indenter in contact with the part at the location where hardness is to be measured (Figure 26.8). Both dial needles will begin to move, pointing upward as you apply the "minor" load. When the tiny needle points directly up, the larger needle should be also pointing upward and within the color bands on the dial face. The large needle seldom points perfectly vertical so the machine provides a means of obtaining true "zero" by adjusting the dial face (Figure 26.10). Never turn back the large needle by reversing the handwheel; this action negates the test.
5. Zero the tester by adjusting the dial face (it rotates) to align the zero mark to the middle of the large needle (Figure 26.10). You are ready to test.
6. Gently trip the load trip lever (located at the top, side, or as a *bar* in front of the screw (Figure 26.11). This action allows application of the "major" load. The large and small needles will rotate

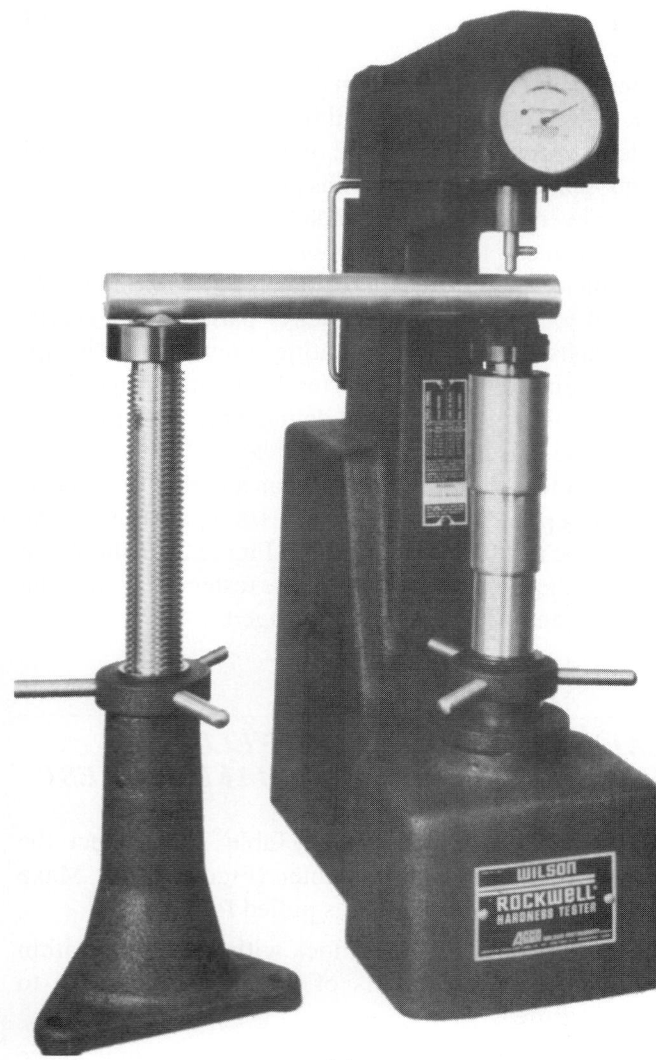

(a)

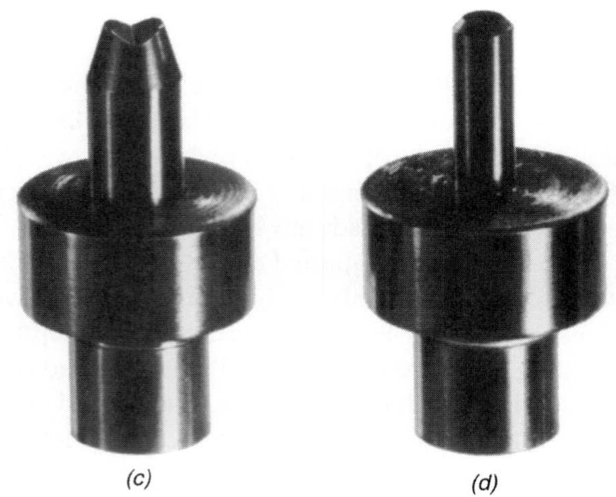

(c) (d)

(e)

(b)

(f)

FIGURE 26.6. The anvils used by the standard and superficial Rockwell hardness test machines provide for testing perpendicular to the axis of the indenter. In addition, and most important, the part being tested must be supported and stable during the application of both minor and major loads. The diamond spot anvil provides a precision test surface for hardness testing small finish machined parts. (*Wilson Instrument Division of ACCO*)

(*a*) THE LOAD LEVEL JACK
(*b*) PLANE ANVIL
(*c*) SHALLOW "V" ANVIL
(*d*) SPOT ANVIL
(*e*) CYLINDRON JR. ANVIL
(*f*) DIAMOND SPOT ANVIL

TABLE 26.4. Approximate equivalent hardness numbers and tensile strengths for Rockwell hardness numbers for steel.

Rockwell C-Scale Hardness Numbers

Rockwell C-scale hardness No.	Vickers hardness No.	Brinell hardness No., 3000-kg load, 10-mm ball: Standard ball	Brinell: Tungsten carbide ball	Rockwell hardness No.: A scale, 60-kg load, Brale indenter	Rockwell: B scale, 100-kg load, 1/16-in. diam ball	Rockwell: D scale, 100-kg load, Brale indenter	Rockwell superficial hardness No., superficial Brale indenter: 15N scale, 15-kg load	30N scale, 30-kg load	45N scale, 45-kg load	Knoop hardness No., 500-g load and greater	Shore Scleroscope hardness No.	Tensile strength (approx), 1000 psi	Rockwell C-scale hardness No.
68	940	...	...	85.6	...	76.9	93.2	84.4	75.4	920	97	...	68
67	900	...	...	85.0	...	76.1	92.9	83.6	74.2	895	95	...	67
66	865	...	...	84.5	...	75.4	92.5	82.8	73.3	870	92	...	66
65	832	...	(739)	83.9	...	74.5	92.2	81.9	72.0	846	91	...	65
64	800	...	(722)	83.4	...	73.8	91.8	81.1	71.0	822	88	...	64
63	772	...	(705)	82.8	...	73.0	91.4	80.1	69.9	799	87	...	63
62	746	...	(688)	82.3	...	72.2	91.1	79.3	68.8	776	85	...	62
61	720	...	(670)	81.8	...	71.5	90.7	78.4	67.7	754	83	...	61
60	697	...	(654)	81.2	...	70.7	90.2	77.5	66.6	732	81	...	60
59	674	...	(634)	80.7	...	69.9	89.8	76.6	65.5	710	80	351	59
58	653	...	615	80.1	...	69.2	89.3	75.7	64.3	690	78	338	58
57	633	...	595	79.6	...	68.5	88.9	74.8	63.2	670	76	325	57
56	613	...	577	79.0	...	67.7	88.3	73.9	62.0	650	75	313	56
55	595	...	560	78.5	...	66.9	87.9	73.0	60.9	630	74	301	55
54	577	...	543	78.0	...	66.1	87.4	72.0	59.8	612	72	292	54
53	560	...	525	77.4	...	65.4	86.9	71.2	58.6	594	71	283	53
52	544	...	512	76.8	...	64.6	86.4	70.2	57.4	576	69	273	52
51	528	...	496	76.3	...	63.8	85.9	69.4	56.1	558	68	264	51
50	513	...	481	75.9	...	63.1	85.5	68.5	55.0	542	67	255	50
49	498	...	469	75.2	...	62.1	85.0	67.6	53.8	526	66	246	49
48	484	...	455	74.7	...	61.4	84.5	66.7	52.5	510	64	238	48
47	471	442	443	74.1	...	60.8	83.9	65.8	51.4	495	63	229	47
46	458	432	432	73.6	...	60.0	83.5	64.8	50.3	480	62	221	46
45	446	421	421	73.1	...	59.2	83.0	64.0	49.0	466	60	215	45
44	434	409	409	72.5	...	58.5	82.5	63.1	47.8	452	58	208	44
43	423	400	400	72.0	...	57.7	82.0	62.2	46.7	438	57	201	43
42	412	390	390	71.5	...	56.9	81.5	61.3	45.5	426	56	194	42
41	402	381	381	70.9	...	56.2	80.9	60.4	44.3	414	55	188	41
40	392	371	371	70.4	...	55.4	80.4	59.5	43.1	402	54	182	40
39	382	362	362	69.9	...	54.6	79.9	58.6	41.9	391	52	177	39
38	372	353	353	69.4	...	53.8	79.4	57.7	40.8	380	51	171	38
37	363	344	344	68.9	...	53.1	78.8	56.8	39.6	370	50	166	37
36	354	336	336	68.4	...	52.3	78.3	55.9	38.4	360	49	161	36
35	345	327	327	67.9	...	51.5	77.7	55.0	37.2	351	48	157	35
34	336	319	319	67.4	...	50.8	77.2	54.2	36.1	342	47	153	34
33	327	311	311	66.8	...	50.0	76.6	53.3	34.9	334	46	149	33
32	318	301	301	66.3	...	49.2	76.1	52.1	33.7	326	44	145	32
31	310	294	294	65.8	...	48.4	75.6	51.3	32.5	318	43	141	31
30	302	286	286	65.3	...	47.7	75.0	50.4	31.3	311	42	138	30
29	294	279	279	64.7	...	47.0	74.5	49.5	30.1	304	41	135	29
28	286	271	271	64.3	...	46.1	73.9	48.6	28.9	297	40	131	28
27	279	264	264	63.8	...	45.2	73.3	47.7	27.8	290	39	128	27
26	272	258	258	63.3	...	44.6	72.8	46.8	26.7	284	38	125	26
25	266	253	253	62.8	...	43.8	72.2	45.9	25.5	278	38	122	25
24	260	247	247	62.4	...	43.1	71.6	45.0	24.3	272	37	119	24
23	254	243	243	62.0	100.0	42.1	71.0	44.0	23.1	266	36	117	23
22	248	237	237	61.5	99.0	41.6	70.5	43.2	22.0	261	35	114	22
21	243	231	231	61.0	98.5	40.9	69.9	42.3	20.7	256	35	112	21

Rockwell B-Scale Hardness Numbers

Rockwell B-scale hardness No.	Vickers hardness No.	Brinell hardness No., 10-mm-diam ball: 500-kg load	Brinell: 3000-kg load	Rockwell hardness No.: A scale, 60-kg load, Brale indenter	Rockwell: C scale, 150-kg load, Brale indenter	Rockwell: F scale, 60-kg load, 1/16-in.-diam ball	Rockwell superficial hardness No., 1/16-in.-diam ball: 15T scale, 15-kg load	30T scale, 30-kg load	45T scale, 45-kg load	Knoop hardness No., 500-g load and greater	Shore Scleroscope hardness No.	Tensile strength (approx), 1000 psi
98	228	189	228	60.2	(19.9)	...	92.5	81.8	70.9	241	34	107
97	222	184	222	59.5	(18.6)	...	92.1	81.1	69.9	236	33	104
96	216	179	216	58.9	(17.2)	...	91.8	80.4	68.9	231	32	102
95	210	175	210	58.3	(15.7)	...	91.5	79.8	67.9	226	...	99
94	205	171	205	57.6	(14.3)	...	91.2	79.1	66.9	221	31	97
93	200	167	200	57.0	(13.0)	...	90.8	78.4	65.9	216	30	94
92	195	163	195	56.4	(11.7)	...	90.5	77.8	64.8	211	...	92
91	190	160	190	55.8	(10.4)	...	90.2	77.1	63.8	206	29	90
90	185	157	185	55.2	(9.2)	...	89.9	76.4	62.8	201	28	88
89	180	154	180	54.6	(8.0)	...	89.5	75.8	61.8	196	27	86
88	176	151	176	54.0	(6.9)	...	89.2	75.1	60.8	192	...	84
87	172	148	172	53.4	(5.8)	...	88.9	74.4	59.8	188	26	82
86	169	145	169	52.8	(4.7)	...	88.6	73.8	58.8	184	26	81
85	165	142	165	52.3	(3.6)	...	88.2	73.1	57.8	180	25	79
84	162	140	162	51.7	(2.5)	...	87.9	72.4	56.8	176	...	78
83	159	137	159	51.1	(1.4)	...	87.6	71.8	55.8	173	24	76
82	156	135	156	50.6	(0.3)	...	87.3	71.1	54.8	170	24	75
81	153	133	153	50.0	...	...	86.9	70.4	53.8	167	...	73
80	150	130	150	49.5	...	...	86.6	69.7	52.8	164	23	72
79	147	128	147	48.9	...	...	86.3	69.1	51.8	161	...	70
78	144	126	144	48.4	...	...	86.0	68.4	50.8	158	22	69
77	141	124	141	47.9	...	...	85.6	67.7	49.8	155	22	68
76	139	122	139	47.3	...	...	85.3	67.1	48.8	152	...	67
75	137	120	137	46.8	...	99.6	85.0	66.4	47.8	150	21	66
74	135	118	135	46.3	...	99.1	84.7	65.7	46.8	148	21	65
73	132	116	132	45.8	...	98.5	84.3	65.1	45.8	145	...	64
72	130	114	130	45.3	...	98.0	84.0	64.4	44.8	143	20	63
71	127	112	127	44.8	...	97.4	83.7	63.7	43.8	141	20	62
70	125	110	125	44.3	...	96.8	83.4	63.1	42.8	139	...	61
69	123	109	123	43.8	...	96.2	83.0	62.4	41.8	137	19	60
68	121	107	121	43.3	...	95.6	82.7	61.7	40.8	135	19	59
67	119	106	119	42.8	...	95.1	82.4	61.0	39.8	133	19	58
66	117	104	117	42.3	...	94.5	82.1	60.4	38.7	131	...	57
65	116	102	116	41.8	...	93.9	81.8	59.7	37.7	129	13	56
64	114	101	114	41.4	...	93.4	81.4	59.0	36.7	127	18	...
63	112	99	112	40.9	...	92.8	81.1	58.4	35.7	125	18	...
62	110	98	110	40.4	...	92.2	80.8	57.7	34.7	124	...	...
61	108	96	108	40.0	...	91.7	80.5	57.0	33.7	122	17	...
60	107	95	107	39.5	...	91.1	80.1	56.4	32.7	120	...	...
59	106	94	106	39.0	...	90.5	79.8	55.7	31.7	118	...	...
58	104	92	104	38.6	...	90.0	79.5	55.0	30.7	117	...	...
57	103	91	103	38.1	...	89.4	79.2	54.4	29.7	115	...	...
56	101	90	101	37.7	...	88.8	78.8	53.7	28.7	114	...	...
55	100	89	100	37.2	...	88.2	78.5	53.0	27.7	112	...	...

FIGURE 26.7. Placing the test block in the machine. (*Wilson Instrument Division of Acco*)

FIGURE 26.9. Specimen being brought into contact with the penetrator. This establishes the minor load. (*Wilson Instrument Division of Acco*)

FIGURE 26.8. Selecting and installing the correct weight. (*Lane Community College*)

FIGURE 26.10. Setting the zero start point. (*Wilson Instrument Division of Acco*)

FIGURE 26.11. Applying the major load by tripping the load lever handle clockwise. (*Wilson Instrument Division of Acco*)

(not to exceed more than 1.5 revolutions) as the indenter penetrates into the test piece.

7. Wait approximately 5 seconds for the needles to stop before *gently* pulling the trip lever forward or counterclockwise to release the major load (Figure 26.11).
8. The indenter will remain in contact with the part being tested. Part hardness is measured by reading the number where the large needle points and the value corresponding to the color for the load range being used to test the part (Figure 26.12).
9. Reverse the handwheel to release the load. Lower the anvil sufficiently to prevent part contact with the indenter.
10. When reading hardness numbers from the tester or from a wall chart, the test ranges (both load and indenter) are color coded to the test range being used (i.e., black, green, and red).

SUPERFICIAL TESTING

After testing sheet metal, examine the underside of the sheet. If the impression of the penetrator can be seen, then the reading is in error and the superficial test should be used. If the impression can still be seen after the superficial test, then a lighter load should be used. A minor load of 3 kg and a major load of 30 kg is recommended for most superficial testing. Superficial testing is also used for case-hardened and nitrided steel having a very thin case.

A brale marked N is needed for superficial testing, as A and C brales are not suitable. Recorded readings should be prefixed by the major load and the letter N when using the brale for superficial testing, for example, 30N78. When using the 1/16 in. ball penetrator, the

FIGURE 26.12. Dial face with reading in Rockwell units after completion of the test. The reading on the dial is RC 55. (*Lane Community College*)

same as that used for the B, F, and G hardness scales, the readings should always be prefixed by the major load and the letter T, for example, 30T85. The 1/16 in. ball penetrator, however, should not be used on material harder than 30T82. Other superficial scales, such as W, X, and Y, should also be prefixed with the major load when recording hardness. See Table 26.5 for superficial test penetrator selection.

The basic **anvils** used for Rockwell testing are shown in Figure 26.6. Flat anvils are used for specimens with flat surfaces and V-type anvils for round specimens. A spot anvil is used when the tester is being checked on a Rockwell test block. The spot anvil should not be used for checking cylindrical surfaces. The diamond spot anvil (Figure 26.6*f*) is similar to the spot anvil, but it has a diamond set into the spot. The diamond is ground and polished to a flat surface. This anvil is used **only** with the superficial tester, and then **only** in conjunction with the steel ball penetrator for testing soft metal.

USING THE BRINELL HARDNESS TESTER

The Brinell hardness test is made by forcing a steel ball, usually 10 millimeters (mm) in diameter, into the test

TABLE 26.5. Superficial tester load and penetrator selection

Scale Symbol	Penetrator	Load (kg)
15N	Brale	15
30N	Brale	30
45N	Brale	45
15T	1/16 in. ball	15
30T	1/16 in. ball	30
45T	1/16 in. ball	45
15W	1/8 in. ball	15
30W	1/8 in. ball	30
45W	1/8 in. ball	45
15X	1/4 in. ball	15
30X	1/4 in. ball	30
45X	1/4 in. ball	45
15Y	1/2 in. ball	15
30Y	1/2 in. ball	30
45Y	1/2 in. ball	45

Source: Wilson Instruction Manual, "Rockwell Hardness Tester Models OUR-a and OUS-a," American Chain & Cable Company, Inc., 1973.

specimen by using a known load weight and measuring the diameter of the resulting impression. The Brinell hardness value is the load divided by the area of the impression, expressed as follows:

BHN = Brinell hardness number, kg/mm^2

D = diameter of the steel ball, mm

P = applied load, kg

d = diameter of the impression, mm

$$\text{BHN} = \frac{P}{(\pi D/2)(D - \sqrt{D^2 - d^2})}$$

A small microscope is used to measure the diameter of the impressions. Various loads are used for testing different materials: 500 kg for soft materials such as copper and aluminum, and 3000 kg for steels and cast irons. For convenience, Table 26.2 gives the Brinell hardness number and corresponding diameters of impression for a 10 mm ball and a load of 3000 kg. The Brinell hardness numbers are obtained by reading the indentation diameter in millimeters on the tested specimen and reading the Brinell number in the table under the appropriate load. Conversion to Rockwell hardness numbers is easily done by reading the appropriate Rockwell numbers across from the Brinell numbers. Tensile strengths are also shown. Just as for the Rockwell tests, the impression of the steel ball must not show on the underside of the specimen, and the same surface preparation is made before testing. Tests should not be made too near the edge of a specimen. Figure 26.13 shows a typical bench type Brinell hardness tester and Figure 26.14 an indentation measuring microscope.

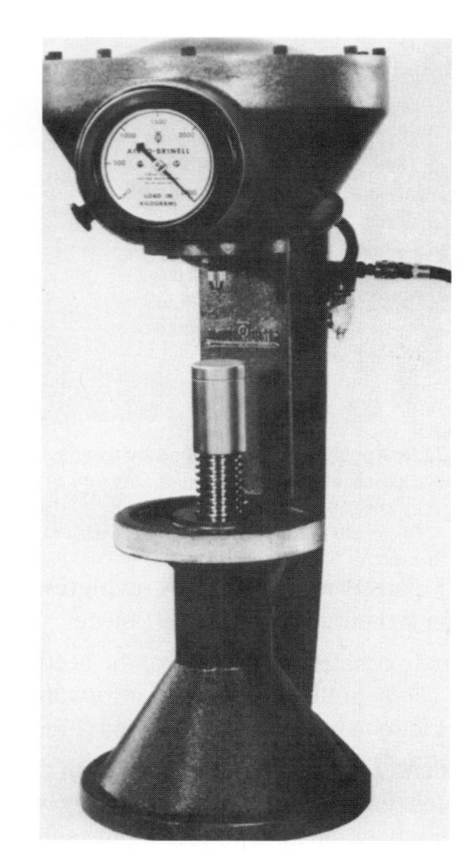

FIGURE 26.13. Typical air-operated Brinell Hardness Tester. (*Tinius Olson Testing Machine Company, Inc.*)

THE TESTING SEQUENCE OF THE BRINELL HARDNESS TESTER

1. The desired load in kilograms is selected on the dial by adjusting the air regulator (Figure 26.15).
2. The specimen is placed on the anvil. Make sure the specimen is clean and free from burrs. It should be smooth enough so that an accurate measurement can be taken of the impression.
3. The specimen is raised to within 5/8 in. of the Brinell ball by turning the handwheel.
4. The load is then applied by pulling out the plunger control (Figure 26.16). Maintain the load for 30 seconds for nonferrous metals and 15 seconds for steel. Release the load (Figure 26.17).
5. Remove the specimen from the tester and measure the diameter of the impression.
6. Determine the Brinell hardness number (BHN) by calculation or by using the table. Soft copper should have a BHN of about 40, soft steel from 150 to 200, and hardened tools from 500 to 600. Fully hardened high carbon steel would have a BHN of 750. A

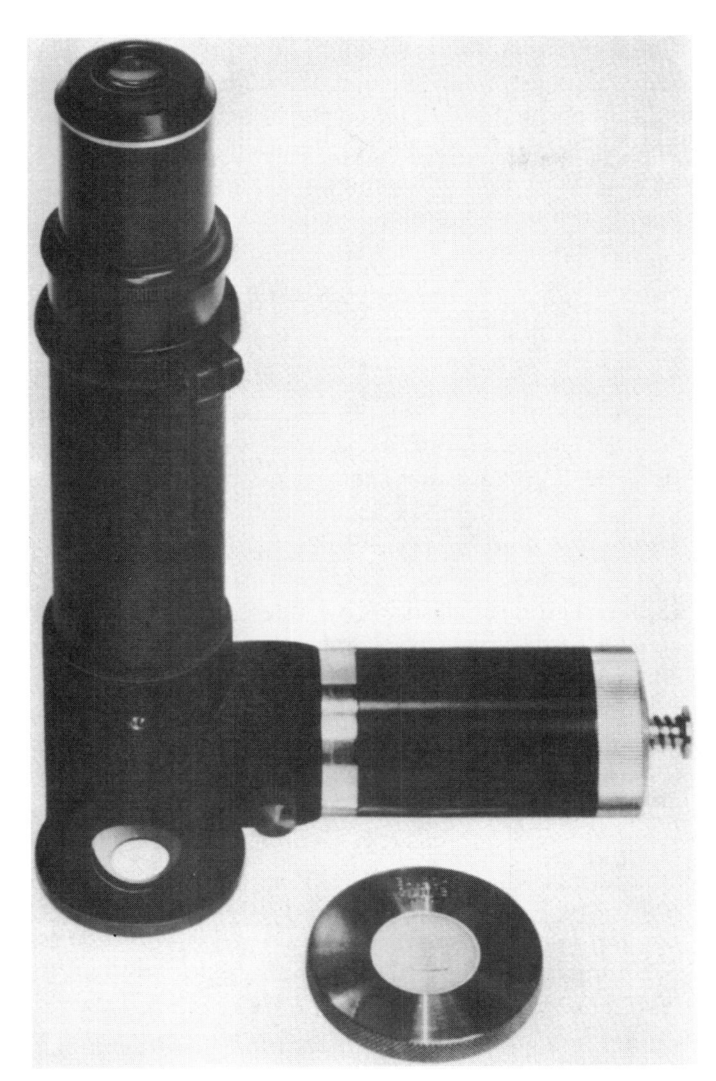

FIGURE 26.14. Microscope used to measure Brinell hardness indentation (*Tinius Olson Testing Machine Company, Inc.*)

Brinell test ball of tungsten carbide should be used for materials above 600 BHN.

Brinell hardness testers work best for testing softer metals and medium-hard steels.

MICROHARDNESS TESTING

Microhardness testing is a precise method of hardness testing on a microscopic scale in comparison to Rockwell and Brinell testers. Microhardness testing allows surface hardness testing of homogeneous microstructure, precise measurement of surface-hardened case depth, and microconstituents of interest to the investigator. Preparation of the microhardness test sample is critical. The surface must be perpendicular to the indenter and highly polished in the same manner as required

FIGURE 26.15. Select load. Operator adjusts the air regulator as shown until the desired Brinell load in kilograms is indicated. (*Tinius Olson Testing Machine Company, Inc.*)

FIGURE 26.16. Apply load. Operator pulls out plunger type of control to apply load smoothly to specimen. (*Tinius Olson Testing Machine Company, Inc.*)

FIGURE 26.17. Release load. As soon as the plunger is depressed the Brinell ball retracts in readiness for the next test. (*Tinius Olson Testing Machine Company, Inc.*)

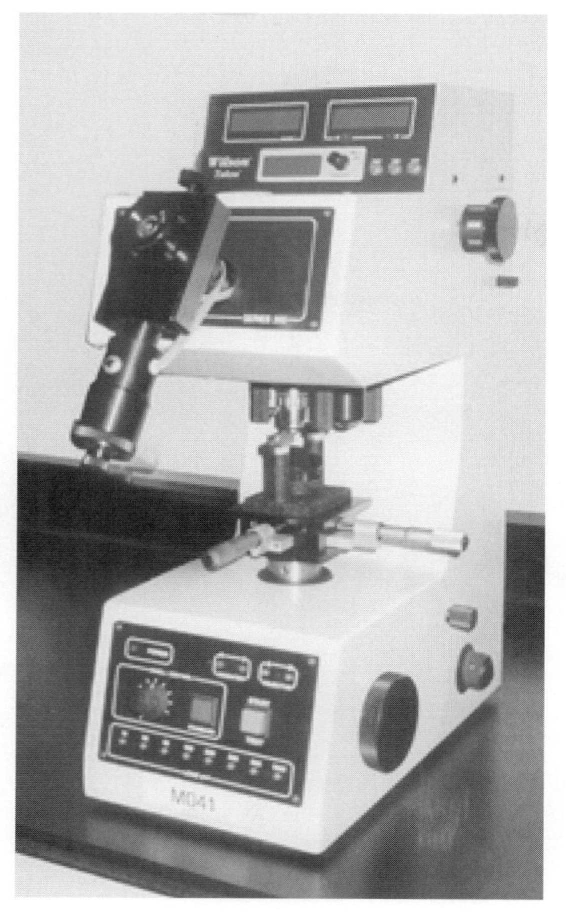

FIGURE 26.18. Wilson Tester MO41 Microhardness Tester. This modern machine performs all conversions from filar units to Vickers or Knoop hardness.

for microstructure examination, except unetched. The microhardness tester is equipped with a metallurgical microscope and one of two types of indenters: a Knoop or Vickers (see Figure 26.18). To use the microhardness tester, precise calibration of both the instrument and the operator of the instrument is required.

CALIBRATION BLOCK

The calibration block used for determining machined hardness variance is usually calibrated in accordance with ASTM Standard E384 (Part C). Hardness indentations on the calibration block are located in five test groups. The certificate of calibration for the test block will have recorded the measured hardness at the location of the test groups 1 through 5 (see Figure 26.19). Each test group consists of five indentations; their average is recorded on the certificate of calibration as the average for the specific group. It is extremely important to not make indentations near or next to these group indentations, since the group indentations are required to be measured as part of the calibration of the person making the hardness measurements.

GENERAL CALIBRATION TEST PROCEDURE FOR MICROHARDNESS TESTING

1. Place a clean microhardness test calibration block in position on the micrometer stage. Read the test block certification.
2. Focus the reticule eyepiece.
3. Focus the microscope by adjusting the coarse and fine focus.
4. Select a "Group" (1, 2, 3, 4, or 5) to measure and adjust the fine focus so as to clearly define the end points of the Knoop indentations.
5. Position one of the certified group indentations of the group in the microscope so that the long diagonal can be measured accurately. Measure the diagonal length.
6. Refer to the calibration certificate for the average length measured for the specific group selected. Continue to practice measuring the group indentations until your measurements equal that indicated on the certification. Eyes take about 15 minutes before becoming accustomed to correctly measuring indentations and tend to become strained during long periods of continuous measuring, so check eye

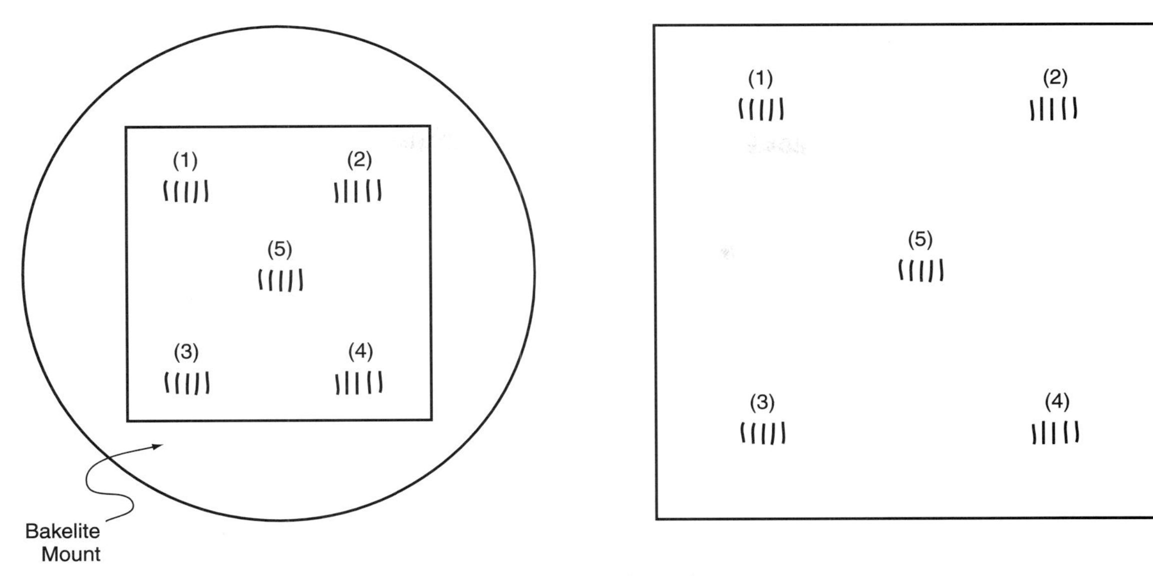

FIGURE 26.19. Schematic of standard calibration blocks showing locations of test groups.

accuracy often by measuring the certified test block indentations.

7. Set the load to 500 grams.

8. Set the time:
steels = 15 seconds; nonferrous = 30 seconds

9. Make an indentation. Locate the indentation in the field of the microscope. Adjust the position of the indentation to the stationary reference measuring line.

10. Make five Knoop indentations; moving the stage micrometer 0.005 in. before each indentation to separate indentations at least two widths apart.

11. Record each indentation.

12. Mathematically average the data; add and divide by 5. The measured hardness should be within the limits of the calibration block. Normally, the measured hardness will not vary more than ±3 Knoop numbers.

13. Refer to the procedure for obtaining a correction factor (CF) to convert measured hardness to actual hardness.

MEASUREMENT PRACTICE: DPH/VICKERS

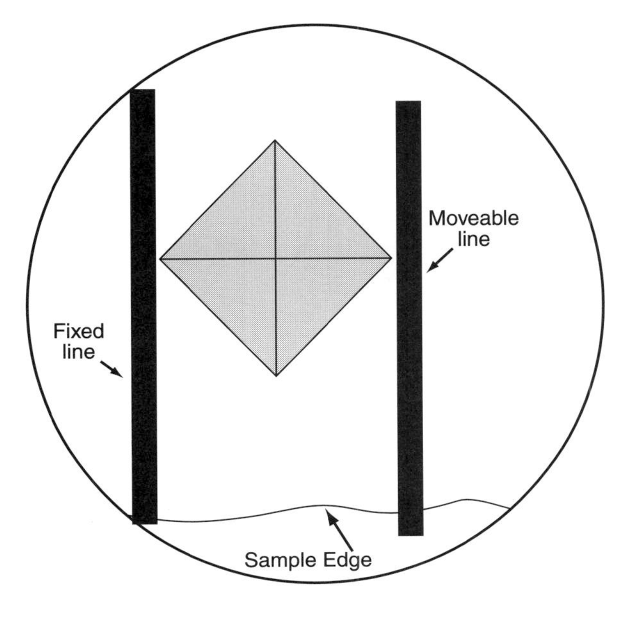

Preferred Method

1. The position of the Vickers DPH is correct.
2. Measurement of both diagonals is made without overlapping the tips: $(d_1 + d_2/2)$.
3. The indentation is acceptable in that the diagonal lengths are within 20 percent.
4. Symmetry also is an indicator of proper sample preparation; a rounded edge may result in faulty data.
5. The surface of the test piece and the resulting indentation are free of stains.
6. The diagonal lines inside the indentation are sharp and perfectly defined, which is an indication of the physical condition of the diamond.

MEASURING THE VICKERS INDENTATION

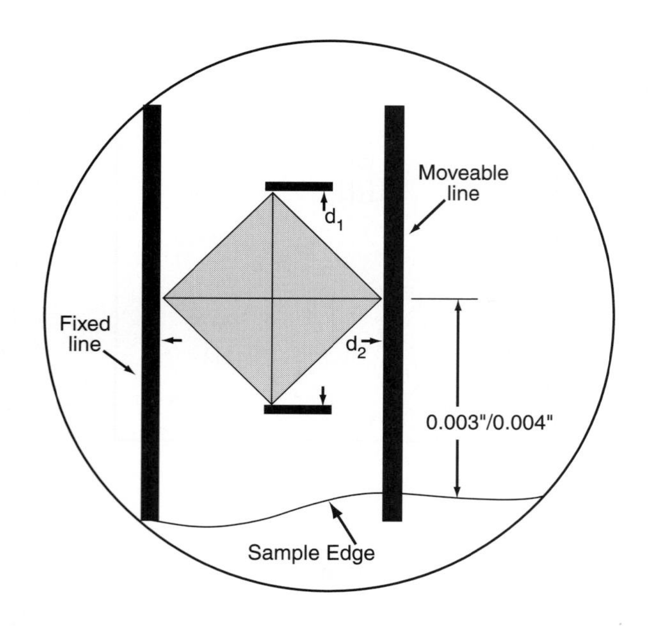

1. The indentation must be made at 0.003 in. from the sample surface or edge to be in compliance with ASTM E384-89 specification and measurement of surface hardness (carburization and/or decarburization).
2. Distance from the edge is measured to the parallel diagonal.
3. Place the points of each diagonal between the thick lines of the microscope and avoid overlapping the edge of the indentation. Test accuracy is dependent upon correct measurement of both diagonals.

IRREGULARITIES IN MEASUREMENT: DPH/VICKERS

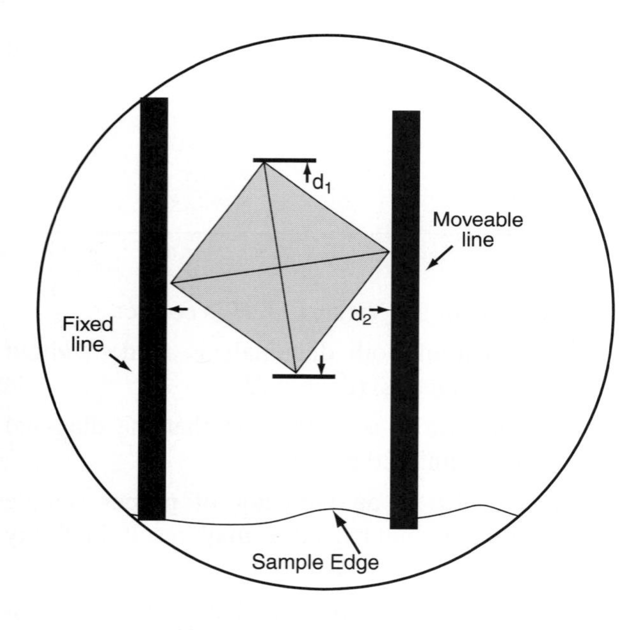

Measuring Unit Not Aligned

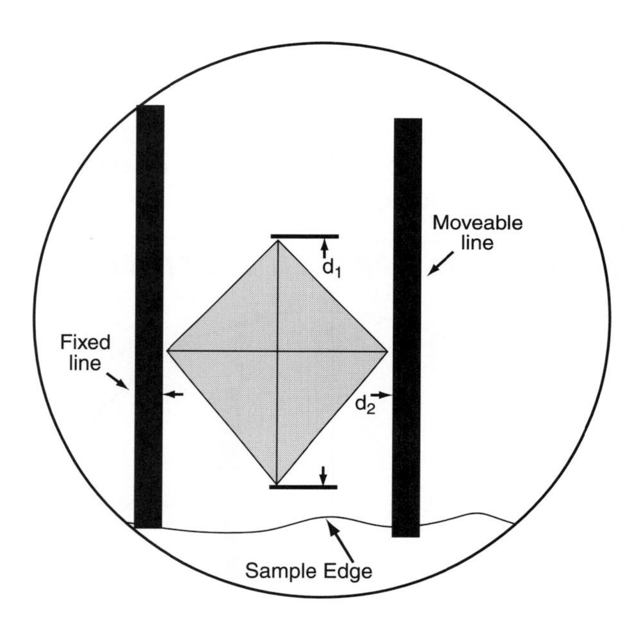

Indentation Too Close To Edge

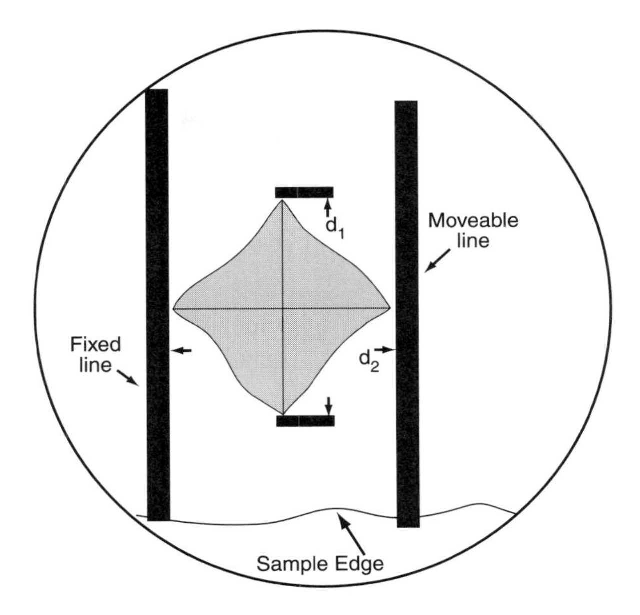

Skewed Indentation
The indentation is skewed due to the elastic recovery of the metal crystal structure. Reading errors can result.

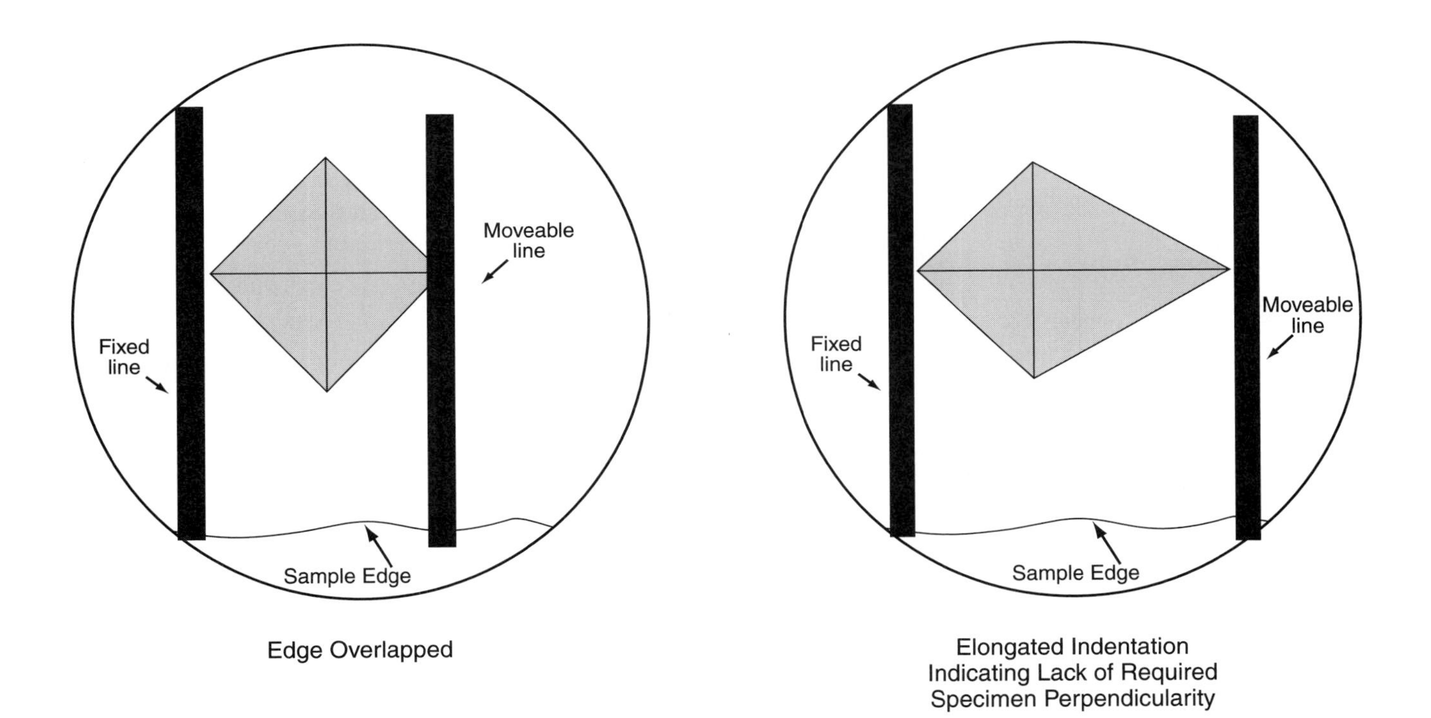

Edge Overlapped

Elongated Indentation Indicating Lack of Required Specimen Perpendicularity

MEASUREMENT PRACTICE: KHN/KNOOP

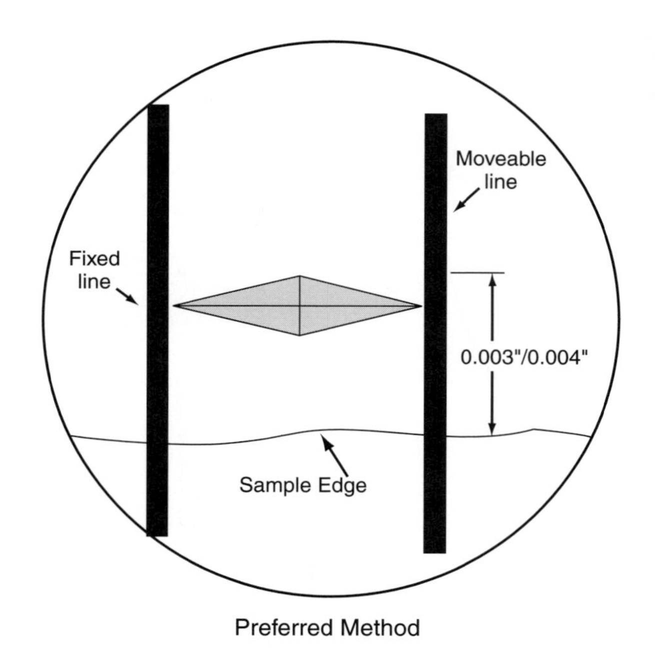

Preferred Method

1 The position of the Knoop is correct.
2. Measurement is made of the full length without overlapping the tips.
3. The indentation is acceptable in that the diagonal lengths are within 20 percent.
4. Symmetry also is an indicator of proper sample preparation; a rounded edge may result in edge irregularities.
5. The surface of the test piece and the resulting indentation are free of stains.
6. The diagonal lines inside the indentation are sharp and perfectly defined, which is an indication of the physical condition of the diamond.

MEASURING THE KNOOP INDENTATION

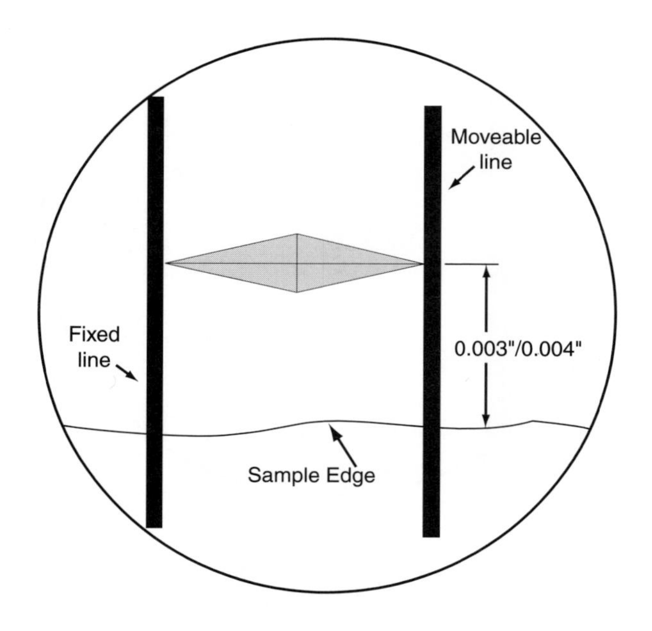

1. The indentation must be made at 0.003 in. from the sample surface or edge to be in compliance with ASTM E384-89 specification and measurement of surface hardness (carburization and/or decarburization).
2. Distance from the edge is measured to the Knoop long diagonal.
3. Place the points of the long diagonal between the thick lines of the microscope and avoid overlapping the edge of the indentation. Test accuracy is dependent upon the precision of indentation placement.

IRREGULARITIES IN MEASUREMENT: KHN/KNOOP

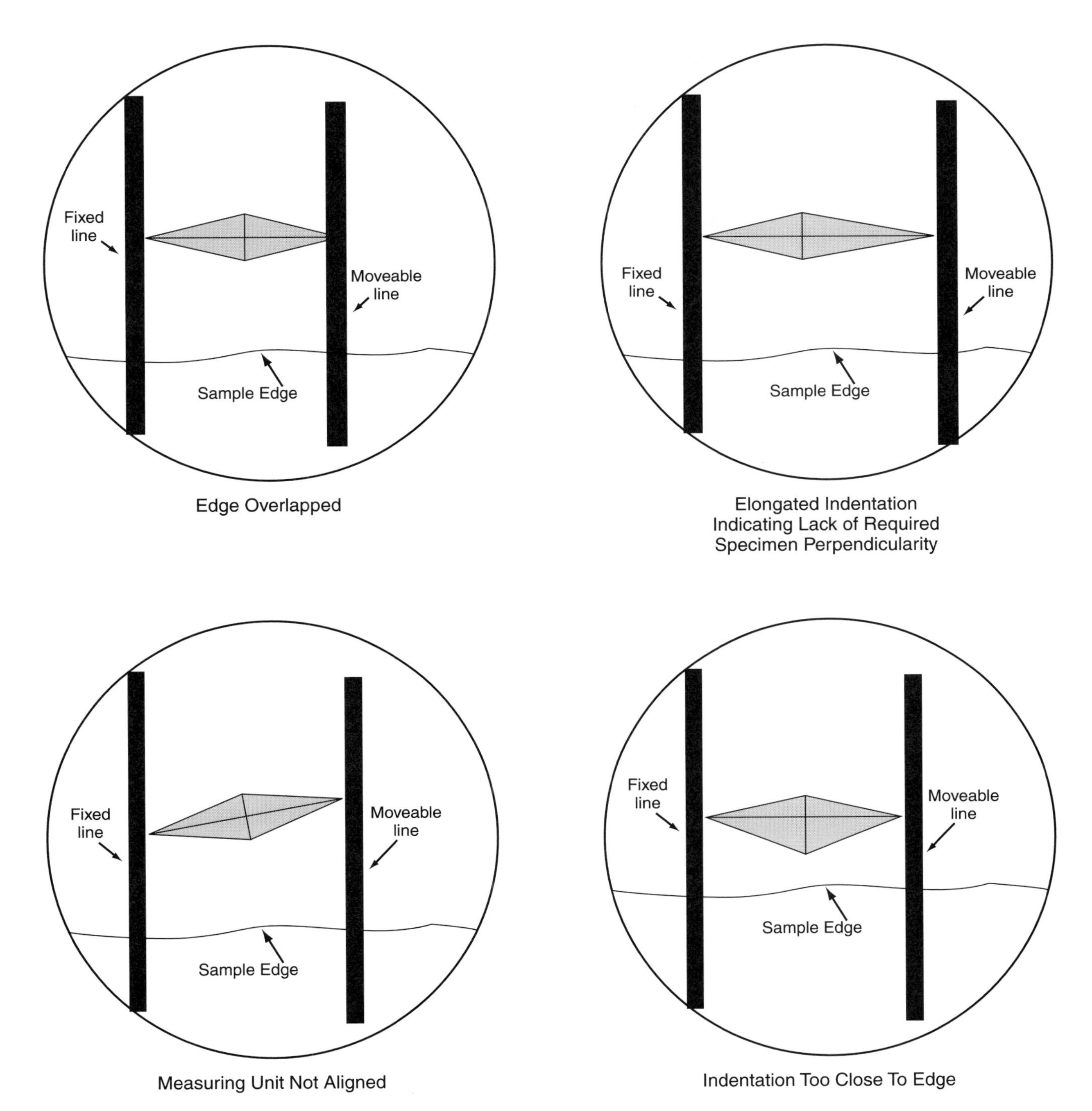

Edge Overlapped

Elongated Indentation Indicating Lack of Required Specimen Perpendicularity

Measuring Unit Not Aligned

Indentation Too Close To Edge

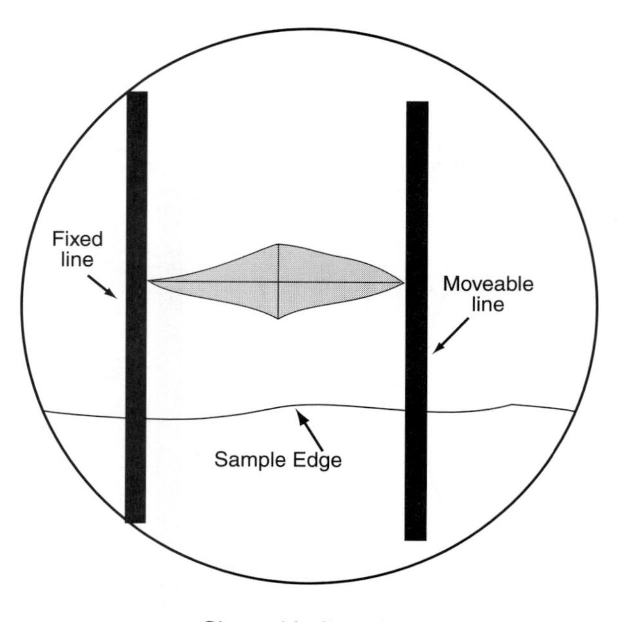

Skewed Indentation
The indentation is skewed due to the elastic recovery of the metal crystal structure. Reading errors can result.

SELF-EVALUATION

1. What one specific category of the property of hardness do the Rockwell and Brinell hardness testers use and measure? How is it measured?
2. State the relationship that exists between hardness and tensile strength.
3. Explain which scale, major load, and penetrator should be used to test a block of tungsten carbide on the Rockwell tester.
4. What is the reason that the steel ball cannot be used on the Rockwell tester to test the harder steels?
5. When testing with the Rockwell superficial tester, is the Brale used the same one that is used on the A, C, and D scales? Explain.
6. The 1/16-in. ball penetrator used for the Rockwell superficial tester is a different one from that used for the B, F, and G scales. True or false?
7. What is the diamond spot anvil used for?
8. How does roughness on the specimen to be tested affect the test results?
9. How does decarburization affect the test results?
10. What does a curved surface do to the test results?
11. On the Brinell tester, what load should be used for testing steel?
12. What size ball penetrator is generally used on a Brinell tester?
13. Explain the steps required for the calibration of a microhardness tester.
14. Explain why the microhardness tester operator must also be calibrated.
15. What is the difference between actual and measured hardness?
16. What happens when the surface being hardness tested is too rough?
17. Why must a ball indenter be used to hardness test aluminum alloys?
18. What indenter and load is required for the Rockwell “C” scale?
19. Explain what happens when a Brinell hardness tester is used to measure the hardness of a carburized steel.

20. What is the difference between the minor load of a standard and superficial Rockwell hardness tester?
21. Explain why large items must be supported during hardness testing.
22. Explain why only one side of a calibration block is used to measure hardness. How far apart must Rockwell indentations be placed?

CASE PROBLEM: FLAME HARDENED WRISTPINS EXHIBIT SOFT SPOTS

A racing engine failed during a top fuel drag race. Upon engine disassembly, one wristpin was found to be shattered. The wristpin is a through hardened 52100 alloy steel. Some of the pieces were found to be soft. Can you verify the actual wristpin hardness? If you find soft spots on the O.D. surface, explain what can cause this problem and how it can be corrected.

WORKSHEET 1

Objectives

1. Learn to use the Rockwell and microhardness testers.
2. Determine actual Rockwell surface hardness and compare these test results with microhardness testing conducted at the same location.

Materials Fractured pieces of wristpin approximately 1 inch square.

Procedure

1. Clean the pieces of wristpin to remove all dirt and oils.
2. Hand grind the I.D. surface flat to support the wristpin during O.D. hardness measurements.
3. Set up the Rockwell tester for a 30N scale hardness test.
4. Calibrate the Rockwell tester using a 30N test block.
5. Perform hardness measurements on the piece of wristpin and record the results. Is there a correction for a curved surface?
6. Locate the soft spot and determine its boundaries.
7. Cross-section through the suspected soft spot and mount this transverse section in bakelite as described in Chapter 27.
8. Grind and polish the mounted section as required for microhardness testing.
9. Calibrate the microhardness tester and yourself.
10. Perform microhardness testing at a depth of 0.002 inch at various locations and record results.

Conclusion Was this testing sufficient to determine the cause of soft spots? Is there a good correlation between the hardness testing methods? What additional test procedure would be appropriate to visually see soft spots on the surface of a hardened steel?

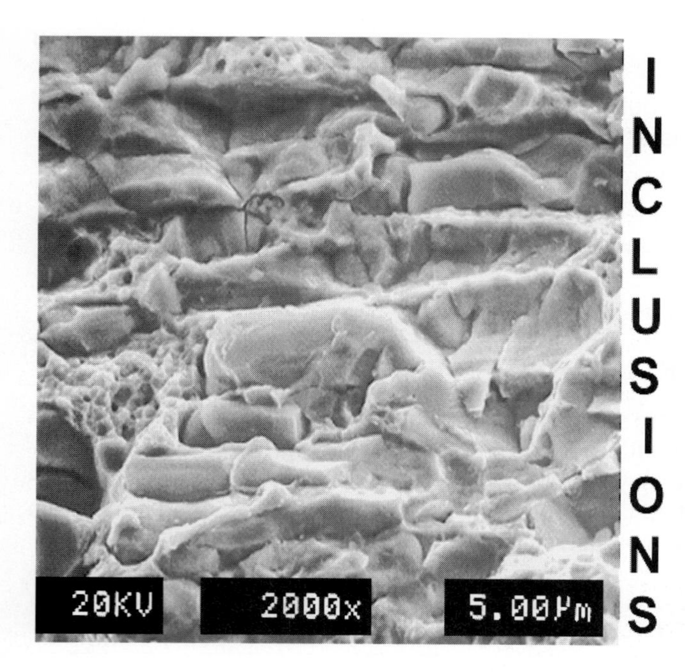

CHAPTER 27

Failure Analysis

One of the most common problems found in industrial equipment and other machinery is mechanical breakdown due to wear, overload, or fatigue. Operational failures sometimes cause a loss of production, loss of work time for employees, and occasional hazardous conditions. Many service problems can be prevented by proper maintenance, selection of materials, and design. This chapter will help you recognize and understand the methods used to determine the cause(s) of material failures and the corrective measures required to eliminate these problems in the future.

The methods and practices of metallurgical failure analysis, chemical analysis fracture surface examination, and optical microcroscopy have changed dramatically over the past thirty years. In former times, wet chemical analysis was the answer to determining alloy conformance to A.I.S.I. chemistry specifications. Wet chemical analysis often took more than three hours and required an experienced B.S. degree chemist to perform testing with the necessary accuracy. During the past thirty years, wet chemical analysis has been systematically replaced by optical emission spectrometry (OES). OES instrument performs alloy qualitative and quantitative chemical analysis in a few seconds with far better repeatability and accuracy. The OES is commonly operated by a technician who learned the use of the instrument while on the job.

In the sixties, the transmission electron microscope (TEM) was the prime instrument for determining fracture mode in material failure analysis. Due to tedious requirements for specimen preparation and difficulty in operating the instrument, the TEM has become outdated and is seldom used.

The replacement instrument of choice is the scanning electron microscope (SEM). Analytical instruments such as the SEM have become commonplace in the world of material failure analysis. Specimen preparation for the SEM is easy in comparison to the TEM. The majority of fracture surfaces are examined directly, the fracture surface is ultrasonically cleaned and/or degreased and then placed directly into the SEM vacuum chamber. When the required level of vacuum is reached, the SEM TV monitor comes on and you're ready to examine the fracture surface. In the rare case where an acetate replica of the fracture surface must be used, the replica is either vapor coated with carbon or gold to make it conductive. The coated replica (acetate intact) is placed directly into the S.E.M. for viewing in the same manner as an actual fracture surface.

Metallographic specimen sectioning requires a water cooled abrasive saw with high cutting speed. The water coolant prevents any heating or burning of the metal. Bakelite is the preferred mounting media used to encapsulate the cut material. The encapsulation, called a mount, holds the specimen intact during grinding, polishing, etching and optical examination. Other mounting media, such as 2-part polymer/resin hardening systems are also used when the size of the cut material is larger than can be accommodated in a 1 1/2 inch diameter mounting press. Metallographic specimen preparation techniques have also improved. A very fast and simple method of metallographic sample preparation is explained in this chapter. This method teaches an easy-to-do and quick method of specimen preparation for the non-expert. More sophisticated methods, procedures, and complicated metallographic specimen preparation equipment are essential in industry and for most senior colleges and universities. These detailed procedures and safety requirements may be found in Appendix 1.

OBJECTIVES **After completing this chapter, you should be able to:**

1. Explain the causes of several industrial problems that lead to failures and list corrective measures for them.
2. Identify and describe the causes of failure in 10 metal parts.
3. Make recommendations for change in material design or in heat treatment for each of the 10 metal parts to correct their service failures.

MATERIAL FAILURE ANALYSIS TECHNIQUES

The primary objective for performing a failure analysis is to determine the cause(s) of the failure; and in doing so, provide product improvement information to prevent future reoccurrence of the failure. Frequently, the importance of contributory causes to the failure must be assessed; new experimental techniques may have been developed, or an unfamiliar field of engineering or science explored. This chapter discusses the techniques used by the metallurgist and/or material scientist to determine the cause(s) of a failure and offers case histories to familiarize the student with problem analysis.

Stages of Material Analysis

The principal stages in a material investigation may comprise the following:

1. Collection of background data or history of the failed item
2. Photo documentation of the as-received condition of the item with labeling and measurement
3. Determination of the type of testing or analysis to be conducted and formation of a protocol for analysis
4. Any nondestructive testing
5. Visual and macroscopic examination of the failed item; fracture surface(s) and other surface features (Often, the areas to be sectioned are marked and identified at this time.)
6. Optical metallographic examination, which includes cutting for this and other testing, mounting, grinding, and polishing, as required for optical microscopic examination of the mounted section in the unetched and etched conditions (X-ray defraction may be performed to determine crystal structure.)
7. Chemical analysis for material verification
8. Mechanical testing to determine mechanical properties
9. Microhardness testing to determine general hardness, case characteristics, and surface effects produced during use and/or by thermal treatment
10. Etching of metallographic section to delineate the microstructure
11. Examination of the fracture surface using the scanning electron microscope (This examination may also include energy dispersive X-ray analysis (EDX). The fracture mode and/or the fracture origin is often identified during this stage of testing.)
12. Data revision, including analysis of fracture mechanics
13. Testing of exemplar samples
14. Engineering analysis and report writing which can include discussion of the failure mode, formation of conclusions, and often, recommendations for product improvement and failure prevention

Although these 14 stages of a failure analysis seem quite complicated, not all of the testing described is required for factual determination of a failure mode. The conscientious metallurgist will select the appropriate testing based on experience and best judgment at the time of initial item examination. The background and service history of a failed item is very important and will influence the selection of proposed tests. Often, the background data are sparse and the metallurgist must rely on the testing to force out information pertaining to original manufacturing processes and service life. Record keeping and item labeling at every stage of analysis are mandatory, especially when compliance to ISO and ANSI requirements are implemented.

EQUIPMENT USED IN FAILURE ANALYSIS

Cameras and video recorders are commonplace when documenting the overall condition of a failed item, especially when litigation is likely. The macroscope—magnification ranges from 5X to 50X—is required to see and select specific details on and around the fracture surface which may help in determining the cause(s) of failure. The optical microscope, equipped with large viewing TV monitors to allow group viewing, is used at many laboratories to examine the cleanliness and inclusion rating of a metal per ASTM E-45. Optical microscopes are also used for plating thickness measurement, for general microstructure analysis, and for determining the microstructure's response to heat treatment (Figure 27.1). Mechanical testing techniques and equipment are discussed in Chapter 3; nondestructive testing techniques and equipment are described in Chapter 20. Chemical analysis can be performed in several ways: by optical emission spectrometry called OES, by "wet" chemical analysis, by X-ray fluorescence, by Auger electron analysis, and by EDX with the scanning electron microscope (Figure 27.2). The scanning electron microscope allows detailed high-resolution examination of the fracture surfaces of electrical conducting and nonconducting metals and nonmetals. Magnifications exceeding 50,000X are possible. Microhardness testers are used to accurately determine hardness in a

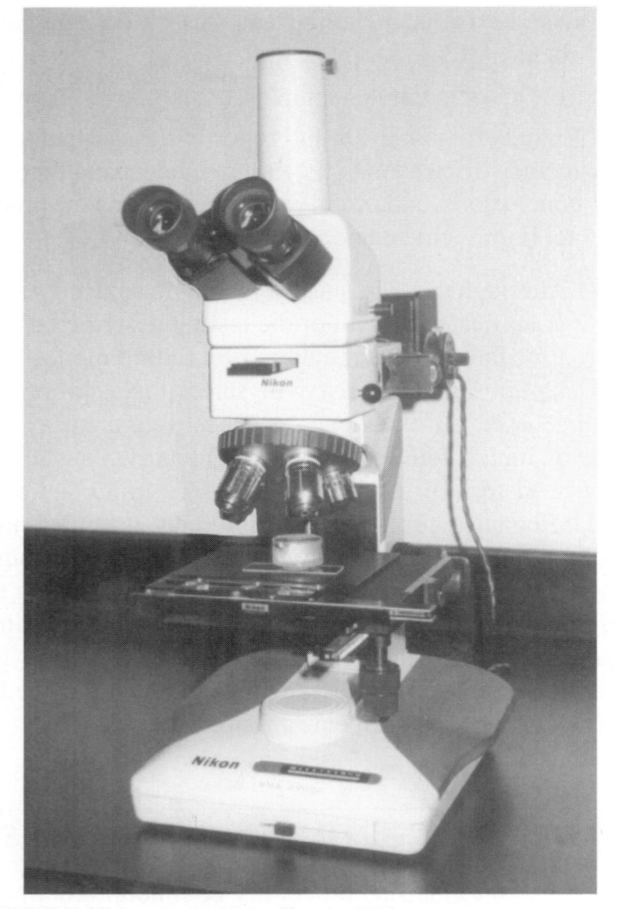

FIGURE 27.1. Optical Metallurgical Microscope

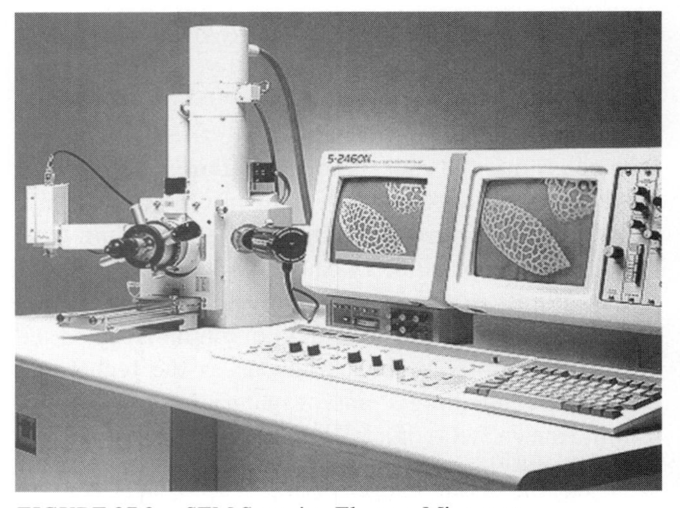

FIGURE 27.2. SEM Scanning Electron Microscope

very small area such as on a cross-sectioned grain. Microhardness testers are equipped with an x-y micrometer stage which allows hardness testing of case microstructures and at specific locations from an edge or surface (see Chapter 26).

SIMPLE METALLOGRAPHIC SAMPLE PREPARATION

The preparation of a metal surface for examination using the optical microscope is a rather simple procedure. In most books, the requirements for the preparation of metallographic samples are quite complicated. The procedure involves cutting, mounting in a media to hold the sample, grinding the surface through progressively finer grit sizes, and finally polishing the surface with progressively finer polishing compounds until a scratch-free surface finish is obtained.

Shortcuts are cost savers which must not be ignored. Therefore, the following procedures are offered for the quick preparation of metal surfaces for optical examination, as-polished and in the etched condition. The materials required are rather simple—grinding papers: 60 grit, 120 grit, 320 grit, 600 grit, and 800 grit; one low nap soft cloth impregnated with 1 micron diamond compound.

The well-kept secret of metallographic sample preparation should only take about 2 minutes maximum per standard sample; a 1.5 in. diameter mount. Mounting is usually done in a heated and pressurized mounting press and takes about 5 to 10 minutes. Most mounting presses are automatic; however, it takes about 4200 psi at 300°F with a little cooling to make the mount. Once the mount is made, the objective is to grind the surface near to or at the level you wish to examine. This is accomplished by wet hand grinding or by using a water-cooled, medium speed rotary wheel with coarse grinding paper, usually 60 or 120 grit is used. The scratch direction will always be from the thumb to the middle finger when holding the mounted sample. Once the scratches are evenly distributed across the surface of the sample, wash off the grit and proceed to 320 grit wet grinding. Grind until all the coarse grit scratches are removed and then proceed to 600 grit wet grinding. If you are using a motorized wheel, reduce speed to slow. When all the former scratches are removed by the 600 grit, proceed to 800 grit wet grinding. The surface of the sample should be almost mirrorlike when the 600 grit scratches have been removed by the 800 grit paper. The last step is to dry the surface of the sample. Polishing is accomplished in one step. Using WD-40 as a spray lubricant, rotary polish the surface of the sample using medium speed and pressure while rotating around the cloth in a clockwise direction. Keep the wheel moist with the WD-40; polishing only takes about 10 seconds to achieve a scratch-free surface on most steels and tool steels. Aluminum and other soft nonferrous alloys take about 20 seconds. All metals are polished using the same wheel and compound. Overpolishing can cause surface smearing and/or work hardening, which is hard to remove. WD-40 can be removed from the surface of the mounted sample by washing with a mild liquid soap. Ultrasonic cleaning also works well to remove oil residue from the surface of the mounted sample.

With a little practice, students will understand the simplicity of the metallographic sample preparation procedure. Elaborate and expensive automatic grinding and polishing machines are not necessary or required. Cross-contamination between different metals does not occur with this sample preparation technique.

Microstructure examination usually begins with optical examination of the as-polished surface to determine inclusion content in a material. The American Society for Testing and Materials (ASTM) provides standards and acceptance criteria for most microstructure examinations common in industry. In addition, the *Metals Handbooks,* by ASM, provides a wealth of microstructure information with detailed procedures for correct sample preparation and evaluation.

Metal failure occurs only in an extremely low percentage of the millions of tons of steel fabricated every year. Those that do occur fall into the following groups.

- Operational failures:
  - Overload
  - Wear
  - Corrosion and stress corrosion
  - Brittle failures
  - Metal fatigue
- Improper design:
  - Sharp corners in high-stress areas
  - Insufficient safety stress factor
  - Incorrect material selection
- Failures caused by thermal treatments:
  - Forging
  - Hardening and tempering
  - Welding
  - Crazing or surface cracks caused by the heat of grinding

OVERLOAD

Overload failures are usually attributed to faulty design, extra loads applied, or an unforeseen machine movement. Shock loads or loads applied above the design limit are quite often the cause of the breakdown of machinery.

Although mechanical engineers always plan for a high safety factor in designs (for instance, the 10 to 1 safety factor above the yield strength that is sometimes used in fasteners), the operators of machinery often tend to use machines above their design limit. Of course, this kind of overstress is due to operator error. Inadequate design can sometimes play a part in overload failures. Improper material selection in the design of the part or improper heat treatment can cause some failures when overload is a factor.

Often a machinist or welder will select a metal bar or piece for a job based on its ultimate tensile strength rather than its yield point. In effect, this is a design error and can ultimately result in breakdown (Figure 27.3). The strength of any material selected for a job should be based on its yield strength plus an adequate safety factor. When a part of a mechanism or structure is stressed beyond its yield point, it has been damaged or ruined.

WEAR

Excessive wear can be caused by continuous overload, but wear is ordinarily a slow process that is related to the friction between two surfaces (Figure 27.4). **Rapid wear** can often be attributed to lack of lubrication or the improper selection of material for the wear surface. Some wear is to be expected, however, and could be called **normal wear.**

Wear is one of the most frequent causes of failure. We find normal wear in machine tooling such as carbide

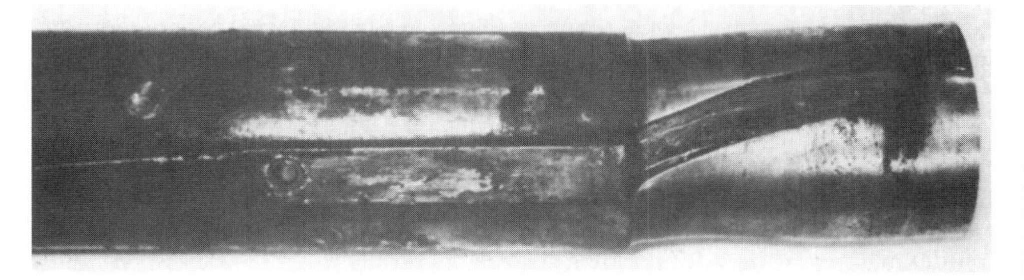

FIGURE 27.3. This CR shaft was stressed beyond its yield point by an overload. This could have been prevented by using a stronger alloy shafting.

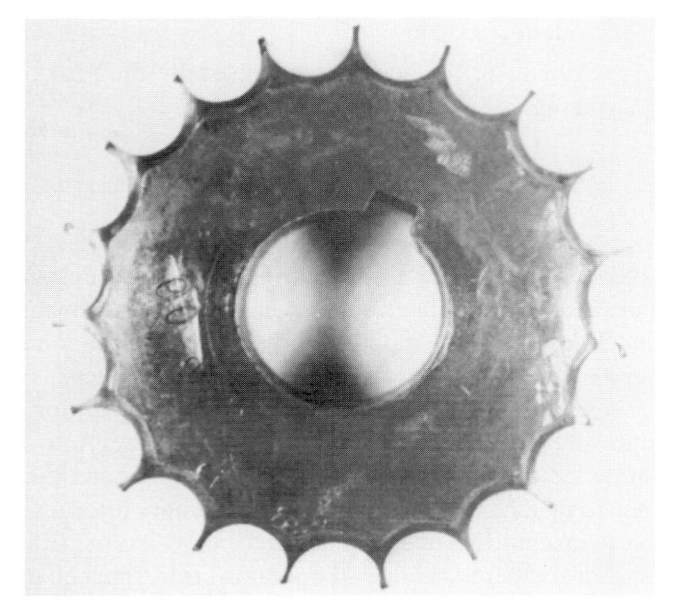

FIGURE 27.4. The teeth on this mild steel roller chain sprocket have worn almost to the point of complete failure. Wear life would be increased considerably if an alloy steel were used for the sprocket, or if the teeth were surface hardened.

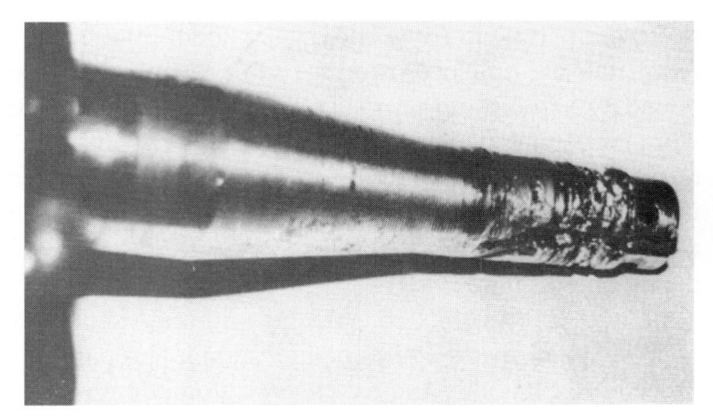

FIGURE 27.5. This type of failure can be hazardous. The spindle bearing seized up due to lack of lubrication and began to twist the spindle. Had the operator of the automobile not stopped very quickly, the wheel would have come off.

and high-speed tools that wear and have to be replaced or resharpened. Parts of automobiles wear until ultimately an overhaul is required (Figure 27.5). Machines are regularly inspected for worn parts, which when found are replaced; this is called preventive maintenance. Often normal wear cannot be prevented; it is simply accepted, but it can be kept to a minimum by the proper use of lubricants. Rapid wear can occur if the load distribution is concentrated in a small area because of the part design or shape. This can be altered by redesign to offer more wear surface. Speeds that are too high can increase friction considerably and cause rapid wear.

Metallic wear is a surface phenomenon, which is caused by the displacement and detachment of surface particles. All surfaces subjected to either rolling or sliding contact show some wear. In some severe cases, the wear surface can become cold welded to the other surface. In fact, some metals are pressure welded together in machines, taking advantage of their tendency to be cold welded. This happens when tiny projections of metal make a direct contact on the other surface and produce friction and heat, causing them to be welded to the opposite surface if the material is soft. Metal is torn off if the material is brittle. Insufficient lubrication and excessive heat are usually the causes of this failure scenario (Figure 27.6).

High-pressure lubricants are often used while pressing two parts together to prevent this sort of welding. Two steel parts such as a steel shaft and a steel bore in a gear or sprocket, if pressed together dry, will virtually always seize or weld and cause the two parts to be ruined for further use (Figure 27.7). In general, soft metals, when forced together, have a greater tendency to cold weld than harder metals. Two extremely hard metals even when dry will have very little tendency to weld together. For this reason, hardened steel bushings and hardened pins are often used in earthmoving machinery to avoid wear. Some soft metals when used together for bearing surfaces (for example, aluminum to aluminum) have a great tendency to weld or seize. Among these metals are aluminum, copper, and austenitic stainless steels.

Cast iron, when sliding on cast iron as is found in machine tools on the ways of lathes or milling machine tables, has less tendency than most metals to seize, because the metal contains graphite flakes that provide some lubrication, although additional lubrication is still necessary.

As a general rule, however, it is not good practice to use the same metal for two bearing surfaces that are in contact. However, if a soft steel pin is used in a soft steel link or arm, it should have a sufficiently loose fit to avoid seizing. In this application it is better practice to use a bronze bushing or other bearing material in the hole and a steel pin, because the steel pin is harder than the bronze and when a heavy load is applied, the small projections of bronze are flattened instead of torn out. Also, the bronze will wear more than the steel and usually only the bushing will need replacing when a repair is needed (Figure 27.8).

Some metals have a tendency to work harden, and although they would gall or seize in their soft condition as they begin sliding together, they start to harden on the surface and minimize the tendency to cold weld. An example of this is in the austenitic manganese steels used in rock-crushing machinery. When these have work hardened sufficiently, they do not tend to cold weld to their own surfaces.

FIGURE 27.6. The babbitted surface of this tractor engine bearing insert has partially melted and torn off. This failure was not due to normal wear but to lack of lubrication.

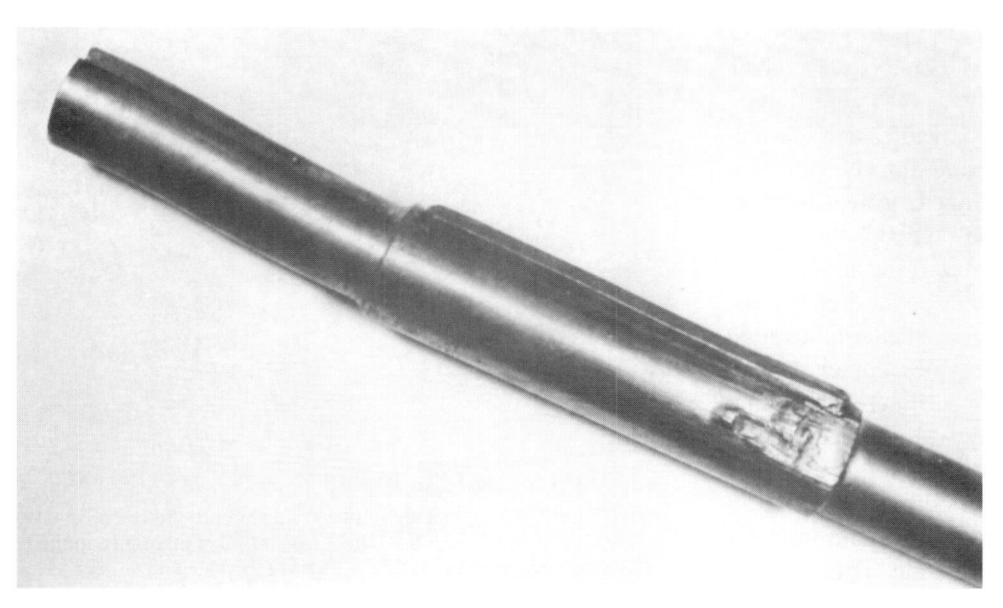

FIGURE 27.7. This shaft had just been made by a machinist and was forced into an interference fit bore for a press fit. No lubrication was used, and it immediately seized and welded to the bore, which was also ruined.

In **abrasive wear,** small particles are torn off the surface of the metal, creating friction. Friction involving abrasive wear is sometimes used or even required in a mechanism, such as on the brakes of an automobile. The materials are designed to minimize wear with the greatest amount of friction in this case. Where friction is not desired, a lubricant is normally used to provide a barrier between the two surfaces. This can be done by heavy lubricating films or lighter boundary lubrication in which there is a residual film.

Erosive wear is often found in areas that are subjected to a flow of particles or gases that impinge on the metal at high velocities. Sand blasting, which is sometimes used to clean parts, utilizes this principle.

Corrosive wear takes place when an acid, caustic, or other corrosive medium comes in contact with metal parts. When lubricants become contaminated with corrosive materials, pitting can occur in such areas as machine bearings.

Surface fatigue is often found on roll or ball bearing races or sleeve bearings where excessive side thrust has been applied to the bearing. It is seen as fine checks (cracks) or spalling (small pieces falling out on the surface).

FIGURE 27.8. The bronze bushing in this arm has seen severe use, is badly worn, and will be replaced, while the steel shaft that turns in the arm shows relatively little wear.

*FIGURE 27.9***a.** These aircraft engine cylinders must withstand high temperatures and wear. The inside of the cylinder wall is porous chromium plated.

PROTECTION AGAINST WEAR

Various methods are used to limit the amount of wear on the part. One common method is simply to harden the part. Also, the part can be surface hardened by **diffusion** of a material, such as carbon or chrome, into the surface of the part. Parts can also be **metallized, hard faced,** or **heat treated.** Other methods of limiting wear are electroplating (especially the use of hard industrial chromium) and anodizing of aluminum. Chromium plate can either be hard or porous. The porous type can hold oil to provide a better lubrication film. Some internal combustion engines are chromium plated in the cylinders and piston rings (Figures 27.9*a* and 27.9*b*). Some nickel plate is used, as well as rhodium, which is very hard and has high heat resistance.

The oxide coating that is formed by anodizing on certain metals such as magnesium, zinc, aluminum, and their alloys is very hard and wear resistant. These oxides are porous enough to form a base for paint or stain to give it further resistance to corrosion.

Some of the types of diffusion surfacing are **carburizing, carbo-nitriding, cyaniding, nitriding, chromizing,** and **siliconizing.** Chromizing consists of the introduction of chromium into the surface layers of the base metal. This is sometimes done by the use of chromium powder and lead baths in which the part is immersed at a relatively high temperature. This, of course, produces a stainless steel on the surface of low carbon steel or an iron-based metal, but it may also be applied to nonferrous materials such as tungsten, molybdenum, cobalt, or nickel to improve corrosion and wear resistance.

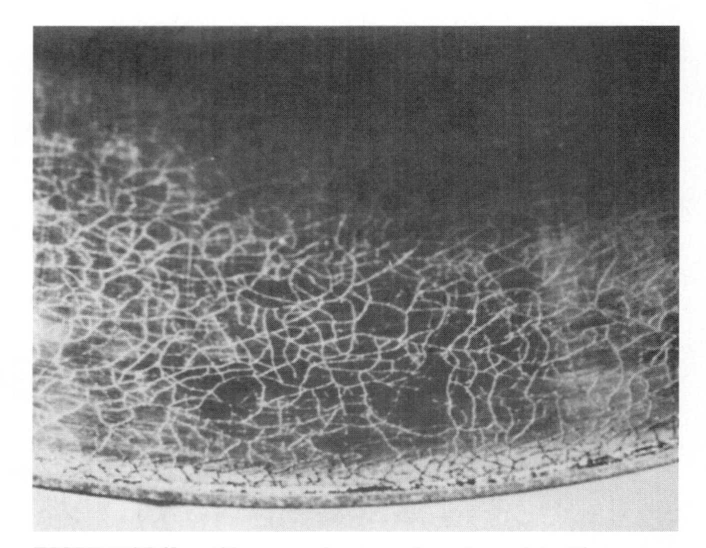

*FIGURE 27.9***b.** Close-up of porous chromium plate. The many grooves or channels will hold the lubricant. This plating is applied in small droplets which, when ground off, produce this effect.

The fusion of silicon, which is called **siliconizing,** consists of impregnating an iron-based material with silicon. This also greatly increases wear resistance. **Hard facing** is put on a metal by the use of several types of welding operations, and it is simply a hard type of metal alloy such as alloying cobalt and tungsten or tungsten carbide that produces an extremely hard surface that is very wear resistant. Metal spraying is used for the purpose of making hard, wear-resistant surfaces and for repairing worn surfaces (Figures 27.10*a* to 27.10*d*). Metallizing is usually done by feeding either a metal powder or a metal wire at a controlled rate through a tool that provides a heat source; the molten particles of metal are forced onto the surface of the base metal at a high velocity. In this process, which is not the same as welding, the liquid metal particles simply flatten out on the base metal and make a mechanical bond with the base metal and the previously deposited material instead of a metallurgical bond, since the cooling is rapid and an oxide film forms over the particle, preventing fusion of the metal particles. Thus, there is only a loose metallic or oxide bond between the particles. This determines to a great extent how strong or how porous the deposited material becomes.

However, some metal-spraying processes are claimed to have at least some metallurgical bonding (fusion). The micrograph in Figure 27.10*d* does seem to show such a bond at the interface of the spray metal and base metal. Welding is also used to build up surfaces for repair, but the stress concentration at the edge of the weld is often a cause of fatigue failure. Metallizing does not produce stress concentration, but undercutting is necessary, which reduces the effective stress area of the part and weakens it to some extent.

***FIGURE 27.10*a.** The undercut that is made ranges from 0.015 to 0.020 in. deep where the buildup is required. Grooves are often made to ensure bonding.

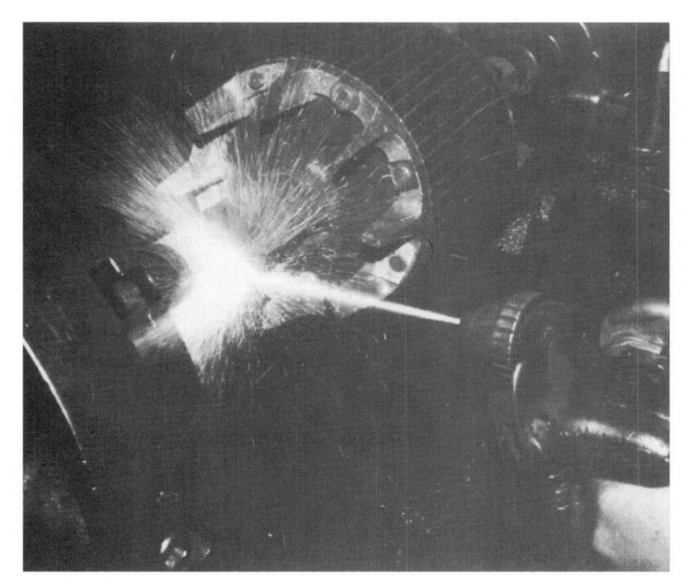

***FIGURE 27.10*b.** Metal being sprayed on prepared surface of shaft. Often a light undercoat of molybdenum is applied first to create a bond that has a physiochemical nature. Steel spray has only an adhesive bond, not a metallic bond as in welding.

CORROSION

When part failures are caused by corrosion alone, the corrosive medium often attacks the grain boundaries in what is known as intergranular corrosion. This weakening of the metal causes sudden failures or cracking along the grain boundaries (intergranular). However, if the part is subjected to a constant stress, a form of failure called stress corrosion often takes place. The rate of corrosion is greatly accelerated by stress reversals. Stress corrosion failure is by both intergranular and

***FIGURE 27.10*c.** When sufficient material has been applied, the surface may be machined to the correct diameter after it has cooled.

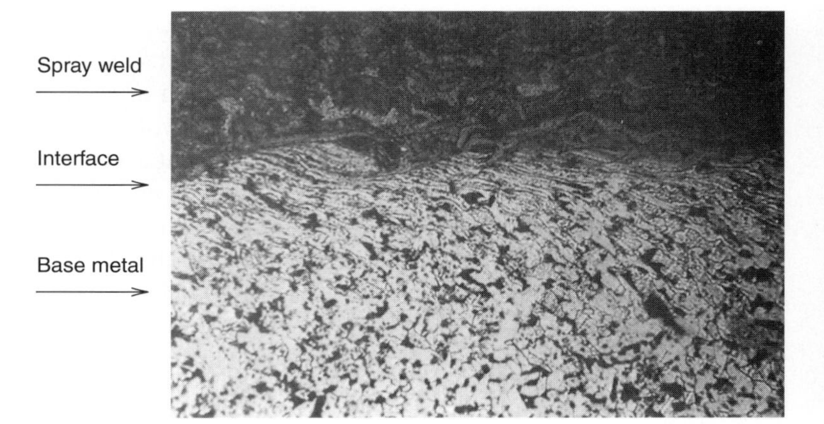

*FIGURE 27.10*d. Spray weld on mild steel surface. The cause of the distorted grains near the surface of the steel was machining (500×).

transgranular cracking, the latter being through the grains along slip planes. This subject was discussed in Chapter 19, "Corrosion in Metals."

BRITTLE FAILURES

A brittle fracture usually shows a crystalline surface on the broken end. When metals are ductile, they will usually deform when stressed beyond the elastic range, but when they reach the limit of plasticity and their ultimate strength, they will break in the remaining section as a brittle fracture. If such deformation is present, it can be assumed that the broken piece was a ductile metal. But if the entire cross section shows a brittle fracture, the metal may or may not be ductile, depending on the speed or suddenness of the shock load, temperature, grain size, and metallurgical structure of the metal. Under certain conditions normally ductile metals can behave as brittle metals. See the section on brittle failures at low temperatures that follows.

The pattern of a break can often reveal how the failure was precipitated. For example, if the break was caused by a sudden shock load such as an explosion, there may be chevron-shaped formations present that point to the origin of fracture. When a stress concentration is present, such as a weld on a structure that is subject to a sudden overload, the fracture is often brittle across the entire break, showing grains, striations, and wave fronts (Figure 27.11*a*). Brittle fractures are often intergranular (along the grain boundaries); this gives the fracture surface a rock candy appearance at high magnification. See Figure 27.11*b*. When grain boundaries are weakened by corrosion, hydrogen, heat damage, or

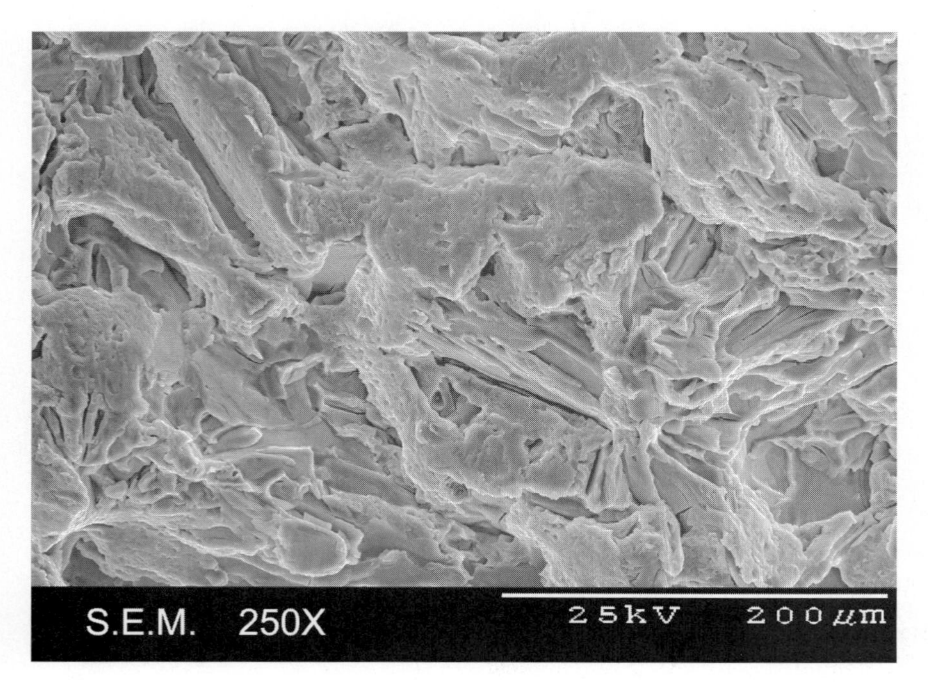

*FIGURE 27.11*a. Brittle fracture in a precipitation hardening stainless steel.

*FIGURE 27.11***b.** Brittle fracture in alloy steel caused by hydrogen embrittlement.

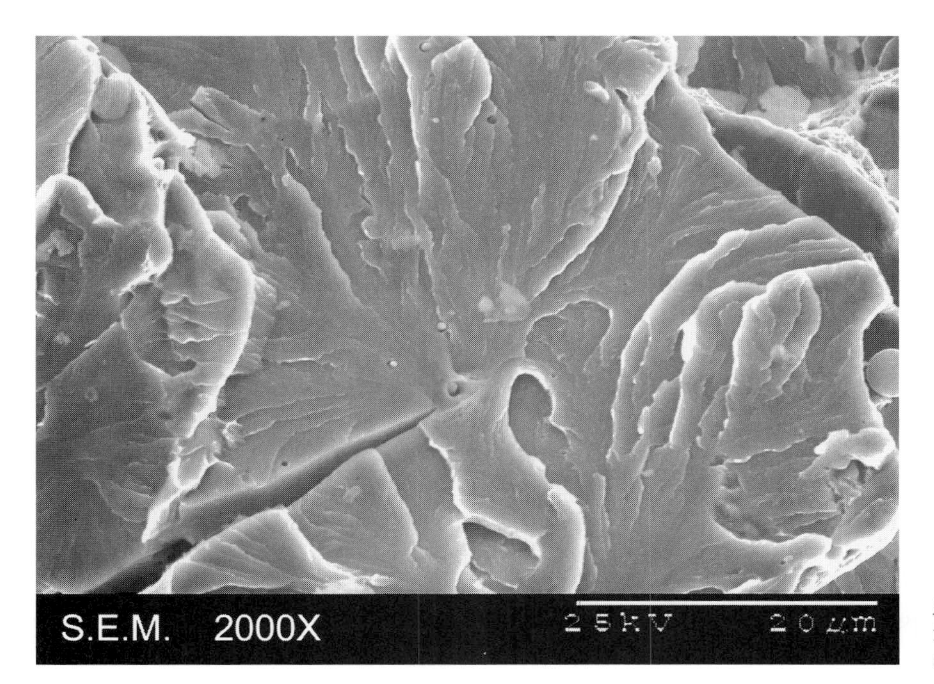

*FIGURE 27.11***c.** Quasi-cleavage brittle fracture in precipitation hardening stainless steel.

impurities, the brittle fracture may be intergranular. Brittle failures can also be transgranular (through the grains). This is called cleavage.

Cleavage fracture is confined to certain crystallographic planes that are found in body-centered cubic or hexagonal close-packed crystal structures. For the most part, metals having other crystalline unit structures do not fail by cleavage unless it is by stress corrosion cracking or by corrosion fatigue. Cleavage should normally leave a flat, smooth surface; however, because metals are polycrystalline with the fracture path randomly oriented through the grains, and because of certain imperfections, certain patterns are formed on the surface. The lines, called river patterns, point to the origin of fracture, and a twist or change of direction of cleavage is marked by a boundary, shown in Figure 27.11*c* between dark and light areas.

Small quantities of hydrogen have a great effect on the ductility of some metals. Hydrogen can get into steels by heating them in an atmosphere or a material

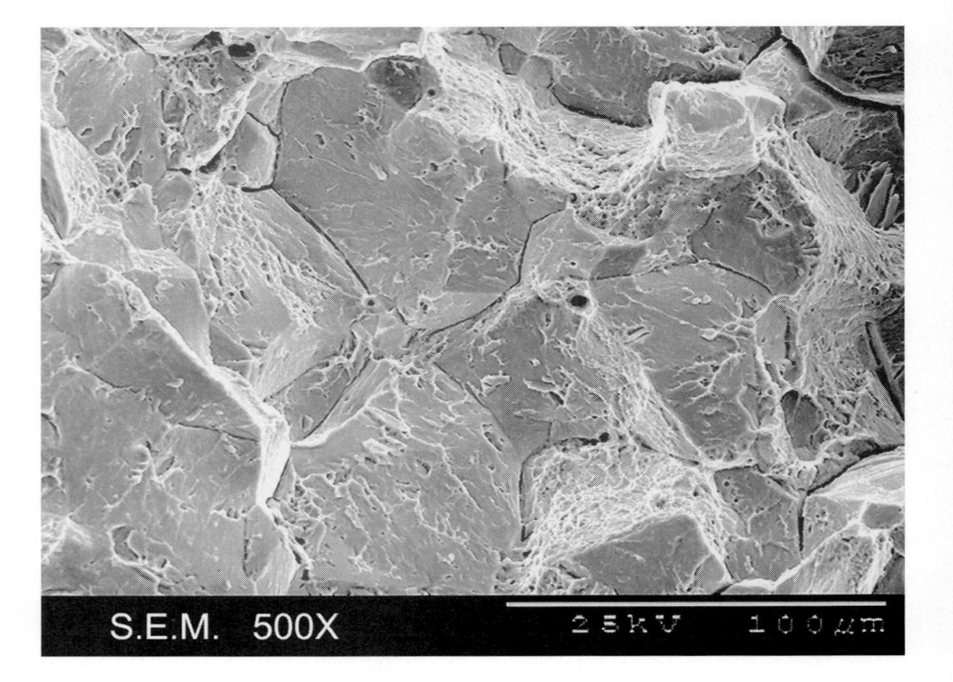

***FIGURE 27.11*d.** Intergranular fracture in normally ductile precipitation hardening stainless steel. This brittle fracture was due to hydrogen embrittlement.

containing hydrogen, such as pickling or cleaning operations, electroplating, cold working, welding in the presence of hydrogen-bearing compounds, or the steel-making process itself. There is a noticeable embrittling effect in steels containing hydrogen, which can be detected in tensile tests and seen in the plastic region of the stress-strain diagram showing a loss in ductility. Typical ductile metals often exhibit severe intergranular cracking as shown in Figure 27.11*d*.

The electroplating of many parts is required to prevent corrosion failure because of their service environment. Steel may be contaminated by electroplating materials that are commonly used for cleaning or pickling operations. These materials cause hydrogen embrittlement by charging the material with hydrogen. Monatomic hydrogen is produced by most pickling or plating operations at the metal-liquid interface, and it seems that single hydrogen atoms can readily diffuse into the metal. Preventive measures can be taken to reduce this accumulation of hydrogen gas on the surface of the metal.

A frequent source of hydrogen embrittlement is found in the welding process. Welding operations, in which hydrogen-bearing compounds such as oil, grease, paint, or water are present, are capable of infusing hydrogen into the molten metal, thus embrittling the weld zone. Special shielding methods are often used to help reduce the amount of hydrogen absorption.

One effective method of removing hydrogen is a "baking" treatment in which the part, or in some cases the welding rod, is heated for long periods of time at temperatures of 250° to 400°F. This treatment promotes the escape of hydrogen from the metal and restores the ductility.

BRITTLE FAILURES AT LOW TEMPERATURES

When body-centered cubic metals are subjected to dynamically applied loads at low temperatures, such as those involved in the impact test, the result is a lowering of ductility with a sharp increase of brittleness. This ductile-to-brittle transition phenomenon is commonly referred to as the transition temperature, transition zone, or brittle range (Figure 27.12). This phenomenon is often exhibited in instantaneous failures of pipelines, storage tanks, ships breaking in half at sea, bridges, and other metal structures that are subjected to low temperatures.

In Chapter 3, "The Physical and Mechanical Properties of Metals," methods of prevention of brittle transition were listed showing that certain alloying elements in steel, such as nickel, tended to lower the transition temperatures of steels. Some steels containing these elements will not fail by embrittlement even at temperatures below −150°F.

Apparently, plain carbon steels have higher transition temperatures than similar alloy steels. Also, coarser grained steels go into the brittle range at higher temperatures. Quenched and tempered fine-grained steels can have even higher ductile-to-brittle transition, sometimes at room temperature. When carbides or ferrite are

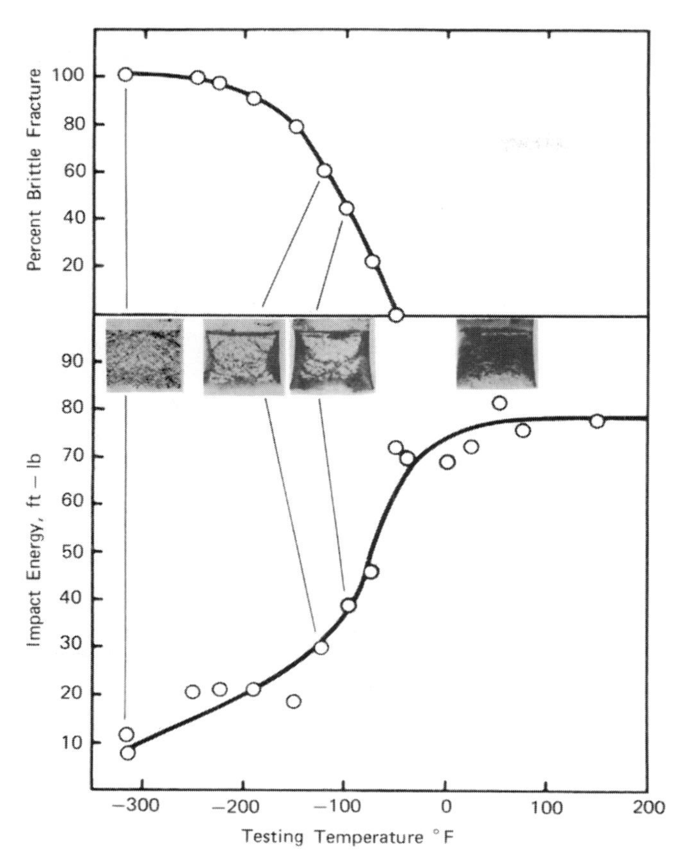

FIGURE 27.12. Appearance of Charpy V-notch fractures that were obtained at a series of tempered martensite of hardness around 30 RC. *(Courtesy Republic Steel Corporation)*

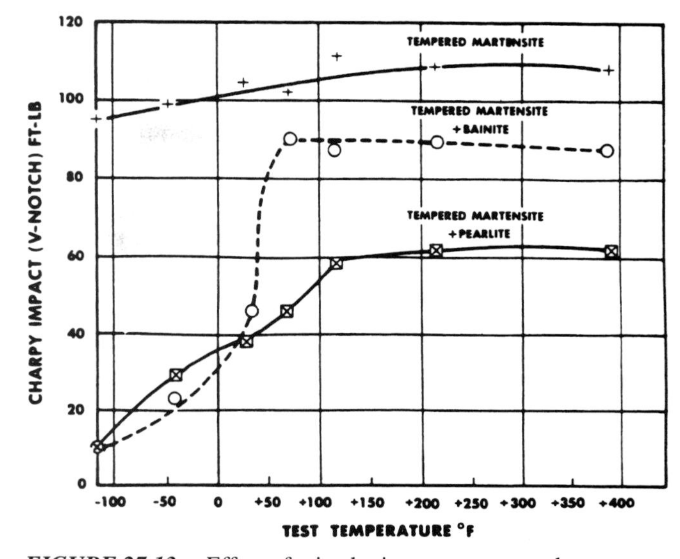

FIGURE 27.13. Effect of mixed microstructures, such as may occur in slack quenching, is to reduce impact resistance and to have a pronounced effect on the ductile to brittle transition. *(Courtesy Republic Steel Corporation)*

outlining grain boundaries, the transition temperature is raised, which greatly reduces the energy required to produce failure. A mixed microstructure caused by improper quenching (slack quenching, between the A_3 and A_1 temperatures) also produces brittleness in steels. Of all the microstructures, ferrite produces the highest transition temperature (Figure 27.13), and pearlite, bainite, and tempered martensite have a slightly lower transition temperature.

SHEAR FAILURES

Shear failures are often found in load-carrying members. In shafts subject to rotational stress this is called torsional shear. Shear failures are not the same as brittle failures, as can be seen in the S.E.M. photomicrograph (Figure 27.14). This fractograph reveals the characteristic elongated oval dimples of shear in a ductile material. Shear is a transgranular fracture (across the grain). When metals are pulled apart in tension, a similar dimple is formed, but sufficiently different so that metallurgists are able to interpret these micrographs to determine the method and cause of failure.

FAILURES CAUSED BY THERMAL TREATMENTS

Quench cracking is probably the greatest single problem related to hardening and tempering. Some of the most common causes of quench cracks are as follows:

1. Overheating during the austenitizing cycle so that normally fine grained steels tend to coarsen. Coarse grained steels increase the depth of hardening and are inherently more prone to quench cracking than fine grained steels.
2. Improper selection of quenchant, that is, quenching too rapidly as with the use of water, brine, or caustic when oil is the proper quenchant for the specific part and type of steel.
3. Improper selection of steel for the design of the part.
4. Time delays between quenching and tempering.
5. Improper design of keyways, holes, sharp changes in section, mass distribution, and other stress raisers.
6. Improper entry of the work into the quenchant with respect to the shape of the part, which results in nonuniform or eccentric cooling.
7. Quench cracks are often filled with oxidation as shown in Figure 27.15*a* or can be caused by inclusions in the metal as shown in Figure 27.15*b*.

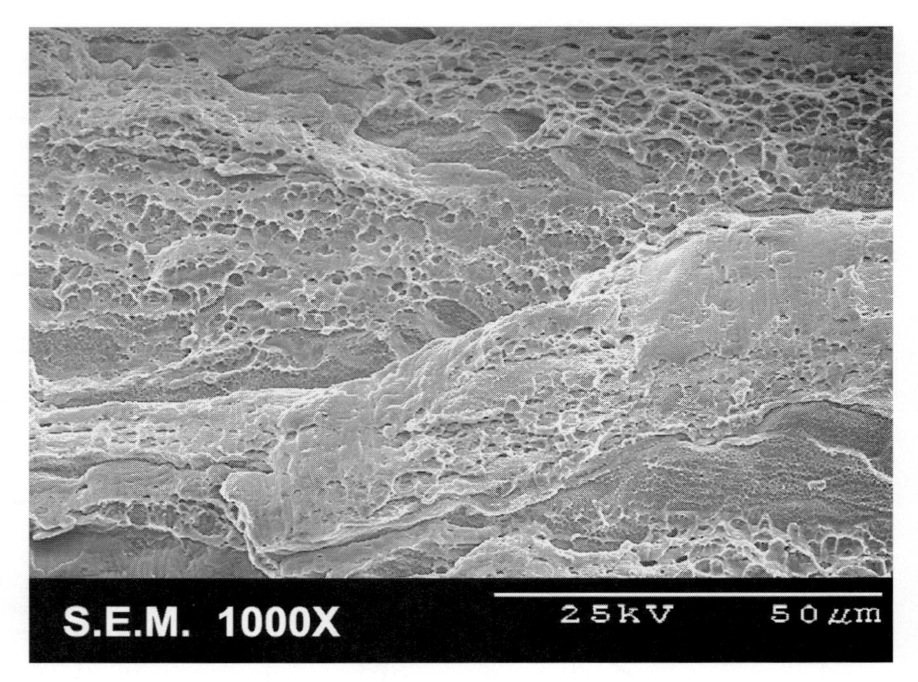

FIGURE 27.14. The dimples on this fracture surface are smeared and elongated which is a typical fracture pattern produced by a shear failure mode.

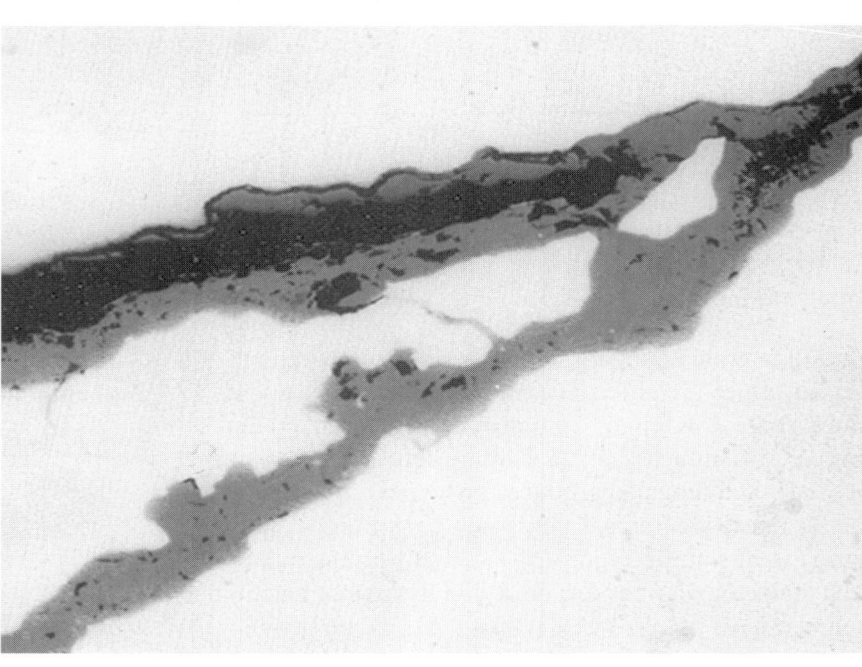

FIGURE 27.15a. The oxide coated crack occurred when this 4140 alloy steel was heated too rapidly to the austenitic temperature.

500× Unetched

FATIGUE

All metals and nonmetals when subjected to repeated thermal and/or mechanical stress can fail by fatigue. Fatigue failure occurs when the endurance limit of a material is exceeded. When a stress or combination of stresses exceeds more than 50 percent of the material's measured elastic limit, the useful stress cyclic life decreases exponentially. Designers often use yield point of a material as the safety limit for fatigue loading, however, materials seldom exceed more than one cycle of stress reversal when loaded above the elastic limit. Yielding means the material has given up and is actively deforming without increase in load. Using the yield point as a stress limit for any mechanical design invites possible catastrophic material failure without forewarning. Actual physical properties of the item being designed, primarily the measured elastic limit as

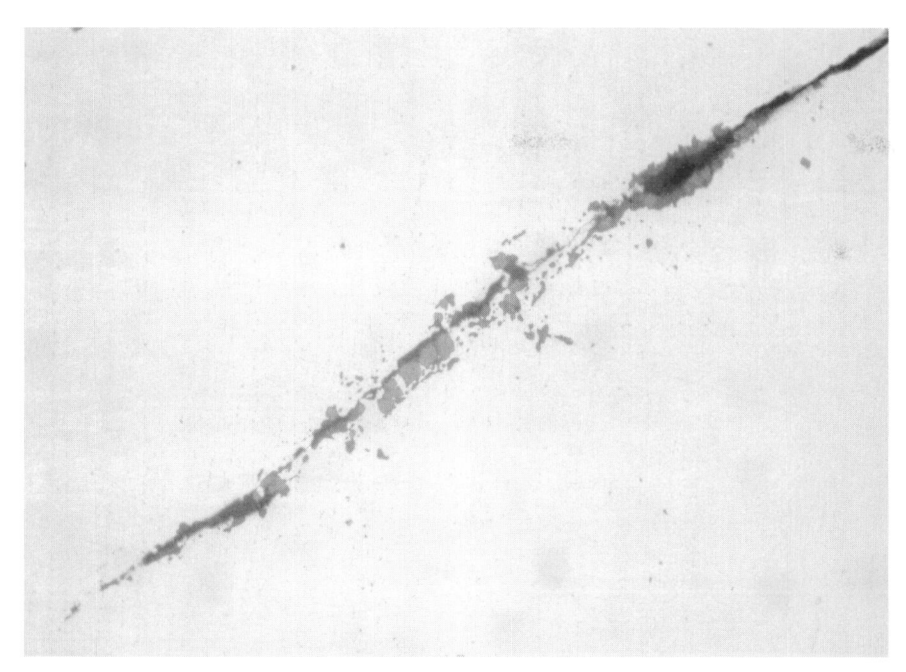

*FIGURE 27.15***b.** This crack occurred due to large oxide inclusions. Note the fragmented condition of the inclusion which occurred during mill hot rolling.

discussed in Chapter 3, must be determined to establish the safe stress level and reduce the risk of fatigue failure.

Fatigue failures can initiate as the result of design deficiencies such as sharp radii, a notch or a crack. Cracks must be a specific minimum size for applied stresses to cause crack propagation. When the windshield of a car is struck by a rock, it may chip and microcrack rather than develop large cracks. Cracks may never develop from the damage unless external pressure is locally applied; then watch the crack grow. By applying pressure in different locations, the crack can be coaxed into changing directions. Likewise, when water enters the crack and freezes during the night, the crack will most always grow. If a crack propagates in a material and the crack is caused by more than one cycle of stress, fatigue failure becomes the primary failure mode rather than failure by sudden overload.

Examination of fracture surfaces directly with the scanning electron microscope has made failure mode recognition much easier. Fatigue failures can develop from location of high stress, but most often fatigue failures develop multiple fracture origins, Figure 27.15, which are similar in appearance (Figure 27.16*b*). Failure occurs when the crack growth reduces the cross-sectional area of the item and overload failure occurs. Although there are many similarities in fatigue failures, comparison of exemplar fracture surfaces is exceedingly helpful when performing a failure analysis. Fatigue fracture surface features can change radically with different metals, Figure 27.16*a*-16*c*.

FRACTURE MODE IDENTIFICATION

The fracture surface of most materials, especially metals, will exhibit the physical signature of the method which caused the fracture. Perhaps you have wondered why accident investigators spend so much time at the crash site of an airplane, recovering all of the broken pieces? The purpose of recovery is to locate the original placement of all items and to thoroughly examine the fracture surfaces for fracture mode identification. "Fracture mode" is the term used to describe the physical features of a fracture created by a specific mechanism such as fatigue or tensile overload. Usually, if all of the items are recovered from the crash site, the metallurgist can characterize each fracture and determine the cause of the accident.

The following series of scanning electron microscope photomicrographs will give you an idea of the characteristic differences associated with different fracture modes. In many cases, a failed material may contain more than one fracture mode, making the metallurgical detective work even more challenging.

The fracture surface of the alloy steel bolt (Figures 27.18*a* through Figure 27.18*d*) is an example of an actual failure analysis using the fractographic information provided in Table 27.1.

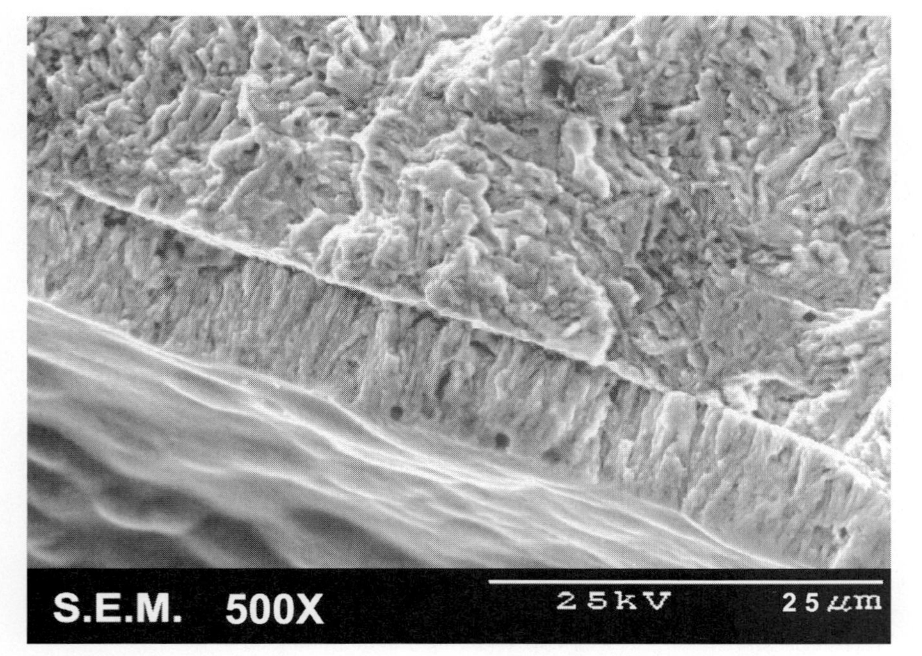

FIGURE 27.16a. The columnar band of grain growth, diagonal in this SEM photomicrograph is fractured cadmium plating. Multiple fatigue fracture origins have propagated at the cadmium plating/base metal interface.

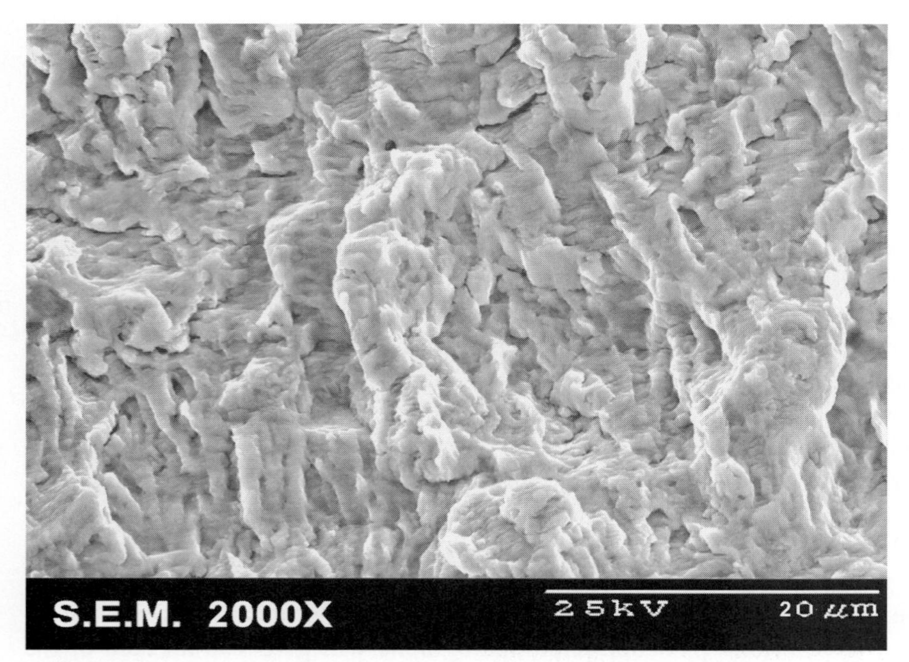

FIGURE 27.16b. Lateral separations in the grain structure are caused by the propagation of a high cycle fatigue crack in an alloy steel (high magnification of a fracture area in Figure 27.16*a*.).

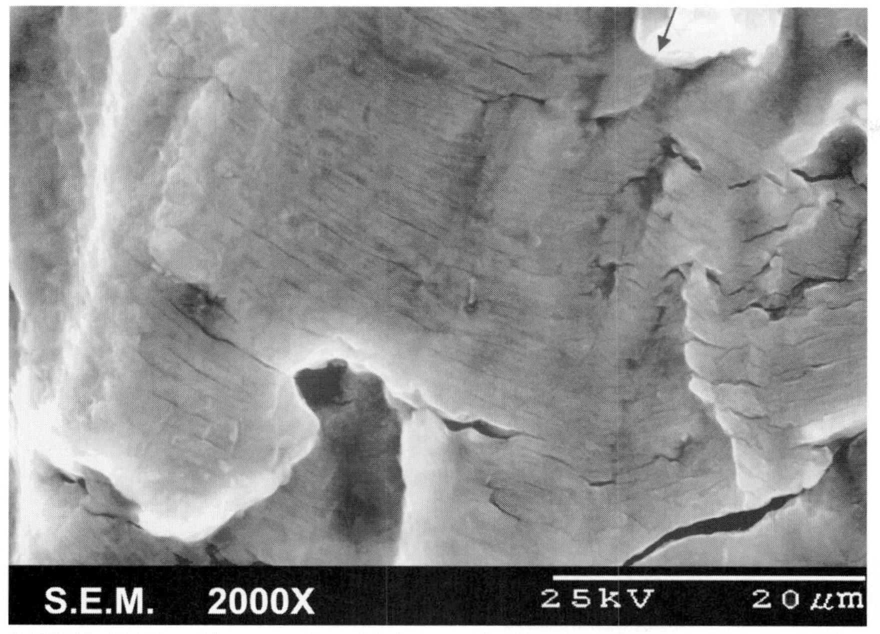

FIGURE 27.16c. The arrow (top right) shows the direction of fatigue crack growth in this aluminum alloy hose collar.

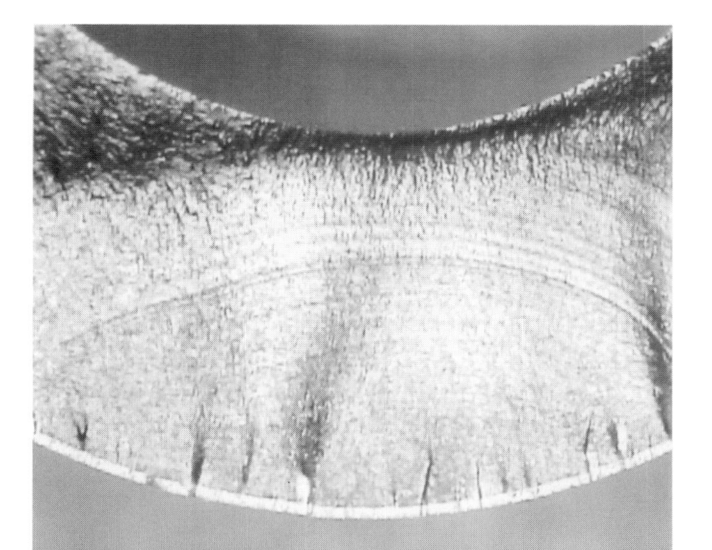

FIGURE 27.16d. This classic "thumb nail/beach mark" fatigue failure occurred in an engine wristpin. The O.D. surface contains vertical ratchet marks at multiple fracture origins due to inherent stress induced microcracks in the hard chrome plating. *(Photo Courtesy of Stork-MMA Labs, Huntington, Beach, CA)*

TABLE 27.1. TYPICAL FRACTURE MODES

FIGURE 27.17a.	TENSILE OVERLOAD - FINE GRAIN SIZE
FIGURE 27.17b.	TENSILE OVERLOAD - COARSE GRAIN SIZE
FIGURE 27.17c.	QUASI-CLEAVAGE - FINE GRAIN SIZE
FIGURE 27.17d.	QUASI-CLEAVAGE - COARSE GRAIN SIZE
FIGURE 27.17e.	SHEAR AND TORSION
FIGURE 27.17f.	STRESS CORROSION CRACKING IN ALUMINUM
FIGURE 27.17g.	SHEAR AND STRESS CORROSION CRACKING
FIGURE 27.17h.	FATIGUE - ALUMINUM
FIGURE 27.17i.	FATIGUE - 15-5PH STEEL
FIGURE 27.17j.	HYDROGEN EMBRITTLEMENT - STEEL
FIGURE 27.17k.	HYDROGEN EMBRITTLEMENT - 17-4PH STEEL
FIGURE 27.17l.	SMASHED METAL

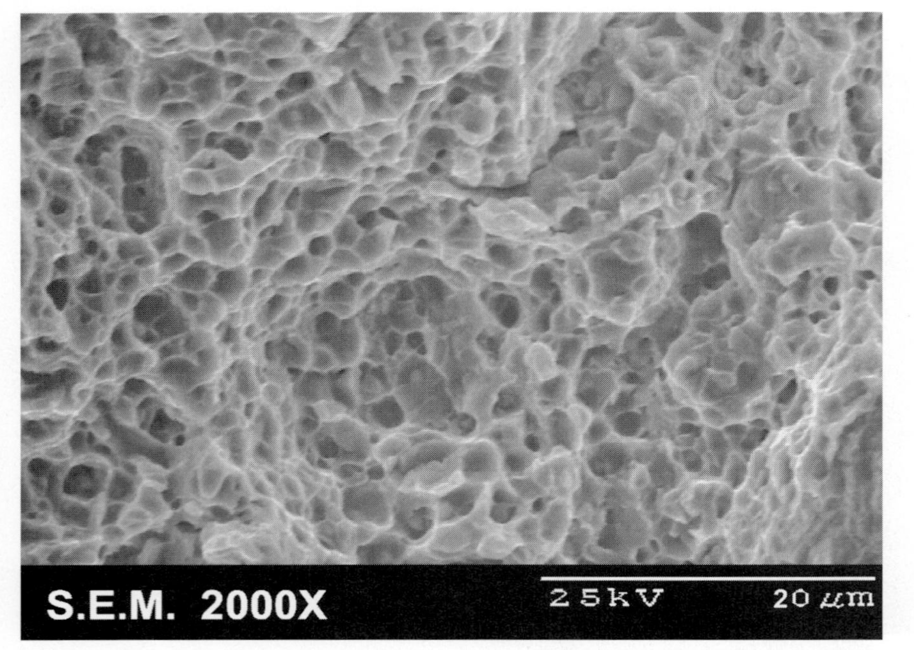

*FIGURE 27.17*a. The photomicrograph shows a dimple fracture surface characteristic of a fine grain size metal.

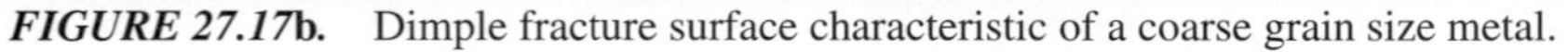

*FIGURE 27.17*b. Dimple fracture surface characteristic of a coarse grain size metal.

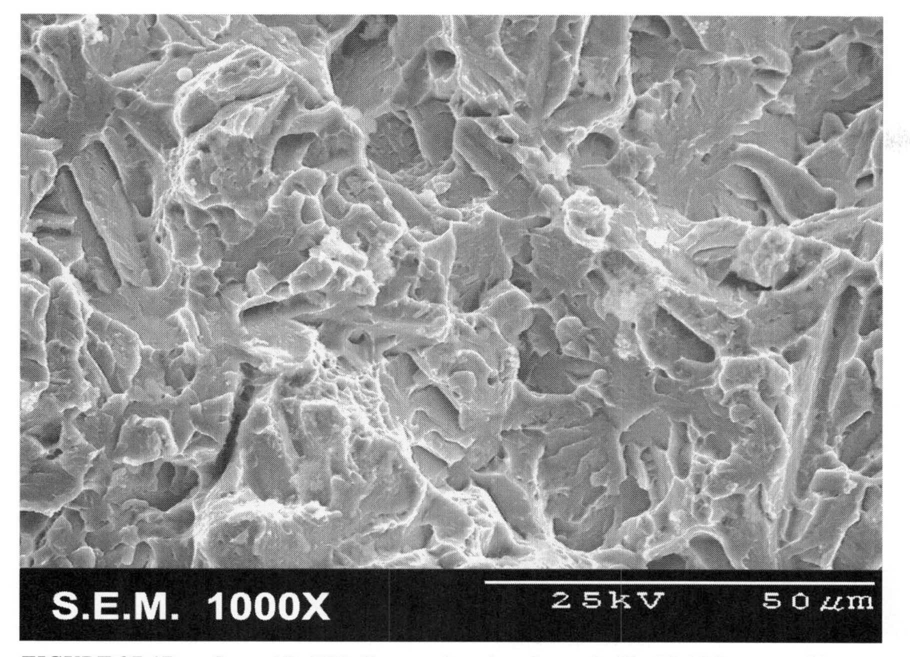

***FIGURE 27.17*c.** Same 17-4PH alloy steel as that shown in Fig 27.17d, except this metal has a very fine grain size.

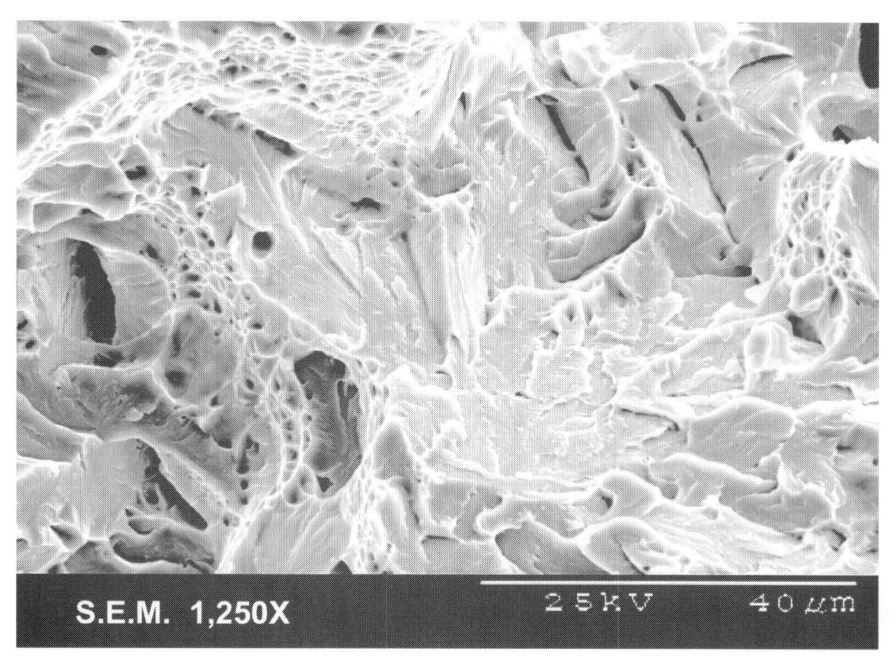

***FIGURE 27.17*d.** A typical coarse grain quasi-cleavage fracture surface in 17-4PH alloy steel as seen with the scanning electron microscope.

FIGURE 27.17e. A shear/torsion mode fracture usually exhibits smeared or flattened surface features. The normally ductile dimples are elongated; compare with Fig 27.17a.

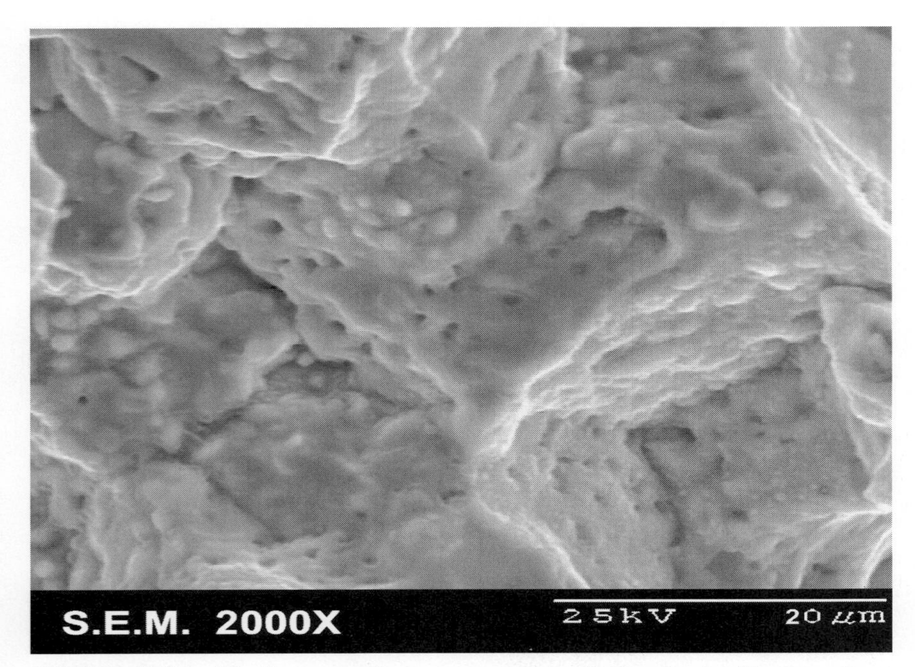

FIGURE 27.17f. The fracture surface appears to be washed or dissolved by the action of corrosion. This aluminum failed by stress corrosion cracking and exhibited a visually brittle appearing fracture surface until examined using the S.E.M.

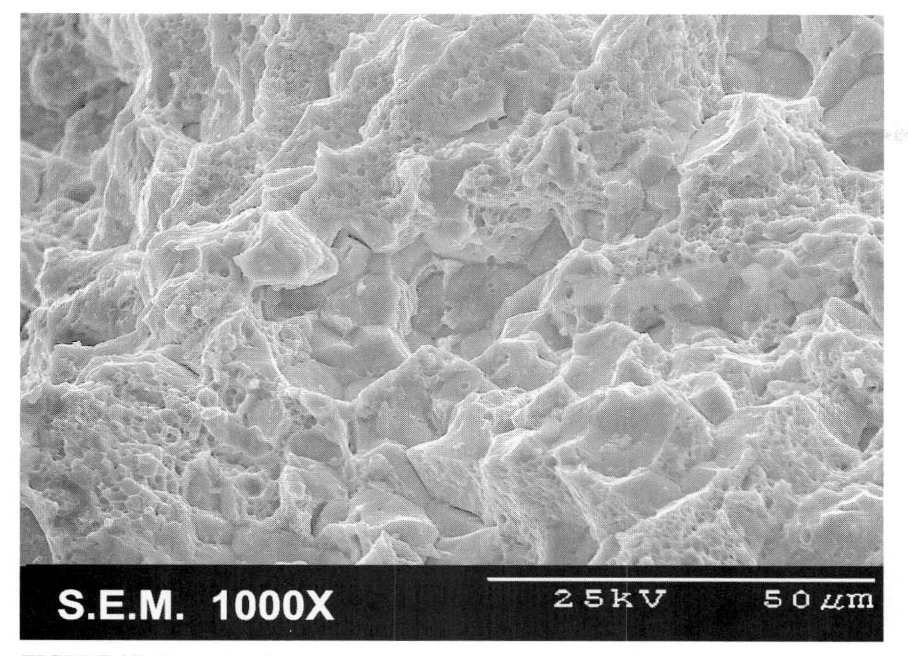

***FIGURE 27.17*g.** This fracture surface contains a mixed failure mode of shear and stress corrosion cracking.

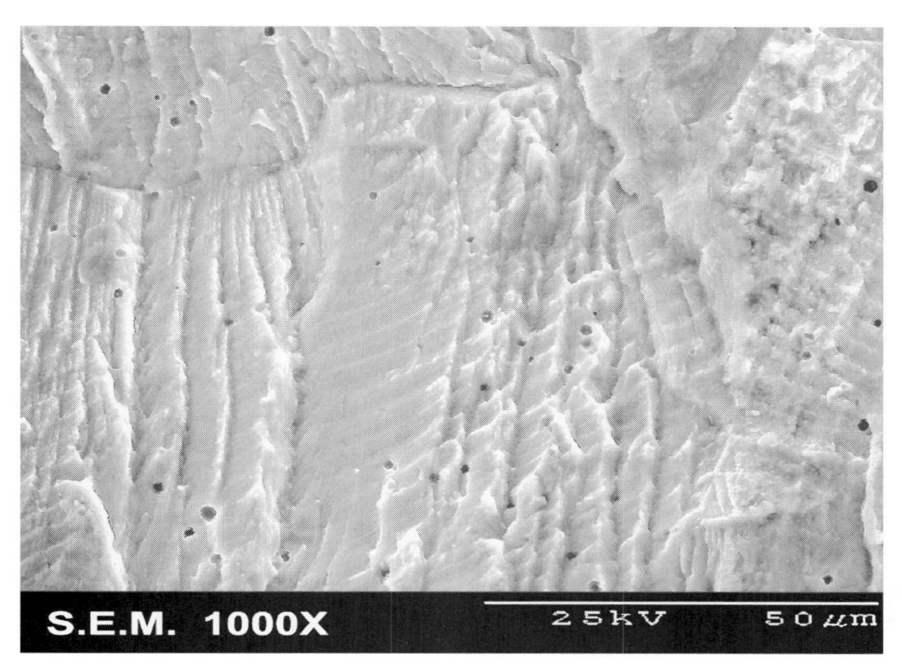

***FIGURE 27.17*h.** Typical high cycle fatigue striations in an aluminum alloy.

FIGURE 27.17i. Classic high cycle fatigue failure in 15-5PH steel.

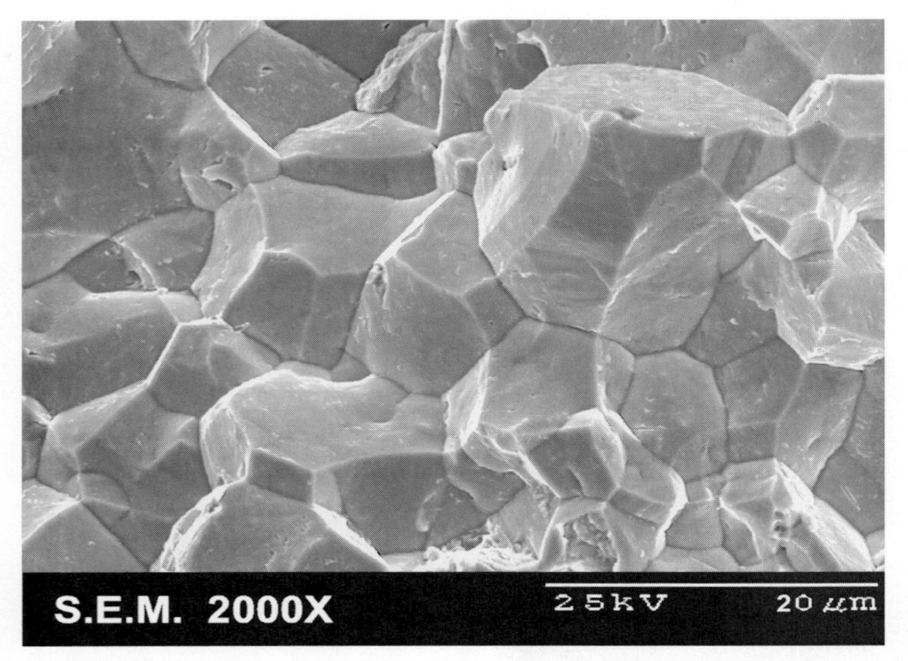

FIGURE 27.17j. Hydrogen embrittlement fracture surface in hardened alloy steel caused by cadmium plating without a bake treatment.

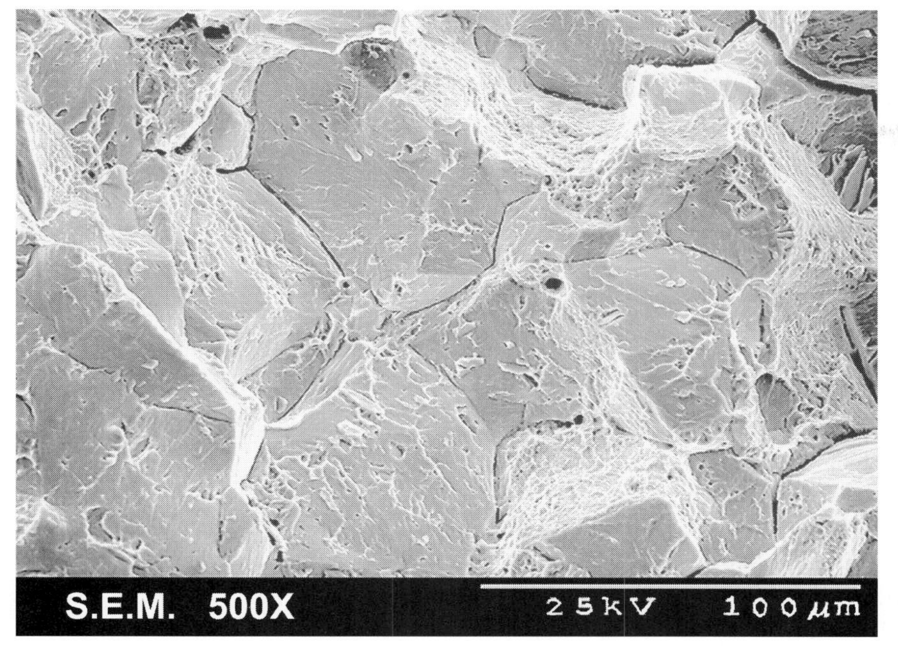

***FIGURE 27.17*k.** Hydrogen embrittlement in 17-4PH steel. Note the intergranular brittle cracking at the grain boundaries.

***FIGURE 27.17*l.** The surface fracture mode features on this surface have been destroyed by plastic deformation.

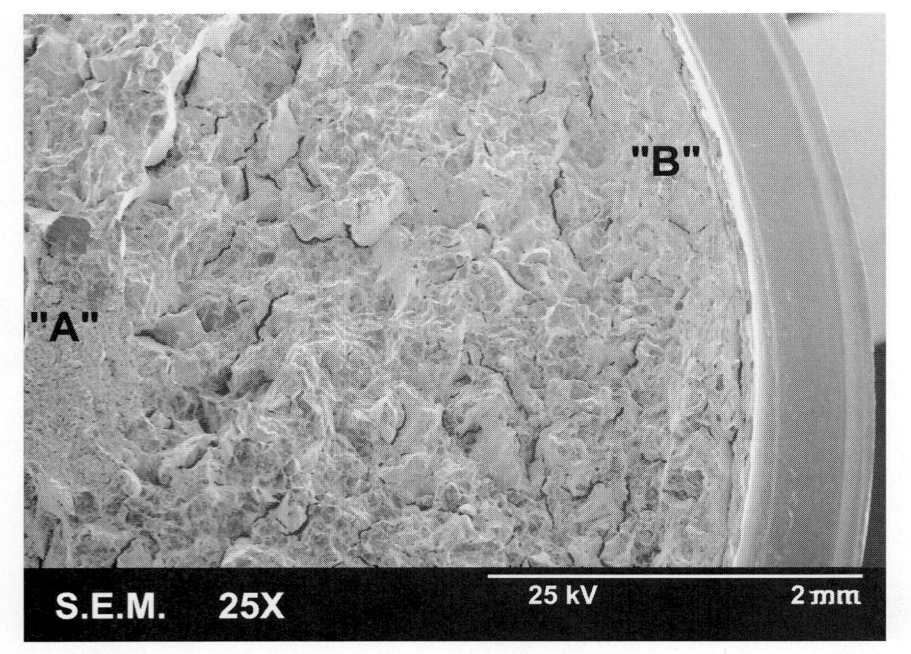

*FIGURE 27.18***a.** This S.E.M. photomicrograph shows the transverse fracture surface of an alloy steel bolt as a result of stress corrosion cracking. The core area "A" exhibits brittle and ductile fracture modes while the fracture surface at area "B" near the thread root, appears dissolved or washed away by the action of corrosion. Note the multitude of secondary microcracks between the center and root areas.

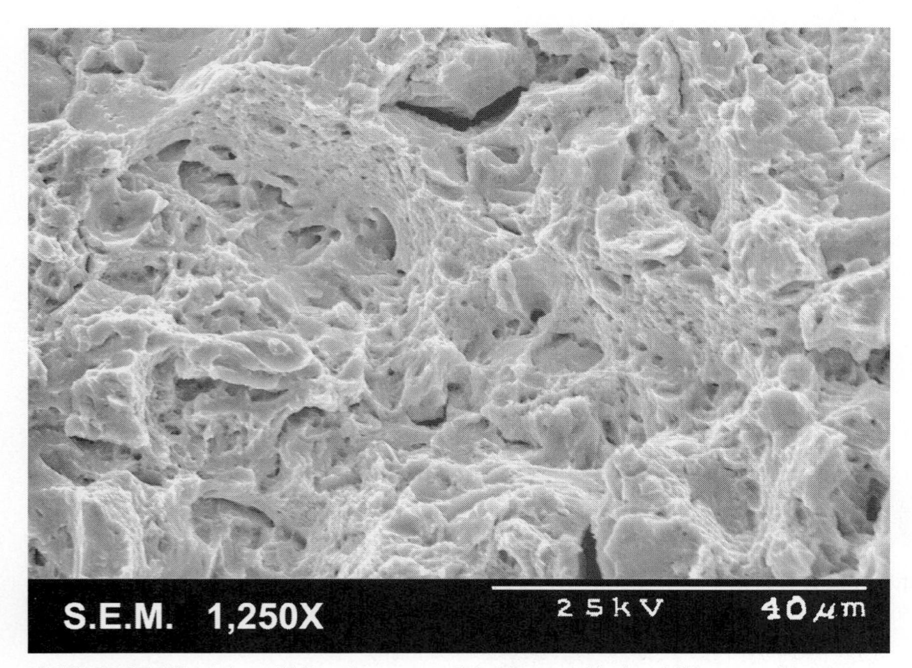

*FIGURE 27.18***b.** Area "A" at the center of the fractured bolt exhibits a combination fracture mode of ductile overload and brittle fracture.

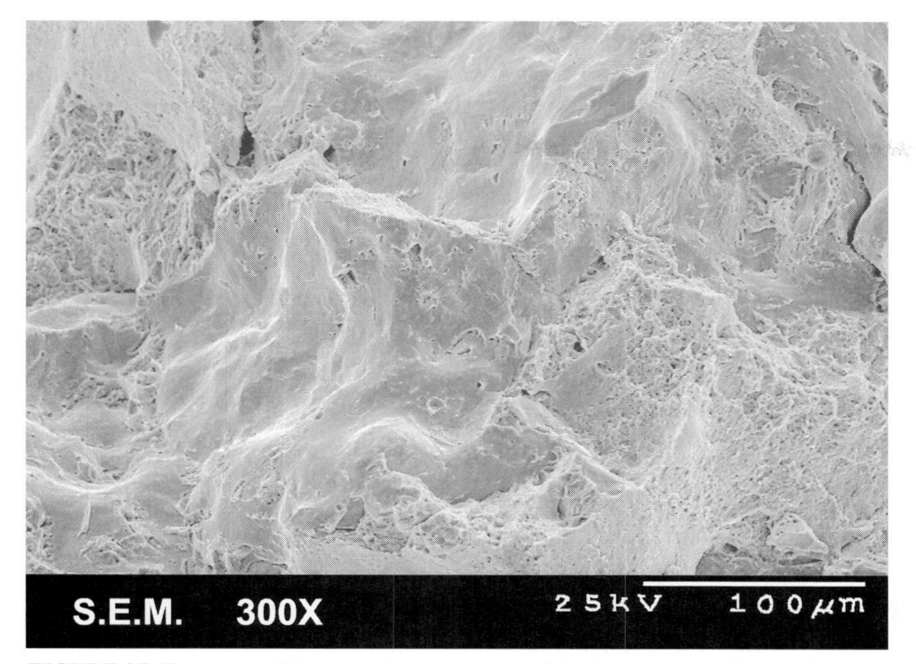

***FIGURE 27.18*c.** Area "B" is near the thread root of the fractured bolt. The washed away appearance of the fracture surface is typically characteristic of a stress corrosion cracking mode.

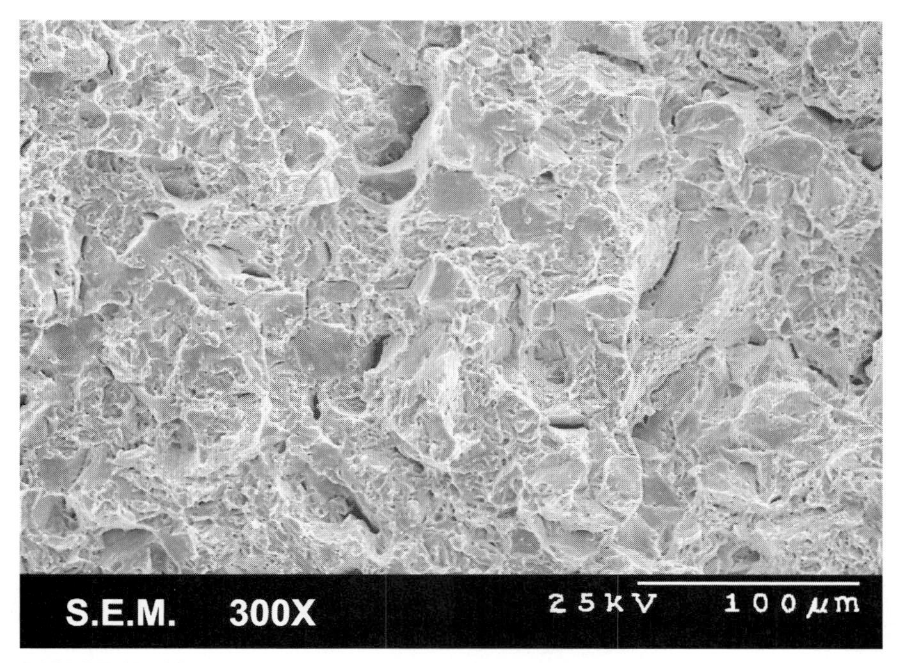

***FIGURE 27.18*d.** This exemplar fracture surface was created by sudden impact overload of a portion of the failed bolt (Fig. 27.18a). The majority of the exemplar fracture surface is a ductile tensile overload with cleaved grains very similar to the center area fracture surface of the bolt area "A" (Figure. 27.18a).

DESIGN CONSIDERATIONS

Parts should be designed and fabricated to utilize directionality of grain flow. Figure 27.19 is an example of a forged shaft that shows slag inclusions running the length of the material. Wrought iron and resulfurized drawn shafting contain inclusions that give the material high-transverse (crosswise) fatigue strength. These metals have a lowered resistance to high torque values, however, tending to split along the stringers of manganese sulfide (Figure 27.20).

Fatigue is caused by a concentration of tensile stress that can often be corrected by a change in design (Figures 27.21*a* and 27.21*b*). Parts that are subject to stress reversals (cyclic stresses) can have their fatigue life extended considerably by raising compressive stresses on the surface (Figure 27.22). This can be done in several ways: by carburizing, nitriding, or surface hardening by induction or flame. To increase fatigue life or the endurance limit, parts are sometimes subjected to shot peening to produce residual compressive stresses on the surface.

A study of individual fatigue problems based on the service conditions and by direct observation of the failure can often lead to a conclusion that explains the cause or causes and suggests some corrective measures. The loading of the part can be high or low for its size; it can have high or low speed or stress reversals caused by misalignment. Vibration is often a cause of fatigue, the frequency and intensity being factors. Occasional overloads are also instrumental in initiating failure.

Fatigue failures are usually characterized by three distinct surfaces or stages:

Stage 1. A smooth surface with wave marks such as seen on clamshells. This area represents a slow progression of the initial crack.

Stage 2. A similar but rougher surface showing coarse wave marks progressing toward the center.

Stage 3. A crystalline area showing the final sudden failure of the part. It is this last portion that prompts the erroneous conjecture that the part has "crystallized."

The symbols and nomenclature related to fatigue failures are illustrated in Figure 27.23.

TYPES OF FATIGUE FAILURES

The symbols and nomeclature used to explain the type of fatigue failure with a specific designation are shown in Figure 27.24. These designations provide a simple explanation for the overall appearance of a fatigue fracture surface. Some of the more common fatigue fracture appearances associated with various bending conditions are illustrated in Figure 27.25. The following cases are examples.

Case 1 One-Way Bending Load

No-Stress Concentration. A small elliptically shaped fatigue crack usually starts at a surface flaw such as a scratch or tool mark. The crack tends to flatten out as it grows. It is caused by the stress at the base of the crack being lower because of the decrease in distance from the edge of the crack to the neutral axis. The degree of overstress in the part is indicated by the amounts of smooth-textured area as compared with the crystalline-textured area of the fracture. A large crystalline area indicates high overstress (Figure 27.25, 1-*b*). A smaller crystalline area indicates a lower overstress, which would require the greater number of cycles necessary to produce failure (Figure 27.25, 1-*a*).

Mild-Stress Concentration. If a distinct stress raiser such as a notch is present, the stress at the base of the

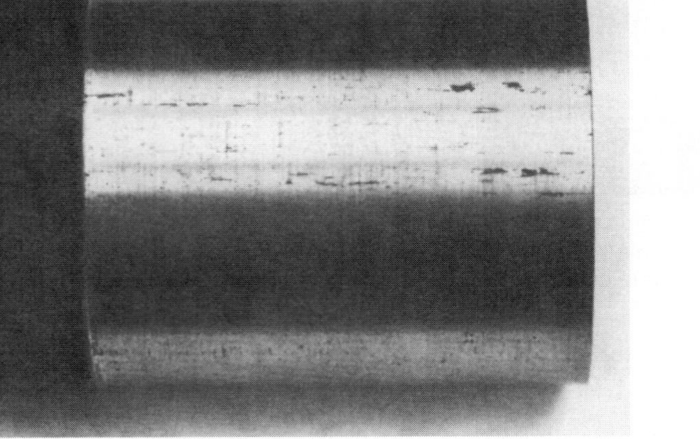

FIGURE 27.19. Wrought iron forging, showing inclusions that run the length of the material.

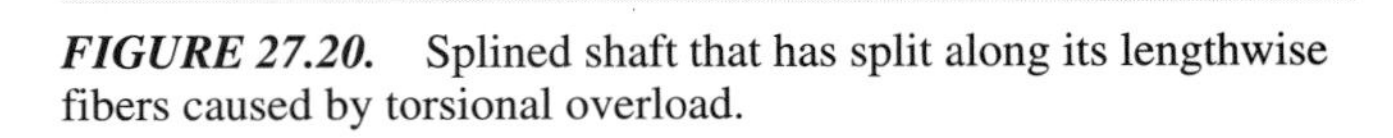

FIGURE 27.20. Splined shaft that has split along its lengthwise fibers caused by torsional overload.

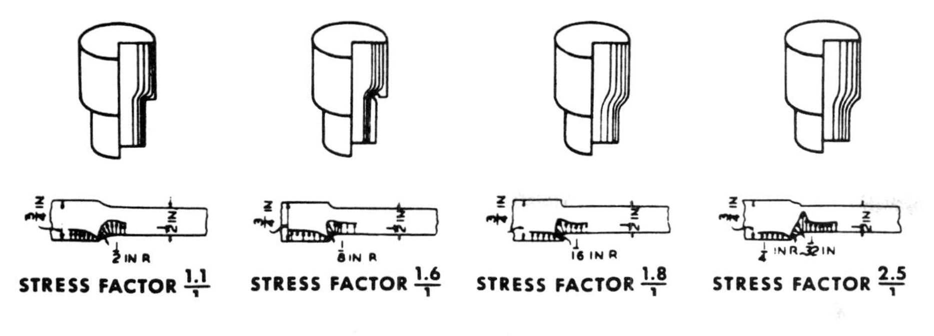

FIGURE 27.21a. Stress factors. *(Courtesy Republic Steel Corporation)*

Design Considerations

CORRECT PRACTICE | **INCORRECT PRACTICE**

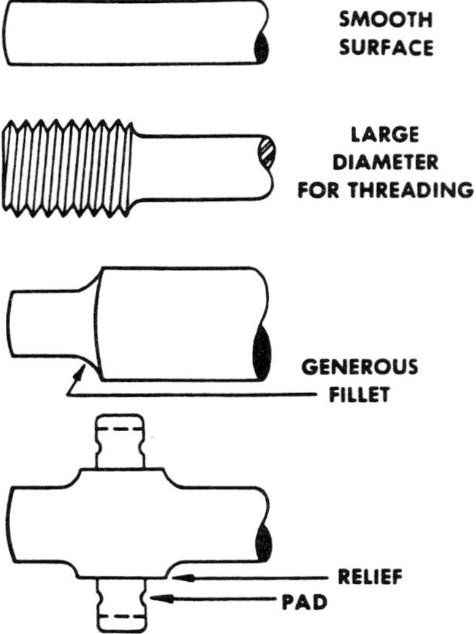

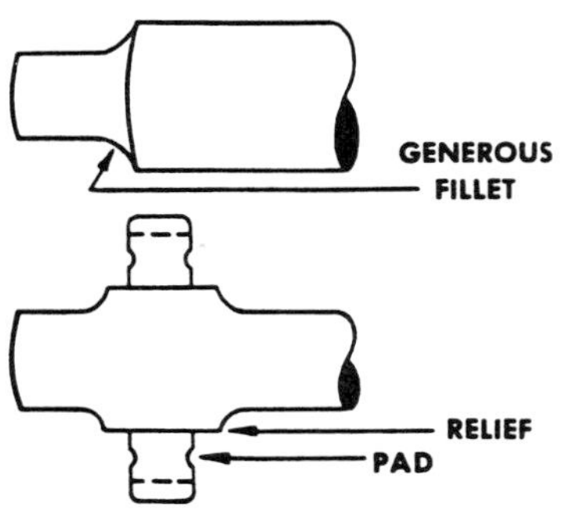

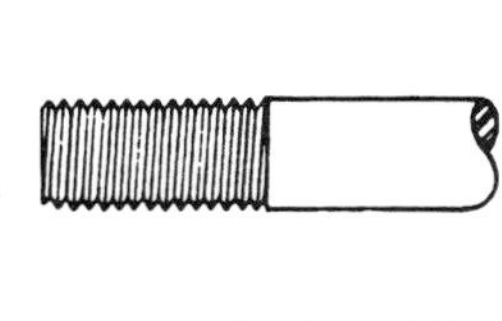

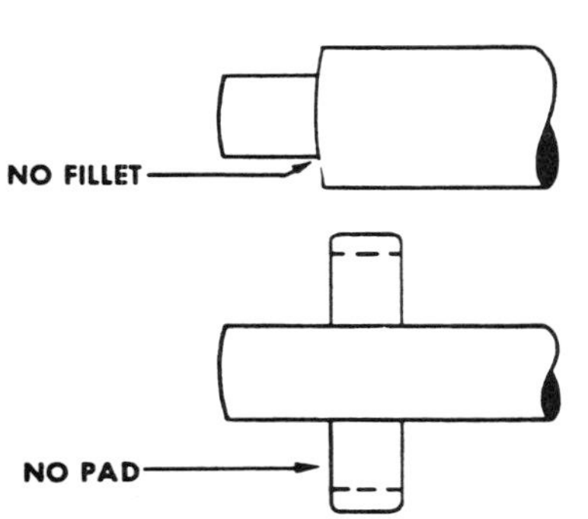

FIGURE 27.21b. Concentration of stress is a function of design. *(Courtesy Republic Steel Corporation)*

Intermittent loading

Compressive stresses induced by shot peening or carburizing tend to cancel out tensile stresses

Tensile stresses are induced here and tend to start fatigue cracks

FIGURE 27.22. Fatigue failures are caused by tensile stresses on areas of stress concentration. Compressive stresses induced by shot peening or carburizing tend to cancel out the tensile stresses and therefore reduce fatigue failures.

FIGURE 27.23. Classic example of fatigue in a shaft. The three distinct areas of fracture can be seen here.

crack would be high, causing the crack to progress rapidly near the surface, and the crack tends to flatten out sooner. The degrees of overstress by the relative areas of smooth and crystalline textures on the fracture surface are shown in Figure 27.25, 1-*c* and 1-*d.*

High-Stress Concentration. The smooth fracture growth can change from concave (as in 1-*b*) to convex as the rate of crack growth circumferentially at the surface exceeds the radial crack growth. In high overstress, the convex texture can occur extremely early in crack formation.

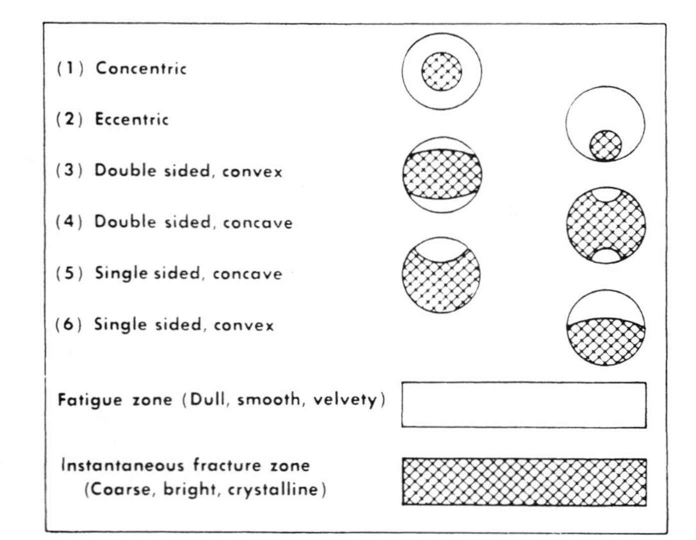

FIGURE 27.24. Symbols and nomenclature established by Bacon are useful for the designation of fracture appearances. *(Courtesy Republic Steel Corporation)*

Case 2 Two-Way Bending Load

No-Stress Concentration. Cracks start almost simultaneously at opposite surfaces when the surfaces are equally stressed. The cracks proceed toward the center at similar rates and result in a fracture that is rather symmetrical (2-*b* of Figure 27.25). In low overstress conditions (2-*a*), cracks may not begin at the same time; consequently the fracture is less symmetrical.

Mild-Stress Concentration. Higher stress concentration and the increased rate of circumferential crack growth cause the fracture to flatten out more quickly. Rapid radial crack growth tends to promote a concave zone with a relatively small radius of curvature. As the relative rates of circumferential and radial crack growth tend to become equalized, the radius of curvature tends to increase.

High-Stress Concentration. In this case, the circumferential crack rate increases rapidly, quickly exceeding the radial crack rate and the radius of curvature changes from concave to convex. The relative areas of smooth and crystalline textures discussed earlier also apply here.

Case 3 Reversed Bending and Rotational Load

No-Stress Concentration. As in Case 2, stress occurs at two extreme surface fibers. Usually the weaker area will fail first. The fracture tends to progress and flatten out from the initial small, concave crack. Eventually the fracture tends to become a straight line. The crack tends to propagate *against* the direction of rotation. With low overstress, the crack growth can proceed well beyond

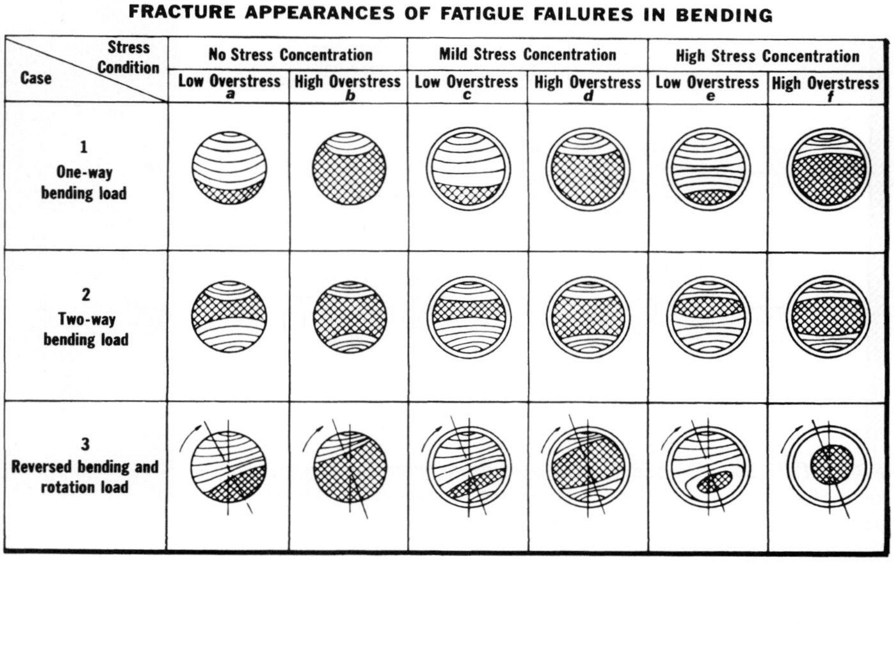

FRACTURE APPEARANCES OF FATIGUE FAILURES IN BENDING

Case \ Stress Condition	No Stress Concentration		Mild Stress Concentration		High Stress Concentration	
	Low Overstress *a*	High Overstress *b*	Low Overstress *c*	High Overstress *d*	Low Overstress *e*	High Overstress *f*
1 One-way bending load						
2 Two-way bending load						
3 Reversed bending and rotation load						

FIGURE 27.25. Fracture appearances of fatigue failures in bending.

the center and promote circumferential growth prior to complete failure. With high overstress, the crack does not proceed as far as shown in 3-*b* of Figure 27.25.

Mild-Stress Concentration. With high overstress, the notch causes early crack formation and rapid crack growth around the periphery, and the crystalline zone is centrally located as shown in 3-*d*. Low overstress tends to start the crack at the weakest point and the crystalline zone is moved away from the point of crack initiation (3-*c*). Extremely tough material will respond to these conditions with a fracture similar to 3-*a*.

High-Stress Concentration. A combination of severe concentration and high overstress, such as a groove machined about the entire periphery with a sharp notch radius, causes cracks to start all around the circumference at the same time. The resultant crystalline failure is somewhat circular in appearance and centrally located (3-*f*). Lower overstress tends to move the point of failure away from the central location (3-*c* of Figure 27.25).

When a shaft is subjected to a torsional load, the maximum shear stress is equal to the maximum tensile stress. However, the corresponding two strengths in steel are not equal, the shear strength being approximately one-half the tensile strength. The shear stress, therefore, will reach the shear strength before the tensile stress will reach the tensile strength; therefore, a shear type of failure will result (Figure 27.26). One reason that transverse (crosswise) cracks are more prevalent than longitudinal cracks is that grinding or machining marks, which accentuate the probability of failure, are oriented in the transverse direction. The quality of surface finish is therefore very important.

Splined shafts almost always produce a characteristic compound fracture. Fatigue cracks originate almost simultaneously at all the spline roots and grow until the shaft ruptures. The use of fibrous type steels tends to increase this problem.

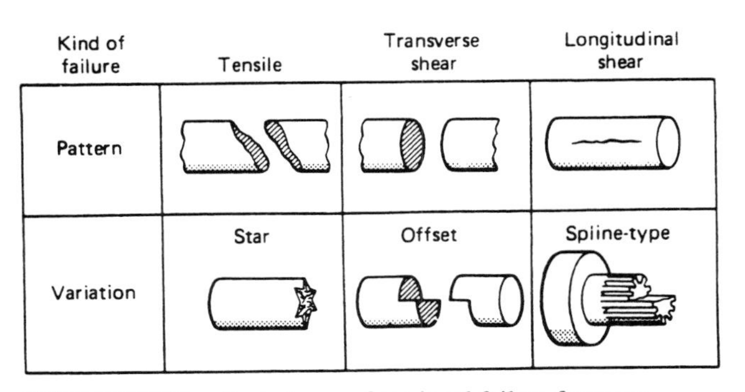

FIGURE 27.26. Basic types of torsional failure fractures.

SOME EXAMPLES OF SERVICE FAILURES

The following illustrations are cases of industrial problems resulting from poor practices, error in design or material selection, or simple overload.

1. *Fatigue failure in helical gear* (Figure 27.27). This pinion gear is part of a heavy reduction drive that was operating on a relatively light load. The smooth clamshell surface extends completely through the gear because the irregular break and the meshing of the gear teeth on both sides of the break prevented any brittle fracture. It was noted from the wear surface on the teeth that there was a slight axial misalignment of the gears and mismating of the teeth because of either an involute error or center distance error. This caused a high-stress tooth contact that resulted in fatigue cracking.
2. *Brittle failure caused by weld* (Figure 27.28). Such brittle failures can be initiated by a stress concentration when there is high overstress. This 3 in. diameter SAE 1040 bar was loaded as a cantilever beam and used as a special fork on a lift truck. A gusset was welded at the top to strengthen it. The short weld across the end of the gusset produced sufficient stress concentration to cause the sudden brittle failure. If welds must be used on a highly stressed member, no weld should be placed in a transverse direction to the outer fiber stresses, that is, across the bar. Welds should be made only lengthwise to avoid

FIGURE 27.27. Fatigue failure in helical gear.

stress concentration. Welds should never be made around a shaft.

3. *Weld on overloaded shaft in a sawmill* (Figures 27.29*a* and 27.29*b*). Millwrights sometimes find it necessary to use any means to keep the machines running, even though they are aware that they are using poor practices. Figure 27.29*a* shows an extreme overload on the shaft and coupling. The drive key has been twisted out of position. Figure 27.29*a* reveals the valiant effort that was made by the millwrights to keep this shaft turning. Welding around a shaft is bad practice because it creates a stress concentration that will rapidly initiate fatigue cracking, especially with high overloads. Here weld has been piled upon weld until the final failure ended the process.

4. *Failure in a tractor engine* (Figures 27.30*a* to 27.30*e*). The bolts in the bearing caps suddenly failed, and the crankshaft pushed the connecting rod

FIGURE 27.28. Brittle failure caused by weld.

***FIGURE 27.29*a.** Extremely overloaded shaft caused this failure, twisting the key out of position.

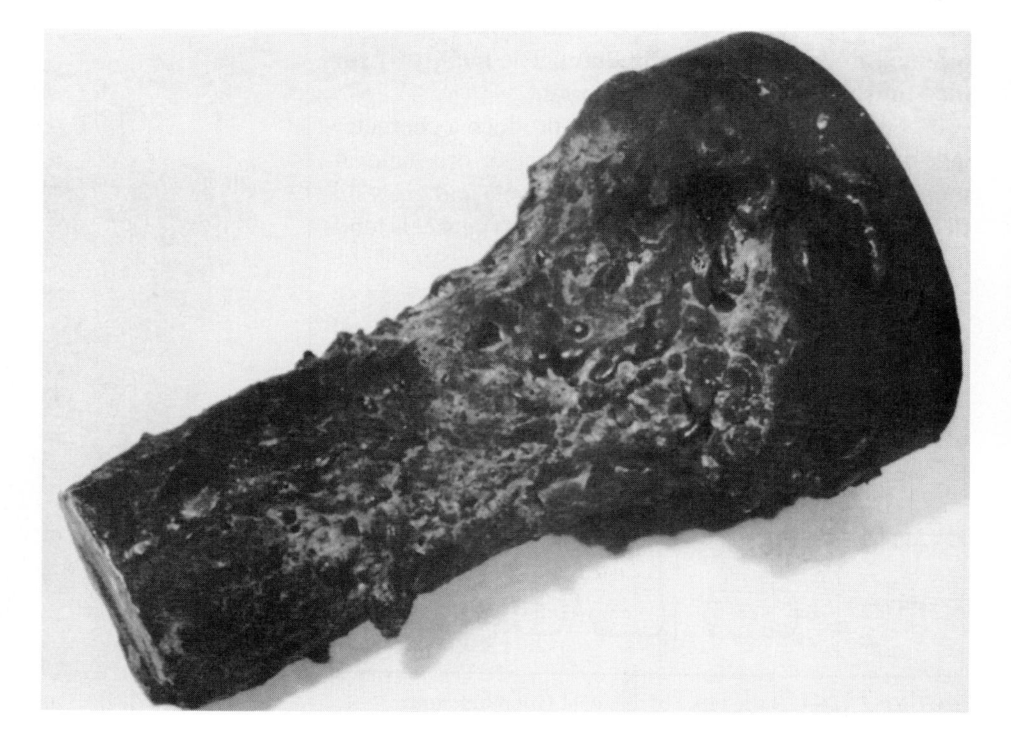

***FIGURE 27.29*b.** Same shaft and coupling as in Figure 27.29*a*, showing repeated welding on the shaft in an effort to keep it in operation.

***FIGURE 27.30*a.** A failure in a tractor engine. A connecting rod came loose from the crankshaft and was pushed out through the side of the block.

***FIGURE 27.30*b.** Close-up showing the hole and connecting rod.

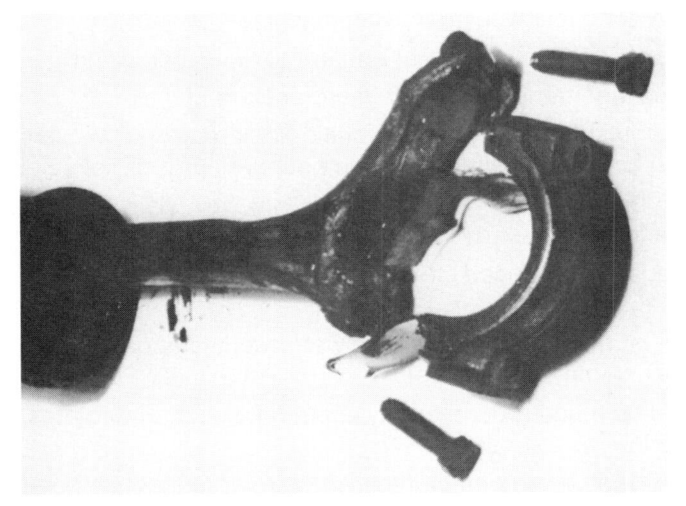

***FIGURE 27.30*c.** Bent connecting rod and cap. Note the necked-down screws, indicating that they were overtorqued.

***FIGURE 27.30*d.** End of one of the cap screws that held the cap on the connecting rod. A typical fatigue pattern is evident.

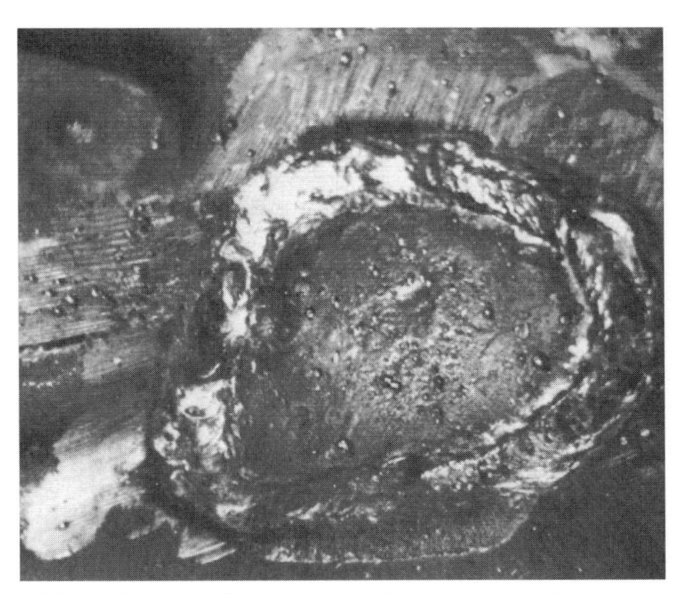

***FIGURE 27.30*e.** The hole in the side of the engine block has been welded with ENi rod. The entire block was cleaned in a dip tank and preheated a few hundred degrees.

through the side of the engine block (Figures 27.30*a* and 27.30*b*). The cap screws that fasten the connecting rod-bearing cap had evidently been overtorqued by the mechanic to the point that they had "necked down" to a smaller diameter (Figure 27.30*c*). This put a much higher stress on the screws, which started a fatigue fracture (Figure 27.30*d*). (Undertorquing bolts can also create fatigue because of the vibration caused by having the bolt stressed with a load value less than that of peak loading on the bolt when in use.) The engine will be repaired by welding in the

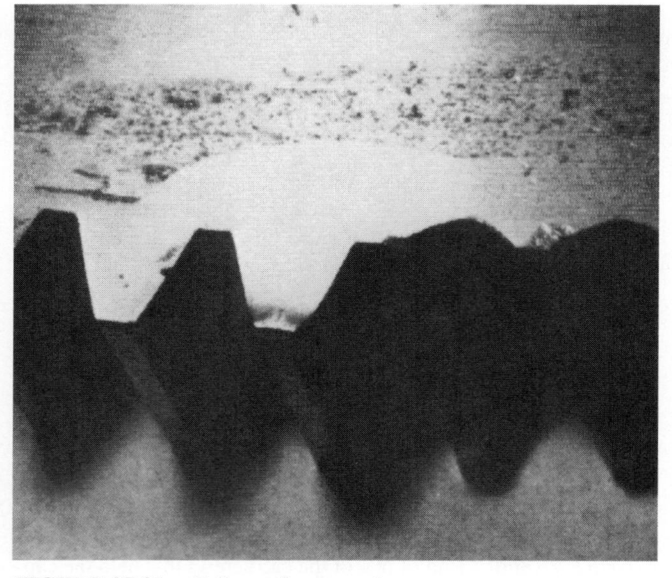

FIGURE 27.31. Failure of gear teeth.

piece of cast iron that was broken out and either by replacing the crankshaft or by metal spraying and grinding the damaged rod-bearing surface on the old one. The connecting rod and insert bearings will also be replaced.

5. *Failure of gear teeth* (Figure 27.31). This large ring gear is a classic example of root cracking and failure of gear teeth by fatigue. Note the sharp corners at the root of the teeth where the crack initiated. Redesign of the gear with a rounded root corrected the problem.

GRINDING PROBLEMS

Grinding may lead to problems resembling heat treating errors. Surface temperatures ranging from 2000°F to 3000°F are generated during grinding. This can cause two undesirable effects on hardened tool steels: development of high internal stresses, causing surface cracks to be formed, and changes in the hardness and metallurgical structure of the surface area.

One common effect of grinding on hardened and tempered tool steels is to reduce the hardness of the surface by gradual tempering; the hardness is lowest at the extreme surface but increases with distance below the surface. The depth of this tempering varies with the amount of depth of cut, the use of coolants, and the type of grinding wheel. If high temperatures are produced locally by the grinding wheel and the surface is immediately quenched by the coolant, a martensite having a Rockwell hardness of C65 to 70 can be formed. This gradient hardness, being much greater than that beneath the surface of the tempered part, sometimes causes very high stresses that contribute greatly to grinding cracks. Sometimes grinding cracks are visible in oblique or angling light, but they can be easily detected when present by the use of magnetic particle or fluorescent particle testing.

When a part is hardened but not tempered before it is ground, it is extremely liable to stress cracking. Faulty grinding procedures can also cause grinding cracks. Improper grinding operations can cause tools that have been properly hardened to fail.

SELF-EVALUATION

1. Is it possible to entirely eliminate wear in moving machinery parts?
2. What is the difference between intergranular and transgranular cracking?
3. In brittle failures, is cleavage transgranular or intergranular?
4. Which metal would be most likely to sustain a brittle fracture at −40°F, plain carbon SAE 1040 with carbide outlining the grain boundary or a nickel steel having 0.30 percent carbon with 5 percent nickel?
5. What is a stress raiser in a metal part? What can it cause?
6. Fatigue failures are caused by induced stress from intermittent loading. What type of stress initiates this cracking?
7. Why should welds not be made across any highly stressed component such as a spring, car or truck frame, or a bumper?
8. Why should a weld not be made around or anywhere on a shaft that rotates, even with moderate to low load stress?
9. How can low-carbon mild steel become brittle next to an area that has been heated to a high temperature and held there for a relatively long period of time?

10. How can a normally ductile steel become brittle when it has been electroplated or when a weld was made on it over paint, oil, or water, or if the welding rod coating has absorbed moisture?
11. When two mild steel surfaces rub together under pressure and without lubricant, what is the probable result?
12. If it is impossible to lubricate the parts in question 11, what can be done to avoid the problem?

CASE PROBLEM: THE SHAFT THAT FAILED

A single diameter shaft, having no steps or shoulders, drives a machine with a fairly heavy load. The shaft rotates at 1200 rpm and extends from the bearing several times its diameter. A vee-pulley is mounted on the shaft. Belts drive the pulley from a motor, causing a considerable side stress. The bearing is kept in place with a snap-ring that is in a narrow groove around the shaft at the edge of the bearing. The shaft has to be replaced several times a year at considerable cost in lost man-hours and material. Every time the shaft is severed, the millwright notices that it is at the location of the snap-ring groove. The break has a smooth surface with tiny grooves and a round section at the center that has a rough crystalline surface as in Figure 27.23. Can you identify the problem and name the type of failure? Name as many changes that could be made to correct the problem.

WORKSHEET

Complete the following worksheet by describing each service failure on the table format that is provided or on a separate sheet of paper. Write in your recommendations for corrective measures or materials. The answers will be found after the Self-Evaluation answers for Chapter 27 in the Appendix.

Objective Identify and describe 10 kinds of industrial service failures and make recommendations that will help to correct the problem.

	Kind of Failure (Description)	Recommendation to Correct the Problem

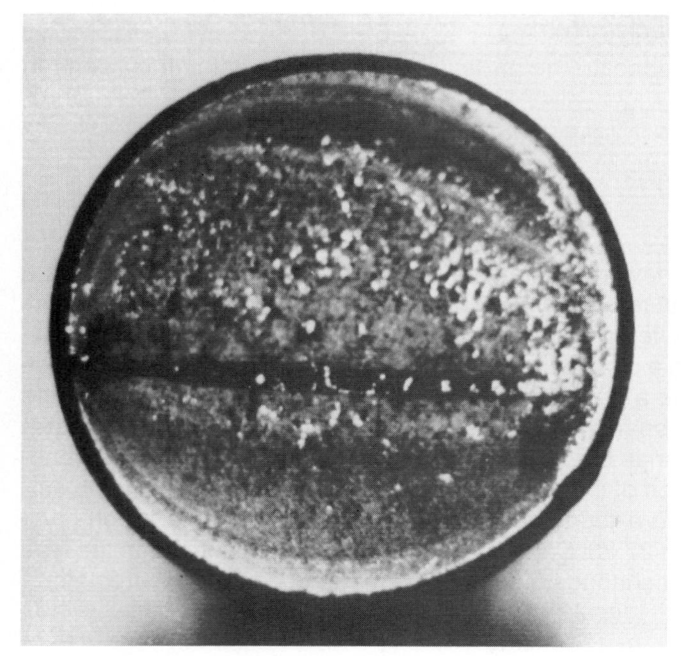

FIGURE 27.32.

FIGURE 27.33.

	Kind of Failure (Description)	Recommendation to Correct the Problem

FIGURE 27.34.

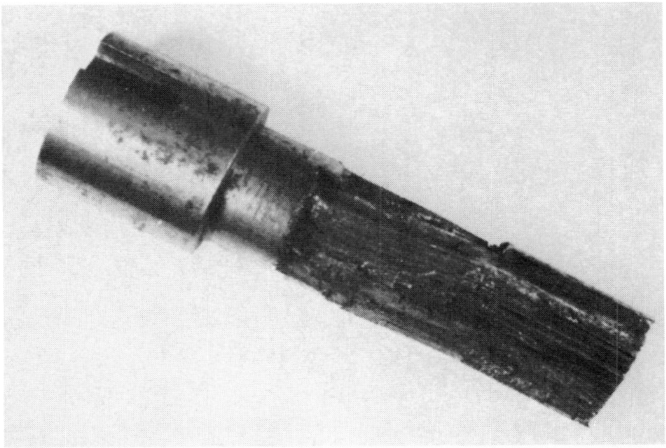

FIGURE 27.35.

	Kind of Failure (Description)	Recommendation to Correct the Problem

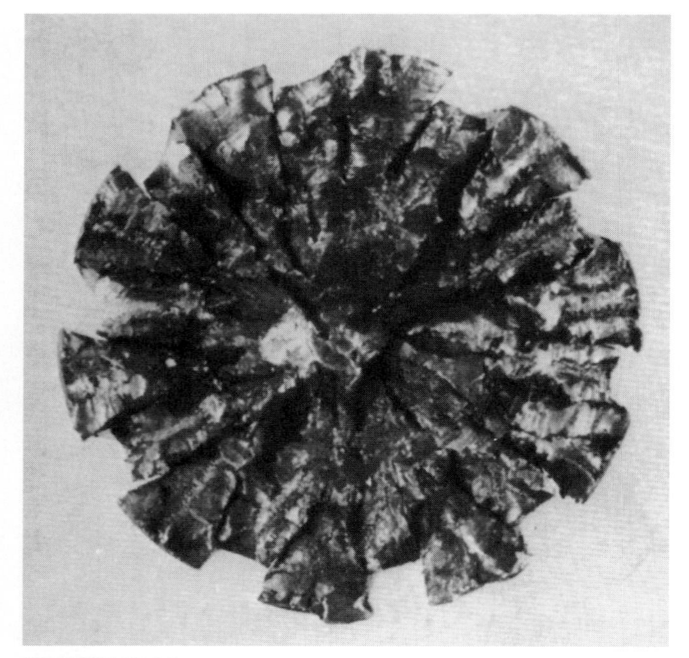

FIGURE 27.36.

FIGURE 27.37.

	Kind of Failure (Description)	Recommendation to Correct the Problem

FIGURE 27.38.

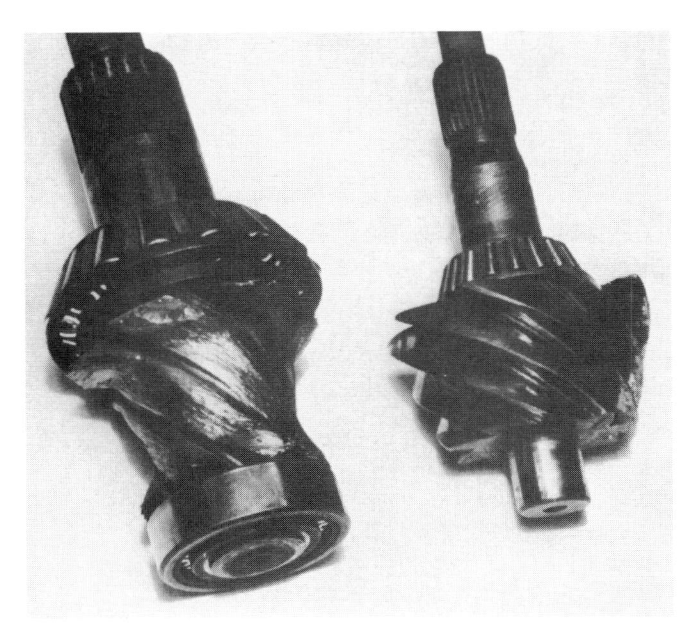

FIGURE 27.39.

	Kind of Failure (Description)	Recommendation to Correct the Problem

FIGURE 27.40.

FIGURE 27.41.

Show your completed worksheet to your instructor for evaluation.

APPENDIX 1

Industrial Method of Specimen Preparation

MICROSCOPIC EXAMINATION OF METALS

Details of the structure of metals are not readily visible to the naked eye, but grain structures in metals may be seen with the aid of the microscope. Metal characteristics, grain size, effects of heat treatment, and carbon content of steels may be determined by studying the micrograph (Figures A1.a through A1.d). The approximate percentage of carbon in steels may be estimated by the percentage of pearlite (dark areas) in annealed carbon steels. For this purpose, a metallurgical microscope (Figure A2) and associated techniques of photomicroscopy are used. The metallurgical reflected light microscope is similar to those used for other purposes, except that it contains an illumination system within the lens system to provide vertical illumination (Figure A3). Some microscopes are also equipped with a reticle and micrometer scale for measuring the magnified image. Another reticle used contains the various grain sizes at 100× magnification for the purpose of comparing or measuring relative grain size. Filters and polarizers are used in the illumination or optical system to reduce glare and improve the definition of grain structures. The magnifying power of the microscope may be determined by multiplying the power of the objective lens by that of the eyepiece. Thus, a 40× objective lens with a 12.5× eyepiece would enlarge the image to 500× (500 diameters). Figure A4 shows a modern image analysis microscope system.

In modern metallurgical laboratories, image analysis systems (Figure A5) may be utilized in the analysis of microstructures which are displayed on a screen. The image can be stored on an optical storage device such as a hard disc. Certain operations such as percentages, particles per unit area, and grain size can be selected and graphed and then put into storage. Results can also be transferred into documents, reports, and standardized forms.

See color inserts for examples of metallography and microstructure.

PREPARATION OF THE SPECIMEN

The specimen should be selected from the area of the piece that needs to be examined and in the proper orientation. That is, if grain flow or distortion is important, a cross section of the part may not show elongated grains; only a slice parallel to the direction of rolling would adequately reveal elongated grains from rolling. Sometimes, more than one specimen is required. A weld is usually cross-sectioned for examination.

Soft materials (under 35 Rc) may be sectioned by sawing, but harder materials must be cut off with an abrasive wheel. Metallurgical cutoff saws with abrasive blades and coolant flow are used for this purpose (Figures A6 and A7). **The specimen must not be overheated,** whether it is hard or soft. The grain structures may become altered with a high cutting temperature.

The specimen should be small, 1/4 to 3/8 in. in width, for ease of preparation. The specimen is usually mounted in plastic by using a mounting press (Figure A8). Thermosetting plastic is often used and is formed by heat and pressure around the specimen (Figure A9). The mold must be kept hot through the curing period (3 to 5 minutes) for thermosetting materials. If thermoplastic materials such as lucite are used, the mold is first heated to soften the plastic and then cooled before the specimen is removed. The time for this operation is about 15 to 20 minutes.

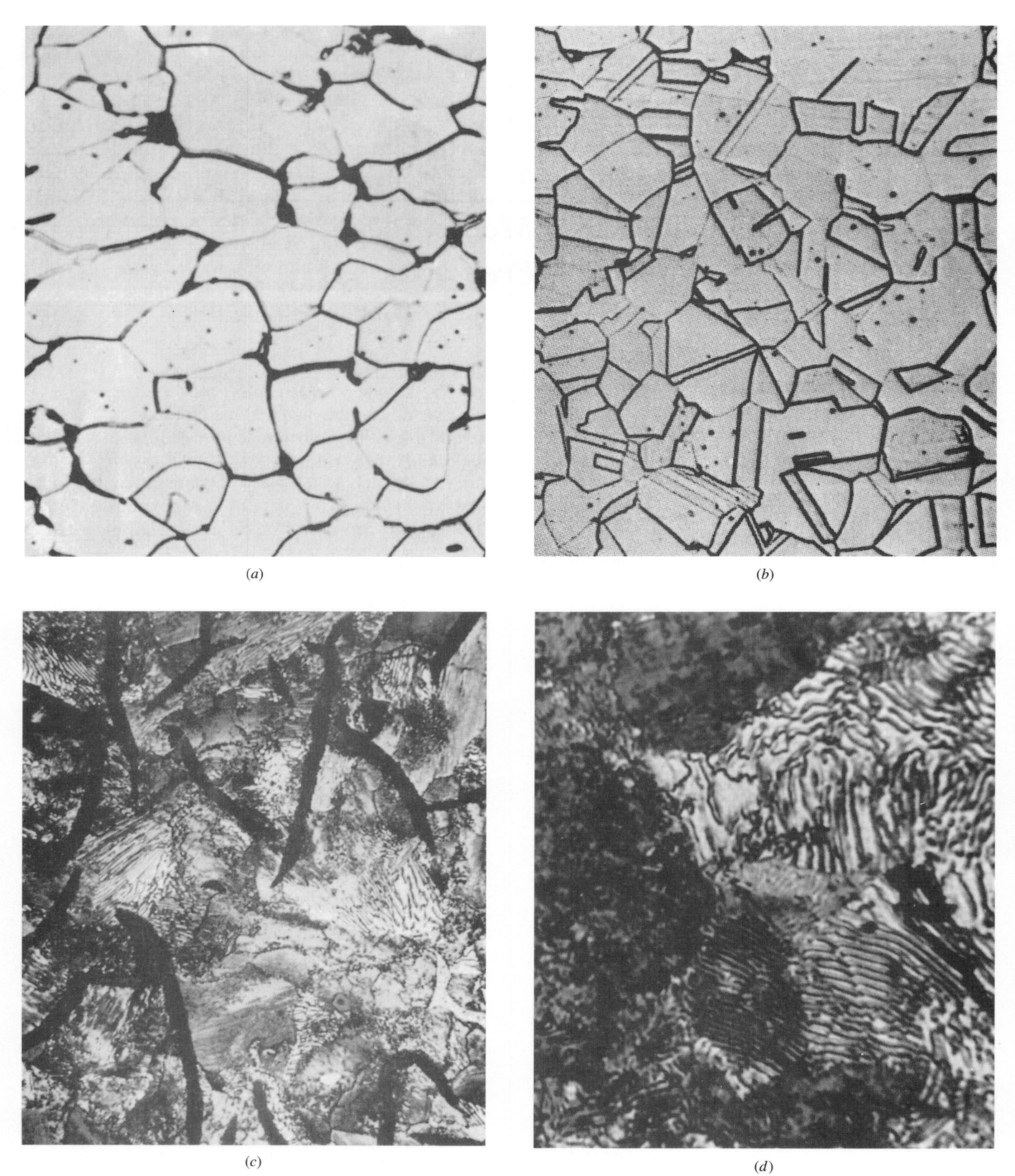

FIGURE A1. Identifying microstructures of various ferrous metals: (*a*) ferrite (annealed low carbon steel), (*b*) austenite, (*c*) gray cast iron (etched), (*d*) martensite (*a through d* are by permission from *Metals Handbook Volume 7, Copyright American Society for Metals, 1972.*)

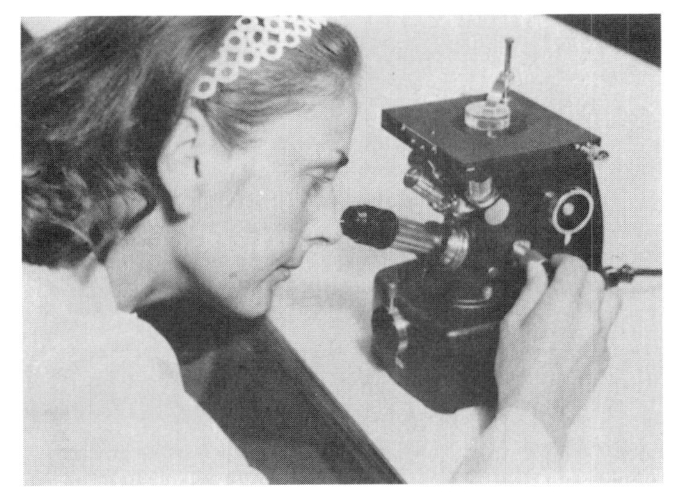

FIGURE A2. Inverted stage microscope. (*Photograph courtesy of Buehler Ltd., Evanston, Illinois*)

FIGURE A4. Modern image analysis microscope system. (*Photo courtesy of Buehler Ltd., Evanston, Illinois.*)

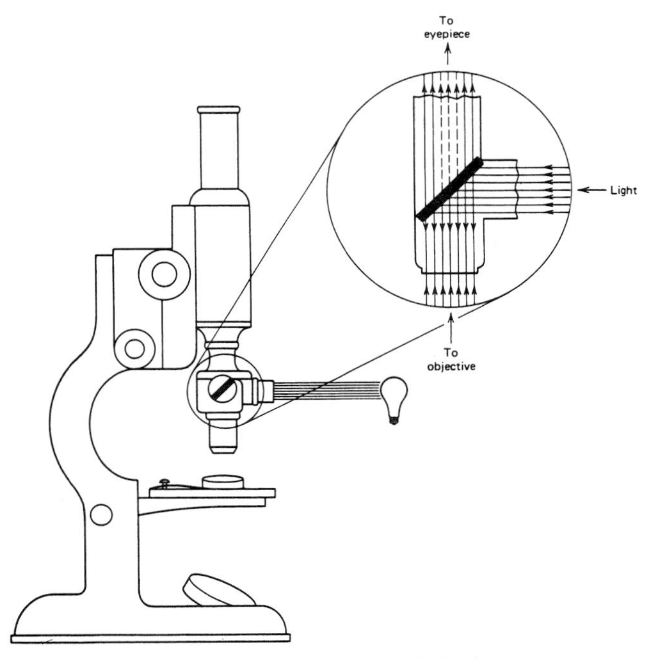

FIGURE A3. Illumination in a metallurgical microscope.

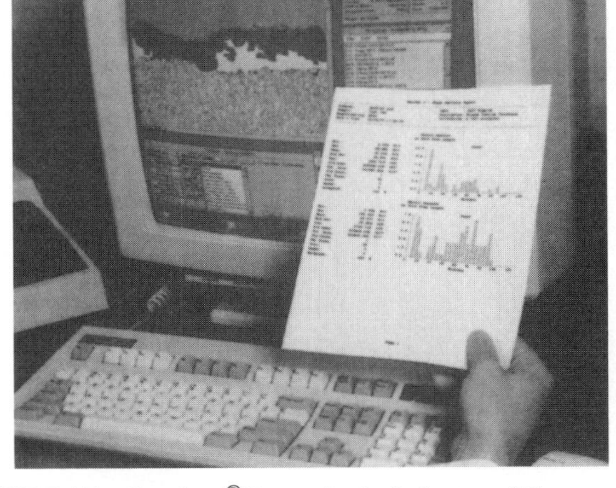

FIGURE A5. Omnimet® Image Analysis System. (*Photograph courtesy of Buehler Ltd., Evanston, Illinois.*)

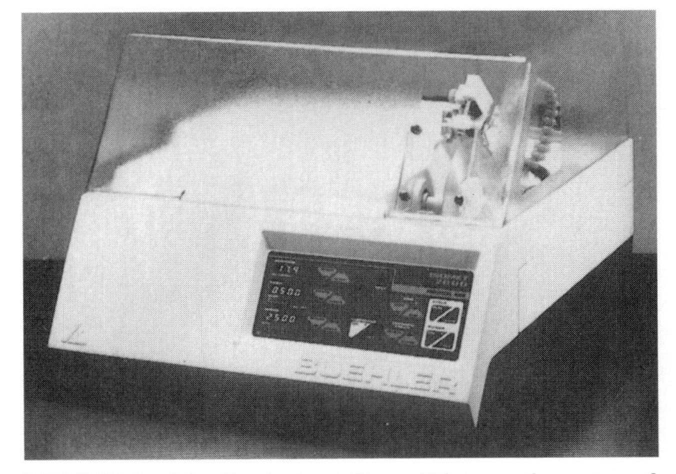

FIGURE A6. Metallurgical cutoff saw. (*Photograph courtesy of Buehler Ltd., Evanston, Illinois*)

POLISHING THE SPECIMEN

Grains and other features of metal cannot be seen unless the specimen is ground and polished to remove all scratches. Various methods of polishing are used, such as electrolytic, rotating, or vibrating polishers. The most common procedure is to first rough grind the face of the specimen on a belt sander and then hand grind on several grades of abrasive paper from 240 to 600 grit. The Handimet® grinder (Figure A9) provides four grinding surfaces upon which water flows constantly. The water

FIGURE A7. Specimen chuck, coolant nozzle, and cut off wheel of Figure A5. (*Photograph courtesy of Buehler Ltd.*)

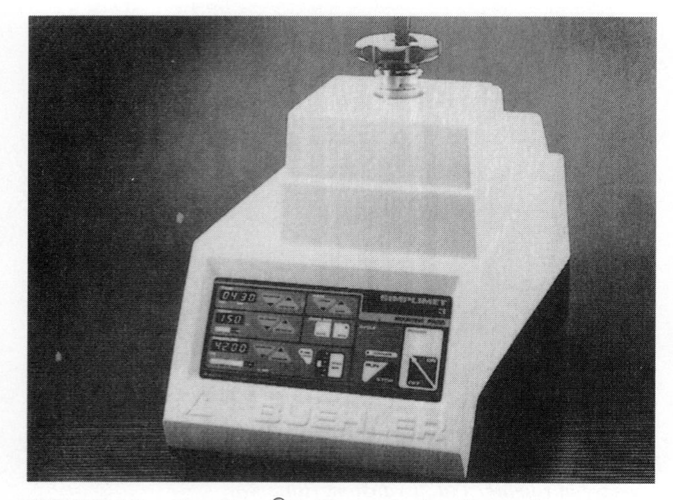

FIGURE A8. Simplimet® 3, Electro-hydraulic specimen mount press. (*Photograph courtesy of Buehler Ltd.*)

FIGURE A9. Small specimen mounted in thermosetting plastic.

FIGURE A10. The Handimet® roll grinder is used to prepare specimens for further polishing with either the rotating or electropolishers. When grinding surfaces become worn, new abrasive surfaces can be drawn in place. (*Photograph courtesy of Buehler Ltd.*)

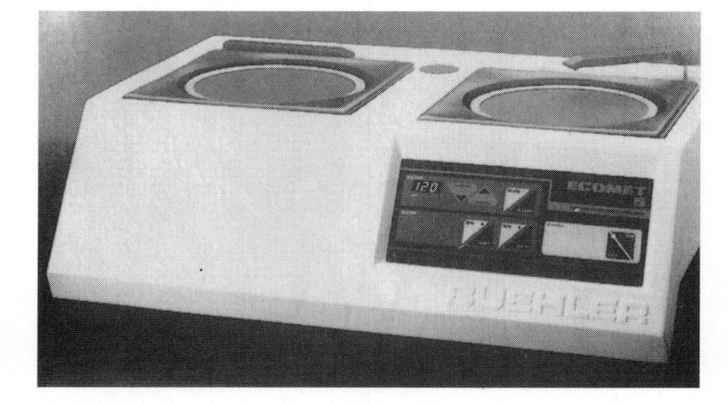

FIGURE A11. The Ecomet® 5, two speed grinder polisher. Electronic circuitry controls speed and water supply. Either abrasive paper or polishing cloths may be used on the wheels. (*Photograph courtesy of Buehler Ltd.*)

removes surface particles and maintains sharp cutting edges. The specimen is first moved back and forth on the coarse grit paper until all the scratches go in one direction **and then the specimen must be thoroughly cleaned before moving to a finer grit.** The second step in grinding should be done so that the new grinding scratches or lines are 90 degrees to the previous lines. This process should continue through all the grits of paper. The specimen is now prefinished and ready for polishing, which can be done in one operation or by rough polishing, followed by finish polishing. This is best done with coarse diamond paste abrasives on nylon cloth, followed by aluminum oxide on a short nap cloth such as billiard cloth.

The specimen may be put in an electropolisher or a vibratory polisher, or a rotating polishing wheel may

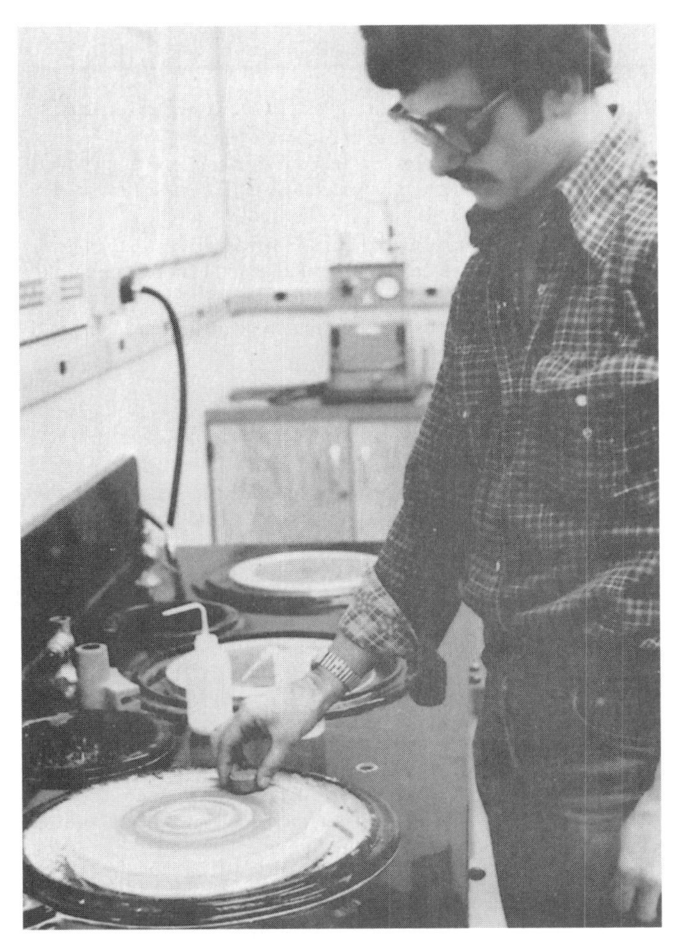

FIGURE A12. Hand polishing a specimen. (*Lane Community College*)

be used (Figure A11). The wheel is covered with a cloth such as billiard felt, and a slurry of finely divided alumina is applied. The specimen is held face down on the wheel and slowly moved around in the opposite direction to the rotation (Figure A12).

When the metal specimen is mirror bright and shows no scratches or lines, it should be cleaned in water. Methyl alcohol should be used for further cleaning and to help the drying operation. A small Pyrex (petri) dish may be used for this purpose, or the specimen may be swabbed with cotton saturated in the alcohol. The alcohol should be dried with a flow of hot air. This can be done with an ordinary hair dryer.

Improper polishing techniques will produce a less than desirable outcome. Insufficient polishing will leave many scratches, which are magnified by the microscope. Too much pressure applied when polishing can deform the surface layer and make grain boundaries and structures indistinct.

ETCHING THE SPECIMEN

Enchants are composed of organic or inorganic acids and alkalis dissolved in alcohol, water, or other solvents. Some common enchants are given in Table A1. The specimen may now be etched for the required time by immersing it face down in a solution contained in a petri dish (Figure A13). An alternate method is to apply the enchant with an eye dropper. If the etching time is too short, the specimen will be underetched, and the grain boundaries and other configurations will be faint and indistinct when viewed in the microscope. If the etching time is to long, the specimen will be overetched and will be very dark, having unusual colors. The time of etching must be controlled very carefully. The etching action is stopped by placing the specimen under a stream of water. Clean the specimen with alcohol and use a hair dryer to finish drying it. Do not rub the polished and etched specimen with a cloth or your finger, as this will alter the surface condition of the metal.

A single sample preparation facility is shown in Figure A14. A specimen holder is mounted on the power head which is rotated at 25 RPM on the platen. The holder may contain 6 specimens. Since the automatic polishing system, with LED indicators monitoring control time, head direction and force, and fluid delivery system, the resultant polished surface is generally superior to hand polishing methods. A polisher/etcher (Figure A15) also takes the guesswork out of both polishing and etching operations. These devices are very useful in metallurgical laboratories. But hand polishing and etching techniques are still applicable in school labs and when a small number of tests are needed.

SAFETY WITH CHEMICALS

Etching reagents, such as acids that are used on prepared and polished specimens of metals to bring out grain structures and other characteristics for viewing under a microscope, can be hazardous if they are mishandled. Since acids or alkaline solutions can cause severe burns to the skin and eyes, care must be exercised in their use. Small amounts of reagents are involved when etching for microscopic study, but much larger amounts are used for macro-etching (for visual study) larger parts, which are often immersed in a heated bath.

TABLE A1. Etching reagents for microscopic examination of metals

Metals	Enchant	Composition	Remarks
Iron and carbon steel	Nital	2 to 5% nitric acid in methyl alcohol	Darkens pearlite in carbon steels. Differentiates ferrite from martensite; reveals ferrite grain boundaries. Shows core depth of nitrided steels. Time: 5 to 60 seconds.
	Picral	4 g picric acid 100 ml methyl alcohol	For annealed and quench-hardened carbon and low alloy steel. Not as good as nital for revealing ferrite grain boundaries. Time: 5 to 120 seconds.
	Hydrochloric and picric acid	5 g hydrochloric acid 1 g picric acid 100 ml methyl alcohol	Reveals austenitic grains in quenched and tempered steels.
Alloy and stainless steels	Ferric chloride and hydrochloric acids	5 g ferric chloride 20 ml hydrochloric acid 100 ml distilled water	Reveals structures in iron-chromium-nickel and steels. Reveals structure of stainless and austenitic nickel steels.
High-speed steels	Hydrochloric and nitric acid	9 ml hydrochloric acid 9 ml nitric acid 100 ml methyl alcohol	Reveals grain size of quenched and tempered high-speed steels. Time: 15 seconds to 5 minutes.
Aluminum and aluminum alloys	Sodium hydroxide	10 g sodium hydroxide 90 ml distilled water	General enchant. Can be used for both micro- and macroetching. Time: 5 seconds.
Magnesium and magnesium alloys	Glycol	75 ml ethylene glycol 24 ml distilled water 1 ml concentrated nitric acid	Can be used to reveal grain structures of most magnesium alloys.
Nickel and nickel alloys	Acetic and nitric acid	50 ml glacial acetic acid 50 ml concentrated nitric acid	Nickel-nickel copper alloys. Time: 5 to 20 seconds.
Copper and copper alloys	Nitric acid	12 to 30% nitric acid	Shows inclusions, porosity, and flow lines. Requires 5 to 20 minutes.
	Kellers concentrated etch	10 ml hydrofluoric acid 15 ml hydrochloric acid 25 ml nitric acid 50 ml water	Store solution in paraffin wax bottle only as the hydrofluoric acid will dissolve glass. Good enchant for copper-bearing alloys.
	Ammonium hydroxide and hydrogen peroxide	5 parts ammonium hydroxide 5 parts hydrogen peroxide 5 parts water	Used for copper and many copper alloys. Peroxide content varies with the copper content of alloy.
	Ammonium persulfate	10 g ammonium persulfate 90 ml water	Good enchant for copper, brass, bronze, and nickel silver.

Face masks and protective clothing (rubberized aprons and rubber gloves) should be worn when using reagents. Rubber gloves and tongs should be used when handling the specimen. If any acid is accidentally spilled on your skin or splashed in your eyes, it should be **immediately** washed off in cold running water. Sodium bicarbonate solutions should always be available for neutralizing acids. Only water should be used for washing out eyes that are burned by acids, and the washing should be done as quickly as possible.

Red fuming nitric acid should not be used for testing purposes. If it is used on a sample of titanium, for example, a very dangerous explosion may follow. Only dilute acids may be used for testing purposes.

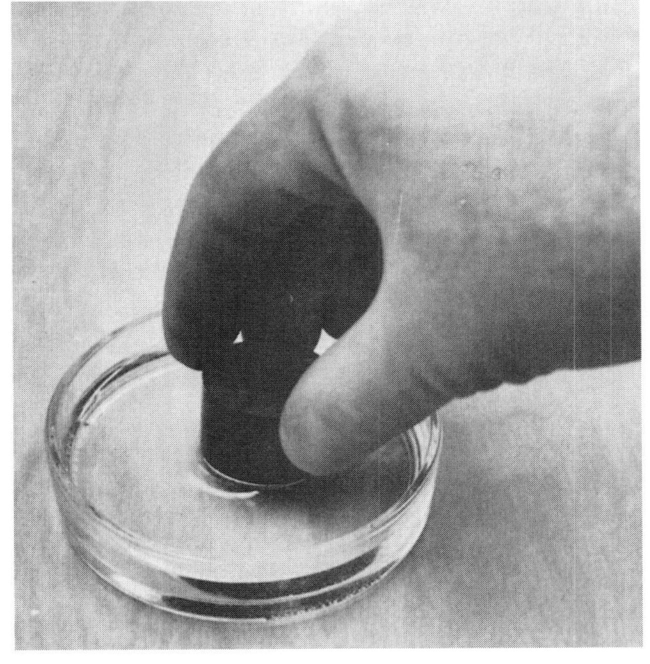

FIGURE A13. Etching the specimen.

FIGURE A14. Automet® 2 Power head with electronic loading system with variable speed and programmable fluid dispensing system. (*Photograph courtesy of Buehler Ltd.*)

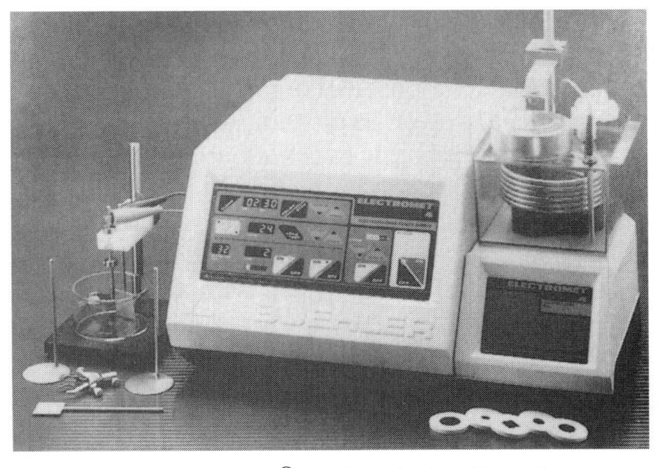

FIGURE A15. Electromet® 4 polisher/etcher. Electrolytic sample preparation system is provided with this equipment for polishing and etching stainless steels, aluminum, titanium, and many copper based alloys. (*Photograph courtesy of Buehler Ltd.*)

A ventilation system with a hood should be installed over the area where acids are used in order to carry away the toxic fumes. An eyewash spray should also be available in the immediate vicinity.

Safety Note. When mixing acids with water or methanol, **always** pour the acid into the solvent. When in doubt about the safe mixing procedures for chemicals, consult a chemist or a comprehensive chemistry handbook.

The Metals Handbooks should be used as additional reference to provide a larger basis of information for specific metallographic sample preparation for metals, composites, and non-metals.

APPENDIX 2

Tables

TABLE 1 Heat Treatments for Various Tool Steels

Carbon Tool Steel
AISI-SAE Type W1

Atlas X-12
Best Carbon
B-F Extra
Blue Label
Carbon
Cutlery
Diamond S
Extra
G. W. Extra
H & R Carbon
Pompton
Regular
Special A. S. V.
Standard

Temp.	R.C.
As-Q	67–68
150° C	64–65
175	63–64
200	61–62
230	60–61
260	59–60
290	57–58
315	53–55
370	50–52
425	46–48
480	40–42
540	33–36

Analysis: C 0.60–1.40
Harden: 770°–815° C
Quench: brine/water

High Carbon High Chromium Oil
AISI-SAE Type D3

Alloy C
Atlas NN
CNS-2
Crocar
H & R K
Hampden
Hicro 200
Huron
Lehigh S
Neor
Republic HC
Superdie
Superior No. 1
Vi-Chrome

Temp.	R.C.
As-Q	64–66
150° C	63–64
175	62–63
200	61–62
230	60–62
315	59–60
370	58–60
510	57–59
540	56–57
550	52–54
565	50–53

Analysis: C 2.25 Cr 12.00
Harden: 925°–980° C
Quench: oil/salt/air

Low Manganese, Cr-W Oil Hardening
AISI-SAE Type 01

Atlan
BTR
B 76
Carpenter O-1
Choyce 77
Fisco Oilhard
Invaro No. 1
Ketos
Kiski
Oil Hardening
Oneida
Ready-Mark
Utex
Veri Best Drill Rod

Temp.	R.C.
As-Q	64–65
150° C	63–64
175	62–63
190	61–62
200	59–61
230	58–59
260	57–58
315	55–56
370	51–53
425	48–50
480	46–47
540	42–43

Analysis: C 0.90 Mn 1.20 Cr .50 W .50
Harden: 790°–815° C
Quench: oil/salt

5% Chromium Air Hardening
AISI-SAE Type A2

Airkool
Airvan
Aircrat
Airtem
Concord
Crest A.H.
Dumore
Econo No. 5
E-Z Die Smooth Cut
Hardnair
Krovan
Pittsburgh
Precision Ground
Simonds Airtrue

Temp.	R.C.
150° C	63–65
150	63–64
175	62–63
200	60–62
260	60–61
315	59–60
370	58–59
425	57–58
480	56–57
540	54–56
565	52–54
590	50–52
650	43–45

Analysis: C 1.00 Cr 5.00 Mo 1.10
Harden: 925°–1025° C
Quench: air

TABLE 1 (continued)

Manganese Oil Hardening
AISI-SAE Type 02

Deward
H Brand
H & R Oil-Hardening
Prescott
Ry-Alloy
Ry-Alloy drill
Republic Arrestite
Simonds 864
S.O.D.
Special Oil-Hardening
Stentor

Temp.	R.C.
As-Q	64–65
150° C	63–64
175	62–63
190	61–62
200	60–61
230	59–60
260	58–59
290	56–58
315	55–56
370	51–53
425	47–49
540	42–43

Analysis: C 0.90 Mn 1.60
Harden: 770°–800° C
Quench: oil/salt

Silicon-Manganese Shock Resisting
AISI-SAW Type S5

Achorn USI Steel
Atsil
Bedco Alloy
BTF Alloy
Champion 255
Chimo
Duro Chip
FB-S5 Tufkut
Fisco Omega
H & R No. 8
Lanark
Ludlum 602
Omega
Rocket
Silicon alloy

Temp.	R.C.
As-Q	61–63
150° C	59–61
175	58–60
200	57–59
230	56–59
260	55–58
315	54–57
370	53–55
425	50–52
480	46–50

Analysis: C 0.55 Si 2.00 Mn .80 Mo .40
Harden: 870°–910° C
Quench: oil/salt

High-Speed
AISI-SAE Type T1

Blue Chip
Blue Streak
Cannon
Clarite
F.C.C. No. 41 M
Gold Anchor
H & R No. 1
H.S.C. 18-4-1
Kutkwik
Record Superior
Red Label
Spartan
Superior Ark H.S.
Supremus

Double Tempered	
Temp.	R.C.
As-Q	63–65
540° C	63–64
550	63–64
565	63–65
580	62–64
590	61–63
620	57–60
650	58–60

Analysis: C 0.75 W 18.00
Harden: 1250°–1290° C
Quench: air/oil/salt

High-Speed Steel
AISI-SAE Type M2

Achorn
Bedco M-2
Blue Streak Moly
Braemow
DBL-2
FB-M2
H & R No. 57
Mocarb
Molite
Moly
Mustang
Rex M2-S
Sixix
Star Mo-M2

Double Tempered	
Temp.	R.C.
As-Q	64–66
540° C	64–66
550	64–66
565	64–66
580	62–64
590	61–63
620	59–61
650	54–57

Analysis: C 0.83 W 6.50 Mo 5.00
Harden: 1175°–1230° C
Quench: air/oil/salt

TABLE 1 (continued)

High Carbon High Chromium Air
AISI-SAE Type D2

Atlan HCC
Atmodie
Atmodie Smoothcut
Airdi 150
Beaver
Darwin No. 1FM
Hyco-1
High Production
Lehigh H
Olympic F M
Ontario
Republic 404
Superior No. 3
White Label

Temp.	R.C.
As-Q	63–64
150° C	62–63
175	61–63
200	60–62
230	60–61
260	59–60
340	58–59
510	58–60
540	57–58
550	56–57
565	54–56
590	48–52

Analysis: C 1.50 Cr 12.00 Mo .80
Harden: 980°–1025° C
Quench: air

Tungsten Chisel
AISI-SAE Type S1

Achorn UBC Steel
Bengal
Brown Label
Bull Dog
Cyclops S 1
Falcon—6
Falcon—4
Ideor
Macco Foolproof
Maxtuff
Par-Exc
Pure-Ore Chiz-Alloy
Seminole Hard
Super Shock

Temp.	R.C.
As-Q	58–61
150° C	58–61
200	57–60
260	56–59
315	56–58
370	54–56
425	52–54
480	50–53
540	47–50
590	44–46
650	40–43

Analysis: C 0.50 Cr 1.35 W 2.50
Harden: 900°–980° C
Quench: oil/salt

Manganese Air Hardening
AISI Types A4, A5, A6

Air 4
Airloy
A.S. No. 121
A.S. Ack Low
Apache
CM Air-Hardening
H & R A-4
Macco AL-4
Milnair 4
Nutherm
Pure-Ore Air-Eez
Tempair
Uni-Die Smooth Cut
Vega Mold Steel

Temp.	R.C.
As-Q	63–64
150° C	61–63
175	61–62
200	60–61
260	58–59
315	55–57
370	54–55
425	52–53
480	50–52

Analysis:
A4 C 1.00 Mn 2.00 Cr 2.00 Mo 1.10
A5 C 1.00 Mn 3.00 Cr 1.00 Mo 1.00
A6 C 0.70 Mn 2.00 Cr 1.00 Mo 1.30
Harden: A4 815°–870° C
A5 790°–845° C
A6 830°–870° C
Quench: air

Tungsten Fast Finishing
AISI Type F2

Atlas XXX
B. F. Fast Finishing
BFS
Celero
Coppco Fast Finishing
Double Special
E-E
F.F.F.
Fast Finishing
Finishing
GR-350
K.W.
Penco No. 20
Special Finishing

Temp.	R.C.
As-Q	66–68
150° C	64–66
175	63–65
200	63–65
230	61–63
260	61–62
315	59–61
370	55–56

Analysis: C 1.30 W 3.75
Harden: 790°–870° C
Quench: brine/water

The above figures are average and for 1-in.-dia rounds, adequately quenched to obtain full hardness.

TABLE 1 (continued)

High-Speed Steel
AISI-SAE Type M30

Como
Electrite Lacomo
H & R Super Molyhi
Super Hi-Mo
Super LMW
Super Motung
8-N2 Cobalt

Temp.	R.C.
As-Q	65–66
540° C	66–67
550	66–67
565	65–66
580	64–65
590	63–64
620	61–62
650	57–58

Analysis: C 0.80 Cr 4.00 Mo 8.25 Co 5.00
Harden: 1200°–1245° C
Quench: air/oil/salt

Stainless Steels
403–410–416 420 440C

Temp.	R.C.	R.C.	R.C.
As-Q	38–45	52–55	58–61
175° C	38–43	52–55	58–60
230	38–43	50–52	56–68
315	38–43	48–50	56–58
425	38–43	48–50	56–58
420	38–43	45–47	56–58
510	38–42	44–46	53–55
525	30–40	42–45	53–55
540	26–35	40–42	52–55
550	25–34	35–38	50–52
620	20–25	33–35	46–48
590	B95–100	28–32	40–43

Harden:
403/410/416–925° C–1010° C
420–980°–1040° C
440C–1010°–1065° C
Quench: air/oil/salt

5% Chromium Hot Work
AISI-SAE Types H11, H12, H13

A. S. Cromat V
Crodi
Cromo-W V
Coppco H. W. No. 1
G. W. 99
H. & R. Hot Work No. 6
H W S
Howard A
Hotform No. 2
Macco M. L.
M. G. R.
Redstone 12
Republic 10-H-W
Thermotem 11

Double Tempered	
Temp.	R.C.
As-Q	54–55
540° C	54–56
550	53–54
565	51–52
570	48–49
580	47–48
590	44–45
605	42–43
620	40–41
635	38–39
650	36–37

Analysis:
H11 C 0.35 Cr 5.00 Mo 1.50 V .40
H12 C 0.35 Cr 5.00 Mo 1.50 W 1.30 V .40
H13 C 0.35 Cr 5.00 Mo 1.25 V 1.00
Harden:
H11 995°–1040° C
H12 995°–1025° C
H13 995°–1025° C
Quench: Air

9% Tungsten Hot Work
AISI-SAE Type H21

Air Hardening No. 30
B.B.
B-44J
C.L.W.
D. C. 66
D.N.V.
E. I. S. 73
F.C.C. No. 14
H & R Hot Work No. 2
Halcolmb H. W.
Macco P-175
Nitung
Nut Piercer
Redstone

Temp.	R.C.
As-Q	52–54
540° C	52–55
565	52–55
590	49–52
620	48–51
635	47–50
660	45–48
675	39–43
705	34–38

Analysis: C 0.35 Cr 3.50 W 9.50
Harden: 1090°–1200° C
Quench: air/oil/salt

Note: Steels listed are representative groups of brands as used in various sections of the United States and are not to be construed as recommendations for specific brands. All analysis percentages are averaged from representative specifications from steel producers.

Source: "Tempering Chart for Tool Steels," *Lindberg, Sola Basic Industries,* Chicago, Illinois, 1973.

TABLE 2 Heat Treatments for Various Production Steels

Alloy Steels (Water Hardening 0.30%C)			
Temp.	**1330**	**2330**	**3130**
205° C	47RC	47RC	47RC
260	44	44	44
315	42	42	42
370	38	38	38
425	35	35	35
480	32	32	32
540	26	26	26
590	22	22	22
650	16	16	16

Heat treatment:
Normalized at 900° C
Water quenched from 880°–815° C
Tempered—2 hours

Alloy Steels (Water Hardening 0.30%C)			
Temp.	**4130**	**5130**	**8640**
205° C	47RC	47RC	47RC
260	45	45	45
315	43	43	43
370	42	42	42
425	38	38	38
480	34	34	34
540	32	32	32
590	26	26	26
650	22	22	22

Heat treatment:
Normalized at 885° C
Water quenched from 880°–855° C
Tempered—2 hours

Alloy Steels (Oil Hardening 0.40%C)		
Temp.	**1340**	**3140**
205° C	57RC	55RC
260	53	52
315	50	49
370	46	47
425	44	41
480	41	37
540	38	33
590	35	30
650	31	26

Heat treatment:
Normalized at 870° C
Oil quenched from 830°–845° C
Tempered—2 hours

Alloy Steels (Oil Hardening 0.40%C)			
Temp.	**4340**	**4640**	**8740**
205° C	55RC	52RC	57RC
260	52	51	53
315	50	50	50
370	48	47	47
425	45	42	44
480	42	40	41
540	39	37	38
590	34	31	35
650	31	27	22

Heat treatment:
Normalized at 870° C
Oil quenched from 830°–855° C
Tempered—2 hours

Alloy Steels (Oil Hardening 0.50%C)			
Temp.	**4150**	**5150**	**6150**
205° C	56RC	57RC	58RC
260	55	55	57
315	43	52	53
370	51	49	50
425	47	45	46
480	46	39	42
540	43	34	40
590	39	31	36
650	35	28	31

Heat treatment:
Normalized at 870° C
Oil quenched from 865°–870° C
Tempered—2 hours

Alloy Steels (Oil Hardening 0.50%C)		
Temp.	**8650**	**8750**
205° C	55RC	56RC
260	54	55
315	52	52
370	49	51
425	45	46
480	41	44
540	37	39
590	32	34
650	28	32

Heat treatment:
Normalized at 870° C
Oil quenched from 815°–845° C
Tempered—2 hours

Source: "Production Steels, Analysis and Heat Treatment," *Lindberg. Sola Basic Industries,* Chicago.

Carbon Steels (Water Hardening 0.30–0.50%C)			
Temp.	**1030**	**1040**	**1050**
205° C	50RC	51RC	52RC
260	45	48	50
315	43	46	46
370	39	42	44
425	31	37	40
480	28	30	37
540	25	27	31
590	22	22	29
650	95RB	94RB	22

Heat treatment:
Normalized at 900° C
Water quenched from 830°–845° C
Tempered—2 hours

Carbon Steels (Water Hardening 0.60–0.95%C)			
Temp.	**1060**	**1080**	**1095**
205° C	56RC	57RC	58RC
260	55	55	57
315	50	50	52
370	42	43	47
425	38	41	43
480	37	40	42
540	35	39	41
590	33	38	40
650	26	32	33

Heat treatment:
Normalized at 885° C
Water quenched from 800°–815° C
Tempered—2 hours

Carbon Steels (Water Hardening 0.40%C)		
Temp.	**1137**	**1141**
205° C	44RC	49RC
260	42	46
315	40	43
370	37	41
425	33	38
480	30	34
540	27	28
590	21	23
650	91RB	94RB

Heat treatment:
Normalized at 900° C
Water quenched from 830°–855° C
Tempered—2 hours

The above figures are average and for 1-in. dia. rounds, adequately quenched to obtain full hardness.

Note: Oil-hardening steels can often be quenched in hot salt or hot oil (martempering) to minimize distortion, cracking hazards, and residual stresses.

TABLE 3 Heat Treatments for Case-Hardening Steels

Case-Hardening Steels

The low carbon steels used for case hardening in pack or gas carburizing must be properly treated to develop desired case and core properties. Various combinations of heating and quenching to refine structure are illustrated below.

	1015		**1018**		**1022**		**C-1117**		**2317**		**2515**		**3120**	
	Co (RB)	**Ca (RC)**	**Co (RC)**	**Ca (RC)**	**Co (RB)**	**Ca (RC)**	**Co (RC)**	**Ca (RC)**	**Co (RC)**	**Ca (RC)**	**Co (RC)**	**Ca (RC)**	**Co (RC)**	**Ca (RC)**
Carburized and oil quenched from 925°C	85	62	14	62	87	62	10	65	34	62	36	62	30	62
Reheated and oil quenched from 800° C	82	61	12	61	85	61	10	64	32	62	35	62	31	62
Double oil quenched from 800° C and 775 ° C	82	60	12	60	85	60	11	63	32	62	35	62	30	63

	4320		**4620**		**4815**		**5120**		**6120**		**8620**	
	Co (RC)	**Ca (RC)**	**Co (RC)**	**Ca (RC)**	**Co (RC)**	**Ca (RC)**	**Co (RC)**	**Ca (RC)**	**Co (RC)**	**Ca (RC)**	**Co (RC)**	**Ca (RC)**
Carburized and oil quenched from 925° C	38	64	28	62	33	62	33	62	35	62	30	62
Reheated and oil quenched from 815° C	38	63	29	64	32	63	29	63	27	63	27	63
Double oil quenched from 815° C and 775° C	38	63	28	63	32	63	30	63	25	62	24	62

Note: All hardness readings are RC unless otherwise indicated.

CA = Case

Co = Core

RB = Rockwell B-Scale

RC = Rockwell C-Scale

The above figures are average and for 1-in. dia. rounds, adequately quenched to obtain full hardness.

Source: "Production Steels, Analysis and Heat Treatment," *Lindberg, Sola Basic Industries,* Chicago, Illinois, 1973.

TABLE 4 Hardness Conversion Table (Approximate Value)

C 150 kg	A 60 kg	15-N 15 kg	B 100 kg	15-T 15 kg	Diamond Pyramid Hardness 10 kg	Knoop Hardness 500 gr & Over	Brinell Hardness 3000 kg	G 150 kg	Tensile Strength Approx.
70	86.5	94.0	—	—	1076	972	—	—	1000 lb per square inch Inexact and only for steel
69	86.0	93.5	—	—	1004	946	—	—	
68	85.5	—	—	—	942	920	—	—	
67	85.0	93.0	—	—	894	895	—	—	
66	84.5	92.5	—	—	854	870	—	—	
65	84.0	92.0	—	—	820	846	—	—	
64	83.5	—	—	—	789	822	—	—	
63	83.0	91.5	—	—	763	799	—	—	
62	82.5	91.0	—	—	739	776	—	—	
61	81.5	90.5	—	—	716	754	—	—	
60	81.0	90.0	—	—	695	732	614	—	
59	80.5	89.5	—	—	675	710	600	—	
58	80.0	—	—	—	655	690	587	—	
57	79.5	89.0	—	—	636	670	573	—	
56	79.0	88.5	—	—	617	650	560	—	—
55	78.5	88.0	—	—	598	630	547	—	301
54	78.0	87.5	—	—	580	612	534	—	291
53	77.5	87.0	—	—	562	594	522	—	282
52	77.0	86.5	—	—	545	576	509	—	273
51	76.5	86.0	—	—	528	558	496	—	264
50	76.0	85.5	—	—	513	542	484	—	255
49	75.5	85.0	—	—	498	526	472	—	246
48	74.5	84.5	—	—	485	510	460	—	237
47	74.0	84.0	—	—	471	495	448	—	229
46	73.5	83.5	—	—	458	480	437	—	221
45	73.0	83.0	—	—	446	466	426	—	214
44	72.5	82.5	—	—	435	452	415	—	207
43	72.0	82.0	—	—	424	438	404	—	200
42	71.5	81.5	—	—	413	426	393	—	194
41	71.0	81.0	—	—	403	414	382	—	188
40	70.5	80.5	—	—	393	402	372	—	182
39	70.0	80.0	—	—	383	391	362	—	177
38	69.5	79.5	—	—	373	380	352	—	171
37	69.0	79.0	—	—	363	370	342	—	166
36	68.5	78.5	—	—	353	360	332	—	162
35	68.0	78.0	—	—	343	351	322	—	157
34	67.5	77.0	—	—	334	342	313	—	153
33	67.0	76.5	—	—	325	334	305	—	148
32	66.5	76.0	—	—	317	326	297	—	144
31	66.0	75.5	—	—	309	318	290	—	140
30	65.5	75.0	—	—	301	311	283	92.0	136
29	65.0	74.5	—	—	293	304	276	91.0	132
28	64.5	74.0	—	—	285	297	270	90.0	129
27	64.0	73.5	—	—	278	290	265	89.0	126
26	63.5	72.5	—	—	271	284	260	88.0	123
25	63.0	72.0	—	—	264	278	255	87.0	120
24	62.5	71.5	—	—	257	272	250	86.0	117
23	62.0	71.0	—	—	251	266	245	84.5	115
22	61.5	70.5	—	—	246	261	240	83.5	112
21	61.5	70.0	100	93.0	241	251	240	82.5	116
20	61.0	69.5	99	92.5	236	246	234	81.0	112
—	60.0	—	98	—	—	241	228	79.0	109
—	59.5	—	97	92.0	—	236	222	77.5	106
—	59.0	—	96	—	—	231	216	76.0	103
—	58.0	—	95	91.5	—	226	210	74.0	101
—	57.5	—	94	—	—	221	205	72.5	98
—	57.0	—	93	91.0	—	216	200	71.0	96

TABLE 4 (continued)

C 150 kg	A 60 kg	15-N 15 kg	B 100 kg	15-T 15 kg	Diamond Pyramid Hardness 10 kg	Knoop Hardness 500 gr & Over	Brinell Hardness 3000 kg	G 150 kg	Tensile Strength Approx.
—	56.5	—	92	90.5	—	211	195	69.0	93
—	56.0	—	91	—	—	206	190	67.5	91
—	55.5	—	90	90.0	—	201	185	66.0	89
—	55.0	—	89	89.5	—	196	180	64.0	87
—	54.0	—	88	—	—	192	176	62.5	85
—	53.5	—	87	89.0	—	188	172	61.0	83
—	53.0	—	86	88.5	—	184	169	59.0	81
—	52.5	—	85	—	—	180	165	57.5	80
—	52.0	—	84	88.0	—	176	162	56.0	78
—	51.0	—	83	87.5	—	173	159	54.0	77
—	50.5	—	82	—	—	170	156	52.5	75
—	50.0	—	81	87.0	—	167	153	51.0	74
—	49.5	—	80	86.5	—	164	150	49.0	72
—	49.0	—	79	—	—	161	147	47.5	—
—	48.5	—	78	86.0	—	158	144	46.0	—
—	48.0	—	77	85.5	—	155	141	44.0	—
—	47.0	—	76	—	—	152	139	42.5	—
—	46.5	—	75	85.0	—	150	137	41.0	—
—	46.0	—	74	—	—	147	135	39.0	—

Note: The values in this table correspond to the values shown in the corresponding joint SAE-ASM-ASTM Committee on Hardness. Conversions as printed in ASTM E 48, Table 3.

Source: "Hardness Conversion Table," *Lindberg, Sola Basic Industries,* Chicago, Illinois, 1973.

TABLE 5 The Periodic Table of the Elements

Light Metals		Heavy Metals										Nonmetals					Inert gases
IA	IIA	←Brittle metals→					←Ductile metals→				Low melting	IIIA	IVA	VA	VIA	VIIA	0
H 1																	He 2
Li 3	Be 4											B 5	C 6	N 7	O 8	F 9	Ne 10
Na 11	Mg 12	IIIB	IVB	VB	VIB	VIIB	VIIIB			IB	IIB	Al 13	Si 14	P 15	S 16	Cl 17	Ar 18
K 19	Ca 20	Sc 21	Ti 22	V 23	Cr 24	Mn 25	Fe 26	Co 27	Ni 28	Cu 29	Zn 30	Ga 31	Ge 32	As 33	Se 34	Br 35	Kr 36
Rb 37	Sr 38	Y 39	Zr 40	Nb 41	Mo 42	Tc 43	Ru 44	Rh 45	Pd 46	Ag 47	Cd 48	In 49	Sn 50	Sb 51	Te 52	I 53	Xe 54
Cs 55	Ba 56	* 57–71	Hf 72	Ta 73	W 74	Re 75	Os 76	Ir 77	Pt 78	Au 79	Hg 80	Tl 81	Pb 82	Bi 83	Po 84	At 85	Rn 86
Fr 87	Ra 88	† 89															
	*	La 57	Ce 58	Pr 59	Nd 60	Pm 61	Sm 62	Eu 63	Gd 64	Tb 65	Dy 66	Ho 67	Er 68	Tm 69	Yb 70	Lu 71	
	†	Ac 89	Th 90	Pa 91	U 92	Np 93	Pu 94	Am 95	Cm 96	Bk 97	Cf 98	Es 99	Fm 100	Md 101	No 102	Lr 103	

In general: it can be stated that the properties of the elements are periodic functions of their atomic numbers. It is in the vertical columns that the greatest similarities between elements exist. There are also some similarities between the A and B groups on either side of the VIII groups. For example, the Scandium group, IIIB, is in some respects similar to the Boron group, IIIA. The most important properties in the study of metallurgy that vary periodically with the atomic numbers are crystal structure, atomic size, electrical and thermal conductivity, and possible oxidation states.

Besides the relationship of the elements in the vertical groups of the transition metals, there are resemblances in several horizontal triads: iron, cobalt, and nickel (all three are ferromagnetic); ruthenium, rhodium, and palladium; and osmium, iridium, and platinum.

TABLE 6 Specific Gravity (Density) and Weights of Some Materials

Material	Specific Gravity g/cm^3	Weight per cu. in.-lb	Weight per cu. ft.-lb
Air	0.001293	0.0000466579	0.080625
Aluminum	2.70	0.097	168.5
Antimony	6.618	0.2390	413.0
Boron	2.535	0.0916	158.2
Brass (approx.)	8.50	0.3067	530.02
Bronze (approx.)	8.75	0.3157	545.6
Cadmium	8.648	0.3123	539.6
Calcium	1.54	0.0556	96.1
Chromium	7.1	0.2562	442.7
Cobalt	8.71	0.3145	543.5
Columbium (Niobium)	8.4	0.3031	523.8
Copper	8.89	0.3210	554.7
Cork	0.24	0.00866	14.9
Gasoline	0.67	2.4177	41.7
Glass (approx.)	2.5	0.0902	155.9
Gold	19.3	0.6969	1204.3
Iridium	22.42	0.8096	1399.0
Iron, Cast (approx.)	7.50	0.2562	442.7
Lead	11.34	0.4096	707.7
Magnesium	1.741	0.0628	108.6
Manganese	7.3	0.2636	455.5
Mercury	13.55	0.4892	845.3
Molybdenum	10.2	0.3683	636.5
Nickel	8.8	0.3178	549.1
Platinum	21.37	0.7717	1333.5
Silver	10.5	0.3789	654.7
Steel (Approx.)	7.85	0.2833	489.4
Tatalum	16.6	0.5998	1035.8
Tin	7.29	0.2633	454.9
Titanium	4.5	0.1621	280.1
Tungsten	19.0	0.6856	1184.7
Uranium	18.7	0.6753	1166.9
Vanadium	5.69	0.2052	354.8
Water	1.0	0.036	62.355
Wood (Approx.)	0.8018	2.89	50.0
Zinc	7.1	0.2562	442.7

TABLE 7 Temperature Conversion Table: Celsius–Fahrenheit

C°	F°	C°	F°	C°	F°	C°	F°	C°	F°	C°	F°	C°	F°
0	32	50	122.0	100	212	600	1112	1100	2012	1600	2912	2100	3812
1	33.8	51	123.8	110	230	610	1130	1110	2030	1610	2930	2110	3830
2	35.6	52	125.6	120	248	620	1148	1120	2048	1620	2948	2120	3848
3	37.4	53	127.4	130	266	630	1166	1130	2066	1630	2966	2130	3866
4	39.2	54	129.2	140	284	640	1184	1140	2084	1640	2984	2140	3884
5	41.0	55	131.0	150	302	650	1202	1150	2102	1650	3002	2150	3902
6	42.8	56	132.8	160	320	660	1220	1160	2120	1660	3020	2160	3920
7	44.6	57	134.6	170	338	670	1238	1170	2138	1670	3038	2170	3938
8	46.4	58	136.4	180	256	680	1256	1180	2156	1680	3056	2180	3956
9	48.2	59	138.2	190	374	690	1274	1190	2174	1690	3074	2190	3974
10	50.0	60	140.0	200	392	700	1292	1200	2192	1700	3092	2200	3992
11	51.8	61	141.8	210	410	710	1310	1210	2210	1710	3110	2210	4010
12	53.6	62	143.6	212	413.6	720	1328	1220	2228	1720	3128	2220	4028
13	55.4	63	145.4	220	428	730	1346	1230	2246	1730	3146	2230	4046
14	57.2	64	147.2	230	446	740	1364	1240	2264	1740	3164	2240	4064
15	59.0	65	149.0	240	464	750	1382	1250	2282	1750	3182	2250	4082
16	60.8	66	150.8	250	482	760	1400	1260	2300	1760	3200	2260	4100
17	62.6	67	152.6	260	500	770	1418	1270	2318	1770	3218	2270	4118
18	64.4	68	154.4	270	518	780	1436	1280	2236	1780	3236	2280	4136
19	66.2	69	156.2	280	536	790	1454	1290	2354	1790	3254	2290	4154
20	68.0	70	158.0	290	554	800	1472	1300	2372	1800	3272	2300	4172
21	69.8	71	159.8	300	572	810	1490	1310	2390	1810	3290	2310	4190
22	71.6	72	161.6	310	590	820	1508	1320	2408	1820	3308	2320	4208
23	73.4	73	163.4	320	608	830	1526	1330	2426	1830	3326	2330	4226
24	75.2	74	165.2	330	626	840	1544	1340	2444	1840	3344	2340	4244
25		75	167.0	340	644	850	1562	1350	2462	1850	3362	2350	4262
26		76	168.8	350	662	860	1580	1360	2480	1860	3380	2360	4280
27	80.6	77	170.6	360	680	870	1598	1370	2498	1870	3398	2370	4298
28	82.4	78	172.4	370	698	880	1616	1380	2516	1880	3416	2380	4316
29	84.2	79	174.2	380	716	890	1634	1390	2534	1890	3434	2390	4334
30	86.0	80	176.0	390	734	900	1652	1400	2552	1900	3452	2400	4352
31	87.8	81	177.8	400	752	910	1670	1410	2570	1910	3470	2410	4370
32	89.6	82	179.6	410	770	920	1688	1420	2588	1920	3488	2420	4388
33	91.4	83	181.4	420	788	930	1706	1430	2606	1930	3506	2430	4406
34	93.2	84	183.2	430	806	940	1724	1440	2624	1940	3524	2440	4424
35	95.0	85	185.0	440	824	950	1742	1450	2642	1950	3542	2450	4442
36	96.8	86	186.8	450	842	960	1760	1460	2660	1960	3560	2460	4460
37	98.6	87	188.6	460	860	970	1778	1470	2678	1970	3578	2470	4478
38	100.4	88	190.4	470	878	980	1796	1480	2696	1980	3596	2480	4496
39	102.2	89	192.2	480	896	990	1814	1490	2714	1990	3614	2490	4514
40	104.0	90	194.0	490	914	1000	1832	1500	2732	2000	3632	2500	4532
41	105.8	91	195.8	500	932	1010	1850	1510	2750	2010	3650	2510	4550
42	107.6	92	197.6	510	950	1020	1868	1520	2768	2020	3668	2520	4568
43	109.4	93	199.4	520	968	1030	1886	1530	2786	2030	3686	2530	4586
44	111.2	94	201.2	530	986	1040	1904	1540	2804	2040	3704	2540	4604
				540	1004								
45	113.0	95	203.0			1050	1922	1550	2822	2050	3722	2550	4622
46	114.8	96	204.8	550	1022	1060	1940	1560	2840	2060	3740	2560	4640
47	116.6	97	206.6	560	1040	1070	1958	1570	2558	2070	3758	2570	4658
48	118.4	98	208.4	570	1058	1080	1976	1580	2876	2080	3776	2580	4676
49	120.2	99	210.2	580	1076	1090	1994	1590	2894	2090	3794	2590	4694
				590	1094								

TABLE 7 (continued)

C°	F°	C°	F°	C°	F°	C°	F°	C°	F°
2600	4712	2700	4892	2800	5072	2900	5252	3000	5432
2610	4730	2710	4910	2810	5090	2910	5270		
2620	4748	2720	4928	2820	5108	2920	5288		
2630	4766	2730	4946	2830	5126	2930	5306		
2640	4784	2740	4964	2840	5144	2940	5324		
2650	4802	2750	4982	2850	5162	2950	5342		
2660	4820	2760	5000	2860	5180	2960	5360		
2670	4838	2770	5018	2870	5198	2970	5378		
2680	4856	2780	5036	2880	5216	2980	5396		
2690	4874	2790	5054	2890	5234	2990	5414		

TABLE 8 Chemical Tests and Reactions for Some Common Metals

Reagents	Carbon steel	Stainless Steels: Types 302, 304	Types 316, 317	Types 304, 321, 347	Types 303, 414, 430F	Aluminum	Magnesium	Nickel	Monel	Inconel
Nital—10% nitric acid in methanol	Blackens surface	No reaction					Cleans surface or turns gray	Slow reaction, turns blue-green	*Slow reaction, turns blue-green	No reaction
Hydrochloric acid	Reacts to clean surface	No reaction		Attacks, releasing gas	Attacks, leaving black smudge, rotten egg odor	Blackens surface	Blackens surface	Cleans surface	Cleans surface	No reaction
Copper sulfate solution	Leaves copper color	No reaction					Reacts, leaving heavy black smudge	No reaction		
Sodium hydroxide	No reaction					Reacts, leaves gray smudge	No reaction			
Sulfuric acid	Cleans surface	Leaves dark surface with green crystals	Slow attack, leaving tan surface	No reaction			Strongly reacts, foams, and leaves clean surface	No reaction		
Zinc chloride solution in water	No reaction						Reacts and turns black	No reaction		
Ferrous chloride	No reaction						Reacts and turns black	No reaction		

**Note.* Monel may be distinguished from nickel (they look alike) by the following test. Place a few drops of nital on the surface of the specimen. It will turn blue-green if it is either monel or nickel. Then place a small piece of soft iron (a nail will do) in the solution on the sample. If there is no reaction, the metal is nickel, but if the iron turns copper colored, copper ions have been transferred from the copper in the metal, and it can be assumed that it is monel, which contains copper. Zinc and tin will react with so many acids to turn them black that chemical tests would be of no value for them. Silver will turn black in sulfuric acid. However, most other metals may be identified by testing for ferromagnetism, density, color, or thermal conductivity.

TABLE 9 Some Common Ores of Industrial Metals

Metal	Ore	Color	Chemical Formula
Aluminum	Bauxite	Gray-white	Al_2O_3 (40 to 60%)
Cadmium	Sphalerite (zinc blende)	Colorless to brown to black	$ZnS + 0.1$ to 0.2% Cd
	Smithsonite	White	$Zn_4Si_2O_7OH \bullet H_2O$
Chromium	Chromite	Green	$FeO \bullet Cr_2O_3$
Cobalt	Cobaltite	Gray to silver-white	$CoAsS$
	Linnaeite (Cobalt pyrite)		Co_3S_4
	Smalt-ite	Bluish white or gray	$CoAs_2$
	Erythrite	Transparent, crimson, or peach-red	$Co_3(AsO_4)_2 \bullet 8H_2O$
Columbium (Niobium)	Columbite (Niobite)	Black	$(FeMn)(CbTa)_2O_6$
Copper	Chalcopyrite	Brass yellow	$CuFeS_2$
	Chalcocite	Lead gray	Cu_2S
	Cuprite	Red	Cu_2O
	Azurite	Blue	$2CuCo_3 \bullet Cu(OH)_2$
	Bornite	Red-brown	$FeS \bullet 2Cu_2S \bullet As_2S_5$
	Malachite	Green	$CuCo_3 \bullet Cu(OH)_2$
	Native copper	Reddish	Cu
Gold	Native (alluvial or in quartz)	Yellow	Au
Indium	Sphalerite	White to yellow	ZnS + low % of In
Iron	Magnetite	Black	Fe_3O_4
	Hematite	Reddish brown	Fe_2O_3
	Limonite	Brown to yellow	$2Fe_2O_3 \bullet 3H_2O$
	Goethite	Brown	$Fe_2O_3 \bullet H_2O$
	Siderite	Blue	$FeCO_3$
	Taconite	Flintlike	Fe_3O_3
Lead	Galena	Black	PbS
Magnesium	Sea water		$Mg(OH)_2$
Manganese	Pyrolusite	Black	MnO_2
	Manganite	Steel-gray to black	$Mn_2O_3 \bullet H_2O$
	Rhodochrosite	Rose-red	$MnCO_3$
Molybdenum	Molybdenite	Blue	MoS_2
Nickel	Niccolite	Pale copper-red	NiAs
	Pentlandite	Bronze-yellow	$(FeNi)_9S_8$
	Garnierite	Bright apple-green	$(MgNi)_3Si_2O_5(OH)_4$
Silver	Argentite	Black	Ag_2S
	Native	White	Ag
Tin	Cassiterite	Brown or black	SnO_2
Titanium	Anatase,	White	
	Brookite, and Rutile	Reddish-brown	TiO_2
Tungsten	Wolframite	Brownish to grayish black	$(FeMn)WO_4$
	Scheelite	Various	$CaWO_4$
	Ferberite		$FeWO_4$
	Huebnerite		$MnWO_4$
Uranium	Uraninite	Black	UO_2
	Pitchblende	Brown to black	U_3O_8
Vanadium	Vanadinite	Brown, yellow, red	$Pb_4(VO_4)_3Cl$
	Carnotite	Yellow	$K_2(UO_2)_2(VO_4)_23H_2O$
Zinc	Franklinite	Black	$ZnFe_2O_4$
	Willemite	Varying	Zn_2SiO_4
	Zincite	Red to orange-yellow	ZnO
Zirconium	Zircon (zirconium silicate sand)	Brown or gray square prisms or transparent	$ZrSiO_4$

APPENDIX 3

Self-Evaluation Answers

Chapter 1
Extracting Metals from Ores

1. Iron ore is essentially iron oxide. When the oxygen is removed, the iron remains. This is done by burning carbon (coke) in the presence of the ore. The carbon monoxide that is formed combines with oxygen in the ore to become carbon dioxide.
2. No. It is either used directly for steel making or for producing refined cast irons.
3. Iron ore, coke, and limestone.
4. No. The carbon content of pig iron (3 to 4.5 percent) renders it too weak and brittle for many uses. Steel with its lower carbon content (0.05 to 2 percent) is more ductile and has a higher tensile strength.
5. Ore dressing concentrates the ore so that a higher percentage of iron is shipped to the smelter.
6. Basically, the carbon must be removed from the pig iron, and also such unwanted elements as sulfur and phosphorus. Some sulfur may be retained by the addition of manganese, which forms the compound, manganese sulfide, which is not so undesirable in the steel as iron sulfide. Free-machining (resulfurized) steels contain manganese sulfide.
7. The electric furnace can produce a highly controlled, high-grade steel product. In areas having little or no pig iron production, but with considerable scrap available, the electric furnace can operate quite well, especially if cheap hydroelectric power is available.
8. "Killed" steel is deoxidized in the furnace, giving it more uniformity and less porosity than "rimmed" steels.
9. These metals tend to gather at the grain boundaries of the steel, making it hot short, that is, tending to split while being hot rolled or forged.
10. Oil, antifreeze, freon, and other soil and atmospheric contaminants must now be disposed of safely.
11. A roasting process. Matte.
12. By the electrolytic process.
13. Bauxite.
14. Cryolite.
15. The oxygen is released as a gas and the metallic aluminum sinks to the bottom of the bath where it is siphoned off and cast into ingots.

Chapter 2
Casting Processes

1. The two types of sand castings are green sand molding and dry sand molding. Moist sand with a small amount of clay and other additives make up the green sand mold. Dry molding sand, often used for making cores, has no clay but contains linseed oil or organic resins.
2. Wood, metal, and wax are common materials used for making patterns. Patterns for sand molding are made of wood or metal and must be tapered so they can be easily removed from the mold. They also must have shrinkage allowance. Patterns may be either simple cope and drag or match plate types.
3. Core sand is rammed into a "core box" or mold and the green cores are baked in an oven to harden them. They are set into the pattern and locked into the sand mold by the core print. The core remains in the mold when the pattern is removed.
4. A "chill" is usually a metal section that is embedded in the mold to increase or control the cooling rate of the casting.
5. The steps in the shell-molding process are:
   a. Heated metal patterns are brought into contact with a prepared sand mix in a dump box.
   b. The adhering sand and pattern are placed in an oven and heated to about 500°F for 2 minutes.
   c. The shell is removed and clamped to its mating half to form a complete mold.

   The sand is mixed with a phenolic resin binder and a silicon release agent is used on the pattern.
6. Molten metal is thrown out of the mold cavity by centrifugal forces when casting pipe, wheels, and other products by this process. The castings are more uniform

and of a higher quality than in gravity casting. Also, two different metals may be cast in layers. Centrifuging is the filling of a mold or molds located near the outside of a rotating body.

7. No. The pattern cannot be removed by separating mold halves as with sand casting since it is a one piece mold. The steps necessary to produce an investment casting are:
   a. The wax pattern is made, including the sprue and riser.
   b. Plaster is cast around the wax or a refractory slurry is used. The pattern is then dipped into the slurry and dried repeatedly until a thick shell has formed on the pattern.
   c. The mold is heated to drive out the wax and the shell is fired to harden it.
   d. The molten metal is poured and, when it has solidified, the mold shell is broken off the casting.
8. Higher production rates and higher precision are possible with permanent molds as compared to sand casting. One disadvantage of permanent molding is that it is suitable for only limited production runs, as the mold deteriorates after a few thousand castings.
9. The molten metal is not poured into the mold as in permanent molding but is injected under very high pressures into the mold cavity.
10. The cold-chamber machine has a separate melting and holding furnace. The hot-chamber machine is characterized by the plunger and gooseneck in the melting pot located on the machine. Two advantages of die casting are:
    a. High production rates are made possible by this process.
    b. High finish and precision are obtained in the finished product.
11. The casting can deteriorate at the grain boundaries which is called intergranular corrosion.
12. By breaking them off. They cannot be reused.

Chapter 3
The Physical and Mechanical Properties of Metals

1. Creep occurs within the elastic range of metals.
2. Creep failures occur over a period of time; the rate of creep increases as the temperature increases.
3. A metal that is ductile above its transition temperature behaves as a brittle metal below the transition temperature and loses much of its toughness.
4. Hardness, strength, and modulus of elasticity increase with a decrease in temperature.
5. Nickel will lower the transition temperature when alloyed with steel.
6. Resistance to penetration may be measured by the Rockwell or Brinell testers; elastic hardness may be measured with a scleroscope and abrasive resistance may be tested in the shop to some degree with a file.
7. Tensile, compressive, and shear.
8. Unit stress $= \frac{\text{load}}{\text{area}} = \frac{40{,}000}{4} = 10{,}000$ psi
9. Ductility is the ability of a metal to deform permanently under a tensile load.
10. Malleability is the ability of a metal to deform permanently under a compressive load.
11. Fatigue strength can be improved by eliminating sharp undercuts, deep tool marks, and other forms of stress concentration.
12. Electrical and thermal conductivity are related. A metal that is a good conductor of electricity is also a good conductor of heat.
13. A pure metal (unalloyed) will conduct best.
14. The rate of thermal expansion is expressed in inches (of dimensional change) per inch (of material length or diameter) per degree Fahrenheit. The value for each material is called the coefficient of thermal expansion.
15. By using an extremely rapid cooling rate.
16. They become superconductors, having no resistance to the passage of electricity and they are repelled by a magnetic field.

Chapter 4
The Crystalline Structure of Metals and Phase Diagrams

1. Atoms are composed of a nucleus consisting of positively charged protons and neutrons (that have no charge). Electrons are negative in charge and revolve in paths called shells. The outer shell or valence electrons are important in determining chemical and physical properties.
2. The electron cloud of free valence electrons creates a mutual attraction for the metal atoms. This free movement accounts for their high electrical and thermal conductivity, plasticity, and elasticity.
3. a. Body-centered cubic (BCC).
   b. Face-centered cubic (FCC).
   c. Hexagonal close-packaged (HCP).
   d. Cubic.
   e. Body-centered tetragonal.
   f. Rhombohedral.
4. Dendrite.
5. No. Each grain grows from its own nucleus in independent orientation; this means that the lattice structure of adjacent grains jams together in a misfit pattern.
6. Check Figure 4 in this chapter for the correct answers.
7. It is a short horizontal section on the line.
8. Lowest transition temperature. Eutectic is the lowest melting temperature. Eutectoid is the lowest transition temperature from one solid phase to another.
9. A_3 shows the beginning of transition from austenite to ferrite on cooling. A_1 shows the completion of austenite

transition to ferrite and pearlite. A_{cm} shows the limit of carbon solubility in austenite.

10. Cementite. It appears dark when in the form of lamellar pearlite but white when massive, such as in grain boundaries or in white cast iron.

Chapter 5
Identification and Selection of Iron Alloys

1. Carbon and alloy steels are designated by the numerical SAE or AISI system.
2. The four basic types of stainless steels are: martensitic (hardenable) and ferritic (nonhardenable), both magnetic and of the 400 series, austenitic (nonmagnetic and nonhardenable, except by work hardening) of the 300 series, and precipitation (hardenable).
3. The identification for each piece would be as follows:
   a. AISI C1020 (CF) is a soft, low carbon steel with a dull metallic luster surface finish. Use the observation test, spark test, and file test for hardness.
   b. AISI B1140 (G and P) is a medium carbon, resulfurized, free-machining steel with a shiny finish. Use the observation test, spark test, and machinability test.
   c. AISI C4140 (G and P) is a chromium-molybdenum alloy, medium carbon content with a polished, shiny finish. Since an alloy steel would be harder than a similar carbon or low carbon content steel, a hardness test should be used, such as the file or scratch test to compare with known samples. The machinability test would be useful as a comparison test.
   d. AISI 8620 (HR) is a tough low carbon steel used for carburizing purposes. The hardness and machinability tests will immediately show the difference from low carbon hot-rolled steel.
   e. AISI B1140 (Ebony) is the same as the resulfurized steel in (b), only the finish is different. The test would be the same as for (b).
   f. AISI C1040 is a medium carbon steel. The spark test would be useful here as well as the hardness and machinability test.
4. A magnetic test can quickly determine whether it is a ferrous metal or perhaps nickel. A chemical test can tell you if it is a stainless steel. If the metal is white in color, a spark test will be needed to determine whether it is nickel casting or one of white cast iron since they are similar in appearance. If a small piece can be broken off, the fracture will show whether it is white or gray cast iron. Gray cast iron will leave a black smudge on the finger. If it is cast steel, it will be more ductile than cast iron and a spark test should reveal a smaller carbon content.
5. O1 refers to an alloy type of oil-hardening (oil quench) tool steel. W1 refers to a water-hardening (water quench) tool steel.
6. a. No.
   b. Hardened tool steel or case-hardened steel.
7. The martensitic and ferritic types are attracted to a magnet while the austenitic types are not.
8. Nickel is a nonferrous metal that has magnetic properties. Some alloy combinations of nonferrous metals make strong permanent magnets, for example, the well-known Alnico magnet, an alloy of aluminum, nickel, and cobalt.
9. Some properties of steel to be kept in mind when ordering or planning for a job would be:
   Strength.
   Machinability.
   Hardenability.
   Weldability (if welding is involved).
   Fatigue resistance.
   Corrosion resistance (especially if the piece is to be exposed to a corrosive atmosphere).
10. Chromium. Nickel.
11. Cast iron is quite brittle and can be bent without breaking only to a very limited extent. Mild steel can undergo considerable distortion and bending without breaking.

Chapter 6
The Manufacturing of Steel Products

1. The ingot is heated to bring it up to rolling temperature, 2200°F, and to give it a uniform temperature throughout its mass.
2. Rolling breaks down the coarse, weak columnar grains in the ingot and they reform into smaller but even grains. This recrystallizing process makes the steel stronger.
3. Mild steel bar contains up to 0.20 percent carbon.
4. Cold-finished steel is stronger than equivalent hot-rolled steel because the grains are permanently deformed and elongated in the direction of rolling. If it is in an annealed condition, the grains are restored to their former state and the metal is no longer much stronger than hot-rolled steel.
5. Hot-rolled steel is covered with black mill scale. Cold-finished steels typically have a bright metallic finish.
6. Hot-formed metals tend to take on a fibrous quality like wood grain. When a tool or a machine part is forged, this grain is shaped with the irregularities instead of being cut through as with machining.
7. Seamless pipe is made from a single billet that is pierced so that it is all one piece with no welds. Butt-welded pipe is made from steel strip and the edges are joined by pressure welding all along its side.
8. Small pipes and tubes are rolled from strip and electric resistance welded. This material does not have to be heated first as with butt-welded pipe. Automatic fusion welding joins pipes of large diameter.
9. The rough cuts should all be made first because of the residual stresses in these products. Forged and hot-rolled bars tend to warp when these stresses are removed by machining processes, and if a finish cut is taken first, it

will warp out of true position and cannot be corrected by more cutting.

10. Steel rod is drawn through a series of increasingly smaller dies. Occasionally this process must be interrupted to anneal the wire before drawing can continue.
11. Rolled steel develops a fibrous quality in its grain structure that is elongated in the direction of rolling. The metal is more likely to crack from bending in one axis than the other.
12. Cold working causes the metal to become stiffer and stronger while adding some hardness.

Chapter 8
I-T/T-T-T Diagrams and Cooling Curves

1. About 50°F above the A_3, or $A_{3,1}$ lines.
2. Martensite is produced by rapid quenching from the austenitizing temperature to the Mf or near room temperature. Time is a major consideration.
3. The Ms temperature is the point at which martensite begins to form, and the Mf is the point where it is at 100 percent transformation.
4. The critical cooling rate is the time necessary to undercool austenite below the M temperature to avoid any transformation occurring at the nose of the S-curve.
5. The microstructure should be partly fine pearlite and partly martensite.
6. An increase in carbon content moves the nose of the S curve to the right.
7. Hardening and tempering.
8. This could be caused by the difference in internal and external cooling rates. Changing to an oil- or air-hardening steel may correct the problem.
9. It can be observed that a cooling curve will always cut through the nose of the S-curve no matter how fast the part is quenched. It is therefore evident that little or no martensite may be produced.
10. A steel that is hardenable and is deep hardening shows an S-curve that is moved to the right.
11. When a part is quenched in oil or air, the cooling rate is slower.
12. The interior and exterior cooling rates are almost the same when low cooling rates are used. When they are drastically different, as in water or brine quenching, cracking and warping of the part can be the result.

Chapter 9
Heat Treating Equipment

1. Electric, gas, oil fired, and pot furnaces.
2. The surface decarburizes or loses surface carbon to the atmosphere as it combines with oxygen to form carbon dioxide. An oxide scale forms on the surface.
3. Dispersion of carbon atoms in the solid solution of austenite may be incomplete and little or no hardening in the quench takes place as a result. Also, the center of a thick section takes more time to come to the austenitizing temperature than does a thin section.
4. Circulation or agitation breaks down the vapor barrier. This action allows the quench to proceed at a more rapid rate and avoids spotty hardening.
5. By furnace.
6. They run from the surface toward the center of the piece. The fractured surfaces usually appear blackened. The surfaces have a fine crystalline structure.
7. a. Overheating.
   b. Wrong quench.
   c. Wrong selection of steel.
   d. Poor design.
   e. Time delays between quench and tempering.
   f. Wrong angle into the quench.
8. a. Controlled atmosphere furnace.
   b. Wrapping the piece in stainless steel foil.
   c. Covering with cast iron chips.
9. a. Tungsten high speed.
   b. Molybdenum high speed.
10. An air-hardening tool steel should be used when distortions must be kept to a minimum.
11. Since the low carbon steel core does not harden when quenched from 1650°F, it remains soft and tough, but the case becomes very hard. No tempering is therefore required as the piece is not brittle all the way through as a fully hardened carbon steel piece would be.
12. A deep case can be made by pack carburizing or by a liquid bath carburizing. A relatively deep case is often applied by nitriding or by similar procedures.
13. No. The base material must contain sufficient carbon to harden by itself without adding more for surface hardening.
14. Three methods of introducing carbon into heated steel are roll, pack, and liquid carburizing.
15. Nitriding.
16. The surface markings are an indication that the grain boundaries have been damaged by overheating and the part cannot be salvaged by any heat treatments.
17. Cold tongs can reduce the surface temperature of the part where they come in contact with it. The steel may not harden in that place.

Chapter 10
Annealing, Stress Relieving, and Normalizing

1. Medium carbon steels that are not uniform, have hardened areas from welding, or prior heat treating need to be normalized before they can be machined. Forgings, castings, and tool steel in the as-rolled condition are normalized before any further heat treatments or machining is done.

2. 1650°F. 100°F above the upper critical limit.
3. The spheroidization temperature is quite close to the lower critical temperature line, about 1300°F.
4. The full anneal brings carbon steel to its softest condition as all the grains are reformed (recrystallized), and any hard carbide structures become soft pearlite as it slowly cools. Stress relieving will only recrystallize distorted ferrite grains and not the hard carbide structures or pearlite grains.
5. Stress relieving should be used on severely cold-worked steels or for weldments.
6. High carbon steels (0.8 to 1.7 percent carbon).
7. Process annealing is used by the sheet and wire industry and is essentially the same as stress relieving.
8. In still air.
9. Very slowly. Packed in insulating material or cooled in a furnace.
10. Since low carbon steels tend to become gummy when spheroidized, the machinability is poorer than in the as-rolled condition. Spheroidization sometimes is desirable when stress relieving weldments on low carbon steels.
11. Recrystallization.
12. The grains will become larger through a phenomenon called grain growth.

Chapter 11
Hardening and Tempering of Steel

1. No hardening would result since 1200°F is less than the lower critical point and no dissolving of carbon has taken place.
2. There would be almost no change. For all practical purposes in the shop these low carbon steels are not considered hardenable.
3. They are shallow hardening and liable to distortion and quench cracking because of the severity of the water quench.
4. Air- and oil-hardening steels are not so subject to distortion and cracking as W1 steels and are deep hardening.
5. 1450°F, 50°F above the upper critical limit.
6. Tempering is done to remove the internal stresses in hard martensite, which is very brittle. The temperature gives the best compromise between hardness and toughness or ductility.
7. Tempering temperature should be specified according to the hardness, strength, and ductility desired. Mechanical properties charts give this data.
8. 525°F. Purple.
9. 600°F. It would be too soft for any cutting tool.
10. Immediately. If you let it set for any length of time, it may crack from internal stresses.
11. Mass of the part, severity of the quench, and hardenability of the material.

Chapter 12
Hardenability of Steels

1. The jominy end-quench hardenability test.
2. A 1 in.-dia specimen about 4 in. long is heated to the quenching temperature and placed in a jet of water so that only one end is cooled without getting the sides wet. When the whole specimen has cooled, flat surfaces are ground on the sides and Rockwell C-scale readings are taken at 1/16 in. intervals.
3. The rate of cooling and hence the microstructure as shown on the I-T diagram are related to the depths of hardening as shown by the jominy end-quench graph.
4. Circulation of the cooling medium in all cases increases depth of hardening.
5. Rc67 is about as hard as any carbon steel will get. Steels with less than 0.83 percent carbon will not get this hard.
6. Coarse pearlite.
7. Austempering is isothermal quenching in the lower bainite region of the S-curve. Since no martensite is formed, a tougher, more ductile microstructure is the result; it is superior to a quenched and tempered product of the same hardness.
8. The best time to temper is immediately after quenching, as soon as the piece is cool enough to be hand-held.
9. The blue brittle tempering range is found between 400°F and 800°F. Only some alloy steels show loss of notch toughness at higher tempering ranges. This is called temper brittleness. It can be avoided by quenching from the tempering temperature.
10. You can predict the hardness that a tempered part will be by consulting a mechanical properties chart for that particular grade of steel.
11. Hardenability (depth hardness) depends on the amount of carbon present, the alloy content, and the grain size.
12. Some tool steels take a longer soaking time than others to effect complete austenitization. If a certain steel is soaked for an insufficient period of time for its specs, it will not get hard when quenched.

Chapter 13
Welding Processes for Iron and Iron Alloys

Part I

1. The weld zone bears a similarity to the ingot as it solidifies into columnar structures. It also compares to a small electric melting furnace. Peening of welds is like hot and cold rolling of metals.
2. Solid phase welding, fusion joining, and liquid-solid phase welding.
3. Diffusion of some of the carbon into the weld melt will take place, causing hard carbides to form when cooling rates are rapid.
4. The weld zone, fusion zone, and heat-affected zone.

5. In the weld zone.
6. Oxyacetylene welding.
7. Martensite or hard carbides can form with high cooling rates in welds.
8. The heating action of subsequent passes tends to normalize and soften the hard zones in the previous pass.
9. Cracking, porosity, and embrittlement (from entrapped gases or formation of martensite).
10. By limiting or inhibiting the formation of martensite.
11. By removing them through tempering, spheroidizing, or annealing.
12. The tough, hard, distorted grains in cold-rolled steel are recrystallized in the heat-affected zone to a softer, larger grain that is not so strong as it was in the original stressed condition.
13. Large masses cause more rapid cooling rates.
14. To protect the weld zone from contaminators of the atmosphere (oxygen and nitrogen), both of which weaken the weld, causing porosity and hard zones.
15. The weld.
16. Lack of cleaning and insufficient preheating of the base metal.

Part II

1. The hardenability of steels, and consequently of welds, and of heat-affected zones is increased with the addition of alloying elements.
2. High postheat (tempering) temperatures tend to promote decomposition of the retained austenite into fine carbides, thus increasing the hardness and brittleness of the weld or heat-affected zone.
3. Complex carbides tend to form, causing brittleness and weakening of the steel in the grain boundaries of the large austenite grains.
4. Vanadium.
5. The brittleness in the HAZ of martensitic stainless steel is caused by the formation of carbides when rapid cooling takes place, but excessive grain growth in the HAZ of the ferritic stainless steel is the cause of brittleness.
6. a. Preheat and postheat are usually needed since the HAZ of tool steel welds tends to become brittle.
   b. The correct filler metal that is similar to the base metal in carbon and alloy content should be used.
   c. Full annealing is necessary after welding if any machining is to be done. This is then followed by hardening and tempering procedures.
7. Standard ASTM specifications.
8. Hot cracks are usually located lengthwise down the center of a weld or in the crater at the end of a weld. Hot cracking can be caused by a too rapid cooling rate or by stresses in the weld zone at high temperatures.
9. Carbon can be picked up in the root pass from oil or grease, causing embrittlement and cracking. All of these materials can cause hydrogen embrittlement in welds.
10. The overlay tends to spall (break off in chunks) when this procedure is followed.
11. No. Prolonged overheating of these steels can alter their microstructures and seriously weaken them and often cause thermal cracking.
12. When a base metal contains medium or high carbon and other alloying elements, and the filler rod is mild steel, a mixing and combining takes place where these elements are picked up and added to the weld melt.

Part III

1. No. All except white cast iron.
2. Repair of casting defects in foundries and of worn or broken castings.
3. They contain too much carbon and silicon.
4. Almost all the welding processes can be used.
5. Gray cast iron has a lower tensile strength than the weld metal (unless the weld is also cast iron) and, when the weld metal contracts while cooling, very high stresses are created in the base metal.
6. The major advantage is that preheat is not used where large castings are involved. The major disadvantage is the weakness of such welds due to microcracking near the junction zone of the weld.
7. Preheating and postheating can eliminate the hard fusion zone. It can be avoided by removing sufficient base metal prior to welding so that machining is done only on the weld itself.
8. The graphite in gray cast iron can be smeared over the surface to be welded by grinding procedures. Weld grooves should be prepared by chipping or sawing.
9. By studding procedures.
10. Gas welding with a cast iron rod raises the malleable iron above the transformation temperature and causes it to revert to white cast iron.
11. Cast iron is somewhat porous and often contains impurities such as oil and water that ruin the weld. These are driven off by preheating. Also preheat makes for a slower cooling rate in the weld which discourages the formation of hard, brittle structure.
12. The casting should be very slowly cooled for several hours or overnight.

Chapter 14
Identification of Nonferrous Metals

1. Advantages: Since aluminum is about one-third lighter than steel, it is used extensively in aircraft. It also forms an oxide on the surface that resists further corrosion.

   Disadvantages: The initial cost is much greater. Higher strength aluminum alloys cannot be welded.

2. The letter "H" following the four-digit number always designates strain on work hardening. The letter "T" refers to heat treatment.
3. Magnesium weighs approximately one-third less than aluminum and is approximately one-quarter the weight of steel. Magnesium will burn in air when finely divided.
4. Copper is most extensively used in the electrical industry because of its low resistance to the passage of current when it is unalloyed with other metals. Copper can be strain hardened or work hardened and certain alloys may be hardened by a solution heat treatment and aging process.
5. Bronze is basically copper and tin. Brass is basically copper and zinc.
6. Nickel is used to electroplate surfaces of metals for corrosion resistance and as an alloying element with steels and nonferrous metals.
7. All three resist deterioration from corrosion.
8. Alloy.
9. Tin, lead, and cadmium.
10. Die cast metals, sometimes called "pot metal."
11. Wrought aluminum is stronger.
12. Large rake angles (12 to 20 degrees back rake). Use of a lubricant. Proper cutting speeds.
13. No. If a fire should start in the chips, the water-based coolant will intensify the burning.
14. Oxygen-free copper is used almost exclusively for these products.
15. Nichrome or chromel.

Chapter 15
Heat Treating of Nonferrous Metals

1. a. 5
   b. 2
   c. 4
   d. 3
   e. 1
2. The solid solution becomes supersaturated and the copper forms into highly dispersed globules of copper aluminide.
3. Solution heat treatment and quench and precipitation heat treatment or aging.
4. Aging causes the copper aluminide particles to act as keys to lock up the slip planes. This lowers the ductility and raises the strength.
5. Pure aluminum is more corrosion resistant than alloyed aluminum; for this reason, a thin layer of pure aluminum is clad to the sheet of alloy.
6. The aluminum is simply heated to about 650°F (343°C) and allowed to cool in air.
7. Nothing. The metal has been ruined and cannot be salvaged.
8. By the process of artificial aging; that is, by heating the quenched part for several hours.
9. A muffle (carbonizing atmosphere) furnace or bright annealing can be used to avoid oxide scale.
10. Alloying elements tend to cause titanium to become HCP (alpha) or BCC (beta). Some alloys are alpha-beta and are highly heat treatable.

Chapter 16
Metallurgy of Welds: Nonferrous Metals

1. It is their great affinity for oxygen that causes the rapid formation of oxides.
2. Because of their high thermal conductivity.
3. Aluminum should be cleaned prior to welding for two reasons: to remove the oxide film and to remove contaminants.
4. Recrystallization and grain growth is increased by strain hardening. These larger, more ductile grains cause the heat-affected zone to have a lower tensile strength than the rest of the base metal.
5. The alloying elements (such as copper in 2024) tend to form compounds and precipitate to the grain boundaries due to the heat of welding. This, of course, produces an exceedingly weak, brittle weld that can be restored to the original strength only by postweld solution heat treatment and aging the entire weldment.
6. Oxygen in copper welds tends to form oxides that migrate to the grain boundaries, thus severely weakening the metal.
7. Quenching is not necessary in these alloys. All that is needed is a simple annealing procedure from 1100 to 1600°F. This step will harden the weldment uniformly and provide a higher corrosion resistance.
8. Nickel alloy electrodes such as monel are used; in addition, austenitic stainless steel, which contains high percentages of nickel, is used for joining dissimilar metals.
9. These metals react violently with oxygen and readily burn to form oxides. They can be welded either in a vacuum chamber or by the TIG process by using an inert gas.
10. By keeping the weld area very clean, avoiding oils and other contaminants; even fingerprints can cause embrittlement.
11. Lead and zinc are low temperature melting metals and tend to volatilize before the copper melts. This causes the alloy to become weak at welding temperatures, causing weld cracking.
12. By capillary action.

Chapter 17
Powder Metallurgy

1. The automotive industry.
2. Porous products such as filters and self-lubricating bearings, friction materials for brake linings, cutting tools,

creep resistant alloys for jet engines, gears, cams, magnets, and metals that are difficult to process, such as tungsten.

3. a. The powder is pressed into a green compact or briquette shape.
   b. The briquette is placed in an oven and sintered at a temperature just under its melting point.
   c. When more precision is required, a resizing operation in a press is carried out.
4. The green compact is "cold welded" by pressure alone, but the bond is very weak. The sintering process forms metallurgical bonds similar to those of fusion welding.
5. Some advantages: small parts can be economically produced by P/M with no scrap loss. Porous materials and metal-nonmetal combinations are possible with this method. All the normal machining, plating, and heat-treating processes can be used on P/M metal. Some disadvantages: P/M parts have lower resistance to corrosion than conventionally formed metals.
6. Density is increased over that of cold compacting and ductility approaches that of rolled metals.
7. Simple die compacting.
8. Porous filters and prelubricated sleeve bearings.
9. 2000 parts per hour.
10. Small parts.

Chapter 18
Precious Metal Processing

1. Precious metals are weighed in troy ounces using an analytical balance.
2. The platinum group metals are platinum (Pt), palladium (Pd), rhodium (Rh), ruthenium (Ru), iridium (Ir), and osmium (Os).
3. Precious metals such as gold and silver are hardened by alloying.
4. Mercury dissolves gold in the ore. When the mercury evaporates, gold dust remains.
5. Mercury has a very low vapor pressure; the fumes are toxic.
6. A sluice box is used to extract gold particles and heavy ore as they pass over the riffles.
7. 41.7% gold is found in a 10k ring.
8. Fineness defines the purity of the metal.
9. Gold alloyed with nickel and copper form white gold. Zinc from 5 to 12 percent is added to modern white gold.
10. A troy ounce is the weight system for precious metals when sold or traded in the market.

Chapter 19
Corrosion of Metals

1. Corrosion may be classified as direct oxidation and galvanic action.
2. An electrochemical process in which the metal is changed to the ionic state in an electrolyte.
3. Metals usually become positive ions while nonmetals become negative ions.
4. For galvanic corrosion, there must be an anode and a cathode electrically connected, and an electrolyte. The anode and cathode may be similar or dissimilar metals and an electric current must flow.
5. In more neutral electrolytes, an accumulation of hydrogen gas is formed on the cathode surface that prevents or slows down corrosion.
6. Oxygen reacts with the layer of hydrogen on the cathode to form water, accepting negative charges of electrons that allow the release of metal ions from the anode and the resultant corrosion.
7. Iron.
8. The current density is greater with the large cathode/small anode couple because of the relatively larger area for depolarization that is available to any oxygen atoms.
9. Metals are least likely to corrode when they are in their pure state or have few contaminants.
10. Anything that can prevent the normally present oxygen from contacting the metal will set up an anode area surrounded by the high oxygen concentration, which is the cathode. The anode becomes activated (stainless steel is normally passivated and cathodic) and is corroded in time.
11. Sacrificial zinc and magnesium anodes are bolted to the ship's hull and replaced when they have deteriorated. Both the bronze propeller and the steel hull become cathodic and the zinc becomes anodic.
12. It protects the steel from rusting and it is also used to increase wear resistance of tools by implanting nitrogen ions into the crystal lattice.

Chapter 20
Nondestructive Testing

1. Nondestructive testing is used to inspect for cracks or other defects in various metal and nonmetal parts without damaging the part for its intended use.
2. Magnetic-particle inspection can only be used for ferromagnetic metals such as iron and steel.
3. The specimen must be magnetized lengthwise in order to find a crosswise crack. A circular magnetic field is needed to find a lengthwise crack.
4. Fluorescent penetrant systems make use of capillary action to draw the penetrating solution into the flaws. The surplus penetrant is removed and a developer is applied to enhance the outline of the defect. The inspection is carried out under black light, which causes the penetrant to fluoresce. This method may be used on almost any dense material.
5. Dye penetrants may be used in remote areas where no power is available for a black light source. Portability, low cost, and ease of application are also advantages.

6. The pulse-echo system and the through transmission system. The pulse-echo system uses one transducer that both sends and receives the pulse. The pulse is reflected and echoed back from the other side or from a flaw. The through transmission system uses a transducer on each side of the test piece. The signal passes through the material and is modified by any flaws present.
7. Internal defects, cracks, porosity, laminations, thicknesses, and weld bonds can be detected with ultrasonic testing systems.
8. X rays have the ability to pass through solids. A solid steel casting, for example, will have an X ray source on one side of the area to be inspected and a photographic plate on the other. A void or crack will appear darker on the negative.
9. Gamma radiation is omnidirectional—that is, it goes in all directions. Therefore, a hollow object may be inspected on its entire surface with one exposure. Radium or another radioactive substance is used to produce gamma rays, and extreme care must be taken with its use.
10. Eddy current techniques are used for sorting materials of various alloy compositions and with differing heat treatments. It can be used for detecting seams or variations in thicknesses, mass, and shape. It is limited to testing only materials that conduct electricity.
11. No. It is also used on plastics composites.
12. Any mechanism can fail. However, some equipment, such as farm machinery, can fail without any great danger to humans as compared to aircraft equipment and therefore need not be periodically tested.

Chapter 21
Composite Materials

1. A composite material is a combination of high strength fibers or particles alloyed or combined usually in lightweight metal.
2. Fiberglass, components of electrical contact switches.
3. Tensile strength, shear strength and hardness.
4. Mechanical and chemical bonding.
5. The abrasion resistance of the hard fibers is much higher than the base metal. The major problem is to get the fiber and the base metal to the same topographic level.
6. A composite containing very fine needle-like particles of usually a hard, strong metal or metalloid.

Chapter 22
Plastics and Elastomers

1. High temperatures and ultraviolet light from sunlight.
2. Thermosetting and thermoplastic.
3. Thermoplastic.
4. Latex from plants such as the pararubber tree is the source of natural rubber.
5. Petrochemicals and acetylene.
6. Neoprene.
7. Ethyl alcohol, tetraethyl lead, ethylene glycol, styrene, polystyrene, polyethylene, ethyl benzene, and nylon.
8. Plastics and other nonmetals are poor conductors of heat, and the temperature rise at the cutting surface overheats the tool, causing failure.
9. Plastics can be joined by using adhesives, with heat, and by ultrasonics.
10. Polymer is a long-chain molecule made up of monomers.
11. Because of their high strength and stability.
12. No. Polystyrene is a cheap plastic having very low strength and is used for disposable products.
13. A composite is a polymer that has been greatly strengthened with a reinforcing material such as glass fiber.
14. It produces foamed plastic.
15. Blow molding is a method of producing hollow plastic products such as bottles.
16. Natural rubber. However, all rubbers will deteriorate to some extent when subjected to ultraviolet light in sunlight for a long period of time.

Chapter 23
Ceramic Materials

1. They usually have ionic or covalent bonds, which are rigid, and have few slip planes to absorb local stresses, which may exceed the bond strength and cause a brittle failure.
2. Ceramic materials have high compression strength but low tensile strength.
3. A refractory material such as clay brick can withstand high temperatures without breaking down or melting. Refractories are used to contain molten metals and for furnace linings.
4. Silica and alumina.
5. False.
6. False.
7. True.
8. The beams are prestressed with steel wires, which impart a higher tensile strength to the concrete.
9. Glass is made by combining molten silica (quartz), lime, and soda. This combination solidifies into the noncrystalline (amorphous) structure of glass.
10. A composite is a material that is combined with another, different material to reinforce or strengthen the resultant mixture.
11. The process is called hydration. It begins with a silica gel that forms on the tiny cement particles, from which fibers begin to grow outward to join with others on adjacent cement particles to form a hard, interlocked mass.
12. Polycrystalline ceramic fibers are used to strengthen metals such as aluminum. Silicon carbide and silicon nitride are being used in gas turbine research and automotive engine parts. Missile radomes, cutting tools, and

wire dies are also applications of these structural ceramic materials.

Chapter 24
Wood and Paper Products

1. The softwoods (coniferous) and the hardwoods (deciduous) are the two classes.
2. Pines, firs, spruces, and hemlock are coniferous trees.
3. Growth takes place in the cambium layer.
4. The age of a tree can be determined by counting its annular growth rings.
5. The log is first sawed into rough slabs in a large circular or band saw called a "head rig." The slabs are then sawed into lumber, which is graded and then dried. The dry boards may then be planed to provide a finished surface.
6. Sawdust and wood chips are by-products that are used for making hardboard, particle board, and fireplace logs.
7. The grains are all in one direction in wood, making it prone to splitting. Wood is simply weak in one axis and strong in the other. Plywood utilizes this strength by crossing each layer or ply alternately and gluing them together.
8. Wood is not a uniform material, having defects such as knots, cracks, pitch pockets, and rot. Boards must be selected according to their intended use.
9. They are made of sawdust and wood chips. They are used for panels in construction, furniture making, and pegboard.
10. A pulp is made by grinding the fibers and soaking them in a vat. The pulp is formed into a mat that is pressed, dried, and squeezed to form a hard, dense sheet.
11. The top and bottom boards where the stress is greatest on a laminated beam are made of clear, selected lumber, while sawed timber may have imperfections such as knots, cracks, and rot in those locations.
12. No. The lifeless core is also the useful part of the tree.

Chapter 25
Adhesives and Industrial Lubricants and Gases

1. The materials used were hides and hooves of animals, casein, starch, and tree gums.
2. Bonded structures resist vibration damage better and are less likely to have corrosion.
3. It should be thoroughly cleaned by sand blasting, by etching, or with a degreaser, and all moisture removed.
4. The bond in these adhesives is catalyzed by the contacting surfaces; the adhesive is anaerobic and will not cure in the presence of air or if it is over 0.002 in. thick.
5. The polysulfides, polyurethanes, and silicones are used as sealants.
6. Hydrogen, acetylene, natural gas, propane, and butane are fuel gases, and nitrogen, helium, and argon are inert gases.
7. Boundary layer or adhesive film lubrication such as the hypoid oils and greases.
8. Liquid asphalts are diluted with solvents or by emulsifying them with water.
9. Cellulosics are not strong adhesives and would fail in use.
10. In order to bond, the adhesive must be within 50 or 60 angstroms of the substrate atoms.
11. Adhesive bonding resists vibration and cracking better than rivets or bolts and reduces corrosion.
12. Epoxy.

Chapter 26
Hardness Testing

1. Resistance to penetration is the one category that is utilized by the Rockwell and Brinell testers. The depth of penetration is measured when the major load is removed on the Rockwell tester and the diameter of the impression is measured to determine a Brinell hardness.
2. As the hardness of a metal increases, the strength increases.
3. The "A" scale and a brale marked "A" with a major load of 60 kgf should be used to test a tungsten carbide block.
4. It would become deformed or flattened and give an incorrect reading.
5. No. The brale used with the Rockwell superficial tester is always marked or prefixed with the letter "N."
6. False. The ball penetrator is the same for all the scales that use the same diameter ball.
7. The diamond spot anvil is used for superficial testing on the Rockwell tester. When used, it does not become indented, as is the case when using the spot anvil.
8. Roughness will give less accurate results than would a smooth surface.
9. The surface "skin" would be softer than the interior of the decarburized part.
10. A curved surface will give inaccurate readings.
11. A 3000 kg weight is used to test steel specimens on the Brinell tester.
12. A 10-mm steel ball is usually used on the Brinell tester.
13. The required steps are as follows:
    a. Place the calibration block into the instrument and locate the desired group.
    b. Measure the group to calibrate your eyes to determine variance.
    c. Perform a hardness test on the calibration block and measure the indentation.
    d. Compensate the measurement and record actual block hardness.
    e. Compare your data with the instrument calibration data sheet.
14. An operator's eyes change focus during the day. Thus, visual dimensional measurements made during the day

will change accordingly. Compensation for minor changes can affect the accuracy of a hardness test.

15. Actual hardness is measured hardness compensated by a variance adjustment.
16. Hardness test will not be repeatable.
17. The grain size of aluminum alloys is very large and the microstructure is very dirty, which prevents the taking of accurate data.
18. A brale indenter and a 150-kilogram load.
19. The ball is indented rather than the carbonized case. An accurate reading is impossible.
20. Standard Hardness Tester of a minor load is 10% of a major load. The minor load for a standard hardness tester is 10 kg. The standard load for a superficial hardness tester is 3 kg.
21. Any lateral load will cause a major fluctuation of the readout from the instrument. In order to reproduce hardness test data with any degree of accuracy, the surface being tested must be perpendicular to the indenter.
22. Indentations create irregularities on the surface, which will not allow proper support for the hardness test.

Chapter 27
Failure Analysis

1. No.
2. Intergranular cracking takes place along the grain boundaries and transgranular cracking is through the grain along slip planes.
3. Transgranular.
4. The plain carbon steel would be the most likely to fracture because it has a higher transition temperature than the nickel steel.
5. A stress raiser is a notch, sharp change in section, or even a scratch which concentrates stress at that point, often initiating a small crack that continues to deepen when stress reversals are present. The final result is total failure of the part.
6. Tensile stress.
7. Welding across a highly stressed part can cause a sudden brittle failure that can be catastrophic.
8. It will almost always initiate a metal fatigue crack and ultimate failure eventually.
9. Excessive grain growth can result from a hot heavy one-pass weld or an oxy-acetylene weld. These large grains in the heat affected zone are weak and brittle and even more so if considerable cold work has been done at that point.
10. The heat affected zone and the weld can both become brittle from the hydrogen in these substances. Even low-hydrogen welding rod can cause hydrogen embrittlement if the coating has absorbed moisture.
11. The two parts will gall and sieze; that is, be cold-welded together.
12. If the parts are made of hardened steel or a work hardening alloy, such as high manganese steel, the parts will not weld together.

Glossary

THE TERMINOLOGY OF METALS

To provide technical clarity and definition for commonly used metallurgical terms, an illustrative glossary section is presented. As with many technologies today, metallurgical processes are often explained with complicated sounding words and terms such as *microstructure* and *phase precipitation.* This glossary is intended to provide you with simple explanations and illustrative descriptions for metallurgical words and terms used in this book and, more importantly, in the everyday world of metals.

A

Acetylene (C_2H_2). A colorless, highly flammable gas with a characteristic odor, used in combination with oxygen in welding (oxy-acetylene) to produce a 6700°F flame.

Acicular. Needle-shape particles usually observed in microstructures of iron, titanium, and many other metals. Acicular particles are often distributed in composite material to increase hardness and wear resistance.

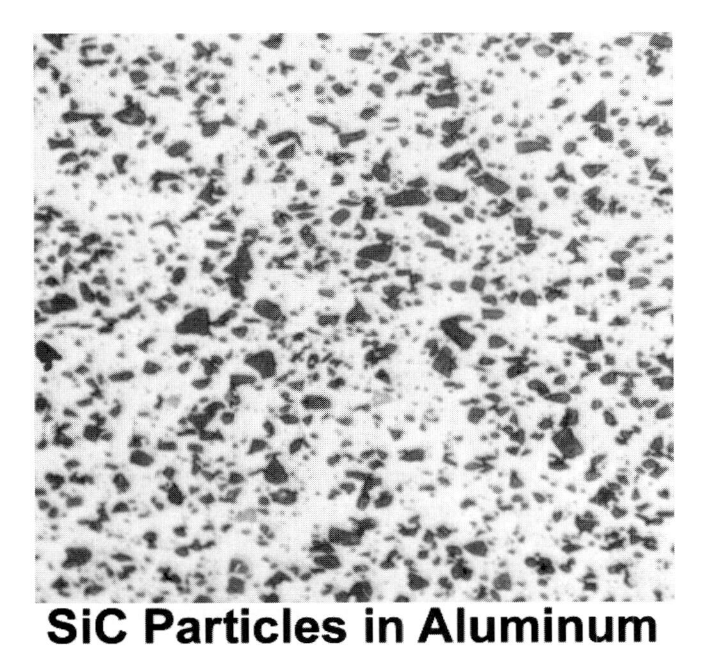

SiC Particles in Aluminum

Acids, bases, and salts. Acids are electrolytes that furnish hydrogen ions. Bases are electrolytes that furnish hydroxyl ions. Salts are electrolytes that furnish neither hydrogen nor hydroxyl ions. Salts are formed by a process called neutralization, the combination of equivalent weights of an acid and a base.

Activation. Changing a passive, nonreactive surface of a metal into a chemically active surface.

Adhesives. Materials or compositions that enable two surfaces to join together. An adhesive is not necessarily a glue, which is considered to be a sticky substance, since many adhesives are not sticky.

Admixture. In welding, the combining of a base metal with a filler metal to alter the characteristics of either.

Aggregate. Small particles as powders that are used for powder metallurgy and are loosely combined to form a whole; also sand and rock as used in concrete.

Aging. A phase precipitation process in metals which follows the laws of thermal solubility. Natural or room temperature aging occurs within a 1-week period following solution heat treatment in aluminum alloys and results in a marked increase in mechanical strength and properties. Artificial aging occurs in a much shorter time period by a controlled heating process dependent upon the type alloy. In aluminum castings, aging is used to promote dimensional stability.

AISI. Abbreviation for *American Iron & Steel Institute.*

Alclad. A thin layer of pure aluminum which is metallurgically bonded to the surface of an aluminum alloy of higher strength for the purpose of corrosion control. Aircraft usually have an alclad surface to prevent salt spray corrosion. Alclad surfaces also provide excellent bonding for paints.

Allotropy. The ability of a material to exist in several crystalline forms.

Alloy. A substance that has metallic properties and is composed of two or more chemical elements of which at least one is a metal.

Amorphous. Noncrystalline, a random orientation of the atomic structure.

Anisotropy. A material exhibiting different physical properties in different directions. Mechanical properties in a billet are usually greater in the rolling direction (longitudinal) than in the transverse direction or perpendicular to the rolling (short transverse and long transverse). Carbon is a great example of an anisotropic material.

Anneal. Soften by thermal processing. A heat treating process used to remove the effects of mechanical deformation and/or prior hardening heat treatment; the result of this process returns the metal to its softest condition and usually results in the formation of an equiaxed grain microstructure.

WORK-HARDENED

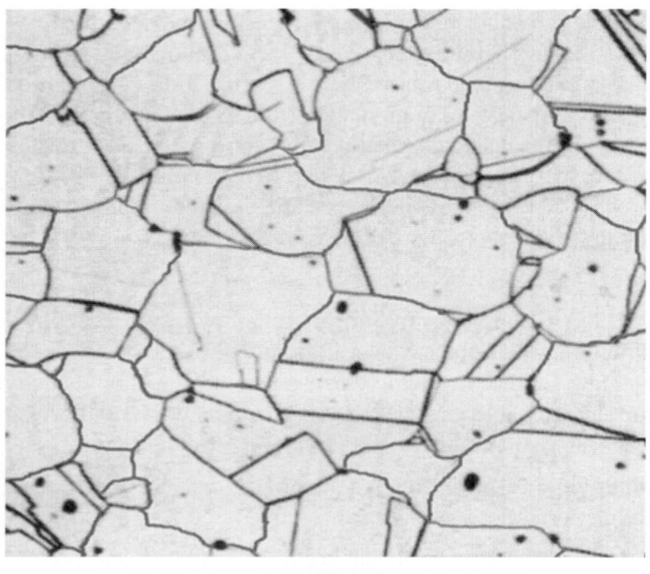

FULL-ANNEALED

Annodizing. The process of subjecting a metal to electrolytic action as the anode of a cell to create a thin layer of controlled oxide growth. The technical term for the oxide produced in boehmite. Annodized coatings are used as a corrosion protective labor and can be colored for decorative purposes.

ASTM. Abbreviation for American Society for Testing and Materials.

Ausforming. A heat treating process consisting of quenching a ferrous alloy from the austenitizing temperature into a salt bath at a temperature just above that which causes formation of martensite and below that which causes the formation of upper bainite. While at temperature, the subject part is forged or hammered, causing plastic deformation of the microstructure, which results in an extremely strong, hard, and highly strengthened bainite microstructure.

Austempering. A heat treating process consisting of quenching a ferrous alloy from the austenitizing temperature into a medium at a temperature just above that which causes formation of martensite and below that which causes formation of upper bainite. Temperature is maintained until the microstructure fully transforms to lower bainite. This microstructure is strong and springlike. Most shovels are heat treated to form a bainitic microstructure.

Austenite. The face-centered cubic form of iron created when pure iron is heated above 1666°F. Austenite is formed by allotropic transformation. Hardenable steels must be heat-treated to form austenite before they can be quench hardened.

Austenitizing. The process of forming austenite by heating a ferrous alloy above the transformation range.

AWS. Abbreviation for *American Welding Society.*

B

Bainite. A microstructure in steel named after E.C. Bain, which forms during controlled cooling (quenching) at a rate faster than required to form pearlite but above the temperature where martensite is formed. The temperature is held for a prescribed time to allow complete transformation of austenite to bainite.

Bainitic hardening. A quench-hardening heat treating process for steels which results principally in the formation of bainite.

Base metal. Also called *parent metal.*

Batch furnace. A furnace used to heat treat a single load at a time as opposed to a continuous feed heat treat furnace. Batch type furnaces are necessary for large parts such as heavy forgings and for heat treatment of fasteners.

Belt or continuous furnace. A continuous feed type furnace which uses a mesh type or cast link belt to carry the parts through the furnace.

Black oxide. A black surface finish put on ferrous metals by immersing into oxidizing salts or a salt solution to provide corrosion protection.

Bluing. A process by which a thin blue film of oxide is formed on the surface of a ferrous alloy for the purposes of improving appearance and corrosion resistance. Most gun barrells are blued.

Boriding. A thermochemical surface-hardening treatment for ferrous alloys which adds borides to the surface at a temperature below the Ac_1 line (lower critical temperature line).

Bright annealing. Annealing in a vacuum or a protective inert atmosphere which prevents surface contamination and/or oxidation discoloration.

Brine quenching. A quench in which brine (saltwater chlorides, carbonates, or cyanides) is used as a quench media.

Brittleness/brittle fracture. The property of materials which exhibit low fracture toughness and tend to shatter under impact loading. Brittleness is the property opposite to plasticity.

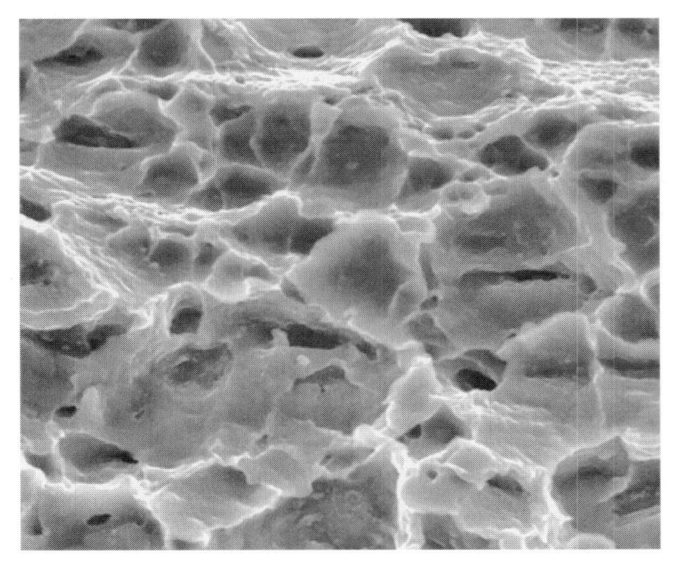

DUCTILE

BRITTLE

BUE. "Built-up edge"; a condition in which the metal being cut sticks to the cutting edge or point of the cutting tool. This condition causes tearing and a rough finish of the base metal. BUE is akin to cold welding.

C

Calorizing. Providing resistance to oxidation to a ferrous alloy by heating in aluminum powder at 1470° to 1830°F.

Carbonitriding. A subcritical temperature surface case hardening process in which both carbon and nitrogen are diffused into the metal. The process can be performed by immersion into proprietary salt baths (melonite) and/or by gaseous exposure in a furnace retort.

Carburizing. A ferrous alloy surface-hardening heat treat process which introduces carbon into the metal at an elevated temperature. Surface carbon concentration and depth of diffusion (case depth) are time-temperature dependent processes. The resulting case hardness is dependent upon the carbon content of the diffused carbon, the quenching media, and the complete transformation from austenite to untempered martensite.

CARBURIZED 9310 / KNOOP

Carburizing flame. A gas flame which will introduce carbon into ferrous metals, that is, by oxy-acetylene welding.

Case hardening. A ferrous alloy heat treat process in which the surface is made considerably harder than the interior or core metal. Some case hardening processes are carburizing, nitriding, carbonitriding, boronizing, ion implanting, cyaniding, flame and induction hardening.

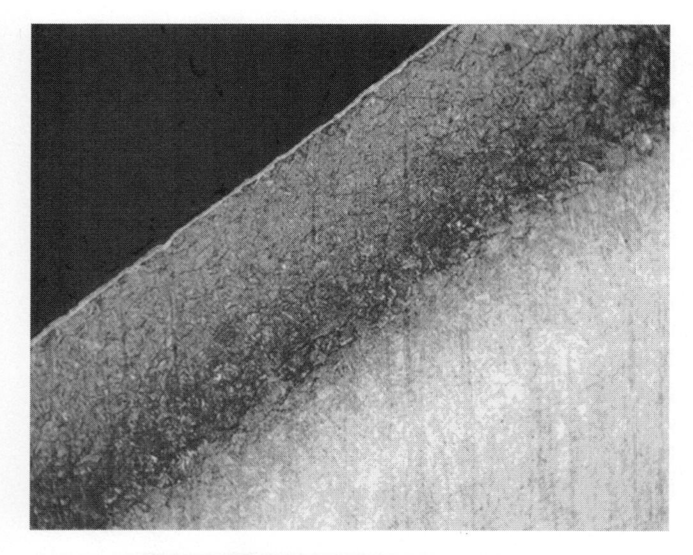

NITRIDED H-13

Casting. A process of producing a metal object by pouring molten metal into a mold.

Cast iron. Iron containing 2 to 4.5 percent carbon, silicon, and other trace elements. Cast iron is a generic name applied to five major alloy groups or cast iron types: *gray cast iron, white cast iron, ductile cast iron, malleable cast iron,* and *compacted graphite iron.*

Catalyst. An agent that induces catalysis, which is a phenomenon observed in chemical reactions in which the reaction between two or more substances is influenced by the presence of a third substance (the catalyst) that usually remains unchanged throughout the reaction.

Cathodic polarization. In an electrolyte in which hydrogen gas can accumulate around the cathode, very little corrosion can take place on the anode. Oxygen acts as a depolarizer and accelerates corrosion when present in the electrolyte.

Caustic. A chemical base material (caustic soda or sodium hydroxide) used to clean aluminum and sometimes used as an etchant in metallography.

Cellulose. A polysaccharide of glucose units that constitutes the chief part of the cell walls of plants. For instance, cotton fiber is over 90 percent cellulose and is the raw material of many manufactured goods such as paper, rayon, and cellophane. In many plant cells, the cellulose wall is strengthened by the addition of lignin, forming ligno-cellulose.

Cementite. Also known as iron carbide, a compound of iron and carbon (Fe_3C).

Centrifuging. Casting of molten metals by using centrifugal forces instead of gravity. The mold or molds are rotated about a center where molten metal is poured and allowed to follow sprues outward and into the mold cavity.

Cladding. The joining of one metal (usually sheet or plate) to another by using heat and pressure or by an explosive force. With this method, a thin sheet of more expensive metal or one less likely to corrode may be applied to a less expensive metal or one more likely to corrode.

Coalescence. The union of particles into larger units. In grains (microstructure), this term refers to the joining of adjacent grains to form a larger grain.

Cold drawing. Reducing the cross section of a metal bar, rod, or wire by drawing the metal through a die. Cold drawing is usually performed at room temperature.

Cold rolling. Reducing the cross section of a metal in a rolling mill for the purpose of producing a smooth hard surface finish. The surface grains of the metal become elongated and work-hardened.

Cold stabilization. The heat treat practice of continuous quenching of steels by allowing no time delay when removing the quenched steel from quench oil and placing the steel into a cold (cryogenic) chamber. See *Cryogenic Quenching.*

Cold working. Plastically deforming a metal at room temperature to introduce strain hardening in and deformation of the microstructure.

Colloid. (a) A substance having particles too small for resolution with an ordinary light microscope and which in suspension or solution fail to settle out. However, the suspension will not pass through a semipermeable membrane. (b) A gelatinous or glutinous substance.

Columnar grain structure. A course grain structure of para-elongated grains formed by unidirectional growth most often observed in castings, weldments, and plated materials.

Composite. A material containing one or more components, usually high strength fibers, particles and/or whiskers, added for strength. Composites may be constructed using a metal/metal, a metal/non-metal, or a non-metal/non-metal combination.

Compressive strength (ultimate). The maximum stress that can be applied to a brittle material in compression without fracture.

Compressive strength (yield). The maximum stress that can be applied to a metal in compression without permanent deformation.

Compressive test. A method of determining mechanical properties of a material subjected to uniaxial compressive load. Data obtained are *ultimate strength, yield strength, compressive set and elastic limit, compressive fracture strength,* and in some cases the test is used as a measure of *malleability.*

Core metal. In case hardening, that portion of the metal which is not part of the case or has not been chemically altered by the case hardening process.

Cores. In metal casting, the hollow parts made by using formed sand shapes strengthened by baking, wire mesh, and/or epoxy. For example, sand cores are used to produce the piston cylinder holes in cast automotive engine blocks rather than to cast a solid block and remove the metal by machining.

Corrosion. A gradual electrochemical attack on a metal caused by galvanic action in an electrolyte such as moisture. The result of metal decomposition by corrosion is the formation of an oxide.

Corrosion embrittlement. The embrittlement caused in certain alloys that are susceptible to intergranular corrosion attack when they are exposed to a corrosive environment.

Corrosion resistance. The ability of some metals to form a protective oxide or passive layer which renders them more resistant to corrosion.

Coupling. Also called galvanic coupling. When two different metals are placed in an electrolyte, making a galvanic corrosion cell in which an electric current is produced.

Creep. (1) High temperature creep—Plastic deformation resulting from sustained loading below the yield point at elevated temperatures. (2) Room temperature creep—Stress relaxation and plastic deformation resulting from sustained loading below the measured yield point.

Critical point. Same as *transformation temperature:* the temperature at which a change in crystal structure, phase, and/or physical properties occurs.

Critical temperature. The preferred term used by metallurgists is *transformation temperature.* The lower A_1 and the upper A_3 temperatures are the boundaries of the transformation range in which ferrite transforms into austenite.

Cryogenic quenching. The process of continuing the quenching of a metal into a cryogenic temperature range of –100°F or cooler. To complete the transformation of austenite to martensite sometimes requires cold stabilization in liquid nitrogen.

Crystalloid. A substance that forms a true solution and is capable of being crystallized.

Crystal unit structure or unit cell. The simplest polyhedron that embodies all the structural characteristics of a crystal and makes up the lattice of a crystal by indefinite repetition.

Cyanogen $(CN)_2$. A colorless gas of characteristic odor. It forms hydrocyanic and cyanic acid when in contact with water. Cyanogen compounds are often used for case hardening.

D

Decarburization. Loss of carbon from the surface of a steel as a result of furnace heating in the presence of an oxidizing environment. Decarburization may be partial or complete.

Deformation. Permanent alteration of the form or shape of a metal as a result of localized overstress past the material's elastic limit.

Dendrite. A crystal characterized by a treelike pattern that is usually formed by the solidification of a metal. Dendrites generally grow inward from the surface of a mold.

Density. The density of a body is the ratio of its mass to its volume. For solids, density is ascertained by measuring the buoyant force upon the specimen of liquid of known density in which it is immersed or by determining the volume of displacement from a specific gravity bottle.

Deoxidizer. A substance that is used to remove oxygen from molten metals—for example, ferrosilicon in steel making.

Dezincification. The dealloying (leaching from solid solution) of zinc which is alloyed with copper as in brass.

Diamagnetism. The phenomenon in which the magnetization in a substance opposes the magnetizing force which induces it. It is opposite to ferromagnetism, which is an attractive phenomenon. Diamagnetism exists to some extent in all substances—some nonferrous metals are strongly diamagnetic.

Die casting. Casting metal into a mold by using pressure instead of gravity or centrifugal force.

Diffusion. The process of intermingling atoms or other particles within a solution. In solids, it is a slow movement of atoms from areas of high concentration toward areas of low concentration. The process may be (a) migration of interstitial atoms such as carbon, (b) movement of vacancies, or (c) direct exchange of atoms to neighboring sites.

Drawing. A term sometimes used for the process of tempering hardened steel. Also used for metal forming in presses and forming wire.

Ductility. The property of a material to deform permanently, or to exhibit plasticity without rupture, while under tension.

Duplex grain size. The simultaneous presence of two grain sizes in substantial amounts, with one grain size appreciably larger than the other.

Duplex microstructure. A two-phase microstructure.

E

Elasticity. The ability of a material to return to its original form after a load has been removed.

Elastic limit. The maximum stress that a material is capable of sustaining without any permanent strain (deformation) remaining upon complete release of the stress.

Elastomer. Any of various elastic substances resembling rubber.

Electrolysis. The producing of chemical changes by passing an electric current through an electrolyte.

Electrolyte. A nonmetallic conductor in which electric current is carried by the movement of ions.

Electron microscope. (1) Transmission electron microscope (TEM)—Used to examine carbon-coated replicas of fracture surfaces at very high magnification. (2) Scanning electron microscope (SEM)—Used to perform real-time high-magnification examination of actual fracture surfaces and/or coated replicas.

Electron microscopy. The study of materials using of an electron microscope.

Electroplating. Coating an object with a thin layer of some metal through electrolytic deposition.

Equilibrium. A condition of balance in which all the forces or processes that are present are counterbalanced by equal and opposite forces or processes where the condition appears to be one of rest rather than of change.

Equilibrium diagram. A graph of the pressure, temperature, and composition limits of phase fields in an alloy system as they exist under conditions of thermodynamic equilibrium; see *phase diagram.*

Etchant. A chemical solution used to delineate a metal or metal matrix composite material to reveal microstructural details.

Eutectic. The alloy composition that freezes at the lowest constant temperature, causing a discrete mixture to form in definite proportions.

Eutectic melting. Melting of localized microscopic areas whose composition corresponds to that of the eutectic in the system.

Eutectoid. The alloy composition that transforms from a high temperature solid into new phases at the lowest constant temperature. In binary (double) alloy systems, it is a mechanical mixture of two phases forming simultaneously from a solid solution as it cools through the eutectoid (A_1 in steels) temperature.

Extrusion. Forcing a solid metal piece (often heated) through a shaped die by using an extremely high force in a way that is similar to squeezing toothpaste from a tube.

F

Fatigue failure. Fracture in a material due to conditions of repeated cyclic overload.

Fatigue strength. A measure of the load-carrying ability of a material subjected to loading which is repeated a definite number of cycles. The maximum alternating stress amplitude sustained by a material, subjected to a specific mean stress, for a specific number of cycles without failure. Fatigue rated materials are never loaded more than 50 percent of the measured elastic limit.

Ferrite. The name given the body-centered cubic crystal structure of pure iron.

Fiber. (a) The directional property of wrought metals that is revealed by a woody appearance when fractured. (b) A preferred orientation of metal crystals after a deformation process such as rolling or drawing. (c) Cellulosic plant cells that are used for manufacturing paper and other products.

Flux. A solid, liquid or gaseous material that is applied to solid or molten metal in order to clean and remove the oxides.

Forging. The shaping of metal by hammering or pressing. Although forging may be used to shape malleable metals in the cold state, the application of heat increases plasticity and permits greater deformation without inducing undue strain in the metal.

Fracture surface. A ruptured surface of a metal which contains the character or fracture mode causing the fracture. Fracture surfaces are often examined using the macroscope, optical microscope, and the scanning electron microscope to determine the actual cause of failure. Fatigue fractures, however, often display a smooth, clam-shell appearance.

Full annealing. An annealing process that uses a predetermined and closely controlled time-temperature cycle to produce a fully recrystallized microstructure. Full annealing of steels is sometimes performed to increase machinability.

Fusion. The merging of two materials while in a molten state.

G

Galvanic series. Also called the electromotive series. A list of metals in an ascending order of electrical or galvanic activity in seawater (electrolyte). The more active metals near the anodic end are most likely to corrode.

Gangue. The commercially undesirable portion of an ore that must be removed before the ore is processed into a metal.

Grain. Individual crystals in metals.

Grain flow. A directional flow of grain structure which occurs as a result of plastic deformation such as in rolling or forging. Upon plastic deformation, an axial crystal orientation results as opposed to an equiaxed or omnidirectional crystal structure observed in a metal casting.

Grain germination. Abnormal coalescence and enlargement of individual grains due to exposure to excessive elevated temperature.

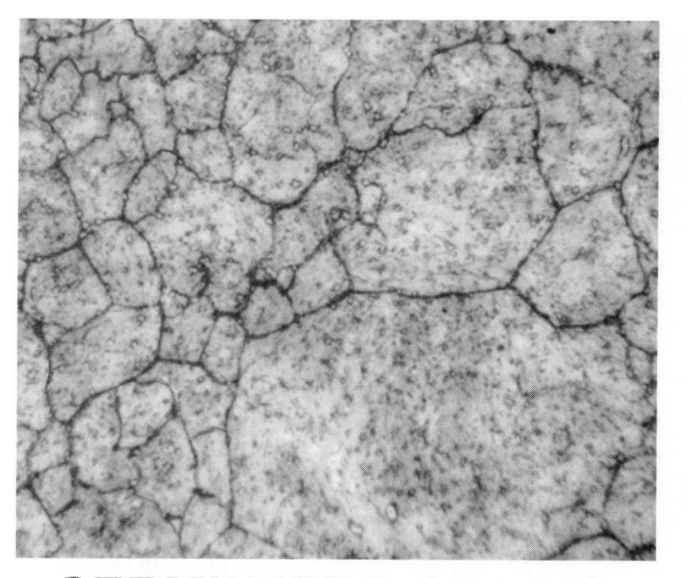

GERMINATED GRAINS

Grain growth. An increase in the average size of the grains in a polycrystalline metal, usually the result of heating at an elevated temperature.

Grain size. A measure of the area or volume of grains in a polycrystalline metal or alloy, usually expressed as an average when individual grain sizes are uniform. In metals containing two or more phases, the measured grain size refers to that of the matrix. The optical microscope is used to compare and measure grain size at 100× for alloy steels and 1000× for tool steels in accordance with the standard grain size chart established by ASTM.

Graphite. One of 14 crystal structures of carbon. Graphite is commonly used in pencils although wrongly referred to as *lead.*

H

Hardenability. The ability of a metal or alloy to be through hardened. The standard of measurement for hardenability is the Jominy end-quench test.

Hardening. The process of increasing the hardness of metals, usually by thermal heat treatment. Hardness in some metals can only be achieved by mechanical deformation or strain/work hardening.

Hard facing. When a hard surface is desired on a soft metal, a hard material (usually another metal, alloy, or composite) is applied to the surface. Some methods include welding, plasma deposition, and electroplating.

Hardness. The ability of a material to resist abrasion, penetration, and deformation.

Heat sink. In castings, heat sinks are used to stabilize the temperature differential between large and small areas of solidifying metal.

Hooke's law. The stress set up within an elastic body that is proportional to the strain to which the body is subjected by the applied load.

Hot rolling. A process of forming metals between rolls in which the metals are heated to temperatures above the transformation range.

Hot-short. Manganese is added to steels to combine with sulfur and prevent hot-shortness.

Hydrogen embrittlement. A marked loss of ductility in steels resulting from absorption of hydrogen at grain boundaries and around grains during plating, acid etching, chemical cleaning, and certain heat treating processes. Hydrogen embrittlement causes severe localized overstress at grain boundaries, creating a brittle, intergranular crystalline–appearing fracture.

HYDROGEN EMBRITTLEMENT

I

Impact test. A test in which small notched specimens are broken in an Izod-Charpy machine. This test determines the notch toughness of a metal.

Inclusions. Particles of impurities that are usually formed during solidification and are usually in the form of silicates, sulfides, and oxides.

Inert gas. Noble gases such as helium or argon that are not reactive with any other elements.

Ingot. A large block of metal that is usually cast in a metal mold and forms the basic material for further rolling and processing.

Interface. The contact area between two materials or mechanical elements.

Intergranular corrosion. A type of galvanic corrosion that progresses along the grain boundaries of an alloy. The grain boundaries become anodic to the grains and deteriorate, usually causing failure of the part.

INTERGRANULAR CRACKING
Photo Courtesy of Stork-MMA Laboratories, Inc.

Intergranular cracking. This is a crack that forms along the grain boundaries and not through the grains.

Interstitial lattice structure. A crystalline lattice containing smaller atoms of a different element within its interstices (voids or holes between the atoms and the lattice).

Ion. An atom or molecule that is electrostatically charged because it has lost or gained one or more valence electrons.

Iron. The term iron always refers to the element Fe and not cast iron, steel, or any other alloy of iron.

Isomerism. The relation of two or more compounds, radicals, or ions that contain the same number of atoms of the same elements but differ in structural arrangement and properties.

Isothermal transformation (I-T). Transformation that takes place at a constant temperature.

J

Jominy end-quench test. A test to determine the hardenability of alloy and tool steels.

K

Kaolin. A fine white clay that is used in ceramics and refractories and is composed mostly of kaolinite, a hydrous silicate of aluminum. Impurities may cause various colors and tints.

Killed steel. Steel that has been deoxidized with agents such as silicon or aluminum compounds to reduce the ingot gas content. This prevents gases from evolving during the solidification period.

L

Lamellar. An alternating platelike structure in metals (as in pearlite).

Lath martensite. Martensite formed partly in steels containing less than 1.0 percent carbon and solely in steels containing less than 0.5 percent carbon as parallel arrays of lath-shape units 0.1 to 0.3 microns thick.

Lattice, space. A term that is used to denote a regular array of points in space—for example, the sites of atoms in a crystal. The points of the three-dimensional space lattice are constructed by the repeated application of the basic translations that carry a unit cell into its neighbor.

Leidenfrost phenomenon. Slow quenching rates associated with a hot vapor blanket surrounding a part being quenched in a liquid medium such as water or oil. The gaseous vapor envelope acts as an insulator, thus dramatically slowing the cooling rate. Inconsistent part hardness and incomplete microstructure transformation are consequences of this phenomenon.

Lignin. A substance related to cellulose that with cellulose forms the woody cell walls of plants and the material that cements them together. Methyl alcohol is derived from lignin in the destructive distillation of wood.

Liquation temperature. The lowest temperature at which partial melting can occur in an alloy that exhibits the greatest possible degree of segregation. See Eutectic melting.

Liquid carburizing. Surface hardening of steel by immersion into a molten bath consisting of cyanides and other salts in which carbon diffuses into the surface of the steel.

Liquid nitriding. A method of surface hardening in which molten nitrogen-bearing, fused-salt baths containing both cyanides and cyanates are exposed to parts at a subcritical temperatures of approximately 1000°F, in which nitrogen readily diffuses into the surface of the steel.

Liquid nitrocarburizing. A nitrocarburizing process during which both carbon and nitrogen are diffused into the surface of a steel utilizing molten liquid salt baths below the lower critical temperature.

Liquid spray quench. Same as *spray quenching.*

Liquidus. The temperature at which freezing begins during cooling and ends during heating under equilibrium conditions, represented by a line on a two-phase diagram.

Localized precipitation. Precipitation from a supersaturated solid solution similar to continuous precipitation, or age-hardening. The precipitate particles form at preferred locations, such as along crystal slip planes, grain boundaries, triple points, or incoherent twin boundaries.

M

Macroscopic. Structural details observed with the naked eye and/or aided with low-power optics at magnifications up to 50×.

Macrostructure. The structure of metals as revealed by macroscopic examination.

Malleability. The ability of a metal to deform permanently without rupture when loaded in compression.

Malleabilzing. Annealing of white cast iron in such a way to cause transformation of carbon to graphite.

Malleable cast iron. Produced by prolonged annealing of white cast iron in which decarburization of graphitization, or both, take place to eliminate some or all of the cementite. The graphite is in the form of temper carbon. If decarburization is the predominate reaction, the product will exhibit a light fracture surface, called "whiteheart malleable cast iron." If the fracture surface is dark, the cast iron is referred to as "blackheart malleable cast iron." Ferritic malleable cast iron has a predominantly ferritic matrix; pearlitic malleable cast iron may contain pearlitic, spheroidite, or tempered martensite depending on the heat treatment and desired hardness.

Maraging. A precipitation hardening treatment applied to a special group of steels containing high nickel content, in which one or more intermetallic compounds precipitate in a matrix of essentially carbon-free martensite.

Martempering. (1) A hardening procedure in which a steel is quenched from the austenitizing temperature into an appropriate medium (usually a salt bath) whose temperature is maintained at the M_s of the steel, held until the temperature is uniform throughout but not long enough to permit bainite to form, and then air cooled. (2) When this procedure is applied to carburized steels, the controlling M_s temperature is that of the case (called marquenching).

Martensite. The crystal structure formed during quench hardening of carbon, alloy, and tool steels. Martensite has an acicular or needlelike microstructure which is formed without diffusion when the steel is rapidly cooled from the austenitizing temperature. Martensite is the hardest transformation product of austenite, and may be in a form called lath or plate martensite.

McQuaid-Ehn test. A metallographic test to reveal prior austenitic grain size in steels.

Meissner effect. When a superconductor is cooled to transition in a magnetic field, the lines of induction are pushed out, as if it exhibited diamagnetism. A superconductor will actually float above a magnet.

Metalloid. A nonmetal that exhibits some, but not all, of the properties of a metal. Examples are sulfur, silicon, carbon, phosphorus, and arsenic.

Metallurgy. The science and study of the behaviors and properties of metals and their extraction from their ores.

Methanol (methyl alcohol, wood alcohol). Produced by the destructive distillation of wood or made synthetically.

Microhardness. The hardness of a material determined by indenting the surface with a Vickers (DPH) or Knoop (KHN) indenter using a very light load. Measurement of the resulting indentation requires the use of a special measuring microscope. Microhardness testing is used to accurately measure the **effective case depth** in case hardening heat treatments.

Microsegregation. Segregation within a grain, crystal, or small particle.

Mill scale. The heavy oxide layer formed during hot fabrication or heat treatment of metals at the steel mill.

Modulus of elasticity. The ratio of the unit stress to the unit deformation (strain) of a structural material is a constant as long as the unit stress is below the elastic limit. Shearing modulus of elasticity is often called the modulus of rigidity.

Monomer. A single molecule or a substance consisting of single molecules. The basic unit in a polymer.

Muffled furnace. A gas- or oil-fired furnace in which the work is separated from the flame by an inner lining or "muffle."

N

Natural aging. Spontaneous aging of a supersaturated solid solution at room temperature. Precipitation of excessive alloy compound, formerly dissolved in solid solution due to high temperature heat treatment; akin to dissolving too much sugar in boiling water and as the water cools, the excess sugar precipitates as a crystal. Aging in metals follows the natural laws of solubility.

Nitriding. A heat treating process in which parts are exposed to a nitrogen atmosphere in a sealed retort at temperatures well below allotropic transformation. With time, nitrogen diffuses to a depth of approximately 0.012″ into the steel, causing the formation of iron-nitride compound. Iron-nitride is very hard, approaching the hardness of diamond.

Nonferrous. Metals other than iron and iron alloys.

Normalizing. Normalizing is a heat treatment used to recrystallize steels following forging.

Notch toughness. The resistance to fracture of a metal specimen having a notch or groove when subjected to a sudden load, usually tested on an Izod-Charpy testing machine.

O

Oil hardening. Quench-hardening treatment involving cooling in oil, as in hardening of carbon steel in an oil bath. Oils are categorized as conventional, fast, martempering, or hot quenching.

Overaging. Aging under conditions of time and temperature greater than those required to obtain maximum change in mechanical properties.

Overheating. Heating a metal or alloy to a temperature that compromises standard mechanical properties. Overheating can create detrimental microstructural artifacts such as *grain germination, excessive phase precipitation, allotropic transformation,* and *eutectic melting.* Machinists often refer to overheating as **burning,** usually caused by a lack of lubrication during surface grinding.

Oxidation. (1) Surface oxidation of a metal caused by thermal exposure in air. (2) A reaction in which there is an increase in valence resulting from a loss of electrons. (3) A corrosion reaction in which the metal combines with oxygen to form an oxide. In steels, an oxidized surface is manifest by discoloration from a straw to a dark blue/black brittle and flaky scale.

Oxidizing flame. A gas flame produced with excess oxygen in the inner flame of an oxy-acetylene torch.

P

Pack carburizing. A method of surface hardening of steel in which parts are packed in a carburizing compound and heated to approximately 1700°F to activate the carbon which diffuses into the steel.

Pearlite. Grains formed in carbon and alloy steels upon slow cooling, consisting of alternate hard iron carbide and soft ferrite layers, thus referred to as a *lamellar microstructure* (see schematic).

Peening. Work hardening the surface of metal by hammering or blasting with shot (small steel balls). Peening introduces compressive stresses on weld surfaces that tend to counteract unwanted tensile stresses.

Permeability. In casting of metals, the term is used to define the porosity of foundry sands in molds and the ability of trapped gases to escape through the sand.

Phase. A portion of an alloy, physically homogeneous throughout, that is separated from the rest of the alloy by distinct bounding surfaces. The following phases occur in the iron-carbon alloy: molten alloy, austenite, ferrite, cementite, and graphite.

Phase diagram. A graphical representation of the temperature and composition limits of phase fields as they actually exist under specific conditions of heating and cooling in an alloy system. Phases can be represented in equilibrium, as an approximation of equilibrium or in a metastable condition.

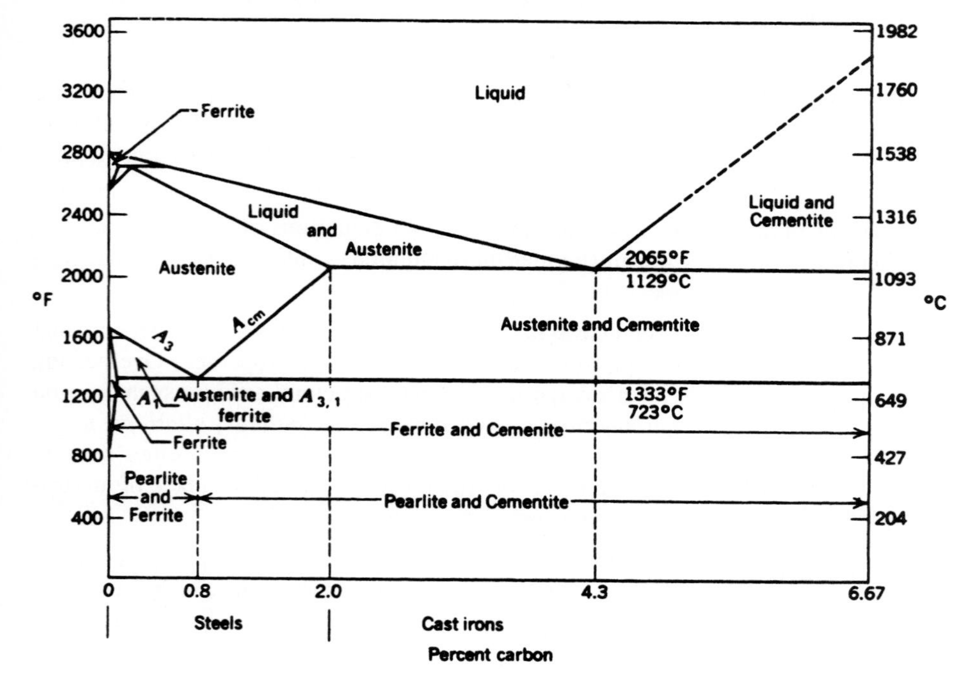

Pig iron. The product of a blast furnace. It is a raw iron that usually contains about 4.5 percent carbon and impurities such as phosphorous, sulfur, and silicon.

Pitting corrosion. Most metals pit corrode when exposed to highly acidic conditions. When corrosion begins, a small amount of positive metal ions are in solution at the metal surface. The positive charged ions attract negative charged ions from the acid solution. Hydrolysis in the metal pit leads to an ever-decreasing pH, gettering more metal into solution and increasing the pit size.

Plasticity. The ability of a metal to sustain permanent deformation or yielding without rupture.

Plate martensite. Rectangular-shape martensite, not to be confused with triangular-shape retained austenite; however, these two phases are often found together. Verification of a microstructure is usually confirmed by X-ray defraction method for determining the crystal structure present.

Poisson's ratio. If a rod of elastic material is stretched with sufficient force, it can be elongated. The unit elongation is known as strain, which may be denoted as *S*. The lateral contraction may be denoted by *C*. The ratio *C/S*, which is a constant for a given material within the elastic limit, is known as Poisson's ratio. The value of 0.30 is generally used for steels.

Polymerization. A reaction in which a complex molecule (a polymer) is formed from a number of simpler molecules that can be alike or unlike.

Powder metallurgy. A process by which metal shapes are pressure formed from metal powder and fused or sintered in a furnace to permanently bind the powder into solid form.

Precipitation hardening. Hardening caused by the precipitation of a constituent or compound from a supersaturated solid solution; sometimes referred to as *age hardening.*

psi. An abbreviation for pounds per square inch.

Q

Quench cracking. Fracture of a metal usually due to thermal overstress when quenching from an elevated temperature.

Quenching. (1) Rapid cooling. (2) The process of rapid cooling of metal alloys for the purpose of hardening; and/or (3) the rapid cooling of a homogenized microstructure to prevent phase precipitation which would occur during slow cooling to room temperature.

R

Recovery. The relaxation and return to the original dimension of the metal after it has been stressed.

Recrystallation. The formation of new, strain-free grains and/or microstructure from that existing in a cold-worked, plastically deformed condition, usually accomplished by heating; see normalizing.

Reducing flame. A gas flame produced with excess fuel in the inner flame.

Refractory. Materials that will resist change of shape, weight, or physical properties at high temperatures. These materials are usually silica, fire clay, diaspore, alumina, and kaolin. They are used for furnace linings.

Retained austenite. The austenitic face-centered cubic crystal structure which results due to incomplete allotropic transformation to martensite when quench-hardening steels.

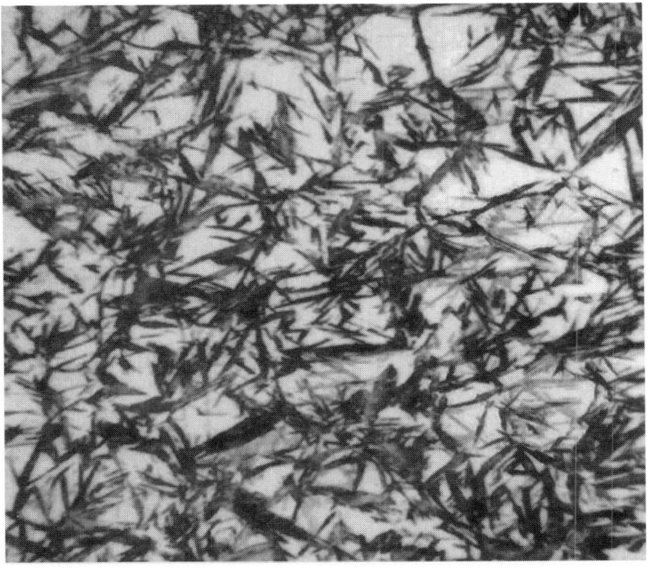

Rimmed steel. A low carbon steel (insufficiently deoxidized) that during solidification releases considerable quantities of gases (mainly carbon monoxide). When the mold top is not capped, a side and bottom rim of several inches forms. The solidified ingot has scattered blow holes and porosity in the center but a relatively thick skin free from blow holes.

(Micromet® Rockwell-type hardness tester, digital model. Photograph courtesy of Buehler Ltd., Evanston, Illinois.)

Rockwell hardness test. A method of indentation hardness testing based on comparing the depth of indentation resulting from application of a minor load versus a major load with a standardized indenter.

rpm. Revolutions per minute.

S

SAE. Abbreviation for *The Society of Automotive Engineers.*

Scale. The surface oxidation on metals that is caused by heating in air or in other oxidizing atmospheres.

Scrap. Materials or metals that have lost their usefulness and are collected for reprocessing.

Sealant. A sealing agent that has some adhesive qualities; it is used to prevent leakage.

Sensitization. The precipitation of chromium carbide at the grain boundaries of an austenitic stainless steel caused by unwanted carbon diffusion and/or slow cooling during the quenching process. The microstructure is sometimes referred to as a sigma phased precipitation.

sfm. Surface feet per minute. Also fpm (feet per minute) and sfpm (surface feet per minute).

Shell molding. A form of gravity casting of metal (usually a high melting temperature metal) in which the mold is made of a thin shell of refractory material.

Sintered. To cause to become a consolidated mass by heating without melting.

Slag (dross). A fused product that occurs in the melting of metals and is composed of oxidized impurities of a metal and a fluxing substance such as limestone. The slag protects the metal from oxidation by the atmosphere since it floats on the surface of the molten metal.

Slip plane. The crystallographic plane in which slip occurs in a crystal.

Smelting. The process of heating ores to a high temperature in the presence of a reducing agent such as carbon (coke) and of a fluxing agent to remove the gangue.

Snap Temper. A precautionary interim tempering heat treatment performed at between 300° and 400°F and applied to high hardenability steels directly after quenching to prevent cracks caused by microstructure transformation and overstress.

Soaking. A prolonged heating of a metal at a predetermined temperature to create a uniform temperature throughout its mass.

Solid solution. A solid crystalline phase containing two or more chemical species in concentrations that may vary between limits imposed by phase equilibrium. Some of the types of solid solutions are *continous, interstitial, substitutional,* and *terminal.*

Solidus. Seen as a line on a two-phase diagram, it represents the temperatures at which freezing ends when cooling or melting begins when heating under equilibrium conditions.

Solubility. The degree to which one substance will dissolve in another.

Solute. A substance that is dissolved in a solution and is present in minor amounts.

Solution heat treatment. A process in which an alloy is heated to a predetermined temperature for a length of time that is suitable to allow a certain constituent to enter into solid solution. The alloy is then cooled quickly to hold the constituent in solution, causing the metal to be in an unstable supersaturated condition. This condition is often followed by age hardening.

Solvent. A substance that is capable of dissolving another substance and that is the major constituent in a solution.

Sorbite. A term that was once used to denote tempered martensite having a microstructure with a granular appearance.

Specific gravity. A numerical value that represents the weight of a given substance as the weight of an equal volume of water. The specific gravity for pure water is taken as 1.000.

Spheroidizing. Heating and cooling to produce a spheroidal or globular form of carbide in steel.

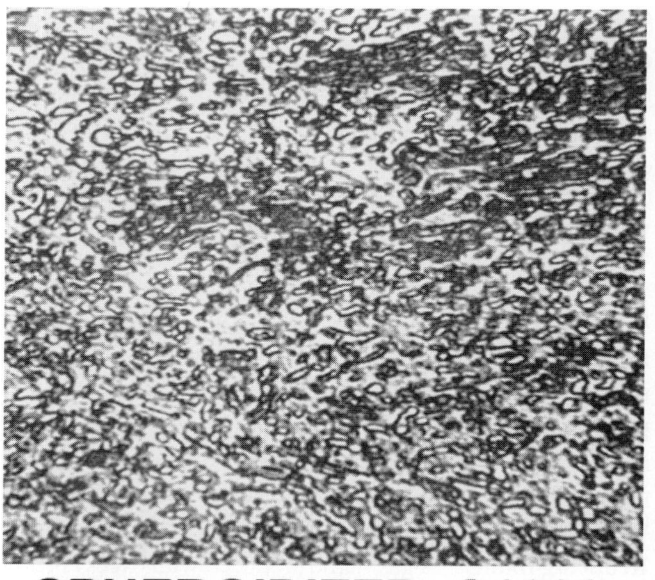

SPHEROIDIZED C1090

Spray quenching. A quenching process using spray nozzles to spray water or other liquids on a part. The quench rate is controlled by the velocity and volume of liquid per unit area per unit of time of impingement.

Stainless steel. An alloy of iron containing at least 11 percent chromium and sometimes nickel that resists almost all forms of rusting and corrosion.

Statistical quality control. The application of statistical techniques for measuring and improving the quality of processes and products.

Steel. An alloy of iron and less than 2 percent carbon plus some impurities and small amounts of alloying elements is known as plain carbon steel. Alloy steels contain substantial amounts of alloying elements such as chromium or nickel besides carbon.

Strain. The unit deformation of a metal when stress is applied.

Strain hardening. An increase in hardness and strength of a metal that has been deformed by cold working or at temperatures lower than the recrystallization range.

Strength. The ability of a metal to resist external forces. This is called tensile, compressive, or shear strength, depending on the load. See **stress.**

Stress. The load per unit of area on a stress-strain diagram. *Tensile stress* refers to an object loaded in tension, denoting the longitudinal force that causes the fibers of a material to elongate. *Compressive stress* refers to a member loaded in compression, which either gives rise to a given reduction in volume or a transverse displacement of material. *Shear stress* refers to a force that lies in a parallel plane. The force tends to cause the plane of the area involved to slide on the adjacent planes. *Torsional stress* is a shearing stress that occurs at any point in a body as the result of an applied torque or torsional load.

Stress corrosion. (Also called corrosion fatigue.) A rapid deterioration of properties resulting from the repeated cyclic stressing of a metal in a corrosive medium.

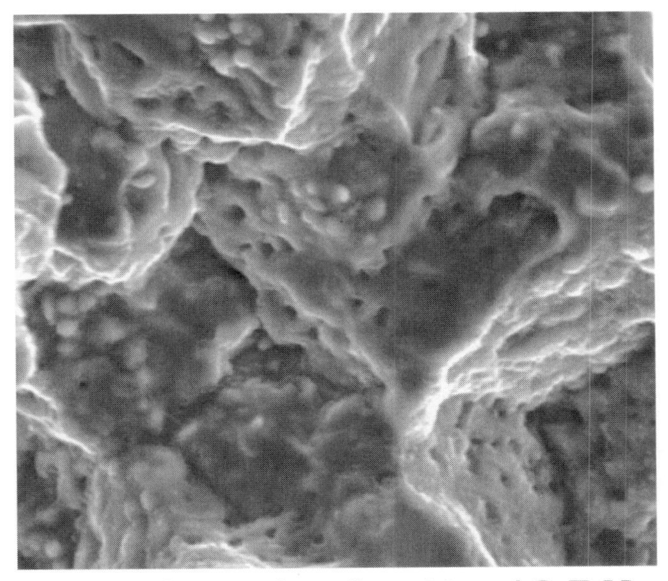

Stress Corrosion Cracking / S.E.M.

Stress raiser. Can be a notch, nick, weld undercut, sharp change in section, machining grooves, or hairline cracks that provide a concentration of stresses when the metal is under tensile stress. Stress raisers pose a particular problem and can cause early failure in members that are subjected to many cycles of stress reversals.

Stress relief anneal. The reduction of residual stresses in a metal part by heating it to a given temperature (in steels, this temperature is 50°F below the last tempering temperature), and holding temperature for a suitable time. This treatment is used to relieve stresses due to machining, cold working, and welding.

T

Temper embrittlement. Embrittlement or *temper brittleness* in some alloy steels caused by tempering in the temperature range of 600° to 1000°F. Temper embrittlement is activated by residual chemical impurities such as arsenic, antimony, phosphorous, and tin, and may be eliminated by tempering outside of the sensitive temperature range.

Tempering. In ferrous metals, the stress relief of steels hardened by quenching for the purpose of toughening them and reducing their brittleness. In nonferrous metals, temper is a condition produced by mechanical treatment such as cold working. An alloy may be cold worked to the hard temper, fully softened to the annealed temper, or two intermediate tempers.

Tensile strength. The maximum load carrying capability of a material based upon a tensile test.

Thermal conductivity. The quantity of heat that is transmitted per unit time, per unit cross section, per unit temperature gradient through a given substance. All materials are in some measure conductors of heat.

Thermal expansion. The increase of the dimension of a material that results from the increased movement of atoms caused by increased temperature.

Thermal stress. Shear stress that is induced in a material due to unequal heating or cooling rates. The difference of expansion and contraction between the interior and exterior surfaces of a metal that is being heated or cooled is an example.

Thermoplastic. Capable of softening or fusing when heated and of hardening again when cooled.

Thermosetting. Capable of becoming permanently rigid when cured by heating; will not soften by reheating.

Tool steel. A special group of steels that are designed for specific uses such as heat-resistant steels that can be heat treated to produce certain properties, mainly hardness and wear resistance.

Torque. A force that tends to produce rotation or torsion. Torque is measured by multiplying the applied force by the distance at which it is acting to the axis of the rotating part.

Torsion. The twisting or wrenching of a body by the exertion of forces tending to turn one end about a longitudinal axis while the other is held fast or turned in the opposite direction.

Toughness. Generally measured in terms of notch toughness, which is the ability of a metal to resist rupture from impact loading when a notch is present. A standard test specimen containing a prepared notch is inserted into the vise of a testing machine. This device, called the Izod-Charpy testing machine, consists of a weight on a swinging arm. The arm or pendulum is released, strikes the specimen, and continues to swing forward. The amount of energy absorbed by the breaking of the specimen is measured by how far the pendulum continues to swing.

Transducer. A device by means of which energy can flow from one or more transmission systems to other transmission systems, such as a mechanical force being converted into electrical energy (or the opposite effect) by means of a piezoelectrical crystal.

Transformation temperature. The temperature(s) at which one phase transforms into another phase; for example, where ferrite or alpha iron transforms into austenite or gamma iron.

Transgranular cracking. Cracking that takes place along crystallographic planes through the grains rather than along the grain boundaries.

Twin. Two portions of a crystal with a definite orientation relationship; one may be regarded as the parent, the other as the twin. The orientation of the twin is a mirror image of the orientation of the parent across a twinning plane or an orientation that can be derived by rotating the twin portion about a twinning axis.

U

Ultimate strength. (Also tensile strength.) The highest strength that a metal exhibits after it begins to deform plastically under load. Rupture of the material occurs either at the peak of its ultimate strength or at a point of further elongation and at a drop in stress load.

V

Valence. The capacity of an atom to combine with other atoms to form a molecule. The inert gases have zero valence; valence is determined by considering the positive and negative atoms as determined by the atoms gaining or losing of valence electrons.

Viscosity. The property in fluids, either liquid or gaseous, that may be described as a resistance to flow; also, the capability of continuous yielding under stress.

Void. A cavity or hole in a substance.

W

Weld metal. The molten area of a weld, either introduced by a filler rod or produced by the fusion of the base metal.

Widmanstatten structure. Widmanstatten structure, also known as a *basketweave microstructure* in titanium alloys, occurs when the metal is heated slightly above the allotropic transformation temperature or *beta transus* temperature.

Work hardening. Also called *strain hardening.* A process in which the grains become distorted and elongated in the direction of working (rolling). This process hardens and strengthens metals but reduces their ductility. Excessive work hardening can cause ultimate brittle failure of the part. Stress relief is often used between periods of working of metals to restore their ductility.

Wrought iron. Contains 1 or 2 percent slag, which is distributed through the iron as threads and fibers, imparting a tough fibrous structure. Usually contains less than 0.1 percent carbon. It is tough, malleable, and relatively soft.

Y

Yield point. The stress at which a marked increase in deformation occurs without an increase in load stress. This phenomenon is seen in mild or medium carbon steel but not in nonferrous metals and other alloy steels.

Yield strength. Observed at the proportional limit of metals and is the stress at which a material deviates from that proportionality of stress to strain to a specified amount.

Index

C

N

O

P

T